Recent Advances in Aerodynamics

Also available from Springer-Verlag

the companion volume on aeroacoustics—

Recent Advances in Aeroacoustics

Proceedings of an International Symposium held at
Stanford University, August 22–26, 1983
Edited by A. Krothapalli and C. A. Smith

Contents: NOISE DUE TO JETS, SHEAR LAYERS, AND TURBULENCE. J.E.Ffowcs Williams: Waves in Turbulent Mixing Layers. C.K.W.Tam: On Broadband Shock Associated Noise of Supersonic Jets. M.S.Howe: On the Absorption of Sound by Turbulence and Other Hydrodynamic Flows. ROTOR ACOUSTICS. A.R.George: Analyses of Broadband Noise Mechanisms of Rotors. F.Farassat: The Evolution of Methods for Noise Predictions of High Speed Rotors and Propellers in the Time Domain. F.H.Schmitz and Y.H.Yu: Helicopter Impulsive Noise: Theoretical and Experimental Status. GENERATION AND PROPAGATION. W.K.Blake and A.Powell: The Development of Contemporary Views of Flow-Tone Generation. N.Rott: A Simple Theory of the Sondhauss Tube. S.M.Candel: A Review of Numerical Methods in Acoustic Wave Propagation. D.G.Crighton: Nonlinear Acoustic Propagation of Broadband Noise. A.H.Nayfeh: Acoustic Propagation in Partially Choked Ducts

Springer-Verlag
New York Berlin Heidelberg Vienna Tokyo

Recent Advances in Aerodynamics

Proceedings of an International Symposium held at Stanford University, August 22–26, 1983

Edited by
Anjaneyulu Krothapalli and Charles A. Smith

With Contributions by
R.K. Agarwal D. Baganoff P.M. Bevilaqua S.S. Davis
E.H. Dowell R.L. Fearn K.Y. Fung U. Ganzer
C.-M. Ho A.K.M.F. Hussain D.A. Johnson
D.R. Kotansky R.E. Kuhn J.T.C. Liu J.G. Marvin
W.J. McCroskey A. Michalke R.H. Miller
S.M. Przybytkowski D. Rockwell T. Sarpkaya
W.R. Sears A.R. Seebass

With 403 Illustrations

Springer-Verlag
New York Berlin Heidelberg Tokyo

Anjaneyulu Krothapalli
Department of Mechanical Engineering
FAMU/FSU College of Engineering
The Florida State University
Tallahassee, FL 32306
U.S.A.

Charles A. Smith
NASA Ames Research Center
Moffett Field, CA 94035
U.S.A.

Library of Congress Cataloging in Publication Data
Main entry under title:
Recent advances in aerodynamics.
Proceedings of the International Symposium on Recent Advances in Aerodynamics and Acoustics, organized by the Joint Institute for Aeronautics and Acoustics.
Includes bibliographies.
1. Aerodynamics—Congresses. 2. Aerodynamic noise—Congresses. I. Krothapalli, A. (Anjaneyulu), 1950– . II. Smith, C. A. (Charles Arthur), 1945– . III. International Symposium on Recent Advances in Aerodynamics and Acoustics (1983 : Stanford University) IV. Joint Institute for Aeronautics and Acoustics.
TL570.R373 1985 629.132′3 84-22117

9 8 7 6 5 4 3 2 1

ISBN-13:978-1-4612-9379-8 e-ISBN-13:978-1-4612-4972-6
DOI. 10.1007/978-1-4612-4972-6

Preface

The Joint Institute for Aeronautics and Acoustics at Stanford University was established in October 1973 to provide an academic environment for long-term cooperative research between Stanford and NASA Ames Research Center. Since its establishment, the Institute has conducted theoretical and experimental work in the areas of aerodynamics, acoustics, fluid mechanics, flight dynamics, guidance and control, and human factors. This research has involved Stanford faculty, research associates, graduate students, and many distinguished visitors in collaborative efforts with the research staff of NASA Ames Research Center.

The occasion of the Institute's tenth anniversary was used to reflect back on where that research has brought us, and to consider where our endeavors should be directed next. Thus, an International Symposium was held to review recent advances in the fields relevant to the activities of the Institute and to discuss the areas of research to be undertaken in the future. This anniversary was also chosen as an opportunity to honor one of the Institute's founders and its director, Professor Krishnamurty Karamcheti. It has been his creative inspiration that has provided the ideal research environment at the Joint Institute.

The International Symposium on Recent Advances in Aerodynamics and Acoustics was held at Stanford University, Stanford, California, U.S.A., August 22–26, 1983. Thirty-five distinguished scientists were invited to present a comprehensive review on the following subject areas: unsteady aerodynamics, jets and shear layers, V/STOL aircraft aerodynamics, rotor dynamics and aerodynamics, aeroacoustics, and rotor acoustics. The proceedings of the symposium are organized in two volumes. One volume contains the papers concerned with aerodynamics and the other volume those papers dealing with aeroacoustics and rotoracoustics.

Acknowledgements

During the preparation for the symposium a number of people, both at NASA Ames Research Center and at Stanford University, contributed in various ways. Here, we mention a few of those who represent the spirit and efforts of many. As honorary chairman of the Symposium, Mr. C. A. Syvertson served invaluably, not only in this capacity but also in several other essential roles before and during the symposium. Mr. C. T. Snyder and Mr. W. D. Deckert were instrumental in helping to organize the outstanding contribution and collaboration of NASA. Mr. David Hickey, through many discussions, provided suggestions and ideas which were incorporated in the running of the Symposium. As a co-chairman of the Symposium, Prof. Leonard Roberts coordinated many of the difficult organizing tasks.

We should like to note that the success of the technical sessions is a credit to the session chairmen, who effectively presided over the presentations and discussions. In this context we would like to extend our gratitude to Dr. Jack Nielsen, Prof. J. E. Ffowcs Williams, Prof. Brian Cantwell, Prof. Harold Levine, Prof. W. C. Reynolds, Dr. Gordon Banerian, Prof. Holt Ashley, and Dr. Craig Simcox.

The expert preparation of the proceedings was done by Ms. Doris Hsia and Ms. Dianne Budnik. It involved an immense effort on their part to get the proceedings in the present form. We give our sincere thanks and appreciation to both of them. We also wish to express our appreciation to Springer-Verlag for their understanding and patience.

An undertaking such as this would not be possible without the generous support of the NASA Ames Research Center. This support is gratefully acknowledged.

Finally, on a personal note we thank Prof. Krishnamurty Karamcheti. His continuous help, encouragement, and personal warmth has always been a source of stimulation for those of us associated with him. It is with a deep sense of appreciation that we dedicate these proceedings to him.

A. Krothapalli
C. A. Smith

Contents

Part IV. Experimental Techniques

Part V. Rotor Dynamics and Aerodynamics

List of Contributors

Agarwal, Ramesh
McDonnel Douglas Corporation
St. Louis, MO 63166

Baganoff, Donald
Department of Aeronautics and Astronautics
Stanford University
Stanford, CA 94305

Bevilaqua, Paul
Rockwell International Corporation
Dept. 71, Bldg. 6
4300 East 5th Ave.
Columbus, OH 43216

Davis, Sanford
NASA Ames Research Center
Mail Stop 227-8
Moffett Field, CA 94035

Dowell, Earl H.
School of Engineering
Duke University
Durham, NC 27706

Fearn, Richard L.
Dept. of Engineering Sciences
University of Florida
Gainesville, FL 32611

Fung, K. Y.
University of Arizona
Tucson, AZ 85721

Ganzer, Uwe
Institut für Luft-und Raumfahrt
Technische Universität Berlin
Marchstr. 14 Sekr. F2 D-1000
Berlin 10, Germany

Ho, Chih-Ming
Dept. of Aerospace Engineering
University of Southern California
Olin Hall 300
Los Angeles, CA 90089–1454

Hussain, A.K.M.F.
Dept. of Mechanical Engineering
University of Houston
Houston, TX 77004

Johnson, Dennis
NASA Ames Research Center
Mail Stop 227-8
Moffett Field, CA 94035

Kotansky, Donald R.
McDonnell Aircraft Company
McDonnell Douglas Corporation
St. Louis, MO 63166

Kuhn, Richard
111 Mistletoe Drive
Newport News, VA 23606

Liu, J.T.C.
Engineering Division
Brown University
Box D
Providence, RI 02912

Marvin, Joseph
NASA Ames Research Center
Mail Stop 229-1
Moffett Field, CA 94035

McCroskey, William J.
NASA Ames Research Center
Mail Stop 202A-1
Moffett Field, CA 94035

Michalke, Alfons
Hermann-Föttinger-Institut für Thermo- und Fluiddynamik
Technische Universität Berlin
Sekr. HF-1
Strasse des 17
Juni 135, D-1000
Berlin 12, West Germany

Miller, Rene H.
Dept. of Aeronautics and Astronautics
Mass. Institute of Technology
Cambridge, MA 02139

Przybytkowski, Stanislaw M.
University of Arizona
Tucson, AZ 85721

Rockwell, Donald
Dept. of Mechanical Engineering and Mechanics
Lehigh University
Bethlehem, PA 18015

Sarpkaya, Turgut
U. S. Navy
Department of Defense
Monterey, CA 93940

Sears, W. R.
Aerospace and Mechanical Engineering Dept.
The University of Arizona
Tucson, AZ 85721

Seebass, A. R.
College of Engineering and Applied Science
AD 1-1 Engineering Center
Campus Box 422
University of Colorado
Boulder, CO 80309

Editors:

Krothapalli, A.
Dept. of Mechanical Engineering
Institute for Engineering
The Florida State University
Tallahassee, FL 32306

Smith, Charles A.
NASA Ames Research Center
Mail Stop 247-1
Moffett Field, CA 94035

Assistant editor:

Hsia, Doris
Joint Institute for Aeronautics and Acoustics
Dept. of Aeronautics and Astronautics
Stanford University
Stanford, CA 94305

Part I
Unsteady Aerodynamics

Advances in the Understanding and Computation of Unsteady Transonic Flow

A. Richard Seebass
University of Colorado
Boulder, CO 80309

K. Y. Fung
and
Stanislaw M. Przybytkowski
University of Arizona
Tucson, AZ 85721

1. Introduction

This paper addresses two topics in unsteady transonic flow. The first is the determination of the aerodynamic response of airfoils to small, unsteady disturbances. The second is the effect of wind tunnel walls on unsteady transonic testing. The results that are presented derive from an investigation of unsteady transonic flow initiated at the University of Arizona in 1977, and that have continued there since then. While these studies were launched under the senior author's direction, the principal contributions were made by the junior authors. Since 1981 the direction of this research has been the responsibility of Dr. K. Y. Fung. These studies are now coming to a close and it is our good fortune to have the opportunity to summarize the contributions they have made to the computation and understanding of unsteady transonic flows.

We are, of course, interested in transonic flight because the efficiency of turbojet powered aircraft is proportional to the Mach number times their lift-to-drag ratio. Until shock waves appear and become reasonably strong, fuel economy increases with increasing Mach numbers. Generally speaking, the most efficient flight is obtained from configurations that are designed to be shock free at remarkably high Mach numbers and lift coefficients, but then operated at still higher Mach numbers and lift coefficients with weak shock waves embedded in the flow. It used to be that speed was more important to the airlines than fuel economy, and even stronger shock waves were tolerated. Thus, when the sun was in a position to create a natural shadowgraph, it was sometimes possible to see a shock wave like that indicated by the shadow on the wing of commercial transports (see Figure 1).

It has been demonstrated mathematically, experimentally, and numerically, that at supercritical Mach numbers, shock-free flows are isolated from one another. Thus, it is a general characteristic of such flows that they have shock waves embedded in them. Small changes in the freestream Mach number or, as illustrated in Figure 2 by the NLR studies of their 7301 airfoil, in the angle of attack can have large effects on the flow field. The most notable effect is on the position and strength of these embedded shock waves. There are two principal characteristics that distinguish unsteady transonic flow from its subcritical counterpart. The first of these is the motion of the embedded shock wave; the second is the delay in the upstream propagation of acoustic signals emanating from positions aft of the supercritical region. Because of the generally large change in the pressure across shock waves, shock wave motion caused by small changes in the flow conditions or geometry can cause large changes in the pressure in the vicinity of the shock foot. Such changes are illustrated by Tijdeman's (1976, 1977) experimental study of an NACA 64006 airfoil with an oscillating quarter chord flap, as shown in Figure 3. The large deviations in the magnitude of the pressure coefficient in the vicinity of the 60% chord point are due to the motion of the shock location; the large departures in the phase lag from those calculated by linear theory are due to the delay in the forward propagation of acoustic signals from the flap caused by the supersonic flow region. Tijdeman (1977) has calculated these delays for this flow and compared them with those for a subcritical Mach number. As can be seen in Figure 4, even a rather small supercritical region can nearly double the time it takes an acoustic signal to reach the leading edge of the airfoil. As a consequence, linear theory, which assumes acoustic propagation through an undisturbed flow field, fails because of the steady disturbance to the uniform flow.

A review of the general features of unsteady transonic flow and a brief discussion of their calculation may be found in Tijdeman and Seebass (1980). Ballhaus and Bridgeman (1980) have reviewed the numerical methods for unsteady transonic flow, and Yang and Batina (1983) have applied these techniques in examining the three-degree-of-freedom response of several airfoils. Thus, no further review than that given above will be attempted here. Rather, we now turn to the two subjects of this paper: the numerical determination of the

Figure 1. Sun illuminated shadowgraph of the shock wave on the wing of a TWA Boeing 707. (From Sobieczky and Seebass, 1984.) "Reproduced, with permission, from the Annual Review of Fluid Mechanics, Vol. 16. © 1984 by Annual Reviews Inc."

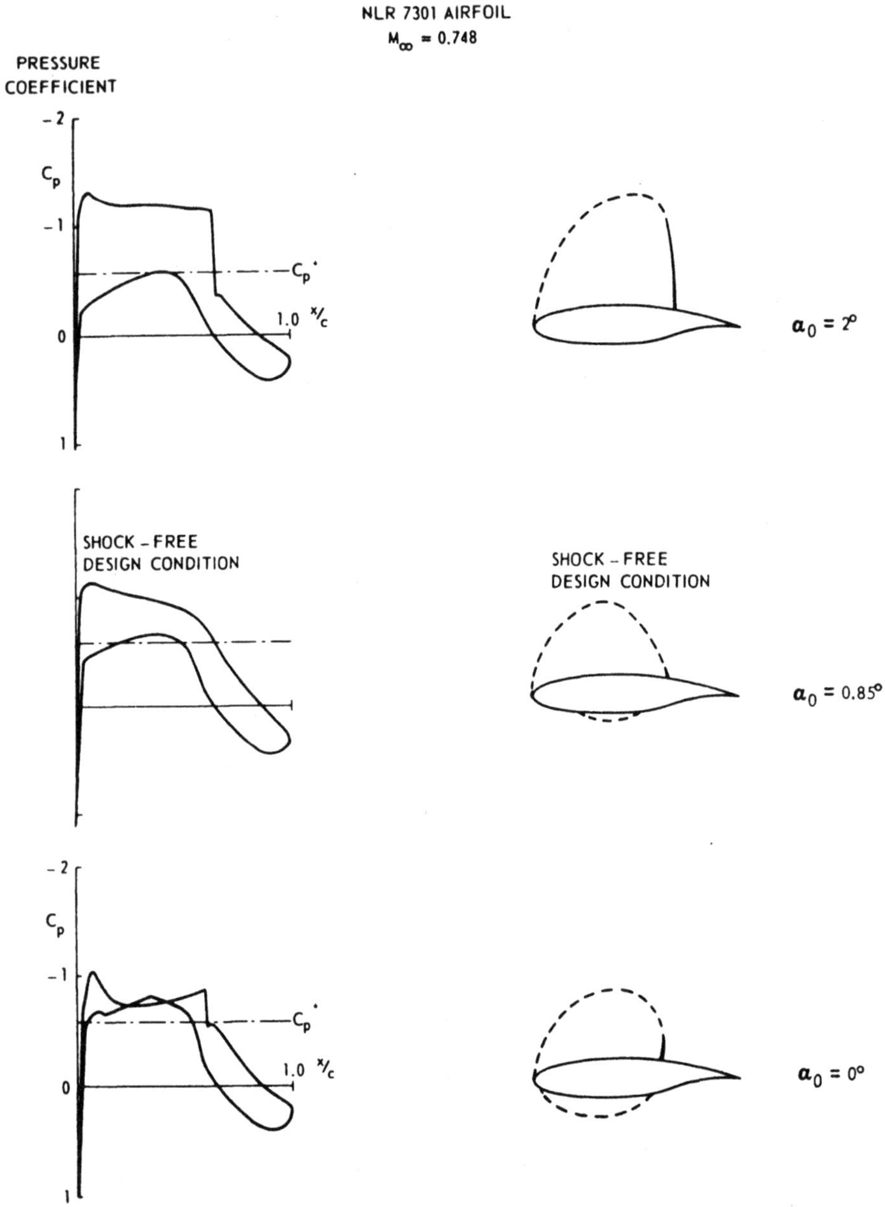

Figure 2. The effect of small changes near the shock-free design point of the NLR 7301 airfoil. (From the studies of Tijdeman (1977) reported in Tijdeman and Seebass, 1980.) "Reproduced, with permission, from the Annual Review of Fluid Mechanics, Vol. 12 © 1980 by Annual Reviews Inc."

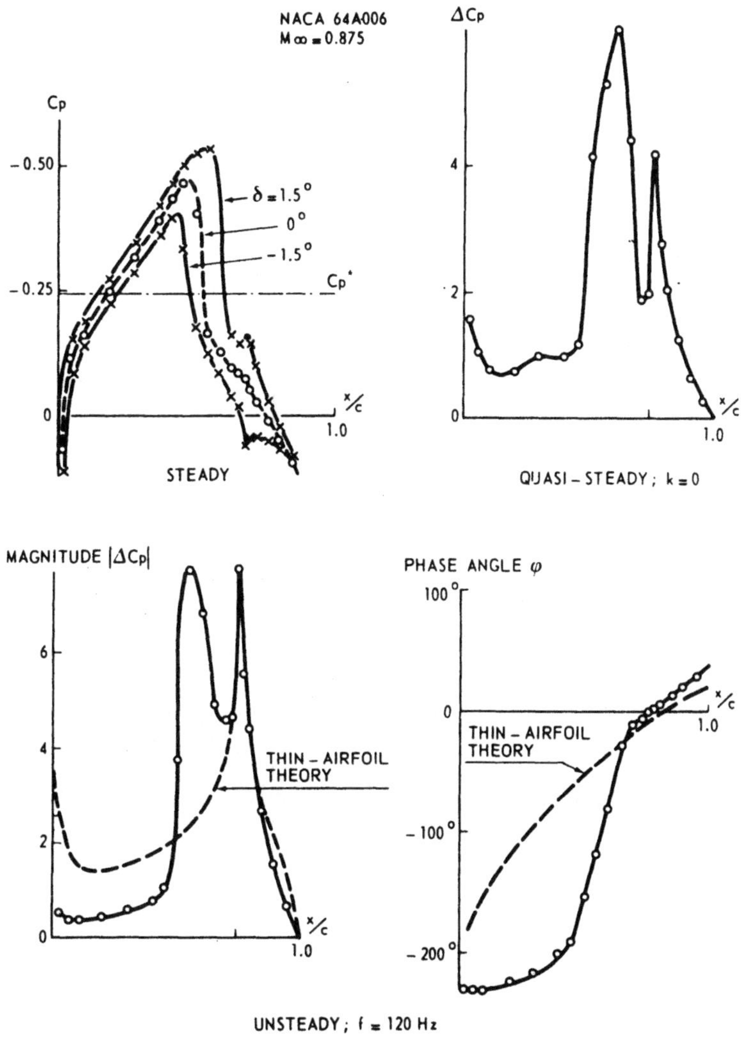

Figure 3. The effect of quarter-chord flap motion on the magnitude and phase lag of the pressure distribution on an NACA 64006 airfoil in the study of Tijdeman (1977). The deflections are 1.5 degrees about 0 and the reduced frequency 0.468. (From Tijdeman and Seebass, 1980.) "Reproduced, with permission, from the Annual Review of Fluid Mechanics, Vol. 12 © 1980 by Annual Reviews Inc."

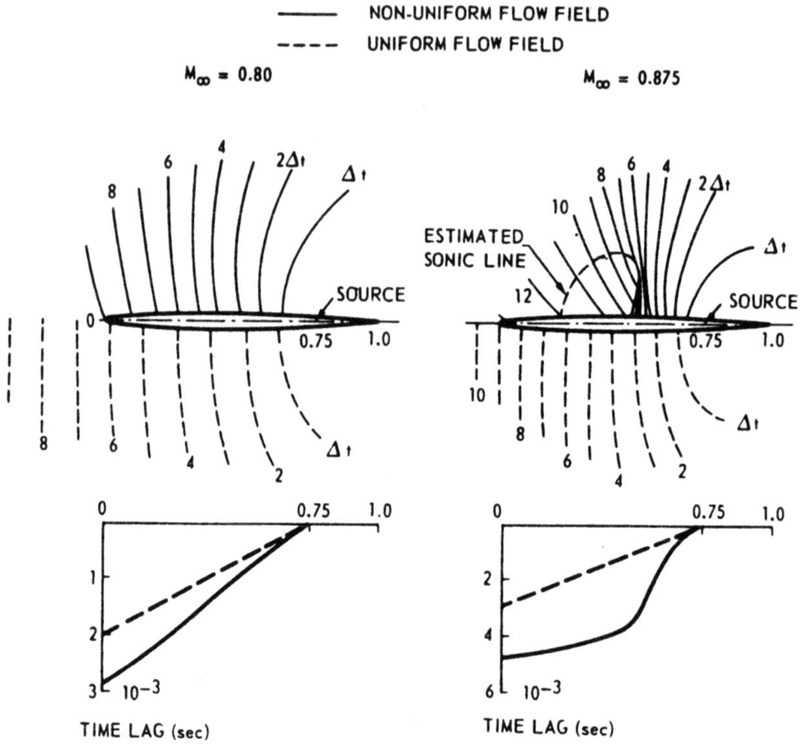

Figure 4. The time delays in the propagation of an acoustic signal caused by an embedded supersonic region as calculated by Tijdeman (1977). (From Tijdeman and Seebass, 1980.) "Reproduced, with permission, from the Annual Review of Fluid Mechanics, Vol. 12 © 1980 by Annual Reviews Inc."

effect of small unsteady motion on unsteady transonic flows; and the effect of wind tunnel walls on such flows.

2. Flutter Boundaries

As noted above, improvements in aircraft efficiency require the design of wings that are nearly shock free and the operation of these wings at high transonic Mach numbers. The flutter boundary of these so-called "*supercritical*" wings may occur at lower Mach numbers than is the case with conventional wings. This is illustrated in Figure 5, from Farmer and Hanson (1976), which shows the dynamic pressure and Mach number at the onset of flutter for two dynamically identical wings. Very small changes in the wing's airfoil sections have made substantial changes in its flutter boundary. The so-called

of the transonic regime. Aircraft wings in flutter extract energy from the passing airstream; this can lead to amplitudes of motion that cause the disintegration of the wing structure. The flutter boundary delineates the conditions where small amplitude disturbances are no longer damped. Obviously the prediction of this flutter boundary depends on an accurate prediction of the aerodynamic response of transonic airfoils and wings.

We have noted the sensitivity of transonic flow fields to rather small changes in the flow conditions. They are, consequently, especially sensitive to viscous effects. An accurate prediction of the steady flow upon which unsteady perturbations are superimposed is, then, essential. The general magnitude of viscous effects can be judged from Figure 6, which depicts the calculated flow past an NACA 64006 airfoil with a one degree flap deflection, with and without a boundary displacement thickness, and NLR experimental results. The effect of the boundary layer is to decrease the overall size of the supersonic zone and to move the shock wave forward on the airfoil. The changes in the steady state due to viscous effects are substantial, and they must be taken into account in any calculation of flutter boundaries. This led us to observe more than five years

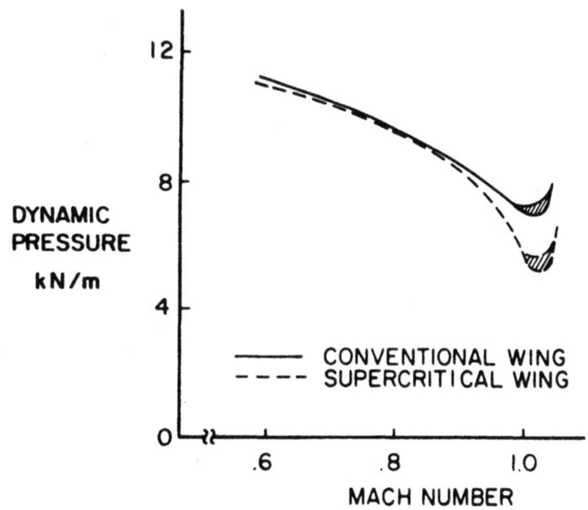

Figure 5. Flutter boundaries for two dynamically equivalent TF-8A wings as determined experimentally by Farmer and Hanson (1976). Courtesy of Springer-Verlag.

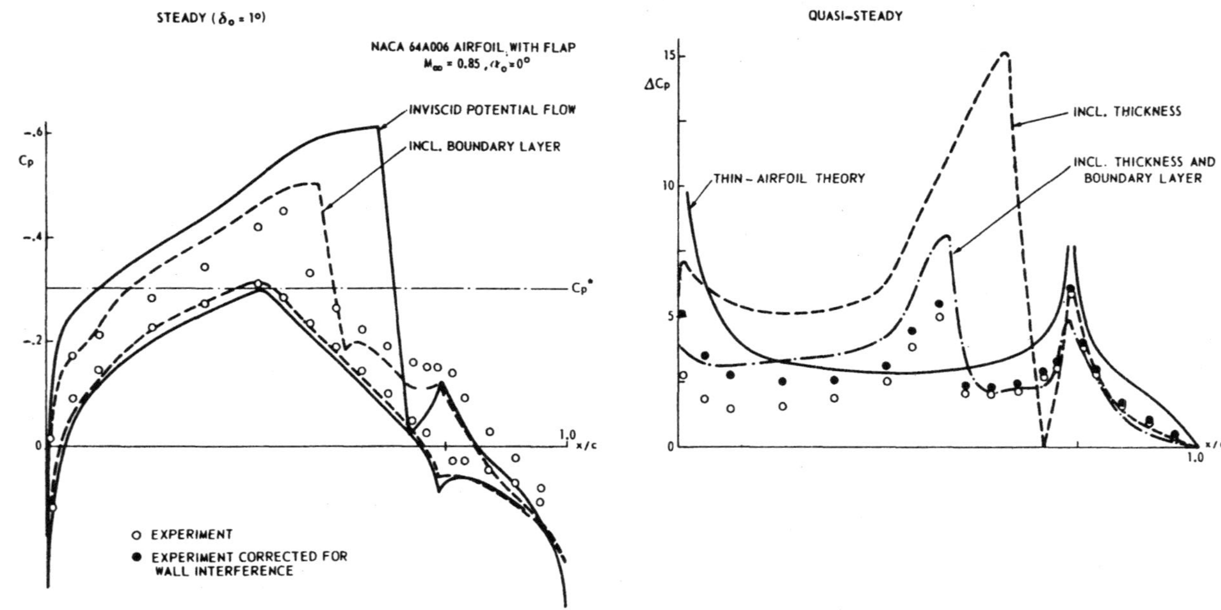

Figure 6. Comparison of inviscid calculations, with and without an added boundary layer displacement thickness, for an NACA 64006 airfoil with a one degree flap deflection, with NLR experimental results. (From Tijdeman and Seebass, 1980.) "Reproduced, with permission, from the Annual Review of Fluid Mechanics, Vol. 12 © 1980 by Annual Reviews Inc."

ago (Fung et al., 1978; Seebass et al., 1978) that an effective algorithm for the computation of the effects of small unsteady perturbations on steady transonic flow would not only have to capture small shock excursions accurately and efficiently, but it would also have to accept either an experimentally or numerically determined steady flow field as input. These two essential ingredients, as well as other needed elements, are now in hand and an accurate and efficient determination of flutter boundaries for unseparated flow past airfoils is now possible. Through strip theory, these results are applicable to high aspect ratio wings of modest sweep. But the method also can be extended to three-dimensional flows.

2.1 Computation Effort

The computational effort required to determine a wing's aerodynamic response to small unsteady motions is proportional to the number of points in an L by M by N mesh and the efficiency, Eff, of the algorithm used. Suppose T time steps are needed to compute a given mode of motion and that this must be done at K reduced frequencies. Then, for each mode of motion, a measure of the computational effort will be

$$\text{Comp. Effort} = K \times T \times L \times M \times N \times \text{Eff}.$$

With L, M, N all the order of 10^2, T typically $5 \cdot 10^2$, and K about 10, then the numerical coefficient of the algorithmic efficiency is $5 \cdot 10^9$. Thus, the algorithms used to determine flutter boundaries in the design process must be efficient in themselves, and they also must resolve unsteady motions without unnecessary time steps. It is obvious that time linearization should be used when it is applicable because a single indicial response will determine the response for the K reduced frequencies with a computational effort proportional to $K \times T$, but with a coefficient that is independent of the product $L \times M \times N$. Indeed, with time linearization, the total computational effort becomes

$$\text{Comp. Effort} = T \times L \times M \times N \times \text{Eff} + \text{const.} \times K \times T$$

where the savings accrue from K being effectively equal to 1. The second term on the right-hand side is negligible when compared to the first term, even in two dimensions.

The accurate and efficient determination of flutter boundaries, as well as exploring the potential for active control to suppress flutter, requires both an algorithm that can resolve small shock motions efficiently, and an algorithm that will create a steady flow field consistent with the prescribed pressures. These two algorithms should be compatible in that the airfoil shape supplied by one algorithm should reproduce the pressures supplied for the steady flow. An algorithm that can capture small shock motions and that avoids numerical instability, even when the shock excursion exceeds one mesh space in a single time step, was first provided by Fung et al. (1978). Since that time the algorithm has been improved, including the incorporation of a far-field boundary condition that avoids the reflection of acoustic waves and time linearization. With time-linearization, Fourier synthesis of the results for an indicial motion provides the magnitude and phase lag predictions for the reduced frequencies of interest. We call this algorithm UTFC (unsteady transonic flow code). An algorithm that determines the airfoil ordinates that will, in the transonic small perturbation approximation, return prescribed pressures, was developed by Fung and Chung (1983). This algorithm is based on a procedure proposed by Tranen (1974) nearly a decade ago. The essential observation here is that the pressure given problem is analogous to an unsteady problem in which the constant that must be added to the perturbation potential is determined by a fictitious unsteady process that relates a pseudo *unsteady pressure* to the failure of the current airfoil coordinates to be consistent with a closed body. As this algorithm is based on the alternate factorization scheme of Ballhaus et al. (1978), AF2, we call this inverse algorithm IAF2. The coupling of these two algorithms yields an effective tool for the aerodynamic input to the determination of flutter boundaries. It is instructive here to review the essential nonnumerical approximations of both algorithms. An accounting of the numerical method in sufficient detail for the reader to reproduce these algorithms from first principles has been provided in a recent review by Fung (1984).

2.2 Governing Equations

The numerical algorithms described here implement the transonic small perturbation approximation. However, the basic input to these algorithms is the pressure field, and this may be determined either experimentally or by using a more sophisticated numerical algorithm. The basic approximations, then, are that the unsteady disturbances are small, that changes in the boundary layer due to the unsteady perturbations are negligible, and that the steady nonlinear flow field is adequately approximated by the steady solution to the small perturbation equations that reproduces the prescribed pressures. The principal limitations are those of unseparated and two-dimensional flow. The latter restriction, at least in principle, can be removed by additional coding effort. The restriction to unseparated flows is a more serious one and one that would be difficult to remove.

In the approximation of small unsteady disturbances to a basic nonlinear steady state, the nondimensional equation governing the perturbation potential φ for disturbances of size δ_0, the ratio of the airfoil's thickness to its chord, in a free-stream of Mach number M_∞ is

$$\frac{k^2 M_\infty^2}{\delta_0}\varphi_{tt} + \frac{2k}{\delta_0}M_\infty^2\varphi_{xt} - \left\{\frac{1-M_\infty^2}{\delta_0} - (\gamma+1)\,M_\infty^2\left[\varphi_x + \frac{\gamma-1}{\gamma+1}k\varphi_t\right]\right\}\varphi_{xx} - \varphi_{yy} = 0. \tag{2.1}$$

Here φ is related to the full potential by $\Phi = UC(x+\delta_0\varphi)$, the other quantities have their usual meanings, and k is the nondimensional reduced frequency obtained by multiplying the circular frequency by the freestream flow speed and dividing by the airfoil's chord. For $k = O(1)$, the nonlinear term in the equation is small compared to the unsteady terms and it may be dropped, giving the linear approximation that applies for high reduced frequency. For $k \to 0$, the unsteady terms may be dropped and we have, by definition, the quasi-steady approximation. The singular limit is k/δ_0 constant as $\delta_0 \to 0$, as pointed out by Lin et al. (1948) many years ago. In this limit the $k^2\varphi_{tt}$ and $k\varphi_t\varphi_{xx}$ terms are small and may be dropped. This corresponds to the assumption that the downstream propagation of an acoustic disturbance is instantaneous.

We are principally interested in small unsteady disturbances to the flow field and may linearize about the basic nonlinear steady state, provided we do so in a way that allows the large changes in pressure due to shock motion to be taken into proper account. Thus, we write the airfoil shape as composed of steady and nonsteady parts:

$$y = \delta Y(x) + \tilde{\delta}\tilde{Y}(x,t).$$

Here we make the assumption that $\tilde{\delta} = o(\delta)$; later we shall be more precise. Thus, we introduce the unsteady disturbance φ to the basic steady disturbance φ^o:

$$\varphi(x,y,t) = \varphi^o(x,y) + (\frac{\tilde{\delta}}{\delta})\tilde{\varphi}(x,y,t) + o(\frac{\tilde{\delta}}{\delta}).$$

With this assumption, the governing equation, the tangency condition on the airfoil, and the continuity of pressure across the wake, have the following approximations, using a priori the fact that the boundary condition (2.4) requires $\delta_0 = \delta^{2/3}$, are:

$$-\frac{2kM_\infty^2}{\delta_0}\tilde{\varphi}_{xt} + \left\{\left[\frac{1-M_\infty^2}{\delta_0} - (\gamma+1)M_\infty^2\varphi_x^0\right]\tilde{\varphi}_x\right\}_x + \tilde{\varphi}_{yy} = 0, \quad (2.2)$$

$$\tilde{\varphi}_y(x,0,t) = \tilde{Y}_x(x,t) + k\tilde{Y}_t(x,t), \qquad 0 \le x \le 1, \qquad (2.3)$$

$$[\![\tilde{\varphi}_x + k\tilde{\varphi}_t]\!] = 0, \quad y = 0, \quad x > 1, \qquad (2.4)$$

where $[\![(\cdots)]\!]$ indicates the jump in $(\cdots)$ across the wake. For the far-field, we may either use the unsteady far-field for a lifting singularity (Fung, 1981), or the non-reflecting boundary condition of Engquist and Majda (1981). The latter condition may be approximated by

$$\tilde{\varphi}_y \pm \left|\frac{1-M_\infty^2}{\delta^{2/3}}\right|^{\frac{1}{2}} \tilde{\varphi}_x = 0 \qquad (2.5)$$

at the upper and lower walls respectively. Numerical experiments in two dimensions have indicated that the linear far-field boundary conditions may be applied somewhat closer to the airfoil or wing than the nonreflecting boundary condition. On the other hand, the linear far-field requires the instantaneous downstream propagation of changes in the circulation. As we have noted above, for moderate values of k the retention of terms of $O(k)$ seems to provide better agreement with the linear theory that applies for $k = O(1)$, and which also retains the terms in equation 2.1. This observation and its ease of implementation may favor the simpler boundary condition (2.5) for two-dimensions. In three dimensions the analog of (2.5) is difficult to implement and, thus, the linear far-field is preferred.

We noted earlier that the singular limit occurs for $k/\delta^{2/3} = O(1)$ and this allows one to drop the k^2 term in equation 2.1. To be consistent with this approximation, we should then drop it everywhere. However, if we drop it in equation 2.2, then we have assumed changes in the circulation that propagate instantaneously downstream. But one of the principal effects of changes in the circulation is the angle of attack induced by the shed vorticity. Thus, retaining the term of $O(k)$ in equation 2.2 might be expected to improve the results for the larger values of k of interest, as it provides downstream propagation at a nondimensional speed $1/k$ corresponding to the circulation being carried downstream at the freestream speed. This was first pointed out and verified by Houwink and van der Vooren (1979), who studied the effects of retaining terms of $O(k)$ in equations 2.3 and 2.4. Thus we retain all such terms here except the $k\varphi_t\varphi_{xx}$ and $k^2\varphi_{tt}$ terms in equation 2.2.

3. Shock Relations

A crucial element in the calculation of unsteady transonic flows is the ability to correctly and efficiently capture shock wave motions. Because the input to this algorithm will be an airfoil shape that reproduces prescribed pressures, the mesh that prescribes this airfoil need be no finer than that necessary to reproduce the given pressure distribution. In addition, with the pressure prescribed, the usual problems associated with the leading edge singularity disappear. Thus, a large number of airfoil points are not necessary and, for small amplitude disturbances, the shock motion during a time period of interest may be less than the mesh spacing on the airfoil. We need to capture this motion during the numerical computations. As we noted earlier (Fung et al., 1978), the shock relations take the form of the conservation of mass across the discontinuity

$$\begin{aligned}\frac{-2kM_\infty^2}{\delta_0}[\![\varphi_x]\!]^2\left(\frac{dx}{dt}\right)_s - \left\{\frac{1-M_\infty^2}{\delta_0} - (\gamma+1)\,M_\infty^2\overline{\varphi}_x\right\}[\![\varphi_x]\!]^2 \\ + [\![\varphi_y]\!]^2 = 0.\end{aligned} \tag{3.1}$$

Here $[\![(\cdots)]\!]$ again indicates the jump in $(\cdots)$ across the discontinuity and $(\overline{\cdots})$ its mean value. In addition we have the retention of irrotationality

$$\left(\frac{dy}{dx}\right)_s = -\frac{[\![\varphi_x]\!]}{[\![\varphi_y]\!]}. \tag{3.2}$$

The first of these equations shows that the shock position, $x_s(y,t)$ can be integrated forward in time knowing the jump in φ_x and φ_y. The second equation describes the instantaneous shape of the shock wave. It is a general characteristic of transonic flows past slender bodies that the shock wave is nearly normal to the flow and that $[\![\varphi_y]\!] \approx 0$. Indeed, away from the shock foot where the shock curvature is singular, one can show that the shock curvature must be the same order as the body curvature and consequently of $O(\delta)$. Thus, consistent with the small perturbation approximation, we can treat the shock as normal to the freestream flow and replace equations 3.1 and 3.2 by the simple relation,

$$\epsilon\frac{d\chi(t)}{dt} = \frac{\gamma+1}{2k}\tilde{\delta}\overline{\tilde{\varphi}}_x\,(x_s^0, 0, t) \tag{3.3}$$

for the shock excursion χ defined by

$$x_s(t) = x_s^0 + \epsilon\chi(t). \tag{3.4}$$

We note that for the reduced frequencies of principal interest, namely $k = 0(\delta^{2/3})$, we must require that

$$\epsilon = O(\frac{\tilde{\delta}}{k}) = O\left(\frac{\tilde{\delta}}{\delta^{2/3}}\right) = O(1). \tag{3.5}$$

This defines the amplitudes of the unsteady motions for which the shock excursions should remain small. The result (3.3) was first given by Landahl (1961) in his monograph.

By expanding $\varphi(x,y,t)$ in a Taylor series about x_s^0, and by requiring that $[\![\varphi]\!] = 0$ on the shock wave, but recognizing that $\tilde{\varphi}$ will change discontinuously at the shock, one may easily show that the change in potential at the old location due to shock motion is given by

$$(\delta/\tilde{\delta})[\![\tilde{\varphi}]\!] = -[\![\varphi_x^0]\!](x_s - x_s^0) \tag{3.6}$$

(Fung et al., 1978). Thus, in the region of shock motion we calculate x_s from equations 3.3 and 3.6. The values of $\tilde{\varphi}$ are determined in this region by analytic continuation from the values outside the region to the steady state shock position. That is, we extrapolate $\tilde{\varphi}$ linearly in the region of shock motion and this determines the shock position. The y location to be used in equations 3.5 and 3.6 remains to be specified; the most important shock motions occur at the shock foot and thus we choose $y = 0$ for this evaluation. The assumption of a normal shock has, of course, eliminated any singular behavior at the shock foot.

The procedures invoked here have been tested by comparison with other numerical computations. Yu et al. (1978) carried out nearly identical calculations for equation 2.1 using shock fitting in two space dimensions and time. These calculations, as well as numerous comparisons with the well-developed algorithm, LTRAN2-HI, introduced by Ballhaus and Steger (1975) have demonstrated the reliability of this approximation.

4. Pressure Given Steady State

The inverse problem of finding the airfoil that corresponds to a given pressure distribution is ill-posed and, consequently, requires iteration. The main task here is to find an algorithm that will produce the input pressure distribution when computed by the procedures outline above. Thus, there is some virtue to using comparable numerical techniques here. To this end, Fung and Chung (1983) modified the approximate factorization scheme, AF2, of Ballhaus et al. (1978) to be an inverse algorithm, IAF2. Because the pressure is given on the airfoil, we know φ there within an arbitrary constant. This constant is determined iteratively through the requirement that there be no mass flux through the airfoil. Thus, on the airfoil

$$\varphi^{n+1} = \varphi^n + C'.$$

Now we may evaluate C' from

$$\frac{2C'}{\Delta Y} = \alpha \left\{ \int_0^1 [\varphi_y(x, 0^+) - \varphi_y(x, 0^-)]\, dx - \beta \right\},$$

where α is a relaxation parameter and β is used to control the airfoil closure and the distribution of its slopes. Fung and Chung point out that the mass fluxes are effectively balanced by a pseudo-unsteady pressure on the airfoil given by

$$\frac{(\varphi^{n+1} - \varphi^{n})}{\Delta y}.$$

Figure 7 shows the input pressure distribution for the 64A010 airfoil studies by Davis and Malcolm (1980). A shock is introduced that satisfies the normal shock jump relations between the downstream extrapolation of the upstream pressure and the upstream extrapolation of the downstream pressure. The upstream pressure (or Mach number) sets the unique position of the shock wave. This steady state pressure then determines the nonlinear steady state and the airfoil shape shown consistent with this steady state.

5. Results

There have been a number of careful and fairly comprehensive experimental studies of two-dimensional unsteady transonic flow including, among others, the pioneering studies of Tijdeman (1977), the pitching airfoil studies of Davis and Malcolm(1980) and Triebstein et al. (1982). In the latter two cases, the chord to wind tunnel height was 1/6.7 and 1/8, respectively, reducing to some extent the wind tunnel wall interference. As numerical values of the steady pressure were available at 19 points from the Davis and Malcolm study, these experiments were used in Fung's and Chung's calculations. Davis and Malcolm studied the 64A010 oscillating in pitch with 0.5 degree half amplitude at $M_\infty = 0.8$. Their steady state pressure distribution (Figure 7) was used as input to IAF2; the steady flow field consistent with this pressure distribution in the small perturbation approximation was then used in the unsteady calculations made with UTFC. A 100 $\times$ 80 grid with 23 points on the airfoil was used in these studies.

The results obtained using the UTFC code for the variation of the amplitudes and the phase lags for the lift and moment with reduced frequency are shown in Figure 8. Let us first compare the numerical results. These were both computed using the airfoil output from IAF2. The algorithms differ in the following details: UTFC is time-linearized, assumes a normal shock, implements shock fitting, and uses the far-field for a lifting element. LTRAN2-HI is nonlinear, does not implement shock fitting, and uses a nonreflecting

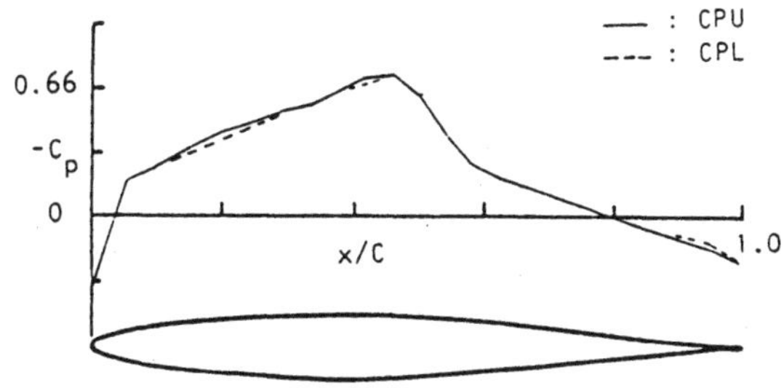

Figure 7. Upper and lower surface pressure distributions from the experiments of Davis and Malcolm (1980) and used in the IAF2 algorithm to generate the airfoil for the unsteady computations using UTFC.

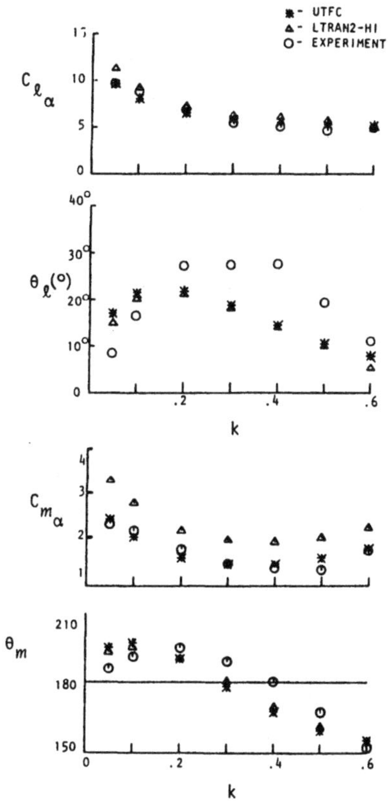

Figure 8. The magnitude and phase lag of the unsteady lift and moment, as a function of reduced frequency, on an NACA 64A010 airfoil oscillating in pitch with 0.5 degree half amplitude at $M_\infty = 0.8$. The experimental results are from Davis and Malcolm (1980); the LTRAN2-HI results were computed using the algorithm described in Hessenius and Goorjian (1982).

boundary condition. The agreement between these results for both the amplitude and phase is remarkably good, although LTRAN2-HI provides somewhat different shock motions and this probably accounts for the discrepancy in the moment amplitudes. If we now compare the UTFC result with the experiments, assuming that the shock fitting procedure provides better results, then both the moment and the lift amplitudes are reproduced by the numerical calculations, which omit any unsteady boundary layer effects. However, the agreement for the phase lag is not nearly as good, indicating either an inadequacy in the procedure detailed here or errors in the experimental results. The only likely source of such errors are unsteady boundary layer effects or the contamination of the results by the wind tunnel walls. As we will show in the next section, this contamination is not only possible, but also quite likely despite the 6.7:1 wind tunnel height to chord ratio. While these discrepancies are small, we have noted earlier that such small deviations can lead to remarkably different flutter boundaries.

If we accept the hypothesis that the disagreement in phase lag between the numerical results and the UTFC calculations stems from the effects of wind tunnel walls, then, with little computational effort it is possible to predict the unsteady aerodynamic behavior of airfoils for conditions where the flow remains attached, provided reliable steady state pressure distributions are available to determine the nonlinear steady state. It would be useful to have similar results for the studies of Triebstein et al. (1982). At just slightly supercritical conditions and a single, high reduced frequency, they report good agreement for the phase lag between their experiments in the ONERA tunnel and small perturbation calculations coupled with an unsteady boundary layer algorithm at an 8:1 wind tunnel height to chord ratio. Their tests were of a finite span model with end plates, and this could have reduced the wind tunnel interference, as would the relatively low Mach number for which this comparison was made.

6. Wind Tunnel Wall Effects

We have been alerted to the effects of wind tunnel walls by the discrepancy between the numerical analysis reported above and reliable experimental results. But we have been aware of these effects for some time and concern for them led to an investigation of far-field boundary conditions that could be applied reasonably close to an airfoil or wing (Fung, 1981). It also launched an investigation by Przybytkowski (1983) into these effects. Here we report on these findings.

The effect of wind tunnel walls in unsteady transonic flows is really a study in acoustics. We are generally dealing with unsteady disturbances that, while propagated through a nonlinear flow field, are very small compared to the ambient pressure, and essentially acoustic. Their interaction with a solid wall is easy enough to understand. But the flow field distortions caused by porosity or by slotted wind tunnel walls cause them to propagate through a flow field that is not easily calculated. Bliss (1983) has shown that for a wide range of conditions, there is at first a linear, and subsequently a quadratic increase, in the flow through a slot in a wind tunnel wall with an increase in pressure drop across the slot. While there have been extensive studies of the modeling of these processes, any boundary conditions we might apply at the wind tunnel walls would remain ad hoc. Thus, we limited our studies to simple boundary conditions. It is intuitively clear that wind tunnel wall effects should be less severe in three dimensions than in two dimensions. Consequently, we felt it essential to study the implementation of various boundary conditions in both two and three dimensions.

6.1 Wall Boundary Conditions

The development of adaptive wall, or smart, wind tunnels pioneered by Sears (1977), Ferri (1973), and Goodyer (1975) will surely make it possible to undertake steady wind tunnel studies with considerably smaller wind tunnel height to chord ratios, and hence to more easily achieve full scale Reynolds numbers. The extent to which this technique will also be useful in unsteady flows will depend on the adaptive mechanism employed in achieving interference free steady flows. To some extent the results presented here provide guidance about the wall mechanisms that will minimize the contamination of unsteady experiments in wind tunnels that have adjusted themselves, or been adjusted, to minimize wall interference in the steady flow. In all cases considered here, the steady flow was that calculated for essentially unbounded conditions. Thus, any contamination of the results given here by wind tunnel walls are due to unsteady reflections.

As we noted above, we limited this study to simple model boundary conditions. These included two far-field conditions, one of which models ideal porous wall conditions, and a solid wall. We can, of course simulate unbounded flows by having a computational region that is large enough that during the time of computation, no acoustic waves have time to reach the boundary and be reflected to the airfoil or wing. The two-dimensional far-field condition that minimizes the linear acoustic energy reflection is that of Engquist and Majda (1981) introduced earlier, viz., equation 2.5; for the full potential.

$$\varphi_n + \sqrt{\kappa^*}\varphi_x = 0, \tag{6.1}$$

where φ_n is the normal outward derivative at the boundary and where

$$k^* = \left| \frac{1 - M_\infty^2}{\delta^{2/3}} - (\gamma + 1)\, M_\infty^2 \varphi_x \right|$$

is the local instantaneous value evaluated at the boundary.

The boundary condition which relates the flow through the wall linearly to the pressure difference across a slot or porous wall is

$$\varphi_n + P\varphi_x = 0, \tag{6.2}$$

which is identical to the linear approximation to equation 6.1 if $P^2 = k^*$. An alternative far-field boundary condition is to supply the values of φ corresponding to a lifting element. In three dimensions this takes the form (Fung, 1983):

$$\varphi = \frac{y}{4\pi} \int_{\frac{-AR}{2}}^{\frac{AR}{2}} \int_{t_0}^{t} \Delta\Gamma\,(\hat{y}, t_0)\, g\,(x, y - \hat{y}, z, t - t_0)\, dt\, d\hat{y}, \tag{6.3}$$

where Γ is the circulation around the wing, AR is the wing's aspect ratio, and

$$g\,(x, y, z, t) = \frac{[1 + x(x^2 + y^2 + z^2)^{\frac{1}{2}}] \cdot \tilde{H}[t + x - (x^2 + y^2 + z^2)^{\frac{1}{2}}]}{y^2 + z^2}$$

Here $\tilde{H}$ represents the Heaviside step function. This linear solution for an elemental vortex sheet assumes the low frequency approximation; that is, a signal propagates instantaneously downstream and, hence, the signal affects the downstream boundary instantaneously after a change in circulation has occurred. We can alter this condition by changing the time constant t_0. This allows us to adjust the phase of the far-field conditions and, consequently, to explore this effect on the results of our numerical studies. Thus we implement (6.3) with $t_0 = nHk/C$ where, for $n = 1$, t_0 is the nondimensional time it takes an acoustic wave to travel a distance H, the tunnel height (and in our study also the width).

7. Results

7.1 Two-Dimensional Results

We begin by noting that in the linear approximation one can calculate the frequencies where reflections from solid wind tunnel walls are in resonance with the airfoil motion. This was done some time ago by Woolston and Runyan (1955) who showed that resonance occurs when

$$k_r = \frac{\pi C}{HM_\infty}\sqrt{1 - M_\infty^2}\,(2n-1), \qquad n = 1, 2, 3, \ldots \tag{7.1}$$

where H is the tunnel height.

For his two-dimensional study Przybytkowski also used the NACA 64A010 studied experimentally by Davis and Malcolm (1980) oscillating in pitch at $M_\infty = 0.8$. The input chosen corresponds to the experimental pressure distribution found for this case and discussed in the previous section. Assuming that this procedure will correctly predict the airfoil's response, then any deviations between the predicted response and that measured are likely to be due to wind tunnel wall effects. In the experiments, the wind tunnel wall was slotted and the tunnel height was 6.7 chords ($H/C = 6.7$).

We begin this study with a linear test case. It is a simple matter to set $\varphi_x^0 = 0$ in our calculations, and thereby calculate a linear result. This Przybytkowski did for $H/C = 14.4$ and $M_\infty = 0.8$. For these conditions equation 7.1 predicts resonance at $k = 0.16$, 0.49, 0.81, etc.; the calculated results for the amplitude, C_l, and the phase, ϑ_l, for unbounded flow and for a solid wall wind tunnel with $H/C = 14.4$ are shown in Figure 9. These results provide the airfoil's aerodynamic response through

$$C_l = C_{l_\alpha} \sin(\omega t - \vartheta_l).$$

We can immediately see the effect of solid wind tunnel walls of half height of 7.2 chords. There are two resonant conditions in the range $0 < k < 0.6$, occurring at roughly $k = 0.12$ and $k = 0.35$. The calculated results preserve the linear theory ratio of 3 between the first two resonant frequencies. The discrepancies between these results and the theory of Woolston and Runyan are due to the neglect of the $k^2\varphi_{tt}$ term in the numerical calculations.

By retaining the φ_x^0 term that we omitted in our fully linear study, we may isolate the effect of the interaction of the acoustic waves on the nonlinear steady state. Since this is also the case studied experimentally by Davis and Malcolm (1980), we can also compare our numerical results with their experimental results. In their experiments the wind tunnel height to chord ratio, H/C, was 6.7. Figure 10 depicts the results for unbounded flow and the solid wall boundary condition with $H/C = 14.4$, repeating the conditions of the fully linear calculation. In comparing Figures 9 and 10 we can see that the nonlinear interaction for the solid wall has increased the amplitude of the response and has left intact the two resonances. We also note that the experimental results, while not markedly different from either of the calculated results, give some evidence of wind tunnel wall effects in the results for the amplitude. In comparing the results for the phase we see that for $k > 0.2$, the experimental results are substantially different from either calculation, and this we attribute to wind tunnel wall effects. We next examined this same case with nonreflecting boundary conditions imposed at the tunnel walls; these results are shown in Figure 11. Note that for the amplitude of the response, the unbounded and nonreflecting boundary conditions give nearly identical results. The wind tunnel interference that we observe here is much more evident in the results for the phase. In examining the phase, we first note that for $k < 0.2$ the two numerical results already give some evidence of the effects of nonlinearity in the boundary conditions, and for $k > 0.2$ it seems likely that the differences between the experimental results and the two calculations are due to wall effects. To explore this possibility further, we implemented the linear far-field conditions with three different delay times t_0. This corresponds to producing a phase difference between the correct linear result at the wall and those

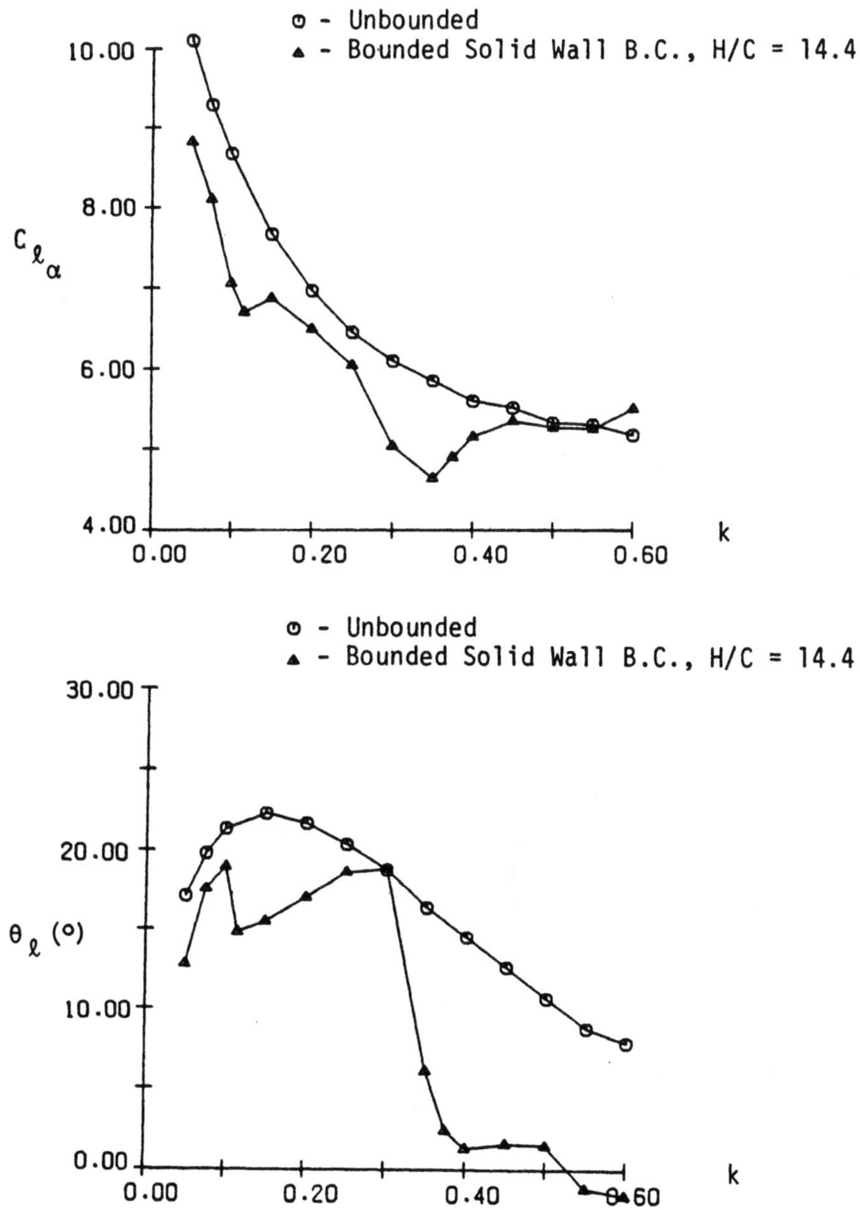

Figure 9. The numerically calculated results, using fully linear theory with small reduced frequency, for the magnitude and phase lag for an NACA 64A010 airfoil oscillating in pitch at $M_\infty = 0.8$ in an unbounded flow, and with solid walls at $H/C = 14.4$.

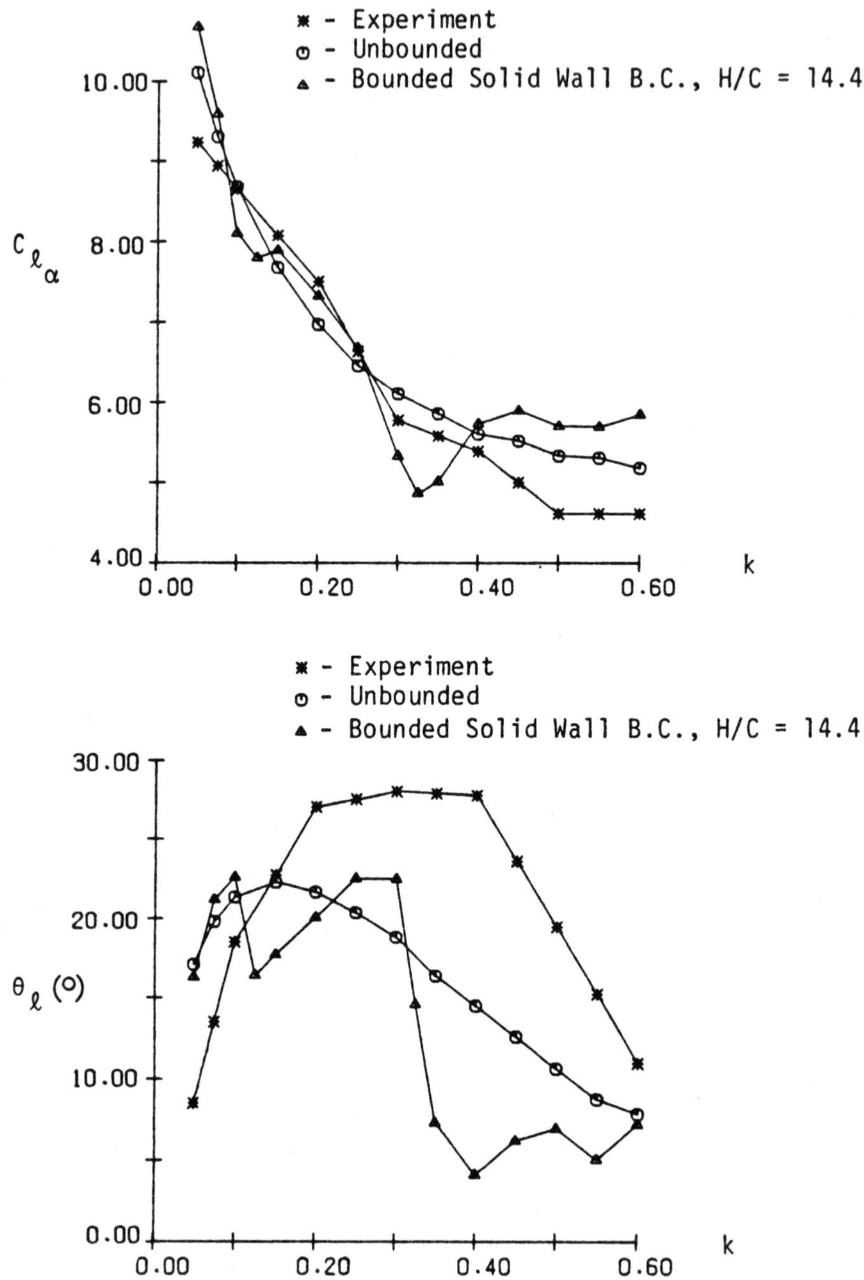

Figure 10. A comparison of the unbounded and bounded time-linearized calculations with the experimental results of Davis and Malcolm (1980) for an NACA 64A010 oscillating in pitch. Again, solid walls are at $H/C = 14.4$.

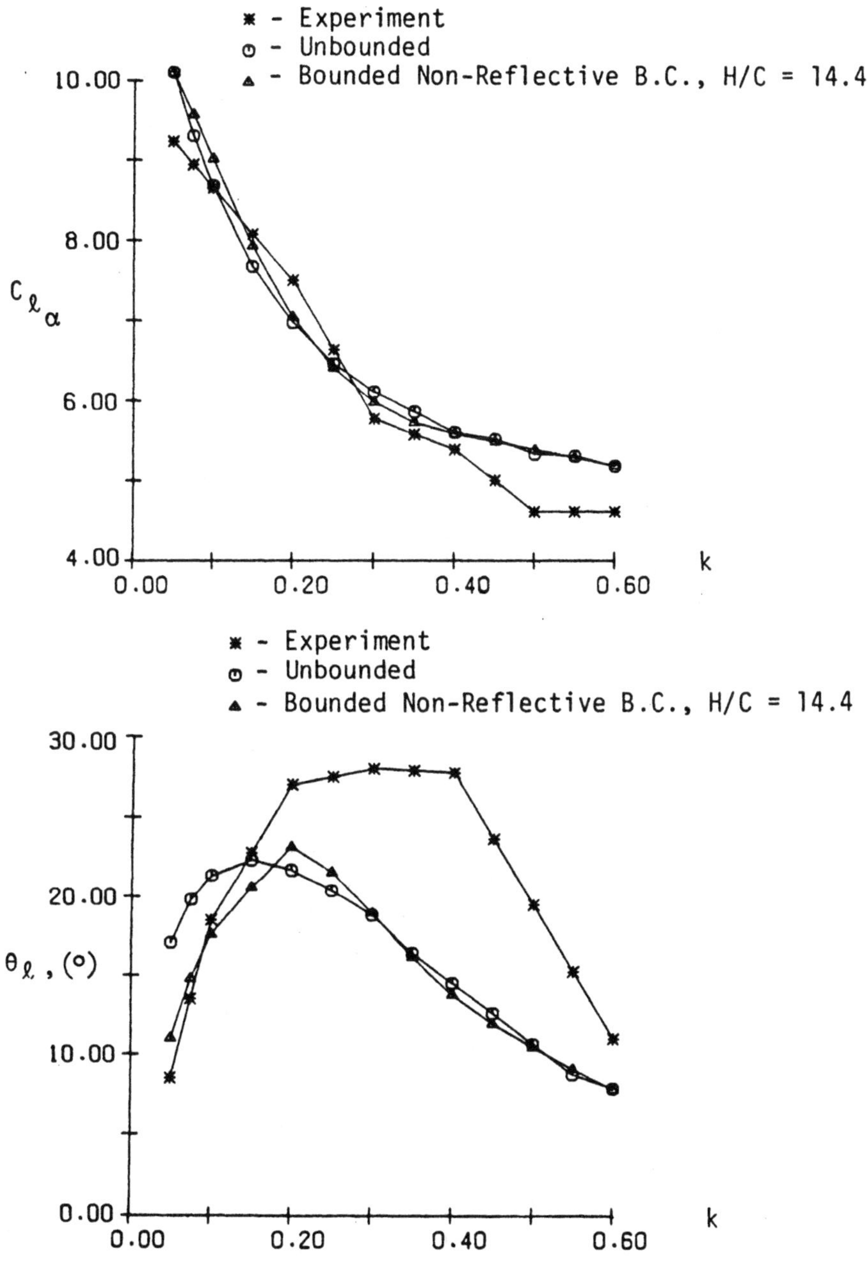

Figure 11. A comparison of the unbounded and bounded time-linearized calculations with the experimental results of Davis and Malcolm (1980) for an NACA 64A010 oscillating in pitch. Nonreflecting walls are at $H/C = 14.4$.

implemented there. The results are shown in Figure 12, where we see that the phases predicted at the higher reduced frequencies are substantially different from those of the unbounded case, and not unlike those of the experiment. The results for $n = 0$ and 1 agree, as they should for a harmonic motion. Thus, we draw the conclusion that wall reflections are especially significant in unsteady transonic flows and a cause for concern, at least in two dimensions.

7.2 Three-Dimensional Results

In order to examine the effects of wind tunnel walls, Przybytkowski (1983) developed an ADI algorithm for unsteady transonic flow inside a rectangular or circular wind tunnel. For the studies reported here, he considered the wind tunnel to be square and of dimension H, and the wing to be rectangular, unswept and of aspect ratio, AR, of 3.8. Thus the wing's span was 3.8 times its chord and the ratio H/AR is simply the wind tunnel height to span ratio. He verified the accuracy of his calculations by comparing his results with those of Bailey and Steger (1973), demonstrating excellent agreement for this simple geometry. No other comparisons were attempted as no cross derivative terms were included. The purpose of his study was to examine the effects of wind tunnel walls on unsteady transonic flows, not to calculate such flows accurately. Before presenting his results regarding wind tunnel walls, it is probably worth examining the effects of dimensionality on the response of unbounded flows. Figure 13 depicts the amplitude and phase for the NACA 64A010 airfoil, and a rectangular wing with this airfoil section and $AR = 3.8$, oscillating in pitch at $M_\infty = 0.8$. We note the decrease in amplitude and the increase in phase lag for the wing with increasing reduced frequency. This is to be expected as the wing tip effects dilute the amplitude response more at higher reduced frequencies, just as they increase the phase lag by allowing an even longer propagation path.

Figures 14 through 16 depict the amplitudes and phase lags of the rectangular wing in the square wind tunnel of varying height H depicted in Figure 14, for solid wall boundary conditions, and for unbounded flow. As the tunnel height to wing span ratio increases from 2.37 to 5.58, we note some diminution of the wall effects, but this is almost offset by the resonance that first appears for $H/AR = 3.37C$ at $k = 0.275$, and that is also evident for $H/AR = 5.58$ at $k = 0.18$. This is consistent with the theoretical result of Woolston

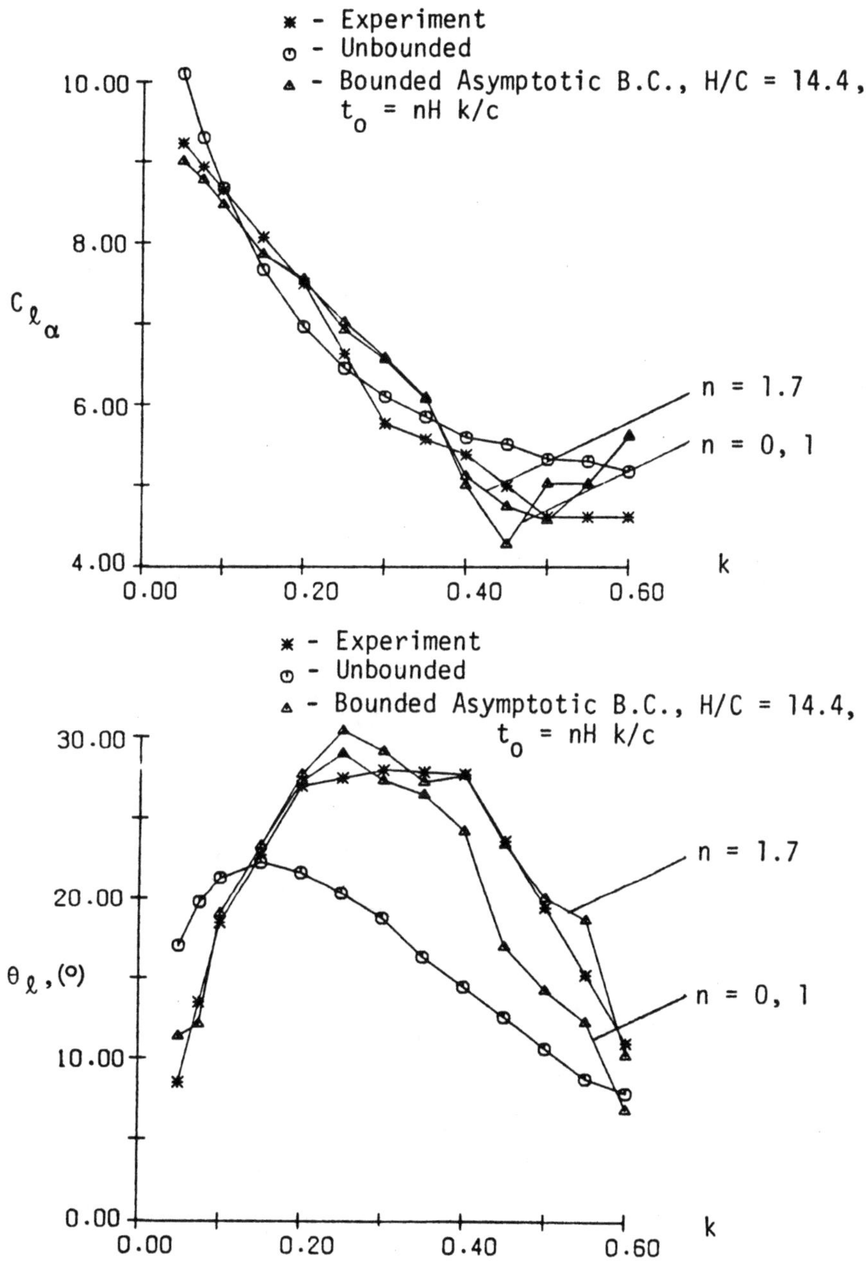

Figure 12. A comparison of the unbounded and bounded time-linearized calculations with the experimental results of Davis and Malcolm (1980) for an NACA 64A010 oscillating in pitch. Asymptotic boundary conditions of different phases are prescribed at $H/C = 14.4$.

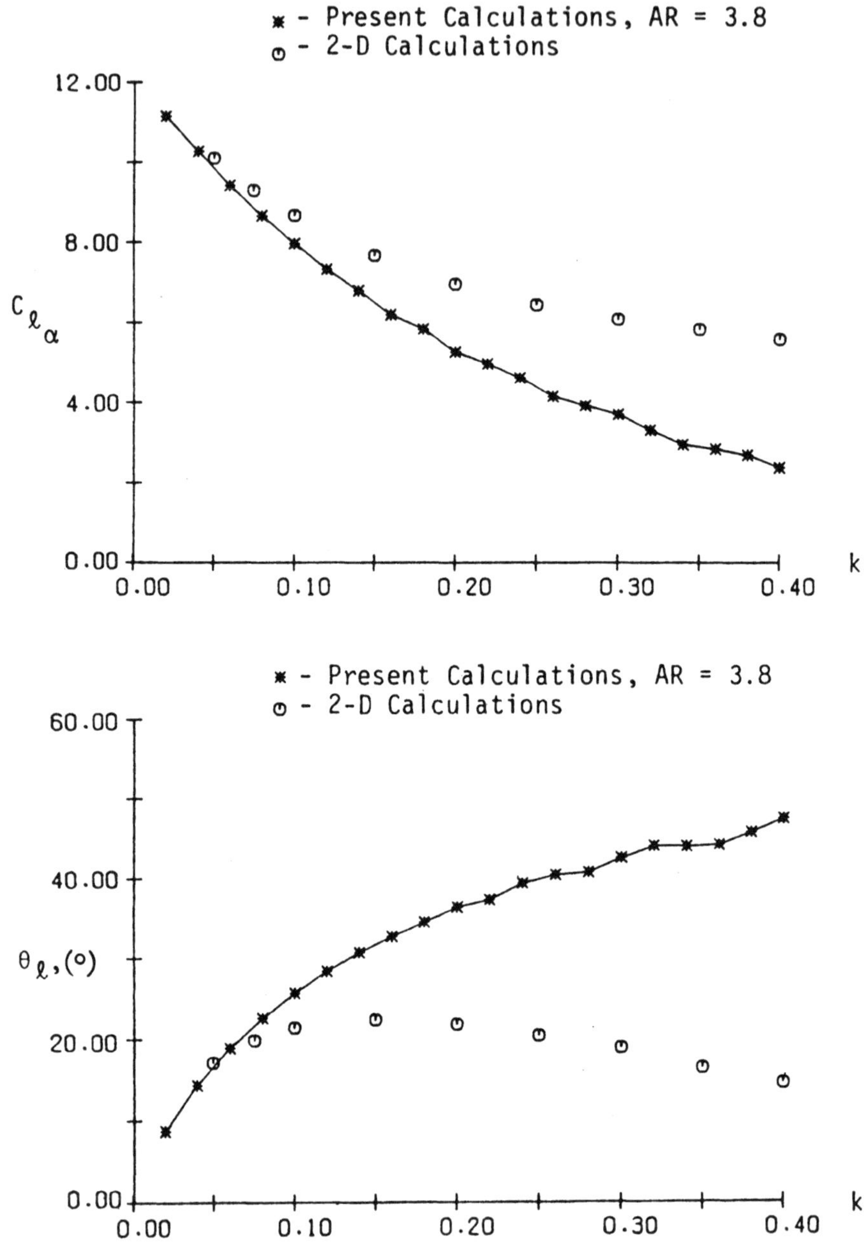

Figure 13. The variation of the amplitude and phase lag with reduced frequency for a rectangular wing with NACA 64A010 sections oscillating in pitch at $M_\infty = 0.8$ in unbounded two-dimensional flow and, for $AR = 3.8$, in unbounded three-dimensional flow.

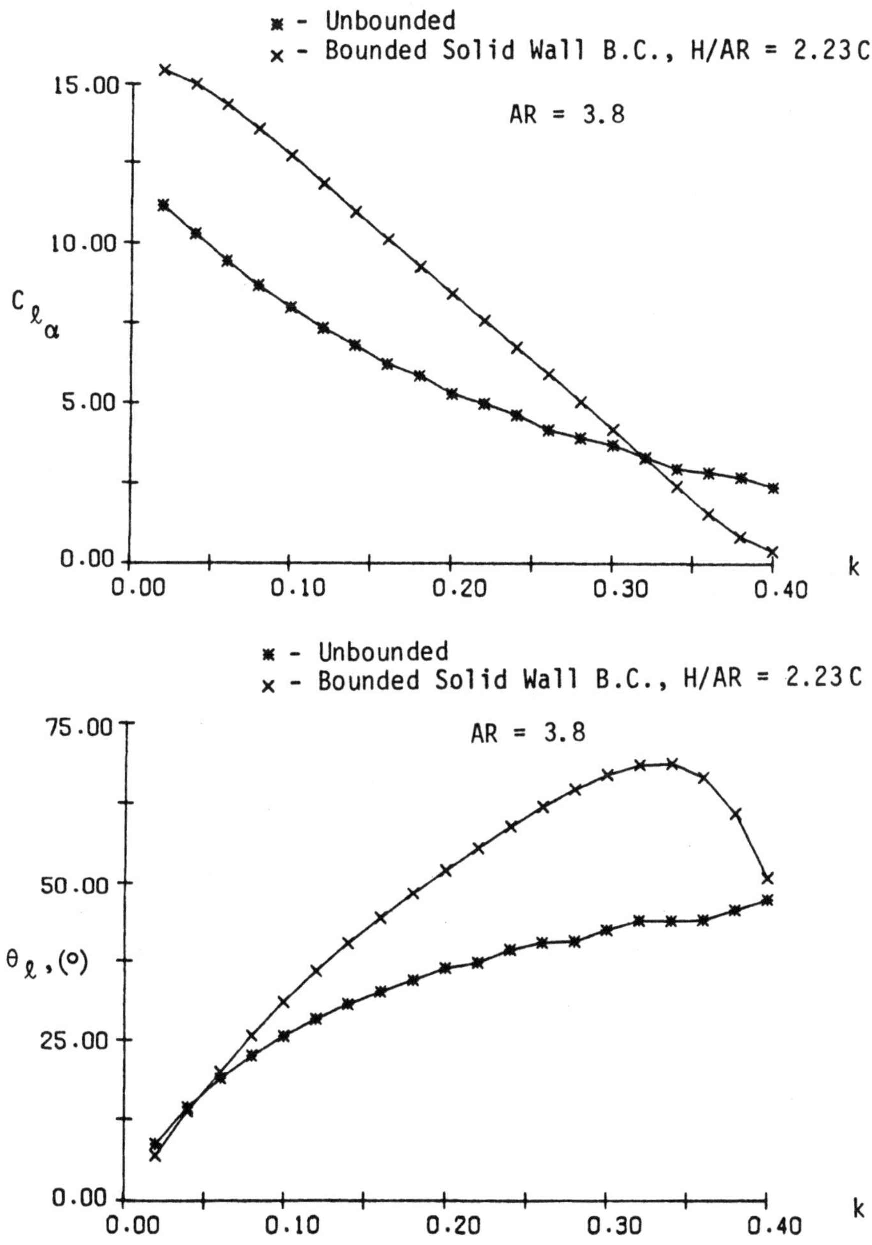

Figure 14. The variation of the amplitude and phase lag with reduced frequency for a rectangular wing with $AR = 3.8$ and oscillating in pitch at $M_\infty = 0.8$ in unbounded flow, and in a solid wall wind tunnel with $H/AR = 2.23C$.

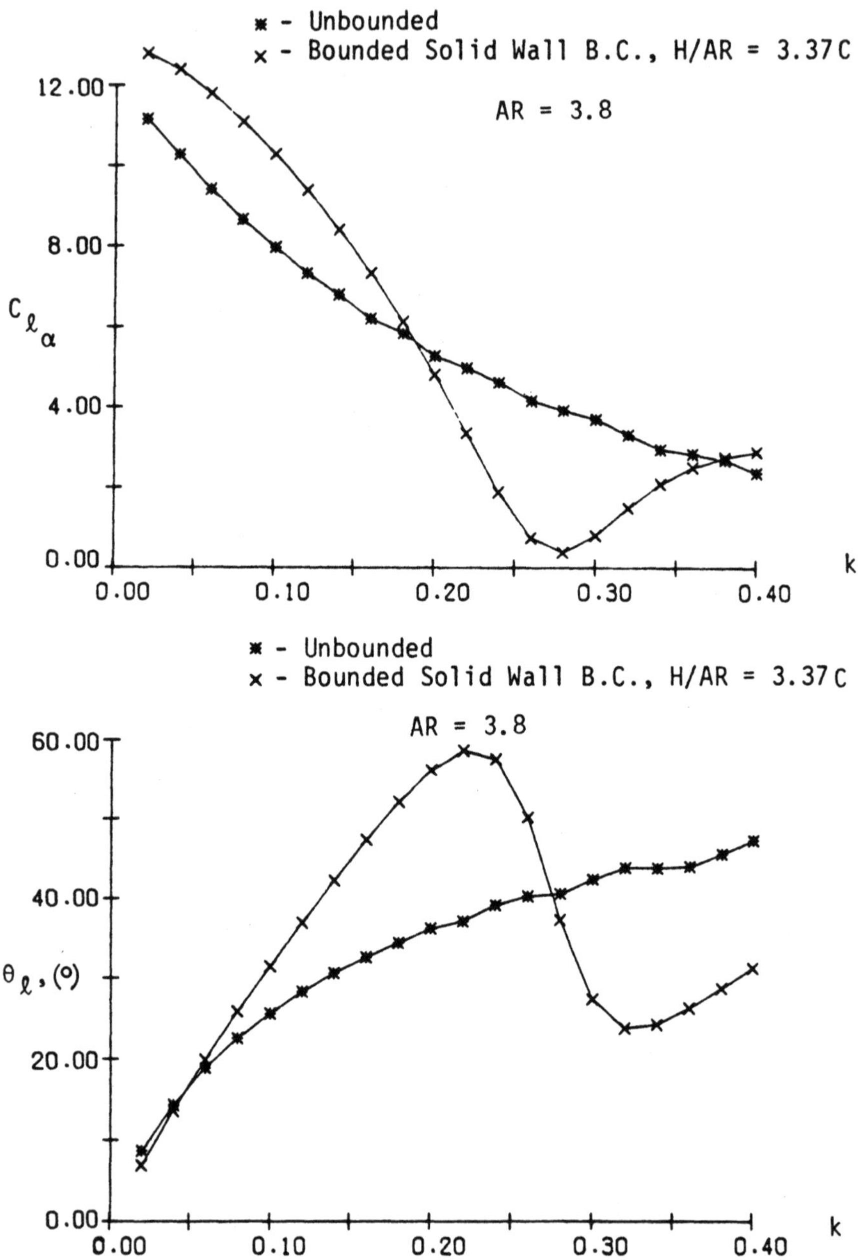

Figure 15. The variation of the amplitude and phase lag with reduced frequency for a rectangular wing with $AR = 3.8$ and oscillating in pitch at $M_\infty = 0.8$ in unbounded flow, and in a solid wall wind tunnel with $H/AR = 3.37C$.

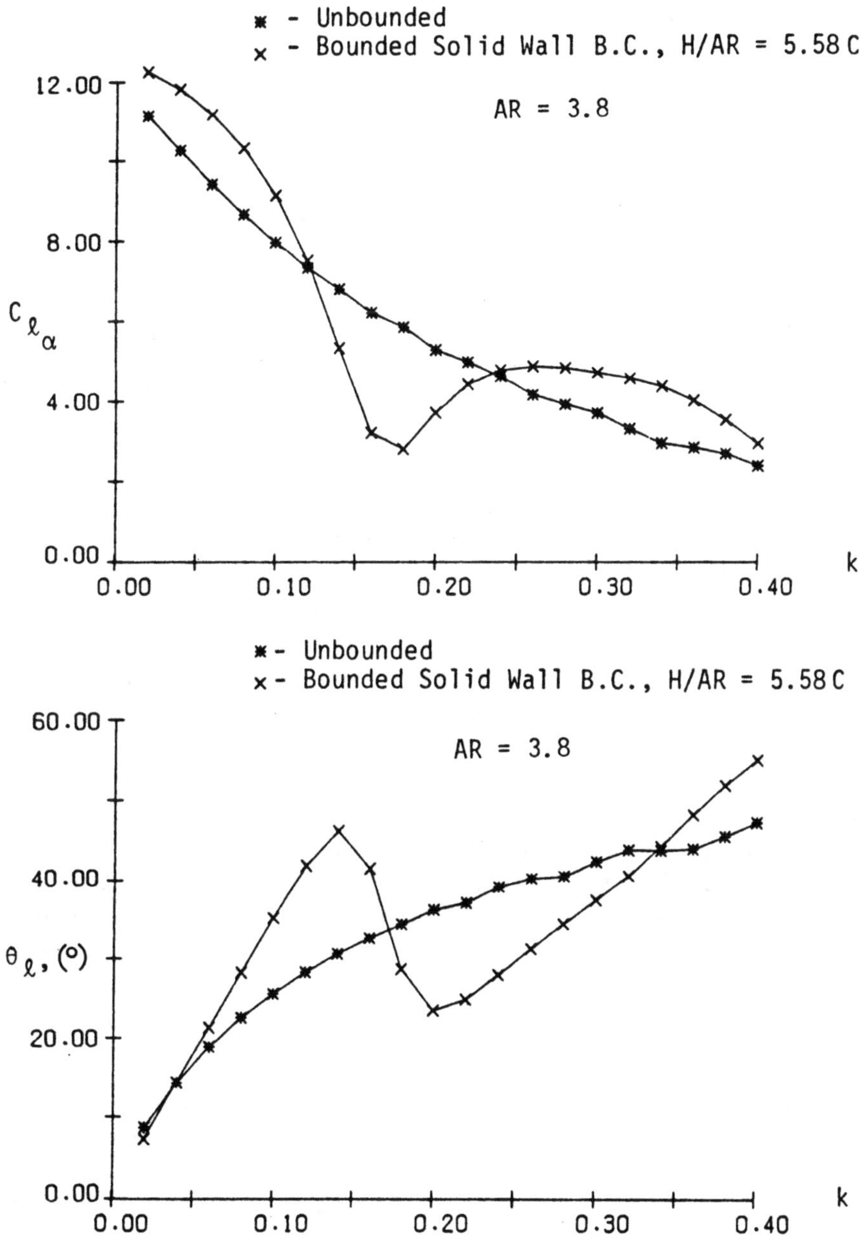

Figure 16. The variation of the amplitude and phase lag with reduced frequency for a rectangular wing with $AR = 3.8$ and oscillating in pitch at $M_\infty = 0.8$ in unbounded flow, and in a solid wall wind tunnel with $H/AR = 5.58C$.

and Runyan (1955) which shows an inverse relation between the reduced frequency at resonance and the wind tunnel height. The resonance amplitudes and the resonant reduced frequencies must both vanish as $H \to \infty$. The effects of solid wind tunnel walls are not even negligible for tunnel height to span ratios of 5.58 and a height to chord ratio of 21.2.

8. Conclusions

We have seen that an accurate and efficient method of calculating an airfoil's aerodynamic response is to use the inverse algorithm IAF2 described here to determine the nonlinear steady state, corresponding to a given pressure distribution, that has been obtained either experimentally or from an advanced numerical algorithm. The unsteady response is then determined using a time-linearized version of LTRAN2-HI that incorporates the shock wave as a moving discontinuity. This decoupling eliminates the need for small time steps due to the highly refined spatial grid near the shock wave needed to accurately capture small shock motions, as well as that needed near the leading edge. The procedure described here provides accurate unsteady pressures using as few as 20 spatial grid points on the airfoil. The procedure is limited to unseparated flows subjected to small unsteady disturbances but, within these restrictions, it should be quite accurate. It is a straightforward task to extend these concepts to three dimensions where the computational savings are more important, such as in the analysis of Borland and Rizzetta (1982), and the ability to prescribe the wing's pressure distribution is highly desirable.

A study of the effects of wind tunnel walls on unsteady transonic flows, whose steady state is free from interference, has shown that the resonances of linear theory remain in the nonlinear flow and can cause substantial discrepancies between unbounded flow and that in the wind tunnel, even for wind tunnel heights that are more than five times the wing's span and twenty times its chord. These findings suggest that the wind tunnel wall devices used in adpative tunnels to minimize the effects of wind tunnel walls on the steady flow will have to be carefully chosen and augmented by acoustic treatment of the walls in order to reduce wall reflections if unsteady testing is to be undertaken.

References

[1] Bailey, F. R. and Steger, J. L. "Relaxation Techniques for Three-Dimensional Transonic Flow About Wings," *AIAA Journal, 11* (1973), 318–25.

[2] Ballhaus, W. F. and Bridgeman, J. O. "Numerical Solution Techniques for Unsteady Transonic Problems," *AGARD Report 679* (1980), 16-1–16-24.

[3] Ballhaus, W. F. and Steger, J. L. "Implicit Approximate-Factorization Schemes for the Low-Frequency Transonic Equation," *NASA Tech. Memo X-73082,* 1975.

[4] Ballhaus, W. F., Jameson, A., and Albert, J. "Implicit Approximate Factorization Schemes for the Efficient Solution of Steady Transonic Flow Problems," *AIAA Journal, 16* (1978), 573–79.

[5] Bliss, D. B. "Aerodynamic Behavior of a Slender Slot in a Wind Tunnel Wall," *AIAA Journal, 20* (1982), 1244–52.

[6] Borland, C. J. and Rizzetta, D. P. "Nonlinear Transonic Flutter Analysis," *AIAA Journal, 20* (1982), 1606–15.

[7] Davis, S. S. and Malcolm, G. N. "Experimental Unsteady Aerodynamics of Conventional and Supercritical Airfoils," *NASA TM-81221,* 1980.

[8] Engquist, B. and Majda, A. "Numerical Radiation Boundary Conditions for Unsteady Transonic Flows," *Journal of Computational Physics, 40* (1981), 91–103.

[9] Farmer, M. G. and Hanson, P. N. "Comparison of Supercritical and Conventional Wing Flutter Characteristics," *Proceedings AIAA/ASME/SAE 17th Structures, Structural Dynamics and Materials Conference,* King of Prussia, PA, New York: Springer-Verlag, 1976, 608–11; also *NASA TMX-72837,* May 1976.

[10] Ferri, A. and Baronti, P. "A Method for Transonic Wind Tunnel Corrections," *AIAA Journal, 11* (1973), 63–66.

[11] Fung, K. Y. "Far Field Boundary Conditions for Unsteady Transonic Flows," *AIAA Journal, 19* (1981), 180–83.

[12] ————. "A Simple Accurate and Effective Algorithm for Unsteady Transonic Flow," to appear in *Recent Advances in Numerical Fluid Mechanics,* ed. W. G. Habaski, England: Pineridge Press, 1984.

[13] Fung, K. Y. and Chung, A. "Computation of Unsteady Transonic Aerodynamic Responses Using a Prescribed Input Steady State Pressure Distribution," *J. Aircraft, 20* (1983), 1058–61.

[14] Fung, K. Y., Yu, N. J., and Seebass, R. "Small Unsteady Perturba-

tions in Transonic Flows," *AIAA Journal, 16* (1978), 815–22.

[15] Goodyer, M. J. "The Self-Streaming Wind Tunnel," *NASA TMX-72699,* 1975.

[16] Hessenius, K. A. and Goorjian, P. M. "Validation of LTRAN2-HI by Comparison With Unsteady Transonic Experiment," *AIAA Journal, 20* (1982), 731–32; also *NASA TM-81037.*

[17] Houwink, R. and van der Vooren, J. "Results of an Improved Version of LTRAN-2 for Computing Unsteady Airloads on Airfoils Oscillating in Transonic Flow," *AIAA Paper No. 79-1553,* 1979.

[18] Landahl, M. T. *Unsteady Transonic Flow,* New York: Pergamon, 1961.

[19] Lin, C. C., Reissner, T., and Tsien, H. S. "On Two-Dimensional Nonsteady Motion of a Slender Body in a Compressible Fluid," *J. Math Phys., 3* (1948), 220–31.

[20] Przybytkowski, S. M. "Effect of Wall Interference in Unsteady Transonic Flows," *Ph.D. thesis,* University of Arizona, 1983.

[21] Sears, W. R. "A Note on Adaptive-Wall Wind Tunnels," *Journal of Applied Mathematics and Physics, 28* (1977), 915–27.

[22] Seebass, A. R. and Fung, K. Y. "Unsteady Transonic Flows: Time-Linearized Calculations," *Symposium on Numerical and Physical Aspects of Aerodynamic Flows.* Ed. T. Cebeci, New York: Springer-Verlag, 1981, 493–505.

[23] Seebass, A. R., Yu, N. J., and Fung, K. Y. "Unsteady Transonic Flow Computations," *AGARD-CP-277* (1978), 11, 1-11, 17.

[24] Sobieczky, H. and Seebass, A. R. "Supercritical Airfoil and Wing Design," *Annual Review of Fluid Mechanics, 16* (1984), 337–63.

[25] Tijdeman, H. "On the Motion of Shock Waves on an Airfoil With Oscillating Flap," *Symp. Transsonicum II.* Eds. K. Oswatitsch and D. Rues, New York: Springer-Verlag, 1976, 43–56.

[26] ———. "Investigations of the Transonic Flow Around Oscillating Airfoils," *Doctoral thesis,* Technische Hogeschool Delft, the Netherlands, 1977.

[27] Tijdeman, H. and Seebass, R. "Transonic Flow Past Oscillating Airfoils," *Annual Review of Fluid Mechanics, 12* (1980), 181–222.

[28] Tranen, T. L. "A Rapid Computer-Aided Transonic Airfoil Design Method," *AIAA Paper No. 74-501,* 1974.

[29] Triebstein, H., Destuynder, R., and Hansen, H. "Investigation of the Unsteady Airloads on a Transport Aircraft Type Airfoil With Two Interchangeable Oscillating Trailing Edge Flaps, at High Transonic Speeds and Reynolds Numbers," *ICAS-82* (1982), 306–15.

[30] Woolston, D. S. and Runyan, H. L. "Some Considerations on the Air Forces on a Wing Oscillating Between Two Walls for Subsonic Compressible Flow," *Journal of the Aeronautical Sciences* (1955), 41–50.

[31] Yang, T. Y. and Batina, J. T. "Transonic Time-Response Analysis of a 3-Degree-of-Freedom Conventional and Supercritical Airfoils," *J. Aircraft, 20* (1983), 703–10.

[32] Yu, N. J., Seebass, A. R., and Ballhaus, W. F. "Implicit Shock-Fitting Scheme for Unsteady Transonic Flow Computations," *AIAA Journal, 16* (1978), 673–78.

Unsteady Transonic Aerodynamics and Aeroelasticity*

Earl H. Dowell†
Duke University
Durham, NC 27706

Abstract

Four principal questions are discussed.

(1) Under what conditions are the aerodynamic forces essentially linear functions of the airfoil motion?

(2) Are there viable alternative methods to finite difference procedures for solving the relevant fluid dynamical equations?

(3) Under those conditions when the aerodynamic forces are *nonlinear* functions of the airfoil motion, what is the significance of the multiple (nonunique) solutions which are sometimes observed?

(4) What are effective, efficient computational procedures for using unsteady transonic aerodynamic computer codes in aeroelastic (e.g., flutter) analyses?

Nomenclature

Sections 2–4

C_L, C_M:	lift, moment coefficients.
$C_{L_\alpha}, C_{M_\alpha}$:	lift, moment curve slopes.
C_p:	pressure coefficient.
c:	airfoil chord.
K:	$= (\gamma + 1)M_\infty^2 \tau/\beta^3$.
k:	$= \omega c/U_\infty$; reduced frequency.
M:	Mach number.
s:	$= (\beta^2 t U_\infty/c)/M_\infty^2$.

*This work is supported by AFOSR Grant 81-0213A. Dr. Anthony Amos is the technical monitor.

†Dean, School of Engineering, Duke University, Durham, NC 27706.

t:	time.
x, y:	spatial coordinates in freestream and vertical directions.
x_s:	shock location.
Δx_s:	shock displacement normalized by the airfoil chord.
α_0, α_1:	mean angle of attack; dynamic angle of attack in degrees.
β:	$= (1 - M_\infty^2)^{1/2}$.
γ:	ratio of specific heats.
ν:	$= kM_\infty^2/\beta^2$.
$\phi^{(0)}, \phi^{(1)}$:	velocity potentials of steady flow and unsteady airfoil motion respectively.
ϕ:	phase angle.
τ:	thickness ratio of airfoil.
ω:	frequency.
Δ:	gradient operator.

Subscripts

∞:	freestream.
L:	local; also lift.
M:	moment.
max:	maximum.
0, 1:	mean, dynamic.
TE:	trailing edge.

Superscripts

c:	shock first forms.
tc:	shock reaches the trailing edge.

Section 5

M:	number of structural modes.
NF:	number of reduced frequencies needed for a flutter analysis.
NR:	number of response levels for a nonlinear flutter analysis.
P:	number of parameters.

T_A: computational time for aerodynamic code to reach a steady state lift value for a prescribed airfoil motion.

T_F: computational time for a simultaneous fluid-structural calculation to complete a transient.

T_{AF}: computational time for aerodynamic code to determine aerodynamic forces for one reduced frequency.

Section 6

$A(\ ), A_L, A_M$: indicial response functions.

a: distance of elastic axis from mid-chord: percent semichord, positive downstream.

b: semichord length.

c: full chord length.

C_L^N: nonlinear lift coefficient.

C_M^N: nonlinear moment coefficient about mid-chord.

C_{Me}^N: nonlinear moment coefficient about elastic axis.

$\overline{D}_L, \overline{D}_M$: components of describing function.

F: output of describing function.

G: structural transfer function.

H: nonlinear aerodynamic transfer function.

$\hat{H}$: aerodynamic describing function.

h: plunging displacement of elastic axis (positive down).

h_c: plunging displacement of mid-chord (positive down).

I_α: moment of inertia per unit span about elastic axis.

k: $= \frac{c\omega}{u}$, reduced frequency.

L: lift force.

M: moment force about mid-chord (positive nose-up).

m: mass per unit span.

M_∞: Mach number of uniform airflow.

R: $= \frac{\omega_h}{\omega_\alpha}$, uncoupled frequency ratio.

r_α: dimensionless radius of gyration about elastic axis (based on semichord); $r_\alpha^2 = r_{cg}^2 + (x_{cg} - a)^2$.

r_{cg}: dimensionless radius of gyration about center of gravity (based on semichord).

S_α: static unbalance.

s: dimensionless variable of Laplace operator; $s = ik$ for harmonic oscillation.

t: time.

U: dimensionless airspeed $\frac{u}{c\omega_\alpha\sqrt{\mu}}$.

u: dimensional airspeed.

x_{cg}: distance of center of gravity from mid-chord; percent semichord, positive downstream.

α: pitching displacement.

ϕ: effective induced angle-of-attack; see equation (6.1).

ϕ_1: amplitude of ϕ oscillation.

μ: mass ratio $\frac{m}{\pi\rho b^2}$.

ω_h, ω_α: uncoupled circular frequency of the airfoil in plunging and in pitch, respectively.

ρ: air density.

τ: dimensionless time $\frac{ut}{c}$.

Superscripts

T: transpose of matrix.

$\hat{}$: quantity associated with describing function.

$'$: $= \frac{d}{dt}$.

$\bar{}$: quantity in the subsidiary domain of Laplace Operator.

1. Introduction

The four questions cited in the abstract are chosen to provide the framework of this paper. This selection was made for several reasons.

- They are fundamental questions which are expected to be of lasting significance.
- Answers to these questions have important consequences for aeroelastic applications of unsteady transonic aerodynamics.
- Recent work has led to at least partial answers.

However, these questions are by no means exhaustive of those of interest. Fortunately others, in this volume and elsewhere, discuss several of these including:

- the important effects of viscosity and flow separation and

- the significant advances which have been made in finite difference solution methods and consequent improved physical understanding of transonic flows.

The four questions are addressed in Sections 2, 3, 4, and 5–6 respectively. Each section may be read relatively independently of the others, and the reader may wish to take advantage of that option.

2. Linear/Nonlinear Behavior in Unsteady Transonic Aerodynamics

2.1 Motivation and General Background

The aeroelastician uses linear dynamic system theory for most aeroelastic analyses. The motivation for doing so is clear. Extensive experience, understanding, and effective computational/experimental procedures have been developed for linear systems. By contrast, although nonlinear methods of analysis and experimentation are available, the results are far more expensive to obtain and also more difficult to interpret. Hence linear models, where applicable, are very powerful, relatively simple, and extremely valuable. Thus, it is highly important to determine the domain of validity of any linear model.

Here our concern is with possible aerodynamic nonlinearities in transonic flow. Of course, aerodynamic nonlinearities may arise in other flow regimes; however, it is in transonic flow where they tend to be most important. Indeed, it is often observed that the transonic flow regime is inherently nonlinear in the governing field equations. However, at any Mach number of any airfoil, if the angle of attack is sufficiently small, the aerodynamic forces and shock motion will be linear in the angle of attack. Moreover, as the frequency of the angle of attack motion increases, the range of angle of attack over which linear behavior persists increases. It is our purpose here to study when linear or nonlinear behavior occurs using as our principal analytical method the low frequency, transonic small disturbance (LTRAN2) procedure of Ballhaus and Goorjian (1977, 1978). Any other present or future nonlinear aerodynamic method could (and should) be used for similar purposes.

It will be helpful to discuss first the shock and its motion which is sometimes a source of confusion. A consequence of any consistent linearization of steady transonic small disturbance aerodynamic theory in the dynamic angle of attack is that a concentrated force or pressure pulse (sometimes called a shock doublet) will appear at the location of the steady state shock (Williams, 1979a, 1979b). The strength of the pressure pulse is equal to the steady state shock pressure jump and its width is proportional to the dynamic angle of attack. By contrast elsewhere on the airfoil chord (away from the shock doublet whose center is at the steady state shock location) the pressure magnitudes (in a transonic linear theory) are proportional to the dynamic angle of attack and become smaller in proportion as the dynamic angle of attack is smaller. Of course this latter behavior is also true in classical theory. The most important (although not the only) distinction between classical, linear theory and transonic, linear theory is the presence of the shock (and its motion) in the latter which creates the concentrated shock force doublet. LTRAN2 and some other transonic computer codes include both the shock and its motion while classical aerodynamic theory includes neither. Some inconsistent transonic methods include the shock's presence, but not its motion.

The behavior described above is seen in a nonlinear dynamic theory as well, when the dynamic and angle of attack becomes small. Consider Figure 1 which was obtained using LTRAN2. It shows the chordwise differential (lower surface minus upper) pressure distribution for an NACA 64A006 airfoil at $M_\infty = 0.86$ for several angles of attack. Here, for simplicity, the reduced frequency is set to zero so there is no distinction (numerically) between steady and dynamic angle of attack. As may be seen for small angles of attack, say $\alpha = 0.125$ deg, 0.25 deg, the pressure distribution has a shock doublet centered at the mean (zero angle of attack) shock location, $x_s/c = 0.584$. The width of the shock doublet is indicated by the vertical lines, the forward one is at the lower surface shock location and the rearward one at the upper surface location. The shock doublet width is proportional to α for the smaller α; however, as α increases to 1 deg the lower surface shock disappears while the upper surface shock moves to the trailing edge and remains there. Also, for the smaller α the shock doublet magnitude is essentially euqal to the pressure jump through the shock at $\alpha = 0$ deg, i.e., 0.43. Away from the shock doublet, the pressures are proportional to α for small α. Finally, note a matter of practical importance. For small

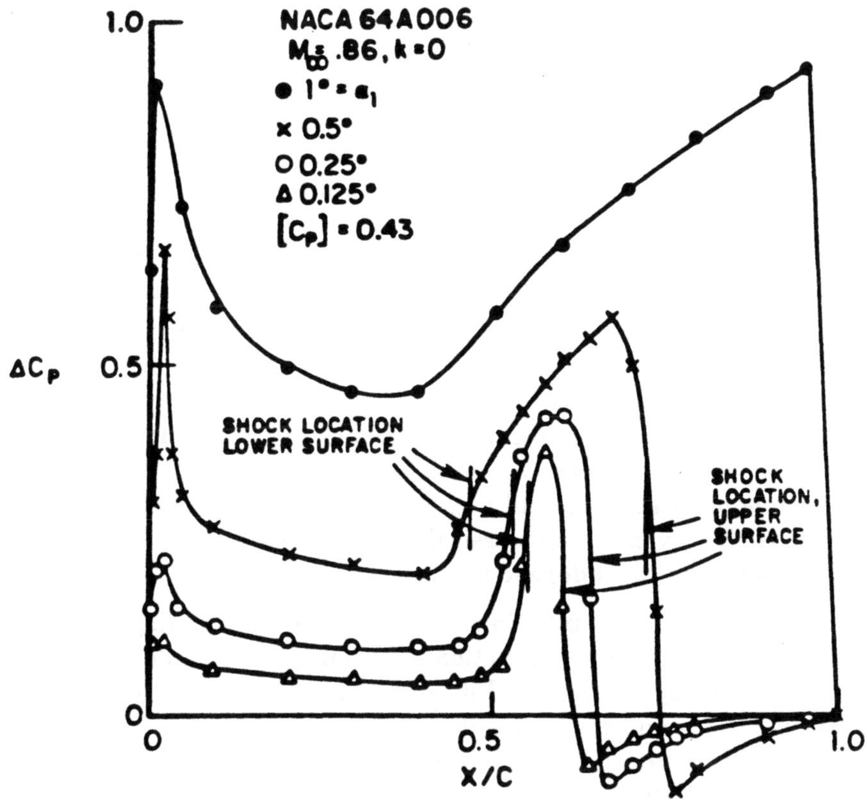

Figure 1. Differential pressure distribution.

α as the shock doublet width narrows, any finite difference scheme nonlinear in α will have a resolution problem as $\alpha \to 0$. By contrast a method a priori linearized in α avoids this difficulty as it computes the shock motion explicitly, e.g., see Williams (1979a, 1979b) and Fung et al. (1978). Also see the discussion of Tijdeman (1977) and Tijdeman and Seebass (1980) for a critical assessment of theory and experiment. The experimental study of Davis and Malcolm (1979) is particularly relevant here as it provides confirmation of the above in broad terms.

2.2 NACA 64A006 Airfoil at $M_\infty = 0.86$ Pitching About its Leading Edge

The following principal issues were studied (Dowell et al., 1983): effect of dynamic angle of attack at various reduced frequencies on dynamic forces and shock motion; boundary for linear/nonlinear behavior; effect of reduced frequency and dynamic amplitude on aerodynamic transfer functions; effect of dynamic angle of attack on steady state forces and shock displacement; and effect of steady-state angle of attack on dynamic forces and shock motion. For the sake of brevity, only the first two issues will be considered here.

2.2.1 Effect of Dynamic Angle of Attack on Dynamic Forces and Shock Motion

It is desirable to assess at what dynamic amplitude nonlinear effects become important in order to determine the relative linear vs nonlinear behavior of lift, pitching moment, and shock motion. Note that the total lift (moment, shock motion) is characterized by $C_L = C_{L_0} + C_{L_1}$, where C_{L_0} is defined to be the lift due to the mean angle of attack, α_0, and C_{L_1} that due to the dynamic angle of attack, α_1, for given α_0. In classical linear theory (but not transonic linear theory) C_{L_1} is independent of α_0.

In Figure 2 lift, pitching moment, and shock displacement amplitudes are shown as a function of dynamic amplitude, α_1, for a reduced frequency of $k = 0.2$. Lift and moment coefficient have their usual definitions and the moment is about the midchord. The shock displacement is normalized by the airfoil chord. Phases are also presented for lift and pitching moment. The shock motion phase was also computed; however, it tended to be less accurately determined. Since it is not needed for our present purposes, it is not shown.

It is seen that lift tends to remain linear to higher dynamic amplitudes than moment which, in turn, tends to remain linear to higher amplitudes than shock motion. Moreover, as will be seen, the larger the reduced frequency, the greater the range of linear behavior. Phase information generally, although not universally, is a more sensitive indicator of departure from linearity than lift, moment, or shock amplitude information. In a strictly linear theory, of course, the phase is independent of the dynamic angle of attack.

It is noted that no measurable higher harmonic content was found in any of the numerical results. The results were virtually sinusoidal signals for lift, moment, and shock motion; hence, determination of magnitude and phase was readily done by any one of several conventional methods. The exception was shock motion phase which is difficult to determine accurately by any method becuase of the relatively coarse finite difference mesh resolution of the shock.

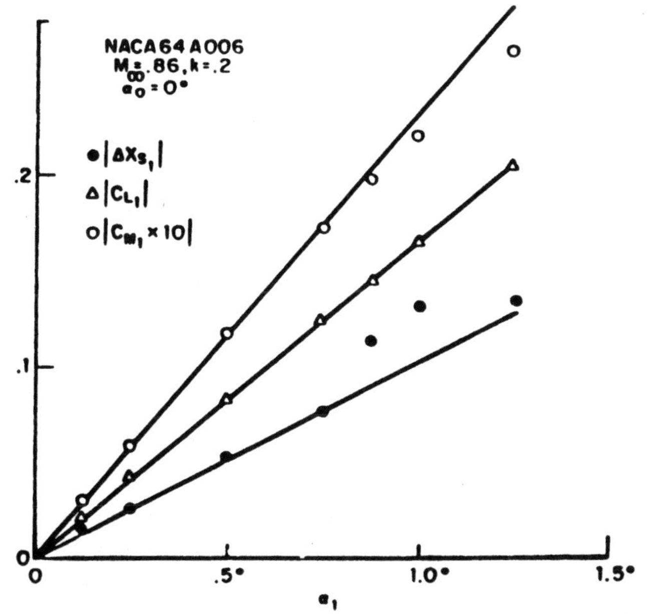

Figure 2a. Effect of dynamic angle of attack on dynamic forces and shock motion: amplitudes.

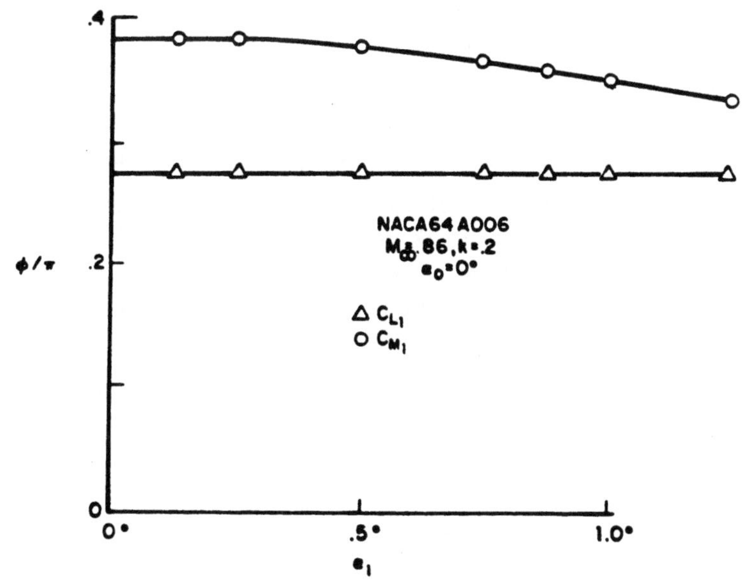

Figure 2b. Effect of dynamic angle of attack on dynamic forces and shock motion: phases.

2.2.2 Boundary for Linear/Nonlinear Behavior

It is highly desirable to provide a criterion by which an aeroelastician may assess when a linear dynamical theory may be used.

Figure 3 has been constructed from Figure 2 and other similar results by identifying the k, α_1 combinations for which the pitching moment deviates by 5% in amplitude or phase from linearity. As expected, at higher k the pitching moment remains linear to larger α_1.

Although Figure 3 provides very useful information, it requires a nonlinear dynamical theory to construct it. A question thus arises: Is there a similar, but perhaps more conservative, criterion which may be used with a linear dynamical theory? The answer is provided by the shock motion. In Figure 4 a similar boundary to that shown in Figure 3 is constructed (again from information such as that provided by Figure 2) based upon shock motion rather than pitching moment. It is observed in Figure 2 that for shock displacement amplitudes of less than 5% the shock motion (as well as lift and pitching moment) behave in a linear fashion. Hence, a 5% shock motion boundary is shown in Figure 4. Note that this boundary could be constructed from a linear dynamical theory. A second boundary (less conservative) based upon the first detectable deviation of shock motion from linearity is also shown. Finally, the boundary from Figure 3 is shown for reference. These results are consistent with those of Ballhaus and Goorjian (1978) who also suggested that shock motions of less than 5% chord correspond to linear behavior.

Thus it is concluded that a simple criterion for departure from nonlinearity based upon shock motion may be used. It can be evaluated by a linear dynamical theory in principle (which enhances its practical utility), although the present results were obtained using a nonlinear, dynamical theory.

A brief digression is in order to explain why the shock motion criterion is extremely useful to the aeroelastician. After the flutter mode is determined from a conventional flutter analysis using linearized but transonic aerodynamics, one may compute the amplitude of the flutter motion which will correspond to a 5% shock motion using the linear transonic aerodynamic model employed in the flutter analysis. This will give the aeroelastician the limit on amplitude for which the linear, flutter calculation is valid. This is very useful information.

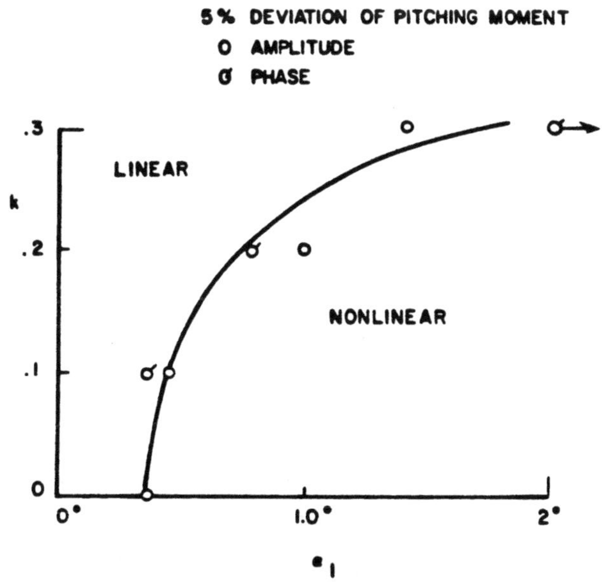

Figure 3. Boundary for linear/nonlinear behavior in terms of reduced frequency and dynamic angle of attack.

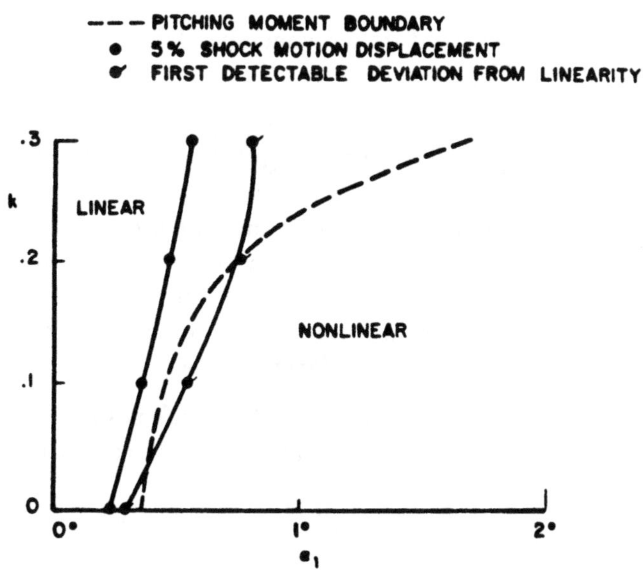

Figure 4. Conservative boundary for linear/nonlinear behavior based upon shock motion amplitude.

2.3 Mach Number Trends

2.3.1 Similarity Law

Here the effects of Mach number are studied systematically for the NACA 64A006 airfoil. We note that a similarity rule holds for low frequency, transonic flow which gives the following results for any family airfoils.

$$C_p = \frac{\tau}{\beta}\overline{C}_p(x/c, s; K, \nu, \alpha/\tau) \tag{2.1}$$

where $\overline{C}_p$ is a universal function of its arguments and

$$\begin{aligned}
\beta &\equiv (1 - M_\infty^2)^{1/2}, \nu \equiv kM_\infty^2/\beta^2 \\
K &\equiv \frac{(\gamma+1)\, M_\infty^2}{\beta^3}, s \equiv \frac{\beta^2 t U_\infty/c}{M_\infty^2} \\
\tau &\equiv \text{thickness ratio of airfoil} \\
\alpha &\equiv \text{angle of attack}
\end{aligned}$$

Equation 2.1 may be further specialized for the case $\alpha \to 0$, by expanding in a Taylor series, i.e.,

$$C_p = \frac{\tau}{\beta}\overline{C}_p(x/c, s; K) + \frac{\alpha}{\beta}\text{Re}\{e^{i\nu s}\overline{C}_{p1}(x/c; K, \nu)\} \tag{2.2}$$

This is the similarity law for dynamic linearization in α, i.e., $\alpha = \alpha_1$. Zero mean angle of attack is assumed for simplicity, $\alpha_0 = 0$, although the result is readily extended. From equation 2.2, it is seen that similarity for the harmonic component requires only that K and ν be the same for two different flows.

Finally, it is noted that since the shock is simply a discontinuity surface of ϕ_x, it satisfies a similarity law expressed by

$$x_s = x_s\,(\beta y/c, s; K, \alpha/\tau, \nu) \tag{2.3}$$

For the limit, $\alpha \to 0$,

$$x_s = x_{s_0}\,(\beta y/c; K) + \alpha/\tau \text{Re}\{e^{i\nu s} x_{s_1}(\beta y/c; K, \nu)\} \tag{2.4}$$

The similarity law given by equation 2.1 was known to Miles (1959). Equations 2.2–2.4 are extensions of his results.

Using the similarity rules, the results for the 64A006 airfoil may be used to obtain results for any other airfoil of the same family, in particular, the 64A010.

From equation 2.4 it may be inferred that the 5% shock motion criterion has the functional form (for a given family of airfoils)

$$\alpha/\tau = F(\nu, K) \tag{2.5}$$

It is interesting to note that Fung et al. (1978) proposed a criterion for the validity of linearization of the form

$$\frac{\alpha}{\tau}/K \ll 0.1 \tag{2.6}$$

Equation 2.6 is clearly a special case of equation 2.5. Using equation 2.5, the data of Figures 3, 4 (and 5, subsequently) may be reinterpreted in terms of similarity variables, and thereby generalized.

2.3.2 Boundary for Linear/Nonlinear Behavior

Using results such as those shown in Figure 2 and invoking the 5% shock displacement criterion, a linear/nonlinear boundary may be constructed in terms of Mach number vs amplitude of airfoil oscillation. Of course, as the shock reaches very near the trailing edge, the 5% criterion would need to be modified. Results are shown in Figure 5 for $k = 0$ and 0.2. Note that for steady flow ($k \equiv 0$) the angle of attack must be very small when $M_\infty = 0.88$ and 0.9 for linear behavior to occur. However, as we have seen before, the 5% shock displacement criterion is conservative. That is, lift and moment tend to remain linear in α to higher α than this criterion would suggest. Nevertheless, the trend should not change using any other reasonable criterion. By contrast, for $k = 0.2$ the linear region is much enlarged. For $M_\infty < M_\infty^c$ or $M_\infty > M_\infty^{tc}$ the linear region is for all practical purposes unbounded. In practice, in this region other physical effects, e.g., viscosity, are likely to come into play before inviscid, small disturbance, transonic theory nonlinearities become important. M_∞^c is the Mach number when the shock first forms and M_∞^{tc} that when it reaches the airfoil trailing edge.

2.3.3 Aerodynamic Transfer Functions

In the linear region it is of interest to display aerodynamic transfer functions vs Mach number. Perhaps the most familiar of

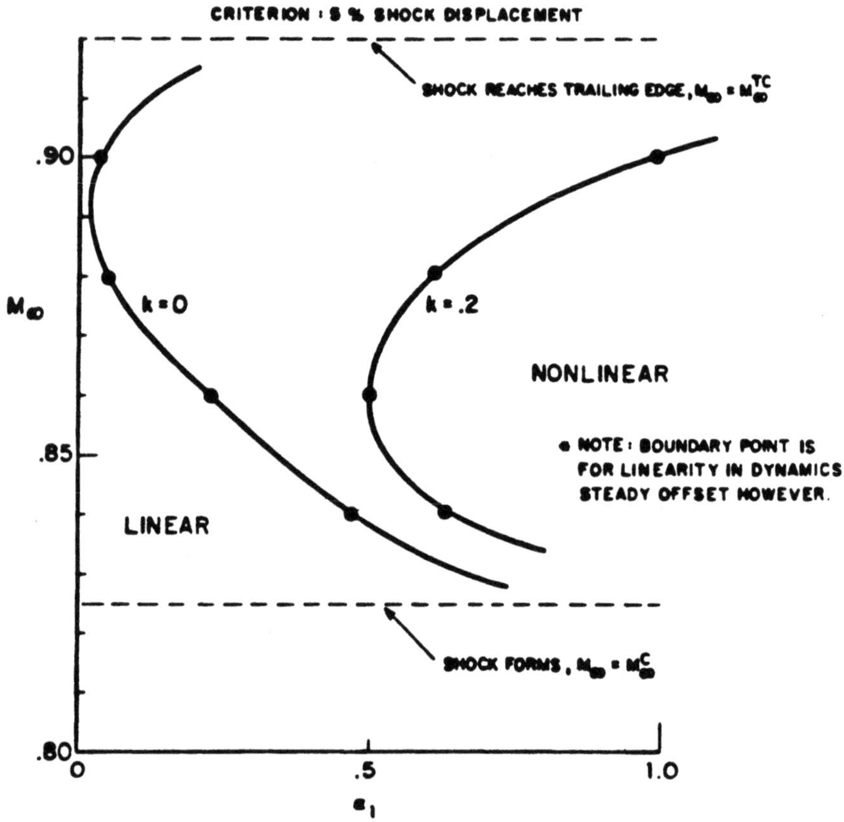

Figure 5. Effect of reduced frequency on boundary for linear/nonlinear behavior: Mach number vs. dynamic angle of attack.

these is lift curve slope, C_{L_1}/α_1. Its amplitude is shown in Figure 6a from LTRAN2 for $k = 0$. Also shown are results from full potential theory, classical subsonic theory, and local linearization. The latter is shown for $M_\infty > M_\infty^{tc}$, i.e., the shock is at the trailing edge. It uses the local trailing edge supersonic Mach number in classical (supersonic) theory. One concludes that for $M_\infty < M_\infty^c$, classical theory gives reasonable results, and for $M_\infty > M_\infty^{tc}$ local linearization gives reasonable results. For $M_\infty^c < M_\infty < M_\infty^{tc}$, LTRAN2 gives markedly different results although it likely fails for $M_\infty = 0.88$, 0.90. Note the difference between transonic small disturbance theory (LTRAN2) which falls well off scale at $M_\infty =$ 0.88 and 0.9 and full potential theory (Bauer et al., 1972).

It should be noted that the full potential results shown in Figure 6a were obtained using a nonconservative finite difference scheme. Full potential results obtained using a quasiconservative finite difference scheme (for technical reasons results were only obtained for $M_\infty < 0.87$) are essentially identical to those of transonic small disturbance theory using a conservative finite difference scheme (LTRAN2). Hence, the difference shown in Figure 6 should be attributed to the distinction between conservative and nonconservative finite differences and not to the distinction between small disturbance and full potential theory. To the extent that the nonconservative finite difference method may be said to have some form of numerical (as opposed to physical) viscosity, the differences may be attributed to the qualitative distinction between inviscid and viscous flow.

In Figure 6b results are shown for $k = 0.2$. For reference, the LTRAN2 results for $k = 0$ are also shown. Again it is seen that the classical subsonic theory and local linearization theory give reasonable results (better than for $k = 0$) for $M_\infty < M_\infty^c$ and $M_\infty > M_\infty^{tc}$, respectively. Moreover, LTRAN2 appears to give reasonable results over the entire Mach number range, although there is no better theory to validate it. Note that from $M_\infty = 0.9$ to 0.92 there is a somewhat abrupt change.

2.4 Conclusions

For $M_\infty < M_\infty^c$, where no shock exists, the aerodynamic forces are linear over a substantial range of angle of attack. This is also true for $M_\infty > M_\infty^{tc}$, i.e., where the shock has moved to the trailing edge. For $M_\infty^c < M_\infty < M_\infty^{tc}$ a boundary of linear/nonlinear behavior may be constructed which shows the angle of attack must be quite small for linear behavior to occur for steady flow. However the region of linear behavior increases substantially for unsteady flow.

In the range $M_\infty^c < M_\infty < M_\infty^{tc}$, transonic small disturbance theory (LTRAN2) and full potential theory appear to fail for steady flow for some narrow band of M_∞ where they substantially overestimate the shock displacement and, hence, the aerodynamic forces. This is tentatively attributed to the absence of viscosity in the theories.

Classical subsonic theory and local linearization are useful approximate tools for unsteady flow provided their limitations are recognized.

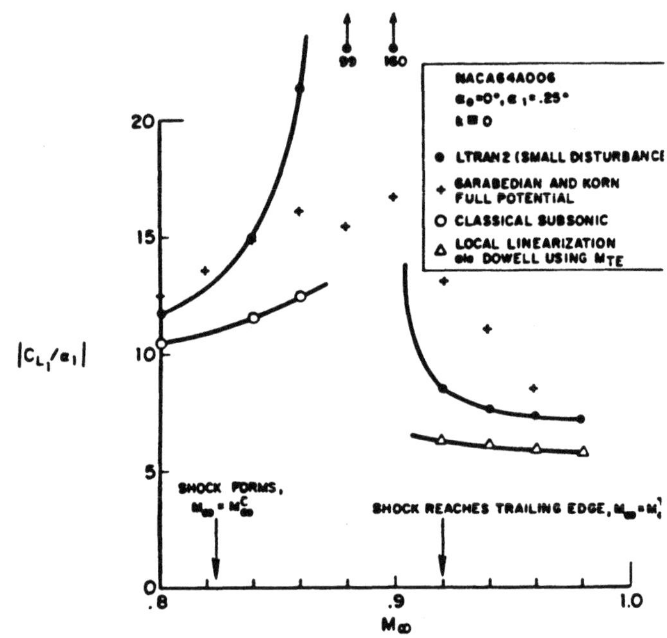

Figure 6a. Effect of Mach number on lift curve slope.

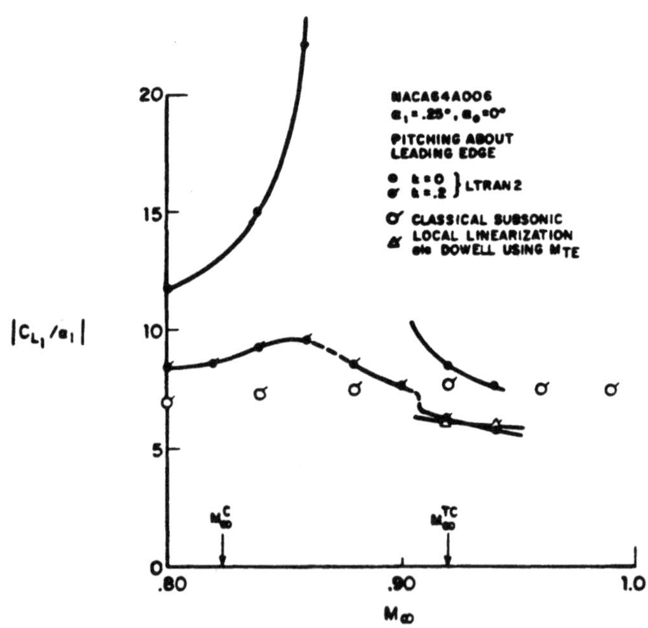

Figure 6b. Effect of Mach number on lift curve slope.

Aerodynamic transfer functions are expected to retain their utility even when nonlinear dynamic effects are important. This is for several reasons, including:

(1) Nonlinear effects diminish with increasing frequency.

(2) At high frequencies, classical linear theory is expected to be reasonably accurate and indeed most inviscid theories will approach classical theory as the frequency becomes larger (Williams, 1979a, 1979b).

(3) The preceding suggests that several theories may be used to provide a composite aerodynamic representation in the frequency domain. For example, one might use BGK for $k = 0$, LTRAN2 for $k = 0.05$–0.2, Williams for $k = 0.2$–1.0, and classical theory (which Williams' theory smoothly approaches) for $k > 1.0$.

A similarity law for low frequency transonic small disturbance theory is available which reduces the number of aerodynamic computations required and generalizes results for one airfoil to an entire family.

Although two-dimensional flows have been discussed here, the general concepts and approach should be useful for three-dimensional flows. In particular, one expects the effect of three-dimensionality to increase the region of linear behavior for transonic flows. For example, the accuracies of transonic small disturbance theory, local linearization, and classical theory should be enhanced by three-dimensional effects.

No transonic method of aerodynamic analysis can be expected to give useful information to the aeroelastician unless the mean steady flow it predicts and uses is accurate. Hence, it is highly desirable to be able to input directly the best steady flow information which is available including that from experiment. The latter would include implicitly viscosity effects on the mean steady flow; in particular it would place the mean shock in the correct position.

The reader may wish to consult the lucid survey article by Tijdeman and Seebass (1980) which provides a context in which to evaluate the present results and conclusions. Also Nixon and colleagues have discussed extensively how the transonic, linear theory may be used in aeroelastic analyses. For example, see Nixon and Kerlick (1980) and Nixon (1981). Finally see the subsequent discussion in Section 5.

3. Viable Alternative Solution Procedures to Finite Difference Methods

Although continuing advances in computer technology will lead to diminishing costs, economics alone will dictate for the next decade a substantial effort to improve the efficiency of finite difference methods and/or consider less expensive alternative solution techniques. Here the latter is discussed drawing largely on the recent work of Hounjet (1981b) and Cockey (1983). Both of these authors have used integral equation methods (IEM) though from rather different points of view. Prior work by Hounjet (1981a) was based upon the Williams-Eckhaus model (Williams, 1979a, 1979b), which also is the point of departure for Cockey. The motivation for considering IEM and a concise description of earlier work is well covered by Hounjet (1981b), Morino (1974), Morino and Tseng (1978), Albano and Rodden (1969), Nixon (1978), Voss (1981), Williams (1978), and Liu (1978).

Both Hounjet and Cockey adopt a transonic small disturbance equation approximation and the associated velocity potential is divided into steady, $\phi^{(0)}$, and unsteady (due to airfoil motion), $\phi^{(1)}$, parts. By assuming (infinitesimally) small airfoil (harmonic) motion the governing field equation for $\phi^{(1)}$ is *linear* with *variable coefficients* which depend upon $\phi^{(0)}$, viz.

$$(1-M_\infty^2)\phi_{xx}^{(1)}+\phi_{yy}^{(1)}-2ikM_\infty^2\phi_x^{(1)}+k^2M_\infty^2\phi^{(1)} = \left[(M_L^2 - M_\infty^2)\,\phi_x^{(1)}\right]_x \tag{3.1}$$

where M_∞ is the free stream Mach number, k is the reduced frequency, $k = \omega c/2u$, in which c denotes the airfoil chord. c is used to make lengths dimensionless. M_L is the local Mach number:

$$M_L^2 = M_\infty^2 + \left[3-(2-\gamma)M_\infty^2\right] M_\infty^2 \phi_x^{(0)} \tag{3.2}$$

Subscripts on $\phi^{(1)}$ denote spatial derivatives.

From this point the approaches of Hounjet and Cockey follow different paths. Note that by setting the right hand side of equation 3.1 to zero, one retrieves classical aerodynamic theory.

3.1 Hounjet

By using the Green's function of classical aerodynamic theory (the LHS of equation 3.3), one obtains an integral equation for the unknown $\phi^{(1)}$. It is

$$\phi^{(1)}(x,y) = \int_0^\infty \Delta\phi^{(1)}(u)\frac{\partial}{\partial y}[E(x-u,y;k,M)]\,du + \int_{-\infty}^{\infty}\int_{-\infty}^{\infty} m(u,v)\,E(x-u,y-v;k,M)\,dudv \tag{3.3}$$

On the right hand side of equation 3.3, the first term is a integral along the airfoil chord (and wake) while the second integral is over the *entire* (but see below) flow field. E represents an elementary source solution which satisfies the radiation condition and the following equation:

$$(1-M_\infty^2)E_{xx} + E_{yy} - 2ikM_\infty^2 E_x + k^2M_\infty^2 E = e^{i\bar{k}x}\delta(x)\,\delta(y) \tag{3.4}$$

where

$$\bar{k} \equiv kM_\infty^2/\beta^2.$$

m is given by

$$m(x,y) = \left[3-(2-\gamma)M_\infty^2\right]M_\infty^2\left[\Phi_x^{(0)}\Phi_x^{(1)}\right] \tag{3.5}$$

In classical (integral equation) aerodynamic theory, of course, the second integral of equation 2.3 is not present because $\Phi_x^{(0)}$ is zero. In transonic IEM the second term may be neglected everywhere in the flow field where $\Phi_x^{(0)} = 0$, i.e., the steady flow field is sensibly uniform. Hence only a relatively small part of the total flow field will contribute to the second integral term and this simplifies the subsequent calculation very substantially. This is the key point which allows the possibility of an efficient computer code. Then the numerical solution of equation 2.3 proceeds as described by Hounjet (1981b).

Numerical results obtained by Hounjet show two principal features:

(1) The accuracy (for a linearization in the dynamic airfoil motion) is the same as that obtained by finite difference procedures.

(2) The computer time is approximately one quarter of that of LTRAN2 (a popular finite difference code).

Recently Hounjet has extended this approach to three-dimensional flow fields.

3.2 Cockey

First rewrite equation 2.1 as

$$(1 - M_\infty^2)\phi_{xx}^{(1)} + \phi_{yy}^{(1)} - 2ikM_\infty^2\phi_x^{(1)} + k^2 M_\infty^2 \phi^{(1)} \\ - (M_L^2 - M_\infty^2)\phi_{xx}^{(1)} - 2M_L \frac{dM_L}{dx}\phi_x^{(1)} = 0 \tag{3.6}$$

If one could determine the Green's function for the LHS of equation 3.6, then an integral equation for $\phi^{(1)}$ will only have an integral along the airfoil chord, shock and wake and none in the flow field per se (except possibly along the shock). Unfortunately obtaining this Green's function is difficult because the last two forms of the LHS of equation 3.6 have variable coefficients.

Thus Cockey modifies equation 3.6 by a local linearization approximation to M_L and $\frac{dM_L}{dx}$ for the purposes of obtaining the Green's function (approximately). The subsequent calculation follows standard techniques except that integrals extend along the shock as well as the airfoil chord (and wake).

Numerical results obtained by Cockey show two principal features:

(1) The accuracy is substantially less than that obtained by finite difference procedures, even though the shock and its movement is taken into account.

(2) However the computer cost is no greater than that associated with classical aerodynamic theory.

(1) is, of course, a disappointing result. However the successful incorporation of the shock into the Cockey model and Hounjet's substantial success suggest a possible way of advantageously combining the features of methods of Hounjet and Cockey.

3.3 A Possible Synthesis of the Hounjet and Cockey Methods

Consider again equation 2.1. Recognizing that in the Cockey method M_L, $\frac{dM_L}{dx}$ are approximated for the purpose of obtaining a

Green's function equation 2.1 is rewritten as follows:

$$\begin{aligned}(1-M_\infty^2)\phi_{xx}^{(1)} + \phi_{yy}^{(1)} - 2ikM_\infty^2\phi_x^{(1)} + k^2M_\infty^2\phi^{(1)} \\ -(M_{LA}^2 - M_\infty^2)\phi_{xx}^{(1)} - 2M_{LA}\frac{dM_{LA}}{dx}\phi_x^{(1)} \\ = (-M_{LA}^2 + M_L^2)\,\phi_{xx}^{(1)} + 2\left(M_L\frac{dM_L}{dx} - M_{LA}\frac{dM_{LA}}{dx}\right)\phi_x^{(1)}\end{aligned} \tag{3.7}$$

Setting the RHS of equation 3.7 to zero, we retrieve the Cockey model. However if the RHS is retained, then Hounjet's method could be used to solve the resulting integral equation where now Cockey's approximate Green's function is used corresponding to the LHS of equation 3.7 rather than the classical Green's function (as Hounjet has used) corresponding to the LHS of equation 2.1. Presumably the advantage of using the hybrid appraoch (i.e., equation 3.7 rather than equation 2.1) is that the RHS of equation 3.7 is usually smaller than the RHS of equation 2.1 and thus the region in the flow field which is included in Hounjet's approach may be smaller and thus the resulting computer code will be more efficient.

We note finally that this approach (Hounjet's original method or the hybrid approach suggested here) is reminiscent of Lighthill's theory of jet noise (Goldstein, 1976) except that here, of course, the RHS in any of its several possible forms is known exactly.

4. Nonuniqueness, Transient Decay Times, and Mean Values for Unsteady Oscillations in Transonic Flow

4.1 Early Work

Kerlick and Nixon (1981) have made the important point that, when using a finite difference, time marching computer code to investigate the lift on an oscillating airfoil in transonic flow, it is necessary to carry the solution sufficiently far forward in time that an essentially steady state solution is obtained. Moreover they offer a method for estimating the transient time before the steady state is reached. They note that, if one stops the time marching solution before the transient is complete and the steady state is reached, then one may reach the incorrect conclusion that a change in the mean lift has occurred due to the oscillating motion of the airfoil when in fact no such change has occurred.

Nevertheless what is perhaps surprising is that for a narrow Mach number range the time for the transient to decay and a steady state to be reached is extraordinarily long. Moreover, for a very narrow range of Mach number a non-zero mean value of lift can occur for an airfoil of symmetrical profile oscillating about a zero angle of attack.

4.2 Recent Work

In Kerlick and Nixon (1981) a NACA 64A006 airfoil was studied using the LTRAN2 computer code for a freestream Mach number of .875 with the airfoil oscillating at a peak angle of attack of .25° and with various reduced frequencies, k. It was shown that typically up to six cycles of airfoil oscillation must be considered for the mean lift to be less than 1% of the corresponding oscillatory lift peak value. Dowell et al. (1983) (aslo using the LTRAN2 computer code) examined various airfoils, 64A006, 64A010, MBB-A3, at various Mach numbers and reduced frequencies. The calculations in Dowell et al. (1983) were carried forward in time through six cycles of airfoil oscillation. As was found by Kerlick and Nixon, the results of Dowell et al. (1983) for the (symmetric) 64A006 airfoil showed the mean or average lift with the airfoil oscillating to be essentially unchanged from its value for no airfoil oscillation (i.e., zero) for **most** Mach numbers studied **after six cycles of oscillation.** However at $M_\infty = .88$ and particularly at $M_\infty = .9$ this was **not** the case. Hence the present author **incorrectly** concluded that a change in mean or average lift had occurred. The correct conclusion is that at $M_\infty = .9, .88$, many cycles of oscillation (> 40) are required for the mean lift to decay to essentially zero.

The results presented in Figure 7 are for $M_\infty = .9$, $k = .2$ and various peak oscillatory angle of attack (Dowell et al., 1983). They are for the lift coefficient; similar results (not shown for brevity) were obtained for moment (about mid-chord) coefficient and shock displacement (normalized by airfoil chord). Both mean steady values and first harmonic are shown **after six cycles of oscillation.** As can be seen the change in mean lift at this Mach number can be as much as 50% of the peak oscillatory lift. For $M = 0.88$ the mean or average components are never more than 10% of the first harmonic oscillatory components, and hence these results are not displayed.

Inspired by the note of Kerlick and Nixon, the case of $k = .2$ and a peak oscillating angle of $.5°$ was carried forward in time through many more cycles of airfoil oscillation for several M_∞. A typical result is shown in Figure 8 for $M_\infty = .9$ in terms of lift vs. angle of attack through 20 cycles of oscillation. Time varies along the (hysterisis) curve. As may be seen after twenty cycles the average lift is still distinctly different from zero and, moreover, is a substantial fraction of the oscillatory lift. The average lift is defined as

$$C_{L_{\text{AVG}}} = \frac{C_L^+ + C_L^-}{2}$$

and the oscillatory lift as

$$C_{L_{\text{OSC}}} = \frac{C_L^+ - C_L^-}{2}$$

where C_L^+, C_L^- are peak values, positive and negative respectively, of any two adjacent peaks.

The essential results can be summarized compactly in Figure 9 where $C_{L_{\text{AVG}}}$ is plotted vs. $1/N$ where N is the cycle number of the angle of attack oscillation. The results are plotted vs. $1/N$ so that the limit as $N \to \infty$ or $1/N \to 0$ may be more readily examined. As may be seen for $M_\infty = .86$ and $.875$ the average lift rapidly declines and is essentially zero for $N > 10$ or $1/N < .1$. (However it should be noted that the oscillatory lift, $C_{L_{\text{OSC}}}$, converges much more rapidly than the average lift. For the sake of brevity, $C_{L_{\text{OSC}}}$ is not shown.) By contrast for $M_\infty = .9$ and $.885$ the average lift persists in measurable values through a much larger number of cycles. For example, for $M_\infty = .9$ at $N = 40$ or $1/N = .025$, there is a clear trend toward zero average lift as $N \to \infty$ or $1/N \to 0$, but at $N = 6$ this would be difficult to perceive. Of course, these results have important practical consequences. A run for $N = 40$ takes 1672 sec of CPU time on an IBM 3033.

Hence at this point the present author concluded that while a non-zero average lift is mathematically possible for a nonlinear aerodynamic system responding to an oscillating angle of attack, no such lift was observed using LTRAN2 for the range of parameters

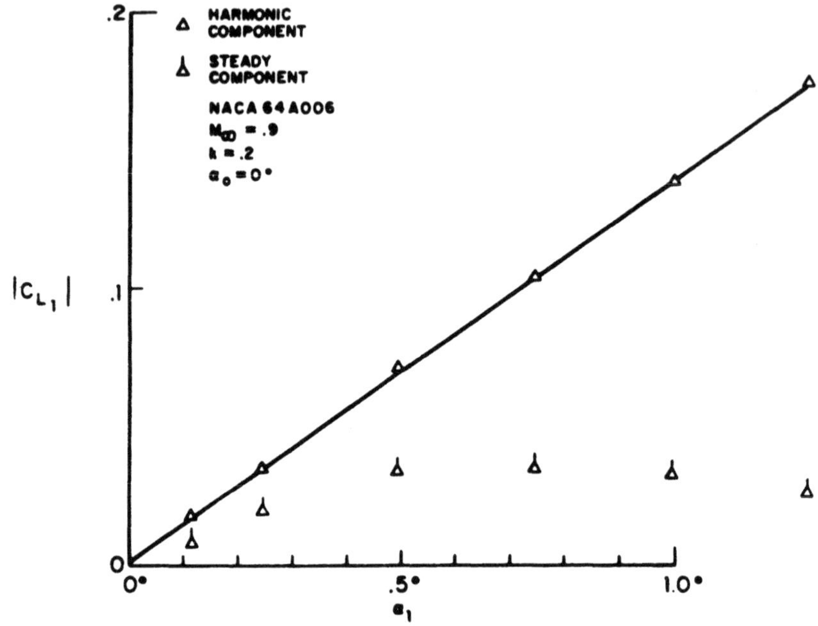

Figure 7. Steady state and first harmonic lift components vs. dynamic angle of attack, NACA 64A006, $M_\infty = .9$, $k = .2$, $\alpha_0 = 0°$.

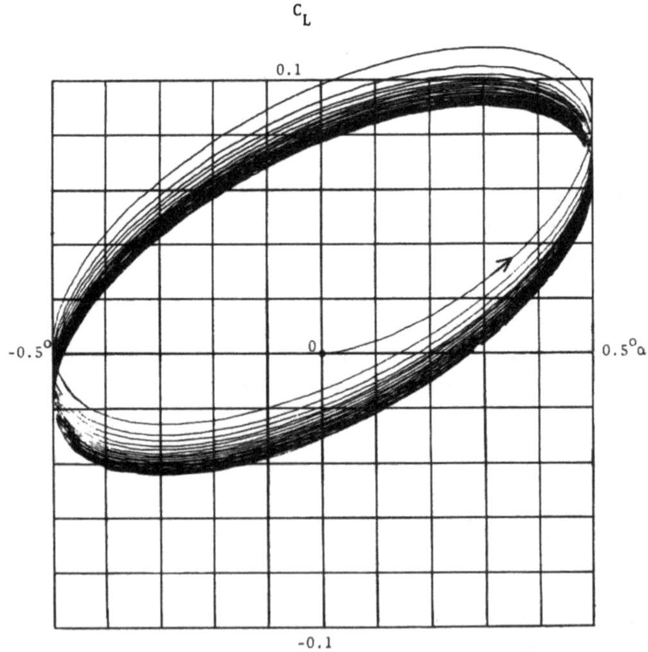

Figure 8. NACA 64A006 airfoil, $M_\infty = 0.9$, $k = 0.2$. Hysteresis curve of lift vs. dynamic angle of attack.

studied. However at some Mach numbers the time for the average lift to decay to zero is extraordinarily long.

Next the present author greatly benefited from a discussion with Dr. Peter Goorjian[19]. He had carried out calculations(Ballhaus and Goorjian, 1977) at $M_\infty = .89$ which indicated a non-zero average lift did occur. Dr. Goorjian's results are more fully discussed in Dowell et al. (1983) where an extended version of the present account is given. Hence the present author and his colleague, Dr. Ueda, also carried out calculations at this M_∞ and the results are shown in Figure 10 (analogous to Figure 8) and Figure 3 (analogous to Figure 9). These results clearly suggest that a non-zero average lift does occur at $M_\infty = .89$.

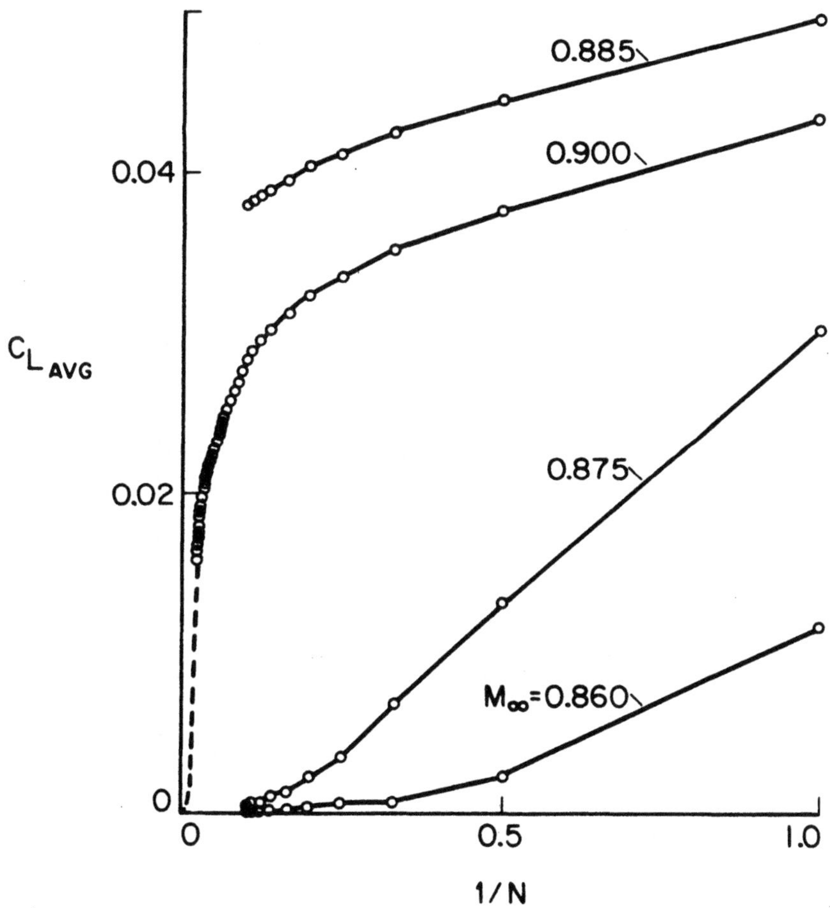

Figure 9. Average lift vs. the inverse of the number of cycles of oscillations.

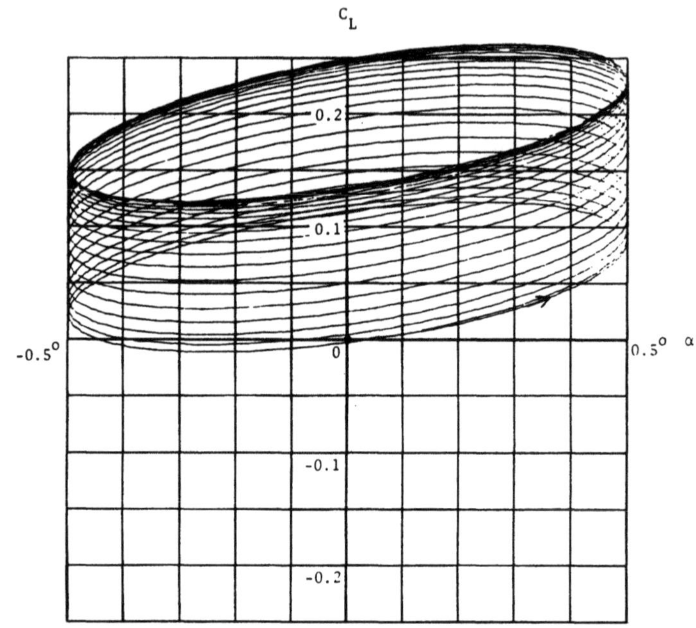

Figure 10a. NACA 64A006 airfoil, $M_\infty = 0.89$, $k = 0.2$, $N = 1$–20 cycles. Hysteresis curve of lift vs. dynamic angle of attack.

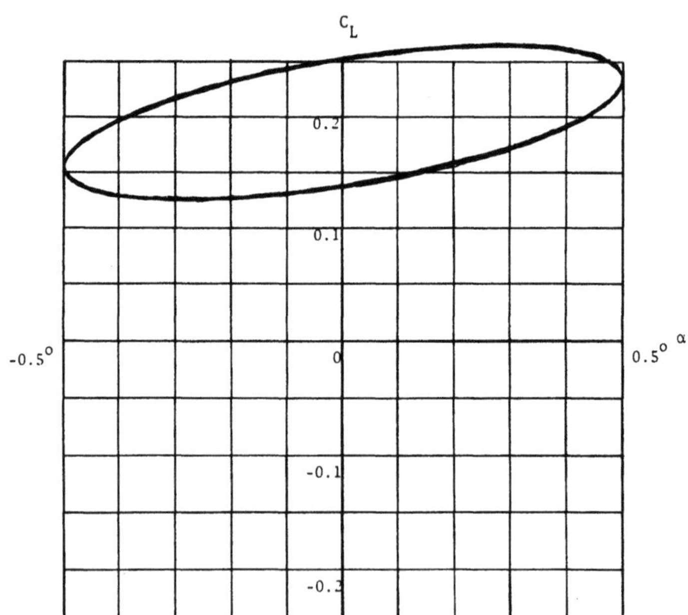

Figure 10b. NACA 64A006 airfoil, $M_\infty = 0.89$, $k = 0.2$, $N = 21$–40 cycles. Hysteresis curve of lift vs. dynamic angle of attack.

As explained by Goorjian, if one starts the airfoil oscillation with an initial negative angle of attack rather than a positive one, the mean lift is correspondingly negative (with the same magnitude) rather than positive. Indeed the entire hysterisis curve is a double mirror image with C_L and α both undergoing sign inversions. Moreover if the oscillating airfoil is now rendered motionless at exactly zero angle of attack, the mean lift persists at some finite, none-zero value. This implies that three solutions for lift occur at this single (zero) angle of attack. This result is fully consistent with that of Steinhoff and Jameson (1981) who obtain nonunique (multivalued) steady flow solutions by direct calculation for the full potential equations.

More recently, Salas (1983) has studied the full potential equations and the Euler equations for steady flow. For similar conditions he has found nonunique solutions for the full potential equations, but not for the Euler equations. However, it is very difficult to **prove** by numerical calculations, the absence of nonunique solutions for all possible conditions of interest.

Questions still remain of course.

- Why does this nonzero average lift occur only over a narrow range of Mach number? Note that the Mach number, $M_\infty \simeq .89$, at which nonunique solutions are observed corresponds to the Mach number for which the range of linear behavior is smallest as $k \to 0$. See Figure 5.
- At what level of mathematical modeling, if any, do nonunique solutions no longer occur?
- Is the result physically significant? In particular, what would be the counterpart, if any, for a viscous fluid model?

5. Effective, Efficient Computational Approaches for Determining Aeroelastic Response Using Unsteady Transonic Aerodynamic Codes

5.1 Various Computational Approaches and Their Relative Merits

The basic issue is whether

- one should run an aerodynamic code for **prescribed** airfoil motions and store the resultant aerodynamic data prior to carrying out a flutter analysis (Option I)

or whether

- one should run an aerodynamic code in conjunction with a structural dynamics code to determine simultaneously the time history of the airfoil and aerodynamic forces (Option II)?

There are, of course, two types of aerodynamic computer codes available, (1) those which calculate in the time domain (Ballhaus and Goorjian, 1977, 1978; Houwink and van der Vooren, 1980; Hessenius and Goorjian, 1982; Borland and Rizzetta, 1982a, 1982b; Borland et al., 1982; Rizzetta and Borland, 1983), and (2) those which calculate in the frequency domain (Williams, 1979a, 1979b; Hounjet, 1981b; Ehlers and Weatherill, 1982). The use of a code of the type (2), of course, precludes the pursuit of option II. However the use of a computer code of type (1) permits the use of option I or II.

Below each type of computer code will be considered in terms of how it may be used most effectively, and for a type (1) code the relative merits of options I and II will be discussed. Until otherwise noted, it will be assumed that we are anticipating a **linear** flutter analysis which only seeks to determine the conditions for the onset of flutter. A type (2) aerodynamic code implicitly assumes this to be the case. Again, as will be discussed later, use of a type (1) aerodynamic code will permit either a linear or a nonlinear flutter analysis to be conducted.

5.2 Type (1) Aerodynamic Code-Time Domain

Assume that flutter solutions are to be found for P parameter combinations (e.g., dynamic pressure values, tip tank masses, control surface stiffnesses) and M structural modes. Moreover assume that T_A is the computational time it takes for the aerodynamic code to reach a steady state lift value for a **prescribed** airfoil motion and T_F is the computational time it takes for a simultaneous fluid—structural time marching calculation to complete a transient. The relative sizes of T_A and T_F depend on airfoil profile and Mach number (for T_A) and structural damping, (T_F). Generally $T_F > T_A$, but there are exceptions. For simplicity, think of the Mach number as fixed, since calculations at several M_∞ will simply increase all computations by the same factor.

Consider now the relative merits of options I and II.

Option I: Generate and Store Aerodynamic Data Prior to Aeroelastic Calculation

The total computational time to generate the aerodynamic forces is

$$M * T_A \tag{5.1}$$

which is independent of P. There is some additional time required for flutter solutions per se, but it is assumed this is negligible compared to the time required to generate the aerodynamic forces.

Option II: Generate Aerodynamic Data and Structural Data Simultaneously

The total computational time will be

$$\mathbf{P} * T_F \tag{5.2}$$

which, of course, is independent of M. Clearly which of the two options is most attractive depends upon wheter $M * T_A \gtrless \mathbf{P} * T_F$. For option II to be more attractive, the number of modes, M, should be somewhat larger than the number of parameters, P. Thus option II will be more attractive in a design verification study while option I will tend to be more attractive in a preliminary design phase.

5.3 Type (2) Aerodynamic Code-Frequency Domain

Here only option I has been used to date, however see the discussion below. Let NF be the number of reduced frequencies needed for the flutter analysis. Let T_{AF} be the time for the aerodynamic code to determine the aerodynamic forces for one frequency. Assume that an aerodynamic influence coefficient approach is used so that the number of modes does not influence the computational time.*

---

*If a relaxation scheme is used rather than direct inversion of the aerodynamic matrix, then another set of issues arises.

Option I:

The computational time to generate the aerodynamic forces is

$$NF * T_{AF} \tag{5.3}$$

Compare equation 5.3 to the computational time associated with time domain aerodynamics. See previous discussion and equation 5.1.

$$NF * T_{AF} \gtrless M * T_A$$

If T_{AF} and T_A are comparable (one might expect competition would tend to make them so), then the method of choice as between equations (5.1) and (5.3) will depend on the number of frequencies, NF, compared to the number of modes, M, needed in the aeroelastic calculations.

Option IA:

Although this approach has not been pursued to date, one could take the frequency domain aerodynamic forces, curve fit them with Pade Approximants or comparable representation(Dowell,1980), use these to deduce a differential equation aerodynamic force representation (Tran and Petot, 1980; McIntosh [31]), and then do a time marching flutter solution. The computational time would still be approximately

$$NF * T_{AF} \tag{5.3a}$$

The comparable computational time using a direct time-marching *aerodynamic* code was (see previous discussion)

$$P * T_F \tag{5.2a}$$

Assuming $T_{AF} = T_F$, one concludes that, if the number of parameters is large compared to the number of reduced frequencies needed, then the frequency domain aerodynamic method tends to be the method of choice over the time domain method which uses option II.

5.4 Summary Comparison

By comparing the estimates, equations 5.1, 5.2, 5.3, 5.3a, one may make an initial judgement as to the method of choice in a given situation.

5.5 Nonlinear Flutter Analysis

5.5.1 General Considerations

If a nonlinear flutter analysis is needed then only the time domain *aerodynamic* method is available, type (1).

Of course, the *aeroelastic* calculation may be done in *either* the time *or* frequency domain. However there is still a trade-off between options I and II. Now, for option I, a further multiplicative factor must be used which is the number of airfoil response levels, NR, which are of interest. For linear flutter analysis only *one* response level is of interest (strictly speaking an infinitesimal response level which approaches zero). Thus the computational times to compare are

Option I: $NR * M * T_A$

Option II: $P * T_F$.

It should be noted moreover that for nonlinear flutter analysis, the number of parameters, P, will tend to be somewhat larger than for linear flutter analysis.

The relative attractiveness of the two options is as before but with a bias shift toward option II, because of the factor NR appearing in option I. The use of option I in a nonlinear flutter analysis does in fact lead to a *flutter* analysis in the *frequency* domain, and the methodology by which that is done is described below in Section 6. Of course this methodology reduces to the classical frequency domain flutter solution method when the airfoil response levels are small. The solution procedures for option II as currently practiced are straightforward and will not be elaborated upon further here. It is also worthy of note here that

- a frequency domain *nonlinear* flutter analysis usually introduces approximations beyond those of a time domain analysis, but
- fortunately, *linear* flutter analysis will often suffice and hence the whole set of questions is frequently moot.

6. Nonlinear Flutter Analysis in the Frequency Domain and Comparison With Time Marching Solutions

The nonlinear effects of transonic aerodynamic forces on the flutter boundary of a typical section airfoil are discussed. The flutter speed dependence on amplitude is obtained by utilizing a novel variation of the describing function method which takes into account the first fundamental harmonic of the nonlinear oscillatory motion. By using an aerodynamic describing function, traditional frequency domain flutter analysis methods may still be used while including (approximately) the effects of aerodynamic nonlinearities. Results from such a flutter analysis are compared with those of brute force time marching solutions. The aerodynamic forces are computed by the LTRAN2 aerodynamic code for a NACA 64A006 airfoil at $M_\infty = 0.86$.

6.1 Motivation and Background

Recent developments in computational aerodynamics have led to renewed interest in the prediction of flutter boundaries of an airfoil in the transonic flow regime (Ballhaus and Goorjian, 1978; Yang et al., 1979; Isogai, 1980). Until recently flutter calculations have either assumed the transonic aerodynamic forces could be approximated as *linear* functions of the airfoil motion so that traditional *linear* flutter analysis methods could be used or, alternatively, taken a brute force approach by simultaneously numerically integrating in time the structural and aerodynamic equations. The latter method does, of course, fully account for aerodynamic nonlinearities.

Ballhaus and Goorjian(1978) calculated the aeroelastic response of a NACA 64A006 airfoil with a single-degree-of-freedom control surface by simultaneously integrating numerically in time the structural equation of motion and also the aerodynamic equations. They used their own code, LTRAN2, for unsteady transonic flow. The indicial method, whereby an aerodynamic impulse function is first calculated by the aerodynamic code and then used in the flutter calculation via a convolution integral, was also studied. The indicial method assumes linearity of the aerodynamic forces with respect to airfoil motion. The flutter of the same airfoil but with two-degrees-of-freedom was analyzed by Yang et al. (1979) with aerodynamic forces obtained by three different methods. These forces were obtained by the time integration method, the indicial method (both of these employed the LTRAN2 code) and the harmonic analysis method in the frequency domain using the UTRANS2 code. The latter method also assumes linearity of the aerodynamic forces. In general, all three methods agree well for the range of parameters studied by Yang. After the flutter boundary was obtained, the response was confirmed near the flutter boundary by simultaneous time integration of the governing structural and aerodynamic equations. Isogai (1980) studied the transonic behavior of the NACA 64A010 airfoil by using his own USTS transonic aerodynamic code which can be applied to supercritical Mach numbers for reduced frequencies, $0 < k < 1.0$. (By contrast, the aerodynamic methods used by Yang were limited to small k.) Isogai used the time integration method for evaluating the aerodynamic forces, but then converted them to linearized harmonic forces for the flutter calculations. See Yang et al. (1981), for further discussion of previous work on flutter calculations including that of other investigators who have used brute force, simultaneous numerical integration of the structural and aerodynamic equations. A discussion of when the aerodynamic forces may be treated as linear in the airfoil motion is given in Dowell et al. (1983) and the present paper. The analysis of Yang et al. (1979) and Isogai (1980) described above assumed linear characteristics for the aerodynamic forces in the flutter calculations. Linearity can be assured if the amplitude of the airfoil oscillation is sufficiently small (Dowell et al., 1983), even though the governing fluid equations are inherently nonlinear for transonic flow fields. Yang et al. (1979) fixed an amplitude of pitching motion at 0.01 radian (0.574°) whereas Isogai (1980) used 0.1 in degrees for the computation of the aerodynamic forces. Dowell et al. (1983) pointed out that increasing the value of the reduced frequency increases the range of amplitude of oscillation for which linear behavior exists in transonic flow. However, the aerodynamic forces often begin to deviate from linear behavior for amplitudes of relatively small value such as 1.0° in pitching motion. Such amplitudes may be attained due to the disturbances an aircraft wing encounters during its flight. It is of importance, therefore, to clarify the aerodynamic nonlinear effect on a flutter boundary, especially when the nonlinear effect may create an aeroelastic softening system, i.e., the flutter speed decreases as the amplitude of oscillation increases. Such softening behavior may cause a dangerous unconservative estimation of a flutter boundary by linear analysis.

Here we study the nonlinear effect of transonic aerodynamic forces on a flutter boundary by utilizing a novel variation of the describing function method (Hsu and Meyer, 1968) which takes into account the first fundamental harmonic of the nonlinear oscillatory motion. By using an aerodynamic describing function, traditional frequency domain flutter analysis methods may still be used while including (approximately) the effects of aerodynamic nonlinearities. Brute force time marching calculations are also presented for comparison purposes.

The method used to calculate the describing functions is briefly this. A step change in angle of attack is specified and the transient aerodynamic force time history (calculated numerically by an appropriate aerodynamic code) is identified as a nonlinear impulse function. The Fourier transform of this impulse function (which in general depends upon the step input level or amplitude) is taken as the aerodynamic describing function (nonlinear transfer function). Calculations have shown that this describing function agrees very well with the one determined by using a harmonic angle of attack input to the aerodynamic code. The latter method of calculation is, of course, much more expensive and time consuming for the range of frequencies needed in flutter analysis. The LTRAN2 computer code is used for determining the aerodynamic forces. However any other nonlinear code could be used in a similar fashion.

6.2 Typical Airfoil Section

A typical airfoil section subjected to transonic flow is considered as shown in Figure 12. Since it can be assumed that the structural deformation is linearly dependent on the aerodynamic load for wings of ordinary modern aircraft during its normal flight, a *linear* structural transfer function is used. The aerodynamic force, however, may depend in a nonlinear manner on the structural deformation in the transonic flow range (Dowell et al., 1983). In order to include the nonlinear effect of large(r) amplitudes of motion on the aerodynamic forces and, hence, on a flutter boundary, we use a *nonlinear* aerodynamic transfer function by employing the describing function method.

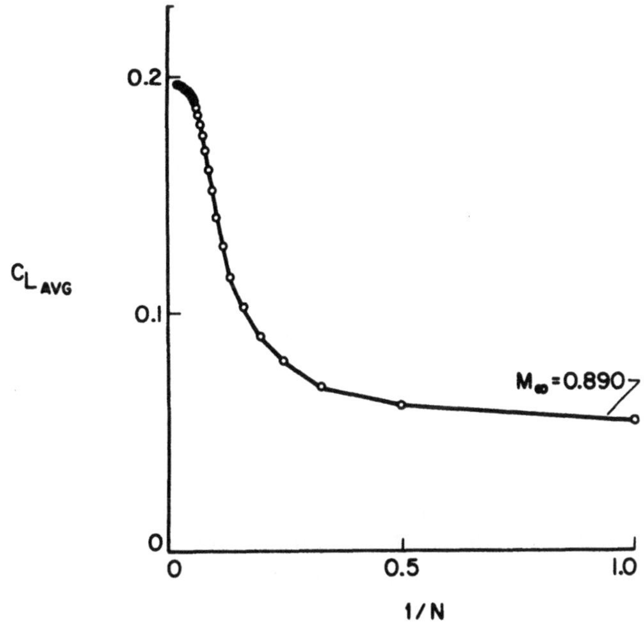

Figure 11. Average lift vs. the inverse of the number of cycles of oscillations. $M_\infty = .89$.

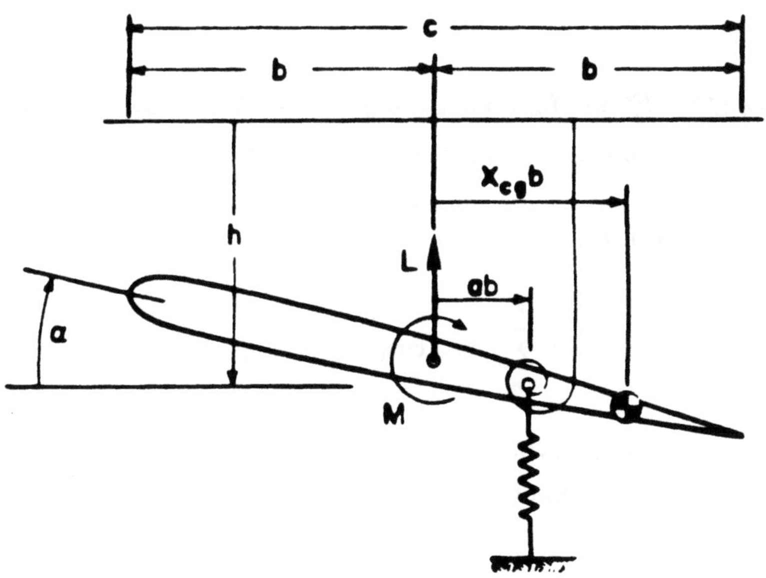

Figure 12. Typical section airfoil.

6.3 Aerodynamic Describing Function

Here we give a summary of the relevant standard describing function results and place the present method in context.

If we assume that the frequency of motion is relatively low, the aerodynamic forces due to the airfoil motion can be approximated as a function of the effective induced angle-of-attack which is given by

$$\phi = \alpha + \frac{\dot{h}_c}{u} \tag{6.1}$$

This quasi-steady approximation is compatible with the low frequency assumption in the LTRAN2 transonic unsteady aerodynamic code which we use in the present flutter calculation. Taking into account the nonlinear effects of the amplitudes of motion, we assume the aerodynamic forces take the form:

$$L = \frac{1}{2}\rho u^2 c C_L^N(\phi, \dot{\phi}), \tag{6.2}$$

$$M = \frac{1}{2}\rho u^2 c^2 C_M^N(\phi, \dot{\phi}), \tag{6.3}$$

where C_L^N, C_M^N are functionals of ϕ, $\dot{\phi}$, i.e., they may, in principle, include the complete time history of ϕ and $\dot{\phi}$.

For general periodic time dependent motion, the effective angle-of-attack ϕ can be expanded in a Fourier series as

$$\phi = \frac{\phi_o}{2} + \sum_n^N [\phi_{I,n} \cos(nk\tau) + \phi_{R,n} \sin(nk\tau)]. \tag{6.4}$$

According to the describing function method, only the first harmonic of ϕ is taken as an input to the aerodynamic force transfer function, i.e.,

$$\phi = \phi_1 \sin k\tau. \tag{6.5}$$

This input motion generates aerodynamic forces through the nonlinear fluid element, call it H, which, in general includes higher harmonics. Thus

$$C_L^N(\phi,\dot{\phi}) = \frac{C_{L_{1,0}}(\phi_1)}{2} + \sum_n^N \left[C_{L_{I,n}}(\phi_1)\cos(nk\tau) + C_{L_{R,n}}(\phi_1)\sin(nk\tau)\right], \quad (6.6)$$

$$C_M^N(\phi,\dot{\phi}) = \frac{C_{M_{1,0}}(\phi_1)}{2} + \sum_n^N \left[C_{M_{I,n}}(\phi_1)\cos(nk\tau) + C_{M_{R,n}}(\phi_1)\sin(nk\tau)\right], \quad (6.7)$$

The describing function method, however, replaces the nonlinear element H by another nonlinear element $\hat{H}$ with the property that it operates on any sinusoidal input, equation 6.5, by passing its fundamental frequency in exactly the same manner as H; however, whatever the input frequency k, $\hat{H}$ generates no higher harmonics. This replacement allows us to write

$$\begin{aligned}\hat{C}_L^N(\phi,\dot{\phi}) &= C_{L_{I,1}}(\phi_1)\cos(k\tau) + C_{L_{R,1}}(\phi_1)\sin(k\tau) \\ &= D_{L_R}(\phi_1)\phi + D_{L_I}(\phi_1)\dot{\phi}/k,\end{aligned} \quad (6.8)$$

$$\begin{aligned}\hat{C}_M^N(\phi,\dot{\phi}) &= C_{M_{I,1}}(\phi_1)\cos(k\tau) + C_{M_{R,1}}(\phi_1)\sin(k\tau) \\ &= D_{M_R}(\phi_1)\phi + D_{M_I}(\phi_1)\dot{\phi}/k,\end{aligned} \quad (6.9)$$

where

$$D_{L_R} = \frac{1}{\pi\phi_1}\int_0^{2\pi} C_L^N(\phi,\dot{\phi})\sin(k\tau)\,d(k\tau), \quad (6.10)$$

$$D_{L_I} = \frac{1}{\pi\phi_1}\int_0^{2\pi} C_L^N(\phi,\dot{\phi})\cos(k\tau)\,d(k\tau), \quad (6.11)$$

$$D_{M_R} = \frac{1}{\pi\phi_1}\int_0^{2\pi} C_M^N(\phi,\dot{\phi})\sin(k\tau)\,d(k\tau), \quad (6.12)$$

$$D_{M_I} = \frac{1}{\pi\phi_1}\int_0^{2\pi} C_M^N(\phi,\dot{\phi})\cos(k\tau)\,d(k\tau), \quad (6.13)$$

Using complex notation for equations 6.8 and 6.9 yields a more compact result, i.e.,

$$\hat{C}_L^N(\phi, \dot{\phi}) = D_{L_1}(\phi_1)\,\phi, \tag{6.14}$$

$$\hat{C}_M^N(\phi, \dot{\phi}) = D_{M_1}(\phi_1)\,\phi, \tag{6.15}$$

where

$$D_{L_1}(\phi_1) = D_{L_R} + iD_{L_I} \tag{6.16}$$

$$D_{M_1}(\phi_1) = D_{M_R} + iD_{M_I} \tag{6.17}$$

In equations 6.14 and 6.15, the coefficients, $\hat{C}_L^N$ and $\hat{C}_M^N$, also have complex values whose real parts correspond to equations 6.8 and 6.9, respectively.

If the amplitude ϕ_1 is fixed, the equations 6.14 and 6.15 take a form identical to that for a linear system. This implies the applicability of the same stability analysis as that for a linear system.

The coefficients defined in equations 6.10–6.13 to construct the describing function can be computed by a time integration code for transonic flow. It is also possible to evaluate them from wind-tunnel experimental data measured on a harmonically (or impulsively) excited airfoil. In the present study, we utilize an *extended nonlinear* version of the indicial method (Ballhaus and Goorjian, 1978) to calculate the aerodynamic coefficients.

Since the describing function assumes the same form as a linear transfer function when the amplitude is fixed, we can regard a typical such element, $\hat{H}_\phi$, which relates any representative aerodynamic forces, F, to airfoil motion, ϕ, as a linear system *with respect to variations in frequency.*

$$F(\tau) = \hat{H}_\phi\,(ik, \phi_1)\,\phi_1 e^{ik\tau} \tag{6.18}$$

This relation corresponds in the subsidiary (effectively frequency) domain of the Laplace operator to the following:

$$\overline{F}(s) = \hat{H}\,(s, \phi_1)\frac{\phi_1}{s - ik} \tag{6.19}$$

If we put $k = 0$, then equation 6.19 represents an indicial response relationship.

$$\overline{\hat{A}}(s, \phi_1) = \hat{H}_\phi(s, \phi_1)\frac{\phi_1}{s} \tag{6.20}$$

By using equation 6.20 we can obtain the describing function, $\hat{H}_\phi(ik, \phi_1)$, from the indicial response to a step input with amplitude ϕ_1. Furthermore, if we neglect the effect of higher harmonics, an assumption already made in the describing function method, then $\hat{H}_\phi(ik, \phi_1)$ can be approximated by using the indicial response $\overline{A}(s, \phi_1)$ of the element H as

$$\hat{H}_\phi(ik, \phi_1) = \overline{A}(ik, \phi_1)\, ik/\phi_1. \tag{6.21}$$

From a linear system, starting from (6.18) one may proceed through (6.19), (6.20) to (6.21) and vice versa by standard mathematical methods. However, as the careful reader will note this is not strictly possible for a nonlinear system, i.e., (6.19) and (6.20) follow from (6.18) only by *analogy* to linear system results. Indeed we may take (6.18) and (6.21), (or (6.20)) as two *independent* definitions of $\hat{H}_\phi(ik, \phi_1)$, the describing function which will be used in the flutter analysis. However by numerical example we will show that, in fact, the two definitions lead to similar results. This is fortunate, because the less obvious definition, (6.21), is far easier to use in practice for generating aerodynamic forces to employ in flutter calculations.

6.4 Working Form of the Aeroelastic System Equations

The governing structural equations of the system are given in nondimensional form by

$$\frac{\pi\mu}{2}\left(\frac{h}{c}\right)'' + \frac{\pi\mu}{2}\left(\frac{S_\alpha}{mc}\right)\alpha'' + \frac{\pi\mu}{2}\left(\frac{c\omega_h}{u}\right)^2\left(\frac{h}{c}\right) = -C_L^N, \tag{6.22}$$

$$\frac{\pi\mu}{2}\left(\frac{S_\alpha}{mc}\right)\left(\frac{h}{c}\right)'' + \frac{\pi\mu}{2}\left(\frac{I_\alpha}{mc^2}\right)\alpha'' + \frac{\pi\mu}{2}\left(\frac{I_\alpha}{mc^2}\right)\left(\frac{c\omega_\alpha}{u}\right)^2\alpha = C_{Me}^N. \tag{6.23}$$

From (6.22) and (6.23) the structural transfer function for the state vector, $[(h/c)\alpha]^T$, is

$$G^{-1}(s,U) = \begin{bmatrix} \frac{\pi}{2}\{\mu s^2 + R^2/U^2\} & \frac{\pi\mu}{4}(x_{cg} - a)\, s^2 \\ \frac{\pi\mu}{4}(x_{cg} - a)\, s^2 & \frac{\pi}{8} r_\alpha^2 \{\mu s^2 + 1/U^2\} \end{bmatrix}. \tag{6.24}$$

As to the aerodynamic describing function, we first assume the indicial response $A(\tau, \phi_1)$ to a step change in ϕ in (6.21) can be expressed in the following form for the lift and moments forces,

$$A_L(\tau,\phi_1) = a_0^L(\phi_1) + \sum_{i=1}^{N} a_i^L(\phi_1)\, e^{b_i^L \tau}, \tag{6.25}$$

$$A_M(\tau,\phi_1) = a_0^M(\phi_1) + \sum_{i=1}^{N} a_i^M(\phi_1)\, e^{b_i^M \tau}, \tag{6.26}$$

where a_o^L and a_o^M are chosen to be identical to the steady state values for $\phi = \phi_1$, since every b_i is chosen to be a negative real number. The coefficients, a_i^L, a_i^M are determined by the least square method for fixed values of the b_i's. The b_i are selected to be in the vicinity of the negative of the k values for which the imaginary parts of the aerodynamic transfer function have extrema. This procedure for selecting the b_i is discussed in detail in Dowell (1980).

After determining the coefficients in equations 6.25 and 6.26, the indicial response functions can be written in the subsidiary domain of the Laplace operator (frequency domain) as,

$$\overline{A}_L(s,\phi_1) = \frac{a_o^L}{s} + \sum_{i=1}^{N} \frac{a_i^L}{s - b_i^L}, \tag{6.27}$$

$$\overline{A}_M(s,\phi_1) = \frac{a_o^M}{s} + \sum_{i=1}^{N} \frac{a_i^M}{s - b_i^M}. \tag{6.28}$$

Then from equation 6.21, the aerodynamic describing function is obtained for the state variable ϕ as

$$\hat{H}_\phi(ik,\phi_1) = \left[\overline{D}_L(ik,\phi_1)\, D_M(ik,\phi_1)\right]^T \tag{6.29}$$

where

$$\overline{D}_L(ik,\phi_1) = ik\overline{A}_L\,(ik,\phi_1)/\phi_1 \tag{6.30}$$
$$\overline{D}_M(ik,\phi_1) = ik\overline{A}_M\,(ik,\phi_1)/\phi_1 \tag{6.31}$$

In order to construct the aerodynamic describing functions so that they are compatible with the structural transfer function, we must transform ϕ and C_M^N to those variables used in the structural equations of motion. The relationships for the state vectors and the moment coefficients are as follows:

$$\overline{\phi}(ik) = \left(ik\;1 - \frac{a}{2}ik\right)\begin{bmatrix}\overline{h/c}\\ \overline{\alpha}\end{bmatrix}, \tag{6.32}$$

$$C_{Me}^N = C_M^N + \frac{a}{2}\,C_L^N. \tag{6.33}$$

As the aerodynamic describing function $\hat{H}(ik,\phi_1)$ is defined by

$$\begin{bmatrix}-\overline{C}_L^N\,(ik,\phi_1)\\ -\overline{C}_{Me}^N\,(ik,\phi_1)\end{bmatrix} = \hat{H}\,(ik,\phi_1)\begin{bmatrix}\overline{h/c}\\ \overline{\alpha}\end{bmatrix} \tag{6.34}$$

it becomes using 6.30–6.34

$$\hat{H}\,(ik,\phi_1) = \begin{bmatrix}A_{11} & A_{12}\\ A_{21} & A_{22}\end{bmatrix} \tag{6.35}$$

where

$$\begin{aligned}
A_{11} &= -\overline{D}_{L_1}\,(ik,\phi_1)\cdot ik\\
A_{12} &= -\overline{D}_{L_1}\,(ik,\phi_1)\left(1 - \frac{a}{2}ik\right)\\
A_{21} &= \overline{D}_{L_1}\,(ik,\phi_1)\frac{a}{2}ik + \overline{D}_{M_1}\,(ik,\phi_1)\cdot ik\\
A_{22} &= \overline{D}_{L_1}\,(ik,\phi_1)\frac{a}{2}\left(1 - \frac{a}{2}ik\right) + \overline{D}_{M_1}\,(ik,\phi_1)\left(1 - \frac{a}{2}ik\right)
\end{aligned}$$

Using the structural transfer function equation 6.24 and the aerodynamic describing function equation 6.35, a self-sustained oscillation of the system shown in Figure 2 is characterized by the equation

$$|G^{-1}(ik, U) - \hat{H}(ik, \phi_1)| = 0. \tag{6.36}$$

Equation 6.36 corresponds to the so-called flutter determinant if the system is linear. For the present nonlinear system, (6.36) allows one to determine the amplitude of the flutter motion as a function of some system parameter, say airspeed, U.

6.5 Extension of the Describing Function

In the earlier discussion of **aerodynamic describing functions**, we assumed the aerodynamic forces can be given as functions of a single variable, ϕ. More rigorously, however, the upwash of the mean camber line of an airfoil is given by

$$-\frac{w}{u} = \phi + \frac{c\dot{\alpha}}{u}(x/c - 0.5) \tag{6.37}$$

where the airfoil is located on $0 \leq x \leq c$. The second term in equation 6.37 includes the effect of the angular velocity of the airfoil motion, $\dot{\alpha}$. If we take into account this aerodynamic effect in the flutter analysis, the aerodynamic describing function must be determined separately for h- and α-motion, or alternatively for ϕ and α. Moreover this procedure makes the describing function method less obvious as to its theoretical basis. Neglecting the effect of $\dot{\alpha}$, however, we encountered fictitious instabilities at high frequencies as well as a decrease of flutter speeds. To eliminate this artifact, an improvement was made to the aerodynamic describing function by adding the component that is derived from the second term of equation 6.37. This is based on the assumption that the $\dot{\alpha}$ effect compared to ϕ is generally secondary for the aerodynamic forces at low reduced frequencies, k, where nonlinear transonic aerodynamic effects are most significant. Thus, the components of the describing function, equation 6.35, are redefined by

$$\begin{aligned} A_{12} &= A_{12} - \overline{D}_{L_2}(ik) \cdot ik \\ A_{22} &= A_{22} + \overline{D}_{L_2}(ik)\frac{a}{2} \cdot ik + \overline{D}_{M_2}(ik) \cdot ik \end{aligned} \tag{6.38}$$

where A_{12} and A_{22} on the right hand side of equation 6.38 are those of equation 6.35. $\overline{D}_{L_2}$ and $\overline{D}_{M_2}$ are the components obtained from an indicial response of the second term of equation (6.37). For brevity, details are omitted.

6.6 Results and Discussions

6.6.1 Aerodynamic Results

All the original aerodynamic data in the following were computed by the LTRAN2 (Ballhaus and Goorjian, 1977) code for a NACA 64A006 airfoil with Mach number set to $M_\infty = 0.86$ where characteristically transonic effects can be observed. The zero angle of attack, steady state shock stands roughly at mid-chord (Dowell et al., 1983). Results were also calculated for a NACA 64A010 airfoil, but these are omitted here.

In this calculation a 113×97 finite difference mesh was employed. The non-dimensional time increment, $\Delta\tau$, for integration was selected at $\pi/12$ to obtain *indicial* responses. The lift and moment forces at every five time increments were used to evaluate the coefficients in equations 6.25 and 6.26. In order to compare the aerodynamic forces with those obtained by the present extended nonlinear *indicial* method, the time integration for the airfoil undergoing *harmonic* excitation of pitching motion about mid-chord axis was also performed with 120 time steps per cycle at various reduced frequencies. The *latter* results (using (6.18)) were compared to those obtained by the *indicial* method using the Laplace transform versions of equations 6.25 and 6.26 which derive from equation 6.21. In general, good agreement was obtained.

The indicial response for lift to step functions of different amplitudes, ϕ_1, are shown in Figure 13. Generally, in this type of indicial response, a spike should appear near $\tau = 0$. However, the low frequency theory of LTRAN2 can not follow a piston-wave-type pressure change because of its infinite propagation rate (Ballhaus and Goorjian, 1978). Hence, in the present calculations the lift coefficient increases gradually from zero at $\tau = 0$.

In the results of Figure 13, the curves prescribed by equation 6.25 are shown after the coefficients were determined by the least square method using 64 time data points for the indicial response at each amplitude. The b_i^L's and b_i^M's, are selected at six values ($N = 6$) which are -0.01875, -0.0375, -0.075, -0.15, -0.3, and -0.6. These results obtained from equation 6.25 are in excellent agreement with the indicial responses computed by the LTRAN2 code, especially for the lower amplitude values of step inputs. Using the coefficients a_i^L, a_i^M, and equation 6.21 we can obtain the elements of the aerodynamic functions, $\overline{D}_{L_1}(ik, \phi_1)$ and $\overline{D}_{M_1}(ik, \phi_1)$. The real parts of the former are shown in Figure 14. They are plotted for reduced frequencies up to 0.3. Although the describing function for higher frequencies can be calculated by equations 6.30 and 6.31, they are no longer meaningful at those frequencies because of the low frequency limitation (Ballhaus and Goorjian, 1977) in LTRAN2.

In Figure 14, the describing functions thus obtained are also compared with the results of the time integration method for simple harmonic motion inputs which use equations 6.10–6.13 (see also 6.18). The agreement between the two methods is generally satisfactory. However, it was seen that the agreement is better for smaller amplitude than larger ones, for lift than for moment, and for real part than for imaginary part (Ueda and Dowell, 1982; 1983). It should be emphasized that the extended nonlinear indicial method has substantially greater simplicity and efficiency in determining the aerodynamic describing function, as compared to the time integration method for simple harmonic motion inputs.

6.6.2 Flutter Results

Some flutter calculations have been done using these aerodynamic describing functions for typical section airfoils. First, the parameters of a typical section airfoil were chosen to compare with the results by Yang et al. (1980). A comparison is made in Ueda and Dowell (1982). The flutter boundary calculated by the present method for an amplitude, ϕ_1, between 0.5° and 1.0° agrees well with that obtained from the *linear* indicial method by Yang et al. (1980).

To investigate the amplitude effect on the flutter boundary, a typical section airfoil corresponding to case B in Isogai (1980) was considered next, although the results cannot be compared directly with those in Isogai (1980) due to the use of a different airfoil profile. The results of the flutter speed as well as of the reduced frequency, bending/torsion amplitude ratio and phase, are shown in Figure 15. Those without the $\dot{\alpha}$ effect are depicted by dashed curves. In this case, the effect of angular velocity on flutter boundaries was very small (Ueda and Dowell, 1982; 1983).

As the aerodynamic describing function method invokes several assumptions, a fully nonlinear time marching solution was computed to verify the above results. The numerical integration scheme adopted for structural equations is the state transion matrix method which Edwards et al. (1982) recommended after examining seven different integrators for a similar calculation (I2 in Edwards et al.,

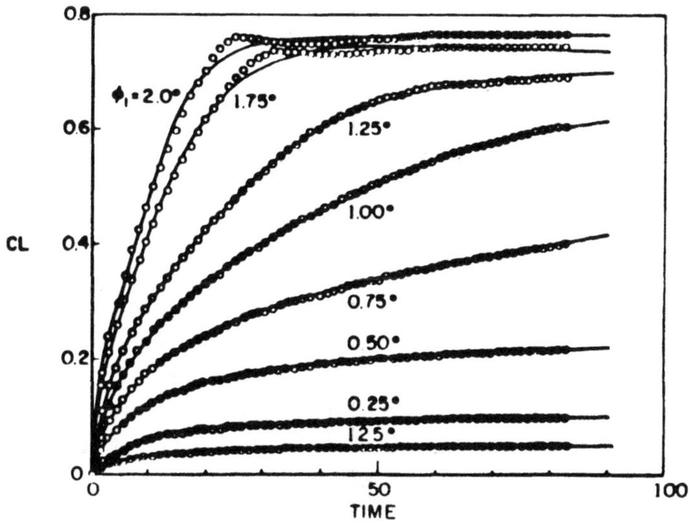

Figure 13. Indicial responses computed by LTRAN2 and exponential curve fit: lift.

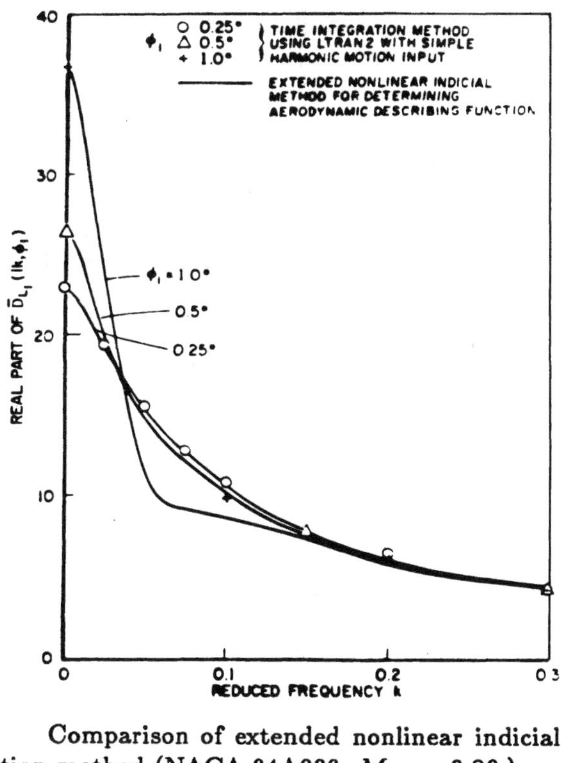

Figure 14. Comparison of extended nonlinear indicial method with time integration method (NACA 64A006, $M_\infty = 0.86$.)

1982). Three time marching calculations with different speed parameters have been carried out for the case shown in Figure 15. The initial state vector was determined from the flutter solution of the describing function method with $\phi_1 = 0.5°$, namely, $U_F = 0.1798$, $k_F = 0.2126$, $|\frac{\bar{h}}{c}/\bar{\alpha}| = 5.446$, and $\phi_{h\alpha} = 4.24°$. If we choose the initial time, $\tau = 0$, as the instant when $\dot{\alpha} = 0$, the flutter solution gives the initial state vector for the time marching as $\underset{\sim}{x} = (0.03164, 0.00583, -0.00049, 0)^T$. The time increment for integration was selected as $\Delta\tau = 0.25$ which, considering the flutter reduced frequency, corresponds roughly to 120 step per cycle. Although the initial state vector is determined from the describing function flutter motion, the time marching is started from a steady-state initial condition of the airfoil at a static angle of attack for the aerodynamic calculations. For example, the initial effective induced angle of attack becomes 0.305 degrees for this case. It should be noted that the second term of equation 6.37 vanishes at the initial upwash since the starting time is set at the instant when the angular velocity becomes zero. The time marching is continued up to $\tau = 250$, which contains one thousand time steps. The variations of the amplitude ϕ_1 of these solutions are shown in Figure 15 and the time histories in Figure 16. At $U = 0.16$, the airfoil shows decaying motion whereas the oscillation is growing at $U = 0.2$. At $U = 0.19$, the oscillation is almost neutrally stable although it is slightly growing. The changes of the peak values in the effective angle of attack of these oscillations are illustrated in Figure 15. The solid line shows the flutter boundary (limit cycle curve) obtained by the describing function method. In this case, the flutter boundary is nearly horizontal at small amplitudes. As can be seen from the figure, the results from the describing function method agree well with those of the time marching solution. Furthermore, the last cycle of the time marching solution at $U = 0.19$ gives the values of $\phi_1 = 0.723°$, $k = 0.212$, $|\frac{\bar{h}}{c}/\bar{\alpha}| = 5.18$, and $\phi_{h\alpha} \simeq 5°$. The agreement of these values with the results in Figure 15 is excellent. It should be noted that the damping in the time marching solutions is attributed to the aerodynamic forces, since we use no structural damping nor artificial damping due to numerical integration schemes. It is known that the transition matrix integrator gives exactly neutral solutions for free structural vibration irrespective of its time step size.

Since the nonlinear effect is most important at relatively *low* reduced frequencies (see Figure 14), the center of gravity was next

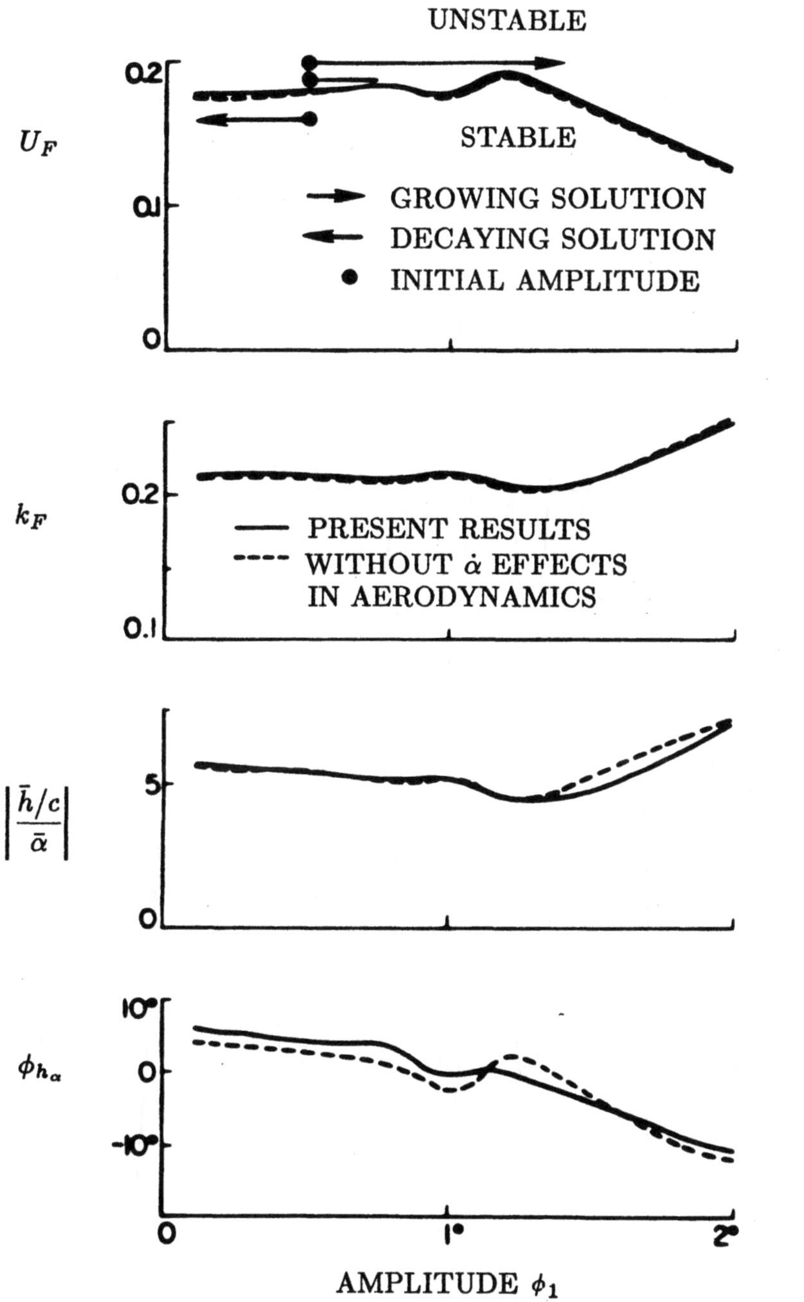

Figure 15. Flutter parameters vs. amplitudes, (NACA 64A006, $M_\infty = 0.86$, $a = -0.3$, $x_{cg} = 0.2$, $r_{cg}^2 = 0.24$, $R = 0.2$, $\mu = 60$).

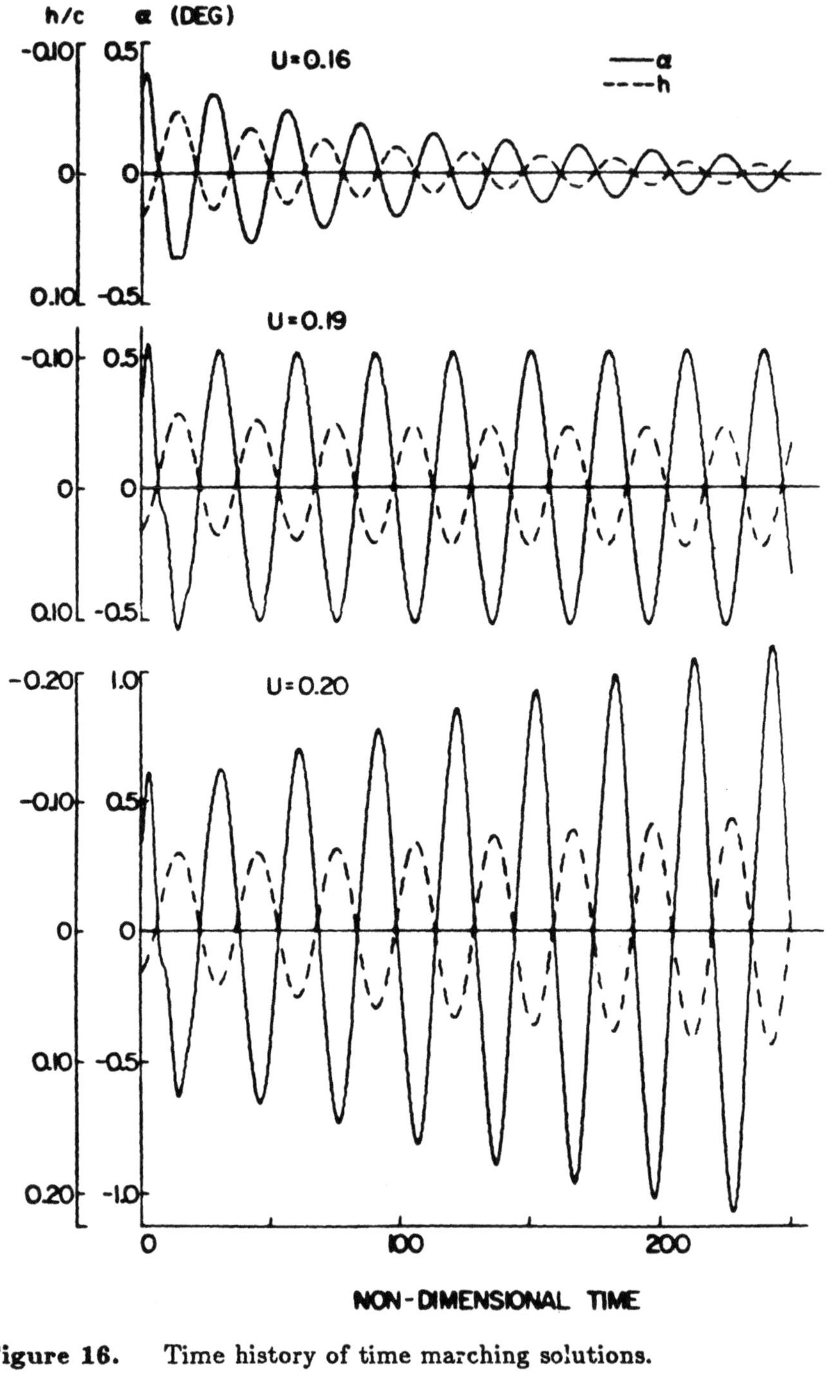

Figure 16. Time history of time marching solutions.

placed at $x_{cg} = -0.25$ and the frequency ratio at $R = 0.1$ in order to obtain a distinctly nonlinear effect. The results are shown in Figure 17. On those portions of the curve where amplitude increases with airspeed, a stable nonlinear limit cycle is predicted. On those portions of the curve where amplitude increases with *decreases* in airspeed, an unstable limit cycle occurs.

Further time marching calculations have been performed to confirm the limit cycle. The initial state vector to start time integrations is varied proportionally to that of the flutter solution of the describing function method with the amplitude, $\phi_1 = -.25°$, which gives $\underset{\sim}{x} = (0.04353, 0.00562, -0.005726, 0)^T$ for $\phi_1 = 0.25°$. This state vector yields the effective induced angle of attack, $-0.0063°$ at $\tau = 0$. Since the reduced frequencies of flutter are about 0.1 for small amplitudes, the time step size of integration was chosen as $\Delta\tau = 0.5$.

As the solid curve of the describing function in Figure 17 anticipates, the limit cycle flutter is also shown by the time marching solutions for small amplitudes. The amplitudes, reduced frequencies, amplitude ratio, and phase angles of these limit cycle oscillations were also calculated from the time history of the solutions and are depicted with open circles in the figure. Convergence to the limit cycle is determined by changing the initial amplitude. For example, the time history with two different initial amplitudes at $U = 0.6$ is shown in Figure 18. Both oscillations converge to the same limit cycle with the amplitude $\phi_1 = 0.355°$ beyond $\tau = 400$. However the average displacement for the h-motion at least up to $\tau = 500$ is different for the two initial conditions. This slow convergence for the average displacement to the neutral position may be attributed to the small frequency ratio, R, which implies weak stiffness of the structure against the h-motion. It should be noted that the *average* translational displacement of an airfoil has no effect on the aerodynamic forces.

The results of the time marching solution can be compared with those obtained by the describing function method in Figure 17. For small amplitudes, the agreement of the results is satisfactory. The reduced frequency of the limit cycle by the time marching solution, however, decreases as the flutter amplitude increases, while the describing function method predicts monotonically increasing reduced frequencies. Further the time marching solution shows stable limit cycle flutter up to the speed of $U = 0.7$ whereas the

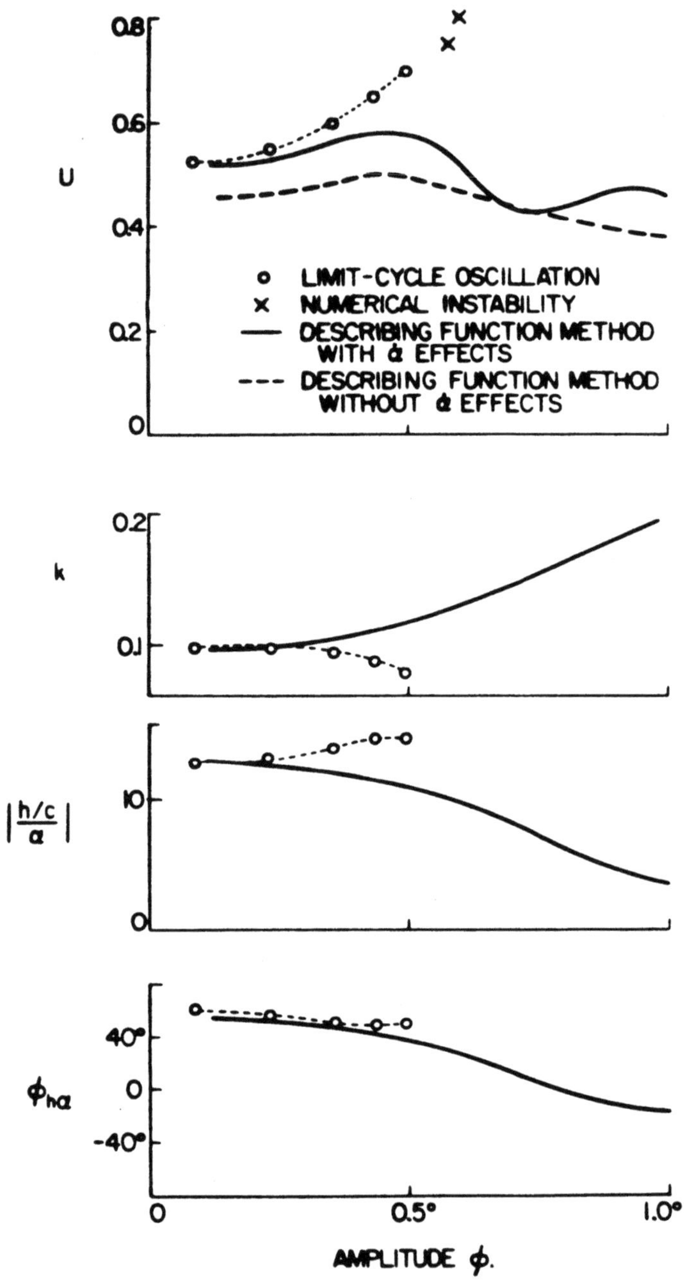

Figure 17. Limit cycle oscillations, (NACA 64A006, $M_\infty = 0.86$, $a = -0.3$, $x_{cg} = 0.2$, $r_{cg}^2 = .24$, $R = 0.2$, $\mu = 60$).

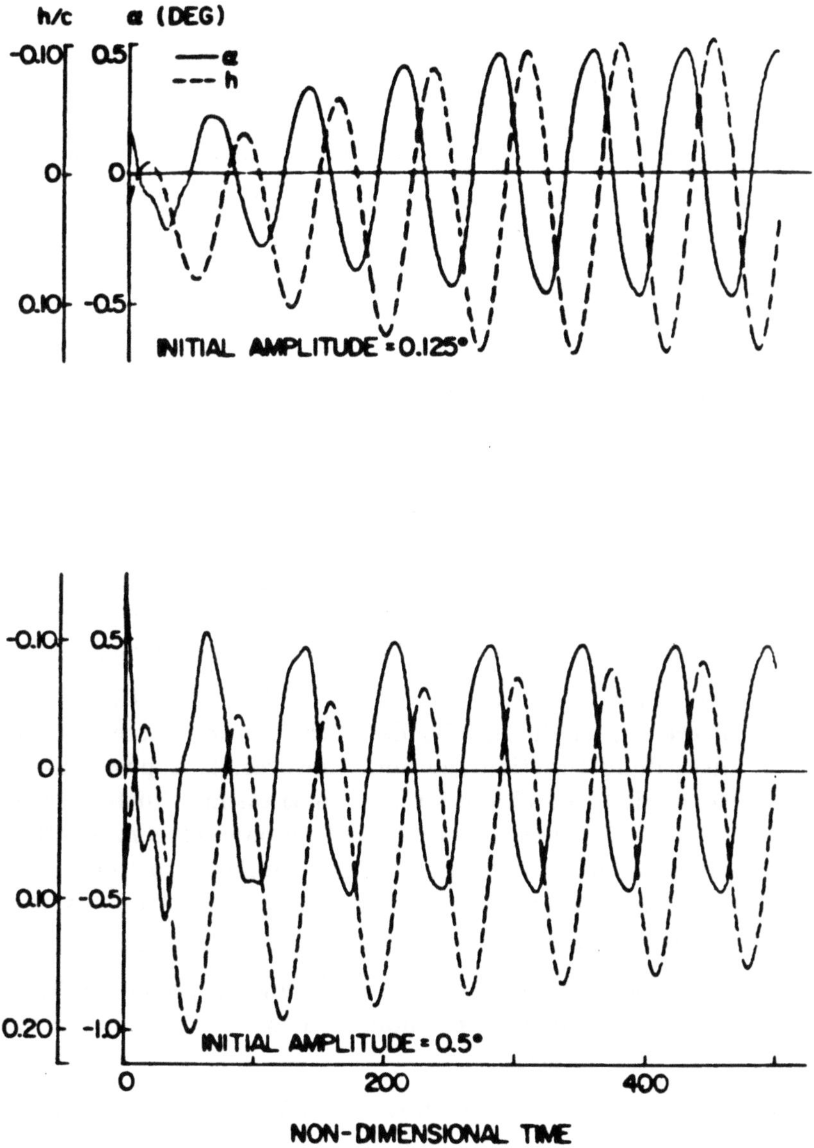

Figure 18. Time history of limit cycle flutter, ($U = 0.6$, $\phi_1 = 0.355°$, $k_F = 0.089$, $|\overline{h}/c/\overline{\alpha}| = 14.4$, $\phi_{h\alpha} = 49.7°$).

describing function method predicts no stable limit cycle solution above $U = 0.58$. At $U = 0.75$ and 0.8, the time marching solutions with the initial amplitude of $\phi_1 = 0.5°$, are terminated by a *numerical* instability of the aerodynamic calculations. The peak values of ϕ just before the occurrence of the instability are plotted with the symbol x. This kind of difficulty was frequently encountered when an initial amplitude of more than 0.5 degrees was used in the time marching calculations. *Possibly* for this reason we failed to detect the divergent unstable limit cycle flutter with larger amplitudes which is predicted by the describing function method.

6.7 Conclusions

An extended nonlinear indicial approach to modeling nonlinear aerodynamic forces for aeroelastic analyses has been developed. The basic approach is based upon describing function ideas.

Flutter boundaries obtained by the describing function method are generally verified by time marching solutions for sufficiently small amplitude flutter motion. Hence the former, less costly method is useful for determining the significance of initial departures from linear behavior. More specific conclusions are listed below.

- Generally the accuracy of the describing function method decreases as the amplitude of the motion increases. The describing function method, however, is a powerful tool to predict the characteristics of transonic flutter since it generally requires a very small amount of computational time for the aerodynamic forces compared to time marching solutions, particularly if a parameter study is to be undertaken.

- The stable nonlinear limit cycle flutter predicted by the describing function method, is also observed in the time marching solutions.

- The component in the upwash distribution due to the angular velocity, $\dot{\alpha}$, of airfoil motion cannot always be neglected even though the aerodynamic code has a low frequency limitation. Sometimes its neglect causes a fictitious flutter instability of the α-motion at high frequencies.

- The $\dot{\alpha}$ effect is properly taken into account by the total describing function decomposition into ϕ and $\dot{\alpha}$ components.

- The nonlinear behavior with the large amplitudes, $\phi > 0.5°$, could not be obtained by the time marching solutions due to a numerical instability in the aerodynamic calculations (even when $\dot{\alpha}$ effects are included).

For an alternative suggestion for achieving the same goals, the reader is directed to Taylor et al. (1980). Also the recent work of Bland and Edwards (1983) should be cited. They deal with the important effects of airfoil shape and thickness (though not nonlinear dynamic effects per se) in a manner similar to that of the present paper. In general the effects of airfoil shape and thickness are more important for flutter than nonlinear dynamics effects per se (Dowell et al., 1983; Bland and Edwards, 1983). Hence *linear* aeroelastic analyses will continue to play a dominant role even in the transonic flow regime.

7. Concluding Remarks

7.1 Some Present Answers

The paper began with four questions. The answers to these will undoubtedly be refined in future years. However, based upon present knowledge, partial answers may be formulated as follows:

(1) For sufficiently small airfoil motions (leading to sufficiently small shock motions, $\approx 5\%$ of airfoil chord or less), the aerodynamic forces will be linear functions of the airfoil motion (Tijdeman, 1977; Tijdeman and Seebass, 1980; Seebass [41]; McCroskey, 1982; Ballhaus and Goorjian, 1977; Fung et al., 1978; Dowell et al., 1983).

(2) At least one viable alternative solution technique to finite difference methods is available, the field panel method of Hounjet (1981b). Further refinements to that approach are expected, possibly including the work of Cockey (1983) as discussed herein and also including the extension to three-dimensional flow fields. The relative merits of finite difference vs. finite element vs. field panel methods remains a subject for future study and, no doubt, vigorous debate.

(3) The nonuniqueness which has been observed under some conditions in transonic small disturbance and full *potential* flow solutions is inherent in the governing field equations themselves and not a numerical artifact of some solution method (Goorjian [19]; Dowell et al. [14]; Steinhoff and Jameson, 1981; Salas, 1983). To date, corresponding nonuniqueness in solutions to the Euler equations has not been observed. However it is not clear at this time whether or not such nonuniqueness in the potential flow equation solutions may have its less (or equally) dramatic counterpart in the solutions to the Euler equations under some conditions. This remains an important topic for future study. Moreover the physical significance, if any, of such solutions remains to be clarified.

(4) While the option of simultaneous time integration of the fluid dynamical and structural dynamical equations of motion to determine aeroelastic response will be attractive for some applications, flutter analysis in the frequency domain will continue to be an important, and at times more attractive, option as well. Methods for generating the frequency domain aerodynamic forces are now available from

- aerodynamic methods which presuppose infinitesimal, harmonic dynamic airfoil motions (Williams, 1979a, 1979b; Hounjet, 1981b; Ehlers and Weatherill, 1982),
- impulse-transfer function ideas (Nixon and Kerlick, 1980; Nixon,1981; Ballhaus and Goorjian,1978; Ueda and Dowell, 1982, 1983) which allow the generation of frequency domain data from a single time history record determined by a time marching aerodynamic code. This approach can be extended (approximately) to large airfoil motions where a nonlinear relationship exists between airfoil motions and aerodynamic forces by using ideas based upon the describing function method (Ueda and Dowell, 1982; 1983).

(5) Simple order of magnitude estimates of the relative computational times for *aeroelastic* analyses can be made and these have been discussed in the text. Linear flutter analysis will continue to play a dominant role even in the transonic flow regime (Ueda and Dowell, 1982, 1983; Yang et al., 1980; Edwards et al., 1982; Bland and Edwards, 1983).

7.2 Future Work

A long list of worthy research topics could be given. Here we focus on a few which have as their common theme improved physical modeling and understanding of the fluid dynamic *and* aeroelastic phenomena of interest.

(1) There is a clear need to understand better the apparent qualitative difference between the solutions of the potential flow equations and those of the Euler equations under those parameter conditions where the former exhibit nonunique solutions. Until this difference is both understood and resolved, it throws into question the whole approach of combining potential flow solution with boundary layer corrections to correct for viscosity under those conditions where nonunique solutions are observed.

(2) A complementary issue is how can more effective (efficient) solution methods be devised for the Euler equations. Subsequent to the pioneering work of Magnus and Yoshihara (1975), little has been done.

One possible approach which has several prospective advantages is to assume that the flow may be treated as a nonlinear mean steady flow plus a small (infinitesimal) linear dynamic pertubation. This technique has been exploited effectively by Williams (1979a, 1979b), Fung and Seebass (1978), Hounjet (1981b), Ehlers and Weatherill (1982), and others for the *potential* flow model. In one of its *limiting* forms this approach has been used by Lighthill (1953), Williams et al. (1977), and others for modeling boundary layer effects in panel flutter (non-lifting) and lifting surface aerodynamics using the Euler equations.

Such an approach will lead to a set of time dependent, *linear*, partial differential equations with variable (spatially dependent) coefficients. These will depend in turn upon the solution of the time-independent, *nonlinear*, partial differential, Euler equations for the mean steady flow. The expected advantages of this approach include the following:

(3) Existing and future steady flow codes may be exploited to maximum advantage to provide input data for the dynamic perturbation equations and their solution. (Also the finite difference grid developed for the mean steady flow may be retained for the dynamic perturbation flow if a finite difference solution method is used for the latter.)

(4) Solution methods for linear equations with variable coefficients may be brought to bear, although these will require extensions

and generalizations. See the work of Hounjet (1981b) and Williams et al. (1977).

(5) Solving the *linear, dynamic* equations will

- in all likelihood meet the requirements of the aeroelastician (within the context of the Euler equations).
- permit inclusion of viscous effects in the mean steady flow modeling and thus indirectly and partially (but not directly and completely) in the dynamic perturbation equations (Lighthill, 1953; Williams et al., 1977).
- allow one to examine the dynamic stability of the mean, steady flow itself and thus contribute to a better understanding of the prospect for nonunique solutions of the Euler equations.

References

[1] Albano, E. and W. P. Rodden. "A Doublet Lattice Method for Calculating Lift Distribution on Oscillating Wings in Subsonic Flows," *AIAA Journal, 7* (1969), 279–85.

[2] Ballhaus, W. F. and Goorjian, P. M. "Implicit Finite Difference Computations of Unsteady Transonic Flows About Airfoils," *AIAA Journal, 15* (1977), 1728–35.

[3] ———. "Efficient Solution of Unsteady Transonic Flows About Airfoils," *Paper 14, AGARD Conference Proceedings No. 226, Unsteady Airload in Separated and Transonic Flows,* 1978a.

[4] ———. "Computation of Unsteady Transonic Flows by the Indicial Method," *AIAA Journal, 16, No. 2* (1978b), 117–24.

[5] Bauer, R., Garabedian, P., and Korn, D. "Supercritical Wing Sections," *Lecture Notes in Economics and Mathematical Systems, 66,* Springer-Verlag, 1972.

[6] Bland, S. R. and Edwards, J. W. "Airfoil Shape and Thickness Effects on Transonic Airloads and Flutter," AIAA SDM Conference in Lake Tahoe, CA, May 1983, *AIAA Paper No. 83-0959.*

[7] Borland, C. J. and Rizzetta, D. P. "Transonic Unsteady Aerodynamics for Aeroelastic Applications, I: Technical Development Summary," *AFWAL TR 80-3107, I,* June 1982a.

[8] ————————————————. "Nonlinear Transonic Flutter Analysis," *AIAA Journal, 20, No. 11* (1982b), 1606–15.

[9] Borland, C. J., Rizzetta, D. P., and Yoshihara, H. "Numerical Solution of Three-Dimensional Unsteady Transonic Flow Over Swept Wings," *AIAA Journal, 20, No. 3* (1982), 340–47.

[10] Cockey, W. D. "Panel Method for Perturbations of Transonic Flows With Finite Shocks," *Ph.D. thesis,* Princeton University, June 1983.

[11] Davis, S. S. and Malcolm, G. "Experiment in Unsteady Transonic Flows," *Proc. of the AIAA/ASME/ASCE 20th Structures, Structural Dynamics and Materials Conference,* St. Louis, MO, April 1979.

[12] Dowell, E. H. "A Simple Method for Converting Frequency-Domain Aerodynamics to the Time Domain," *NASA TM 81844,* 1980.

[13] Dowell, E. H., Bland, S. R., and Williams, M. H. "Linear/Nonlinear Behavior in Unsteady Transonic Aerodynamics," *AIAA Journal, 21* (1983), 38–46.

[14] Dowell, E. H., Ueda, T., and Goorjian, P. M. "Transient Decay Times and Mean Values of Unsteady Oscillations in Transonic Flow," *AIAA Journal, 21, No. 12* (1983), 1762–64.

[15] Edwards, J. W., et al. "Time-Marching Transonic Flutter Solutions Including Angle-of-Attack Effects," *AIAA Paper 82-0685,* presented at the 23rd SDM Conference, New Orleans, LA, May 1982.

[16] Ehlers, F. E. and Weatherill, W. H. "A Harmonic Analysis Method for Unsteady Transonic Flow and its Application to the Flutter of Airfoils," *NASA CR-3537,* 1982.

[17] Fung, K. Y., Yu, N. J., and Seebass, R. "Small Unsteady Perturbations in Transonic Flows," *AIAA Journal, 16* (1978), 815–22.

[18] Goldstein, M. E. *Aeroacoustics,* New York: McGraw-Hill, 1976.

[19] Goorjian, P. M. Private communication, NASA Ames Research Center.

[20] Hessenius, K. A. and Goorjian, P. M. "Validation of LTRAN2-HI by Comparison With Unsteady Transonic Experiment," *AIAA Journal, 20, No. 5* (1982), 731–32.

[21] Hounjet, M.H.L. "A Transonic Panel Method to Determine Loads on Oscillating Airfoils With Shocks," *AIAA Journal, 19* (1981a), 559–66.

[22] ——————————. "A Field Panel Method for the Calculation of Inviscid Transonic Flow About Thin Oscillating Airfoils With Shocks," *NLR MP 81043 U,* National Aerospace Laboratory, Netherlands. Presented at the International Symposium on Aeroelasticity, Nuremburg, October 5–7, 1981b, Germany.

[23] Houwink, R. and van der Vooren, J. "Improved Version of LTRAN2 for Unsteady Transonic Flow Computations," *AIAA Journal, 18, No. 8* (1980), 1008–10.

[24] Hsu, J. C. and Meyer, A. U. *Modern Control Principles and Applications,* New York: McGraw-Hill, 1968.

[25] Isogai, K. "Numerical Study of Transonic Flutter of a Two-Dimensional Airfoil," *Technical Report of National Aerospace Laboratory,* Japan, *NAL-TR-617T,* 1980.

[26] Kerlick, G. D. and Nixon, D. "Mean Values of Unsteady Oscillations in Transonic Flow Calculations," *AIAA Journal, 19, No. 11* (1981), 1496–98.

[27] Lighthill, M. J. "On Boundary Layers and Upstream Influence, II; Supersonic Flows Without Separation," *Proceedings of the Royal Society, A217* (1953), 478–507.

[28] Liu, D. D. "A Lifting Surface Theory Based on an Unsteady Linearized Transonic Flow Model," *AIAA Paper 78-501,* 1978.

[29] Magnus, R. J. and Yoshihara, H. "Calculation of Transonic Flow Over an Oscillating Airfoil," *AIAA Paper 75-98,* Jan. 1975.

[30] McCroskey, W. J. "Unsteady Airfoils," *Annual Review of Fluid Mechanics, 14* (1982), 285–311.

[31] McIntosh, S. private communication.

[32] Miles, J. W. *The Potential Theory of Unsteady Supersonic Flow,* Cambridge: Cambridge University Press, 1959, 4–13.

[33] Morino, L. "A General Theory of Unsteady Compressible Potential Aerodynamics," *NASA CR-2464,* December 1974.

[34] Morino, L. and K. Tseng. "Time-Domain Green's Function Method for Three-Dimensional Nonlinear Subsonic Flows," *AIAA Paper 78-1204,* 1978.

[35] Murman, E. M. "Analysis of Embedded Shock Waves Calculated by Relaxation Methods," *Proceedings of the AIAA CFD Conference,* July 1973, 27–40.

[36] Nixon, D. "Calculation of Unsteady Transonic Flows Using the Integral Equation Method," *AIAA Paper 78-13,* January 1978.

[37] ———. "On the Derivation of Universal Indicial Functions," *AIAA Paper 81-0328,* Jan. 1981.

[38] Nixon, D. and Kerlick, G. D. "Calculation of Unsteady Transonic Pressure Distributions by Indicial Methods," *Nielsen Engineering and Research Paper 117,* 1980.

[39] Rizzetta, D. P. and Borland, C. J. "Unsteady Transonic Flow Over Wings Including Inviscid/Viscous Interactions," *AIAA Journal, 21,*

No. 3 (1983), 363–71.

[40] Salas, M. D., Jameson, A., and Melnik, R. E. "A Comparative Study of the Nonuniqueness Problem of the Potential Equation," presented at the 6th AIAA CFD Conference, Danvers, Mass., July 13–15, 1983.

[41] Seebass, R. "Advances in the Understanding and Computation of Unsteady Transonic Flows," *Recent Advances in Aerodynamics,* ed. A. Krothapalli and C. Smith, 1984.

[42] Steinhoff, J. and Jameson, A. "Multiple Solutions of the Transonic Potential Flow Equations," *AIAA Paper No. 81-1019,* AIAA Computational Fluid Dynamics Conference, Palo Alto, CA, June 1981.

[43] Taylor, R. F., Bogner, F. K., and Stanley, E. C. "A Stability Prediction Method for Nonlinear Aeroelasticity," *AIAA Paper 80-0797,* presented at the AIAA/ASME/AHS/ASCE Conference, Seattle, WA, May 12–14, 1980.

[44] Tijdeman, H. "Investigation of the Transonic Flow Around Oscillating Airfoils," *Ph.D. thesis,* Delft University, 1977.

[45] Tijdeman, H. and Seebass, R. "Transonic Flow Past Oscillating Airfoils," *Annual Review of Fluid Mechanics, 12* (1980), 181–222.

[46] Tran, C. T. and Petot, M. "Semi-empirical Model for the Dynamic Stall of Airfoils in View of the Application of Responses of a Helicopter Blade in Forward Flight," *Paper 48, Proceedings,* 6th European Rotorcraft and Powered-lift Aircraft Forum, Bristol, England, 1980.

[47] Ueda, T. and Dowell, E. H. "Flutter Analysis Using Nonlinear Aerodynamic Forces," *AIAA Paper 82-0728,* 1982.

[48] ————————. "Describing Function Flutter Analysis for Transonic Flow: Extension and Comparison With Time Marching Analysis," *AIAA Paper 83-0958,* 1983.

[49] Voss, R. "Time-Linearized Calculation of Two-Dimensional Unsteady Transonic Flow at Small Disturbances," *DFVLR FB-81-01,* 1981.

[50] Williams, M. H. "Unsteady Thin Airfoil Theory for Transonic Flows With Embedded Shocks," Department of Mechanical and Aerospace Engineering, *Report No. 1376,* Princeton University, May 1978; also *AIAA Journal, 18* (1980), 615–24; also see "Unsteady Airloads in Supercritical Transonic Flows," *Proceedings of the AIAA/ASME/ASCE 20th Structures, Structural Dynamics and Materials Conference,* St. Louis, MO, April 1979b.

[51] ————————. "The Linearization of Transonic Flows Containing Shocks," *AIAA Journal, 17* (1979a), 394–97.

[52] Williams, M. H., et al. "Aerodynamic Effects of Inviscid Parallel Shear Flows," *AIAA Journal, 15, No. 8* (1977), 1159–66.

[53] Yang, T. Y., Guruswamy, P., and Striz, A. G. "Aeroelastic Response Analysis of Two-Dimensional, Single and Two Degree of Freedom Airfoils in Low Frequency, Small-Disturbance Unsteady Transonic Flow," *AFFDL-TR-79-3077,* 1979.

[54] ———————————. "Application of Transonic Codes to Flutter Analysis of Conventional and Supercritical Airfoils," AIAA SDM Conference in Atlanta, GA, *AIAA Paper 81-0603,* 1981.

[55] Yang, T. Y., et al. "Flutter Analysis of a NACA 64A006 Airfoil in Small Disturbance Transonic Flow," *J. Aircraft, 17, No. 4* (1980), 225–32.

Modeling of Turbulent Separated Flows for Aerodynamic Applications

Joseph G. Marvin
NASA Ames Research Center
Moffett Field, CA 94035

Abstract

A review is given of the advances made over the past decade in modeling steady, high speed, compressible separated flows through numerical simulations resulting from solutions of the mass-averaged Navier-Stokes equations. Emphasis is placed on bench-mark flows that represent simplified (but realistic) aerodynamic phenomena. These include impinging shock waves, compression corners, glancing shock waves, trailing edge regions, and supersonic high angle-of-attack flows. A critical assessment of modeling capabilities is provided by comparing the numerical simulations with experiment. The importance of combining experiment, numerical algorithm, grid, and turbulence model to effectively develop this potentially powerful simulation technique is stressed.

1. Introduction

Although the separation that occurs in many aerodynamic flows can have a profound influence on vehicle performance, it remains one of the least understood and most difficult problems in fluid dynamics. Over the past decade, two primary factors have operated to intensify interest in understanding turbulent separation: the imposition on the vehicle designer of higher performance standards, and an increased possibility of predicting separation by applying recent advances in computational fluid dynamics.

A potentially powerful approach to predicting turbulent separated flows is to solve directly the Reynolds-averaged Navier-Stokes equations (Chapman, 1979). For practical reasons, such an approach is favored over direct simulation of the time-dependent, unaveraged Navier-Stokes equations because the three-dimensional, widely varying scales of turbulence present impossible requirements for even the largest and fastest computers (Chapman, 1981). A significant amount of research has been under way at Ames Research Center to develop the technology required to solve separated flows of practical interest within the framework of the Reynolds-averaged Navier-Stokes equations. An obvious advantage of such an approach is that the entire viscous and inviscid portions of the flow are captured simultaneously, and the potential exits for focusing directly on turbulence modeling, which is an important pacing item for the successful development of computational fluid dynamics. A disadvantage is the long computing time and large storage limitations of current computers, which has hampered attempts to focus directly on turbulence modeling without considering numerical resolution and accuracy. As it now stands, the competing elements of turbulence modeling, numerical resolution, and accuracy must all be considered in any evaluation of our ability to compute flows with separation (Marvin, 1982). This will be particularly true for three-dimensional flows, which are the interesting ones from the viewpoint of applications.

The purpose of this article is to review the advances made over the past decade in modeling separation in practical aerodynamic flows. As the paper develops, the problems remaining will become obvious, as will the need for future study. Nevertheless, it will also become apparent that great strides have been made and that the potential of numerical modeling has not diminished. In order to keep the scope of the paper within reasonable proportions, attention will be directed to steady, high-speed, compressible flows of particular interest to the author and his colleagues. Some of the practical situations of current interest are shown in the photographs of Figure 1. Figure 1a shows a shadowgraph of the ascent configuration of the space shuttle and the multiple impinging shock waves that exist. Figure 1b shows an oil-flow pattern of the region on a lifting surface where separation occurs when a control surface is deflected. Figure 1c shows the transonic flow over an airfoil where a strong shock wave develops. And Figure 1d shows the space shuttle oribiter at high angle of attack and at a supersonic speed where separation dominates the leeside flow. Such problems obviously involve many complications and the approach to their solution has been attempted in simplified stages, which this author has referred to as a building-block approach (Marvin, 1980).

The article begins with a section that develops the governing equations, presents a short discussion of the technique developed to

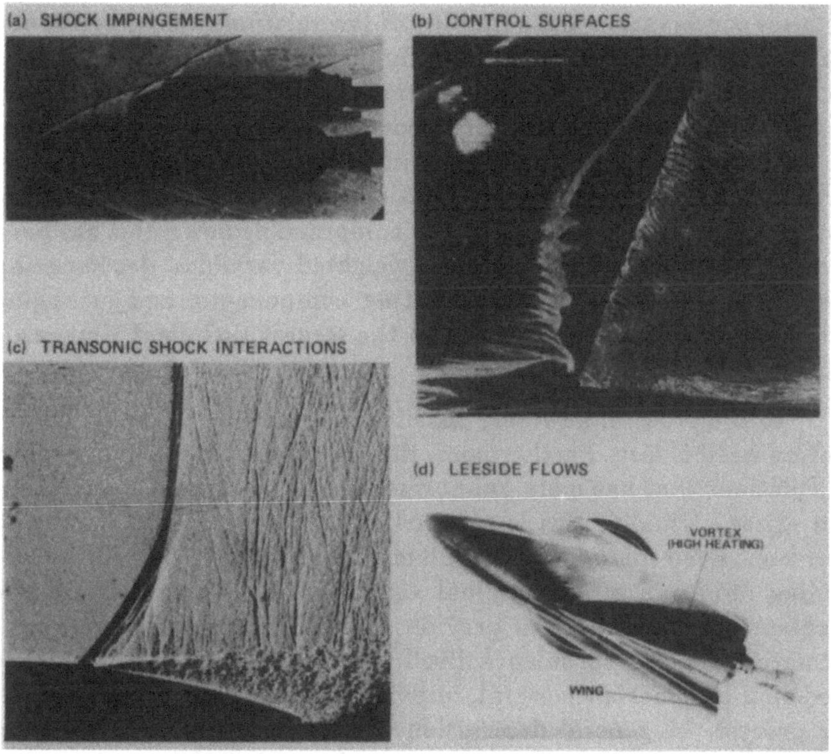

Figure 1. Photographs showing aerodynamic flows with separation: (a) shock impingements; (b) deflected control surfaces; (c) transonic airfoils; (d) leeside flows.

solve those equations, and introduces the various turbulence models under development. Subsequently, examples of solutions for some building-block flows are presented and critically assessed by comparing the results of computations and experiments.

2. Flow Modeling

Modeling of turbulent separated flows is a combination of numerical modeling of the discretized form of the governing equations and the requirement for providing an adequate model of the turbulent correlations in the governing equations.

2.1 Governing Equations

The time-dependent Navier-Stokes equations, supplemented

by mass conservation and suitable gas-law relationships, describe the turbulent motion of a continuum fluid. Solutions to the equations for turbulent flows of practical interest are virtually impossible using today's computers, because turbulence is three-dimensional and has an enormous range of length and time scales. The difficulty can be circumvented by rewriting the equations for another set of variables, obtained by suitable averaging. For compressible flows, this has been accomplished by introducing mass-weighted variables, decomposing them into their mean and fluctuating components, and averaging over a time that is long relative to the largest turbulent time-scale (Rubesin, 1973).

In the process, however, physical information on the turbulent motion itself is lost. Furthermore, the formalism results in a new set of equations that has more unknowns, and an equation-closure problem arises. Usually, this is referred to as the turbulence-modeling problem. Even introducing supplemental equations, derived by obtaining moments of the original equations, does not alleviate the problem, but does help to provide a means to introduce more information on the turbulence itself. Necessarily then, turbulence modeling becomes an integral, important part of our overall modeling process. A general description of various turbulence-modeling approaches used for applications in aerodynamic flows was presented by Marvin (1982).

Solutions to most of the complex aerodynamic flows discussed herein use eddy-viscosity turbulence models. The governing equations in mass-average variables and supplemental equations used in some of the eddy-viscosity models are written for plane flow in vector form as follows:

$$\frac{\partial U}{\partial t} + \frac{\partial F}{\partial x} + \frac{\partial G}{\partial y} = H \tag{2.1}$$

$$U = \begin{bmatrix} \rho \\ \rho u \\ \rho v \\ \rho e \\ \rho k \\ \rho s \end{bmatrix}, \qquad F = \begin{bmatrix} \rho u \\ \rho u^2 + \sigma_{xx} \\ \rho uv - \tau_{xy} \\ u(\rho e + \sigma_{xx}) - v\tau_{xy} + q_{Tx} \\ \rho uk + q_{ky} \\ \rho us + q_{sx} \end{bmatrix} \tag{2.2a}$$

$$G = \begin{bmatrix} \rho v \\ \rho uv - \tau_{xy} \\ \rho v^2 + \sigma_{yx} \\ v(\rho e + \sigma_{yy}) - u\tau_{xy} - q_{Ty} \\ \rho vk + q_{ky} \\ \rho vs + q_{sy} \end{bmatrix}, \qquad H = \begin{bmatrix} 0 \\ 0 \\ 0 \\ 0 \\ H_k \\ H_s \end{bmatrix} \tag{2.2b}$$

The last two equations are the supplemental equations providing the velocity $(k)^{1/2}$ and length scale s required in higher-order eddy-viscosity models. In the column vectors, q_{Tx} and q_{Ty} are the laminar-plus-turbulent heat-flux vectors; σ_{xx}, σ_{yy} are the laminar-plus-turbulent normal stresses; τ_{xy} is the laminar-plus-turbulent shear stress; and q_{ky}, q_{sx}, and q_{sy} are flux vectors associated with the turbulence field variables.

The stress terms and flux vectors are

$$\begin{aligned} \sigma_{xx} &= p + \frac{2}{3}\rho k - \tau_{xx}, \\ \tau_{xy} &= \mu_T\left(\frac{\partial u}{\partial y} + \frac{\partial v}{\partial x}\right), \\ \tau_{xx} &= \frac{2}{3}\mu_T\left(2\frac{\partial u}{\partial x} - \frac{\partial v}{\partial y}\right), \\ \mu_T &= (\mu + \rho\epsilon), \\ q_{Tx} &= k_T\frac{\partial T}{\partial x}, \\ q_{kx} &= -\mu_k\frac{\partial k}{\partial x}, \\ q_{sx} &= -\mu_x\frac{\partial s}{\partial x} \end{aligned} \tag{2.3}$$

where p is the hydrostatic pressure; $\frac{2}{3}\rho k$ is the pressure associated with the turbulence; k_T is the thermal conductivity, including the turbulent diffusivity; and $\rho\epsilon$ is the turbulent eddy viscosity. The functional forms of the source functions H depend on the choice of the turbulence model.

2.2 Solution Methods and Turbulence Models

The methods available for solving equation (2.1), along with the various turbulence models, are introduced in historical order so that the unfamiliar reader will be able to see what the technological developments have been and how they arrived at their present state.

Development of methods for solving the mass-weighted form of the Navier-Stokes equations began after MacCormack (1971) used an explicit time-marching scheme to solve the laminar form of the equations. In this second-order-accurate method the equations are discretized and advanced in time such that

$$U^{n+1} = L(\Delta t)\, U^{n}_{i,j} \tag{2.4}$$

The $L(\Delta t)$ term is replaced by a sequence of time-split, one-dimensional operators, for example,

$$L(\Delta t) = L_x\left(\frac{\Delta t}{2}\right) L_y(\Delta t)\, L_x\left(\frac{\Delta t}{2}\right) \tag{2.5}$$

where L_x solves the parts of equation (2.1) given by

$$\frac{\partial U}{\partial t} + \frac{\partial G}{\partial x} = 0$$

and L_y solves the part given by

$$\frac{\partial U}{\partial t} + \frac{\partial F}{\partial y} = 0 \tag{2.6}$$

The operators are advanced in time to a steady state, if one exists, according to a predictor-corrector sequence of steps. A numerical stability criterion exists that limits the time-step used to advance the solution. Typically, in high-Reynolds-number turbulent flows the limiting time-step occurs in computational sweeps normal to the surface. It is given by

$$\Delta t_y \le \frac{\Delta y}{|v| + c + \left\{\frac{V_1}{\Delta x} + \frac{V_2}{\Delta y}\right\}} \tag{2.7}$$

where c is the sound speed and V_1 and V_2 represent viscous terms. The Δy step interval has to be very small to resolve the wall region of a turbulent boundary layer, and this time-step limit presents severe limitations which result in long computing times. Nevertheless, many solutions of shock-separated flows were reported using this method in the mid-1970s.

Given the severe time-step restriction of the method and computer storage limitations, most investigators chose simple zero-equation eddy-viscosity models that use mean-flow information to close the governing equation. These two-layer eddy-viscosity models employed Prandtl's mixing-layer hypothesis in the inner layer,

$$\epsilon_{\text{inner}} = l^2 \left| \frac{\partial u}{\partial y} + \frac{\partial v}{\partial x} \right| \tag{2.8}$$

where

$$\begin{aligned} l &= 0.4y\,(1 - \exp^{y/A}) \\ A &= A^+ \mu_w / (\tau_w/\rho)^{1/2} \\ A^+ &= 26 \end{aligned}$$

In the outer region, either a mixing-length value was chosen, based on some length scale such as boundary-layer thickness, for example,

$$l = l_{\text{max}} \tag{2.9}$$

or Clauser's eddy-viscosity formulation was chosen with an intermittency factor, for example

$$\epsilon_{\text{outer}} = 0.0168 u_{\text{max}} \delta_i^* / [1 + 5.5(y/\delta)^6] \tag{2.10}$$

where u_{max} is the maximum velocity achieved in the boundary layer and δ_i^* is the kinematic displacement thickness. The turbulent heat flux is modeled through a turbulent Prandtl number. To date, this latter aspect of modeling has not been altered. As will be shown later, solutions with these formulations fail to give satisfactory predictions, although they qualitatively reproduce many experimentally observed features. Most of the shortcomings were earlier blamed on turbulence modeling, but not many of the studies reported effects of grid dependence or numerical smoothing which in retrospect may have been as important as the turbulence model.

Even though computing times were excessive (several hours on a CDC 7600 computer), attempts were made to modify the turbulence model and some improvement in the solutions to complex separated-flow problems was demonstrated. Two approaches are worth noting. One used experimental data to guide modifications to the mixing-length constants in the turbulence model (Marvin et al., 1975), and the other attempted to relax the outer eddy viscosity to account for the fact that turbulence does not adjust immediately to rapid changes in the mean flow (Shang et al., 1976; Baldwin and Rose, 1975); for example

$$\rho\epsilon = \rho\epsilon_o + [\rho\epsilon_{eq} - \rho\epsilon_o]\left(1 - \exp^{\alpha(x - x_o/\delta_o)}\right) \tag{2.11}$$

where $(\rho\epsilon)_o$ and δ_o are undisturbed values ahead of the interaction region, $(\rho\epsilon)_{eq}$ is the usual unmodified value given by equation (2.10), and α is a relaxation length obtained by a best-fit comparison of final computed results with experiment. It is obvious that both attempts rely heavily on experimental data over a wide range of conditions which limits their generality. However, these studies illustrated the potential of the numerical simulations and encouraged development of faster computing methods and better turbulence models.

At this point, the numerical algorithm development research branched. MacCormack (1976) developed his more efficient explicit hybrid method, and Beam and Warming (1978) developed their factored implicit scheme. Also, turbulence-modeling improvements using higher-order eddy-viscosity models followed in the wake of the hybrid-method development, and improvements to algebraic eddy-viscosity models, mostly from a computational compatibility standpoint, followed in the wake of the factored-implicit scheme.

The time-step efficiency of the MacCormack explicit method was improved by combining the advantages of implicit numerical stability with physical insight of the wave-propagating property of the fluid. Conceptually, this was accomplished by further splitting of the y-operator, L_y, into hyperbolic and parabolic parts,

$$L_y(\Delta t) = L_{yh}(\Delta t)\, L_{yp}(\Delta t) \tag{2.12}$$

The hyperbolic operator contains the convective and pressure terms in the column vector G such that

$$\frac{\partial U}{\partial t} + \frac{\partial G_h}{\partial y} = 0 \tag{2.13}$$

In the prediction solution to G_h, pressures and velocities are obtained by the method of characteristics in a manner that eliminated the speed of sound from the time-step limit such that

$$\Delta t_y = \frac{\Delta t}{|v|} \tag{2.14}$$

Since the finest portion of the mesh is usually confined to the wall-bounded region where v is small, the stability bound of the allowable time-step is much less restrictive than that given by equation (2.7). The corrector step is applied as before. The parabolic operator L_{yp} is treated implicitly and, therefore, unconditionally stable with regard to time advances. The programming for the hybrid method is complicated by the necessity of using characteristic relations in the prediction step for the hyperbolic operator. However, decreases in computing times by an order of magnitude or more relative to the purely explicit method were achieved. Such decreases encouraged some investigators to apply higher-order eddy-viscosity models (e.g., see Viegas and Horstman, 1979), and others to move forward in the computations of three-dimensional flows (Hung and MacCormack, 1978).

Higher-order eddy-viscosity turbulence models were introduced into the hybrid method by expanding the column vectors to include the turbulent kinetic-energy and length-scale equations in equation (2.2). The one-equation model from Rubesin (1976), the two-equation model from Jones and Launder (1971), and the two-equation model from Wilcox and Rubesin (1980) have been examined for a range of different problems. The full equations describing the implementation of these models in the hybrid algorithm are given in Viegas and Horstman (1979). Modeling constants developed for incompressible flows are usually used without modifications. Authors have reported mixed results, but conclude overall that the higher-order models produce improvements.

Concurrently, development of implicit methods was undertaken. For our purposes, the factored-implicit scheme of Beam and Warming (1978) will be briefly described. The method is an extension of their earlier development of an inviscid-flow solver, and, for convenience, the essential elements of the method will be discussed in that context. Time-differencing of equation (2.1), where F and G contain only inviscid terms, is accomplished by the unconditionally stable scheme given by

$$U^{n+1} = U^n + \frac{\Delta t}{2}\left[\left(\frac{\partial U}{\partial t}\right)^n + \left(\frac{\partial U}{\partial t}\right)^{n+1}\right] + 0\,(\Delta t)^3 \tag{2.15}$$

where

$$\frac{\partial U}{\partial t} = -\left(\frac{\partial F}{\partial x} + \frac{\partial G}{\partial y}\right)$$

In this form, however, the system of equations is nonlinear and contains a large system of algebraic equations; as a result, the advantage of unconditional stability might not result in solution times significantly smaller than the times for explicit schemes. However, they linearized the equations while maintaining temporal accuracy by a Taylor-series expansion of the nonlinear terms. For example, they let

$$F^{n+1} = F^n + \left(\frac{\partial F}{\partial U}\right)^n (U^{n+1} - U^n) + 0(\Delta t)^2 \tag{2.16}$$

Substituting this expression and a similar one for G, writing the resulting in a Delta form $\Delta U^n = U^{n+1} - U^n$, and employing spatial factorization, the final form of the equation was written as

$$\left(I + \frac{\Delta t}{2}\frac{\partial A^n}{\partial x}\right)\left(I + \frac{\Delta t}{2}\frac{\partial B^n}{\partial y}\right)\Delta U^n = -\Delta t\left(\frac{\partial F}{\partial x} + \frac{\partial G}{\partial y}\right)^n \tag{2.17}$$

The solution is marched in time to a steady state, if one exists, through a three-step sequence, as follows:

$$\begin{aligned}
\left(I + \frac{\Delta t}{2}\frac{\partial A^n}{\partial x}\right)\Delta U^* &= -\Delta t\left(\frac{\partial F}{\partial x} + \frac{\partial G}{\partial y}\right)^n \\
\left(I + \frac{\Delta t}{2}\frac{\partial B^n}{\partial y}\right)\Delta U^n &= \Delta U^* \\
U^{n+1} &= U^n + \Delta U^n
\end{aligned} \tag{2.18}$$

Results from this procedure compare favorably with those of the hybrid method for the same test problems. Refinements to this method and other implicit solvers have been developed on a continuing basis; see for example, Briley and McDonald (1977) and Coakley (1983). MacCormack (1982) has recently reported a new mixed, explicit-implicit scheme which reduces the computation times and the complex programming problems associated with his hybrid method.

Solutions to separated-flow problems using the implicit scheme developed by Beam and Warming (1978) have usually employed zero-equation turbulence models and the thin-layer approximation to the full equations; for example, see Baldwin and Lomax (1978). The thin-layer approximation neglects derivatives of the viscous stresses in the flow direction. Baldwin and Lomax (1978) argue that this is computationally acceptable for even large separated flow regions because the accuracy of these derivatives in the discretized form of the full equations is questionable since the aspect ratio of computational cells in the near-wall viscous regions is usually very much less than unity for grids used to resolve turbulent layers. Briley and McDonald (1977) and Coakley (1983), however, have employed higher-order, two-equation models and the full equations.

One aspect of zero-equation turbulence-model improvement, still presently employed in the thin-layer implicit Navier-Stokes codes under development at Ames, is that carried out by Baldwin and Lomax (1978). The development of the model was initiated to circumvent a shortcoming of the Clauser outer-eddy-viscosity formulation (equation (2.10)), arising because in many instances the inviscid regions in complex flows have a nonuniform velocity field, and determination of the viscous-layer edge needed to evaluate δ_i^* in the model becomes difficult. The outer eddy viscosity is redefined as

$$\epsilon_{\text{outer}} = 0.0168\, C_1 F_{\text{wake}} \left[1 + 5.5 \left(\frac{0.3y}{y_{\max}} \right)^6 \right]^{-1} \tag{2.19}$$

where

$$F_{\text{wake}} = \left\{ \begin{matrix} y_{\max} F_{\max} \\ \text{or} \\ C_{wk} y_{\max} U_{\text{diff}} / F_{\max} \end{matrix} \right\} \text{the smaller} \tag{2.20}$$

The values of $F_{\max}$ and $y_{\max}$ are determined from

$$F(y) = y \left[\left(\frac{\partial u}{\partial x} - \frac{\partial v}{\partial y} \right)^2 \right]^{1/2} \left[1 - \exp^{-y^+/A^+} \right] \tag{2.21}$$

In wakes, the exponential part of $F(y)$ is set to zero. The $F_{\max}$ term is the maximum value of the function and $y_{\max}$ is the corresponding value of Y at $F_{\max}$; U_{diff} is the difference between the maximum and minimum total velocity at a fixed x-station. The constant C_1 was determined to have a value of 1.6 by ensuring that the resulting skin friction computed for a flat plate was equivalent to the value obtained from the original Cebeci-Smith model formulation. In order to have a correct value of eddy viscosity for a far-wake, C_{wk} was taken to be 0.25. For two test problems involving shock-wave interaction, the model gave results that were improved relative to those of the simple two-layer zero-equation model and more or less comparable to those achieved with the relaxation formulation given by equation (2.11). However, recent studies suggest that a certain degree of caution be exercised in applying this model. It requires modification of constants for Mach-number changes, the function $F(y)$ is not always a smoothly varying one, and the choice of $F_{\max}$ is problem-dependent. (See for example Degani and Schiff (1983) and Visbal and Knight (1983)).

3. Experimental Requirements

The emergence of methods for computing complex, turbulent separated flows places stringent requirements on experiments used to assess the development of the methods. In addition to the traditional role of providing basic understanding of the controlling mechanisms, they must also provide guidance for modeling approximations and provide sufficient detail so that accurate checks on computational output can be made. A synergistic framework for advancing computational aerodynamics consisting of closely coordinated experiments and computations was described by Marvin (1982). The continued necessity for data required to support the development of research-, pilot-, and production-type computer codes was emphasized in that work and will not be repeated here.

At the present stage of their development, computer codes used to solve separated-flow problems that employ the mass-averaged Navier-Stokes equations are probably best classified as research codes. Furthermore, the experimental data used to assess their development vary in completeness and accuracy because the flows are complicated by the presence of shock waves or separation or both, and because many investigators used instrumentation techniques that were themselves in developmental stages.

Nevertheless, a series of building-block or bench-mark flows has been developed that can assist in the development of computational methods (Marvin, 1982). Those used for the problems discussed in this paper are given in Tables 1–6. They represent a cross section of simple, but practical, aerodynamic flows. The tables provide the unfamiliar reader with ample bibliographic sources for further study. In addition to bibliographic citations, information is given on test conditions, grid size, and type of turbulence model employed. Grid size alone is not the only criterion for assessing computational resolution, however, no grid stretching and special refinement in regions of rapid flow changes are important techniques commonly used by most investigators. But the sizes provide some measure for comparison between various computations. Likewise, the turbulence models used are only broadly categorized, because they usually differ in detail as a result of programming decisions made by the various investigators. Some experiments conducted before 1981 are noted in the tables; they were reviewed by an independent evaluation committee and ascertained to contain the most comprehensive data sets for code validation (Kline et al., 1981).

4. Results and Discussion

The modeling of the complex separated flows introduced earlier will now be critically reviewed. The physical characteristics of the flows, as determined by experiment, will be introduced and then comparisons of the results of computations and experiments will be presented to illustrate how well these physical characteristics can be simulated computationally.

4.1 Impinging Oblique Shock Waves

Sketches showing the important features of two-dimensional oblique shock-wave interactions are shown in Figure 2. For a purely inviscid flow the uniform upstream flow processed by the incoming shock wave is uniformly turned toward the surface and then straightened again by the reflected shock. The corresponding surface-pressure signature is shown. Analytic expressions are available to predict this rather simple situation. The presence of a boundary layer confounds the problem, and the resulting flow-field characteristics depend on the strength of the incoming shock wave.

TABLE 1. BENCHMARK FLOWS FOR EVALUATING NAVIER-STOKES COMPUTATIONS: IMPINGING OBLIQUE SHOCK WAVES

FLOW	EXPERIMENTS	M_∞	Re x 10^{-6}	COMPUTATIONS	GRID x, y, ϕ	MODEL[a]
2-d	HOLDEN (1972)	8.5	22	BALDWIN AND MacCORMACK (1974)	40, 32	0; 2
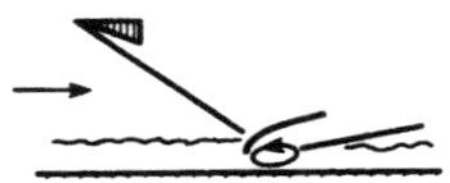	REDDA AND MURPHY (1973)	3	57	BALDWIN AND ROSE (1975)	40, 32	0 (MOD.)
				BALDWIN AND LOMAX (1978)	40, 32	0 (MOD.)
				VIEGAS AND HORSTMAN (1979)	40, 35	0; 2
	KUSSOY et al. (1975)[b]	7.2	13	MARVIN et al. (1975)	40, 78	0; 0 (MOD.)
				COAKLEY et al. (1977)	29, 45	0; 1; 2
CLOSED SEPARATION				VIEGAS AND HORSTMAN (1982)	89, 50	2
3-d	KUSSOY et al. (1980)[b]	2.2	36	KUSSOY et al. (1980)	47, 20, 20	0
				VIEGAS AND HORSTMAN (1982)	47, 20, 20	0; 2
OPEN SEPARATION						
3-d	BROSH et al. (1983)	3	18	BROSH et al. (1983)	45, 34, 38	0 (MOD.)
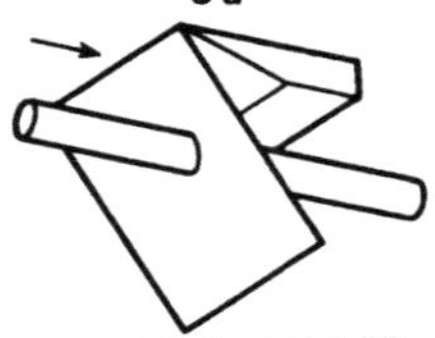						
OPEN SEPARATION						

[a]ZERO-, ONE-, AND TWO-EQUATION MODELS

[b]SELECTED FOR AFOSR/HTTM STANFORD CONFERENCE (KLINE et al., 1981)

TABLE 2. BENCHMARK FLOWS FOR EVALUATING NAVIER-STOKES COMPUTATIONS: COMPRESSION CORNER

FLOW	EXPERIMENTS	M_∞	Re x 10^{-6}	COMPUTATIONS	GRID x, y, z	MODEL[a]
2-d 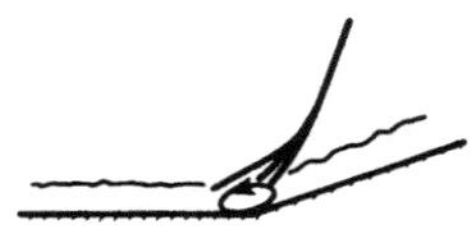	SETTLES et al. (1976, 1979)[b]	2.9	63–200	HORSTMAN et al. (1977)	50, 32	0; 1
				VIEGAS AND HORSTMAN (1979)	50, 32	0; 1; 2
				VIEGAS AND HORSTMAN (1982)	80, 36	2
					160, 40	2
				HUNG (1982)	77, 46	0 (MOD.)
	HOLDEN (1972)	8.7	22	HUNG AND MacCORMACK (1977)	63, 31	0; 0 (MOD.)
	LAW (1974)	3	10	SHANG AND HANKEY (1975)	64, 30	0; 0 (MOD.)
CLOSED SEPARATION				HUNG AND MacCORMACK (1977)	63, 31	0; 0 (MOD.)
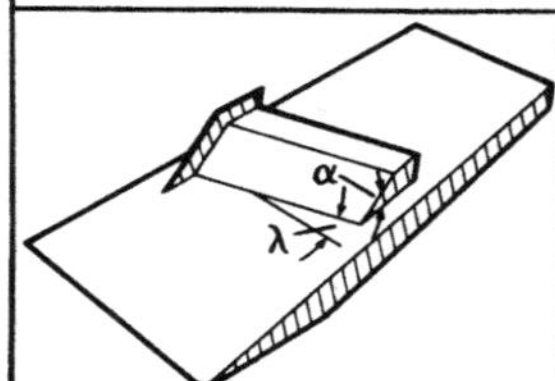 OPEN SEPARATION	TENG AND SETTLES (1982)	2.9	6	HORSTMAN (PVT. COMM.)	40, 35, 20	0; 2

[a]ZERO-, ONE-, AND TWO-EQUATION MODELS
[b]SELECTED FOR AFOSR/HTTM STANFORD CONFERENCE (KLINE et al., 1981)

TABLE 3. BENCHMARK FLOWS FOR EVALUATING NAVIER-STOKES COMPUTATIONS: GLANCING SHOCK WAVES

FLOW	EXPERIMENTS	M_∞	Re x 10^{-6}	COMPUTATIONS	GRID x, y, z	MODEL[a]
3-d 	PEAKE (1976)[b]	2	10	HORSTMAN AND HUNG (1979)	21, 28, 31	0 (MOD.)
				VIEGAS AND HORSTMAN (1982)	21, 28, 31	0 (MOD.)
				HUNG AND MacCORMACK (1978)	32, 36, 32	0 (MOD.)
	OSKAM et al. (1975)	3	5	HORSTMAN AND HUNG (1979)	21, 28, 31	0 (MOD.)
	WEST et al. (1972)	3	0.8	SHANG et al. (1979)	17, 33, 33	0 (MOD.)
OPEN SEPARATION						
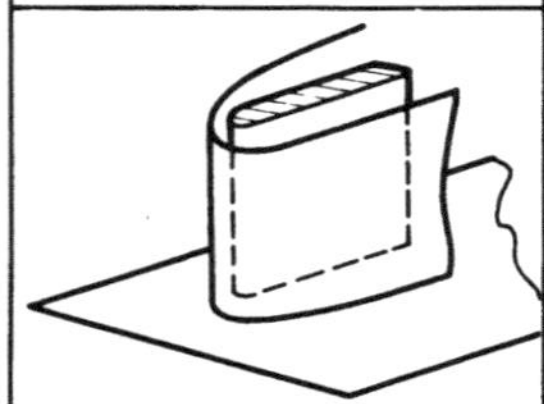	DOLLING AND BOGDONOFF (1982)	3	0.8	HUNG AND KORDULLA (1983)	40, 32, 32	0 (MOD.)
OPEN SEPARATION						

[a]ZERO-, ONE-, AND TWO-EQUATION MODELS

[b]SELECTED FOR AFOSR/HTTM STANFORD CONFERENCE (KLINE et al., 1981)

TABLE 4. BENCHMARK FLOWS FOR EVALUATING NAVIER-STOKES COMPUTATIONS: IMPINGING NORMAL SHOCK WAVES

FLOW	EXPERIMENTS	M_∞	Re x 10^{-6}	COMPUTATIONS	GRID x, y	MODEL[a]
2-d	SEDDON (1960)	1.5	0.6	VIEGAS AND HORSTMAN (1979)	40, 35	0; 1; 2
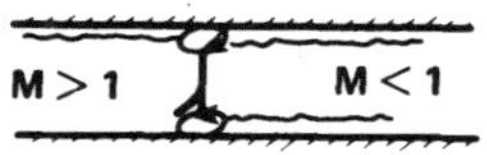				VIEGAS AND RUBESIN (1983)	20, 35	2 (MOD.)
	MATEER et al. (1976, 1979)[b]	1.3-1.5	10-200	VIEGAS AND HORSTMAN (1979)	40, 35	0; 1; 2
				VIEGAS AND HORSTMAN (1982)	38, 40	2
					76, 40	2
				McDONALD (1982)	31, 41	2
CLOSED SEPARATION STRAIGHT WALLS	OM et al. (1982)	1.3-1.5	0.5-1	OM et al. (1982)	170, 40	2
2-d	DELERY (1983)[b]	1.3-1.5	2-4	CAMBIER et al. (1981)	61, 21	0
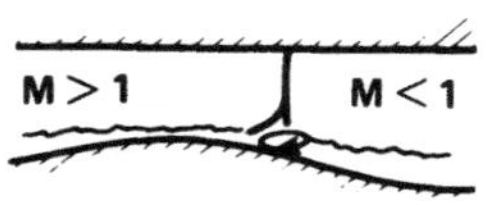				HORSTMAN (PVT. COMM.)		0; 2
	SALMON et al. (1981)	1.3-1.4	4	LIOU et al. (1981)	80, 50	2
CLOSED SEPARATION CURVED/DIVERGED WALL(S)						

[a]ZERO-, ONE-, AND TWO-EQUATION MODELS

[b]SELECTED FOR AFOSR/HTTM STANFORD CONFERENCE (KLINE et al., 1981)

TABLE 5. BENCHMARK FLOWS FOR EVALUATING NAVIER-STOKES COMPUTATIONS: TRAILING-EDGE FLOWS

FLOW	EXPERIMENTS	M_∞	Re x 10^{-6}	COMPUTATIONS	GRID x, y	MODEL[a]
→ $M_\infty < 1$ 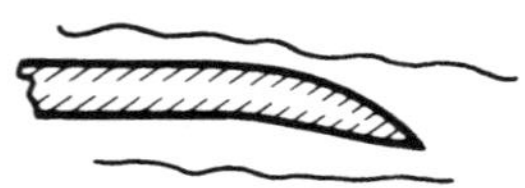	RAMAPRIAN et al. (1981) VISWANATH et al. (1980)[b] VISWANATH AND BROWN (1982)	0 0.4 0.4	1 47 26	MARVIN (1982) VISWANATH et al. (1980) HORSTMAN (1983)	50, 40 60, 50 79, 82	2 0; 2 2

[a]ZERO-, ONE-, AND TWO-EQUATION MODELS

[b]SELECTED FOR AFOSR/HTTM STANFORD CONFERENCE (KLINE et al., 1981)

TABLE 6. BENCHMARK FLOWS FOR EVALUATING NAVIER-STOKES COMPUTATIONS: SUPERSONIC – HIGH ANGLE OF ATTACK

FLOW	EXPERIMENTS	M_∞	$Re \times 10^{-6}$	COMPUTATIONS	GRID x, y, ϕ	MODEL[a]
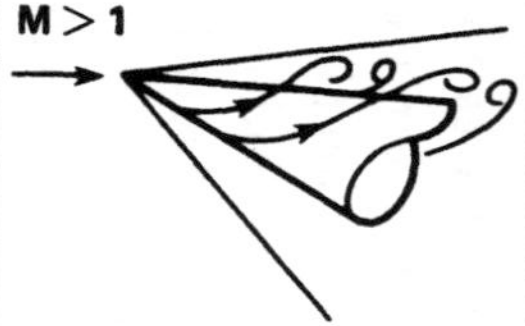 OPEN SEPARATION	RAINBIRD (1968)[b]	1.8	25	RAKICH et al. (1982)	VAR., 50, 47	0
				DEGANNI AND SCHIFF (1983)	0, 50, 144	0 (MOD.)
	BANNINK et al. (1978)	3	7	DEGANNI AND SCHIFF (1983)	0, 50, 144	0 (MOD.)
	McRAE et al. (1980)	1.6	25	McRAE et al. (1980)	0, 30, 36	0 (MOD.)
	KAYSER AND STUREK (1978)	3	7	SCHIFF AND STEGER (1980)	VAR., 50, 36	0 (MOD.)

[a]ZERO-, ONE-, AND TWO-EQUATION MODELS

[b]SELECTED FOR AFOSR/HTTM STANFORD CONFERENCE (KLINE et al., 1981)

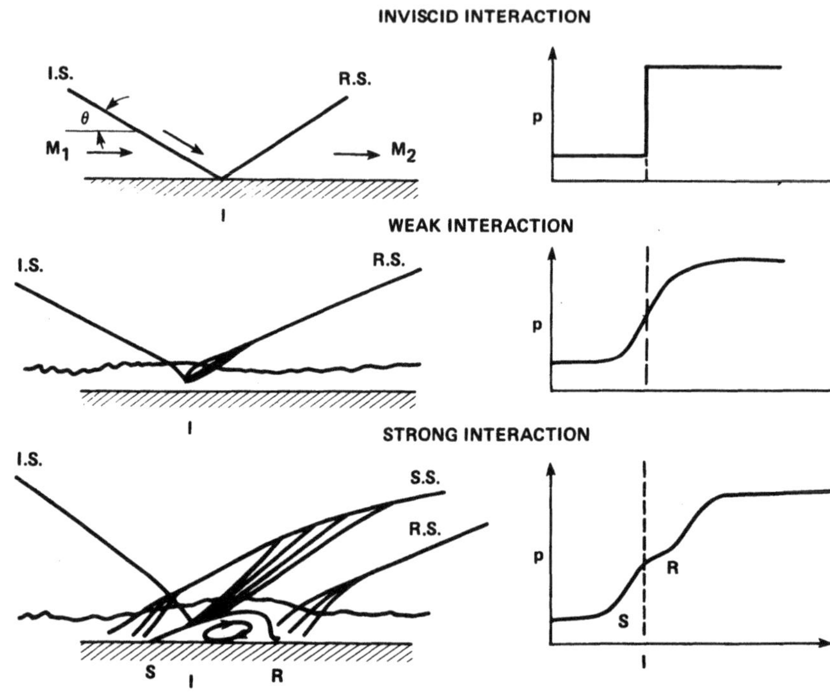

Figure 2. Physical characteristics of two-dimensional oblique shock-wave interactions.

In the weak interaction, the shock wave penetrates the turbulent boundary layer and turns more steeply toward the surface as it encounters the lower speeds within the viscous layer. It reflects from the viscous layer through a series of compression waves that coalesce into a reflected shock wave. A uniformly increasing surface-pressure signature is found, whose overall rise is nearly equivalent to the inviscid jump.

In the strong interaction, the shock wave also penetrates the viscous layer, but that layer cannot overcome the pressure rise, and separation takes place. The viscous layer is turned above the separation through a series of compression waves that coalesce into what is called a separation shock which is later weakened by expansion waves emanating from the viscous flow accelerating over the separation bubble. Downstream, where the bubble terminates, a series of compression waves coalesce into a reflected shock where the flow aligns itself with the surface. The corresponding surface pressure is characterized by a smooth pressure rise and an inflection region characteristic of separation. Also, it is assumed that the separation is closed by a dividing streamline that separates the mass entrained in the region from the outer flow and that a recirculating region is present. In actuality, the turbulent-flow probably leads to unsteadiness within this separated region, but how much influence this has on the mean characteristics is not understood at this time and further study is warranted. Above the separated region an island of very high peak pressure exists near the bifurcation associated with the intersection of the incoming and separation shocks. The extent (scale) of the interaction depends on the boundary-layer thickness, flow Reynolds number, and Mach number.

One of the first considerations in computing such flows is the ability of the computation to resolve shock waves. As reported by Metha and Lomax (1982), the solution methods discussed previously are all capable of capturing shock waves. However, the degree of shock sharpness depends on the numerical method and computational mesh. An example, taken from Coakley (1983), which illustrates what can be achieved with a reasonably good numerical method and a uniform mesh, is shown in Figure 3. Pressures along the solid surface and at a location about midway up in the mesh above the surface are shown for the case of an oblique wave inclined at 29° at a free-stream Mach number of 2.9. Similar results would be displayed in pressure distributions normal to the surface as the shock wave was traversed. The mesh used in this example is typical

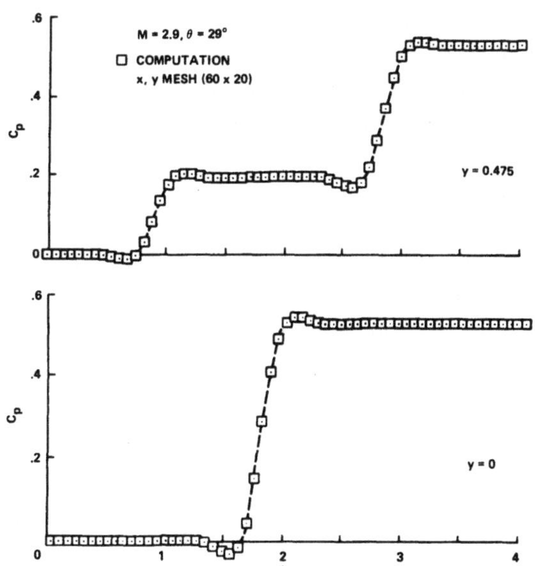

Figure 3. Modeling of an inviscid oblique shock-wave interaction.

of the mesh dimensions used in the Navier-Stokes codes out in the inviscid regions of the flow. The point to note is that the numerical method requires at least several mesh points to capture the pressure jump associated with the waves. From results such as these, it is easy to deduce that for solutions to the strong-interaction problems, in which separation and reflected-shocks occur, mesh choice will have an influence on how well the flow is modeled and further that a certain amount of shock *"smearing"* will always occur in practice. What seems to be missing in studies reported in the literature on shock-separated flow problems is an assessment of this effect on the results.

Many of the first computations of separated turbulent flows were directed toward solving the two-dimensional, strong impinging-shock interaction problem (see Table 1). Turbulence modeling was reported to have a strong influence on the results. An illustrative example is shown next. The bench mark experimental flow of Kussoy and Hortsman (1975) was computed with an explicit numerical method. The experimental apparatus was axisymmetric and thus eliminated three-dimensional effects now known to be present in other *two-dimensional* experiments.

Pressure contours from the experiment and two computations are shown in Figure 4. The experimental contours show the presence of the incident-, separation-, and reflected-shock waves as evidenced by the closely spaced contour levels. An island of very high pressure exists above the separation near the intersection of the incident and separation shocks. The computations were made with zero-equation eddy-viscosity models and the equations were solved down to the wall; the baseline computation used the mixing-length formulation given by equations (2.8) and (2.9), and the modified mixing-length model was determined from data analysis (Marvin et al., 1975). The grid was chosen to allow good shock capture in the outer regions, and in the viscous region a fine mesh was placed near the wall to resolve the turbulent boundary layer. The eddy viscosity from the baseline model is too high in the interaction region and as a consequence the computation only predicts the existence of a reflected shock wave.

On the other hand, the modified model, which results in lower eddy viscosities, gave a better simulation of the experimental flow. In addition to the reflected shock wave, the presence of a separation shock is evident, but it appears to be weaker and smeared compared with the experiment. This deficiency in the calculations is probably a

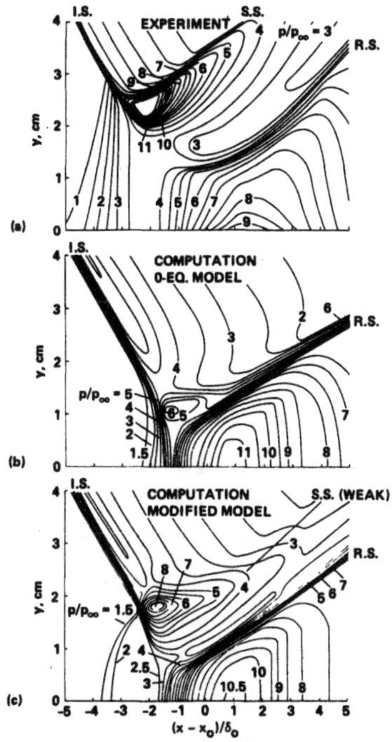

Figure 4. Modeling of a strong oblique shock-wave interaction: (a) experimental isobars; (b) computation using a 0-eq. turbulence model; (c) computation using a turbulence model modified by experimental information.

result of two things: the grid, which is still probably not fine enough to resolve the flow in the region of the island of high pressure, and the modified turbulence model, which still gives a small separation-bubble height relative to the experimental one. Surface skin friction and heat transfer were not accurately predicted within the separated zone, although the model modification did improve the results. In this instance, the model modification was experiment-dependent and, therefore, not extendable to the other conditions of Mach number and Reynolds number.

Although advances in numerical methods that improved computational efficiency provided the opportunity for investigating improvements in turbulence modeling, there has not yet been a significant advance in our ability to predict the flow detail within the separated region. What is known is that zero-equation eddy-viscosity models developed for attached flows must be modified or abandoned in favor of other approaches to provide a physically plausible representation of the flow, and that the model must provide some mechanism for altering the effective viscosity in the interaction zone. Two approaches have provided some improvement: modifying the zero-equation model eddy viscosity (Baldwin and Lomax, 1978) and using two-equation eddy-viscosity models (Viegas and Horstman, 1979).

The former approach, which is advantageous from the viewpoint of computational efficiency, has been used extensively in three-dimensional computations where computer storage and speed make application of higher-order models less attractive.

Results of a recent study by Brosh et al. (1983) of a three-dimensional shock interaction are worth examining because they illustrate current limitations. The flow field is sketched in Figure 5. A plane shock impinges on a cylinder aligned with free-stream flow. Separation occurs on the windward surface because of shock interaction, and on the leeward surface because, in part, of the cross flow imposed by the windward portion of the free stream being processed by the oblique shock. On the windward plane of symmetry the shock interaction is similar to that depicted in Figure 2, but the separation is not closed, and the flow within it is not a result of recirculation fed by downstream flow reattachment. (There has been some speculation that such open separations may be modeled appropriately with zero-equation eddy-viscosity models.)

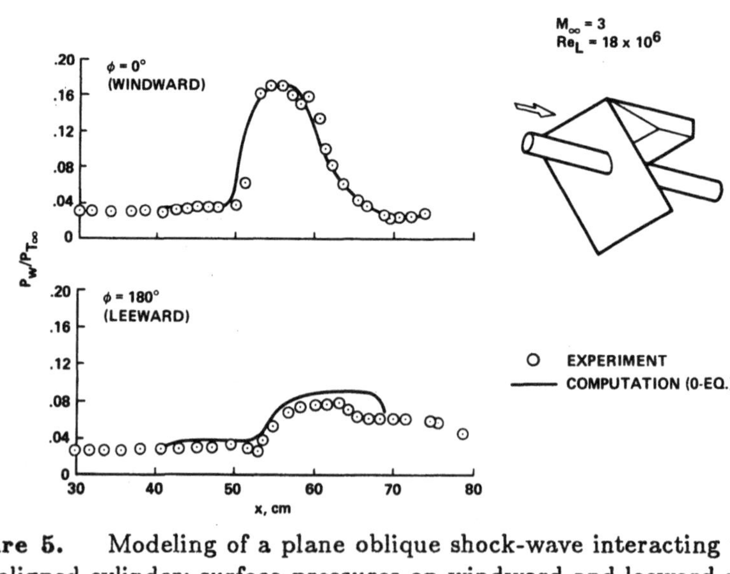

Figure 5. Modeling of a plane oblique shock-wave interacting with a flow-aligned cylinder: surface pressures on windward and leeward planes.

A cursory examination of the computed results indicates that many of the features observed experimentally are simulated, for example, surface-pressures distributions (Figure 5) and the initial separation line. More detailed examination, however, shows deficiencies that result from both turbulence modeling and grid resolution. In Figure 6, the windward plane flow field, determined by flow-field surveys, is sketched, and comparisons with static-pressure profiles are shown. Grid resolution in the region outside the viscous zone leads to significant shock smearing, and no separation shock is predicted.

In Figure 7, the surface skin-friction directions from the computations are compared with a photograph of oil-flow patterns on a Mylar sheet that had been placed around the cylinder and then *unwrapped* and photographed after the test. On the windward plane ($\phi = 0$), a single separated line is predicted, whereas a double separation line is evident in the experiment. It is likely that the deficiencies of the computation are caused by the combination of a poor turbulence model, which gives an effective viscosity that is too high, and poor numerical resolution of the shock system, which causes a local weakening of the shock strength. As the flow proceeds around to the leeward side, a single line of separation is predicted, whereas a double line of separation is measured. As we will see in a later section, the turbulence model of Baldwin and Lomax (1978) is unlikely to predict secondary separations without modification and, in addition, the azimuthal grid spacing was probably too coarse. Hence, grid resolution and turbulence modeling must both be improved before definitive conclusions can be reached on the modeling of three-dimensional, impinging-shock, separated flows. These particular calculations took two hours on a Cray 1-S computer, so finer grid resolution that could help resolve this issue is costly and has not been carried out.

4.2 Supersonic Compression Corner

The physical characteristics and corresponding wall pressures for a two-dimensional compression corner are sketched in Figure 8. For the inviscid flow situation, a single shock forms and the pressure rises abruptly to the level predicted by wedge-flow relations. The presence of a boundary layer complicates the flow, as depicted for

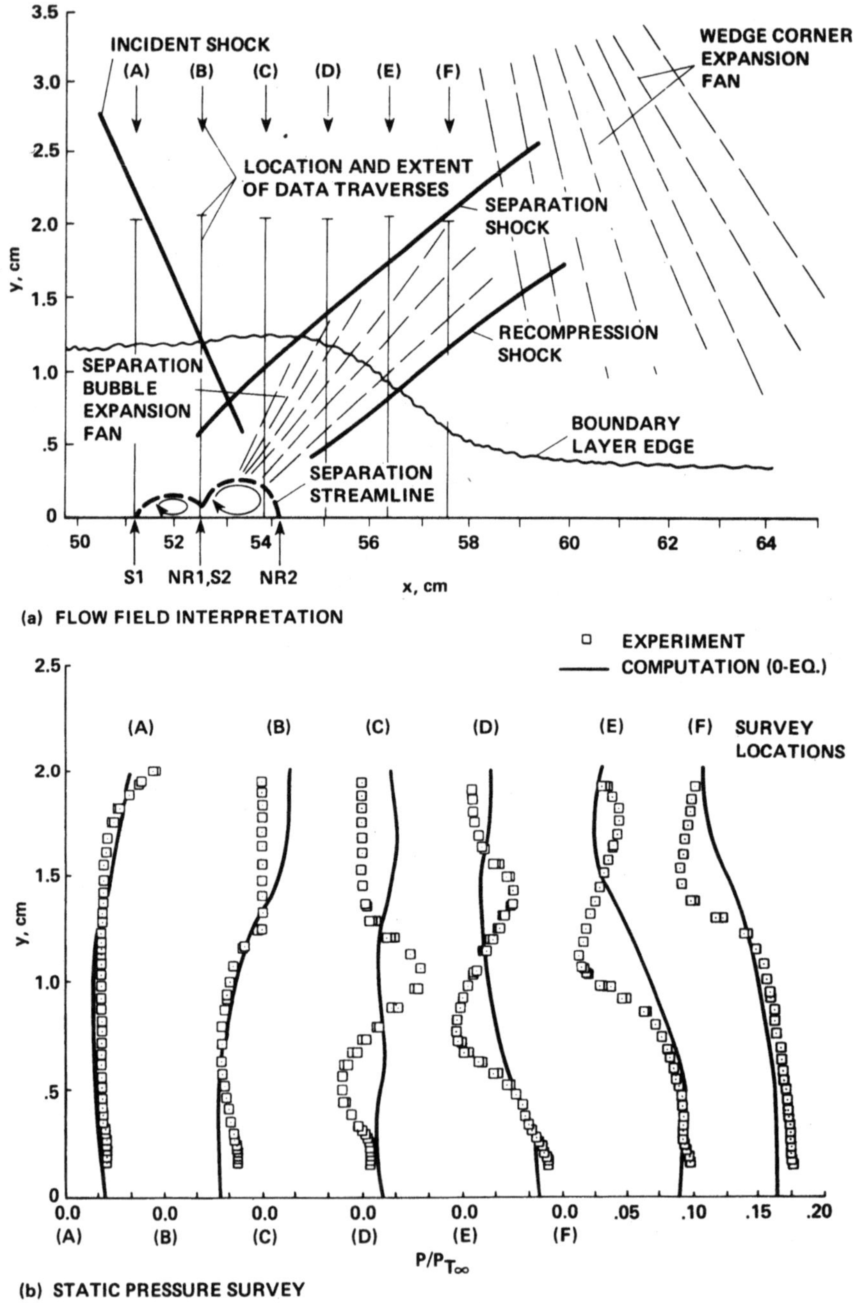

Figure 6. Modeling of a plane strong oblique shock-wave interacting with a flow-aligned cylinder: (a) windward symmetry-plane flow field interpretation from experiment; (b) comparison of experimental and computational static pressure surveys in the windward symmetry-plane.

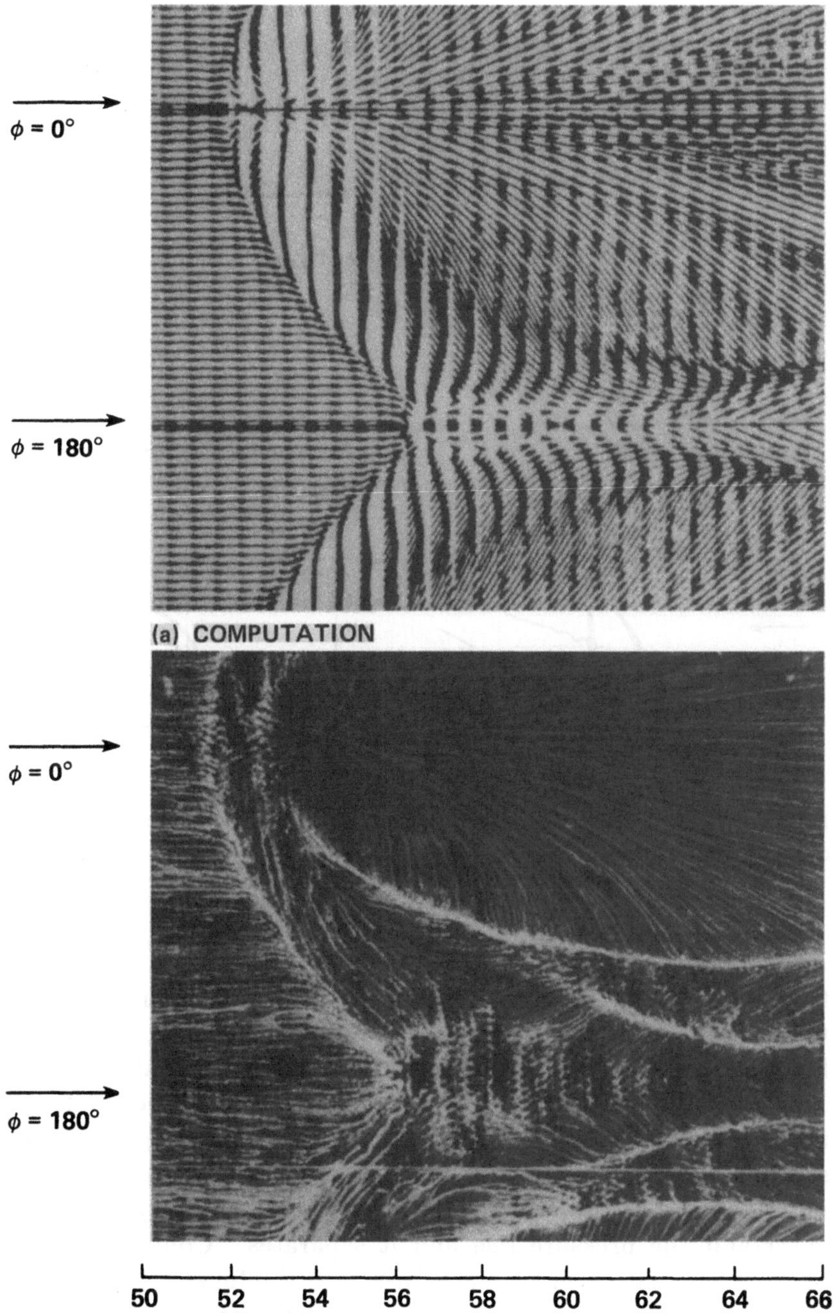

Figure 7. Modeling of a plane strong oblique shock-wave interacting with a flow-aligned cylinder: comparison of experimental and computational skin friction lines.

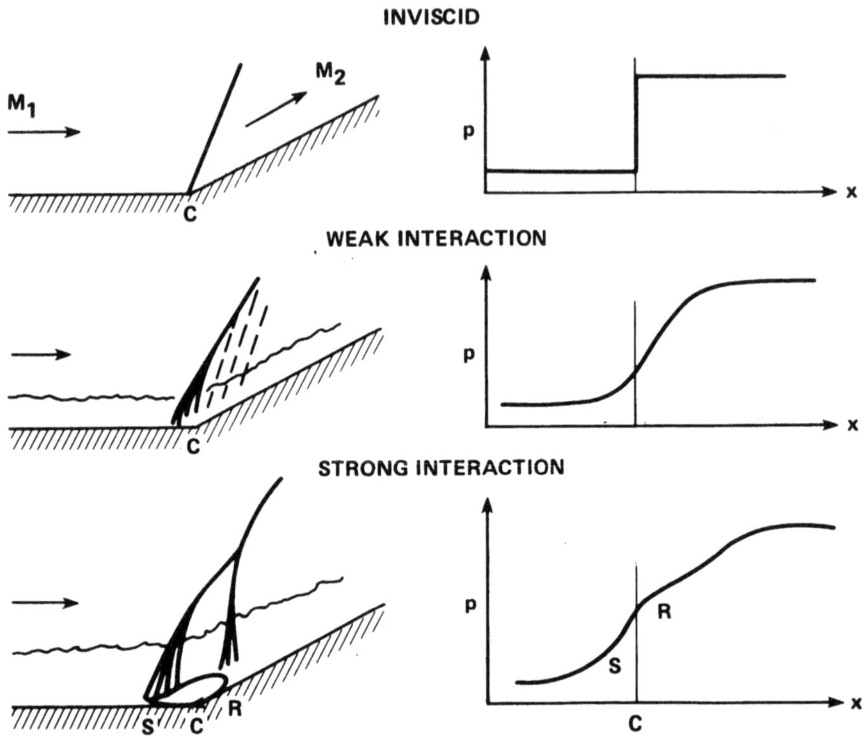

Figure 8. Physical characteristics of two-dimensional compression-corner interactions.

two situations, the weak and strong interactions. In the weak interaction, a series of compression waves forms within the boundary layer as it encounters the pressure rise and they coalesce with the shock formed in the inviscid flow, which is required to turn the flow in the direction of the ramp. The corresponding pressure rise shows a smoothing of the pressure at the beginning and end of the interaction. For the strong interaction, the boundary layer cannot withstand the pressure rise and it separates. Compression waves that coalesce into a shock wave form near the forward portion of the separation bubble as the outer viscous flow negotiates the pressure rise.

Experimentally, the separation shock-angle is found to be independent of the corner angle. If the separation is large enough and the free-stream Mach number high enough, a second shock will form downstream when the flow over the separated region reattaches and turns in the direction of the ramp. The separation and recompression shocks coalesce with the outer shock wave. The corresponding pressure rise shows inflection over the separated region and the upstream influence is more pronounced than in the weak case. Conceptually, the flow in the closed separated region is divided from the outer flow, and mass is entrained and recirculated through the reattachment process. However, as we shall see, there is experimental evidence of unsteadiness in this process. The characteristic scale of the interaction depends on the boundary-layer thickness and free-stream Mach number.

Computations of this complex flow have been reported, as indicated in Table 2. Different numerical methods and turbulence models have been employed. A comparison of two of the more recent computations with experiment is shown in Figure 9. Two cases are shown, one near incipient separation (weak interaction) and one with separation (strong interaction). In one computation, an implicit algorithm and the thin-layer form of the equations were used with the modified zero-equation model of Baldwin and Lomax (1978) which was described earlier. In the other, the MacCormack hybrid algorithm and the full equations were used with the two-equation turbulence model of Wilcox and Rubesin (1980). Metha and Lomax (1982) stated that these different numerical schemes should yield similar results, since comparable grids are used and care in carrying out the computation is exercised. Accepting that premise, the differences between these calculations mainly reflect differences owing to turbulence modeling.

In both the weak and strong cases, the pressures predicted using either model agree reasonably well with the data, and this reflects the common observation that the pressure rise can be estimated, for engineering purposes, using any of the eddy-viscosity models. However, differences occur in the viscous regions. The modified zero-equation model predicts skin-friction values that are much too low downstream of the weak interaction, and this manifests itself more critically in the strong-interaction case by predicting reattachment too far downstream and velocity profiles that do not compare well with experiment. On the other hand, velocity profiles and shape factors in the downstream region are predicted better by the two-

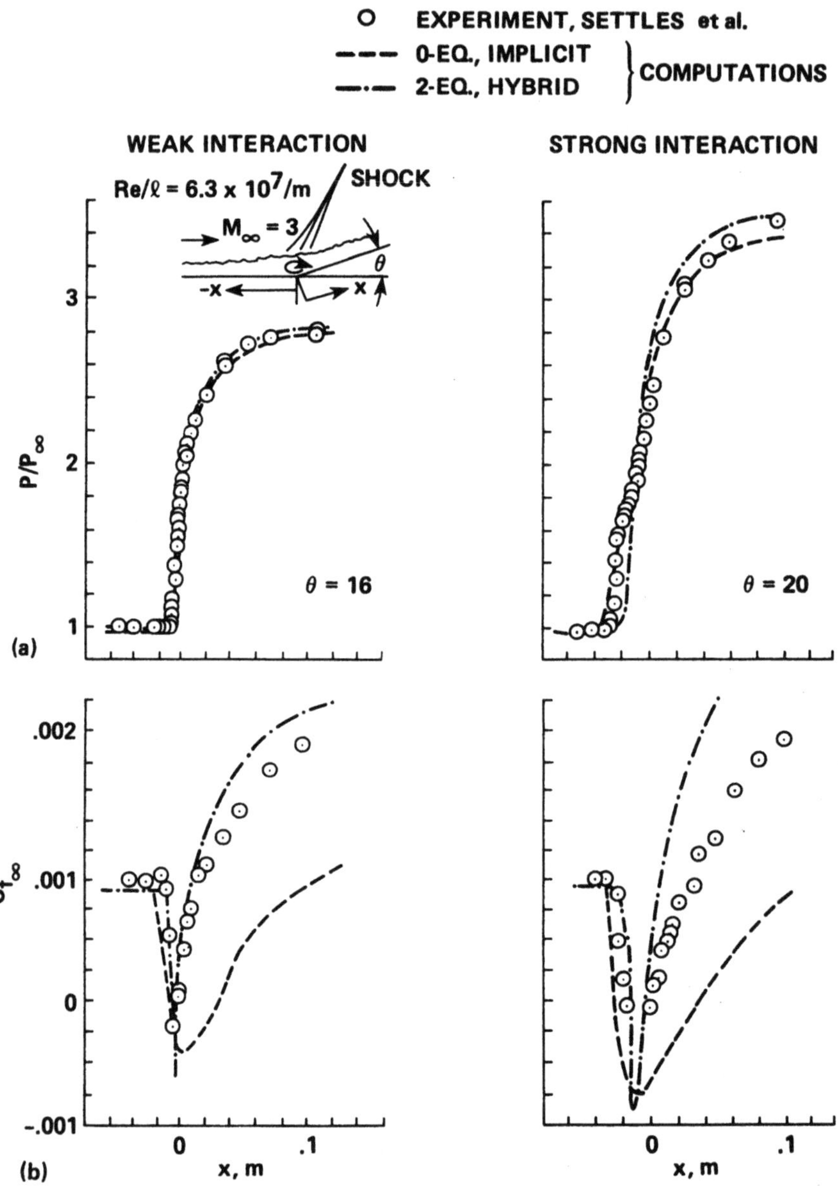

Figure 9. Modeling of a compression-corner interactions using different algorithms and different turbulence models: (a) pressure distributions; (b) skin friction distributions.

equation model, even for the strong-interaction case in which skin friction is somewhat overpredicted (Marvin, 1982). It is thought that the failure of the two-equation model to predict the skin friction resides in the low-Reynolds-number modeling terms developed to allow integration to the wall, but this must be investigated further and in light of the experimental observations on unsteadiness, which is discussed next.

Unsteady pressures have been measured on compression corners by Dolling and Or (1983). Results from a 20° compression corner test are shown in Figure 10. Normalized mean pressures, $\bar{p}_w/\bar{p}_{w_o}$, their rms fluctuations, σ_w, and an intermittency factor, γ, are shown for positions upstream of the corner. The intermittency factor represents the fraction of time that $\bar{p}_w > \bar{p}_{w_o} + 3\sigma\bar{p}_{w_o}$ (i.e., the time that the instantaneous pressure is greater than that of the undisturbed turbulent boundary layer). The peak rms fluctuations occur ahead of separation in the region of the initial pressure rise. The intermittency reaches a value of 1 near the point of maximum fluctuations and just ahead of the mean separation point. It was deduced that these measurements most probably indicate a separation shock movement of about one boundary-layer thickness.

Such unsteadiness could be caused by unsteady mass entrainment in the recirculating zone as a result of scale changes within the turbulent structure. None of the computations reported have indicated unsteady motion of this sort, and if it is caused by time-varying turbulent structure changes, turbulence models based on mass-averaged variables will not be appropriate for modeling the unsteady details. Much, therefore, remains to be learned about modeling for these shock interacting flows. At this time, only mean pressures can be predicted with reasonable confidence, as can the trends of separation and the reattachment location movement with changing Reynolds number based on the incoming boundary-layer thickness (see Figure 11).

Three-dimensional compression corner flows are also now under study; see for example Teng and Settles (1982). Interesting classifications of these flows on the basis of conical and cylindrical upstream influence have been postulated. Although no calculations have been reported, one of the author's colleagues, C. C. Horstman, has had recent success in predicting the flows with conical upstream influence. These results should be available shortly.

4.3 Glancing Shock Wave

Control surfaces on vehicles or missiles can produce shock waves that sweep across adjacent boundary layers. Some bench mark experiments depicting the essential features of these flows are available for verifying computations (see Table 3). Geometries for two of these are sketched in Figure 12 along with surface skin-friction lines and shock-wave structures which help to describe the general physical characteristics of the flows.

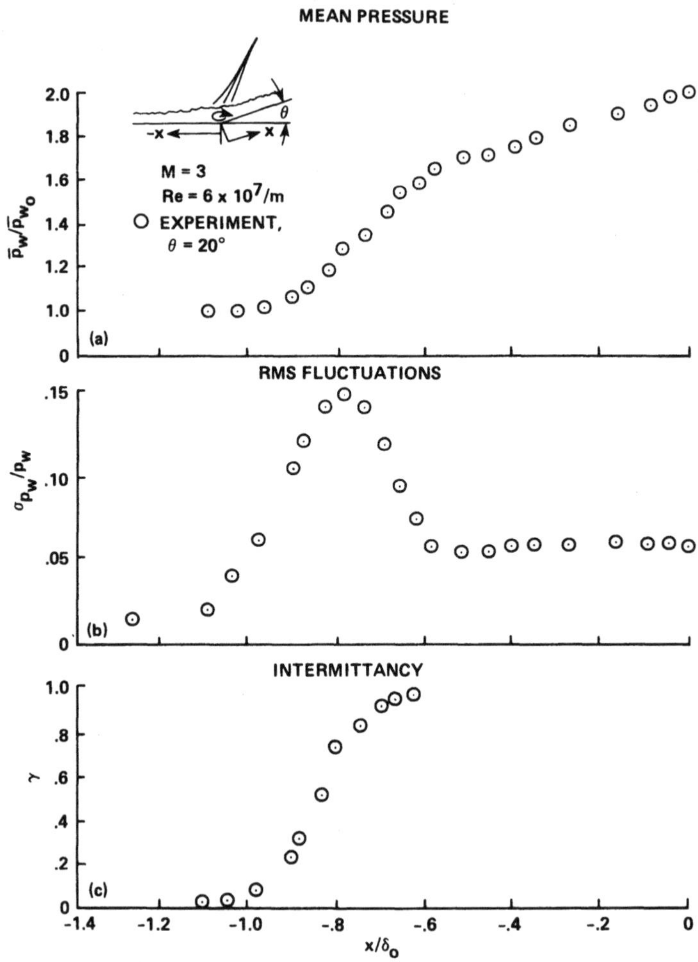

Figure 10. Effects of unsteadiness on the physical structure of a compression-corner interaction: (a) mean surface pressure; (b) rms of fluctuating pressure; (c) intermittency relative to undisturbed flow.

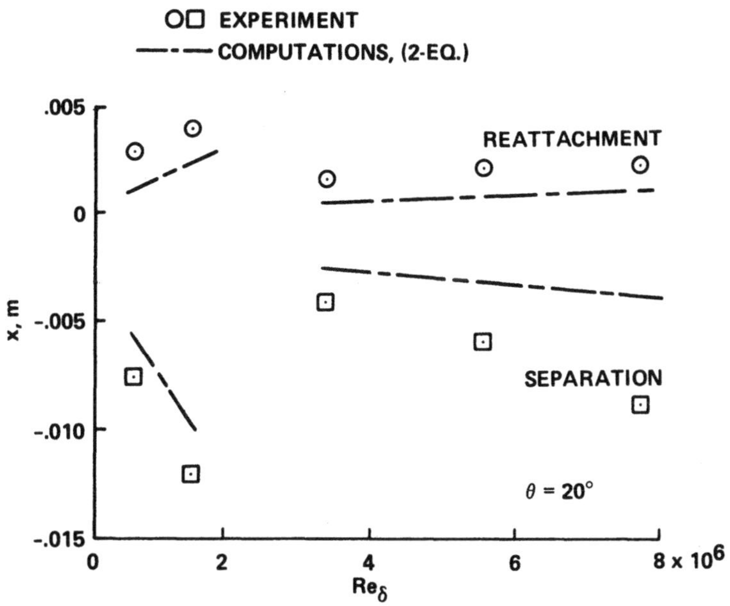

Figure 11. Modeling of separation and reattachment points for a compression corner.

The sharp leading-edge shock generator can result in both weak and strong interactions. In the weak case, the shock interacts with the incoming boundary layer and causes simple flow-turning, with the lower momentum fluid near the wall undergoing larger turning than the higher momentum fluid at the boundary-layer edge. Far from the generator leading edge, the shock pattern formed by the component of the Mach number normal to the shock wave might appear as a weak shock, as sketched in Figure 12. In the strong interaction, the boundary layer cannot overcome the pressure gradient, and a separation line forms ahead of the shock wave and a reattachment line forms downstream. Skin-friction lines accompanying such characteristics are sketched in the figure (see Peake and Tobak, 1980). In this case, the component of the Mach number normal to the shock wave is larger, and the interaction is stronger and a shock wave with the characteristic lambda foot emanating from the compression waves formed near the separation line. In contrast to the two-dimensional, normal-shock-wave case, the flow in the separation region is not closed and continued recirculation of the shock-processed fluid does not occur. In this sense, the swept shock flows are probably more steady than the two-dimensional flows. Furthermore, the flow relief owing to the third dimension causes the boundary layers to separate sooner and to have correspondingly larger upstream influence than the two-dimensional flows. The scale of these interactions is determined mainly by the incoming boundary-layer thickness and Mach number.

In the case of the blunt leading edge, a bow shock wave is formed and a strong interaction takes place. Separation and reattachment lines form ahead and downstream of the bow shock wave. A horseshoe vortex forms as a result of the presence of the blunt generating surface, and it streams around it. The shock wave in the plane of symmetry can form a lambda foot near the separation line for the strongest interactions and an inviscid shear layer emanates from the bifurcation point. The scale of the interaction is determined by the bluntness of the generator, because the shock standoff position and the horseshoe vortex scale are proportional to it.

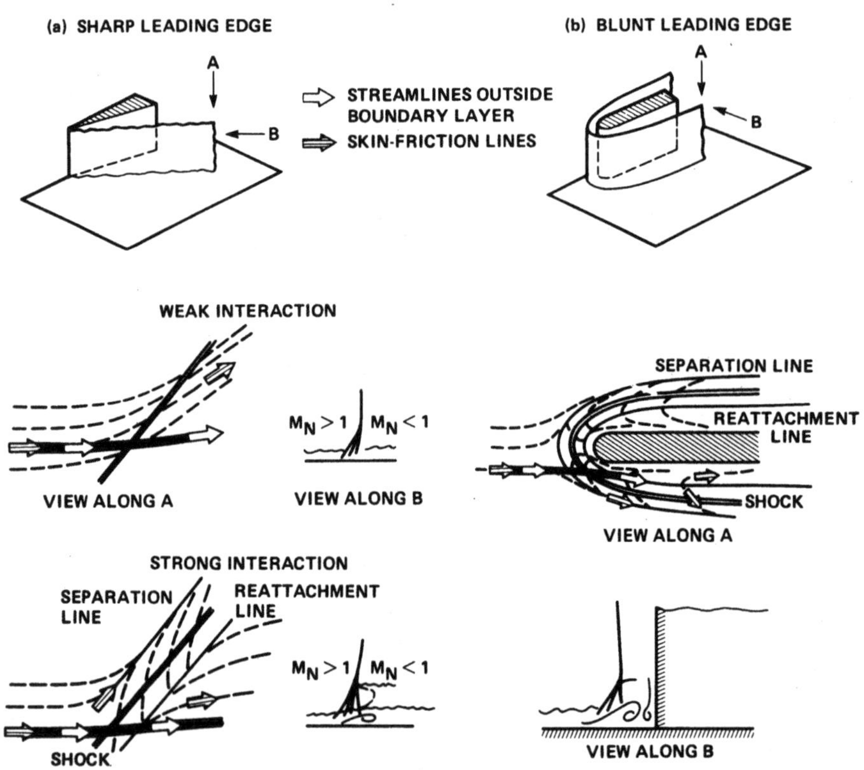

Figure 12. Physical characteristics of glancing shock-wave interactions: (a) sharp leading edge; (b) blunt leading edge.

Surprisingly, numerical simulations of these glancing shock-wave flows using rather coarse grids and a simple turbulence model provide adequate predictions of experimental data in contrast to the impinging-shock-wave and corner-flow results discussed in previous sections. To illustrate this point for the sharp-generator case, typical comparisons of computation and experiment are shown in Figures 13–15. The computations by Horstman and Hung (1979) were made with the MacCormack hybrid method along with a two-layer, zero-equation, mixing-length eddy-viscosity model (equations (2.8) and (2.10)), modified by Hung and MacCormack (1978) to account approximately for the flow in the corner formed at the intersection of the generator and the plate.

The axial variations of pressure and skin friction (Figure 13) and the spanwise variation of pressure and heat transfer (Figure 14) agree with the measurements except in the corner where modeling is undoubtedly incorrect. Differences in the axial variations at the farthest downstream location are caused by locating the computational boundary there. Although not shown here, agreement with mean-velocity profiles is also good. Similarly good comparisons of surface and flow-field quantities have been reported for wedge angles to 12° and Mach numbers to 6.

Surface skin-friction lines from the computations are shown in Figure 15. Locations of the main features of this strong-interaction case are noted. The separation and reattachment lines were determined by examining cross-flow velocity vector plots oriented in a plane normal to the center of the vortex formed by the interaction. They correspond closely to the converging and diverging lines usually associated with the separation and reattachment locations (Peake and Tobak, 1980).

Several factors are believed responsible for the good agreement between computations, in which coarse grids and a simple mixing-length turbulent model are used and experiment. First of all, the normal component of the Mach number is not large and therefore the shock-wave structure is not so difficult for capture. (In the example shown $M_N = 1.3$.) Secondly, the separated-flow region is not closed and highly turbulent fluid is not recirculating. And, as a consequence of the latter, the flow within the separated region is probably more steady than that within a two-dimensional separated region.

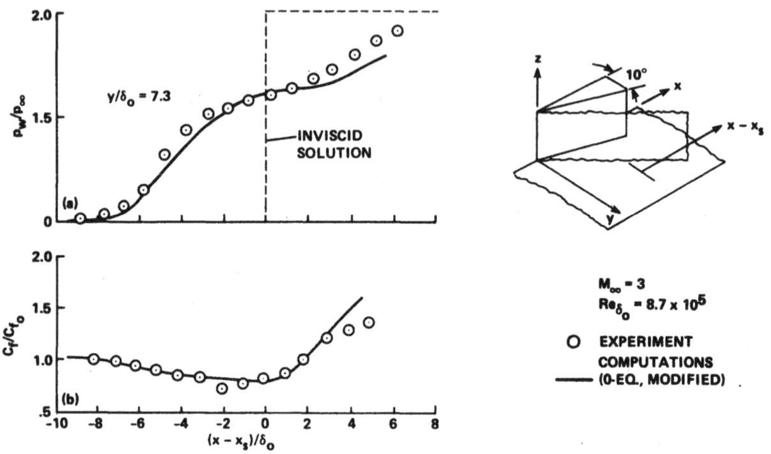

Figure 13. Modeling of a glancing shock-wave interaction from a sharp leading-edge wedge: (a) pressures along streamwise direction; (b) skin friction along streamwise direction.

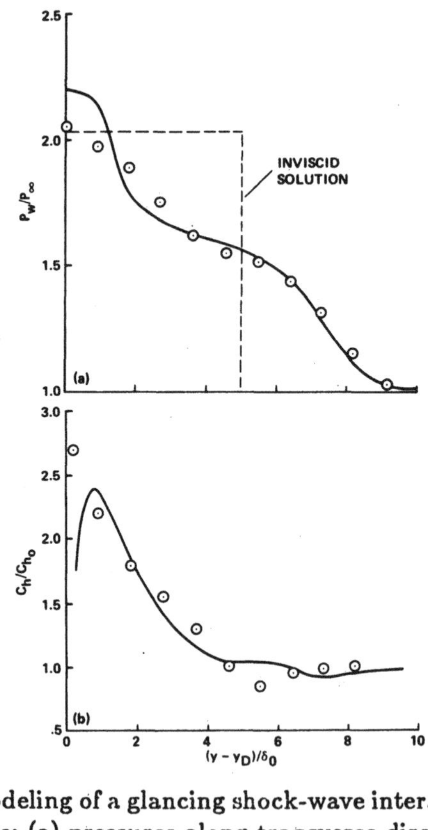

Figure 14. Modeling of a glancing shock-wave interaction from a sharp leading-edge wedge: (a) pressures along transverse direction; (b) skin friction along transverse direction.

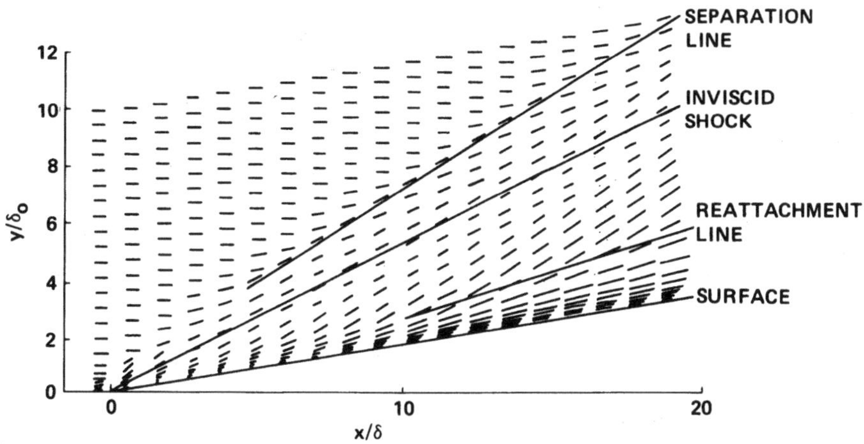

Figure 15. Skin-friction line directions computed for a glancing shock wave from a sharp leading-edge wedge.

The strong interactions resulting from a blunt generator have been recently computed by Hung and Kordulla (1983). The computations were made using a finite-volume version of the newest implicit-explicit method of MacCormack (1982) and the zero-equation turbulence model of Baldwin and Lomax (1978) modified in the same manner as the sharp-generator case to account for the presence of the generator wall. Some example comparisons of these computations with the experimental data of Dolling and Bogdonoff (1982) are shown in Figures 16 and 17. Surface pressures along the flat plate and along the generator surface are shown. It can be inferred from these comparisons that the scale of the interaction, including its upstream influence on the oncoming flow and its height relative to the oncoming boundary-layer thickness are probably being predicted quite well, although no flow-field data are available to verify such a conclusion.

The predicted particle paths which represent streamlines in the plane of symmetry are shown in Figure 18 to illustrate the resolution of the flow-detail within the horseshoe vortex. Although not readily apparent in this plot (because of the scale of the figure), there is a secondary vortex formed at the junction between the blunt generator and the plate (Hung and Kordulla, 1983). The separation region formed by the horseshoe vortex is open and the vortex streams around the blunt generator.

Most of the features observed in oil-flow photographs taken during the experiment also compare favorably, at least qualitatively, with these computations. Again, it may appear surprising that the computations are doing so well, considering the grid resolution and simplicity of the turbulence model. However, the scale of the interaction is set mainly by the blunt leading edge of the generator, in contrast to the three-dimensional, impinging-shock case (Figures 5 and 6) in which no physical scale other than boundary-layer thickness is present.

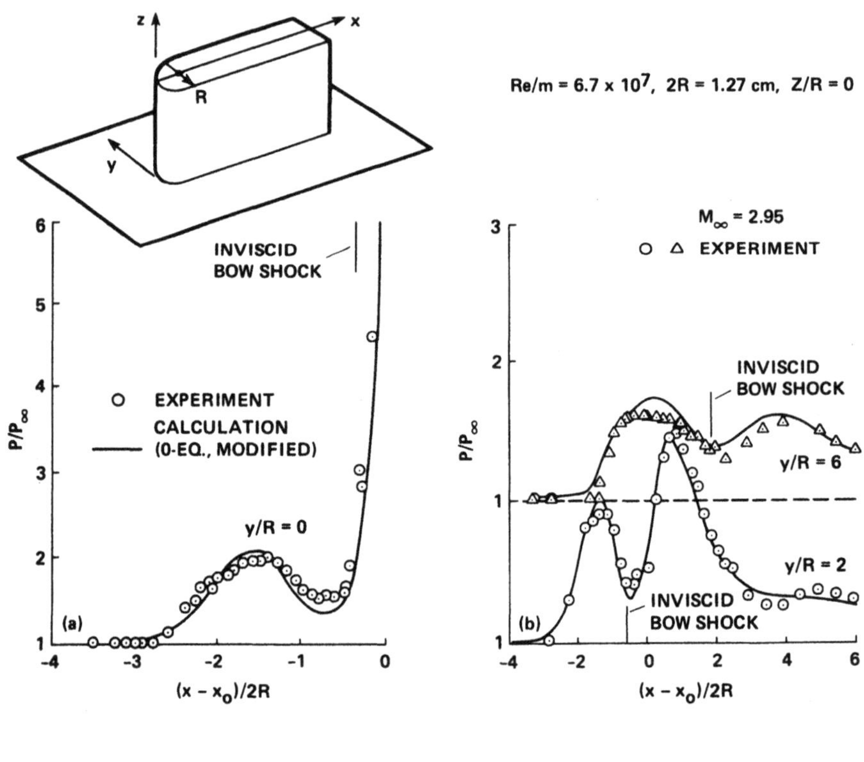

Figure 16. Modeling of a glancing shock-wave interaction from a blunt-plate generator: (a) pressures on the flat plate in the plane of symmetry; (b) pressures on the flat plate off the plane of symmetry.

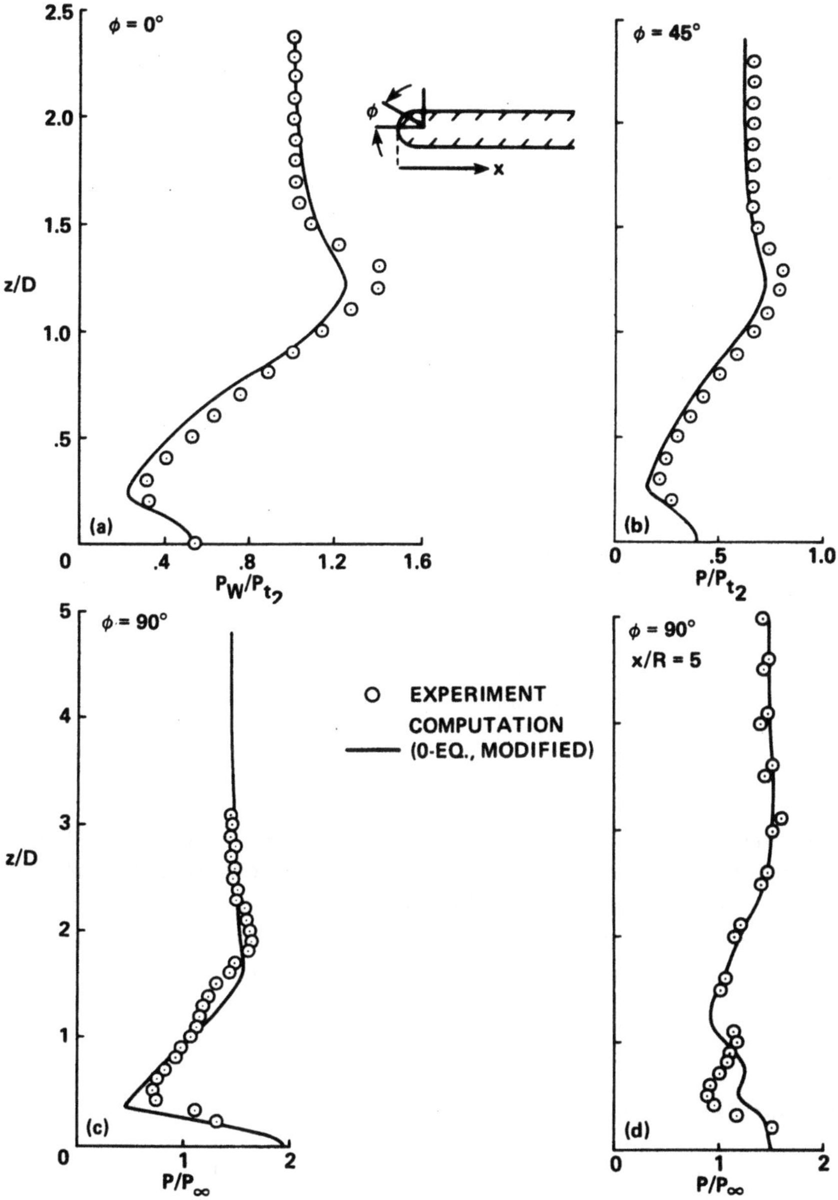

Figure 17. Modeling of a glancing shock-wave interaction from a blunt-plate generator: (a) pressures on the blunt-plate generator, $\phi = 0°$; (b) $\phi = 45°$; (c) $\phi = 90°$; (d) $\phi = 90°$, $X/D = 2.5$.

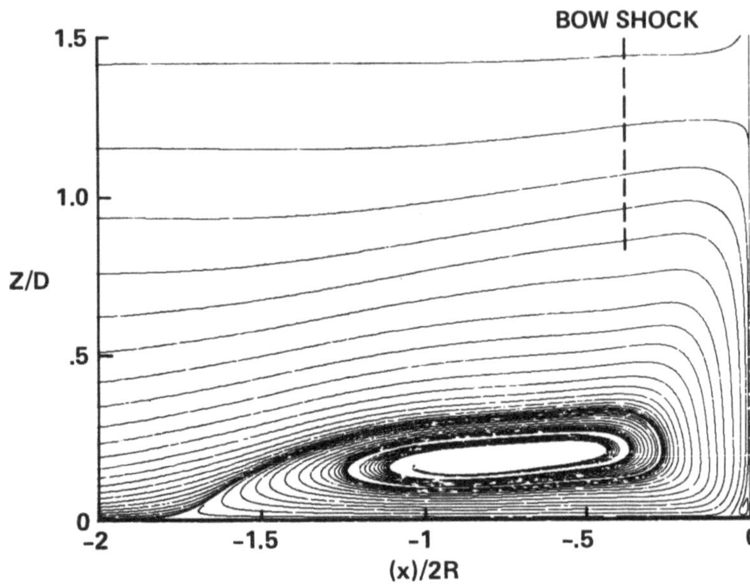

Figure 18. Particle paths that depict streamlines in the plane of symmetry ahead of the generator from computations of a glancing shock-wave interaction from a blunt-plate generator.

Further study of this blunt generator case is needed to determine whether important quantities, such as heat transfer or skin friction, can be predicted. It should also be mentioned that an unsteady shock-wave structure was found experimentally and that no such unsteadiness was found in the computation.

4.4 Normal-Shock-Wave Interaction

Understanding the normal-shock-wave problem is important for the development of supercritical-wing technology. In this paper we will focus our attention on studies that have attempted to isolate the flow in the vicinity of the shock wave and in which the elliptic nature of transonic flow does not have to be considered (see Table 4). Some of the physical characteristics are depicted in Figure 19. The Schlieren photographs and Mach contours from the two-dimensional experiment of East (1976) are shown. At the lowest Mach number, a weak interaction develops and very little change in

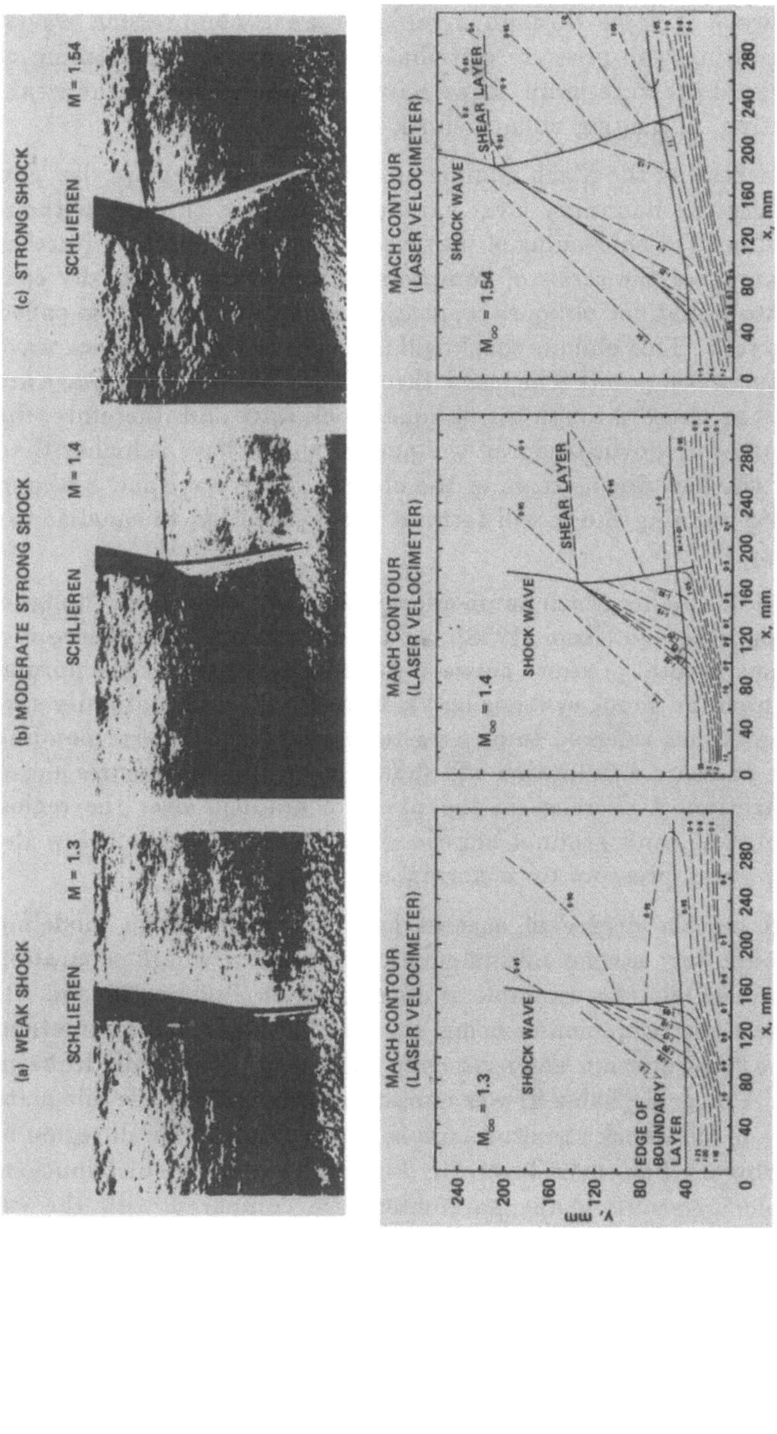

Figure 19. Physical characteristics of normal shock-wave interactions: (a) weak shock; (b) moderately strong shock; (c) strong shock.

the normal-shock-wave structure occurs. A thickening of the subsonic layer takes place during the movement of the viscous-layer from the supersonic to the subsonic regions. A small foot to the normal shock wave appears through a series of weak compression waves. The resulting wall-pressure distribution appears as a smoothing of the inviscid pressure jump, as we have seen previously for the weak-interacting, impinging oblique-shock flows.

Increasing the Mach number strengthens the pressure rise, and eventually the boundary layer can no longer pass through without separating. The thickening of the viscous layer occurs sooner (farther upstream) and the series of compression waves can eventually coalesce into a distinct oblique, separation shock forming the so-called lambda foot. This oblique shock will intersect the normal shock wave at a bifurcation point. The losses through the normal shock wave are larger than those through the oblique shock wave and, therefore, the static pressure downstream of the normal shock wave is higher than that of the flow downstream of the oblique shock wave and a second rearward-running shock will form at the bifurcation to equalize the disparity.

At the higher Mach numbers, existence of a supersonic *"tongue"* has been observed (Kooi, 1978). At the bifuraction point there is a difference in total pressure between the flow processed by the normal and compound shock systems and a shear layer (a discontinuity surface sometimes referred to as a vortex layer) forms. Corresponding surface-pressure distribution will show a steep rise in pressures ahead of separation, a decrease in the pressure gradient over the region of separation, and gradual increase to a level somewhat below the inviscid jump pressure for a normal shock.

A certain degree of success has been achieved in modeling the moderately strong normal-shock interaction where separation is rather small. An example is illustrated in Figures 20 and 21. Computations were made using the MacCormack hybrid method and the two-equation eddy-viscosity model of Wilcox and Rubesin (1980). The grid (Table 4) was chosen in order to provide adequate capture of the shock structure and to resolve the near-wall region of the turbulent boundary layer. In Figure 20, pressure-distributions and velocity-profile shape parameters are compared with the experiment reported by Om et al. (1982) for a range of Mach number and Reynolds number. The experiment was performed in an axisymmetric test section so that three-dimensional effects could be

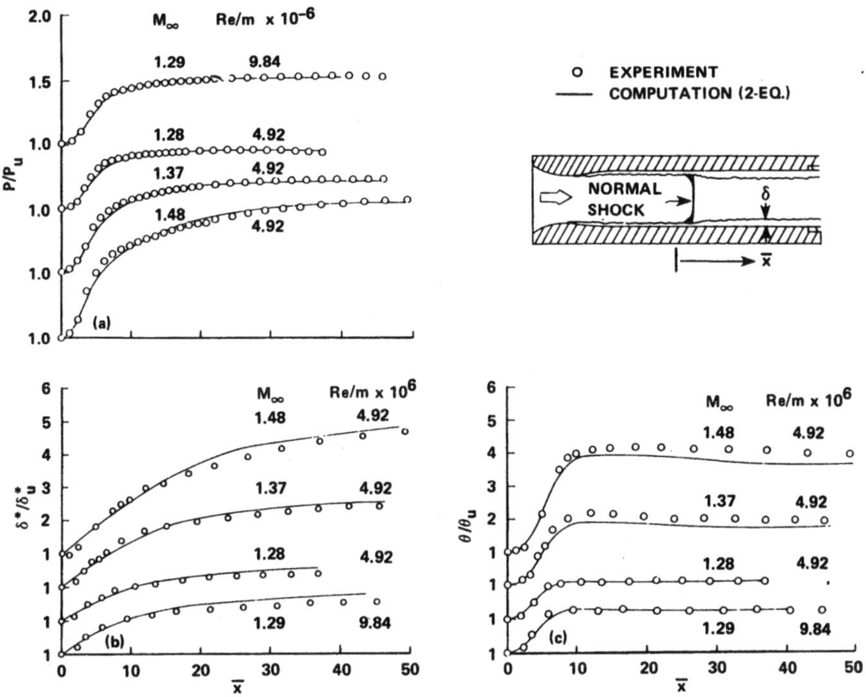

Figure 20. Modeling of a moderately strong normal-shock wave interaction for various Mach and Reynolds numbers: (a) surface pressures; (b) displacement thicknesses; (c) momentum thicknesses.

eliminated; therefore, a high degree of confidence can be placed in the experimental trends that are observed.

The effects of Mach number and Reynolds number are predicted by the computations, except possibly in the immediate vicinity of the shock at the higher Mach numbers. Mach contours are compared in Figure 21 for the highest Mach number case. For the most part the shock structure is also predicted by the computations. The shock is weakened because of viscous-layer thickening near the separation, and a series of compression waves coalesces into the normal shock. A smaller region of supersonic flow is predicted. One would not expect to capture any discontinuity surface in total pressure that would lead to a so-called vortex layer because the grid is obviously too coarse. The extent of separation in the prediction is somewhat smaller than that of the experiment.

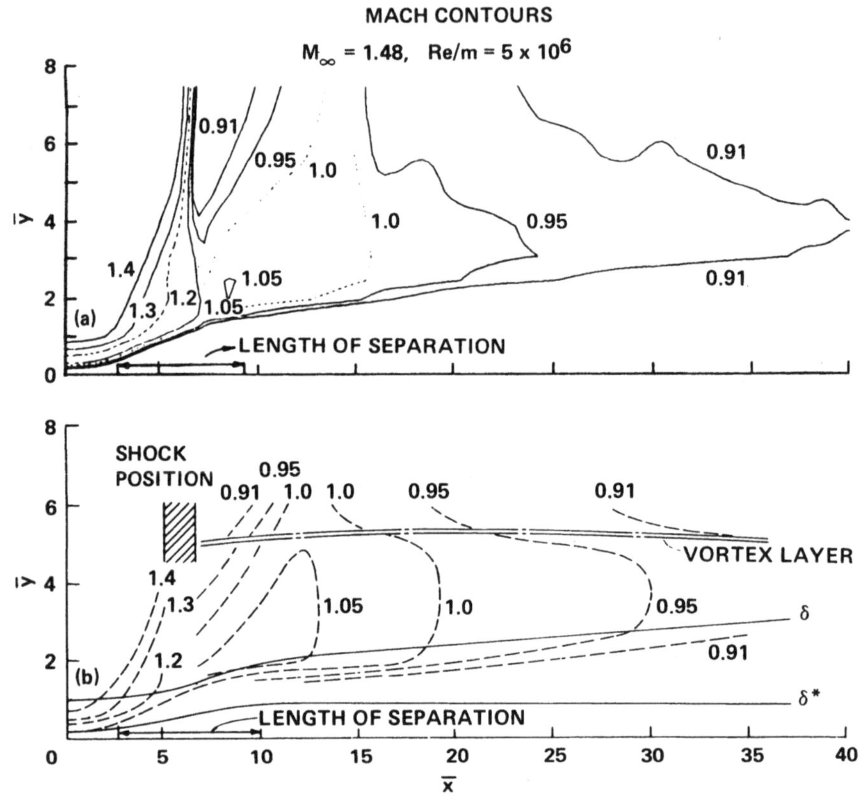

Figure 21. Modeling of a moderately strong normal-shock wave interaction: (a) Mach contours from the computation; (b) Mach contours from the experiment.

Studies have shown that the choice of turbulence model has an influence on the predictions (Viegas and Horstman, 1979). An illustrative example is shown in Figure 22. Although the turbulence model has little influence on the prediction of the overall pressure rise, models that use information on the turbulent kinetic energy changes through the shock wave to form the velocity scale of the eddy viscosity provide much better estimates of the skin friction. We can also note that trends with Reynolds number over a wide, practical range are predicted with those higher-order eddy-viscosity models. However, even these higher-order models have to be applied with caution when wall skin friction or heat transfer is being predicted, because the low-Reynolds-number functions, required when integrating the equation system from a wall boundary out into the flow field have not always been developed adequately.

The reader is referred to a very recent paper by Viegas and Rubesin (1983) in which that aspect of higher-order eddy-viscosity modeling for the moderately strong, normal-shock problem is studied. Figure 23 summarizes the main points from that study. When integrating from the wall boundary, only the Wilcox-Rubesin model gives adequate skin-friction predictions. In developing this model's low Reynolds-number functions, particular attention was given to ensure that modeling was adequate for attached, large adverse-pressure-gradient flows, and evidently the model can also perform adequately in moderately strong normal-shock interactions in which small separation occurs.

On the other hand, the model of Jones and Launder (1971), with its original formulation of the low Reynolds-number terms, and one developed by Chien (1982) to minimize computational stiffness encountered when applying the model of Jones and Launder, do not have the same degree of success. However, they did provide adequate predictions of surface-pressure and velocity-profile shapes. Wall functions were developed by Viegas and Rubesin (1983) for all the models to eliminate the need for integration to the wall. Successful prediction of the skin friction was achieved with all models, as shown in the second part of Figure 23.

In addition to developing wall functions for the two-equation models, the study of Viegas and Rubesin (1983) also showed that the computer code became more robust and converged faster. Together with the savings in grid points near the wall and the advantage of robustness, computational times using wall functions were decreased by nearly one order of magnitude over those using integration to the wall boundary.

Although the axisymmetric bench mark experimental flows have the advantage of minimizing three-dimensional effects, they are limited to moderately strong interactions because the flow is confined and separation extent is limited. Therefore, one must exercise caution in generalizing these results for two-dimensional situations, in which for the same free-stream Mach number, separation may be considerably larger. In those cases, predictions from computations are not as good. To illustrate this aspect, unpublished computations by C. C. Horstman of Ames Research Center for the experiment reported by Delery (1983) are presented in the next figures.

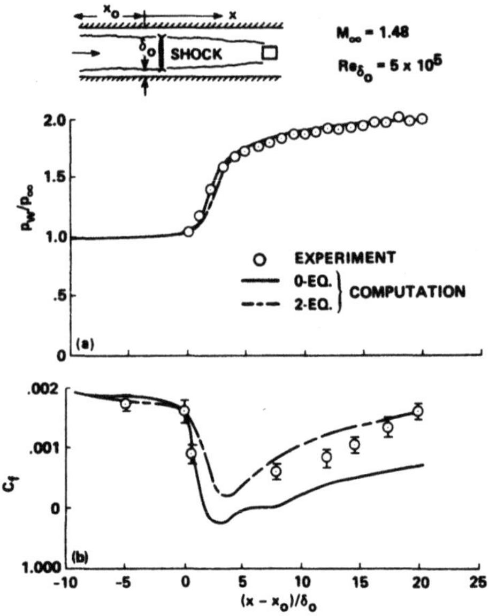

Figure 22. Effect of turbulence modeling on moderately strong normal shock-wave interactions: (a) surface pressure; (b) skin friction.

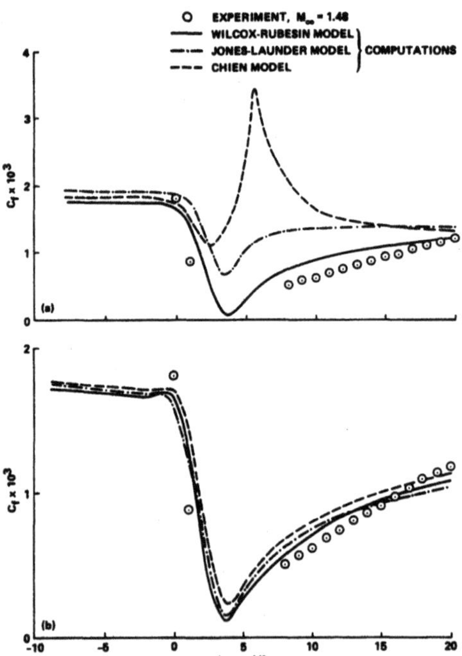

Figure 23. Effects of low-Reynolds-number-term treatment in 2-eq. models used to predict moderately strong normal shock-wave interactions: (a) integration to the wall; (b) wall-function treatment.

In the experiment by Delery, a region of supersonic flow was achieved in an asymmetric channel formed by having a bump on one wall of a rectangular test section. In addition to forming a lambda shock foot, a separated region developed which closed downstream of the junction formed by the bump and the channel wall. Although the flow was choked across the channel, the significant viscous interaction effects only occurred in the bump-wall side. The computations were made using the new implicit-explicit method of MacCormack (1982) along with the two-equation turbulence model of Jones and Launder (1971). Both walls were treated viscously, but the grid resolution was rather coarse on the far wall where interaction effects were small. The equations were integrated to the wall.

An interferogram taken of the flow above the bump-wall is shown in Figure 24. Mach contours determined from the interferogram are also shown. They can be compared with the computed contours using two different turbulence models. The unmodified model of Jones and Launder (1971), with the low-Reynolds-number formulation of Chien (1982), predicts a region of separation smaller

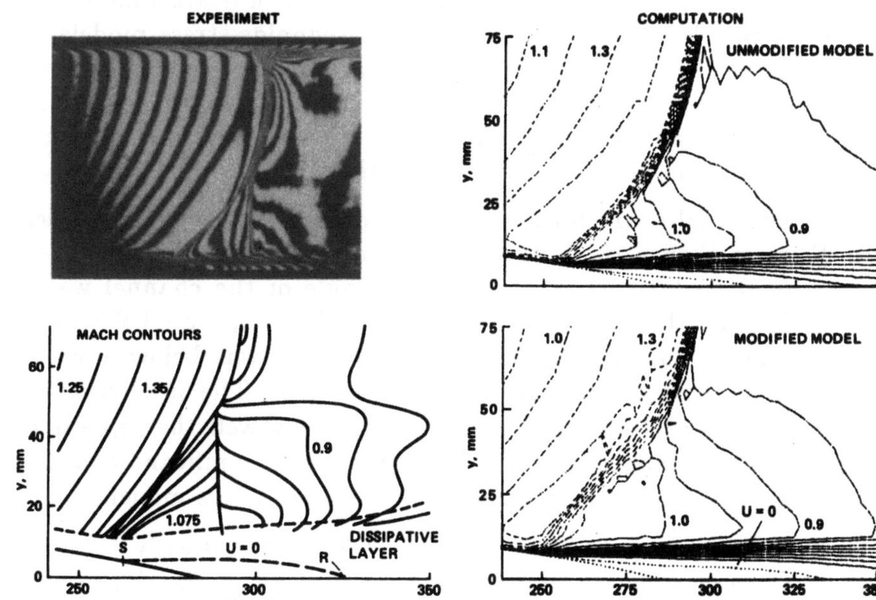

Figure 24. Modeling of strong normal shock-wave interactions: comparison of shock structure from experiment and computation.

than that found experimentally. As a result, the shock structure also differs in that the computed lambda foot of the shock is weaker and the zone of supersonic flow smaller. As mentioned previously, the low-Reynolds-number functions of the turbulence model may be affecting these calculations, but at the time they were made, that weakness of this model had not been reported. Therefore, Horstman made another computation using an ad hoc modification to the model that had provided some improvements in other separated-flow computations (see Horstman, 1983) to see if the correct flow field could be predicted. The results, shown in Figure 25, provide a better comparison for the Mach contours and extent of separation. It is worth noting that these transonic flows also have unsteady aspects that may influence our ability to model the separated region.

The shear layer that develops during this strong normal-shock-wave interaction behaves like the one that develops downstream of a rearward-facing step (Seegmiller et al., 1978; Delery, 1983). Other studies (Driver and Seegmiller, 1982; Driver et al., 1983) of such a flow, which eliminates uncertainties in separation location and the complicating presence of unsteady shocks, indicate that eddy-viscosity models do not work as well as Reynolds-stress models in predicting the flow within the separated region. Such models remove the assumption that the stresses respond immediately to changes in the strain rate and therefore constitute a more plausible physical description in the case of strong interactions.

Some results which exemplify the main thrust of these studies are shown in Figure 26. In the experiment, the size of the separated zone was altered by varying the upper angle of the channel walls, which changed the reattachment location. The separated flow was also unsteady, but the characteristic frequency was below any expected turbulence frequencies by a factor of 3. The prediction of the reattachment location is a measure of how well the separated zone is being calculated, and one can note that significantly better results are achieved with an algebraic Reynolds-stress model.

In these computations by Sindir (1982), the steady form of the equations was solved and the wall functions were developed and implemented to eliminate the low-Reynolds-number terms needed for integration to the wall. It was recognized early in Sindir's study that the scale equation used in the original algebraic stress model was the weakest aspect of the model formulation. Therefore, when

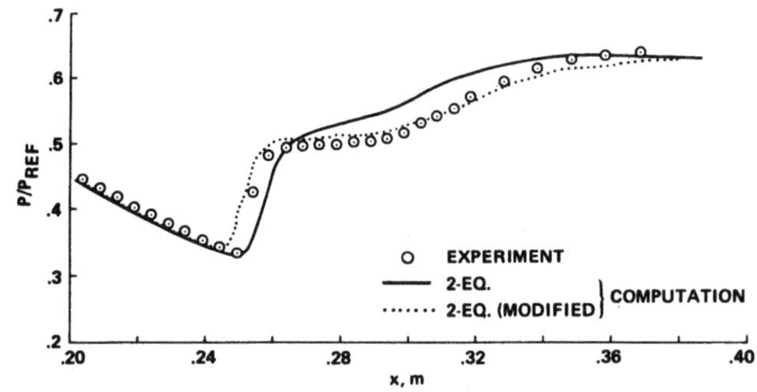

(a) surface pressures.

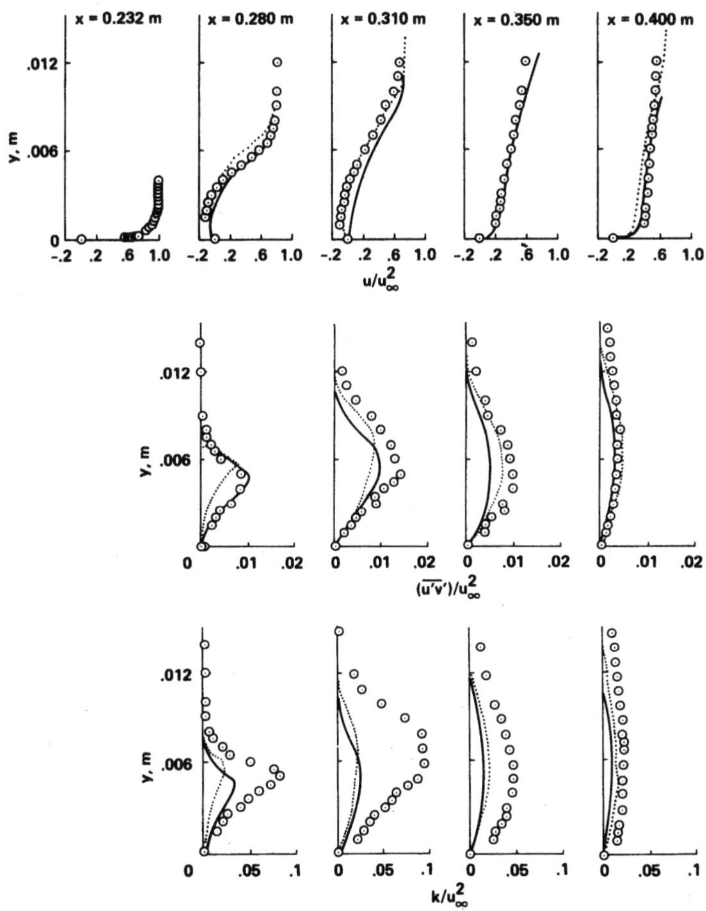

(b) Velocity and turbulence profiles.

Figure 25. Modeling of strong normal shock-wave interactions.

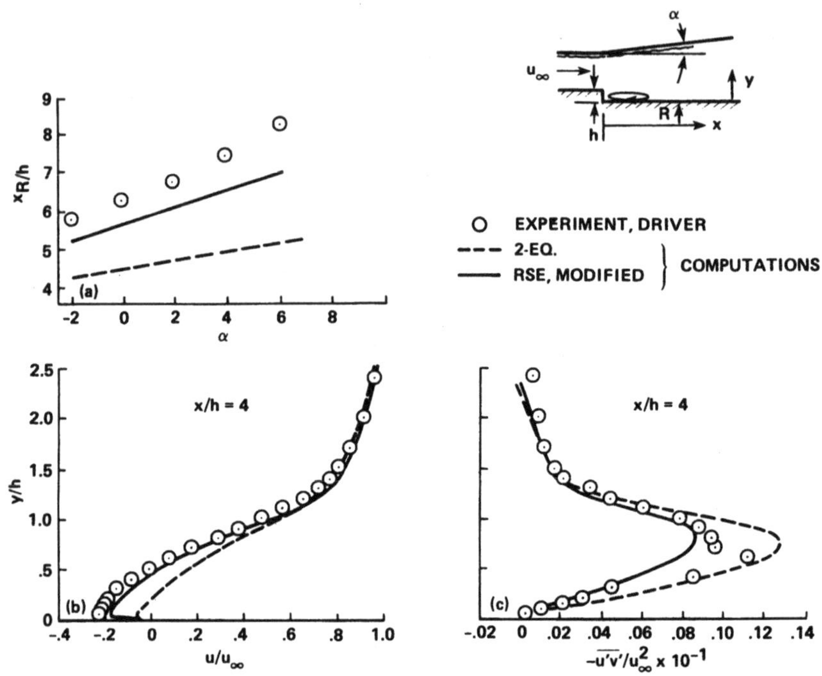

Figure 26. Step-flow studies undertaken to provide modeling-guidance for large separated flow regions: (a) reattachment length; (b) velocity; (c) shear stress profiles.

the original stress-model formulation failed to predict the experiment, the production term was modified. The change causes the dissipation to increase and shear stresses to decrease with the net effect yielding an increase on reattachment length or larger separated zones. Within the separated zone, improved velocity and shear-stress profiles are also achieved. Such a model has not yet been implemented in shock-separated flows because of computational efficiency considerations. However, with the compressible-flow wall functions developed by Viegas and Rubesin (1983) future attempts can be anticipated.

4.5 Trailing-Edge Flow Interactions

Supercritical-wing technology development also depends on an understanding of the flow at the trailing edge because of its global influence on wing lift and drag. A series of two-dimensional bench mark flows, as shown in Table 5, has been under investigation to provide modeling guidance. The experimental flows range from attached incompressible flows with no pressure gradient to high subsonic speed compressible flows with adverse pressure gradients leading to small separation. For incompressible flows, viscous-inviscid interactions have not been important, and modeling studies have shown that two-equation eddy-viscosity models are adequate to resolve the flow in the near-wake region (Marvin, 1982). For the higher speed flows, viscous-inviscid interactions are important and some additional discussion is warranted herein.

Some of the important physical characteristics of these higher speed flows are depicted in Figure 27. A spark shadowgraph and meanflow characteristics, determined from laser velocimeter measurements of Viswanath and Brown (1982), are shown. There is a rapid

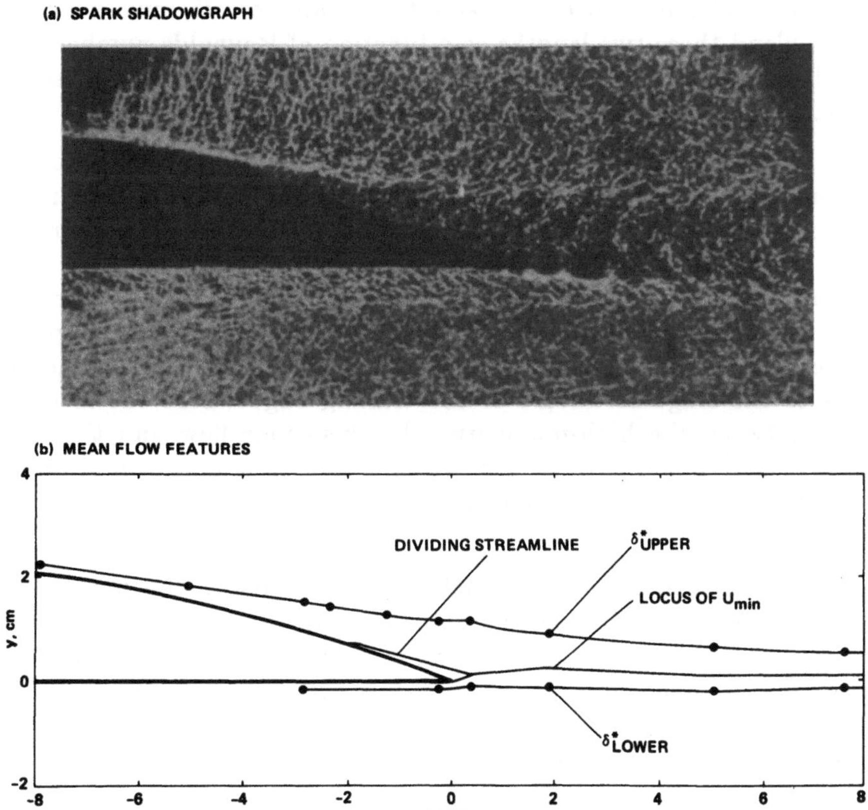

Figure 27. Physical characteristics of a trailing-edge flow with small separation: (a) spark shadowgraph; (b) mean flow features from laser velocimeter experiment.

thickening of the upper surface displacement thickness as the trailing edge is approached. This displacement effect causes changes in streamline curvature and significant influence on the outer inviscid flow. Such effects must be accounted for if correct drag and lift characteristics are to be predicted. Two viscous layers of different thickness merge in the near-wake, and the locus of $U_{\min}$ does not occur along the imaginary geometric line separating the upper and lower surfaces.

Turbulence modeling that reflects the mixing of two viscous regions of different characteristic scales must be addressed. In this regard, the incompressible experiments have been extremely useful. They have verified that two-equation models accomplish this mixing of two different length scales quite adequately, and they have a definite advantage over zero-equation models which must heuristically blend these two lengths as a function of Reynolds number and angle of attack. When pressure gradients are large enough, small separation can occur as in this example.

The flow direction in this separated region has been found to be intermittent (Viswanath and Brown, 1982). Another feature of the flows, observed in the short-duration shadowgraph exposure, is the existence of distinct counterclockwise vortical structures that originate at the trailing edge, grow, and merge with one another downstream. Such structures have also been observed in high-speed, unseparated, asymmetric trailing-edge flows. Their occurrence is due to the singular nature of the trailing edge where the interaction between the high-momentum lower-surface flow and the low-momentum upper-surface flow occurs. It is not known if these structures are the primary causes of the unsteadiness in the small separated zone.

Some success has been achieved in modeling the mean flow within complex trailing-edge regions (Horstman, 1983). An example of comparisons between experiment and computations using the two-equation turbulence model of Wilcox and Rubesin (1980) are shown in Figures 28 and 29 for the case with small separation. Grid resolution was fine enough to resolve the expected interaction between the viscous and inviscid regions, to provide proper integration of the near-wall, low-Reynolds-number terms in the modeling equations, and to provide adequate numerical transition between the no-slip wall-boundary condition and the near-wake flow without resorting to grid alignment with the experimentally determined $u_{\min}$ streamline.

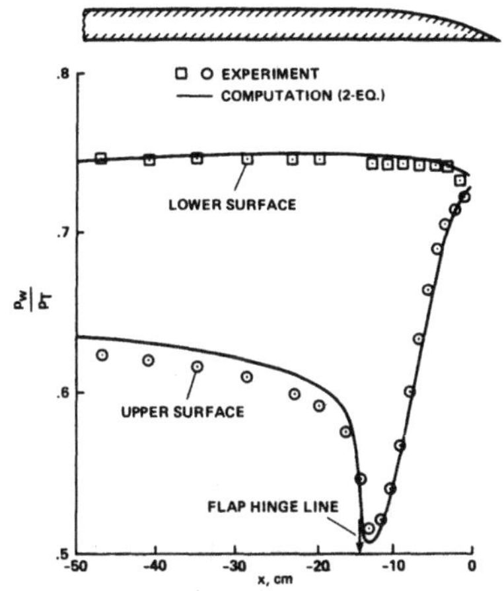

Figure 28. Modeling of a trailing-edge flow with small separation: surface pressures.

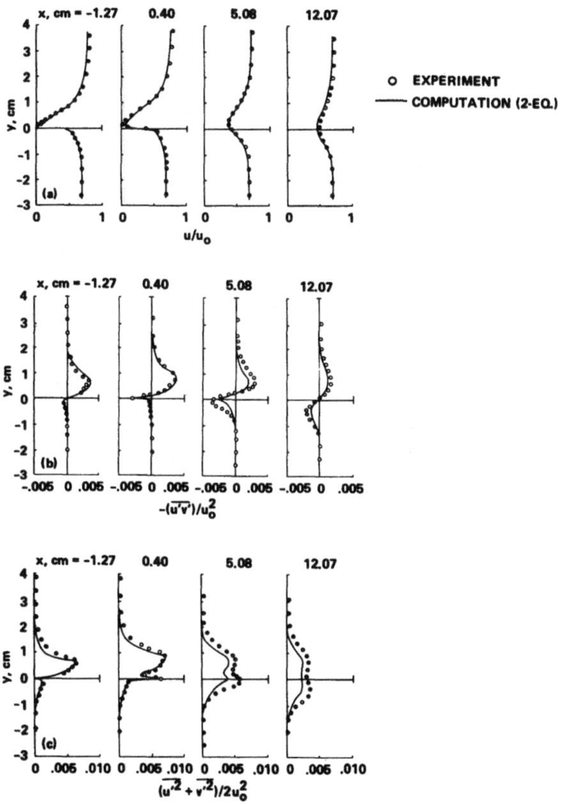

Figure 29. Modeling of a trailing-edge flow with small separation: (a) velocity profiles; (b) shear-stress profiles; (c) Kinetic energy profiles.

In Figure 28, the predicted pressure distribution is shown to agree adequately with the experimental one. Although the location of separation was different in the computation and experiment, values of the viscous displacement and momentum thickness also showed excellent agreement. Corresponding mean-velocity and turbulence profiles are shown in Figure 29. These comparisons also indicate good agreement. The intermittent nature of the separated flow and the vortical structures were not predicted.

Whether these aspects could ever by predicted within the framework of the mass-averaged Navier-Stokes equations depends on whether the characteristic shedding times are much longer than the characteristic times associated with the turbulent structure. Although no estimates of the vortical shedding frequencies were made, a characteristic frequency of pressure fluctuations beneath the separated zone was measured and found to be about 9 kHz. This frequency is of the order of the characteristic frequencies within the attached turbulent layer and hence it is unlikely that computations would resolve these structures. The importance of resolving these unsteady features has not been fully explored and further study of this aspect may be warranted.

4.6 Cross-Flow Separation at Supersonic Speeds

Aircraft maneuverability requirements and space transportation vehicle reentry attitudes require an understanding of flows over bodies at high angle of attack where separation can occur on their leesides. For our discussion, we will limit consideration to steady separations on simple shapes for which some modeling successes for fully turbulent flows have been demonstrated. Bench mark flows are shown in Table 6.

Some typical physical characteristics of these flows can be explained with the aid of Figure 30, where cross-flow streamline patterns and surface skin-friction directions are illustrated. The ratio of angle of attack to cone half-angle, $\overline{\alpha}$, is often used to categorize these simple flows. When $\overline{\alpha}$ is less than 1, inviscid theory, coupled with boundary-layer techniques, is adequate for predicting these flows. As $\overline{\alpha}$ nears or exceeds 1, separation occurs near the leeside generator. As $\overline{\alpha}$ increases, separation moves farther from the leeside generator. Secondary, and even tertiary, separation can manifest itself. The converging and diverging skin-friction direction represents separation and attachment lines, respectively. Depending on the flow Mach number and angle of attack, cross-flow shock waves can also occur; they too can cause separation (such a circumstance is not illustrated).

The separations that occur are open in the sense that there is no recirculation of downstream fluid to forward locations. And, for supersonic flows, the downstream influence on the upstream flow is felt only through the subsonic region of the boundary layers, which can be quite small for turbulent flows. In such situations, the mass-averaged Navier-Stokes equations can be put in parabolic form by neglecting derivatives of the shear stresses in the marching direction, and providing some special procedures in the subsonic region of the flow (see for example Schiff and Steger, 1980; Rakich et al., 1982). Such approximations provide considerable improvement in computational efficiency because the solutions are marched in space coordinates only, time being superfluous. Considerable success has been shown for this approach, using implicit procedures when $\overline{\alpha}$ is less than 1, and when the flows are attached. In these cases, adequate turbulence modeling is provided by zero-equation eddy-viscosity models.

When separation occurs, viscous effects determine the leeside flow structure. Numerical resolution and turbulence modeling both become important. To illustrate the effect of the choice of turbulence model, the data from the bench mark flow published by Rainbird (1968) at a given axial location are compared with several computations in Figure 31. The surface shear-stress angle directions ω_s are defined relative to the conical generator. Therefore, when the angle is positive, the flow is toward the leeward plane of symmetry and away from it when the angles are negative. Thus, this flow shows primary separation at the first location of a sign change in ω (i.e., 123°), and secondary separation at the next sign change location (154°).

The computations shown in Figure 31 were performed using two different algorithms that solve the parabolized form of the Navier-Stokes equations. They differ mainly in the manner in which the subsonic region of the flow is treated. Rakich et al. (1982) solve for a portion of the pressure term in that region, to enable the solution to march more efficiently in regions where departure solutions may occur, and they actually march the solution from the cone apex.

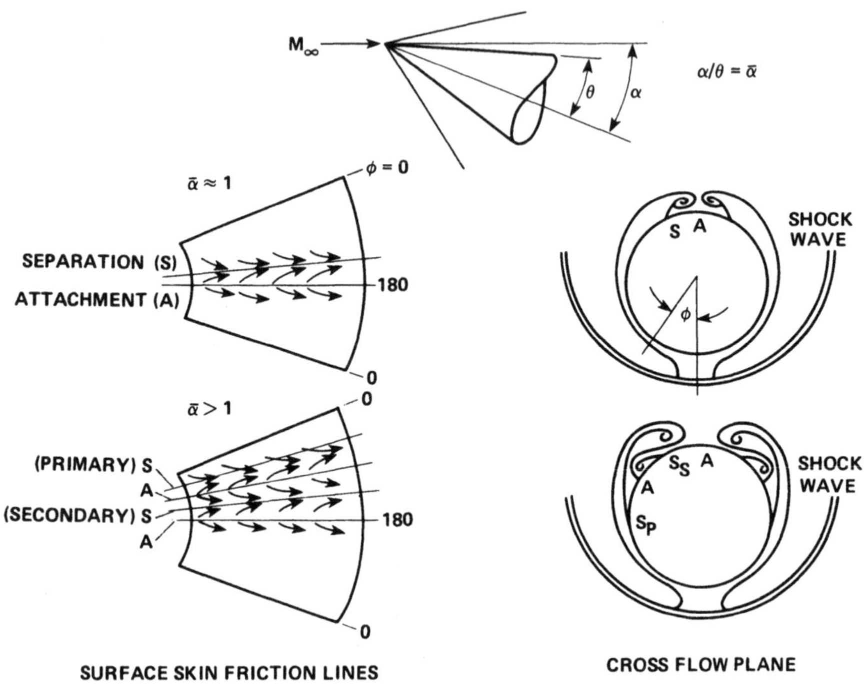

Figure 30. Physical characteristics of supersonic flow over sharp cones at high angle of attack.

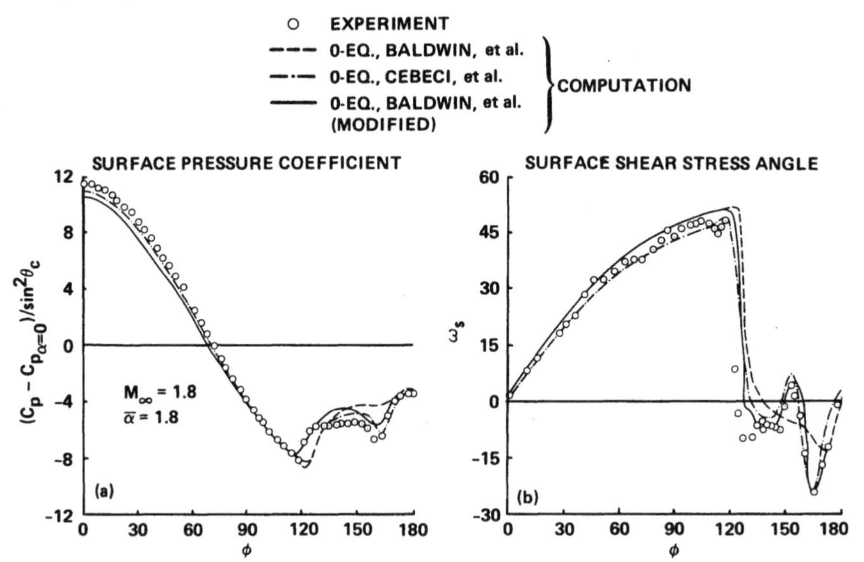

Figure 31. Modeling of supersonic flow over cones at high angle of attack from computations using the parabolized Navier-Stokes equations: (a) surface pressure; (b) surface shear-stress angle.

Degani used the approach of Schiff and Steger (1980) in which the pressure within the subsonic region is assumed to be constant and equivalent to that in the supersonic region, and the flow is treated as if it were conical. Both methods used implicit procedures to advance the solutions. Although no thorough comparisons of these two methods have been reported, example computations using comparable grids on similar shapes at lower angles of attack, where the flow is attached, would suggest the two methods yield essentially the same solutions. The comparison in Figure 31 tends to substantiate this for separated flows as well.

The differences in results between the computations in Figure 31 probably show the effects of different turbulence modeling. First, we observe that the computations of Rakich who used the unmodified zero-equation model (equations (2.8) and (2.10)) does a reasonably good job of predicting both pressure distribution and shear-stress direction. On the other hand, Degani's solution using the model of Baldwin and Lomax (1978) fails to predict the leeside flow, giving a simple primary separation with reattachment near the leeside generator. Degani found that the maximum of the vorticity function, equation (2.21), used to establish the length scale, did not have a single maximum. Choosing the y associated with maximum farthest from the body introduced artificially high values of viscosity and erroneous prediction of the flow. Modifying the computational logic to ensure that the length scale was determined at the location of the first maximum resulted in much better prediction. In most respects, the results of the modified model compare with the zero-equation computations of Rakich.

A further illustration of the influence of modeling on the computations is shown in the top portion of Figure 32a where velocity vectors in the cross-flow plane (taken from Degani and Schiff, 1983) are shown. Proper modeling results in the correct flow field with the primary, secondary, and tertiary vortices. However, another facet of flow modeling—numerical resolution—must also be considered. This facet is shown in Figure 32b. The computations are for another cone and another set of conditions. Using the modified turbulence model, Degani showed that when the cross-flow grid is too coarse, $\Delta\phi = 5°$, a single vortex associated with primary separation is found. For the finer grid ($\Delta\phi = 2°$) the primary, secondary, and tertiary vortices are found. The latter corresponds better with the measured results from the experiment, and it is concluded that the fine-resolution computations fairly represent the real flow field. The obvious conclusion from such studies is that grid resolution and turbulence modeling in the initial stages of cross-flow development are both very important.

Further work remains to refine our ability to model these flows with large separations. Removing the ambiguity in defining the proper length scale in the eddy-viscosity model would be a fruitful area for study, as well as establishing some criteria for proper grid resolution.

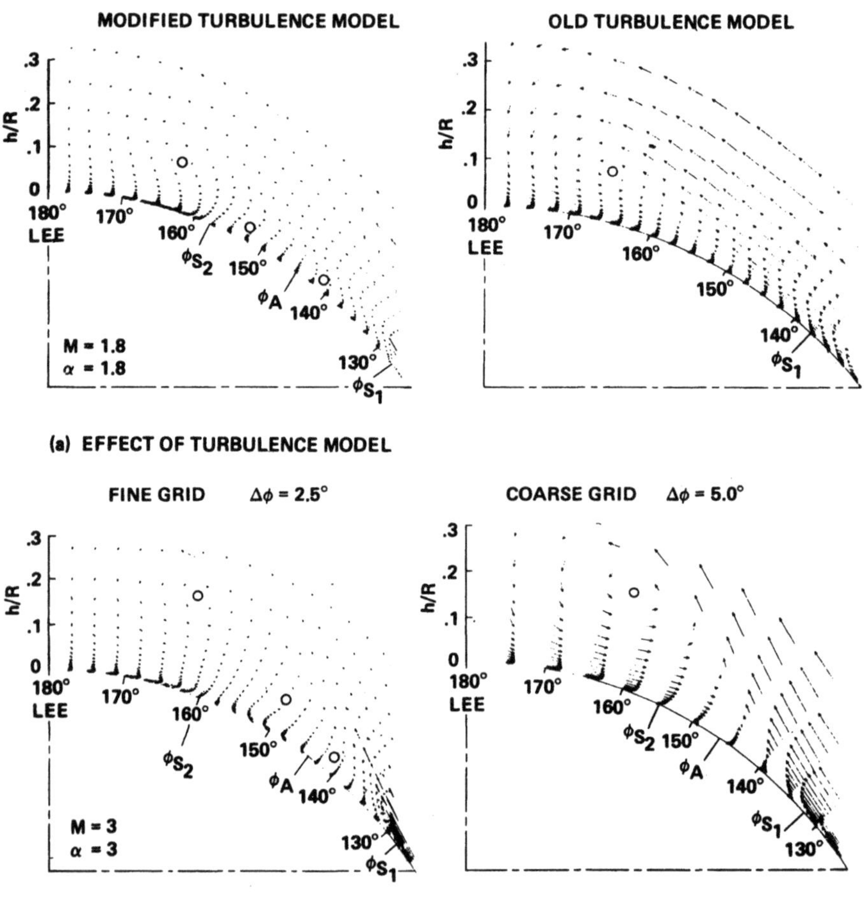

Figure 32. Modeling of supersonic flow over cones at high angle of attack from computations using the parabolized Navier-Stokes equations: (a) effect of turbulence model; (b) effect of azimuthal grid resolution using the modified turbulence model.

5. Concluding Remarks

The status of flow modeling in which numerical simulations of the mass-averaged Navier-Stokes equations are used to compute high-speed, compressible, turbulent separated flows was reviewed. Emphasis was placed on bench mark flows that represent simplified, but realistic aerodynamic phenomena. These included impinging shock waves, compression corners, glancing shock waves, trailing-edge regions, and supersonic, high-angle-of-attack flows. In each case, comparison with experiment provided an assessment of modeling capabilities and shortcomings. Consideration was given to show the importance of combining experiments, numerical algorithm, grid, and turbulence model to develop effectively this potentially powerful technique for solving separated flow problems.

The mass-averaged Navier-Stokes computer codes in use today are still in their developmental stages. They represent a compromise between the choice of numerical algorithm, grid, and turbulence model. The compromise is dictated by constraints of numerical efficiency and the lack of an adequate turbulence model. Provided that adequate safeguards are used to ensure numerical resolution, it is apparent that the computations employing eddy-viscosity turbulence models can give a qualitatively good representation of many two- and three-dimensional, complex aerodynamic flows involving shock waves and separation. Although flow details within separated regions cannot be predicted with complete confidence, the solutions can now provide a bridge for connecting computations on either side of embedded separated regions.

Of the work remaining in developing these codes into predictive tools, proper physical modeling remains paramount. The challenges of better numerical accuracy and resolution along with better turbulence modeling are areas for further exploration. With regard to the latter, it is clear that some distinct advantages are gained by employing higher-order turbulence models. Without inordinate increases in computational times (25% increases are typical), two-equation models provide unambiguous, albeit approximate, determinations of the length and velocity scales needed to define an effective viscosity, and they provide inherent means to allow turbulence to adjust itself appropriately to rapid changes in the mean flow.

Whether they can be improved to provide completely adequate

modeling or whether they must give way to Reynolds stress modeling is a debatable issue, the resolution of which requires additional study. More has to be done to determine the causes and effects of flow unsteadiness and its importance in modeling both two- and three-dimensional flows. And, more has to be done experimentally to define flow-field structures and critical parameters, to gain further understanding of modeling, and to provide well-documented bench mark tests against which progress can be gauged.

References

[1] Baldwin, B. S. and MacCormack, R. W. "Numerical Solution of a Strong Shock Wave With a Hypersonic Turbulent Boundary-Layer," *AIAA Paper 74-558,* 1974.

[2] Baldwin, B. S. and Rose, W. C. "Calculations of Shock-Separated Turbulent Boundary Layers," *NASA SP-347,* 1975.

[3] Baldwin, B. S. and Lomax, H. "Thin Layer Approximation and Algebraic Model for Separated Turbulent Flows," *AIAA Paper 78-257,* 1978.

[4] Bannink, W. J. and Nebbeling, C. "Measurement of the Supersonic Flow Field Past a Slender Cone at High Angles of Attack," High Angle of Attack Aerodynamics, *AGARD CP-247, Paper 22,* 1978.

[5] Beam, R. M., and Warming, R. F. "An Implicit Factored Scheme for the Compressible Navier-Stokes Equations," *AIAA J., 16, No. 4* (1978), 393–402.

[6] Briley, W. R. and McDonald, H. "Solution of the Multidimensional Compressible Navier-Stokes Equations by a Generalized Implicit Method," *J. Comput. Phys., 24, No. 4* (1977), 372.

[7] Brosh, A., Kussoy, M. K., and Hung, C. M. "An Experimental and Numerical Investigation of the Impingement of an Oblique Shock Wave on a Body of Revolution," *AIAA Paper 83-1757,* 1983.

[8] Cambier, L., Ghazzi, W., Veuillot, J. D., and Viviand, H. "Une Approche por Domaines Pour Le Calcul D'ecoulements Compressibles," 5 'eme Coll. Int. Sur Les M'ethodes de Calcul Scientifique et Technique, *INRIA TP ONERA No. 1981-143,* 1981.

[9] Chapman, D. R. "Dryden Lectureship in Research, Computational Aerodynamics Development and Outlook," *AIAA J., 17, No. 12* (1979), 1293–1313.

[10] ——————. "Trends and Pacing Items in Computational Aerodynamics," Seventh International Conference on Numerical Methods

in Fluid Dynamics, *Lecture Notes in Physics, New York: Springer-Verlag, 1981.*

[11] Chein, K. Y. "Predictions of Channel Boundary-Layer Flows With a Low-Reynolds-Number Turbulence Model," *AIAA J., 20* (1982), 33–38.

[12] Coakley, T. J. "Implicit Upwind Methods for the Compressible Navier-Stokes Equations," *AIAA Paper 83-1953,* 1983.

[13] Coakley, T. J., Viegas, J. R., and Horstman, C. C. "Evaluation of Turbulence Models for Three Primary Types of Shock Separated Boundary Layers," *AIAA Paper 77-692,* 1977.

[14] Degani, D. and Schiff, L. B. "Computation of Supersonic Viscous Flows Around Pointed Bodies at Large Incidence," *AIAA Paper 83-0034,* 1983.

[15] Delery, J. M. "Experimental Investigation of Turbulent Properties in Transonic Shock/Boundary Layer Interaction," *AIAA J., 21, No. 2* (1983), 180–85.

[16] Dolling, D. J. and Or, C. T. "Unsteadiness of Shock Wave Structure in Attached and Separated Compression Ramp Flow Fields," *AIAA Paper 83-1715,* 1983.

[17] Dolling, D. S. and Bogdonoff, S. M. "Blunt Fin-Induced Shock Wave/Turbulent Boundary Layer Interaction," *AIAA J., 20, No. 12* (1982), 1674–80.

[18] Driver, D. M. and Seegmiller, H. L. "Features of a Reattaching Turbulent Shear Layer Subject to an Adverse Pressure Gradient," *AIAA Paper 82-1029,* 1982.

[19] Driver, D. M., Seegmiller, H. L., and Marvin, J. G. "Unsteady Behavior of a Reattaching Shear Layer," *AIAA Paper 83-1712,* 1983.

[20] East, L. F. "The Application of a Laser Anemometer to the Investigation of Shock-Wave Boundary Layer Interactions," *AGARD CP 193, Applications of Nonintrusive Instrumentation in Fluid Flow Research, Paper 5,* 1976.

[21] Holden, M. S. "Shock Wave-Turbulent Boundary Layer Interaction in Hypersonic Flow," *AIAA Paper 72-74,* 1972.

[22] Horstman, C. C. "Numerical Simulation of Turbulent Trailing Edge Flows," Second Symposium on Numerical and Physical Aspects of Aerodynamic flows, Long Beach, CA, 1983.

[23] Horstman, C. C. and Hung, C. M. "Computation of Three-Dimensional Turbulent Separated Flows at Supersonic Speeds," *AIAA Paper 79-002,* 1979.

[24] Horstman, C. C., et al. "Reynolds Number Effects on Shock Wave Turbulent Boundary-Layer Interaction," *AIAA J., 15, No. 8* (1977), 1152–58.

[25] Hung, C. M. The 1980–81 AFSOR/HTTM-Stanford Conference on Complex Turbulent Flows: *Comparison of Computation and Experiment; Vol. III. Comparison of Computation With Experiment and Computers Summary Reports.* Ed. S. J. Kline, B. J. Cantwell, and G. M. Lilley, 1982, 1372–74.

[26] Hung, C. M. and MacCormack, R. W. "Numerical Simulation of Supersonic and Hypersonic Turbulent Compression Corner Flows," *AIAA J., 15, No. 3* (1977), 410–16.

[27] ———. "A Numerical Solution of Three-Dimensional Shock Wave and Turbulent Boundary Layer Interaction," *AIAA J., 16* (1978), 1090–96.

[28] Hung, C. M. and Kordulla, W. "A Time Split Finite Volume Algorithm for Three-Dimensional Flow Field Simulations," *AIAA Paper 83-1957,* 1983.

[29] Jones, W. P. and Launder, B. E. "The Prediction of Laminarization With a Two-Equation Model of Turbulence," *International Journal of Heat and Mass Transfer, 15,* Pergamon Press, 1971.

[30] Kayser, L. D. and Sturek, W. B. "Experimental Measurements in the Turbulent Boundary Layer of a Yawed, Spinning Ogive-Cylinder Body of Revolution at $M = 3$; Part II. Data Tabulation," *ARBRL-MR-02813,* U. S. Army Ballistic Research Laboratory, ARRADCOM, Aberdeen Proving Ground, MD, 1978.

[31] Kline, S. J., Cantwell, B. J., and Lilley, G. M., eds. The 1980–81 AFSOR/HTTM-Stanford Conference on Complex Turbulent Flows: *Comparison of Computation and Experiment; Vol. I. Objectives, Evaluation of Data, Specifications of Test Cases, Discussion, and Position Papers.* Stanford: Mechanical Engineering Department, 1981.

[32] Kooi, J. W. "Influence of Free-Stream Mach Number on Transonic Normal Shock Wave/Turbulent Boundary Layer Interaction," *NLR MD 78013 U,* 1978.

[33] Kussoy, M. I. and Horstman, C. C. "An Experimental Documentation of a Hypersonic Shock-Wave Turbulent Boundary Layer Interaction Flow—With and Without Separation," *NASA TM X-62412,* 1975.

[34] Kussoy, M. I., Horstman, C. C., and Viegas, J. R. "An Experimental and Numerical Investigation of a 3-D Shock Separated Turbulent Boundary Layer," *AIAA J., 18, No. 12* (1980), 1477–84.

[35] Law, C. H. "Supersonic Turbulent Boundary-Layer Separation,"

AIAA J., 12, No. 6 (1974), 794–97.

[36] Liou, M. S., Coakley, T. J., and Bergman, M. Y. "Numerical Simulation of Transonic Flow in Diffusers," *AIAA Paper 81-1632,* 1981.

[37] MacCormack, R. W. *Numerical Solution of the Interaction of a Shock Wave With a Laminar Boundary Layer, Lecture Notes in Physics, 8,* New York: Springer-Verlag, 1971, 151–63.

[38] ——————. "An Efficient Numerical Method for Solving the Time-Dependent Compressible Navier-Stokes Equations at High Reynolds Number," *NASA TM X-73129,* 1976.

[39] ——————. "A Numerical Method for Solving the Equations of Compressible Viscous Flow," *AIAA J., 20, No. 9* (1982), 1275–81.

[40] McDonald, H. The 1980–81 AFSOR/HTTM-Stanford Conference on Complex Turbulent Flows: *Comparison of Computation and Experiment; Vol. III. Comparison of Computation With Experiment and Computors Summary Reports.* Ed. S. J. Kline, B. J. Cantwell, and G. M. Lilley, Stanford: Mechanical Engineering Department, 1982, 1424–27.

[41] McRae, D. W., Peake, D. J., and Fisher, D. F. "A Computational and Experimental Study of High Reynolds Number Viscous/Inviscid Interaction About a Cone at High Angle of Attack," *AIAA Paper 80-1422,* 1980.

[42] Marvin, J. G. "Advancing Computational Aerodynamics Through Wind-Tunnel Experimentation," AGARD Fluid Dynamics Panel Meeting on Integration of Computers and Wind Tunnel Testing, Chattanooga, TN, Sept. 24–25, 1980.

[43] ——————. "Turbulence Modeling for Computational Aerodynamics," *AIAA Paper 82-0164,* 1982.

[44] Marvin, J. G., et al. "An Experimental and Numerical Investigation of Shock-Wave-Induced Turbulent Boundary Layer Separation at Hypersonic Speeds," *AGARD-CCP 168,* 1975.

[45] Mateer, G. G. and Viegas, J. R. "Effect of Mach and Reynolds Numbers on a Normal Shock/Turbulence Boundary Layer Interaction," *AIAA Paper 79-1502,* 1979.

[46] Mateer, G. G., Brosh, A., and Viegas, J. R. "A Normal Shock-Wave Turbulent Boundary Layer Interaction at Transonic Speeds," *AIAA Paper 76-161,* 1976.

[47] Metha, V. and Lomax, H. "Reynolds-Averaged Navier-Stokes Computations of Transonic Aerodynamics," *Progress in Astronautics and Aerodynamics, 82,* ed. D. Nixon, New York: AIAA, 1982, 297–375.

[48] Om, D., Childs, M. E., and Viegas, J. R. "An Experimental Investi-

gation and a Numerical Prediction of a Transonic Normal Shock Wave/Turbulent Boundary Layer Interaction," *AIAA Paper 82-0990,* 1982.

[49] Oskam, B., Bogdonoff, S. M., and Vas, I. E. "Study of Three-Dimensional Flow Fields Generated by the Interaction of a Skewed Shock Wave With a Turbulent Boundary Layer," *AFFDL-TR 75-21,* Air Force Flight Dynamics Laboratory, Wright-Patterson Air Force Space, Ohio, 1975.

[50] Peake, D. J. "Three-Dimensional Swept Shock/Turbulent Boundary Layer Separation With Control of Air Injection," *Aeronautical Report L. R., 592,* National Research Council, Canada, 1976.

[51] Peake, D. J. and Tobak, M. "Three-Dimensional Interaction and Vortical Flows With Emphasis on High Speeds," *NASA TM-81169,* 1980.

[52] Rainbird, W. J. "Turbulent Boundary Layer Growth and Separation on a Yawed Cone," *AIAA J., 6* (1968), 2410; also *AGARD CP 30,* 1968.

[53] Rakich, J. V., Davis, R. T., and Barnett, M. "Simulation of Large Turbulent Structures With the Parabolic Navier-Stokes Equations," Eighth International Conference on Numerical Methods in Fluid Dynamics, Aachen, W. Germany, 1982.

[54] Ramaprian, B. R., Patel, V. C., and Sastry, M. S. "Turbulence Wake Development Behind Streamlined Bodies," *IIHR Report 231,* Iowa Institute of Hydraulic Research, Univ. of Iowa, Iowa City, Iowa, 1981.

[55] Redda, D. C. and Murphy, J. D. "Shock Wave Turbulent Boundary-Layer Interaction in Rectangular Channels; Part II. The Influence of Sidewall Boundary-Layers on Incipient Separation and Scale of Interaction," *AIAA J., 11, No. 10* (1973), 1367–68.

[56] Rubesin, M. W. "A One-Equation Model of Turbulence for Use With the Compressible Navier-Stokes Equations," *NASA TM X-73128,* 1976.

[57] Rubesin, M. W. and Rose, W. C. "The Turbulent Mean Flow Reynolds-Stress and Heat Flux Equations in Mass-Averaged Dependent Variables," *NASA TM X-62248,* 1973.

[58] Salmon, J. T., Bogar, T. J., and Sejben, M. "Laser Velocimeter Measurements in Unsteady, Separated, Transonic Diffuser Flows," *AIAA Paper 81-1197,* 1981.

[59] Schiff, L. B. and Steger, J. L. "Numerical Simulation of Steady Supersonic Viscous Flows," *AIAA J., 18, No. 12* (1980), 1421–30.

[60] Schiff, L. B. and Sturek, W. B. "Numerical Simulation of Steady

Supersonic Flow Over an Ogive-Cylinder/Boattail Body," *AIAA Paper 80-0066*, 1980.

[61] Seddon, J. "The Flow Produced by Interaction of a Turbulent Boundary Layer With a Normal Shock Wave of Strength Sufficient to Cause Separation," *ARC R&M No. 3502*, March 1960.

[62] Seegmiller, H. L., Marvin, J. G., and Levy, L. L., Jr. "Steady and Unsteady Transonic Flows," *AIAA J., 16, No. 12* (1978), 1262–70.

[63] Settles, G. S., Vas, I. E., and Bogdonoff, S. M. "Details of a Shock-Separated Turbulent Boundary Layer at a Compression Corner," *AIAA J., 14* (1976), 1709–15.

[64] Settles, G. S., Fitzpatrick, T. J., and Bogdonoff, S. M. "Detailed Study of Attached and Separated Compression Corner Flow Fields in High Reynolds Number Supersonic Flow," *AIAA J., 15, No. 4* (1979), 1152–58.

[65] Shang, J. S. and Hankey, W. L., Jr. "Numerical Solution of the Navier-Stokes Equations for a Compression Ramp," *AIAA J., 13, No. 10* (1975), 1368–74.

[66] Shang, J. S., Hankey, W. L., Jr., and Law, H. C. "Numerical Simulation of Shock-Wave-Turbulent Boundary Layer Interaction," *AIAA J., 14, No. 10* (1976), 1451.

[67] Shang, J. S., Hankey, W. L., Jr., and Petty, J. S. "Three-Dimensional Supersonic Interacting Turbulent Flow Along a Corner," *AIAA J., 10, No. 5* (1979), 652–56.

[68] Sindir, M. "Numerical Study of Separating and Reattaching Flows in a Backward-Facing Step Geometry," *Ph.D. thesis*, Mechanical Engineering Department, Univ. of Calif. at Davis, Davis, CA, 1982.

[69] Teng, H. Y. and Settles, G. S. "Cylindrical and Conical Upstream Influence Regimes of 3-D Shock/Turbulent Boundary Layer Interactions," *AIAA Paper 82-0987*, 1982.

[70] Viegas, J. R. and Horstman, C. C. "Comparison of Multiequation Turbulence Models for Several Shock Boundary Layer Interaction Flows," *AIAA J., 17, No. 8* (1979), 811-20.

[71] ——————————. The 1980–81 AFSOR/HTTM-Stanford Conference on Complex Turbulent Flows: *Comparison of Computation and Experiment; Vol. III. Comparison of Computation With Experiment and Computors Summary Reports.* Ed. S. J. Kline, B. J. Cantwell, and G. M. Lilley, Stanford: Mechanical Engineering Department, 1982, 1535–39.

[72] Viegas, J. R. and Rubesin, M. W. "Wall Function Boundary Conditions in the Solution of the Navier-Stokes Equations for Complex Turbulent Flows," *AIAA Paper 83-1694*, 1983.

[73] Visbal, M. and Knight, D. "Evaluation of the Baldwin-Lomax Turbulence Model for Two-Dimensional Shock-Wave Boundary Layer Interactions," *AIAA Paper 83-1697,* 1983.

[74] Viswanath, P. R. and Brown, J. L. "Separated Trailing-Edge Flow at a Transonic Mach Number," *AIAA Paper 82-0348,* 1982.

[75] Viswanath, P. R., et al. "Trailing Edge Flows at High Reynolds Number," *AIAA J., 18, No. 9* (1980), 1959–65.

[76] West, J. E. and Korkegi, R. H. "Supersonic Interaction in the Corner of Intersecting Wedges and High Reynolds Number," *AIAA J., 10, No. 5* (1972), 652–56.

[77] Wilcox, D. C. and Rubesin, M. W. "Progress in Turbulence Modeling for Complex Flow Fields Including the Effects of Compressibility," *NASA TP-1517,* 1980.

An Alternative Look at the Unsteady Separation Phenomenon

Chih-Ming Ho
University of Southern California
Los Angeles, CA 90089

1. Introduction

When a solid surface is in unsteady motion relative to the fluid, the flow sometimes breaks away from the surface and forms a thick layer with high level velocity fluctuations. This phenomenon is called unsteady separation and several interesting but nontrivial problems are associated with it. First, what is the proper criterion to define the onset of separation? Second, what is the physical process producing the unsteady separation? Third, how are the aerodynamic properties on a lifting surface affected by the unsteady separation?

During the past thirty years, these problems have been investigated by numerous researchers. The studies concerning the separation criteria were reviewed by Williams (1977) and Telionis (1981). McCroskey (1982) surveyed the progress of the dynamic stall investigations. It is not the purpose of this paper to provide another review of the incompressible laminar unsteady separation problem. Instead, we suggest looking at the unsteady separation from a new point of view. A recent work by Didden and Ho (1984), shows that the onset of the unsteady separation can be identified as the initiation of a local shear layer. Therefore, we will re-examine the aforementioned three problems following the idea of a spatially developing local shear layer.

2. Criteria of Unsteady Separation

In a steady separated boundary layer, the criterion of separation suggested by Prandtl (1904) is the vanishing of the wall shear stress. This condition has been experimentally and theoretically proven to be correct in predicting the phenomenon. In an unsteady flow, the situation is much more complicated. Sears (1956), Rott (1956), and Moore (1958) pointed out that the zero wall shear stress is not necessarily accompanied by the breakaway of the flow from the solid surface, although flow reversal will occur. A recent experiment of the pitching airfoil (McAlister and Carr, 1978) serves as a good example to demonstrate this point. At a certain phase of the oscillation cycle, reversed flow develops at the trailing edge of the airfoil. The reversed flow propagates upstream along the surface and remains as a thin layer. Intrusion from the viscous region into the inviscid flow does not occur during this part of the cycle, and no symptoms of separation can be detected. When the reversed flow reaches the leading edge of the airfoil, a large vortex is produced there and the viscous region thickens considerably. Apparently, flow reversal and zero wall shear stress cannot properly indicate the initiation of an unsteady separation. Thirty years ago, Moore (1958), Rott (1956), and Sears (1956) (MRS) realized this point. They proposed an alternative criterion. The MRS criterion says that the unsteady separation should occur at the zero shear stress location in a coordinate convected with the separation velocity. This criterion does indicate the singular behavior of the boundary layer and will reduce to Prandtl's separation condition in the steady situation. However, the main drawback is that the speed of the separation point is not known *a priori*. Due to this problem, many other criteria, have been proposed. For example, the ejection of vorticity; the abrupt thickening of the boundary layer or the occurrence of large transverse velocity was used to indicate the initiation of the unsteady separation.

Despard and Miller (1971) experimentally investigated the separation in an oscillating mean flow. The amplitude of the free stream velocity fluctuation was about 20% of the time averaged value. They suggested another criterion which is "*the separation commences with the initial occurrence of zero velocity or reverse flow at some point in the velocity profile throughout the entire cycle of oscillation*". Tsahalis and Telionis (1974) followed a similar numerical technique developed by Sears and Telionis (1971) and studied the separation in an oscillating flow on a stationary plate. They found that the criterion suggested by Despard and Miller (1971) could be verified under certain but not all situations.

The MRS criterion mentions the concept that the stagnation point in the coordinate convected with the separation point is away from the solid surface, and therefore a saddle point should occur there. Koromilas and Telionis (1980) investigated the separation on several unsteady flow configurations. In the case of a rotating cylinder, they visually identified a saddle point away from the cylinder surface (Figure 1).

A positive proof of the MRS criterion still remains unavailable. This is due to the difficulty of measuring the separated velocity near the wall. The unsteadiness increases its difficulties, because the extra temporal variation of a 2-D or 3-D velocity field needs multiple probes and large data storage. These complications result in a very limited data base of measured unsteadily separated velocity fields.

The unsteady separation point can move either upstream or downstream in different flow configurations. Flow reversal occurs in the upstream moving separation case, but not in the downstream moving separation situation (Figure 2). Didden and Ho(1984) studied the unsteady separation in a boundary layer produced by an impinging jet (Figure 3). The primary vortex shedding from the jet nozzle induces a secondary vortex from the wall. The boundary abruptly thickens to form a bulge (Figure 4). The separation point moves downstream in this flow. No flow reversal occurs in the velocity field which can then be investigated in detail by hot-wires. The result *quantitatively* validates the MRS criterion; the velocity profile with zero shear stress away from the wall was identified (Figure 5). The local velocity at the zero shear stress point is about 40% of the jet exit velocity and equals the convection speed of the bulge of the separated boundary layer (Figure 4). We note that the singularity in the boundary layer analysis is reflected by the presence of the bulge in the laboratory (Van Dommelen and Shen, 1982; Cebeci, 1983). Hence, the separation speed is the convection speed of the singularity which is noted from the data of Didden and Ho (1984). The other proposed criteria, e.g., ejection of vorticity, etc., can be detected in the vicinity but not exactly at the same point which satisfies the MRS criterion. The streamline pattern computed from the velocity components shows no saddle point at the position upstream of the separation point (Figure 6a). At $r/d = 1.1$, the separation occurs, and the saddle point appears (Figure 6b). This confirms Koromilas and Telionis' (1980) visualization experiment and Walker's (1978) numerical results. At the same location, a closed streamline forms near the wall and indicates the formation of the secondary vortex.

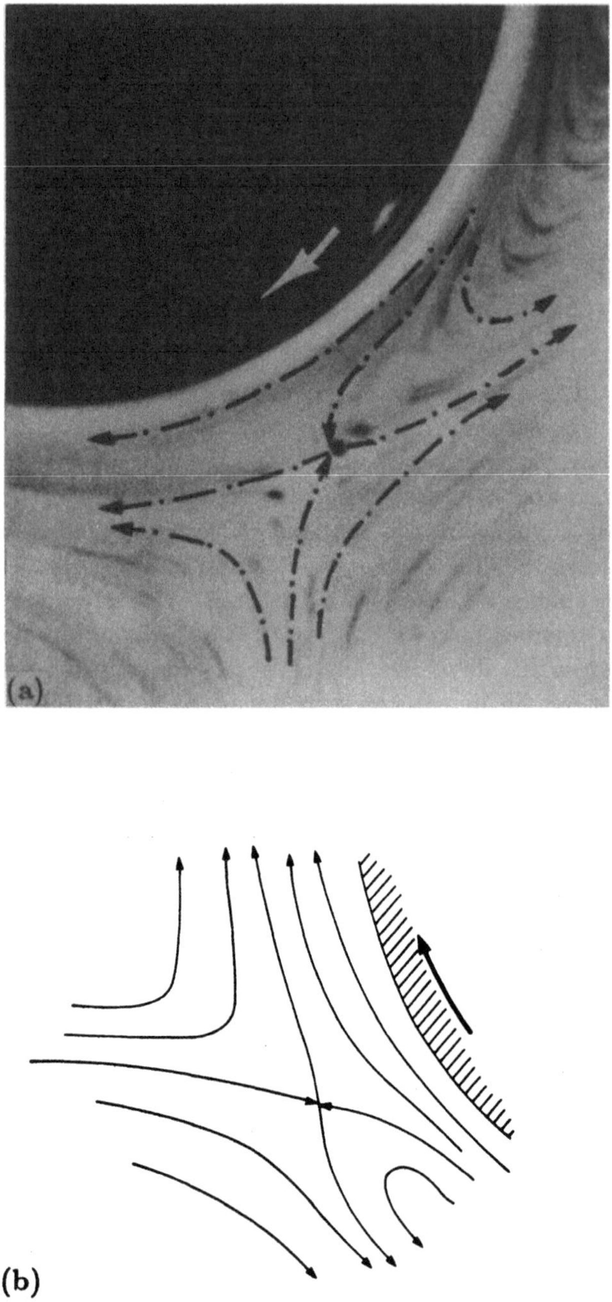

Figure 1. Saddle-point pattern for an upstream-moving wall: (a) visualization picture, (b) streamline obtained from (a) (from Koromilas and Telionis, 1980).

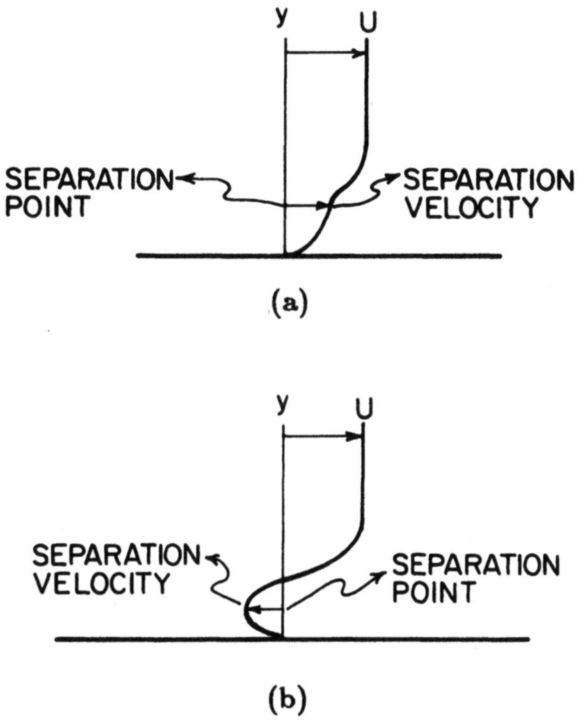

Figure 2. (a) Velocity profile for downstream moving separation. (b) Velocity profile for upstream moving separation.

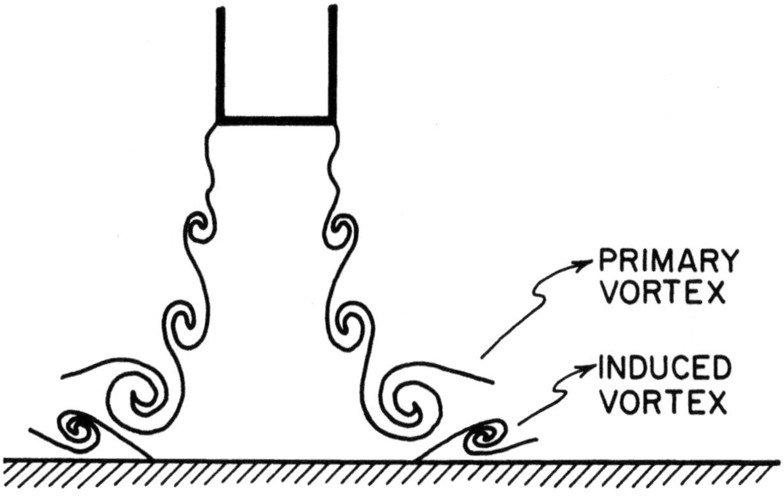

Figure 3. Unsteady separation in an impinging jet.

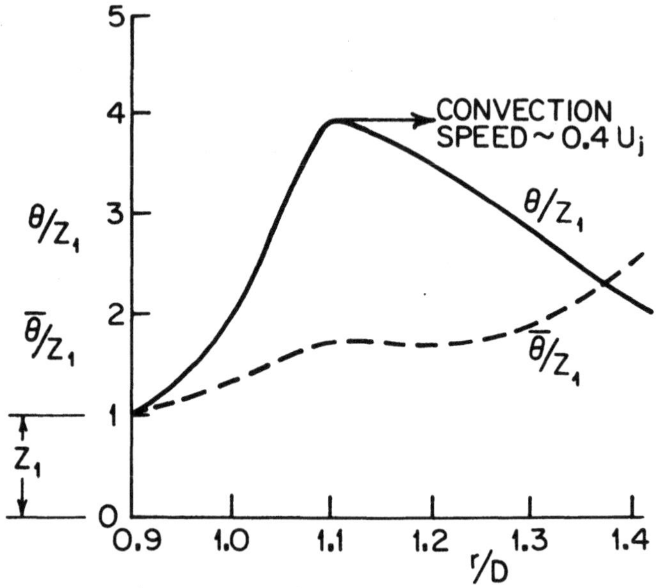

Figure 4. Phase averaged momentum thickness (θ/Z_1) at $t/T = 0.1$ and time average momentum thickness ($\bar{\theta}/Z_1$),
U_j—jet exit speed T—period of passing primary vortices
D—diameter of jet r—radial distance from stagnation point.

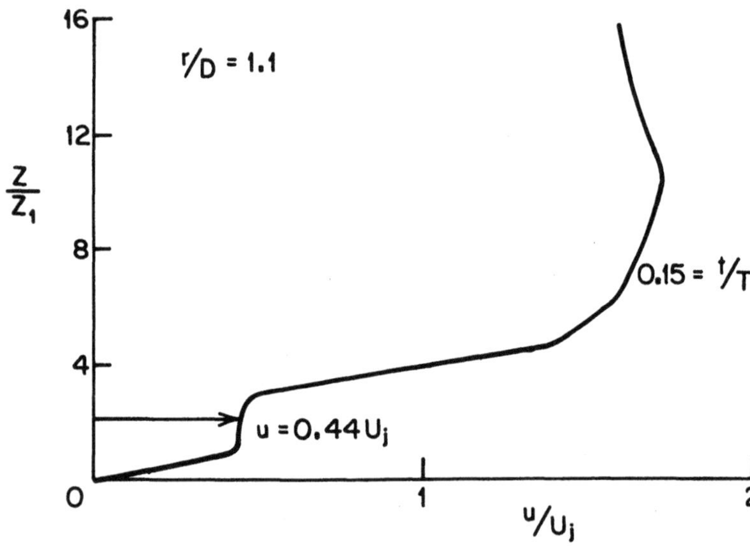

Figure 5. Phase averaged velocity profile (u/U_j) at the separation point.

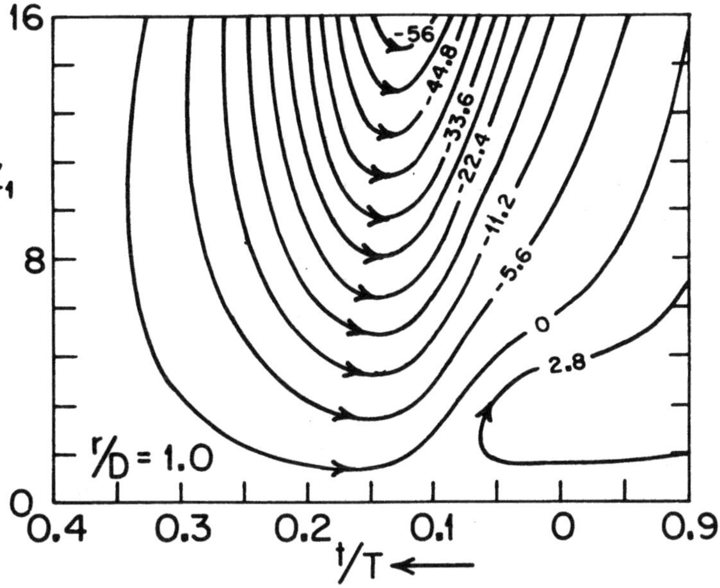

(a)

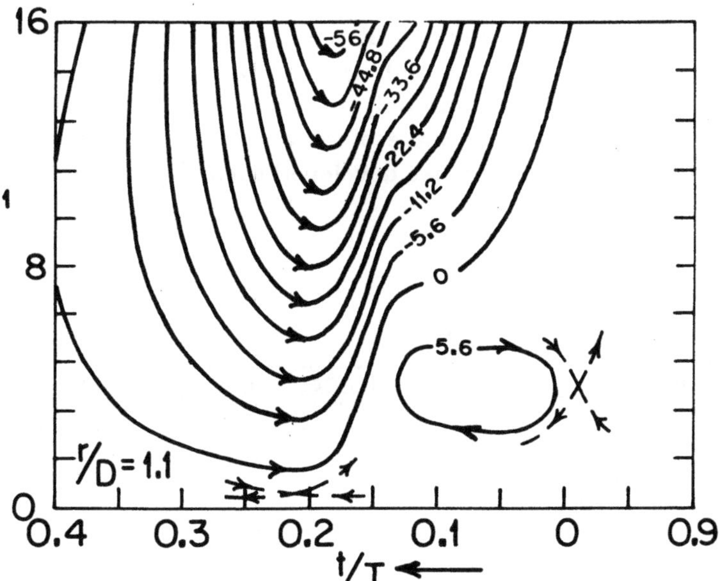

(b)

Figure 6. Phase averaged streamlines: (a) upstream from separation point $r/D = 1.0$, (b) at separation point $r/D = 1.1$.

3. The Local Shear Layer

In Figure 5 the velocity profile shows that a local shear layer starts at the separation point. The local shear layer is created by the inviscid-viscous interaction (Figure 7) near the wall. The primary vortex in the *inviscid* region produces a fast moving stream while it approaches the wall. The accelerating flow causes *an unsteady adverse wall pressure gradient* which retards the flow in the *viscous* layer. Hence, *a shear layer* occurs at the interface of the inviscid and viscous regions. The shear layer is unstable and rolls up into vortical structures and becomes the secondary vortex. The local shear layer has been observed not only in the impinging flow but also in almost all separated flows (Jones et al., 1981, Van Dommelen and Shen, 1982). Although detailed documentation of the shear layer is not available in other flows, we believe that the local shear layer is a generic flow module of the separated flows.

4. The Catastrophically Separated Region

Downstream of the initial separation position, the separated layer is thick and has a high level of velocity fluctuations. The flow in this region is commonly referred to as the catastrophically separated region or the wake of separation. Actually this region can be viewed as the further developed shear layer which originated

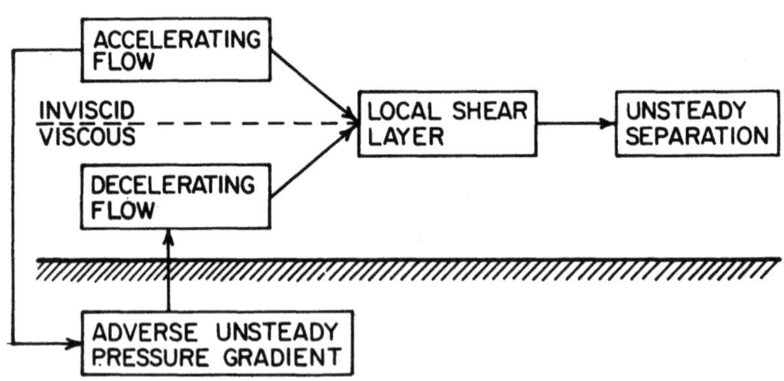

Figure 7. Inviscid-viscous interaction.

from the separation point. Coherent vortical structures form at this point and convect with the free stream flow. Due to velocity fluctuations produced near the initial separation region, the vortices form with irregular periods and are usually three-dimensional. The flow in this region appears to be fairly erratic from visualizations, and is therefore called catastrophic separation. We should obtain very fruitful results if we study this flow region based upon the local shear layer concept. The knowledge acquired from a similar but extensively investigated free shear layer (Ho and Huerre, 1984) will be very helpful for this study.

In the unsteady separation, the local shear layer is externally forced at a certain frequency. Ho and Huang (1982) demonstrated that the evolution of coherent structures in a free shear layer can be greatly altered by forcing near the most amplified frequency. Jones et al. (1981) studied the catastrophically separated region in an unsteady water channel. The forcing frequency was varied from subharmonics to harmonics of the most amplified frequency of the local shear layer. The spectra of velocity fluctuations were measured at several streamwise locations. Amplifications of the forcing waves were detected. The amplifications of the harmonics seemed to be more dominant than that of the subharmonics. The presence of high level subharmonics is an indication of vortex merging. Based upon this experiment, it appears that the solid wall inhibits the amalgamation of the vortices. The experiment by Jones et al. (1981) showed that the local shear layer is a promising approach to study the fully separated region. Obviously much more work along this line of thought is needed.

5. Unsteady Separation on Airfoils

The occurrence of unsteady separation on airfoils or turbine blades causes sudden changes of the lift, drag, and momentum. The performance of airplanes or turbines are severely affected. Considerable amounts of visualization and integral aerodynamic properties (McCroskey, 1981) have been documented in the past. The detailed separation process, however, is not quite clear, because difficulties in measuring separated velocity fields around a moving surface prohibits progress in this direction.

During the unsteady movement of the airfoil, the excursion of the aerodynamic properties from that of the static case depends on the operational parameters. The difference can be very pronounced if a large separation vortex occurs at the leading edge. An example is shown in Figure 8. Several aspects of this test condition are very enlightening. A thin layer reversed flow is first observed at the trailing edge and then progresses toward the leading edge. During this portion of the phase angle, the boundary remains very thin, the drag (Figure 8b) is small and the lift continues to increase. The separation characteristics are not present even when the flow reversal exists. When the reversed flow reaches the leading edge, a separation vortex bursts at that point and convects downstream. We should note a very interesting feature: that the lift *continues to increase* until the vortex reaches about half chord. In other words, the suction on the upper surface of the wing still grows at the initial duration of the separation, the sudden decrease of the suction occurs when the leading edge is in the wake of the separation. A similar situation was found in the impinging jet experiment. High level *suction* is produced by the induced counter-rotating vortex (Figure 9) which evolves from the local shear layer (Didden and Ho, 1984). At a later time, the vortex convects downstream, where high level *pressure* is found in the wake region of the separation. It will be very interesting to find out the flow structure which causes this phenomenon on the unsteady airfoil. A pair of a separation vortex and an induced vortex has been clearly shown on a stationary airfoil in an accelerating mean flow (Figure 10, Freymuth et al., 1983) but the induced vortex has never been observed on pitching airfoils by the hydrogen bubble technique. It, however, does not rule out the possible existence of the induced vortex. This is because whether or not it can be seen is very sensitive to the type of visualization technique used. Nevertheless, the physical mechanism which causes the sharp rise in the lift at the initial separation phase is very important and deserves further investigation.

6. Conclusion

In order to properly describe it, a considerable amount of effort has been devoted to the problem of the instant and the origin of the unsteady flow breaking away from the surface. With help from the advanced instrumentation and numerical techniques, the MRS criterion has been validated in several flows. The progress has been quite appreciable in the past decade.

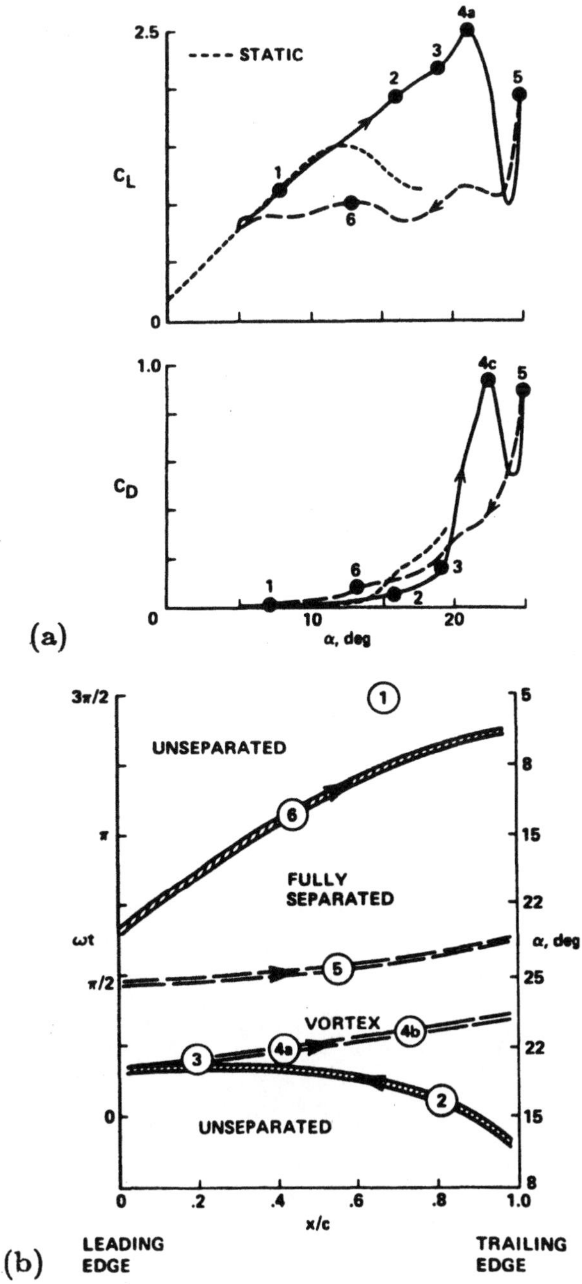

Figure 8. Dynamic stall on the VR-7 airfoil at $M = 0.25$, $\alpha = 15° + 10° \sin \omega t$ and 0.1 reduced frequency. (a) Loci of dynamic stall events on the upper surface of the airfoil. (b) Lift and drag coefficients of the pitching airfoil (from McCroskey, 1981).

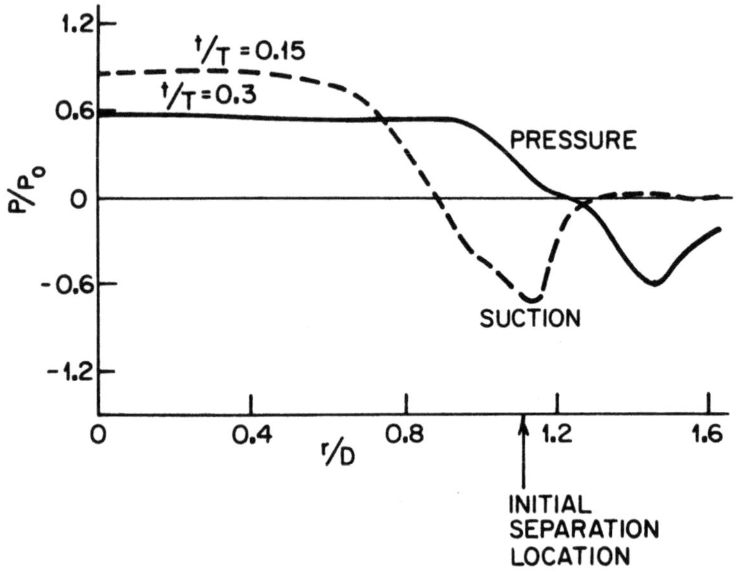

Figure 9. Phase averaged surface pressure traces at the instant of separation $t/T = 0.15$ and after a short time interval $t/T = 0.3$, P_o—stagnation pressure.

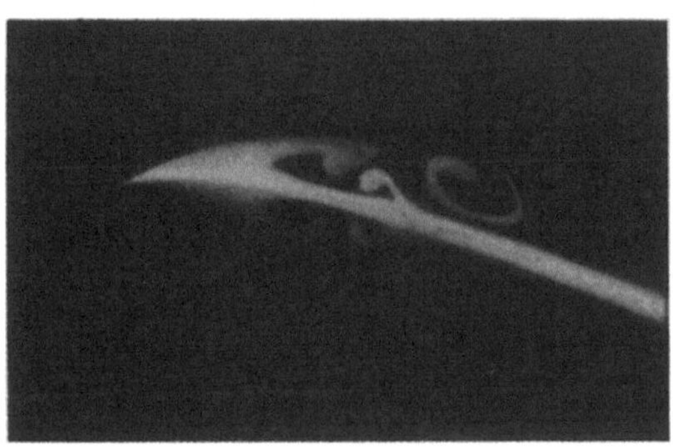

Figure 10. Visualization of a stationary airfoil in a flow with constant acceleration (from Freymuth et al., 1983).

Comparatively speaking, the inviscid-viscous-interaction mechanism of producing the unsteady separation and the flow structures near the separation region are much less understood. Didden and Ho's (1984) experiment demonstrates that a detailed map of the velocity and surface pressure fields is necessary to accomplish this task. Although the process is painstaking, it is very rewarding. The understanding of the basic producing mechanism of the unsteady separation could lead to an effective technique of active control of the unsteady separation.

Acknowledgement

The author would like to thank Professor Norbert Didden for his help. This work was supported by the AFOSR under Contract No. F49620-82-K-0019.

References

[1] Cebeci, T. "Unsteady Separation," *Numerical and Physical Aspects of Aerodynamic Flows." Ed. T. Cebeci, New York: Springer-Verlag, 1982, 265–77.*

[2] Despard, R. A. and Miller, J. A. "Separation in Oscillating Laminar Boundary-Layer Flows," *J. Fluid Mech., 47* (1971), 21–31.

[3] Didden, N. and Ho, C. M. "Unsteady Separation in the Boundary Layer Produced by an Impinging Jet," *1984,* (to be published).

[4] Freymuth, P., Bank, W., and Palmer, M. "Visualization of Accelerating Flow Around an Airfoil at High Angle of Attack," *Z. für Flug, 7* (1983), 392–400.

[5] Ho, C. M. and Huang, L. S. "Subharmonics and Vortex Merging in Mixing Layers," *J. Fluid Mech., 119* (1982), 443–73.

[6] Ho, C. M. and Huerre, P. "Perturbed Free Shear Layers," *Ann. Rev. Fluid Mech., 16* (1984), 356–424.

[7] Jones, G. S., Telionis, D. P., and Barbi, C. "Separation and Wake Interaction of a Pulsating Laminar Flow," *AIAA Paper No. 81-0053,* 1981.

[8] Koromilas, C. A. and Telionis, D. P. "Unsteady Laminar Separation: an Experimental Study," *J. Fluid Mech., 97* (1980), 347–84.

[9] McAlister, K. W. and Carr, L. W. "Water Tunnel Visualization of Dynamic Stall," in *Nonsteady Fluid Dynamics*. Ed. D. E. Crow and J. A. Miller, New York: Pub. ASME, 1978, 103–10.

[10] McCroskey, W. J. "The Phenomenon of Dynamic Stall," *NASA TM-81264*, 1981.

[11] ——————. "Unsteady Airfoils," *Ann. Rev. Fluid Mech., 14* (1982), 285–311.

[12] Moore, F. K. "On the Separation of the Unsteady Laminar Boundary Layer," in *Boundary-Layer Research*. Ed. H. G. Görtler, Berlin: Springer, 1958, 296–310.

[13] Prandtl, L. "Über Flüssigkeitsbewegung bei sehr Kleiner Reibung," *Proc. III Int. Math. Congr.*, Heidelberg, 1904, 484–91.

[14] Rott, N. "Unsteady Viscous Flow in the Vicinity of a Stagnation Point," *Q. Appl. Math., 13* (1956), 444-51.

[15] Sears, W. R. "Some Recent Development in Airfoil Theory," *J. Aerosp. Sci., 23* (1956), 490–99.

[16] Sears, W. R. and Telionis, D. P. "Unsteady Boundary Layer Separation," IUTAM Symposium on Recent Res. on Unsteady Boundary Layers 1. Ed. E. A. Eichelbrenner, 1971, 404–41.

[17] Telionis, D. P. *Unsteady Viscous Flows*, New York: Springer-Verlag, 1981.

[18] Tsahalis, D. Th. and Telionis, D. P. "Oscillating Laminar Boundary Layers and Unsteady Separation," *AIAA J., 12* (1974), 1469–76.

[19] Van Dommelen, L. L. and Shen, S. F. "The Genesis of Separation," *Numerical and Physical Aspects of Aerodynamic Flows*, ed. T. Cebeci, New York: Springer-Verlag, 1982, 293–311.

[20] Walker, J.D.A. "The Boundary Layer Due to Rectilinear Vortex," *Proc. R. Soc. Lond. A, 359* (1978), 167–88.

[21] Williams, J. C., III, "Incompressible Boundary-Layer Separation," *Ann. Rev. Fluid Mech., 9* (1977), 113–44.

Part II
Jets and Shear Layers

Vortex-Edge Interactions

Donald Rockwell
Lehigh University
Bethlehem, PA 18015

1. Introduction

Vorticity fields in the form of coherent vortices that impinge upon leading-edges (and surfaces) produce unsteady loading and noise generation in a number of applications including helicopter rotors, marine propellors, as well as a variety of flap, cavity, tube bundle, and valve/gate configurations. In addition, the interaction of these vorticity fields with the leading-edge provides the initial conditions for evolution of the downstream boundary layer; indeed, the very nature of the transition process can be expected to be a function of the interaction at the edge.

Considerable insight into this class of flows has evolved from study of highly coherent shear layer-edge interactions, largely of the jet-edge or edgetone type. Research eras of Powell (1961, 1965) and Karamcheti et al. (1969), as well as more recent investigations summarized by Rockwell (1983), have provided a rich heritage that leads us to the broader range of unsteady shear layer-surface interactions; they involve incident vorticity fields having a variety of concentration, orientation, and degree of coherence.

Even for the well-posed case of a planar acoustic wave incident upon an edge and triggering downstream travelling instability waves (Goldstein, 1982, 1983; Howe, 1981; Shapiro, 1977), the flow structure in the leading-edge region takes on a complex form somewhat analogous to the triple-deck structure of trailing-edge flows (Goldstein, 1982, 1983). On the other hand, gust-type flows involve fluctuating and mean vorticity fields approaching the edge; consequently, our description of the leading-edge region becomes even more complex since we must account for distortion of these fields as they approach the edge, as well as their interaction at and downstream of the edge. Description of these flows is eased, to some extent, if we recognize that the distortion or interaction at the edge is rapid in comparison with the characteristic time of the freely evolving vorticity field (Goldstein, 1978; Hunt, 1977); in fact, a linearized inviscid analysis of this rapid distortion or interaction can lead to a remarkably accurate description of the radiated sound field (Goldstein, 1979).

Much has been advanced in the theoretical approaches of the foregoing, as well as those summarized by Rockwell (1983); however, details of the flow structure in the leading-edge region remain unclarified. In striving towards resolution of the three-dimensional turbulent distortion problem, we first consider a series of quasi-two-dimensional edge interactions shown in Figure 1: a linear disturbance field characterized by a single frequency corresponding to the most rapidly amplified disturbance in the mean shear flow; a nonlinear disturbance field involving vortex coalescence and exhibiting a number of spectral components; and a simpler nonlinear disturbance in the form of a single vortex having a single predominant frequency and small amplitude higher harmonics. From an experimental standpoint, this single vortex is easiest to generate and control; its relatively high disturbance amplitude induces surface pressure fluctuations that one can measure accurately. Herein we focus on

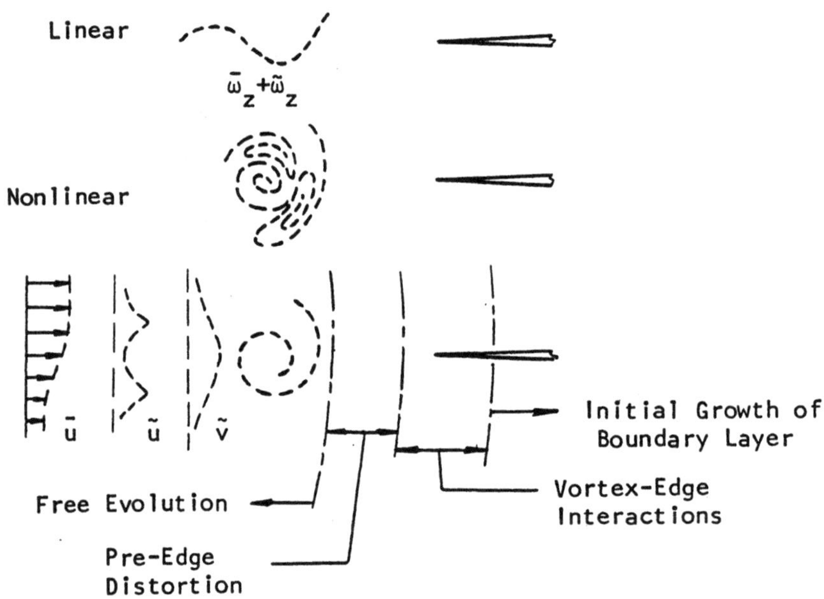

Figure 1. Shear layer edge interactions showing: (a) linear disturbance; (b) interacting vortices; (c) single vortex at single predominant frequency.

this particular type of interaction, keeping an eye towards the extreme cases of linear disturbance-edge and coalesced (multiple) coherent structure-edge interactions. In addition, consideration will be given to combinations of single (noncoalescing) vortices approaching the edge. These vortex patterns show several well-defined spectral components, and their interaction with the edge can produce rapid onset of vortex coalescence. Of prime interest in all these cases will be the conversion from these types of incident vorticity fields to surface pressure fluctuations at the edge.

2. Observations and Challenges

Figure 2 shows several types of vortex-edge interactions; all involve deformation of the incident vortex and secondary shedding of vorticity. For a given edge configuration, we expect the orientation, scale, and degree of concentration of the incident vorticity field (i.e., vortex(ices)) to play a dominant role in determining the detailed flow structure of the interaction region. Moreover, since the secondary shedding of vorticity arises from wall viscous effects, one must suitably account for them as well. We now examine several interaction mechanisms arising from vortex formation in mixing-layer and jet flows, along with the corresponding unsteady pressure fields/forces in the leading-edge region.

2.1 Mixing-Layer-Edge Interactions

Growth of an unstable disturbance in a free mixing-layer leads to vortex formation and eventual coalescence. Focussing on the formation of the initial, single vortex (Figure 2), it is possible to approximate its vorticity distribution by the nonlinear inviscid model of Stuart (1967). Of course, the structure of this incident vortex will depend on its means of generation, i.e., upstream history. Stuart's model allows the concentration of vorticity to be adjusted by varying a constant α; $\alpha = 1.0$ corresponds to a point vortex and $\alpha \ll 1$ represents highly distributed vorticity. Figure 2 shows typical distributions of longitudinal u and transverse v velocity fluctuations for a vorticity concentration $\alpha = 0.7$. Here, $\tilde{v}_{rms}$ represents the

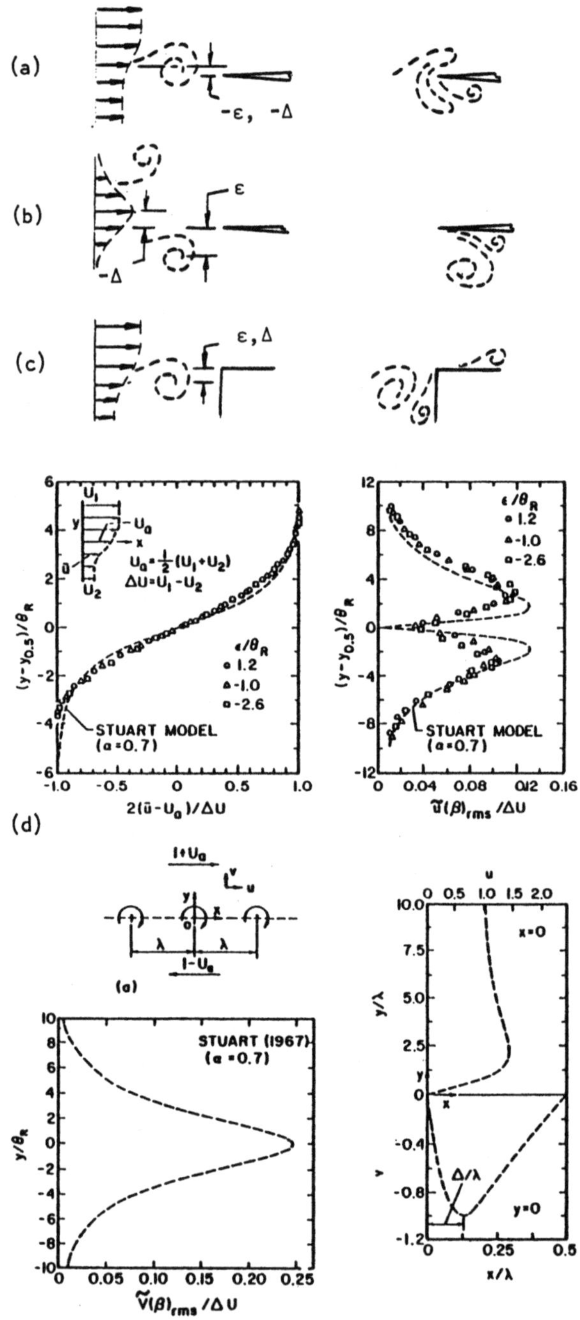

Figure 2. Vortex-edge interaction patterns and simplified incident and shed vortex combinations; and structure of Stuart (1967) vortex.
Courtesy of Cambridge University Press.

rms of the v fluctuation, and u and v represent the instantaneous values of the velocity fluctuations. It turns out that the strong gradient of v in the y direction, as well as the occurrence of maximum v at $x/\lambda = 0.125$ have important consequences for the unsteady loading at the leading edge (Ziada and Rockwell, 1982). These considerations are, of course, necessary for each type of incident vortex; generalization of the interaction mechanism without specific knowledge of the degree of vorticity concentration is inappropriate.

In the event that the center of the incident vortex is slightly above the leading-edge, there is pronouced shedding of vorticity of opposite sense (Figure 3). The pressure fields on the upper ($\tilde{p}_u$) and lower ($\tilde{p}_l$) sides of the leading-edge are shown in the form of an envelope of symmetrical solid lines and a shaded area. They represent, respectively, the rms and instantaneous values of the fluctuating pressure. The normalizing length and time scales are λ and T, corresponding to the wavelength between, and period of, the incident vortices. It is evident that this interaction mechanism produces a large pressure amplitude at the leading edge, which rapidly decreases with streamwise distance. The key role of secondary vortex shedding

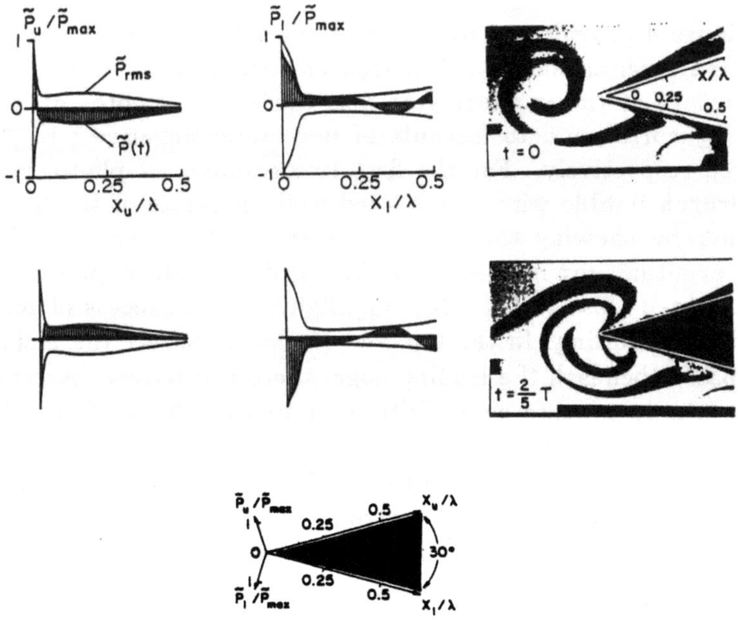

Figure 3. Instantaneous pressure fields corresponding to incident mixing layer vortex-edge interaction $Re_{\theta_R} = 230$ (Kaykayoglu and Rockwell, 1985a). Courtesy of Cambridge University Press.

from the leading edge is underscored by the pressure distribution on the lower surface at $t = 2/5T$; at this instant, secondary shedding has commenced, and the leading-edge pressure is maximum negative. Just downstream of the edge, the pressure distribution takes on a wave-like character corresponding to downstream convection of the shed vorticity. In contrast, the upper surface pressure distribution, which is primarily due to the distortion of the incident vortex, is nearly in phase over a distance of half a wavelength. Comparing the instantaneous distributions along the upper and lower surfaces, one observes that they tend to be approximately π out of phase for $x/\lambda \lesssim 0.25$.

In cases where the vortex approaches the edge lower or higher than in Figure 3, the character of the pressure fields changes and the leading-edge pressure takes on lower values. In fact, by integrating the instantaneous pressure distributions in the region near the leading edge, and properly accounting for phase variation, it is possible to demonstrate that the relative force at the edge follows, in an approximate sense, the relative magnitude of the leading-edge pressure (Kaykayoglu and Rockwell, 1984a)!

Figure 4 provides an overview of several types of vortex-edge interaction mechanisms as a function of vortex offset; in all cases, the structure of the incident vortex is the same. Photos at times t_B and t_E correspond to instants of maximum negative and positive force respectively. For the first two columns of photographs, the hydrogen bubble wire was located well upstream of the leading edge, thereby showing the cumulative effect of the vortex roll-up and its eventual impingement; in the third column of photos, the wire is located close to the edge, highlighting the process of secondary vortex shedding. In the top row of photos, where the incident vortex passes beneath the leading edge, there is a barely discernible shed vortex; it appears as a white glow in the photo of the third column. The incident vortex retains its identity as it passes along the edge. In the case where the center of the incident vortex is just above the leading edge, there is strong secondary vortex shedding that dominates the underside of the edge. Along the upper surface, the passage of the incident vortex is evident, though there is no continuation of the vortex roll-up process. Finally, in the event that the incident vortex is well above the leading-edge, the shed vorticity, evidenced by the concentrated white glow on the underside of the edge, is clearly less pronounced than in the aforementioned case.

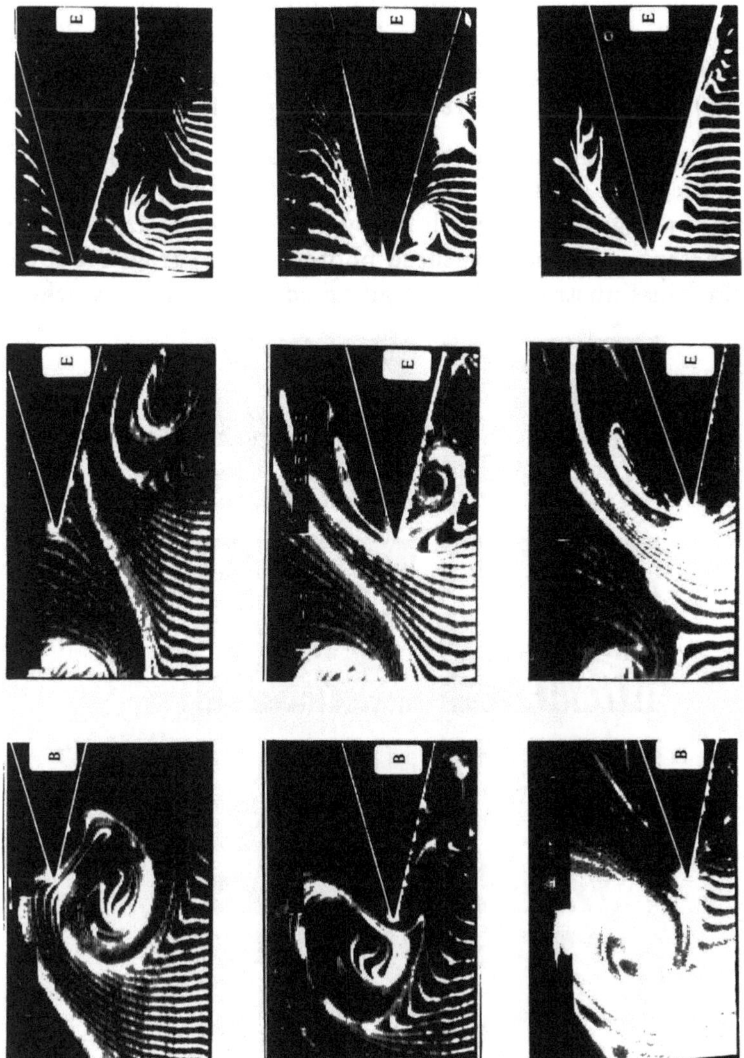

Figure 4. Vortex-edge interaction patterns for cases where incident (mixing-layer) vortex is below, approximately at, and above location of leading edge. $Re_{\theta_R} = 230$ (Ziada and Rockwell, 1982). Courtesy of Cambridge Univ. Press.

These observations bring forth the need for studies of the detailed structure of the incident and shed vortices at the edge. For example, as the incident vortex encounters and is severed at the leading-edge (i.e., middle row of photos of Figure 4), what is the evolution of its structure immediately downstream of the edge? Can it be simulated as a wave-like viscous layer (adjacent to the edge surface) embedded in a field of frozen vorticity (external to the viscous layer) defined by the vortex incident upon the leading edge?

The fact that the incident and shed vortex interactions are closely interwined, at least for certain ranges of vortex-edge offset, is brought forth in Figure 5, in which the symbols A, B, C, D represent successive instants of time. In these photos, the hydrogen bubble wire is located just upstream of the leading-edge. For the case where

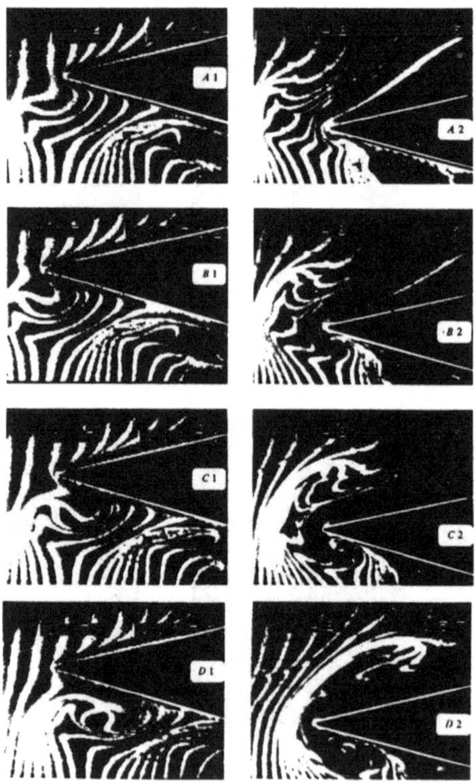

Case 1 Case 2

Figure 5. Vortex-edge patterns for cases where mixing-layer vortex is below (case 1) and above (case 2) leading edge. Hydrogen bubble wire located just upstream of leading edge $Re_{\theta_R} = 230$ (Ziada and Rockwell, 1982). Courtesy of Cambridge Univ. Press.

the major share of the vortex passes beneath the edge, secondary shedding is not evident. In contrast, when the incident vortex is well above the edge, it produces an induced (reverse) flow on the upper surface which enhances the process of secondary shedding. This feeding of fluid adjacent to the upper surface into the vortex shed from the lower surface is, in turn, linked to the large pressure amplitudes at the edge (Figure 3). Consequently, simulations of this interaction process must account for *sweeping* of fluid about the edge rather than a simple vortex sheet springing from its tip.

With regard to analysis of these interactions, it is evident from these visualizations of Figures 3, 4, and 5 that one must accommodate: an arbitrary distribution of vorticity of the incident vortex; and a means of accounting for nonlinear and viscous effects at the edge. Currently available simulations involving frozen vorticity distributions (Widnall and Wolf, 1980), linearized distributions of vorticity (Rogler, 1978), and second order theories accounting for incident gust distortion (Goldstein and Atassi, 1976) have, in their own right, provided a considerable insight. The degree to which these inviscid simulations can approximate the actual nonlinear, viscous description of the interaction awaits investigation. A full Navier-Stokes simulation of this general class of problems is well underway (Ohring, 1983). Of course, another possibility is to simulate the incident vortex with a large number of discrete vortices (see reviews of Clements and Maull, 1975; Rockwell, 1983; Aref, 1983). A major point here is to what extent the cluster of discrete vortices can be adjusted to simulate vorticity distributions of practical interest.

2.2 Jet-Edge Interactions

We now move from single, well-defined concentrations of vorticity to more complex combinations of them. In the case of an oscillating jet, for example, impinging upon the leading edge, the incident concentrations of vorticity have an alternating sense (Figure 2b). Depending upon the development of the incident unsteady shear layer, i.e., degree of concentration of vorticity, the interaction patterns may take on quite different forms. In the following, we consider incident jets of varying maturity and spectral content, and the consequent pressure fields at the leading edge.

Figure 6 (top photo) shows, for the case of zero edge offset, that the relatively immature vortices incident upon the edge quickly roll up after passing it. Moreover, they give rise to vortices of opposite sense due to wall viscous effects. When the offset of the edge relative to the jet centerline is increased (middle photo), similar vortex formation and interaction occurs along the upper face, but is dramatically reduced along the lower surface. At the largest value of offset (lower photo), there is a drastic decrease in the fluctuation amplitude of the incident jet, and, therefore, a retarded vortex evolution on the upper and lower surfaces. In this case, there are no induced vortices of opposite sense produced by the incident vortex-edge boundary layer interaction. A number of issues deserve further attention: enhancement of the rate of growth of the free-shear-layer vortex as it interacts with the leading-edge; rate of growth of the instability wave along the surface of the edge whose vorticity has both like and opposite senses to that of the incident shear layer; and the flow structure at the tip of the edge that gives rise to this wave.

If we examine the unsteady pressure field at the leading-edge for the case where the incident jet possesses a single (predominant) oscillation frequency, then the distribution is as shown in Figure 7. Here the transverse velocity fluctuation of the jet is larger than in Figure 6, the vortex formation process immediately downstream of the edge is correspondingly enhanced, and there is secondary shedding of vorticity of opposite sense, revealed by the injection of dye at the leading edge (not shown here). Remarkable is the fact that the fluctuating pressure drops rapidly as the tip of the edge is approached, at least for this type of interaction mechanism. Moreover, it shows a maximum just downstream of the tip due to rapid onset of the illustrated vortex patterns. However, our findings also indicate that small offset of the edge can substantially alter the distribution of pressure amplitude. This observation suggests that the transverse locations of the local maxima and minima of the mean and fluctuating vorticity distributions, relative to the edge, is of central importance in determining the edge pressure field and requires further study. A major difficulty in experimental work is maintaining the same incident vorticity field as one varies the relative position of the leading-edge because of the feedback coupling, i.e., a lower amplitude pressure field at the edge induces smaller vorticity fluctuations in the upstream shear layer. Appropriate numerical simulation could circumvent this dilemma by prescribing an invariant vorticity field incident upon the edge irrespective of its location/orientation.

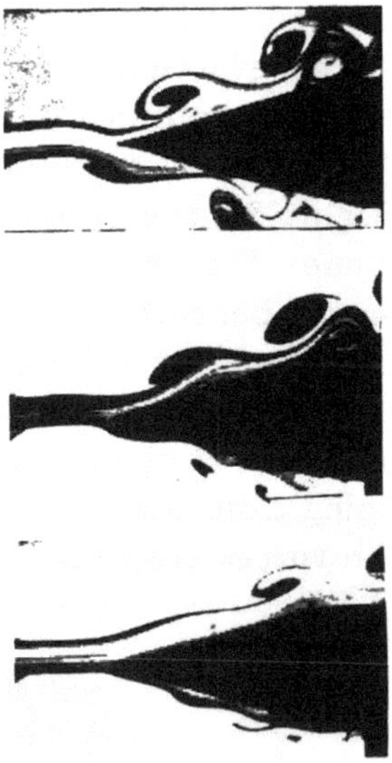

Figure 6. Effect of edge interaction on enhancement of jet vortices and induced boundary layer vortices for: (a) wedge at centerline ($\Delta/w = 0$); (b) wedge offset one-quarter nozzle width from centerline ($\Delta/w = 0.25$); and (c) wedge offset of three-eighths nozzle width ($\Delta/w = 0.375$). $Re_w = 600$; $L/w = 3.25$ (Kaykayoglu and Rockwell, 1985b). Courtesy of Cambridge University Press.

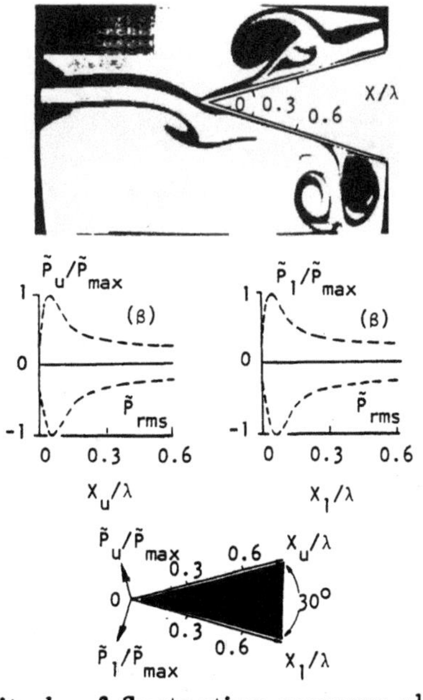

Figure 7. Amplitude of fluctuating pressure along upper and lower sides of edge for jet oscillation at single, predominant frequency. $Re_w = 600$; $L/w = 4.5$ (Kaykayoglu and Rockwell, 1985b). Courtesy of Cambridge University Press.

Downstream of the leading-edge, the counter-rotating vortex pairs shown in Figure 7 can eventually interact, or coalesce, with each other. Figure 8 shows a number of these counter-rotating pairs; let us focus, for the moment, on those pairs along the upper surface of the edge. In Figure 8a, the upstream pair is about to nest within the downstream one. Figure 8b reveals that the upstream counter-rotating pair passes beneath the downstream vortex that originally arose from free-shear-layer instability. Further scenes show completion of this process; it produces very large scale vortices, as depicted in Figure 8c (not part of this same sequence). The two vortices originally stemming from the unstable free shear layer have coalesced and allowed the vortex erupting from the boundary layer

Figure 8. Secondary instability downstream of leading-edge $Re_w = 300$; $L/w = 4.75$ (Lucas and Rockwell, 1983). Courtesy of Cambridge University Press.

to pass beneath them. This type of multiple vortex interaction of Figure 8 invites consideration of the concept of strange attractors (Aref, 1983), which has received considerable attention for interactions in free space. Depending upon the initial conditions, chaotic vortex motion may set in. Here we have an adjacent solid boundary in the form of an edge, and the corresponding image vortices. What are the conditions for regular and irregular trajectories of the vortices?

An interesting feature of the vortex-vortex interaction of Figure 8 is that it tends to move upstream with increasing Reynolds number, or with increasing (impingement) length from the origin of the free shear layer. Its character may change somewhat, but the important point is that the spectral content of the leading-edge pressure field shows an increasing number of discrete components as the vortex-vortex interaction pattern moves closer to the edge.

Figure 9 shows such a case. The spectral content of the incident shear layer shows three predominant frequency components rather than a single one as in Figure 8. Correspondingly, the pressure amplitudes near the leading-edge show the same three spectral components $\beta/3$, $2\beta/3$, β; $\beta/3$ is the frequency at which the stem of the jet oscillates, β is frequency of vortex formation in the jet shear layer, and $2\beta/3$ their difference frequency. For all components, the same normalizing pressure is employed: the maximum amplitude at zero edge offset ($\Delta/w = 0$). It is evident that the shape and amplitude of the pressure distributions is strongly dependent upon the frequency component under consideration. Clearly, if one simply considers the total rms energy of the pressure fluctuations, without undertaking spectral analysis, then proper physical insight is not gained. For example, if we wish to interpret the validity of a leading-edge Kutta condition in terms of pressure amplitude at/near the edge for practical, multi-frequency component flows, it may vary with the frequency component under consideration!

Finally, in Figure 10, we witness the effects of variations of Reynolds number on the interaction process near the leading edge. At $Re = 250$, a single frequency component dominates the fluctuating pressure fields; at higher values of $Re = 600$ and 900, the frequency content of the incident shear layer becomes increasingly richer, and the induced pressure fields correspondingly so, in accordance with our observations of Figure 9. However, there are certain characteristics of the vortex interaction patterns that persist over the

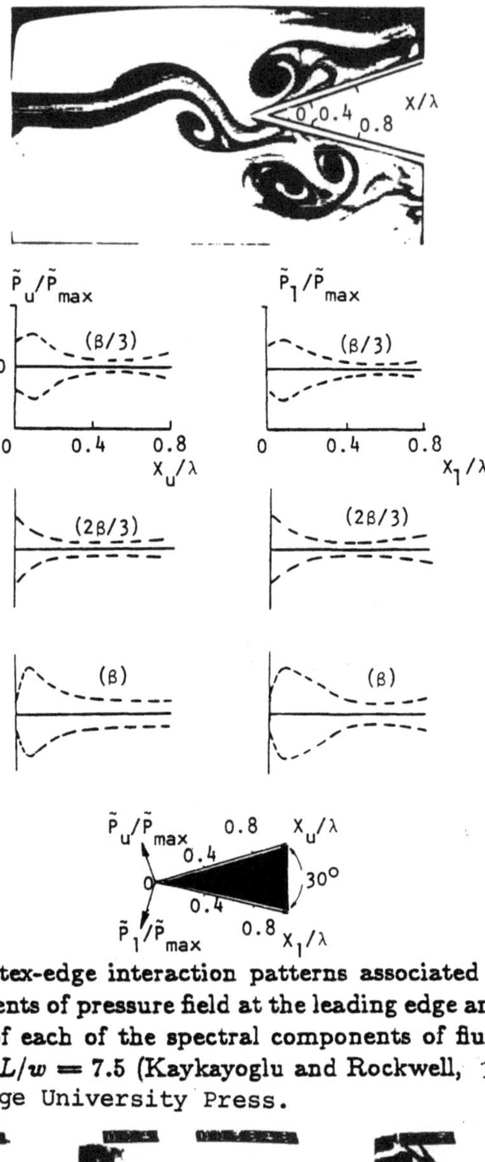

Figure 9. Vortex-edge interaction patterns associated with multiple frequency components of pressure field at the leading edge and corresponding distributions of each of the spectral components of fluctuating pressure. $Re_w = 600$; $L/w = 7.5$ (Kaykayoglu and Rockwell, 1985b). Courtesy of Cambridge University Press.

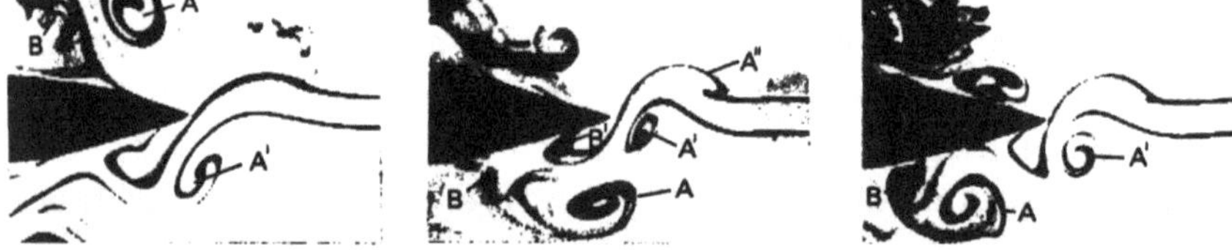

Figure 10. Effect of Reynolds number on vortex-edge interaction patterns for $Re_w = 250$; $Re_w = 600$; and $Re_w = 900$ from left to right (Lucas and Rockwell, 1983). Courtesy of Cambridge University Press.

entire range of Reynolds number, namely the frequency of appearance of the large scale vortices near the edge. The vortex interaction pattern adjusts itself to accommodate this large scale vortex formation which, in turn, modulates upstream regions of the flow at low frequency (Lucas and Rockwell, 1983). Of course, consideration of these vortex patterns is essential for not only for an understanding of the leading-edge conversion, but also for insight into the upstream influence of the edge region (Powell, 1962).

At higher Reynolds number we expect certain features of the coherent vortex-edge interactions to persist, especially in conjunction with strong feedback coupling such as occurs in the impingement of an underexpanded jet. Indeed, the Schlieren photo (Figure 11) of Krothapalli (1983) strongly suggests organized patterns of vortical structure-surface interaction. Moreover, by examining sequential photos of a case where there are strong transverse undulations of the jet (Figure 12), we see a rapid amplification of the undulation amplitude as the surface is approached, followed by the apparent generation of large scale structures at the surface. Obviously, resolution of the detailed mechanics of the interaction process is difficult at these high speeds; it deserves further investigation with an eye towards cross-spectral analysis of surface pressure fields to determine the convective character of the structures of Figures 11 and 12.

2.3 Mixing-Layer-Edge Interactions in Bounded Flows

Shear layers that reattach to a downstream boundary or wall form a bounded recirculation zone. At the impingement or reattachment location, vortex-corner or -wall interactions yield distortion of the incoming concentrations of vorticity, as well as secondary vortices shed from the edge (Figure 2c). As in the foregoing cases, the mechanism of interaction and induced pressure field are strong functions of the vortex structure and its relative position with respect to the corner. Figure 13 shows the case where the major share of incident vorticity is swept down along the front face of the cavity, producing a secondary vortex of opposite sense. Upon its impingement, the incident vortex produces maximum negative pressure. Moreover, appearance of the secondary vortex produces a corresponding peak of negative pressure. The occurrence of a peak on the top face is due to secondary vortex shedding arising from modulation of the boundary layer by the incident vortex.

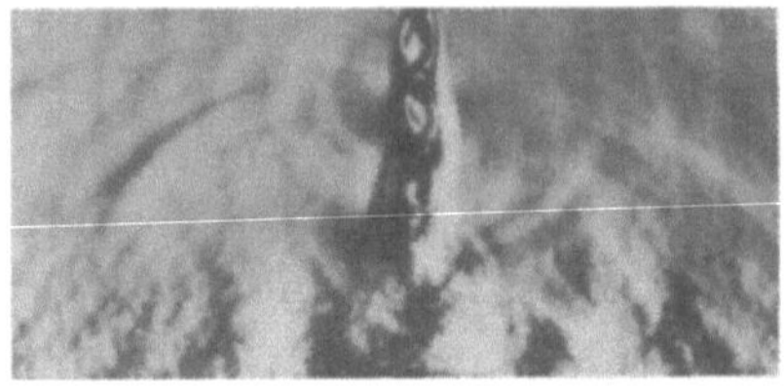

Figure 11. Patterns of coherent structure-surface interaction due to planar underexpanded jet impinging upon a plate. Pressure ratio = 4.09; impingement length L to jet diameter ratio D, $L/D = 14$ (Krothapalli, 1983). "Copyright © the American Institute of Aeronautics and Astronautics; reprinted with permission of the AIAA "

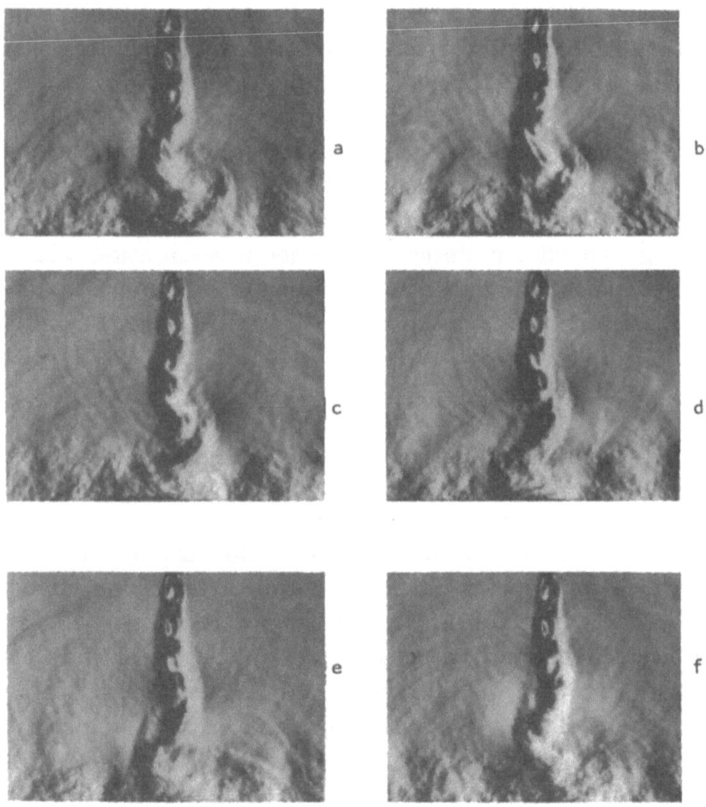

Figure 12. Transverse undulations of a planar underexpanded jet and associated flow-surface interaction (Pressure ratio = 5.07, Krothapalli, 1983). "Copyright © the American Institute of Aeronautics and Astronautics; reprinted with permission of the AIAA."

Concerning the phasing between pressure fluctuations on the front and top surfaces, the corresponding pressure maxima are π out-of-phase, suggesting a dipole-like behavior.

At longer length scales or higher Reynolds numbers than in Figure 13, there is substantial amplitude modulation of the pressure field at the corner. This is due to variations in structure and location of the incident vortex as shown in Figure 14. By examining the photos covering two oscillation periods, it is evident that vortex patterns along the front face of the cavity vary quite substantially. This difference stems from subtle modulations of the structure and location of the incident vortex (compare photos at $t = 0$ and $t = T$). In the former case, there results well-defined formation of a counter-rotating vortex pair along the face, while in the latter, there is complete absence of such formation. These types of patterns are repetitive over many cycles of oscillation. Time-averaged spectra at the impingement surface show clearly defined peaks related to this amplitude modulation, which in turn is connected to nonlinear interaction of instability waves in the approach shear layer (Knisely and Rockwell, 1982).

At still longer length scales and higher Reynolds numbers, the modulation patterns become even more complex, but still exhibit quasi-ordered behavior. There are alterations between two of the several possible types of vortex-corner interactions when the pressure at the edge shows a pronounced predominant frequency and its subharmonic. However, if there are alterations between more than two of these vortex interaction patterns,the surface pressures show at least two subharmonics in addition to the predominant frequency. In some instances as many as four subharmonics are present. Again, all these components are related to each other through nonlinear interaction of the approach shear layer.

Since these types of flows show a tendency towards conservation of mass within the recirculation zone, i.e., a balance between entrainment of the separated shear layer and return flow at reattachment, one should further examine the possible coupling between the dynamics of this recirculation zone, the shear layer instability, and the vortex interaction mechanisms at the corner.

2.4 Oscillating-Edge-Vortex Interactions

In the case of oscillations of a leading edge in the presence of

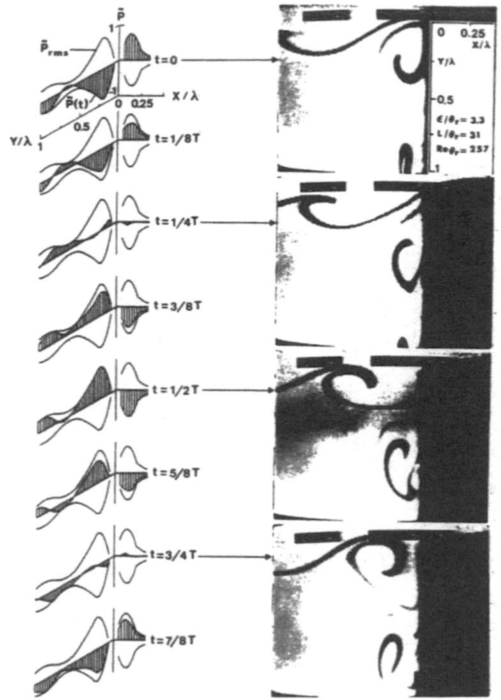

Figure 13. Instantaneous pressure fields corresponding to vortex-corner interaction $Re_{\theta_R} = 257$; $L/\theta_R = 31$ (Tang and Rockwell, 1983). Courtesy of Cambridge Univ. Press.

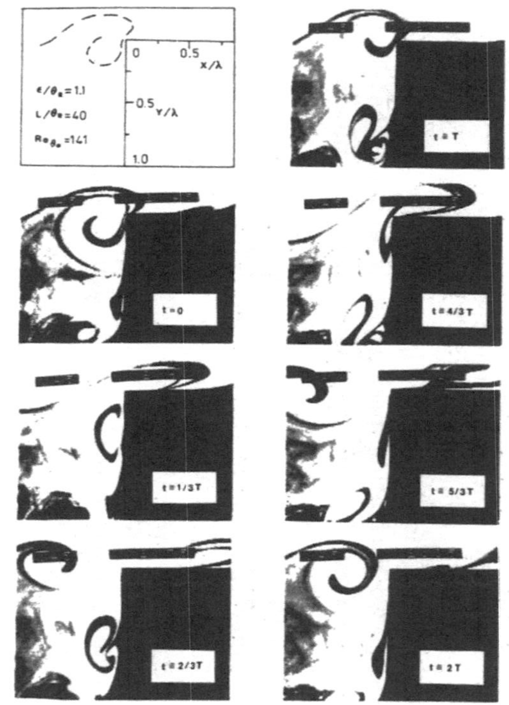

Figure 14. Amplitude modulation (at constant frequency of vortex-corner interaction $Re_{\theta_R} = 257$; $L/\theta_R = 40$) (Tang and Rockwell, 1983). Courtesy of Cambridge Univ. Press.

incident vortical structures, one must consider the dimensionless frequency and structure of the incident vortices. In addition, the phase between the incident vortices and the edge motion is of central importance. The left column of photos in Figure 15 illustrates the case where the ratio of edge frequency (f_{edge}) to natural vortex shedding frequency from the upstream trailing-edge (f_v) is: $f_{\text{edge}}/f_v = 0.6$. For the right column of photos, $f_{\text{edge}}/f_v = 0.4$. This difference in excitation frequency produces a substantial phase shift between the respective incident vortex-leading edge interactions which, in turn, gives quite different incident vortex-shed vortex patterns. For the case $f_{\text{edge}}/f_v = 0.6$, the top two photos of the left column show that upward motion of the wedge, in combination with the incident vortex, produces a strong induced flow in the downward direction that determines the character of the shed vortex. The bottom two photos show that this is followed by movement of the incident vortex above the edge as its tip moves in the downward direction. This observation contrasts with that for the stationary edge (Figure 3), where the secondary vortex shedding is determined solely by the incident vortex-edge interaction.

The top two photos in the right column of Figure 15 show the onset of secondary shedding as the wedge starts its movement in the upward direction. The process of secondary vortex shedding appears to lead that of its higher excitation frequency counterpart described above. Moreover, subsequent photos show that the phase relation between the wedge motion and the incident vortex is such that the major share of it is swept below, rather than above, the oscillating leading edge. Also apparent is a secondary instability in the form of a smaller scale incident vortex riding upon the back of the major incident vortex; it eventually appears as a vortex on the upper surface of the edge (see top photo of right column). This secondary incident vortex is due to the fact that the excitation frequency is well away from the natural shedding frequency of the upstream trailing edge (i.e., $f_{\text{edge}}/f_v = 0.4$). It may well be associated with nonlinear wave interaction in the approach shear layer.

Clearly, there are many interesting facets to pursue in this area of oscillating leading edges. A major issue which pervades all such interactions is the degree to which the flow field induced by the wedge motion (as opposed to that induced by the incident vortex) dominates the leading-edge pressure field. Of course, at high

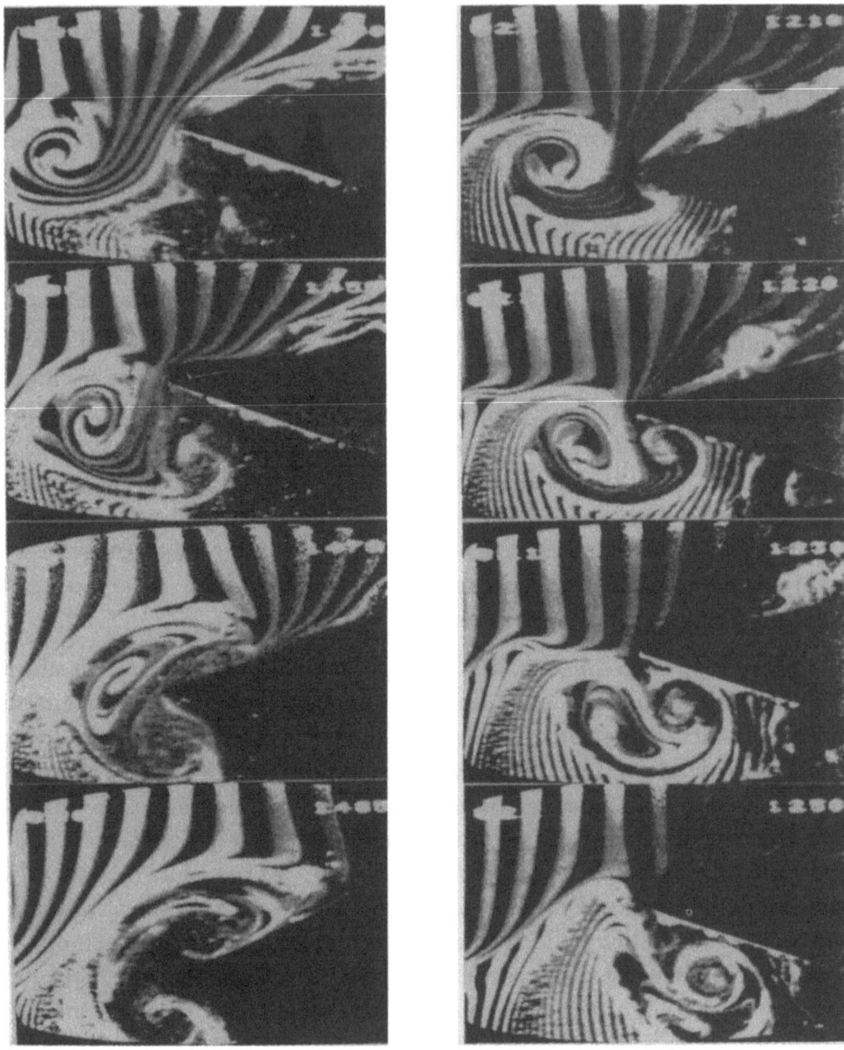

Figure 15. Comparison of interaction patterns of vortex incident upon oscillating leading edge at successive instants of time (top to bottom) for: $f_{edge}/f_v = 0.6$ (left column); and $f_{edge}/f_v = 0.4$ (right column). Phase between oscillating edge and incident vortex is substantially different for these two excitation conditions (Kaykayoglu and Rockwell, 1985b). Courtesy of Cambridge University Press.

amplitudes, and in an appropriate frequency range, relative to those of the incident vorticity field, one expects the field induced by the edge motion to dominate. What are the ranges of dimensionless amplitudes and frequencies for which this limiting condition is valid? In general, however, the photos of Figure 15 show that the incident vortex, shed vortex, and edge motion are all intimately related.

3. Conclusions

In the foregoing, we have witnessed a number of challenging aspects of vorticity field-leading edge interaction, posing interesting research possibilities for both the experimenter and theoretician. In a general sense, one may state that the central and unresolved issue is: to relate the vorticity concentration of a vortex, as well as combinations of them, to the flow distortion and unsteady pressure field downstream edge. This involves characterization of the classical instability/turbulence parameters immediately upstream of, at, and downstream of the edge region, with a careful eye towards possible influence of three-dimensional effects. On the analytical side, linearized inviscid theories have already proven to be acceptable in certain respects, but a full simulation of the problem properly accommodating nonlinear and viscous effects is essential to provide guidance for such approximations.

Acknowledgements

The author is indebted to Messrs. Ruhi Kaykayoglu and Michael Lucas for their collaboration in preparation of the manuscript; their contributions are evident from the citations in the main text. These investigations would not have been possible without the support of the National Science Foundation in Washington, DC and the Volkswagen Foundation of Hannover, West Germany, over the past several years, and more recently, from the Office of Naval Research.

References

[1] Clements, R. and Maull, D. "The Representation of Sheets of Vorticity by Discrete Vortices," *Progress in Aerospace Sciences, 16* (1975), 129–46.

[2] Goldstein, M. E. "Characteristics of the Unsteady Motion on Transversely Sheared Mean Flows," *Journal of Fluid Mechanics, 84, Pt. 2* (1978), 305–29.

[3] ——————. "Scattering and Distortion of the Unsteady Motion on Transversely Sheared Mean Flows," *Journal of Fluid Mechanics, 91, Pt. 4* (1979), 601–32.

[4] ——————. "Generation of Instability Waves at a Leading-Edge," *AIAA Paper No. 82-9236;* presented at AIAA, ASME 3rd Joint Thermophysics, Fluids, Plasma and Heat Transfer Conference, June 7–11, 1982.

[5] ——————. "The Evolution of Tollmien-Schlichting Waves Near a Leading Edge," *Journal of Fluid Mechanics, 127* (1983), 59–81.

[6] Goldstein, M. E. and Atassi, H. "A Complete Second-Order Theory for the Unsteady Flow by an Airfoil due to a Periodic Gust," *Journal of Fluid Mechanics, 74, Pt. 4* (1976), 741–65.

[7] Hunt, J.C.R. "A Review of the Theory of Rapidly Distorted Flow and Its Application," *Fluid Dynamics Transactions, 9* (1977), 122–52.

[8] Karamcheti, K., et al. "Some Features of an Edge-Tone Flow Field," *Basic Aerodynamic Noise Research, NASA Sp 207* (1969), 275–304.

[9] Kaykayoglu, R. and Rockwell, D. "Unsteady Pressure Fields at a Leading-Edge due to an Incident Vortex," Journal of Fluid Mechanics, 156 (1985), 454.

[10] ——————————. "Vortices of a Jet Interacting With a Wedge: Unsteady Pressure Fields," Journal of Fluid Mechanics, (1985).

[11] ——————————. "Incident Vortex-Oscillating Edge Interactions," Journal of Fluid Mechanics, (1985).

[12] Knisely, C. and Rockwell, D. "Self-Sustained Low Frequency Components in an Impinging Shear Layer," *Journal of Fluid Mechanics, 116* (1982), 157–186.

[13] Krothapalli, A. "On Discrete Tones Generated by an Impinging Underexpended Rectangular Jet," *AIAA Paper No. 83-0729;* AIAA 8th Aeroacoustics Conference, April 11–13, 1983, Atlanta, GA.

[14] Lucas, M. and Rockwell, D. "Self-Excited Jet: Low Frequency Modulation and Multiple Frequencies," Journal of Fluid Mechanics, 147 (1984), 340.

[15] Ohring, S. Private communication, Numerical Mechanics Branch, Naval Ship Research and Development Center, Carderock, MD., 1983.

[16] Powell, A. "On the Edgetone," *Journal of the Acoustical Society of America, 33, No. 4* (1961), 359–409.

[17] ————. "Vortex Action in Edgetones," *Journal of the Acoustical Society of America, 34, No. 2* (1962), 163–66.

[18] ————. "Advances in Aeroacoustics," *Rapports du 5ᵉ Congrès International d'Acoustique, 11,* Confèrences Gènèrales, Liege, 1965.

[19] Rockwell, D. "Oscillations of Impinging Shear Layers," invited lecture at the AIAA 20th Aerospace Sciences Meeting, Orlando, FL, January 11–14, 1982, *Paper 82-0047;* also *AIAA Journal, 21, No. 5* (1983), 645–64.

[20] Rockwell, D. and Knisely, C. "The Organized Nature of Flow Impingement Upon a Corner," *Journal of Fluid Mechanics, 93, Pt. 3* (1979), 413–32.

[21] Rogler, A. "The Interaction Between Vortex-Array Representations of Downstream Turbulence and Semi-Infinite Flat Plates," *Journal of Fluid Mechanics, 87, No. 3* (1978), 583–606.

[22] Shapiro, P. J. "The Influence of Sound Upon Laminar Boundary Layer Instability," Mass. Institute of Tech., Acoustics and Vibration Laboratory, *A/V-83458-83560-1,* Sept. 1977.

[23] Stuart, J. T. "On Finite Amplitude Oscillations in Laminar Mixing Layers," *Journal of Fluid Mechanics, 29* (1967), 417–40.

[24] Tang, Y. P. and Rockwell, D. "Instantaneous Pressure Fields at a Corner Associated With Vortex Impingement," *Journal of Fluid Mechanics, 126* (1983), 187–204.

[25] Widnall, S. and Wolf, T. L. "Effect of Tip Vortex Structure on Helicopter Noise due to Blade-Vortex Interaction," *Journal of Aircraft, 17* (1980), 705–11.

[26] Ziada, S. and Rockwell, D. "Vortex-Leading-Edge Interaction," *Journal of Fluid Mechanics, 118* (1982), 79–107.

Large-Scale Organized Motions in Jets and Shear Layers

A.K.M.F. Hussain
Department of Mechanical Engineering
University of Houston
Houston, TX 77004

1. Introduction

Turbulence research has undergone a major change in recent years as a result of the discovery of large-scale organized motions, popularly called "coherent structures". This discovery has brought about a reexamination and, in the opinion of many, even a redefinition of the fundamental concepts in turbulence. Also, this has rekindled hopes for significant advances in the understanding, and perhaps modeling, of turbulent shear flows. So profound is the impact of this discovery that virtually every turbulence researcher is pursuing coherent structures in one form or another. I will summarize some of the key questions in this topic and review a few recent results from our laboratory.

There is little debate over the impact of coherent structures on the current widespread activity in turbulence research or over the need for serious and continued investigations of them, but there is over who discovered them first. Many feel that the presence of coherent structures in turbulent shear flows has been well known for decades. Perhaps this is not a totally unjustified claim. In particular, organized large-scale motions in transitional flows have been known for over a century. For example, rolled up vortices in the near field of circular jets were known to Rayleigh and Laconte; vortex pairing was shown by Brown (1935) and Anderson (1954), among others; exactly a century ago, Reynolds (1883) discovered *slugs*, and perhaps also *puffs*, which were extensively investigated by Lindgren (1959) and Rotta (1956); *spots* were discovered by Emmons (1951) in 1951; organized *rolls* in turbulent Taylor-Couette flow were discovered by Pai in the thirties (Corrsin, 1983); etc. Even in the fully-developed turbulent flows like the wake, large-scale organized structures were suggested by Townsend (1956), Grant (1958), Keffer (1965), Payne and Lumley (1967), etc. The presence of organized events in jet turbulence was recognized by Mollo-Christensen (1967) and Bradshaw et al. (1964).

There are some (for example, Corrsin (1983)) who feel that the role and significance of coherent structures were known long before Brown-Roshko-Winant-Browand. Lumley (1982) contends, "responsible researchers have always described turbulence as having short and long-range order and structure." Corrsin further suggests that coherent structures were obviously implied by the mixing length hypothesis, presumed to explain momentum transport across a shear region.

While these assertions cannot be ignored, there is little doubt in my mind that the dominant presence and significance of coherent structures were neither recognized nor emphasized until recently, principally through the investigations of Crow, Champagne, Kline, Reynolds, Brown, Roshko, Browand, Laufer, Kaplan, Coles, and others. The discovery of coherent structures indeed represents a revolution in turbulence started in the West Coast of the United States, a revolution which clearly has reached the shores of all continents.

The intense and widespread interest in coherent structures has been spurred by the expectations that they are quasi-deterministic and their evolution is mathematically tractable. It is also assumed by many that coherent structures can account for virtually all that turbulent flow does; that is, they dominate entrainment and mixing, momentum transport, production of turbulence, aerodynamic noise generation, etc. It therefore follows that the essence of the turbulent shear flow can be represented in terms of coherent structures, thus raising hopes for successful modeling of turbulent shear flows. These are highly attractive propositions but need careful scrutiny. Based on our detailed data in a number of shear flows, I took the position that too much was claimed or promised about what the coherent structures approach would do, and that some of these claims must be moderated (Hussain, 1980; 1981; 1983a).

Regardless of whether these expectations are realistic or not, coherent structures are highly interesting and need to be vigorously studied. They are central to the understanding of the physics of turbulent shear flows as these structures appear to be the characteristic features of (perhaps all) shear flows. It stands to reason that viable turbulence models must explicitly incorporate coherent structures. Control of these structures may produce technologically interesting control of turbulence phenomena, for example, entrainment and mixing, heat and mass transfer, drag and aerodynamic noise.

1.1 What Are Coherent Structures?

In spite of their being the focus of widespread investigations, there is no consensus on what is meant by coherent structures. In fact, current perceptions not only vary widely but are even in conflict with one another. Even though scientific concepts are typically left understood rather than precisely defined (Liepmann, 1983), I felt that a clear discussion of the flow physics was not possible without a definition of coherent structures.

A coherent structure is defined as *a large-scale turbulent fluid mass with spatially phase-correlated vorticity* (Hussain, 1980; 1981). First of all, a coherent structure has to be turbulent: a laminar vortex is a degenerate or highly special case. Therefore, underlying the three-dimensional, random, vorticity fluctuations inherent to turbulence, there is a (large-scale) vorticity which is instantaneously phase-correlated over the spatial extent of the structure. We will use the alias, *coherent vorticity*, to denote "spatially phase-correlated large-scale instantaneous vorticity". Coherent vorticity, therefore, is the fundamental property characterizing a coherent structure. The instantaneous boundary of coherent vorticity in three-dimensional space denotes the boundary of the structure. Fluctuating motions superimposed on coherent structures are called **incoherent turbulence.**

Many have attempted to identify coherent structures via correlation of velocity, pressure and/or *intermittency* signals. This is not acceptable in general. The flapping of a laminar jet or wake due to external pressure disturbances produce high correlations of velocity or pressure without any coherent structure being necessarily involved. Turbulence intermittency signal (defined on the basis of small-scale vorticity fluctuations) in general may detect a coherent structure in a free shear flow (Hussain et al., 1980; Cantwell and Coles, 1983) but is not useful when patches of incoherent turbulence are also involved. For example, the intermittency signal is always nonzero in a turbulent pipe or channel flow and is thus useless in detecting coherent structures in these flows. Velocity or pressure correlations are also produced by irrotational (hence nonturbulent) fluctuating motions. Correlations of velocity and/or pressure extend beyond the boundary of the structure and are thus unsuitable for coherent structure eduction. Thus, it is inescapable that coherent vorticity is the only reliable identifier of a coherent structure. By my definition, not all large-scale organized motions are coherent structures. One example is the impulsively started jet studied by Cantwell (1983).

Coherent structures transport significant amounts of mass, momentum and energy without themselves being necessarily highly energetic. They are highly dominant in transitional flows, in resonant situations (like *shear layer tone, jet tone* phenomena), in flows under controlled excitation, and in the *wall layer* of a turbulent boundary layer. But, they are not necessarily predominant in the vast majority of turbulent flow situations where incoherent turbulence becomes energetic, for example, in fully-developed states of turbulent flows. Even though the qualifier *large-scale* is not typically used, the current interest in coherent structures concerns those which are large-scale, i.e., of sizes comparable to the length scale of the flow—denoted by the thickness of the shear region. Thus, the Reynolds numbers involved are large, and the dynamics of coherent structures is essentially *inviscid*. (Motion at the *Kolmogoroff scale* η is the most coherent as significant variations in velocity, vorticity or pressure can not occur within this scale. But η does not enter into any discussion of coherent structures unless one addresses *dissipation* or *molecular diffusion*.) Energetic coherent motion can occur at scales which are large compared to η, yet smaller than the flow length scale. Examples include *hairpin eddies*, longitudinal *wall vortices, typical eddies, pockets, ribs*, etc. For consistency, we call them *coherent substructures*. In the parlance of classical turbulence, their size is of the order of the *Taylor microscale*, while coherent structure size is of the order of *integral length scale*.

All coherent structure interactions like tearing and (complete, partial, and fractional) pairing are nonlinear so that the result of an interaction is the creation of a new structure (with a new scale). In this sense, the classical concept of energy *cascade* and equilibrium coexistence of different eddies at the same spatial point does not apply to coherent structures. Remember, each coherent structure has an independent territory and boundary. However, energy cascade and equilibrium may still be applicable to incoherent turbulence, especially at the smaller scales. There can be especial cases of equilibrium between coherent structures and incoherent turbulence when externally driven as in the case of turbulent Taylor-Couette flow, convection in a box, etc. These spatially constrained flows can produce equilibrium coherent structures which are highly interesting candidates for detailed study of topological properties of the structure as well as *turbulence production* via rapid distortion theory.

The coherent structures are seldom periodic unless periodically driven. Even then, the periodicity decreases with increasing downstream distances as a result of increasing jitter. Coherent structures can presumably occur periodically in batches in fully-developed turbulent flows (Townsend, 1979; Mumford, 1983). No explanation is yet available for this periodic occurrence. However, there should be little doubt that these batches occur randomly.

The survival distance of a coherent structure is not as long as commonly assumed. Our experience indicates that the life expectancy or survival distance of a coherent structure decreases with increasing Reynolds number. When the Reynolds number is sufficiently high, the survival distance becomes comparable to the size of the structure. The structure undergoes continual changes as a result of tearing, and complete, partial and fractional pairings. These evolutionary changes may be viewed as progressive stages of instabilities of the flow (discussed later). Based on these, it seems that little will be gained by viewing coherent structures as *solitons*.

It may be worthwhile to comment in passing on the relevance of coherent structures to *strange attractors* (Lanford, 1982; Swinney, 1983). Strange attractors, which predict well time evolutions of a variety of natural phenomena of *low degrees of freedom*, seem to explain initial stages of transition to *chaos* (or turbulence!) in closed flow systems like Taylor-Couette flow, box convection, etc. It has been speculated (Ruelle and Takens, 1971) that strange attractors may explain turbulence, which has *infinite degrees of freedom* even in closed systems. There is no evidence yet that strange attractors can even account for the initial stages of transition in open systems like shear layers, boundary layers, pipe or channel flows, etc.; an effort is underway in our laboratory to experimentally address this question. Far less likely it is that fully-developed turbulence has the characteristics of a strange attractor. The principal problem appears to lie with the fact that strange attractors can explain time-evolution of a dynamical system; I am unaware of any effort to apply these to spatially evolving flow systems. I expect that *coherent structures behave as strange attractors*; these structures, being organized and large-scale, should have a much lower dimension than fully-developed turbulence.

1.2 Flow Visualization

The question of the persistence of coherent structures brings up an interesting feature of modern turbulence research viz., flow visualization. Flow visualization played important roles in early studies of viscous flows but became rarer because of increasing emphasis on quantitative data and because of availability of inexpensive (constant-temperature) anemometers and their simple operation. However, persistent campaigns of Kline at Stanford, notwithstanding many influential skeptics, brought flow visualization new respectability. It was indeed flow visualization which played a key role in the discovery of coherent structures (see for example, Kline et al. (1967), Hama and Nutant (1963), Crow and Champagne (1971), Brown and Roshko (1974)). In fact, many recent significant contributions (for example, Bradshaw et al. (1964), Corino and Brodkey (1969), Winant and Browand(1974), Dimotakis and Brown(1976), Chandrsuda et al. (1978), Browand and Laufer (1975), Yule (1978), Falco (1977), Lau (1979), Gad-el-Hak et al. (1981), Oster et al. (1977), Head and Bandyopadhyay (1981), Perry and Lim (1978), Perry et al. (1980), Smith (1983)) have resulted from flow visualization studies. On the other hand, it is not difficult to find many significant investigations which involved no flow visualization (for example, Townsend (1956), Grant (1958), Keffer (1965), Payne and Lumley (1967), Kaplan and Laufer (1968), Kovasznay et al. (1970), Willmarth and Lu (1971), Wallace et al. (1972), Blackwelder and Kaplan (1976), Bruun (1977), Lumley (1981)). While flow visualization may become a gimmick for showing cute fluid phenomena, it can play a crucial role in understanding flow behavior. It can be said safely that flow visualization is now an integral part of coherent structure investigations. It not only provides an understanding of the flow physics but even helps provide guidance for appropriate measurement schemes for coherent structure eduction.

There are a number of constraints of flow visualization. First, one can find many irrelevant events in flow visualization just like one can find so in random signals (Lumley, 1981). In fact, observations based on flow visualization must be carefully scrutinized and verified by hard data; flow visualization often provides more information than one can digest or process. Furthermore, the local events or processes are often obscured by markers introduced or created upstream. One eminent example is the flow physics in the most critical region of a jet, viz. 3–10 diameters from the exit. Even though this region is characterized by the highest entrainment, turbulence level, Reynolds stress and noise production, almost nothing is known about the flow physics in this region. Dye or smoke introduced at the nozzle becomes too diffuse in this region to reveal any local instantaneous dynamics. The breakdown process in this region is extremely complex and needs to be explained. The organized motions in the self-perserving regions of jets, pipes, channels, boundary layers, and wakes are hard to visualize for the same reasons.

Flow visualization can often be misleading. For example, indefinite persistence of vortices shed from a cylinder was assumed on the basis of persistence of dye or smoke introduced upstream of or at the cylinder. Even in a laminar wake, it should have been obvious from consideration of diffusion of vorticity that such persistence is unlikely. That is, since the Schmidt number is rather high, the boundary of smoke lumps in the wake should have little to do with the vorticity contour of the structure; vorticity diffused away far beyond the boundary of smoke or dye. When the shed vortices are turbulent, the persistence of an individual structure is much less than expected from flow visualization. Because of turbulent diffusivity, the peak vorticity decays rather rapidly and some cancellation of opposite-signed vorticity from structures on opposite sides of the mid-plane is to be expected. Cantwell and Coles (1983) indeed suggest annihilation of circulation due to migration of vorticity between opposite-signed vortices in a plane wake. Furthermore, by marking the flow with smoke-wires locally in the far-wake, where there are otherwise periodic lumps of smoke when smoke is released at the cylinder, Nagib (1983) has shown that no vortical structures are indeed involved. This is a strong confirmation of what is to be clearly expected; that is, sufficiently away from the point of tagging, the smoke or dye boundaries have little to do with coherent structure boundaries.

A new development which combines both flow visualization and quantitative measurement is **image processing** (Meynart, 1983; Dimotakis et al., 1981). Interpretations of flow dynamics on the basis of dye or smoke boundaries via image processing of course suffer from the same limitations as mentioned regarding flow visualization. However, image processing involving particle motions is free from this constraint and promises to be a highly attractive and powerful research tool of the future. An effort is underway in our laboratory to develop *image displacement velocimetry* with illumination with a pulsed laser. This new technique will greatly advance measurement capability in coherent structures, complex turbulent flows, chemically reacting flows, etc.

2. Formation of Coherent Structures

The formation mechanism of coherent structures, like their dynamical role and significance, is an area of considerable speculation and controversy. The initial formation resulting from instability of initial laminar states is conceptually understandable and even mathematically tractable. However, their formation further downstream, in particular in fully-turbulent states, is less clear. Many believe that coherent structures in the fully-developed turbulent regions are legacies of initial transition (Lin, 1983). There are reasonable justifications for this point of view, provided that initial structures persist for a sufficiently long distance. Also, since subsequent evolutions are dependent on the interactions of the initial structures, the initial instability perhaps retains an indelible imprint on the structures far downstream. This was the argument used to explain the lingering effects of the initial condition in free turbulent shear flows (Hussain, 1977).

It is important to emphasize that coherent structures also result from initially fully-turbulent states. In fact, an axisymmetric mixing layer originating from a fully-turbulent initial state was found to be more organized than when the layer was initially laminar (Clark and Hussain, 1979). The topological details of the preferred mode coherent structure of an initially *fully-turbulent* axisymmetric mixing layer have been reported by Hussain and Zaman (1982). (It will be discussed later.) There are also clear demonstrations of formation of coherent structures from a fully-turbulent wake (Nagib, 1983; Taneda, 1959) even though the coherent structure details have not been educed. Our investigations show that the detailed coherent structure in the plane wake is quite different from that suggested previously (Hayakawa and Hussain, 1983).

There is sufficient evidence that the self-preserving jet, wake or mixing layer consists of characteristic coherent structures. In the case of the mixing layer, the evolution involves successive pairings so as to keep pace with the local length scale, even though the pairing is not likely to be of the complete type (Clark and Hussain, 1979). In the case of jets and wakes, the evolution is far from clear. It is reasonable to suggest that structures continually grow, saturate and then decay. The flow may be regarded as a superposition of many instability modes which are everpresent. The flow evolves in the presence of these competing modes and undergoes successive instabilities, each state of the flow responding to the most unstable mode of the everpresent disturbances. This scenario is attractive from stability concepts, but appears very simplistic because the flow is expected to achieve a defined state before another instability sets in. Visualization suggests that the flow is continually in an unstable state, the evolving interactions continually producing new structures. That is, the structures carry with them their own mechanisms of energizing themselves and promoting the interactions that produce new structures presumably scaling with the local flow scales (Hussain, 1983b).

2.1 Instability of the Time-Mean Profile

A number of investigations have pursued the instability of the time-average profile in the fully-developed turbulent state (Goldstein, 1983; Gaster et al., 1983). Even though some of the computed results show impressive agreement with data (Wygnanski, 1983), it is my contention that such an instability analysis is not meaningful on account of a number of reasons. First, the instability analysis of the time-average profile would be acceptable if the instantaneous profile deviates from the time-average profile only slightly. I do not think that the highly nonlinear instability of a turbulent shear flow can be modeled by a linear theory. Second, such an instability analysis would be meaningful if the turbulence time scale is considerably smaller than the time scale of the disturbance; that is, if the shear-flow turbulence were indeed purely fine-scale. In fact, the streamwise and temporal variations of the instantaneous velocity field are due to large-scale structures whose length and time scales are precisely the scales at which the instability is sought. In other words, the disturbance never sees the time-mean profile during its evolution. Third, even if the profile shape were the same at all times but it merely changed in magnitude with time, the instability study of the steady profile will not be very helpful because the instability of a time-dependent profile can be quite different from that of the same profile when steady (Davis, 1976; von Kerczek, 1982) depending on the ratio of time scales of flow unsteadiness and instability wave. Finally, no one has shown that the instability of the instantaneous flow to the leading order is the same as that of the time-average profile. Of course, since virtually infinity of instantaneous profiles are involved, how this is to be established is not clear.

The events in a turbulent shear flow are clearly both highly time-dependent and nonlinear. Thus, both effects need to be considered. In addition, nonparallel effects, when present, must be taken into account (Crighton and Gaster, 1976; Plaschko, 1979). At present, theories can handle either weakly nonlinear or weakly nonparallel effects. (Recently, a theory has included both weakly nonlinear and weakly nonparallel effects (Plaschko and Hussain, 1983), but subject to a fairly restrictive assumption.) Considering the complexity, it appears that the direct numerical simulation or even a vortex simulation may indeed be the only hope for a realistic model for turbulent shear flows.

3. Eduction of Coherent Structures

Eduction means measurement of the spatial distributions of various properties over the spatial extent of the coherent structure in selected cross-sectional planes. These may include *coherent vorticity, coherent Reynolds stress, incoherent Reynolds stress, incoherent turbulence intensities, coherent strain rate, coherent normal* and *shear productions*, etc. These properties provide the topological distributions of structure properties (Hussain, 1980, 1981, 1983a). Even though many investigations have addressed flows with coherent structures, the number studying the spatial details of structure properties has been very few (Cantwell et al., 1978; Hussain, 1980, 1981, 1983a; Hussain et al., 1980; Browand and Weidman, 1976; Nishioka et al., 1981; Riley and Metcalfe, 1980; Knight and Murray, 1980; Cantwell and Coles, 1983; Corcos and Sherman, 1983; Corcos and Lin, 1983). Without these spatial details, investigations of these structures are incomplete. In fact, I accept only those as coherent structure studies which address the topological details of the structures.

A coherent structure is typically buried under incoherent turbulence; eduction requires sifting out the coherent component. An average of a large number of appropriately phase-aligned structures (of the same mode and same parameter size) appears to be the only way to define a structure. We denote the *spatially aligned ensemble average* of identical structures at a particular phase as the **phase average**. (Thus, phase average does *not* denote average over all phases.) The incoherent or phase-random component will have no contribution to the phase average. Thus, whatever survives the phase average after appropriate spatial alignment is the coherent structure, and the departure of each realization from the phase average is incoherent turbulence.

3.1 Preferred Mode

By **mode** of a structure, we mean its spatial configuration: for example, toroidal, helical, bihelical, roller, hairpin, etc. Different flows may (mostly indeed do) consist of different modes of coherent structures. On the other hand, a variety of modes may occur in the same flow. If a particular mode predominates—both in frequency and in dynamical significance—we call it the **preferred mode** of the flow (Hussain and Zaman, 1981). Thus, the preferred mode may change from one flow to another (for example, toroidal vortex in the circular jet near field, spiral vortex in the far field of a circular jet, spanwise rolls in the plane mixing layer, hairpin vortex in boundary layers and wakes) as well as from one region to another in the same flow. Even for structures of a particular mode, there is in general a large *dispersion* in the characteristic structure parameters like shape, size, orientation, strength, lateral displacement, etc.; these together denote the *parameter size* of a structure. If a few parameter sizes dominate, the corresponding structures are then called the dominant preferred modes. That is, in the multidimensional function space of the characteristic parameters, there will be isolated peaks in the probability density. The structures corresponding to these peaks dominate the flow physics. Thus, it is adequate to educe only these dominant preferred modes, denoted as *preferred modes* for simplicity. The parameter values corresponding to each peak denote the *preferred parameter size* of the structure. The dispersion in the parameter size typically increases with increasing downstream distance in a flow. This dispersion can be reduced under controlled excitation or when there is resonance and can be virtually eliminated in an overconstrained system like the Taylor-Couette flow or spiral turbulence.

Eduction of coherent structures in a flow involves first, identification of the mode and second, selection of the parameter size for a given mode. Only structures of the same mode and same parameter size should be accepted for phase averaging. However, since the structure results in general from phase averaging, the mode and parameter size are not known a priori, and eduction is obviously a tedious and iterative process. It is the researcher's challenge to shorten this process.

When the flow consists of a unique mode, the eduction is simplified as mode identification is then unnecessary. Further simplification can be achieved via small-amplitude periodic excitation. This allows simple phase-locked ensemble averaging via periodic sampling of the signal from a single sensor and moving the sensor to different spatial points over the extent of the structure at the frozen phase. Even though the periodic excitation can be used as a clock for triggering eduction, the inherent jitter, which increases with increasing distance x from the origin of structure induction, makes the eduction progressively ineffective with increasing x (Cantwell and Coles, 1983). Use of a local detection triggered on a signal obtained at the measurement station, even in case of periodic excitation, can significantly reduce jitter (Hussain and Zaman, 1981).

What is the relevance of the excited structure to the natural structure? It was claimed by us previously that for the preferred mode, structures induced via small-amplitude excitation must be the same as natural structures as the excitation merely paces the formation of the structures. Since the geometry of the flow is the same and the evolution of the structures is governed by the same equations, induced and natural structures should be the same. We have subsequently confirmed this speculation via detailed measurements: the topological properties of structures induced via controlled excitation have been found to be the same as those of natural structures. However, our results show that the structure is quite different under high-amplitude excitation. Under excitation, the periodic thickening and thinning of the exit boundary layer releases vorticity nonuniformly. If the excitation amplitude is too high, the initial vorticity oscillation amplitude will be considerably higher than that during natural roll-up. In the case of an axisymmetric jet, the preferred mode excitation typically involves a frequency lower than the initial instability frequency f_s. If the excitation frequency is $1/N$ times f_s (i.e., N structures naturally roll up during each period of excitation) and if the amplitude is sufficiently high, then

N initial structures will be rolled up together (Zaman and Hussain, 1980). It should also be noted that the evolution of periodically induced structures is different from that of an impulsively induced isolated structure. In the former case, all structures are identical while in the latter case the induced structures are interspersed among a variety of natural structures. The evolutions of the structures in the two cases are different because of the different environments and structure interactions.

The eduction of natural structures, including those in fully-developed turbulent flows, whether initially excited or not, is a sophisticated art and must involve pattern recognition, as structures then typically do not have a characteristic *footprint* in the signals. The structure passage can be qualitatively inferred from smoothed velocity signals from a linear (trans-verse) array of sensors (say, single hot-wires); a coincidence of characteristic inflexions or features in the smoothed signals from the array may suggest passage of a large-scale structure. Of course, short-time cross-correlation is an effective way to unmask an underlying coincidence. The detection of large-scale structures is unambiguous when based on instantaneous vorticity maps in convenient planes. This is the procedure followed by us in a number of flows. From a transverse linear array (i.e., rake) of X-wires across the shearing region covering the thickness of the shear flow, vorticity Ω maps in the (y, t) plane can be obtained after some smoothing of the signals in both y and t; here y is the transverse coordinate and t is time. Peak values Ω_p in these maps of the $\Omega(y, t)$ contours indicate the structure centers, and the streamwise and transverse extents indicate the structure cross section. The location of the peak vorticity of all accepted structures must occur at the same y; structures which have wandered away transversely are rejected. From a variety of such structures, only those of a certain size (determined by the short-time correlation of $\Omega(t)$ between two selected y-locations) and of a certain strength (denoted by the peak value Ω_p of $\Omega(y, t)$) should be accepted. The straightforward way to align successive realizations is to align on Ω_p. This is the scheme applied in the eduction of coherent structures in the far field of an axisymmetric jet (Tso, 1983) and in mixing layers and wakes (Hayakawa and Hussain (in preparation)). Further improvement in eduction is possible via stricter alignment criteria. First, the alignment can be based on spatial correlation R of Ω in t at all y. This involves iterative determination of R over the extent of each accepted realization with the ensemble average. Second, realizations producing a value of R below a specified value can be discarded as being *freak*.

The approach just outlined is useful only for two-dimensional structures. In each flow, of course, the mode of the structure must be determined first via auxilliary sensors away from the rake. When the structure is three-dimensional like the hairpin vortex, detection must be three-dimensional, requiring multiple rakes. While, ideally, spanwise alignment is required, practical consideration suggests acceptance of only those structures of the selected mode and strength which have the same relative spanwise displacement.

It is necessary to reemphasize that eduction of a three-dimensional structure must require a three-dimensional array of sensors. It is not surprising that different detection schemes based on a point detection in a boundary layer have produced different structure signatures (Sato, 1983; Kunen et al., 1983). Furthermore, the educed structure has been found to depend on the threshold level of even the same detection scheme.

Capturing an isolated structure and then smoothing the properties spatially to obtain their contours is an attractive simplification to eduction via phase averaging, but is not rigorous as there is no way to determine the spatial variation of the smoothing scale. Such smoothing will eliminate local spatial fronts or gradients (thus dilute peak values of strain rate), which will be retained via phase averaging. Also, this method of eduction cannot provide the vital incoherent turbulence measures like $\langle u_r^2\rangle^{1/2}$, $\langle v_r^2\rangle^{1/2}$, $\langle u_r v_r\rangle$, and production.

The coherent structures approach to a turbulent flow is helpful only if the flow has a *very few* preferred modes. It is then sufficient to educe these modes as these can be assumed to dominate the flow. On the other hand, if a flow has a variety of coherent structures, that is, if the multi-dimensional probability density function (pdf) in the parameter space has no strong, isolated peaks, then the coherent structures approach is not very helpful. Then, the eduction would involve classification of structures into various constituent modes, classification of all structures of a given mode into subclasses identified by parameter sizes, obtaining phase average properties for each subclass, and stating the relative frequency of occurrence of each subclass. This effort is clearly prohibitive. Furthermore, even if all the details of the variety of constituent coherent structures were known, their inclusion in a theory will be cumbersome and may require statistical approaches. Such a statistical theory of coherent structures will suffer from limitations similar to those of the classical turbulence theories.

4. Some Specific Topics

4.1 Negative Production

In most turbulent shear flows, the mean velocity gradient and the mean momentum transport by turbulence (i.e., the Reynolds stress, $-\overline{uv}$) retain the same sign across the flow or switch sign together so that their product, i.e., the *shear production*, remains unchanged in sign. (In coordinates aligned with the freestream velocity, the time-average production in high-speed shear flows is essentially shear production as normal production is comparatively negligible.) The momentum transport is then in accord with the gradient transport hypothesis; i.e., the turbulent momentum transport is down the mean momentum gradient. However, in the case of nonsymmetric flows like the wall jet or a turbulent channel with the two sides of unequal roughnesses, the zeros of the mean velocity gradient and the Reynolds stress $-\overline{uv}$ do not coincide; consequently, there is a (small) region of *negative production* where the mean momentum transport by turbulence is counter to the mean momentum gradient $\partial U/\partial y$. Even though this does not violate any basic principle and should not be particularly surprising, this has been the subject of some excitement, controversies, and investigations (Beguier et al., 1977; Hinze, 1970; Hanjalic and Launder, 1972). It is clear that turbulent heat or mass transfer in shear flows with nonsymmetric mean temperature or mean concentration profiles should also have regions of counter-gradient heat or mass transport.

Our studies of coherent structures have provided specific examples of negative turbulence production (Hussain and Zaman, 1980; 1982) and direct explanations for negative production in terms of coherent structures. While there may be other possibilities, these two scenarios are likely to be the most dominant events contributing to negative production. It will be shown that control of coherent structures via excitation can generate negative production in a flow which shows no negative production in the absence of excitation.

The first of these relates to the orientation of the large-scale coherent structure in a shear flow. Consider the large two-dimensional, spanwise rolls in a mixing layer with a velocity profile as in Figure 1a. If the structure were of a circular cross section, it is clear that the coherent Reynolds stress $\langle u_c v_c \rangle$ will consist of four regions of equal strength in a clover-leaf type distribution with alternate signs (Figure 1b). This is a simple consequence of kinematics and is irrespective of whether the flow is vortical or even turbulent. At any transverse location, the time average Reynolds stress $-\overline{uv}$ would be zero. Thus, nonzero production requires that the cross section be noncircular and be inclined with the flow direction. For convenience, consider an elliptic cross section of the structure, which is inclined with an acute angle with x in configuration A (Figure 1c) or an obtuse angle with x in configuration B (Figure 1d). From the closed streamline pattern, it is clear that the configuration B will produce much more positive uv than negative uv. The configuration A will produce just the opposite, i.e., primarily co-gradient Reynolds stress. Thus, an average over B will contribute primarily counter-gradient momentum transport. At a given location, if structures alternate equally between configurations A and B, the net momentum transport will again be zero. However, if the configuration B mostly occurs in a fixed region of space, production will be negative in that region. It is clear that the configuration that will be typically expected to cause net transfer of energy to the coherent motion (i.e., positive production) is opposite to what would be expected based on the shear. On the other hand, the configurations aligned with the shear contribute to negative production.

The second scenario involves planetary motions of two vortices around each other during the pairing process. Let us identify four successive phases of the pairing process by the configurations C, D, E, and F (Figures 2a–2d). Ideally, these four configurations can cyclically repeat themselves indefinitely. In reality, if the Reynolds number is high enough, merger typically takes place abruptly soon after the phase F. (In a low Reynolds number axisymmetric jet, the leap frog motion of two adjacent vortex rings can continue for a few cycles before merger (Reynolds and Bouchard, 1981).) Detailed coherent Reynolds stress measurements show that the phase F produces significant counter-gradient Reynolds stress over the lower-speed half of the layer thickness and thus must cause significant negative production. Note that with respect to the mean shear, the overall motion due to the two pairing vortices together in phase F

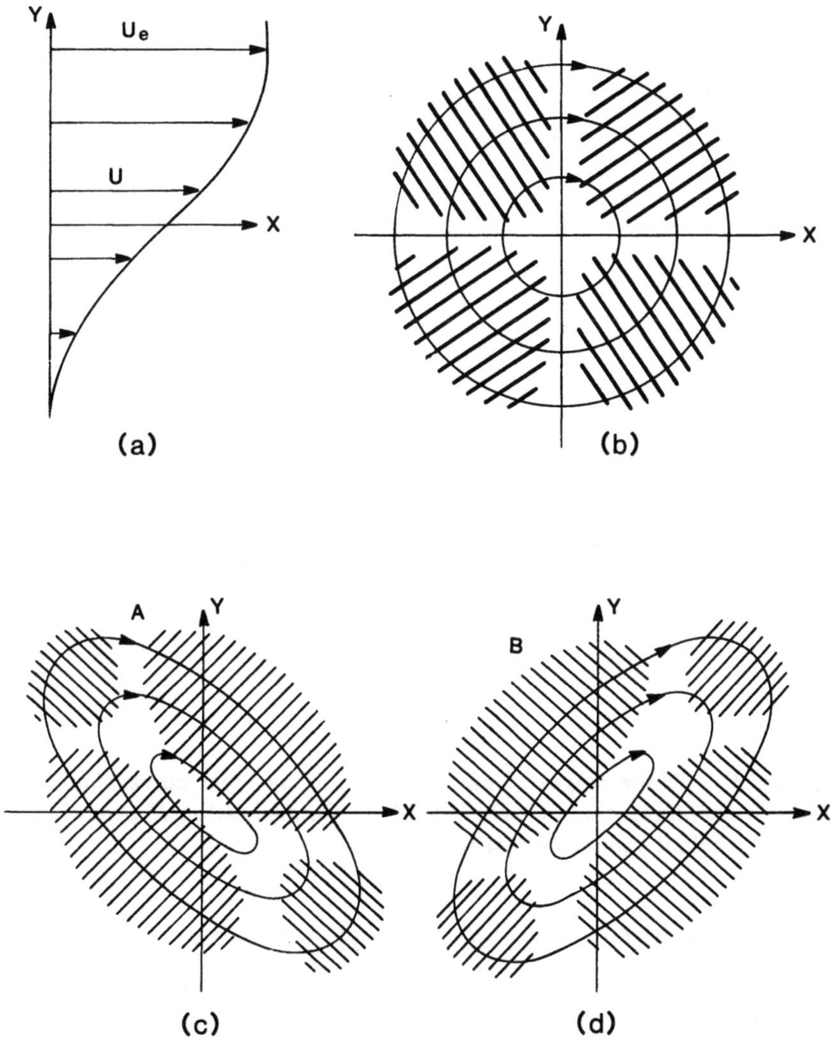

Figure 1. Explanation for negative production in a mixing layer due to a single coherent structure. (a) mean velocity profile; (b) coherent Reynolds stress contours associated with a spanwise roll of circular cross section; (c) and (d) coherent Reynolds stress contours for structures of elliptic cross sections inclined to the flow. ///// denotes $-uv$, \\\\\ denotes uv. Configuration B produces net counter-gradient momentum transport.

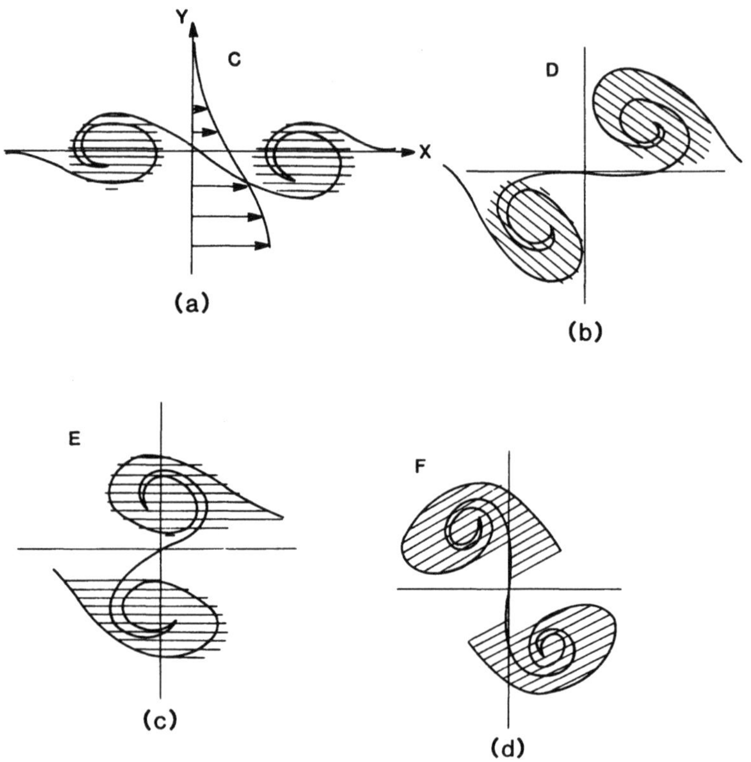

Figure 2. Four stages of pairing in a mixing layer corresponding to successive configurations C, D, E, and F. Configuration F produces net counter-gradient momentum transport.

is not quite different from that of the single structure in phase B.

Since either of these negative-production configurations is expected in general to occur randomly in space and time, their presence will not be strongly felt in a time average. However, if either of these configurations can be made to occur repetitively at a fixed location, then even the time average will show counter-gradient momentum transport by turbulence and hence local *negative production* at that location. Via appropriate excitation, we have been able to induce at fixed locations both stable pairing (i.e., pairing occurring repeatedly at that location) and configuration B of a single structure in plane and axisymmetric mixing layers. We have shown that this spatial localization produces time-mean negative production. In some flow situations, the excitation may be naturally available either via (settling chamber) resonance or via upstream feedback from structures themselves. But, negative production due to spatial localization of phase B of a single structure or phase F of pairing requires much stronger forcing than in general available naturally.

Two points need mention. The negative coherent Reynolds stress is derived from closed structure streamlines drawn in a frame advected downstream with the structure convection velocity U_c. Hence, the Reynolds stress contours in Figure 1 are of $\langle (u_c - U_c)\, v_c \rangle$, while the time-average Reynolds stress measured with a stationary probe in a laboratory is $\overline{(u-U)(v-V)}$. If we denote these two by $\langle r \rangle$ and $\overline{R}$, respectively, it follows that $\overline{\langle r \rangle} = (U - U_c)V + \overline{R} - \overline{u_r v_r}$, since $u = U + \tilde{u}_c + u_r = u_c + u_r$ and $v = V + \tilde{v}_c + v_r = v_c + v_r$. Thus, contours in Figure 1 cannot directly explain the laboratory measurement of $\overline{R}$ without the superposition of contours of UV, $U_c V$ and $\overline{u_r v_r}$ on the contours of $\langle r \rangle$.

The second point refers to turbulence production which is denoted by the product of the Reynolds stress $r_{ij} = -u_{ri} u_{rj}$ and the strain rate $S_{ij} = (\partial u_i/\partial x_j + \partial u_j/\partial x_i)/2$. Note that the subscript r denotes incoherent (random) components and does not imply summation. The coherent production is denoted by $\langle P \rangle = \langle r_{ij} \rangle \langle S_{ij} \rangle$ while the time-average production is denoted by $\overline{P} = \overline{r_{ij}} \cdot \overline{S_{ij}}$. Either kind of production can be divided into *normal* and *shear productions*, i.e., $\langle P \rangle = \langle P_n \rangle + \langle P_s \rangle$. Most laboratory measurements address $\overline{P}_s$ only. It should be clear that only $\langle P \rangle$ is invariant under coordinate rotation but neither $\langle P_n \rangle$ nor $\langle P_s \rangle$ is; similarly, $\overline{P}$ is invariant under coordinate rotation but $\overline{P}_s$ or $\overline{P}_n$ is not. Thus, production can be dominated by normal or shear stresses depending on the coordinates chosen. At a point in the flow, $\overline{P}$ will equal $\overline{P}_n = \overline{r}_{11}\overline{S}_{11} + \overline{r}_{22}\overline{S}_{22}$ when the coordinates are aligned either with the principal axes of the stress tensor $\overline{r}_{ij}$ at that point (so that $\overline{r}_{12} = 0$) or with the principal axes of $\overline{S}_{ij}$ at that point (so that $\overline{S}_{12} = 0$). On the other hand, $\overline{P}$ will equal $\overline{P}_s = 2\overline{r}_{12}\overline{S}_{12}$ when the coordinates are aligned with axes along which the mean normal strain rates are zeros (which is always possible for incompressible flow). Note that the corresponding axes in which $\overline{P}$ becomes equal to $\overline{P}_s$ do not exist for the stress tensor as the (kinematic) normal turbulence stresses $-\langle u_r^2 \rangle$, $-\langle v_r^2 \rangle$ are always nonzeroes along any axes. The same arguments can also be applied to $\langle P \rangle$, $\langle P_n \rangle$, and $\langle P_s \rangle$. It is typical that experimentalists align the x-axis with the mean flow direction and thus negative Reynolds stress and negative production relate only to these axes. Furthermore, *production* is typically used to denote mean shear production. This is somewhat justified as mean normal production in these coordinates is negligible in most shear flows. However, $\langle P_n \rangle$ can be comparable to $\langle P_s \rangle$; that is, coherent normal and shear productions can be comparable. These aspects will be discussed in further detail in Hussain (1983b).

4.2 Coherent Structures and Jet Noise

Since the acoustic analogy model proposed by Lighthill in the early fifties, progress in the understanding of jet noise phenomena has been both scant and sporadic. Coherent structures have rekindled some interest in jet noise phenomena because of the prospect for a more deterministic representation for the Lighthill's quadrupole source term in the aerodynamic noise theory. There is now a general consensus that coherent structures are important in aerodynamic noise generation in a jet. However, neither the noise production mechanism nor the significance of coherent structures in jet noise is known yet. A new noise production mechanism was proposed by me, which involves coherent substructures rather than coherent structures themselves (discussed later).

Foremost among the mechanisms proposed is the effect of instability waves. These are likely to be important only in supersonic jets as subsonic travelling waves are inefficient radiators of noise. On the other hand, interactions of coherent structures can produce significant noise. This prompted Laufer to propose in 1973 that pairing was the principal mechanism for jet noise production. This proposition, although found support in many quarters, was considered by me to be unlikely.

First of all, most practical jets, being turbulent at the exit, seldom involve pairing. The initially turbulent mixing layer rolls up but typically at the *jet column mode*. When a jet is initially laminar, most pairing activity occurs in the *shear layer mode* and is complete within the first one or two diameters while most noise originates from near the end of the potential core. In this early region of pairing, there is significant sound radiation, yet this is less dominant than that produced between 4 to 8 diameters. These facts suggest that pairing—whether induced or occurring naturally—can produce significant sound; but *cannot* be the dominant cause of noise in practical jets.

Our proposition is that it is the *breakdown process* of the initial toroidal structures into *substructures* near the end of the potential core and their *interactions* that produce most noise (Hussain and Zaman, 1981). In this sense, the noise production mechanism involves coherent substructures, rather than coherent structures. It is quite likely that the breakdown process involves the "cut-and-connect mechanism" via which substructures result from the initial toroidal structure. Precursor to the breakdown process is the evolution of the spanwise lobes resulting from the Widnall instability. However, the details of the breakdown process remain elusive. Flow visualization with dye or smoke injected at the nozzle has been ineffective in revealing the precise details of the breakdown process due to rapid smearing of the dye by turbulence. The random, three-dimensional flow in the noise producing region has frustrated efforts to unravel the dynamics via quantitative measurements. Thus, even though this region is the most crucial region of the jet, being characterized by intense entrainment and mixing, and high Reynolds stress and aerodynamic noise production, the phenomenon involved is unknown. It is not obvious that measurement or flow visualization schemes can be devised to reveal adequately the local flow physics. Two approaches appear attractive in addressing this problem. One is to study the structures in the breakdown region via correlation of azimuthal vorticity maps with two radial rakes of X-wires. The second is to trigger a photosensitive chemical reaction so that the product is of a different color; by illuminating at convenient locations and planes, the local flow dynamics can be unveiled from the topology of the product color.

Some aspects of jet noise will be discussed here briefly; these are: broadband noise amplification and suppression, initial condition, and feedback effects.

Broadband noise amplification, independently discovered by Bechert and Pfizenmaier (1975) and Moore (1977), is a matter of curiosity to jet noise researchers. At sufficiently high Reynolds numbers, it is seen that pure-tone excitation can produce an increase in the far-field noise, the increase being virtually uniform over the entire spectral range. Broadband amplification is of great technological significance as it may hold the key to *excess noise* (Ffowcs-Williams et al., 1972; Crighton, 1980), which has frustrated efforts to predict noise of practical jets. This phenomenon has captured a lot of attention because not only is it surprising, it may be the explanation for the excess or core noise, which is the difference (about 10 dB) between scaled up model data and practical jet engine data. Since jet engines abound with many internal noise sources (combustion, separation, vortex-structure interactions, etc.) which are not totally predictable and can vary from one engine to another, and which can produce broadband amplification, comparison of noise data from two practical rigs or of noise data from a practical jet with scaled up model data may be moot as the extent of broadband amplification can vary considerably.

Kibens (1980) has suggested that interaction between shear layer and jet column modes of pairing (Zaman and Hussain, 1977) produces an inherent jitter in shear layer mode of pairing which produces broadband amplification. This kind of explanation is also supported by Ffowcs-Williams and Kempton (1978) and Crighton (1981).

We do not subscribe to this point of view. First of all, broadband noise amplification seems to occur when the exit boundary layer is turbulent. When it is laminar, *broadband suppression* occurs (see below). Even though an initially turbulent jet may have shear layer type roll up, most jets studied in the laboratory for jet noise phenomena do not involve any shear layer mode of pairing. A piece of technological research in our laboratory with the *whistler nozzle* showed that *broadband noise amplification* (by about 10 dB) in the far field is associated with *broadband turbulence amplification* in the noise producing region. These amplifications occurred via self-excitation at Strouhal numbers St_D in the range 0.35–0.6, representing the "preferred modes" of the jet. Since no pairing is involved, it is clear that broadband noise amplification does not necessarily have to involve pairing.

To reiterate, the primary noise production mechanism in a jet proposed by me is the breakdown mechanism, i.e., the evolution of the azimuthal substructures (perhaps via a cut-and-connect mechanism) and their interactions (Hussain, 1983a). Controlled excitation can delay the initiation of the breakdown process (Hussain and Zaman, 1980). It appears that breakdown must occur by the end of the potential core. Therefore, an axisymmetric excitation results in the breakdown process to occur more rapidly, and thus must cause enhanced noise production.

When the initial shear layer is laminar, excitation at the most unstable frequency produces *turbulence suppression*. This was demonstrated by Zaman and Hussain (1981). They also speculated that there might be a concomitant far-field noise reduction, a phenomenon established by Hussain and Hasan (1983). There are alternative interpretations for turbulence and noise suppressions. Kibens (1980) suggested that when pairing is spatially stabilized via *stable shear layer mode of pairing* (Zaman and Hussain, 1977), turbulent kinetic energy is locked up in the subharmonics associated with pairing, thus producing a broadband noise suppression. The explanation of the turbulence suppression mechanism proposed by Zaman and Hussain (1981) is quite different. They showed that the suppression was due to the fact that excitation at the most unstable frequency produced (fastest amplification and) earliest saturation and violent breakdown. When unexcited, these structures persist for longer distances and grow to larger sizes before breakdown. Thus, excitation at $St_{\theta_e} \simeq 0.017$ arrests the instability growth and causes reduction of turbulence and hence noise.

The phenomenon of broadband noise suppression has been experimentally investigated via acoustic excitation through a thin slit all around the lip of a 4 cm low-subsonic axisymmetric jet (Hussain and Hasan, 1983). Figure 3a shows the comparison of the unexcited and excited noise spectra at the emission angle (from the jet axis) of $\phi = 45°$; the data are similar at other ϕ. These data are obtained with a microphone located at 60 diameters from jet exit. The noise suppression due to excitation is broadband. The variation of the OASPL (overall sound pressure level) as a function of the shear layer St_{θ_e} is shown in Figure 3b. When the spectral peak at the excitation frequency is removed, the OASPL resulting from integration of the area under the spectrum then is a measure of the aeroacoustic noise. Figure 3c shows that broadband noise suppression occurs over the St_{θ_e} range 0.01–0.02. Hot-wire measurements within the shear layer of the jet revealed that the excitation produced decreases in the shear layer thickness, turbulence levels and Reynolds stress below the unexcited values, as to be expected on the basis of Zaman and Hussain's (1981) results.

From these data, a number of significant observations can be made, the excitation that produces turbulence suppression can also produce noise suppression. The spectral range of noise suppression agrees with that for turbulence suppression in the noise producing region. What is striking is that excitation can even produce a net suppression of noise even when the excitation sound is not subtracted. We have not found in the literature any evidence of total noise suppression via excitation. The second interesting point is that there are no far-field noise peaks at $f/2$, $f/4$, $f/8$, etc. (Figure 3a). According to Kibens' data, when excitation produces broadband noise suppression, pairing is localized so that energy is locked up in the subharmonics. Our data showing broadband noise suppression without any noticeable peaks at subharmonics once again confirm the stand taken by me that pairing is not central to jet noise.

Hot-wire measurements in the noise-producing region revealed broadband turbulence suppression for the excitation cases which produce braodband far-field noise suppression (Hasan, 1983). Also broadband far-field noise amplification was found to be accompanied by broadband turbulence amplification; this is also the observation in the case of whistler jet (see later).

Thus, we conclude that far-field *broadband noise amplification/suppression* is a direct consequence of near-field *broadband turbulence amplification/suppression.*

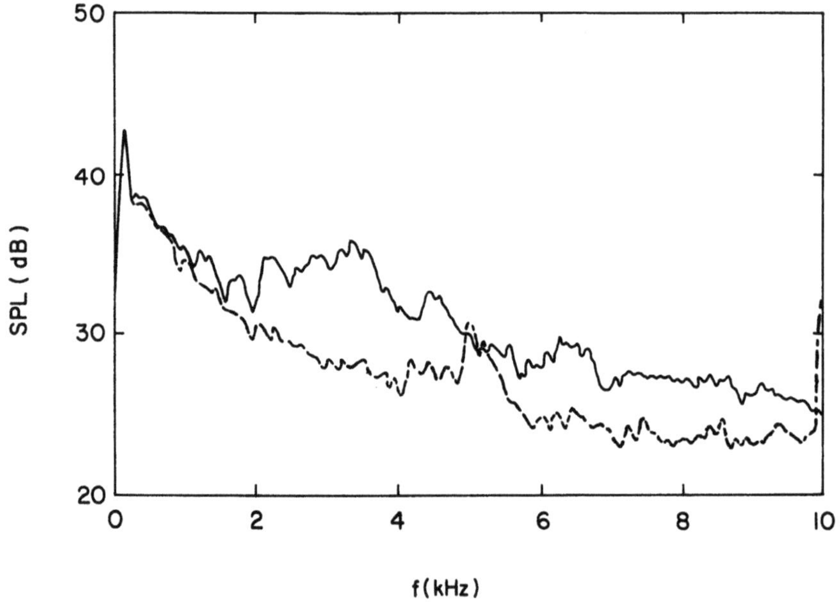

Figure 3a. Far-field sound spectrum at emission angle $\phi = 45°$ at a radial distance $r = 57\,D$ from jet exit of diameter $D = 4$ cm. Solid line for unexcited jet; dotted line for excitation at 10 kHZ corresponding to $St_{\theta e} = 0.0155$.

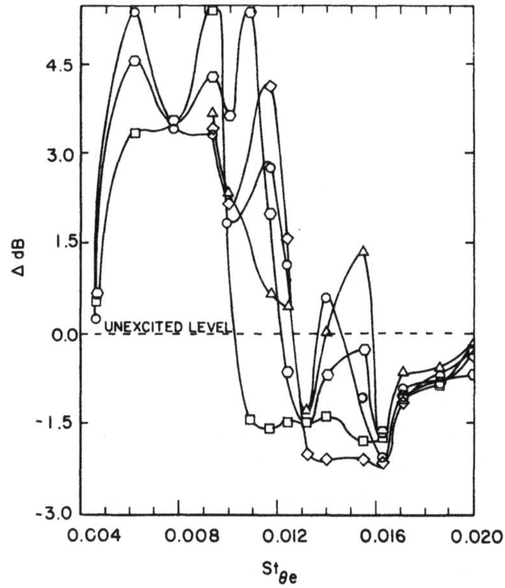

Figure 3b. Relative change in far-field ($r = 57\,D$) OASPL (in dB) with $St_{\theta e}$ at $M = 0.15$, $D = 4$ cm emission angle ϕ values are: □, 45°, ⬡, 60°, Δ, 75°, o, 90°, , 105°.

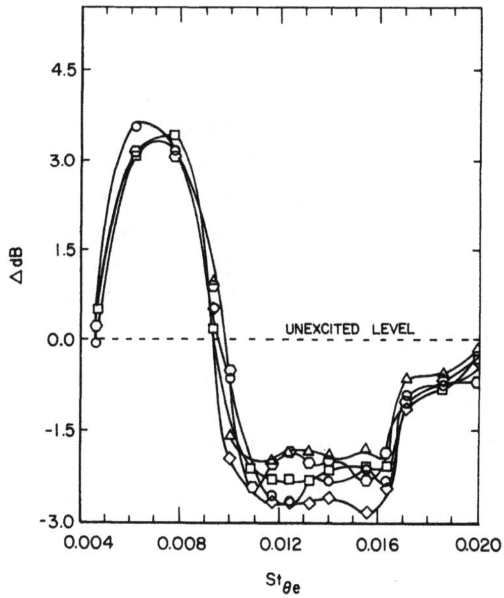

Figure 3c. Broadband far-field ($r = 57\,D$) noise suppression at $M = 0.15$ with the excitation noise subtracted out; $D = 4$ cm.

There is significant dependence of the far-field jet noise OASPL and spectra on both the Reynolds number and the Mach number. Summarizing all excited jet study, Crighton (1972) concluded that broadband noise amplification will occur for $Re_D > 10^5$ and suppression at lower Re_D. Data from a series of experiments in our laboratory are consistent with this claim (Hussain, 1980). Data show collapse of far-field noise spectra for $M \gtrsim 0.25$. It is my opinion that these dependencies on M and Re_D are indirect and the primary controlling factor is the *initial condition*. Most laboratory jets are usually laminar (if carefully designed) for $M \lesssim 0.25$ and $Re_D \lesssim 10^5$, otherwise turbulent. Of course, the use of appropriate *trips* can make the exit boundary layer fully turbulent below these limits. Researchers in the past have paid little attention to the initial condition, and much less to the tripping process, and assumed the presence of a *trip wire* to guarantee turbulent exit boundary layer. In the case of some studies, there is doubt about whether the tripped boundary layer even reattached to the wall. The proper choice of a trip and criteria for identification of the initial condition have been discussed in detail elsewhere (Hussain, 1980; 1981; 1983a).

An associated question is the concept of *Reynolds number similarity* or *asymptotic invariance*. Even though this is perhaps the most sacred concept in turbulence, there has never been a rigorous test of this hypothesis. The underlying concept is that as Reynolds number is increased, the fine scales progressively decrease to allow larger dissipation. On the other hand, the large scales are determined by the geometry and remain essentially unchanged for a given geometry. Since large scales control the energetics, it follows that the flow physics cannot be sensitive to Re_D provided that Re_D is large enough. How large should Re_D be for asymptotic invariance to be valid? Based on our extensive studies, we found (Hussain, 1980) that Re_D dependence is not likely to be significant above $Re_D \simeq 10^5$. It is clear that Reynolds number similarity in jets is more valid when the exit boundary layer is turbulent. When untripped, the exit boundary layer being typically turbulent for $Re_D \gtrsim 10^5$, the similarity should hold above this Re_D. Sufficiently far downstream, similarity should hold anyway as has been experimentally verified. On the other hand, within the axisymmetric shear layer of a large jet, the time average characteristics have been found to be quite different at different Re_D. Clearly, for this region, Re_D is not meaningful, but Re_θ is. Ease of observation has motivated many researchers to explore coherent structures at very low Re_D's. A caution is in order against straightforward extrapolation of the observed phenomena to higher Re_D's.

4.3 Preferred Mode and Pairing in Axisymmetric Jet

The axisymmetric jet evolving from a contraction nozzle has two characteristic length scales: the exit boundary layer thickness (say, momentum thickness θ_e) and the jet diameter D. Associated with these two scales, there are two distinct modes of instability as well as two distinct modes of structure interaction like pairing. These were first investigated by Zaman and Hussain (1977) who called these as "shear-layer mode" and "jet column mode". Even though there were some objections initially to such classification, these terms are now fairly well accepted; see Kibens (1980), for example.

Sufficiently near the lip of a nozzle with a top-hat exit profile (i.e., $\theta_e/D \ll 1$) the axisymmetric mixing layer behavior is not dissimilar from that of the plane layer. This is true typically for at least up to $x \simeq 0.5\ D$. In this region, θ_e is the appropriate length scale; since the instability frequency f_s scales uniquely with θ_e, one can take the initial instability wavelength λ_e as an alternative length scale for the initial region. Farther downstream, the mixing layer momentum thickness θ becomes of the order of D, and the effect of the mixing layer spanwise curvature can no longer be ignored. In this region, D is the appropriate length scale. The shear layer mode is identified by the value of $St_{\theta e} = f\theta_e/U_e$. The jet column mode involves essentially the entire cross section and is identified by $St_D = fD/U_e$.

What are the values of $St_{\theta e}$ and St_D for axisymmetric jets? According to the instability theory (Michalke, 1965), $St_{\theta e}$ should be 0.017, as a shear layer instability receives maximum amplification rate at this value. This has been verified experimentally (Freymuth, 1966; Miksad, 1972). When the shear layer is initially turbulent, the $St_{\theta e}$ value is hard to determine as the roll up takes a long length and the roll up frequency is not very clear from the velocity spectra. The value appears to be smaller than 0.017, probably (about) half. However, farther downstream, the structure passage frequency can be determined with a fair certainty from the high-speed edge signal. At any location x from the origin, the structure passage frequency has been found to vary as x^{-1}, while the shear layer momentum thickness θ is proportional to x. Thus $\mathrm{St}_\theta = f\theta/U_e$ should be a constant. We have found this constant to be 0.024. Thus, at any location in a fully-turbulent mixing layer, the average structure passage frequency is 0.024 U_e/θ.

It is necessary to emphasize here that, contrary to the widely-held belief, the Strouhal number $\mathrm{St}_{\theta e}$ of the natural instability has been found by us (Zaman and Hussain, 1980; Hussain and Zaman, 1978; Husain and Hussain, 1983) to be consistently lower than the most unstable frequency predicted theoretically (Michalke, 1965) and verified experimentally. The explanation is straightforward when it is recognized that the natural instability frequency $\mathrm{St}_{\theta e} \simeq 0.012$ corresponds to the mode with the *maximum amplification*, while $\mathrm{St}_{\theta e} \simeq 0.017$ represents the mode with the *maximum amplification rate*. This subtle difference explains why the *shear layer tone* is maximized at $\mathrm{St}_{\theta e} \simeq 0.012$ while the phenomenon of *turbulence suppression* is maximized at $\mathrm{St}_{\theta e} \simeq 0.017$. This appears to be a direct evidence of feedback of vortex roll up in triggering successive instability waves in a free shear layer. The instability of laminar shear layers in axisymmetric jets has a mild dependence on the jet diameter (Husain and Hussain, 1983), suggesting that the preferred mode (Hussain and Zaman, 1981) of the jet has an additional effect, obviously via feedback. These two, therefore, are indications of feedback playing a role in jet flows (further discussed later).

Crow and Champagne (1971) suggested that the preferred mode should be identified by the frequency that produces the maximum amplification of jet centerline turbulence u'_c. They identified the preferred mode to be $\mathrm{St}_D \simeq 0.30$. On the other hand, Zaman and Hussain (1980) showed that the maximum amplification of centerline turbulence level u'_c occurs for excitation at $\mathrm{St}_D \simeq 0.85$, independent of the initial condition. Crow and Champagne missed this effect because they limited their investigation to lower St_D. Zaman and Hussain defined **preferred mode** *as that frequency at which the fundamental amplitude* u'_f *receives maximum amplification.* Thus, when redefined on the basis of u'_f, $\mathrm{St}_D \simeq 0.30$ is still the preferred mode of the axisymmetric jet. The exact value of the preferred mode is not critical as the jet turbulence spectrum is really broad-band and does not have a sharp peak. Furthermore, the measured frequency not only depends on x but much more so on the transverse position across the thickness of the shear layer (Hussain and Clark, 1981). A large variation (about threefold) in the structure passage frequency across a shear layer has been reported by a number of investigators (for example, Lau, 1979; Hussain and Clark, 1981) but never explained previously. This variation appears to be attributable to tilting of individual structures (Hussain and Zaman, 1981) as well as complex events like tearing and partial and fractional pairings (Hussain and Clark, 1981).

In an initially laminar axisymmetric jet, the rolled up structures pair up as these advect downstream. The jet column mode will naturally evolve from the shear layer mode via successive pairings provided that $f_s/f_j = 2^N$. In laboratory jets N can be 1, 2, or 3 depending on one, two, or three pairings. We have not found N to exceed 3. However, we have found that the jet column mode instability is independent of the shear layer instability as the former occurs independent of the shear layer being initially laminar or turbulent. If the excitation amplitude is large and excitation is at a period which is an integral multiple of shear layer instability period, then an arbitrary number will amalgamate together (Zaman and Hussain, 1977; 1980). This was subsequently studied in detail by Ho and Nosseir (1981) and called "collective interaction".

There are circumstances when excitation produces *stable pairing*, i.e., successive pairings occur at the same location at regular intervals. Such a stable pairing, which produces large turbulence, is very helpful in understanding the dynamics of the pairing process via phase-locked measurements. As discussed earlier, a specific phase of stable pairing contributes to negative production.

Why does St_D reported in the literature vary over a range (Hussain and Zaman, 1981)? The evolution of the structure depends on the initial condition and Reynolds number; the latter effect is actually included in the former. The preferred mode determined by us in various jets appear to be virtually independent of Re_D. The numbers reported in the literature vary to some extent because they are not very precisely recorded or reported, and their measurement is not consistent with the definition. First, the determination of the preferred mode in an unexcited jet is not typically meaningful as all jets are essentially *driven* because of various acoustic modes and disturbances which manifest as *free-stream turbulence* (Hussain, 1980; 1981; 1983a; Hussain and Zaman, 1981). If these disturbances fall within the *receptivity band* (Morkovin, 1983), these will be amplified depending on the amplitude and frequency. Second, even when excited, the determination of the "preferred mode" should be based on the growth of the fundamental amplitude u'_f (i.e., at the frequency of excitation) rather than the total u'_c, as the nonlinearity makes the growths of u'_c and u'_f distinctly different.

Thus, identification of the preferred mode is meaningful only when determined via small-amplitude pure-tone excitation, but at an amplitude above the background disturbance or "free-stream turbulence". It is necessary to emphasize here that because of the residual (mostly acoustic) disturbance modes, which do not vary continuously with speed, many experimentalists have reported steps in the f vs. $U^{3/2}$ relation for the shear layer or jet instability frequency f. I think that these steps are spurious and must be due to *lock-in* with the various acoustic modes of the tunnel or the laboratory environment. This controversy can be conclusively settled by determining via controlled excitation the frequency f_m which produces the maximum growth rate at each U_e. In the resulting f_m vs. $U_e^{3/2}$ plot, such steps disappear. Thus, extreme care is necessary for producing a disturbance-free environment, especially when instability experiments are involved.

We will mention in passing that the initial instability of various axisymmetric jets in our laboratory has been found to be of the axisymmetric mode, while investigations elsewhere have suggested helical mode of initial instability (Drubka and Nagib, 1981). Of course, the coherent structures in the fully-developed region of the jet are found to be predominantly of the helical mode (Tso, 1983), not inconsistent with the instability mode of the bell-shaped profile of the fully developed jet flow (Batchelor and Gill, 1962). The upstream feedback of the helical coherent structures in the self-preserving region can trigger the helical mode of instability, especially in a very *clean* jet facility. We do not feel that laboratory jets could be clean enough for this feedback to dominate initial instability.

4.4 Elliptic Jet

The rather different characteristics of circular and plane jets—both unexcited and excited—suggested to us that the elliptic jet behavior should be intermediate between circular and plane configurations. Because of the two length scales along the two orthogonal planes of symmetry and possible unequal instability growth rates and spreading rates along these two planes, the elliptic jet near field was expected to be different from both circular and plane jets.

Crighton's (1973) study also indicated some interest in elliptic jets. Unless especial precaution is taken, the exit boundary layer thickness of an elliptic jet shows a large azimuthal variation. This variation would suggest, on the basis of shear layer stability theory, that the instability frequency will vary around the lip. However, this would appear unlikely as the roll-up of the shear layer is expected to occur simultaneously all around the lip. All of these motivated us to investigate elliptic jets. Since an elliptic jet of a very high aspect ratio should behave like a plane jet, we limited our investigation to elliptic jets of moderate aspect ratios.

Elliptic jets of aspect ratios 2 : 1 and 4 : 1, both having the same exit area, have been studied via hot-wire measurements and flow visualization. The exit velocity U_e was 30 m/sec, corresponding to the exit Reynolds number $\mathrm{Re}_{De}(\equiv U_e D_e/\nu)$ of 10^5. Here $D_e(\equiv 2\sqrt{ab})$ is the equivalent diameter, and a and b are the major and minor semiaxes of the elliptic cross section at exit; by aspect ratio we mean the ratio a/b. The exit boundary layer momentum thickness θ_e was found to vary by about 50%; however, the instability frequency and roll-up was the same all around the perimeter at the lip, scaling on θ_e along the minor axis. Time average measurements downstream in the jet showed that the cross section, while remaining centered on the jet axis, switched axes. Controlled excitation revealed many peculiarities of the elliptic jet response in contrast with the circular and plane jet responses. We felt that it was essential to separate the effects of the ellipticity of the jet from that of the variation of θ_e along the perimeter, because either of these (i.e., ellipticity of jet or variation of θ_e) could produce significant deviations of the elliptic jet characteristics from circular or plane jet behavior. Careful iterative efforts via recontouring the geometry eventually resulted in elliptic nozzles with constant θ_e all around the exit perimeter. The exit boundary layers were acceptably laminar (i.e., nominally laminar (Hussain, 1980)) with the shape factor agreeing with the Blasius profile value of 2.59 (within 3%). The fluctuation intensity profiles peaked to about 1.5% at $y \simeq \delta^*$ (the displacement thickness), before decreasing asymptotically to the free-stream value of lower than 0.2%. The maximum variation of the exit momentum thickness θ_e at four locations, i.e., along both ends of major and minor axes, was 2% (of 0.124 mm) for the 2 : 1 jet and 3% (of 0.083 mm) for the 4 : 1 jet. The Strouhal number $\mathrm{St}_{\theta e}(\equiv f\theta_e/U_e)$ for the natural instability was around 0.011.

For small-amplitude excitation (at $\mathrm{St}_{\theta e} \simeq 0.011$), the growth rate of the fundamental instability wave along the major axis was higher than along the minor axis for the 2 : 1 jet. For the 4 : 1 jet, the growth rates were about the same along both axes. These data are in contrast with the theoretical predictions (Crighton, 1973). Detailed hot-wire measurements revealed that the jet cross section remained elliptic for at least $x = 100D_e$. However, the axes switch twice within this distance; the locations of the switchings depend on the jet aspect ratio and the initial condition. For further details, see Husain [50].

The jet was excited via longitudinal mode resonance of the jet settling chamber cavity, induced by a loudspeaker attached to the chamber. Sinusoidal surging was held constant at 2% of the exit velocity. Figure 4 shows the streamwise variation along the centerline of the longitudinal fluctuation intensity u'_c for the 4 : 1 elliptic jet at various Strouhal numbers $\mathrm{St}_{De}(\equiv fD_e/U_e)$. Note that for the unexcited jet, the peak intensity occurs at $x/D_e \simeq 5$. With increasing St_{De}, this peak decreases and an earlier peak appears at $x/D_e \simeq 2$ and becomes the strongest at $\mathrm{St}_{De} \approx 0.85$. Thus, if the definition of Crow and Champagne (1971) were followed, $\mathrm{St}_{De} \approx 0.85$ would be called the "preferred mode". If the amplitude of the fundamental $u'_f(x)$ is plotted, it is seen that its maximum occurs at $\mathrm{St}_{De} \approx 0.4$. At higher or lower St_{De} the maximum of u'_f is lower. In fact, the u'_f peak is very low at $\mathrm{St}_{De} \approx 0.85$. Thus, following Zaman and Hussain's (1980) definition, $\mathrm{St}_{De} \approx 0.4$ is the "preferred mode" of the elliptic jet. Spectral data at $\mathrm{St}_{De} = 0.4$ show that the fundamental u'_f saturates near the peak of u'_c due to the development of higher harmonics, but there is no subharmonic—consistent with the fact that the excitation corresponds to the *terminal Strouhal number.* With increasing x, the spectral peaks progressively become submerged in the evolving broadband turbulence. The reason for the fundamental u'_f being the largest at $\mathrm{St}_{De} \simeq 0.4$ and the total intensity u'_c being the largest at $\mathrm{St}_{De} \simeq 0.85$ is that the latter is associated with vortex pairing in the jet column mode. The excitation in this case being at twice the preferred mode value of St_{De}, the jet column mode undergoes only one stage of pairing before the flow settles down to the *terminal Strouhal number*. Because of the appropriate phasing, this produces *stable pairing*, i.e., successive pairings always occur at the same spatial location ($x/D_e \simeq 2$). The resulting high value of $u'_c(x)$ is mostly due to the subharmonic content. If this pairing were to occur at random locations, as it happens in an unexcited jet, the u'_c peak would be weaker. Detailed spectral data show that the subharmonic (i.e., $f/2$ component) indeed dominates the total signal (Figure 5). Flow visualization in an excited elliptic water jet in a large water tank also confirmed the occurrence of stable pairing; any phase of the pairing event could be spatially frozen with strobe lighting.

Similar are the data with 2 : 1 elliptic jet. Remember that Zaman and Hussain (1980) found from a number of jets that $\mathrm{St}_D \simeq 0.3$ and $\mathrm{St}_D \simeq 0.85$ corresponded to the preferred mode and jet column pairing mode of the circular jet, respectively. Thus, it seems that the newly defined equivalent diameter D_e is the appropriate length scale of the elliptic jet. And when scaled with D_e, the elliptic jet behavior is surprisingly similar to the axisymmetric jet. In this sense, the centerline decay characteristics of elliptic jets are also similar to circular jet decay characteristics, even though the spreading characteristics are different, partly because of the switching of axes in the elliptic jet.

Figure 6 shows the streamwise variation of the jet half-width B (which is the transverse distance between the jet centerline and the location of the half mean-velocity point) for the 4 : 1 elliptic jet. Note that when unexcited, the axes switch at $x/D_e \simeq 16$. Until this first switching point, the spread rate is negative along the initial major axis and positive along the initial minor axis. Excitation significantly alters the jet spreading; the axes now switch at $x/D_e \simeq 2$. However, unlike the unexcited jet, the excited jet cross section becomes progressively more elliptic (i.e., aspect ratio becomes higher) with increasing x up to $20D_e$. Thus, at this last station, excitation not only significantly increases the ellipticity but also rotates the elliptic cross section by 90°.

The increase in spreading of the elliptic jet due to excitation causes true enhanced mixing. Flow visualization shows that excitation indeed produced enhanced small-scale mixing. The maximum increase in the jet cross section (by about 200%) occurs at the location of maximum turbulence level. This enhanced mixing is different from that in *bifurcating and blooming jets* where vortices wander away from the axis of the jet, producing the effect of apparent mixing (Lee and Reynolds, 1982). Excitation at the preferred mode produces enhanced mixing, but this enhancement is not as large as that due to stable pairing. The enhanced mixing in an elliptic jet via excitation at the stable pairing mode is more than that achievable in

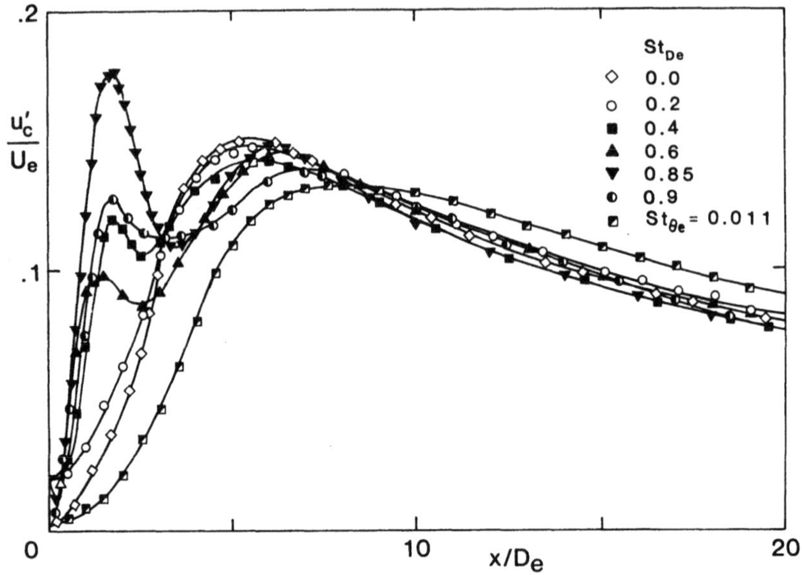

Figure 4. Downstream variation of the centerline longitudinal fluctuation intensity u'_c/U_e for the 4 : 1 elliptic jet for 2% excitation at different St_{De}, compared with the unexcited case ($\mathrm{St}_{De} \simeq 0.0$). Jet exit velocity is 30 m/sec. Note that excitation at $\mathrm{St}_{\theta e} = 0.011$ produces a turbulence reduction for $x \lesssim 8\,D$.

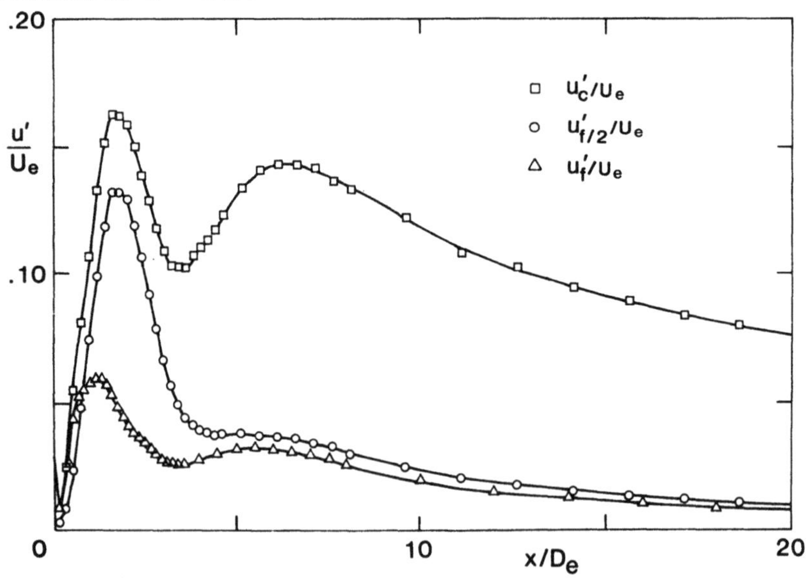

Figure 5. Evolutions along the centerline of the rms total (u'_c), fundamental (u'_f) and subharmonic ($u'_{f/2}$) for excitation at $\mathrm{St}_{De} = 0.85$ of the 4 : 1 elliptic jet.

an axisymmetric jet via excitation in the preferred mode or stable pairing mode, or that achievable in a whistler nozzle. An excited elliptic jet issuing from an orifice plate-type nozzle (investigated by Dr. O. Islam in our laboratory) produced a far more dramatic increase in mixing. It would appear that an elliptic jet with a whistler nozzle will produce a very profound effect on jet mixing and noise production.

These results suggest exciting possibilities for the excited elliptic jets in many technological applications.

A number of interesting coherent structure interactions have been observed in the excited elliptic jet.

When excited at the natural roll-up frequency (i.e., corresponding to $\mathrm{St}_{\theta e} \simeq 0.011$), the vortices in the elliptic jet undergo a peculiar mode of pairing: an elliptical vortex tilts about the major axis and its two sides pair up with two different vortices on either side in the fashion shown in Figure 7 (which is a cross-sectional view passing through the minor axis). That is, a half of each vortex pairs with the next. Note that excitation at this $\mathrm{St}_{\theta e}$ is also associated with suppression (Figure 4).

There is also another mode when, after one stage of pairing, alternate vortices evolve differently: two adjacent vortices undergo another pairing and an explosive breakdown, while the next vortex escapes pairing and persists for a long distance. Thus, after first-stage pairing, every third vortex escapes further pairing and retains identity for a long distance.

The difference between the pairing activity along the two axes of the elliptic jet suggested to us that the radiated far-field noise will also be different along these two planes. Detailed data (Bridges, 1983) show that the difference is indeed significant when the exit Mach number M is below about 0.2 but is negligible at higher M (Figure 8a). This appears to be a simple consequence of the initial condition, which is laminar at lower M. When initially laminar, pairing of the rolled up vortices produce significant noise. When initially turbulent, there is no shear-layer mode pairing, at least at the jet sizes studied. Thus, most noise is then probably produced via the breakdown process, which is likely to be the same at different speeds. Figure 8b shows that the difference between noise radiation characteristics along the major and minor axes of an elliptic jet disappears when the exit boundary layer is turbulent (achieved by tripping), even at low Mach numbers.

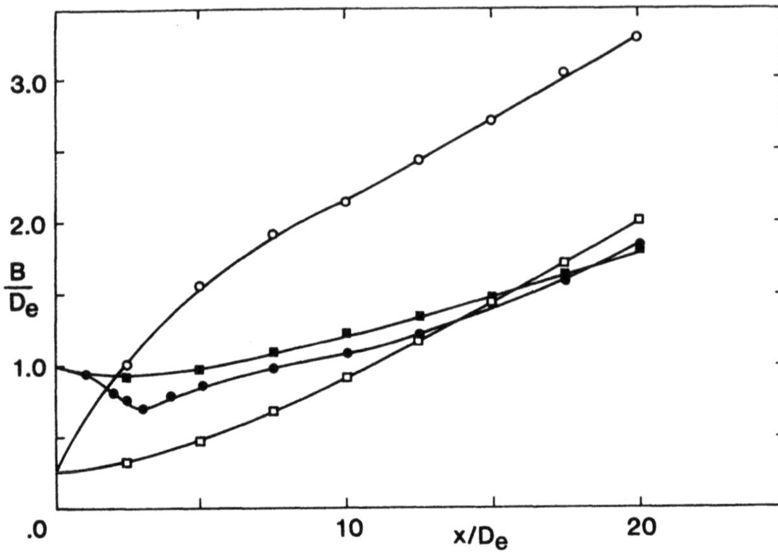

Figure 6. Variation of the jet width B along major and minor axes of a 4 : 1 elliptic jet for forcing ($u'_e/U_e = 0.15$) at the preferred mode ($\mathrm{St}_{D_e} \simeq 0.4$); □, along minor axis, unexcited; ■, along major axis unexcited; ○, along minor axis, excited; •, along minor axis, excited. The unexcited jet switches axes of elliptic cross section at $x \simeq 16D_e$; excitation shifts the switching location to $x \simeq 2D_e$.

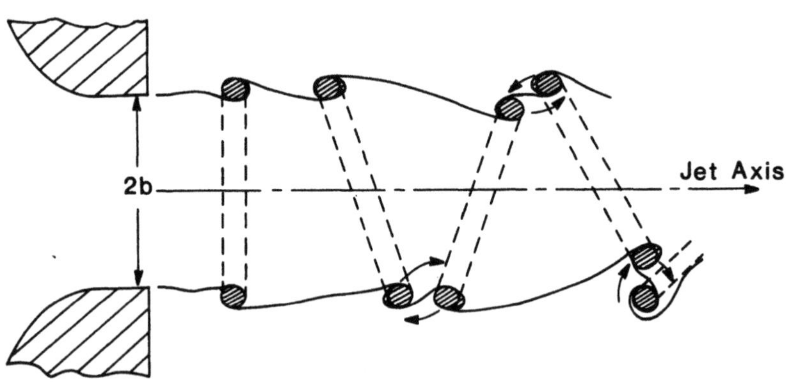

Figure 7. Shared pairing of elliptic vortex rings along the minor axis, when mildly excited at $\mathrm{St}_{\theta_e} \simeq 0.011$. This is a schematic in the plane containing the minor axis and the centerline.

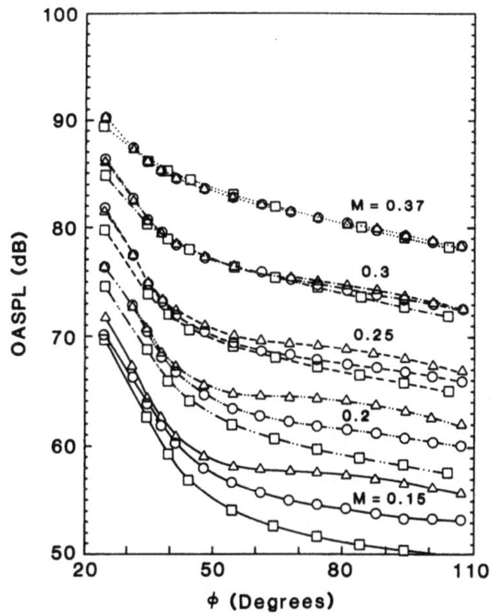

Figure 8a. OASPL vs. emission angle ϕ in a 2 : 1 elliptic jet at various Mach numbers M. ○, data along major axis; Δ, data along minor axis; □, data for a circular jet of $D = 4$ cm. Data were taken at a radial distance $r = 30D$ from jet exit.

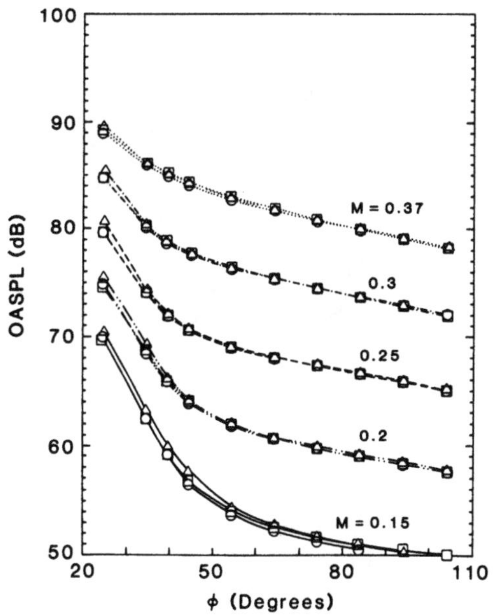

Figure 8b. OASPL vs. ϕ data of Figure 8a but with the exit boundary layer tripped.

4.5 Whistler Jet

Shear-Layer Tone Phenomenon:

Measurements in an initially laminar free shear layer showed that the instability frequency depended on the downstream distance of the hot-wire probe. It was subsequently established that the probe itself triggered self-sustained *resonance* of the shear layer. The phenomenon, controlled by the intrinsic *instability* characteristics of the shear layer, was termed "shear-layer tone" (Hussain and Zaman, 1978). Even though the frequency jumps are quite similar to the *jet tone* phenomenon (Karamcheti et al., 1969), the shear-layer tone is inherently different as the latter involves vortices of one sign only. The phenomenon was characterized by taking data with a wedge placed in a shear layer aligned in the spanwise direction. The frequency variation with the characteristic velocity U_e of the shear layer or with the lip-wedge distance h can be collapsed for each *stage* of the phenomenon if the shear layer initial momentum thickness θ_e is used as the *length scale*, thus confirming the role of the shear layer instability in the phenomenon. Figure 9 shows the collapse of data

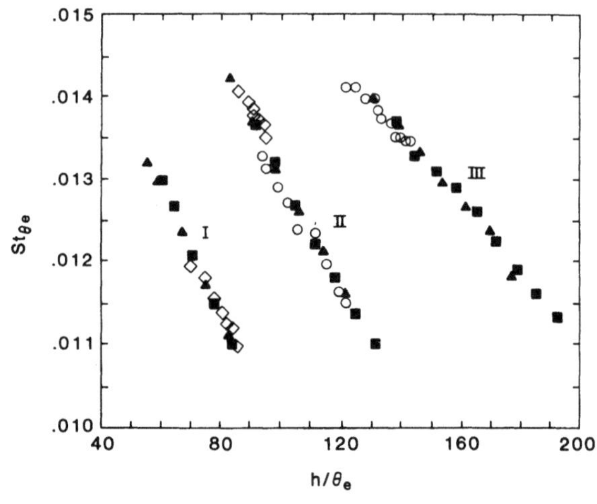

Figure 9. Shear-layer tone Strouhal number $St_{\theta e}$ as a function of h/θ for stages I, II, III Data are obtained by placing a wedge spanwise in a plane shear layer at a distance h downstream from the lip. Data correspond for four different experiments: ▲, variable h at $U_e = 43.3$ m/sec; ■, variable h at $U_e = 29.3$ m/sec; ◇, variable U_e at $h = 1.22$ cm; o, variable U_e with $h = 1.83$ cm.

for four different flow conditions. Note that apart from the data collapse, the average $St_{\theta e}$ range is essentially the same in all stages (I, II, and III). The shear layer tone is strongest at $St_{\theta e} \approx 0.012$, which is considerably lower than the most unstable frequency, i.e., $St_{\theta e} \simeq 0.017$; this difference, explained earlier in this paper, has been observed in independent measurements in our laboratory.

The whistler nozzle is a passive device consisting of a constant diameter tailpipe attached to the downstream end of a circular nozzle and a circular pipe collar that can slide over the pipe. For a given pipe diameter D, exit velocity U_e, and step height h, as the collar is pulled out, the jet produces a loud pure-tone sound. With increasing collar length L_c, the sound first disappears and then appears again. This self-excitation of a *whistler jet* may be called *whistler tone*. When the whistler tone is in action, there is a large increase in the jet turbulence and an associated enhanced decay of the centerline mean velocity (Figures 10a, 10b). The enhanced mixing by whistler excitation suggests many possible technological applications for the whistler nozzle. An axial traverse of a hot-wire inside the pipe shows that the whistler operation is associated with an organ-pipe resonance of the pipe nozzle.

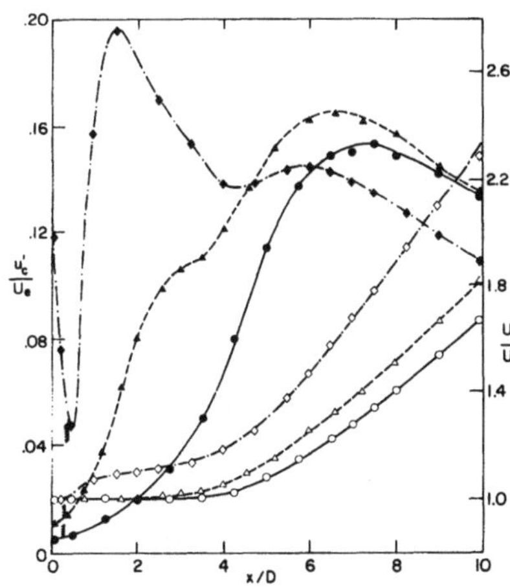

Figure 10a. Variation along the centerline of mean velocity U_c (open symbols) and rms longitudinal fluctuation u'_c (solid symbols) due to whistler tone excitation at different excitation amplitudes. $D = 2.54$ cm; $L_p = 30.48$ cm; $U_e = 36$ m/sec. Values of collar length, whistler tone frequency, and exit excitation amplitude u'_e/U_e are: o, (0.812 cm, 0, 0.7%); Δ, (0.83 cm, 516 Hz, 1.1%); $\Diamond$, (0.00 cm, 504 Hz, 12%).

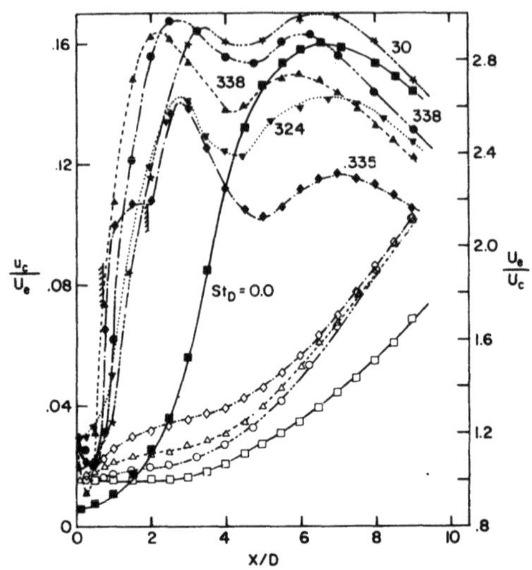

Figure 10b. Variation along the centerline of mean velocity U_c (open symbols) and rms longitudinal fluctuation intensity u_c' (solid symbols) due to whistler tone excitation at the fixed amplitude $u_e'/U_e = 3\%$ for the same D, L, and U_e as in Figure 10a. The stage of excitation and step heights (i.e., difference between radii of pipe and collar) are: ■, unexcited, 0.3175 cm; Δ, I, 0.3175 cm; ο, I, 0.635 cm; ◇, II, 0.3175 cm. data denoted by ▼ were taken by loudspeaker excitation of the pipe without the collar; * denotes excited contraction nozzle data (without pipe) of Zaman and Hussain (1980). Note that the St_D values (indicated in figure) of whistler excitation correspond to the preferred mode.

We have also provided an explanation for the whistler nozzle phenomenon. The whistler excitation results from the *coupling of two independent resonance mechanisms:* shear-layer tone resulting from the impingement of the pipe-exit shear layer on the collar lip and *organ-pipe resonance* of the pipe nozzle. If the shear-layer tone is crucial to driving the organ-pipe resonance, then an obstruction at the location of the collar lip in the absence of the collar should also produce the phenomenon. Indeed, the event was reproduced in pipe-ring and pipe-hole configurations (Hasan, 1983). Unlike the shear layer tone and jet tone phenomena where successive stages overlap, whistler stages are separated by *dead zones* where conditions for both resonance mechanisms cannot be met. Also, unlike the shear-layer and jet tones, the whistler frequency cannot be varied continuously by changing the speed. Because the phenomenon involves coupling of two independent mechanisms, frequency variations defy a simple nondimensional representation for the entire range of its operation. Reasonable collapse of the data is achieved, however, when the pipe-exit momentum thickness is used as the length scale, thus emphasizing the role of the shear-layer tone in whistler excitation (Figure 11). Note the agreement with the data obtained in a free shear layer. It appears from our data that the phenomenon is the strongest when the self-excitation frequency matches the *preferred mode* of the jet (Figure 10a), thus suggesting upstream feedback from the dominant large-scale coherent structures in the jet.

Measurements of the far-field noise of a whistler jet show that the effect of the (nonlinear) self-excitation is rather dramatic both on the OASPL and on the spectra (Hasan, 1983). Figure 12 shows the directivity pattern measured at a distance of 60 D for three different cases ($D = 3.81$ cm); the variation of OASPL with ϕ is highly repeatable. The nature of the broadband sound amplification is shown in Figure 13; the Mach number dependence of broadband amplification is not significant. Note that even though the directivity patterns are quite irregular, the broadband amplification is virtually uniform across the spectral range and monotonically increases (slightly) with the emission angle ϕ, reaching about 10 dB at $\phi = 90°$.

Once again, there is no vortex pairing associated with the whistler excitation with a turbulent exit boundary layer. The far-field acoustic spectrum shows no subharmonic (verified through careful examinations of the local spectral range). The variation of the centerline velocity spectrum is shown in Figure 14; again, there is no subharmonic peak. It is interesting to note that even at this strong forcing, turbulence is spectrally the same as in the unexcited jet beyond six diameters from the exit.

4.6 Natural Structures in Axisymmetric and Plane Mixing Layers

4.6.1 Axisymmetric Mixing Layer

Earlier studies of the mixing layer were carried out by us with the help of a controlled excitation, which provided a simple phase reference for structure eduction. While this provided the convenience of phase-locked measurements with a single (X-wire) sensor with the measurements repeated at different spatial locations at the selected phase, a number of questions remained unanswered. For example, what is the relevance of the forced structure to the natural structure?

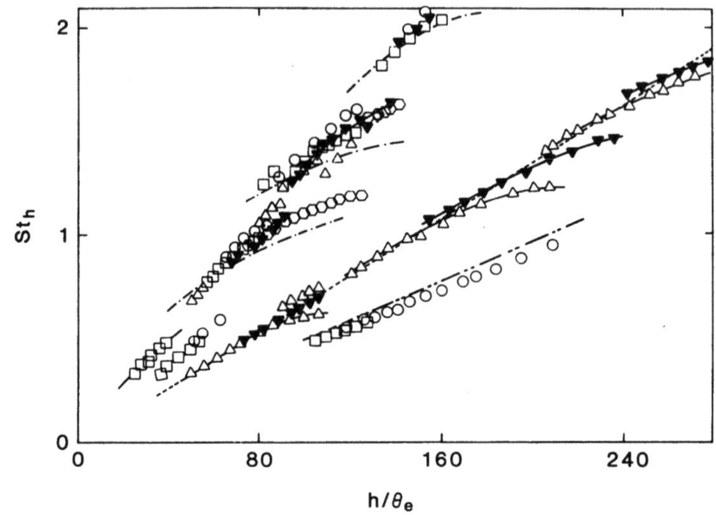

Figure 11. Variation of the Strouhal number $St_h(\equiv fh/U_e)$ with h/θ_e for pipe-hole (PH) and pipe-ring (PR) configurations. The exit velocity and corresponding configurations are: ⬡, 10 m/sec, PR; △, 24 m/sec, PR; □, 36 m/sec, PR; ▽, 24 m/sec, PH; ○, 36 m/sec, PH; — · — · — shear-layer tone data of Hussain and Zaman (1978); — ·· — ·· — whistler nozzle excitation.

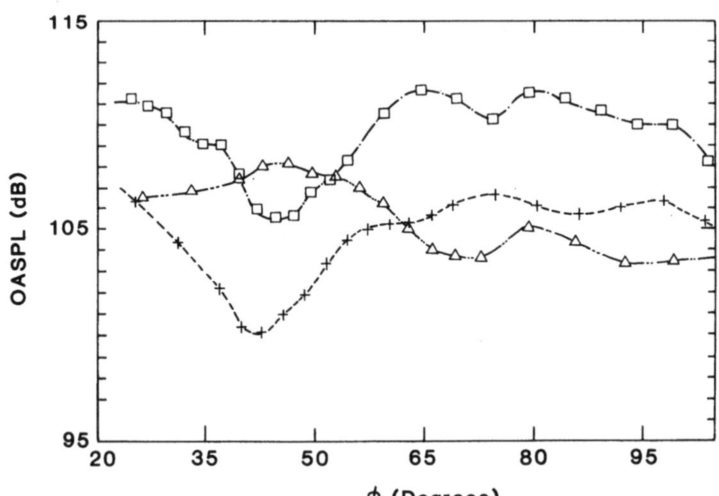

Figure 12. OASPL vs. emission angle ϕ for whistler excited jet for $D = 3.81$ cm pipe diameter. The whistler nozzle pipe length, Mach number and St_D values are: □, 5 cm, 0.37, 0.5; +, 10 cm, 0.37, 0.35; △, 30 cm, 0.37, 0.4.

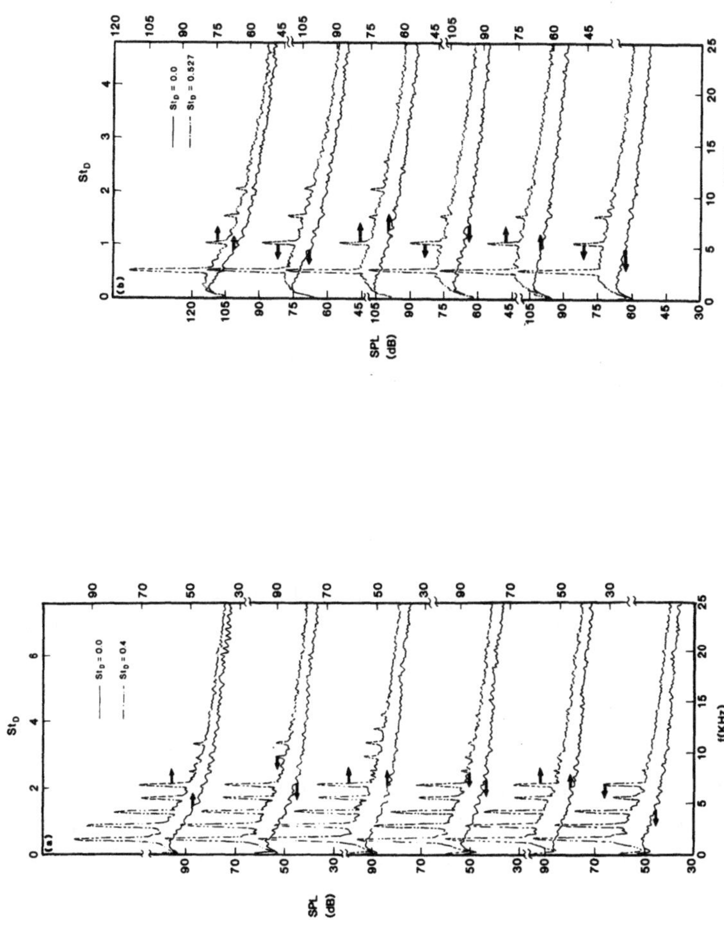

Figure 13. Whistler tone excited and unexcited (without collar) far-field sound spectra for $D = 3.81$ cm. The pipe length and Mach numbers are: (a) $L_p = 30.5$ cm, $M = 0.37$; (b) $L_p = 5$ cm, $M = 0.57$.

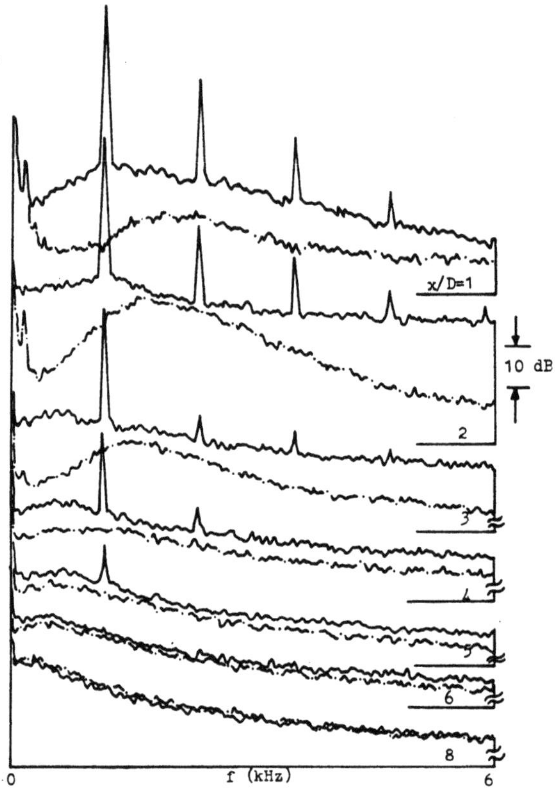

Figure 14. Evolution of the u-spectrum on the whistler jet centerline for $L_p = 30$ cm, $M = 0.37$; — · — · — , unexcited (without collar); ——— , $St_D = 0.4$.

Even though we had claimed that the structure induced by small-amplitude excitation must be the same as natural structures, this could not be conclusively established without educing a natural structure. We were interested in finding a simple eduction scheme based on triggering at a single detection probe. Eduction based on a rake of X-wires is obviously a more accurate approach and is being pursued currently in our laboratory.

A detailed effort to educe the structure proved successful (Zaman and Hussain, 1982). What constitutes the success of an eduction scheme? If a well-defined structure results then the eduction is successful. What is the most successful eduction scheme? It is the scheme which produces the highest peak coherent vorticity in the educed structure. It was observed that eduction was successful when triggered on the peaks of a reference u_{r1}-signal (longitudinal fluctuating velocity) obtained from the high-speed edge of the mixing layer. (Subscripts 1 and 2 denote the high-speed and low-speed sides, respectively, of a mixing layer.) These peaks clearly occur at the instants of passage of large-scale organized structures. The eduction becomes refined with increasing threshold level for detection of the u_{r1}-peaks. However, the experimental time necessary to achieve stable ensemble average increases correspondingly. It was found that a threshold level of $u_{r1} \geq 2\sigma$ (where σ is the rms value of u_{r1}) was adequate for the eduction. Since the structure passage also involved clear negative u_{r1}-peaks at the high-speed edge, attempts were made to trigger on negative u_{r1}-peaks. This was found to be a poorer choice because negative peaks of the high-speed edge u_{r1}-signal occur away from the structure center. Since smearing increases with increasing time separation from the trigger, eduction based on the negative u_{r1}-peaks should produce a large jitter at the structure center. The eduction is sensitive to the transverse location of the detection probe, the optimum location being the high-speed edge (i.e., $U/U_e \simeq 0.99$ point). The use of the transverse velocity v_{r1}-signal does not produce an improvement in the eduction, nor does a joint criterion based on both u_{r1} and v_{r1} signals.

Positive or negative peaks in $u_{r2}(t)$ or $v_{r2}(t)$ signals on the zero-speed edge of the mixing layer are not helpful in triggering eduction. There is also very little correlation between the positive spikes in the high-speed and zero-speed side signals. Since the isolated spikes in the zero-speed side signals must be *footprints* of some highly energetic events, it seems that the structures do not always span the entire thickness of the axisymmetric mixing layer—contrary to the suggestions of Bruun (1977), Yule (1978), and Lau (1979). The explanation appears to lie with tearing and partial and fractional pairings of the coherent structures observed via flow visualization (Hussain and Clark, 1981). Large radial variation of the structure passage frequency as well as of the *convection velocity* across the layer, which has been well documented in many studies, is consistent with this suggestion. Most probably because of this reason, eduction based on simultaneous occurrence of peaks in the signals obtained from the two sides of the mixing layer—a technique successfully employed by Browand and Weidman (1976) in a low Reynolds number two-stream plane mixing layer—failed in the axisymmetric mixing layer.

Figure 15 shows the radial variations of the educed spanwise vorticity at $x = 3D$ as a function of time with respect to the trigger which is denoted by the plus sign. Vorticity values in Figures 15–17 have been nondimensionalized by the preferred mode frequency. The abscissa increases with decreasing time, hence increasing x. Since this is the non-pairing *preferred mode* structure, the implied application of the Taylor hypothesis is well justified (Zaman and Hussain, 1981). Figure 16 shows the same data as in Figure 15 and obtained with the same detection scheme, but with mild forcing (at 0.1% excitation) at the preferred mode. Note that the peak coherent vorticity is nearly the same. Excitation reduces the jitter so that the adjacent educed structure has a higher peak vorticity than in the unexcited case (Figure 15). Thus, for small-amplitude excitation, the induced structure is the same as the natural structure.

What is the sensitivity of the structure to excitation amplitude? Figures 17a and 17b show the corresponding contours of azimuthal vorticity for 0.5% and 1.5% excitations, respectively; both excitations are at the preferred mode. It is clear that higher excitation amplitudes produce stronger (denoted by the peak vorticity) structures, perhaps because the vorticity perturbation at the lip is larger at larger excitation amplitude. Note that even though the peak vorticity increases with higher excitation, the total *circulation* per wavelength remains unchanged.

The same eduction criterion ($u_{r1} \geq 2\sigma$) was applied to educe natural structures in axisymmetric mixing layers with different Reynolds numbers and initial conditions. No significant Re_D-dependence of the natural large-scale structures was observed over the range $5 \times 10^4 \leq \mathrm{Re}_D \leq 8 \times 10^5$. A weak dependence on the initial condition was observed. The structure was found to be more smeared when the exit boundary layer was laminar compared to the case when it was turbulent. This is consistent with observations based on flow visualization (Hussain and Clark, 1981).

In a separate study with 2% excitation (Hussain and Zaman, 1981), it was found that structure properties were virtually independent of the initial condition but very mildly dependent on the Re_D. With increasing Re_D, the structures were slightly stronger and were associated with larger coherent Reynolds stress, even though the incoherent Reynolds stress remained the same. These small differences between the effects of Re_D and initial condition depending on whether structures are excited or natural are not necessarily in contradiction, but a clear explanation is not yet available.

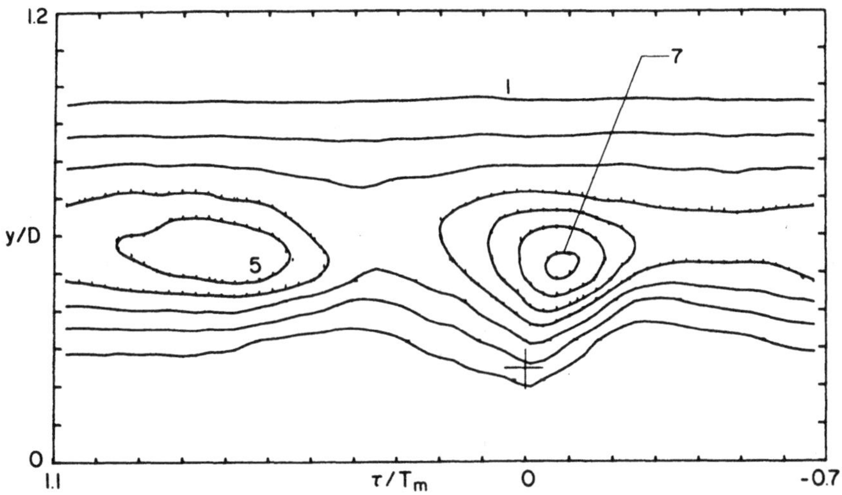

Figure 15. Contours of azimulhal vorticity, nondimensionalized by the preferred mode frequency of natural structures. Data were taken at $x/D = 3$ in a 7.62 cm air jet at $\mathrm{Re}_D = 110,000$.

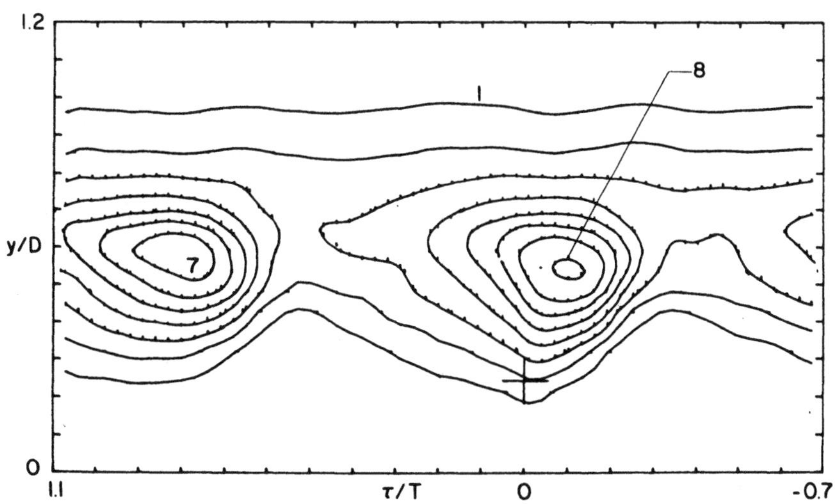

Figure 16. Same data as in Figure 15 but with 0.1% excitation at the preferred mode.

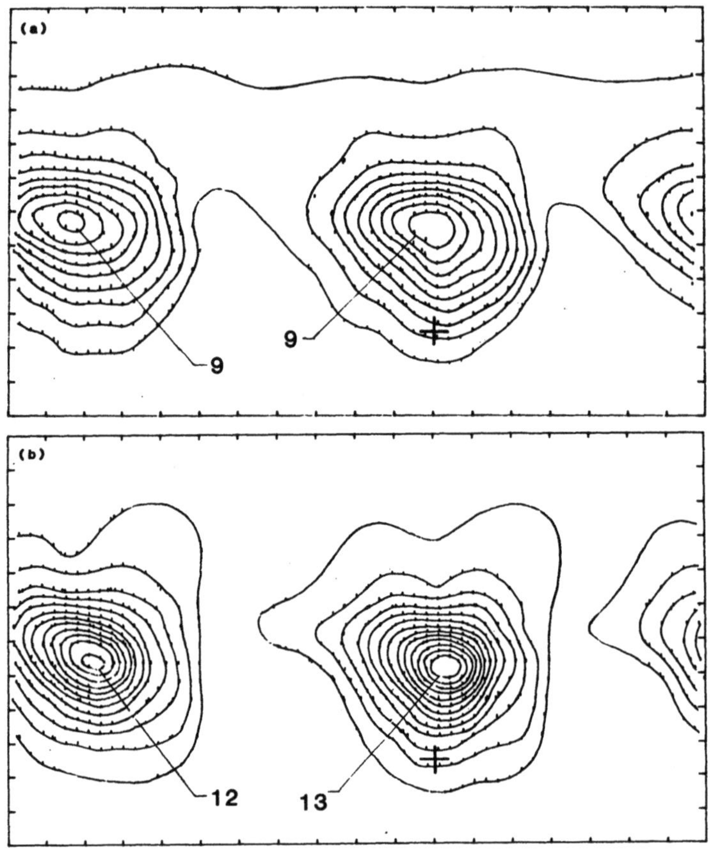

Figure 17. Same data as in Figure 15 but with: (a) 0.5% excitation and (b) 1.5% excitation.

4.6.2 The Plane Mixing Layer

The time-average measures of the axisymmetric mixing layer have been found to be strong functions of the initial condition and the initial fluctuation characteristics (Hussain and Zedan, 1978; Husain and Hussain, 1979), perhaps because of the limited streamwise extent of the mixing layer (as well as a feedback effect). Data in the plane mixing layer also suggested a strong dependence on the initial condition (Liepmann and Laufer, 1947; Wygnanski and Fiedler, 1970; Batt, 1975). Since asymptotic and local invariances suggest that a mixing layer should be independent of the initial condition sufficiently far downstream, I felt that many mixing layer facilities were not sufficiently long and the coherent structures were perhaps organized via feedback due to impingement of the structures on steps or walls downstream. With this in mind, a large plane mixing layer facility was built (Kleis and Hussain, 1979). The layer was studied for streamwise distances of $19,350\,\theta_e$ and $8700\,\theta_e$ for laminar and fully-turbulent exit boundary layers, respectively. Contrary to published data in the literature, the mixing layer was found to achieve a *universal asymptotic state* independent of the initial condition. In the asymptotic state, the peak values of u'/U_e, v'/U_e, w'/U_e and $-\overline{uv}/U_e^2$ were 0.18, 0.131, 0.145, and 0.011, respectively. The *virtual orgin* was upstream of the separation point and the asymptotic value of $d\delta_\omega/dx$ was 0.133 for both layers. Here, δ_ω is the *vorticity layer thickness.*

It was felt interesting to investigate the evolution of the natural coherent structure in the self-preserving region of the mixing layer. In order to eliminate any possible effect of the initial instability, it was decided to trip the boundary layer. Furthermore, mean data suggested achievement of self-preservation much earlier (i.e., at $x \simeq 500\,\theta_e$) in an initially turbulent case than in an initially laminar case. An appropriate three-dimensional trip was placed 10 cm upstream of the separation lip so that the boundary layer at the lip was *fully-developed turbulent* by all criteria (Hussain, 1980; 1981; 1983a).

The initially fully-turbulent shear layer undergoes its own instability and roll up into discrete coherent structures which then successively amalgamate as they evolve downstream. The structures have been detected for up to $x = 3\,m$ or $5000\,\theta_e$; the structure passage frequency f_m decreases with increasing x, being inversely proportional to x. Thus, the local Strouhal number St_θ ($\equiv f_m\,\theta/U_e$) is a constant (because θ increases linearly with x), as shown in Figure 18 which also shows the variation of θ with x.

The linearity of $\theta(x)$ suggests achievement of self-preservation starting from $x \simeq 500\,\theta_e$. The slope of this line, i.e., $d\theta/dx$, denotes the zero-speed side entrainment V_E/U_e; its value of 0.032 thus provides an accurate measurement of the entrainment velocity. Note that for two different flow conditions, St_θ has a value of about 0.024, which gives the universal structure passage frequency in the self-preserving region of any mixing layer.

The coherent structures in this plane mixing layer have been educed at different stages of their development via a conditional sampling triggered on the u_{r1}-peaks of a high-speed side reference signal. The technique explained in the preceding subsection was found to be the optimum for this flow also. In contrast with the axisymmetric case, signals from the high-speed and zero-speed sides of the plane mixing layer are better correlated. Since the zero-speed side *flow reversal,* associated with the passage of large-scale structures, must be similar in both configurations, poorer correlation in the axisymmetric case can not be due to the effect of flow reversal on the hot-wire signals. Even though the zero-speed side signature is relatively stronger for the plane mixing layer, eduction triggered on a signal from this side is less successful than when based on the high-speed side signal.

Figures 19a–19d show the contours of coherent spanwise vorticity at $x/\theta_e = 150$, 300, 600, and 4800, respectively; vorticity has been nondimensionalized by the average structure passage frequency f_m. Note that the initially turbulent layer has begun to roll up at $x = 150\,\theta_e$, the roll-up being completed somewhere between $x = 300\,\theta_e$ and $600\,\theta_e$. The contour details at $x = 600\,\theta_e$, $2000\,\theta_e$ (not shown), and $4800\,\theta_e$, are identical within the experimental uncertainty. Contours of other coherent structure properties also indicate similar agreement. Thus, the mixing layer coherent structure appears to reach an asymptotic state at $x \approx 500\,\theta_e$, consistent with the achievement of self-preservation based on the mean profiles. Figures 20a–20d show the contours at $x/\theta_e = 4800$ of the coherent longitudinal velocity $\langle u_c \rangle / U_e$, the coherent transverse velocity $\langle v_c \rangle / U_e$, the incoherent turbulence intensity $\langle u_r^2 \rangle^{1/2} / U_e$, and coherent shear production $\langle P \rangle$. Initially, the phase-average incoherent Reynolds stress near the exit is much larger than the coherent Reynolds stress. With the progression of the roll-up process both increase, the latter increasing faster. These reach constant but comparable values at $x \gtrsim 500\,\theta_e$. Note that production is minimum at the structure center (Hussain, 1980; 1981; 1983a). The peaks of coherent vorticity and coherent production are about 50% higher than the corresponding time-average peak values. Because the detection scheme is based on the high-speed edge signal only, a certain amount of smearing is bound to occur. However, it is quite unlikely that the peak of coherent vorticity will be an order of magnitude higher than the peak time-average vorticity. These data suggest that in the self-preserving state of the mixing layer, incoherent turbulence is comparably important as the coherent structures.

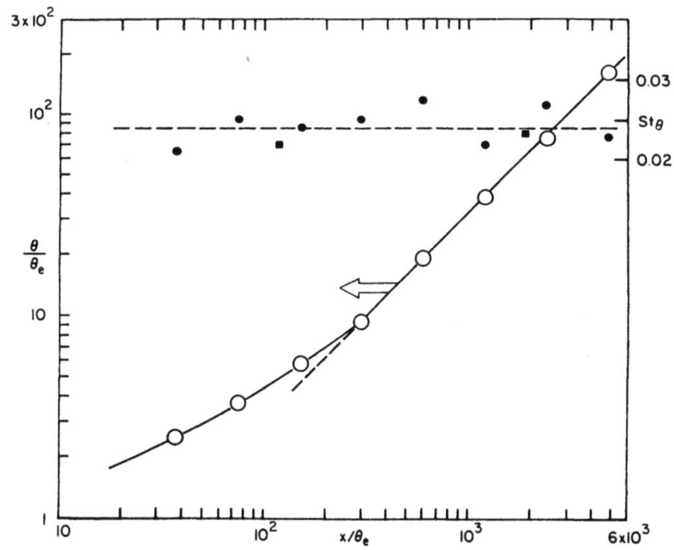

Figure 18. Streamwise variation of momentum thickness θ (open circles) and Strouhal number $St_\theta = f_m\theta/U_e$ (solid circles) for an initially fully-turbulent plane mixing layer at $U_e = 12$ m/sec. Solid squares represent data for $U_e = 9$ m/sec.

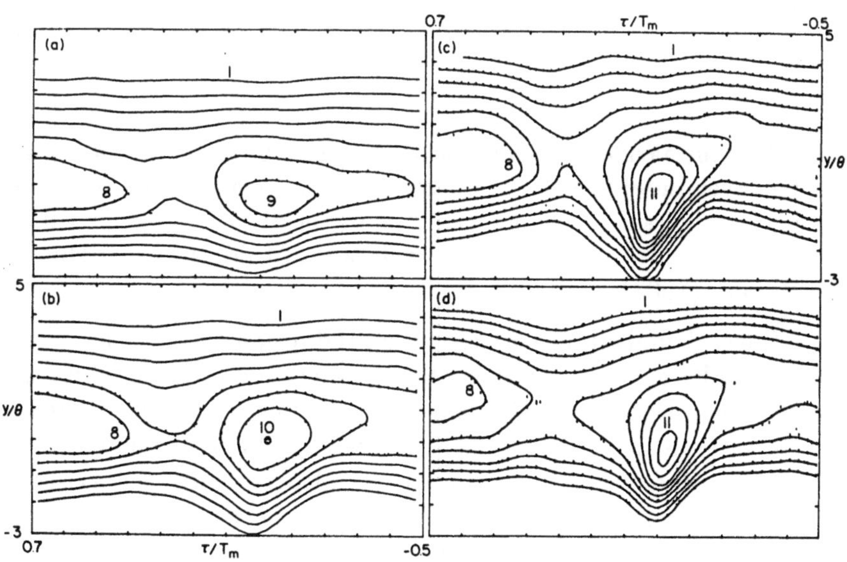

Figure 19. Contours of spanwise coherent vorticity (nondimensionalized by average structure passage frequency f_m) in the initially fully-turbulent plane mixing layer at $U_e = 12$ m/sec. (a) $x = 150\,\theta_e$; (b) $x = 300\,\theta_e$; (c) $x = 600\,\theta_e$; (d) $x = 4800\,\theta_e$.

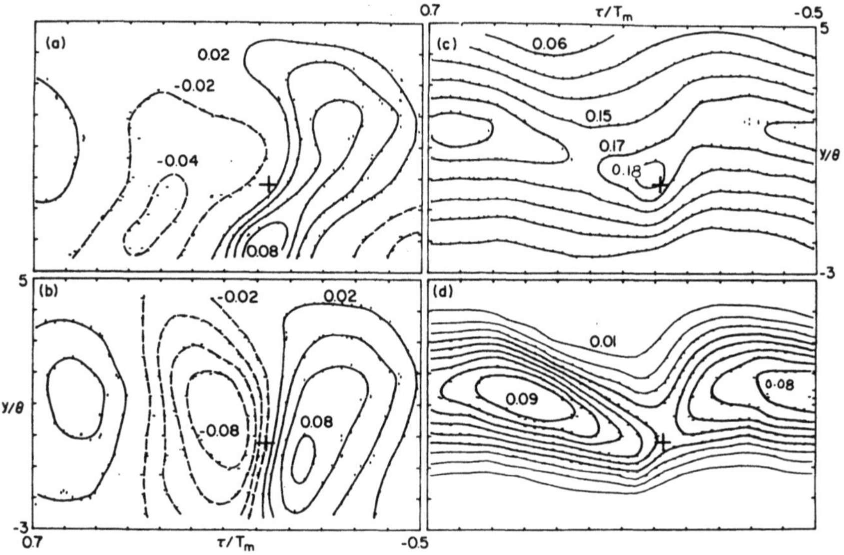

Figure 20. Coherent structure properties in the initially fully turbulent mixing layer at $x = 4800\,\theta_e$ and $U_e = 12$ m/sec. (a) longitudinal coherent velocity $\langle u_c \rangle / U_e$; (b) transverse coherent velocity $\langle v_c \rangle / U_e$; (c) longitudinal incoherent turbulence intensity $\langle u_r^2 \rangle^{1/2} / U_e$ with contour levels: 0.06, 0.09, 0.12, 0.15, 0.17, 0.18; (d) coherent shear production $-\langle P \rangle$; successive contour levels are at the increment of 0.01.

To summarize, coherent structures can be detected in a plane mixing layer indefinitely. The average structure passage frequency f_m has a universal non-dimensional value of $\mathrm{St}_\theta = f_m \theta / U_e \simeq 0.024$ at any distance from the origin. For details of this study, see Hussain and Zaman (1982).

Careful analysis of vorticity and production contours associated with the coherent structures in the plane mixing layer suggests that the *braid* connecting the structure must consist of longitudinal vortices of alternating vorticity signs (Hussain, 1983b). Flow visualization in our and other laboratories has also revealed the existence of these vortices which we call "ribs". These ribs are continually stretched between two corotating large-scale spanwise vortex rolls. Such stretching is thus at the heart of both production and entrainment. New turbulence resulting from the stretching of the ribs is advected away and deposited to the rolls. Thus, the coherent structures carry with them the mechanisms of their own survival. These and other related results will be discussed in detail in Hussain (1983b).

It is suggested that vortex stretching is the principal mechanism of production and entrainment in all turbulent flows.

5. Concluding Comments

It appears that the only meaningful way to define coherent structures is in terms of large-scale instantaneous vorticity. This also helps one to understand transport phenomena in turbulent shear flows as mostly *induction-driven* rather than *gradient-driven*. It seems that vortex interactions, particularly vortex stretching, is the principal mechanism of entrainment and production in all turbulent flows. Thus, vorticity dynamics is at the heart of turbulence phenomena: small scale and large scale.

There is little doubt that the concept of coherent structures is here to stay. What is in doubt is the predominance of these structures in all turbulent shear flows. These structures are highly dominant in initial transitional flows, in forced or resonant flows, or in the wall layer of a turbulent boundary layer. In the fully-developed turbulent states, incoherent turbulence is also comparably important and cannot be ignored. Thus, a theory based on coherent structures alone is not likely to be viable as a general theory of turbulent shear flows.

Preparation of this manuscript has been possible through the financial support of an NSF grant (MEA-811676) monitored by Dr. G. K. Lea and of an ONR grant (N00014-76-C-0128) monitored by Dr. M. M. Reischman. Their continuing encouragement and financial support over the past many years of our active research in coherent structures are deeply appreciated. The ideas presented here evolved through extensive interactions with a number of my colleagues, in particular, Drs. K.B.M.Q. Zaman, S. J. Kleis, A. R. Clark, L.S.G. Kovasznay, Z. D. Husain, J. Tso, M.A.Z. Hasan, H. S. Husain, and others. Some of their direct contributions are evident from the references. The author is grateful to Dr. M. Nallasamy for reviewing this manuscript.

References

[1] Anderson, A.B.C. *J. Acous. Soc. Am., 26* (1954), 21.

[2] Batchelor, G. K. and Gill, A. E. *J. Fluid Mech., 14* (1962), 529.

[3] Batt, R. G. *AIAA J., 13* (1975), 245.

[4] Bechert, D. and Pfizemaier, E. *J. Sound Vib., 43* (1975), 581.

[5] Beguier, C., et al. *Lect. Notes Phys., 76 (1977),* 22.

[6] Blackwelder, R. F. and Kaplan, R. E. *J. Fluid Mech., 76* (1976), 89.

[7] Bradshaw, P., Ferriss, D. H., and Johnson, R. F. *J. Fluid Mech., 19* (1964), 591.

[8] Bridges, J. E. *M. S. thesis,* Univ. of Houston, 1983.

[9] Browand, F. K. and Laufer, J. *Turb. Liquids,* Univ. Missouri-Rolla, 1975, 333.

[10] Browand, F. K. and Weidman, P. D. *J. Fluid Mech., 76* (1976), 127.

[11] Brown, G. B. *Phys. Soc., 47* (1935), 703.

[12] Brown, G. L. and Roshko, A. *J. Fluid Mech., 64* (1974), 775.

[13] Bruun, H. H. *J. Fluid Mech., 64* (1977), 775.

[14] Cantwell, B. J. Private communication, 1983.

[15] Cantwell, B. J. and Coles, D. E. Private communication, 1983.

[16] Cantwell, B. J., Coles, D., and Dimotakis. *J. Fluid Mech., 87* (1978), 641.

[17] Chandrasuda, C., et al. *J. Fluid Mech., 85* (1978), 693.

[18] Clark, A. R. and Hussain, A.K.M.F. *Turb. Shear Flow* (1979), Lond., 230.

[19] Corcos, G. M. and Lin, S. J. *J. Fluid Mech.,* (1983), submitted.

[20] Corcos, G. M. and Sherman, F. S. *J. Fluid Mech.,* (1983), submitted.

[21] Corino, E. R. and Brodkey, R. S. *J. Fluid Mech., 37* (1969), 1.

[22] Corrsin, S. Private communication, 1983.

[23] Crighton, D. G. *J. Fluid Mech., 56* (1972), 683.

[24] ——————. *J. Fluid Mech., 59* (1973), 665.

[25] ——————. *Lect. Notes Phys., 136* (1980), 340.

[26] ——————. *J. Fluid Mech., 106* (1981), 261.

[27] Crighton, D. G. and Gaster, M. *J. Fluid Mech., 77* (1976), 397.

[28] Crow, S. C. and Champagne, F. H. *J. Fluid Mech., 48* (1971), 547.

[29] Davis, S. H. *Ann. Rev. Fluid Mech., 8* (1976), 57.

[30] Dimotakis, P. E. and Brown, G. L. *J. Fluid Mech.*, *78* (1976), 535.

[31] Dimotakis, P. E., Debussy, F. D., and Koochesfahani, M. M. *Phys. Fluids*, *24* (1981), 996.

[32] Drubka, R. E. and Nagib, H. M. *IIT Rept. No. R81-2*, 1981.

[33] Emmons, H. W. *J. Aero. Sci.*, *18* (1951), 490.

[34] Falco, R. *Phys. Fluids Suppl.* *20* (1977), S214.

[35] Ffowcs-Williams, J. E. and Kempton, A. J. *J. Fluid Mech.*, *84* (1978), 673.

[36] Ffowcs-Williams, J. E., et al. *ARC Current Paper No. 1195*, 1972.

[37] Freymuth, P. *J. Fluid Mech.*, *25* (1966), 683.

[38] Gad-el-Hak, M., Blackwelder, R. F., and Riley, J. J. *J. Fluid Mech.*, *110* (1981), 73.

[39] Gaster, M., Kit, E., and Wygnanski, I. *J. Fluid Mech.*, (1983), to appear.

[40] Goldstein, M. Private communication, 1983.

[41] Grant, H. L. *J. Fluid Mech.*, *4* (1958), 149.

[42] Hama, F. R. and Nutant, J. *Proc. Ht. Tr. Fl. Mech. Inst.*, Stanford Univ. (1963), 77.

[43] Hanjalic, K. and Launder, B. E. *J. Fluid Mech.*, *51* (1972), 301.

[44] Hasan, M.A.Z. *Ph.D. thesis*, Univ. of Houston, 1983.

[45] Hayakawa, M. and Hussain, A.K.M.F. 1983, in preparation.

[46] Head, M. R. and Bandyopadhyay, P. *J. Fluid Mech.*, *107* (1981), 297.

[47] Hinze, J. O. *Dutch J. Appl. Sci. Res.*, *22* (1970), 163.

[48] Ho, C. M. and Nosseir, N. S. *J. Fluid Mech.*, *105* (1981), 119.

[49] Hunt, J.C.R. Private communication, 1982.

[50] Husain, H. S. *Ph.D. thesis*, in preparation.

[51] Husain, Z. D. and Hussain, A.K.M.F. *AIAA J.*, *17* (1979), 48.

[52] ——————————. *AIAA J.*, 1983, to appear.

[53] Hussain, A.K.M.F. *Lect. Notes Phys.*, *75* (1977), 103.

[54] ——————. *Lect. Notes Phys.*, *136* (1980), 252.

[55] ——————. "Surveys in Fluid Mechanics." Ed. R. Narasimha and S. M. Deshpande, *Indian Acad. Sci.*, (1981), 35.

[56] ——————. *Phys. Fluids*, *26* (1983a), 2816.

[57] ——————. IUTAM Symposium on Turbulence and Chaotic Phenomena in Fluids, Kyoto, Japan, Sept. 1983b, to appear.

[58] Hussain, A.K.M.F. and Clark, A. R. *J. Fluid Mech.*, *104* (1981), 263.

[59] Hussain, A.K.M.F. and Hasan, M.A.Z. 1983, (submitted for publication).

[60] Hussain, A.K.M.F. and Zaman, K.B.M.Q. *J. Fluid Mech.*, *87* (1978), 349.

[61] ———. *J. Fluid Mech.*, *101* (1980), 493.

[62] ———. *J. Fluid Mech.*, *110* (1981), 39.

[63] ———. *Report FM-14*, Univ. of Houston, 1982.

[64] Hussain, A.K.M.F. and Zedan, M. F. *Phys. Fluids*, *21* (1978), 1100.

[65] Hussain, A.K.M.F., Kleis, S. J., and Sokolov, M. *J. Fluid Mech.*, *98* (1980), 97.

[66] Kaplan, R. E. and Laufer, J. *Proc. 12th Int. Congr. Mech.*, Stanford, 1968, 236.

[67] Karamcheti, K., et al. *NASA Rept. SP-207* (1969), 275.

[68] Keffer, J. F. *J. Fluid Mech.*, *22* (1965), 135.

[69] Kibens, V. *AIAA J.*, *18* (1980), 434.

[70] Kleis, S. J. and Hussain, A.K.M.F. *Bull. Am. Phys. Soc.*, *24* (1979), 1132.

[71] Kline, S. J., et al. *J. Fluid Mech.*, *30* (1967), 741.

[72] Knight, D. D. and Murray, B. T. *Lect. Notes Phys.*, *136* (1980), 62.

[73] Kovasznay, L.S.G., Kibens, V., and Blackwelder, R. F. *J. Fluid Mech.*, *41* (1970), 283.

[74] Kunen, J.M.G., Ooms, G., and Vink, P.J.J. *Symp. Turb.*, Univ. of Missouri-Rolla, 1983, to appear.

[75] Lanford, O., III *Ann. Rev. Fluid Mech.*, *14* (1982), 347.

[76] Lau, J. C. *Proc. Roy. Soc. Lond.*, *A268* (1979), 547.

[77] Lee, M. J. and Reynolds, W. C. *Bull. Am. Phys. Soc.*, *27* (1982), 1185.

[78] Liepmann, H. W. Private communication, 1983.

[79] Liepmann, H. W. and Laufer, J. *NACA Rept. TN-1257*, 1947.

[80] Lin, C. C. Private communication, 1983.

[81] Lindgren, E. R. *Arkiv. Fys.* *12* (1959), 1.

[82] Lumley, J. L. In *Transition and Turbulence*, ed. R. E. Meyer, 1981, 215.

[83] ———. Private communication, 1982.

[84] Meynart, R. *Phys. Fluids, 26* (1983), 2074.

[85] Michalke, A. *J. Fluid Mech., 23* (1965), 521.

[86] Miksad, R. W. *J. Fluid Mech., 56* (1972), 695.

[87] Mollo-Christensen, E. *J. Appl. Mech., 89* (1967), 1.

[88] Moore, C. J. *J. Fluid Mech., 80* (1977), 321.

[89] Morkovin, M. Private communication, 1983.

[90] Mumford, J. C. *J. Fluid Mech.,* (1983), to appear.

[91] Nagib, H. M. Private communication, 1983.

[92] Nishioka, M., Asai, M., and Iida, S. In *Transition and Turbulence,* ed. R. E. Meyer, 1981, 113.

[93] Oster, D., et al. *Lect. Notes Phys., 75* (1977), 48.

[94] Payne, F. R. and Lumley, J. L. *Phys. Fluids Suppl., 10* (1967), S194.

[95] Perry, A. E. and Lim, T. T. *J. Fluid Mech., 88* (1978), 451.

[96] Perry, A. E., Lim, T. T., and Chong, M. S. *J. Fluid Mech., 101* (1980), 243.

[97] Plaschko, P. *J. Fluid Mech., 92* (1979), 209.

[98] Plaschko, P. and Hussain, A.K.M.F. 1983, (to be submitted).

[99] Reynolds, O. *Phil. Trans., 174* (1883), 935.

[100] Reynolds, W. C. and Bouchard, E. E. In *Unsteady Turbulent Shear Flows,* 1981, 402.

[101] Riley, J. J. and Metcalfe, R. *AIAA Paper No. 80-0274,* 1980.

[102] Rotta, J. *Ing. Arch., 24* (1956), 258.

[103] Ruelle, D. and Takens, F. *Commun. Math. Phys., 20* (1971), 167.

[104] Sato, H. Private communication, 1983.

[105] Smith, C. R. Private communication, 1983.

[106] Swinney, H. L. Private communication, 1983.

[107] Taneda, A. *J. Phys. Soc. Jpn., 14* (1959), 843.

[108] Townsend, A. A. *The Structure of Turbulent Shear Flow,* Cambridge: Cambridge Univ. Press, 1956.

[109] ———. *J. Fluid Mech., 95* (1979), 515.

[110] Tso, J. *Ph.D. thesis,* The Johns Hopkins Univ., 1983.

[111] von Kerczek, C. H. *J. Fluid Mech., 116* (1982), 91.

[112] Wallace, J. M., Eckelmann, H., and Brodkey, R. S. *J. Fluid Mech., 54* (1972), 39.

[113] Willmarth, W. W. and Lu, S. S. *J. Fluid Mech.*, *55* (1971), 481.

[114] Winant, C. E. and Browand, F. K. *J. Fluid Mech.*, *63* (1974), 237.

[115] Wygnanski, I. Private communication, 1983.

[116] Wygnanski, I. and Fiedler, H. E. *J. Fluid Mech.*, *41* (1970), 327.

[117] Yule, A. J. *J. Fluid Mech.*, *89* (1978), 413.

[118] Zaman, K.B.M.Q. and Hussain, A.K.M.F. *Turb. Shear Flow*, Penn. St. Univ., 1977, 11.23–11.31.

[119] ———. *J. Fluid Mech.*, *101* (1980), 449.

[120] ———. *J. Fluid Mech.*, *103* (1981), 133.

[121] ———. *Report FM-13*, Univ. of Houston, 1982.

Results of Jet Instability Theory

Alfons Michalke
Hermann-Föttinger-Institut für Thermo- und Fluiddynamik
Technische Universität Berlin

Summary

Theoretical results concerning the instability of axisymmetric jets are reviewed. For inviscid parallel jet flow the various parameters affecting jet instability as shear layer thickness, Mach number, temperature ratio, and external flow velocity are discussed. Furthermore, viscous and nonlinear effects are considered. Finally, the influences of flow divergence and of nozzle-jet interaction are discussed.

1. Introduction

The instability of jets is a phenomenon which is known since long time. In 1858 Leconte [21] observed the gaslight flickering in response to music. Nine years later, Tyndall (1867) found that the flame was unnecessary, the fluid of the jet itself being sensitive to sound. In 1879, Lord Rayleigh [43] investigated theoretically a simple model of a circular jet and found instability for axisymmetric disturbancies.

When jets became important for the propulsion of aircrafts, the instability of jets has found growing interest with respect to the laminar-turbulent transition. A typical picture of the jet instability is shown in Figure 1, taken from Wille (1963). Smoke is introduced into the wall boundary layer of an axisymmetric nozzle. Downstream of the exit plane the shear layer of the circular jet becomes unstable and rolls up to almost axisymmetric ring vortices which can pair and merge, and finally become turbulent.

Beyond the scope of the laminar-turbulent transition, the jet instability got additional importance, when Crow and Champagne (1971) found that the strongly-coherent structures observed in turbulent jets showed properties which are related to the jet instability problem. Michalke (1971) found that the measured phase velocity of [10] agreed quite well with that derived from stability theory, if the

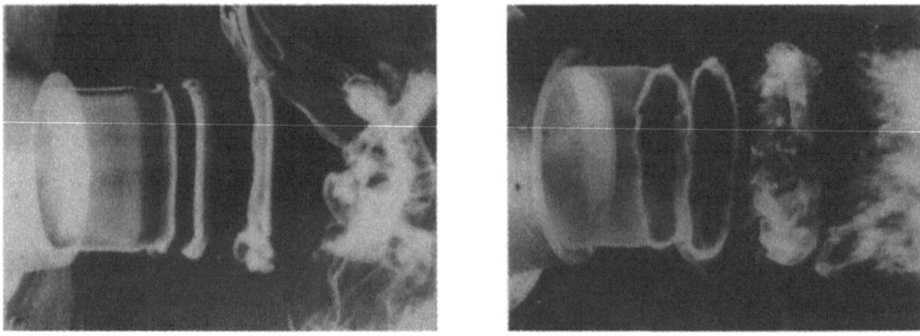

Figure 1. Jet instability visualized by smoke. Jet velocity 2 *m/s*, nozzle diameter 10 *cm*. (from [48])

turbulent mean velocity profile was used in the stability calculation. After that many investigations of jet turbulence have shown that the coherent structures of jet turbulence are a consequence of the instability of the turbulent jet shear layer.

The aim of the present paper is to review the results of jet instability theory. The results which are presented are restricted to axisymmetric jets. Only in some cases, results derived for plane shear layers are included.

In the sense of the classical hydrodynamic stability theory, first investigations were base on temporally growing disturbances (Schade 1962, Batchelor and Gill 1962, Michalke and Schade 1963, Lessen et al. 1965, Gill 1965, Kambe 1969). However, from a physical point of view, spatially growing disturbances are more appropriate, as suggested by Watson (1962) and Gaster (1962). In fact, results of stability theory with use of spatially growing disturbances by Michalke (1965) showed better agreement with experimental results. Hence only this type of disturbance will be considered.

The different parameters affecting jet instability are discussed separately, in order to show their individual influences, although in actual jets more than one parameter like the Mach number, temperature ratio, Reynolds number and others will be simultaneously important. In the main part, the jet instability in a parallel-flow-approximation is discussed. Non-parallel flow effects are treated in Section 3, while the nozzle-jet interaction is discussed in Section 4.

2. Instability of Parallel Axisymmetric Jet Flow

An axisymmetric jet flow originates, if a fluid under the action of an over-pressure is flowing through a circular aperture in a room containing a fluid at rest. The aperture can be the final cross section of a nozzle or of a pipe. We assume that there is no swirl in the flow, and that the fluid of the jet is the same as the ambient fluid. If the Reynolds number of the flow is large, then close to the jet origin the axial velocity component is large compared to the radial component, i.e., the jet flow is approximately parallel. In order to investigate the jet instability, it is therefore reasonable to assume that the undisturbed basic jet flow is parallel, i.e., the jet velocity vector has only an axial component $\overline{U}$.

Since the basic jet flow has to be axisymmetric, it is furthermore reasonable to use a cylindrical coordinate system (x, r, ϕ) with the x-axis being the jet axis. The continuity equation is satisfied, if we have $\overline{U} = \overline{U}(r)$, and the basic density distribution is $\overline{\rho} = \overline{\rho}(r)$. In the parallel jet flow the pressure $\overline{p} = p_\infty$ has to be constant for large Reynolds numbers where p_∞ is the pressure in the ambient fluid which is mostly stagnant. We assume additionally that we have a thermally ideal gas with constant heat capacities, and that the Prandtl number is unity. Then for large Reynolds numbers we can assume that the disturbance motion is not essentially influenced by viscosity, heat conduction and dissipation. Hence the corresponding equations for the problem are the continuity and Euler equations and the conservation law of entropy.

2.1 The Basic Jet Flow

Let us first talk about the basic jet flow. As already mentioned, in the parallel flow approximation the pressure is $\overline{p} = p_\infty$ and the axial velocity component is $\overline{U} = \overline{U}(r)$. A suitable normalisation of the velocity profile is $\overline{U}(0) = U_0$ and $\overline{U}(R) = U_0/2$. The centerline velocity U_0 and the jet radius R are used for normalization.

For compressible flow of a homogeneous fluid the density $\overline{\rho}(r)$ and the absolute temperature $\overline{T}(r)$ are generally not constant. For an ideal gas the equation of state yields for constant pressure

$$\overline{T}(r)/T_0 = [\overline{\rho}(r)/\rho_0]^{-1} \tag{2.1}$$

where again the index 0 refers to the centerline conditions. It is convenient to relate the temperature to the velocity by means of the Busemann-Crocco-law which is valid for a boundary layer flow with a Prandtl number $Pr = 1$ and constant pressure:

$$\overline{T}/T_0 = T_\infty/T_0 + (1 - T_\infty/T_0)(\overline{U}/U_0) + (\gamma - 1)\, M^2\, (\overline{U}/U_0)(1 - \overline{U}/U_0)/2 \tag{2.2}$$

where $T_\infty = \overline{T}(\infty)$ is the ambient temperature, γ is the ratio of the specific heat capacities and $M = U_0/a_0$ is the jet Mach number with a_0 being the sound speed at the jet temperature T_0. The parallel flow approximation is valid even for a supersonic jet provided the jet is ideally expanded.

An essentail parameter of the jet instability is the ratio of the jet thickness to the jet shear layer thickness, since the instability is caused by the induction of the vorticity contained in the jet shear layer. The thinner the shear layer, the higher the vorticity for fixed U_0. The thickness of the shear layer can be expressed by the momentum boundary layer thickness θ which has frequently used in the past. It is mostly defined as

$$\theta = \int_0^\infty (\overline{U}/U_0)(1 - \overline{U}/U_0)\, dr. \tag{2.3}$$

Then the ratio R/θ is a jet parameter which can be used to characterize the jet velocity profile. Since in an actual jet the ratio R/θ is decreasing in downstream direction x, the jet parameter R/θ can be used to characterize the jet velocity profiles at different axial positions x.

In the past various velocity profiles $\overline{U}(r)$ have been used for the investigation of the jet instability. Here only two different types will be discussed:

Profile 1:

$$\overline{U}/U_0 = 0.5[1 + tanh[b_1(1 - r/R)]] \quad \text{where } b_1 = 0.5\, R/\theta \gg 1 \tag{2.4}$$

Profile 2:

$$\overline{U}/U_0 = 0.5[1 + tanh[b_2(R/r - r/R)]] \quad \text{where } b_2 = 0.25\, R/\theta. \tag{2.5}$$

Profile 1 has been used by Michalke (1971) and Morris (1976), but it is suitable only for large values R/θ, since $\overline{U}(0) = U_0$ is true only for large values b_1. For $\theta \to 0$ it yields the plug flow jet profile induced by a cylindrical vortex sheet. Profile 2 has been applied by Michalke (1971), Chan and Leong (1973), Morris (1976, 1983), Crighton and Gaster (1976), Plaschko (1979, 1983) and Michalke and Hermann (1982). It has not the restriction of Profile 1 and can also be used for jet profiles close to the end of the potential core of the jet.

There are slight differences in the velocity profiles and, more pronounced, in the vorticity distributions, even if the jet parameter R/θ is the same for different profiles.

Since the jet parameter R/θ is related to the maximum vorticity, it depends for an actual jet on the axial position x. Crighton and Gaster (1976) and Plaschko (1979) used for Profile 2

$$\theta/R = 0.06\,(x/D) + 0.04 \tag{2.6}$$

where D is the diameter of the nozzle exit. However, the relation may depend on the initial conditions at the nozzle exit. Therefore θ/R as function of x/D may be different for different jets.

2.2 Linearized Disturbance Equations for Inviscid and Nonconducting Flow

The disturbance equation can be derived, if disturbances c'_x, c'_r, c'_ϕ, p', ρ' are introduced in the continuity-, Euler-, and entropy conservation equations. For small disturbances the equations can be linearized around the basic jet flow. Furthermore, we can assume spatially growing, wavy disturbances of the type

$$[c'_x, c'_r, c'_\phi, p', \rho'] = [u(r), v(r), w(r), p(r), \rho(r)] \exp[i(\alpha x + m\phi - \omega t)] \tag{2.7}$$

Here the radian frequency ω and the integer azimuthal wavenumber m are real, while $\alpha = \alpha_r + i\alpha_i$ is generally complex. α_r is the axial wavenumber and $-\alpha_i > 0$ is the spatial growth rate. With Equation (2.7) the disturbance equations can be reduced to a single one for the pressure amplitude $p(r)$:

$$\frac{d^2p}{dr^2} + \left(\frac{1}{r} - \frac{1}{W}\frac{dW}{dr}\right)\frac{dp}{dr} - \left[\alpha^2(1 - M^2W) + (m/r)^2\right]p = 0 \tag{2.8}$$

where

$$W(r) = \left[(\overline{U}(r) - \omega/\alpha)/U_0\right]^2 / \left[\overline{T}(r)/T_0\right]. \tag{2.9}$$

The other amplitude functions are related to $p(r)$ by

$$\overline{\rho}v = i\frac{dp}{dr}/(\alpha\overline{U} - \omega) \tag{2.10}$$

$$\overline{\rho}u = -\frac{dp}{dr}\frac{d\overline{U}}{dr}/(\alpha\overline{U} - \omega)^2 - p/(\overline{U} - \omega/\alpha) \tag{2.11}$$

$$\overline{\rho}w = -(m/r)p/(\alpha\overline{U} - \omega) \tag{2.12}$$

$$\rho = -\frac{1}{\overline{\rho}}\frac{d\overline{\rho}}{dr}\frac{dp}{dr}/(\alpha\overline{U} - \omega)^2 + p/\overline{a}^2 \tag{2.13}$$

Here $\overline{a}$ is the local sound speed of the basic jet flow with

$$\overline{a}^2 = a_0^2(\overline{T}/T_0). \tag{2.14}$$

For $r \rightarrow 0$ (jet axis) and $r \rightarrow \infty$ (ambient fluid) the quantity W of Equation (2.9) approaches constant values. Then the asymptotic solutions of the differential equation (2.8) are given by the modified Bessel functions I_m and K_m of order m. The boundary conditions for the pressure disturbance require that $p(0)$ has to be bounded and $p(\infty)$ has to be zero. Hence for $r \rightarrow 0$

$$p(r) \simeq I_m\left[\alpha\sqrt{1 - M^2W(0)}\,r\right] \tag{2.15}$$

and for $r \rightarrow \infty$

$$p(r) \simeq K_m\left[\alpha\sqrt{1 - M^2W(\infty)}\,r\right] \tag{2.16}$$

where the real part of the argument of K_m has to be positive.

Hence an eigenvalue problem for the complex eigenvalue α has to be solved for given $\overline{U}(r)/U_0$, $\overline{T}(r)/T_0$, M, ω, and m. For this reason, the differential equation (2.8) must be integrated numerically for a chosen value α. It has to be varied up till the boundary condition (2.15) and (2.16) are satisfied. In this way, only eigenvalues of amplified disturbances ($-\alpha_i > 0$) can be found. For damped disturbances ($\alpha_i > 0$) it is necessary to consider the differential equation (2.8) in the complex r-plane. It was shown by Tam (1975) that in this case the singularity $W = 0$ of (2.8) lies above the real r-axis with a branch cut below it. Then, for $\alpha_i > 0$ the integration path along the real r-axis has to avoid a crossing of that branch cut and to take a path above that singularity (see Section 3). However, the physically interesting eigenfunctions along the real r-axis can then be discontinuous at that branch cut.

An analytical approximation of the eigenvalue problem has been derived by Haertig (1981). For a special basic velocity profile and constant jet temperature $\overline{T} \equiv T_\infty$ he transformed the differential equation (2.8) with respect to the variables. The solution is then expanded around the neutral eigenvalue of the equivalent plane shear layer (tanh-profile) with respect to the dimensionless eigenvalue and several parameters. Thus an approximate eigenvalue relation depending on ω, m, R/θ, and M is obtained, which is very precise for large values R/θ.

2.3 Instability Properties of Incompressible, Isothermal, Inviscid Jet Flow

Let us first discuss results of jet instability for inviscid (equation (2.8)), incompressible ($M = 0$) and isothermal ($T_0 = T_\infty$) flow. The simplest case of a circular jet flow is the plug flow with a shear layer thickness $\theta = 0$ which is generated by a cylindrical vortex sheet of radius R. For temporally growing disturbances the problem has been discussed by Lord Rayleigh (1879) and by Batchelor and Gill (1962). Lessen, Fox, and Zien (1965) and Gill (1965) included compressibility. For spatially growing disturbances results were obtained by Michalke (1970) and by Crow and Champagne (1971). Typically for the plug flow is that it is unstable for all frequencies. A surprising result was that in a certain frequency range additional modes exist which may be called "irregular", since they have a non-zero growth rate as the frequency tends to zero. Their wavenumbers α_r are very small and vanish at a certain frequency. Hence the axial phase velocity $c_{ph} = \omega/\alpha_r$ is very large and can become infinite. The physical meaning of these irregular modes is quite uncertain. However, the irregular modes are not a consequence of the discontinuous plug profile, but they do exist for continuous jet velocity profiles, too. For finite jet parameter R/θ and at small frequencies the axisymmetric disturbance ($m = 0$) behaves like that of the plug flow. In the following only the regular modes will be discussed except in Section 2.5.

The main difference between the plug flow and a jet with a non-zero shear layer thickness θ is that unstable disturbances exist only in a finite frequency range. For fixed θ and $R \to \infty$ the results of the circular jet tend to those of a plane shear layer. The Haertig approximation (Haertig 1981) yields in this case the eigenvalues of the plane tanh-velocity profile (Michalke 1965) with an accuracy of three decimals.

For large R/θ the growth rate $-\alpha_i\theta$ and the range of unstable frequencies for $m = 0$ and Profile 1 does not change very much with R/θ. This indicates that the shear layer thickness θ is the characteristic length scale of the problem.

The influence of the jet parameter R/θ becomes more important for $R/\theta < 12$ indicating that the jet radius R can equivalently be used as length scale. In Figure 2 the growth rates of the disturbances with $m = 0$ and $m = 1$ are compared for Profile 2. The values of R/θ are 10, 5, and 2.5 which correspond due to equation (2.6) to axial positions $x/D = 1$, 2.67, and 6, respectively. Obviously, the growth rates decrease with decreasing R/θ, i.e., the jet flow becomes less unstable in downstream direction. While for $R/\theta = 10$ the growth rate for $m = 0$ is slightly larger than that for $m = 1$, the reverse is true farther downstream for $R/\theta < 5$. Nevertheless for $\omega\theta/U_0 < 0.1$ always the disturbance with $m = 1$ is more unstable. In Figure 3 the corresponding phase velocities are shown. We see that for these values R/θ the phase velocity c_{ph} does not exceed the jet velocity U_0, but that of the axisymmetric disturbance increases with decreasing jet parameter R/θ.

As already mentioned, the results can also be normalized with the jet radius R. In Figure 4 the dimensionless frequencies, i.e., the Strouhal numbers $St = \omega R/(U_0\pi)$ for both the maximum growth rate and the neutral frequency are plotted as function of θ/R, which is due to equation (2.6) proportional to the axial distance x/D, for $m = 0$ and $m = 1$ and profile 2. Close to the nozzle, where θ/R is small, the local instability properties change strongly. High Strouhal number disturbances are unstable only within a small region of θ/R. Disturbances with Strouhal numbers smaller than 0.35, however, are unstable in the whole potential core region of the jet. For Profile 2 and $\theta/R > 0.1$ the maximum growth rate for $m = 1$ belongs to a smaller Strouhal number as compared with $m = 0$.

It should be noticed that different profiles yield slightly different results, even if the value of the jet parameter R/θ is the same.

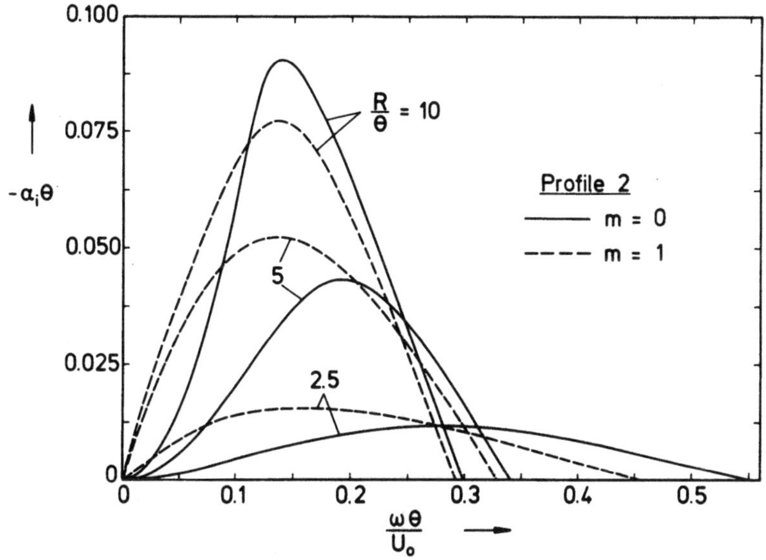

Figure 2. Spatial growth rate $-\alpha_i$ of the axisymmetric ($m = 0$) and of the first azimuthal ($m = 1$) disturbances vs. frequency ω for various values R/θ. Profile 2, isothermal, incompressible flow. (from [32])
Courtesy of Cambridge University Press.

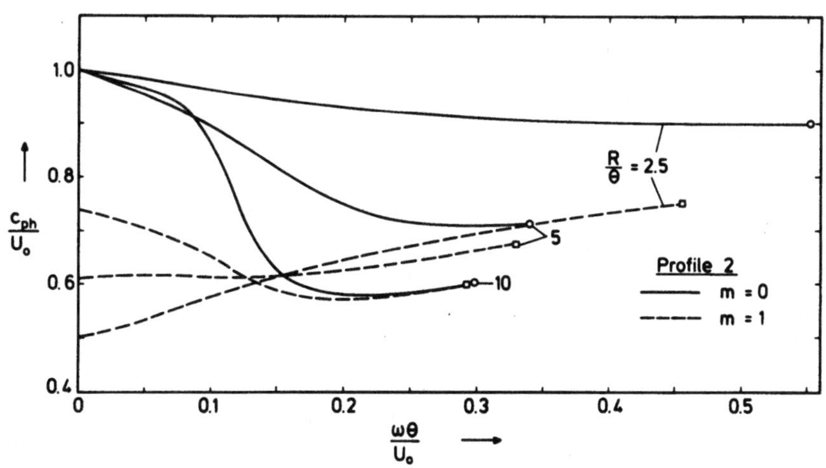

Figure 3. Phase velocity c_{ph} of the axisymmetric ($m = 0$) and of the first azimutahl ($m = 1$) disturbances vs. frequency ω for various values R/θ. Profile 2, isothermal, incompressible flow. (from [32])
Courtesy of Cambridge University Press.

Especially for higher frequencies differences occur. For instance, the results for Profile 2 by Michalke (1971) and for a Gaussian-type profile used by Mattingly and Chang (1974) showed not negligible deviations.

2.4 Mach Number Effect

So far, the inviscid instability problem for an incompressible and isothermal jet has been discussed. The influence of the jet Mach number M shall now be considered, if the jet temperature T_0 is equal to the ambient temperature T_∞. In this case the jet Mach number M appears explicitly in the disturbance equation (2.8). Additionally, the Mach number influences the instability via the basic temperature distribution $\overline{T}(r)$ which is now according to equation (2.2) no longer constant.

Furthermore, the basic jet velocity profile will also depend on the Mach number leading to a thicker jet shear layer with increasing Mach number. But, in order to investigate the influence of the Mach number on the jet instability, it is reasonable to compare always the same jet velocity profiles (as in the incompressible case) with any Mach number dependence thrown in the relation between the jet parameter R/θ and x/D.

The influence of the jet Mach number on the axisymmetric ($m = 0$) and first azimuthal ($m = 1$) disturbance is shown in Figure 5. In the results of [29] for Profile 2, however, $R/\theta_c = 6.25$ was kept constant. θ_c is the compressible momentum thickness which depends on the Mach number via the density distribution and is therefore not suitable to characterize the velocity shear layer. One can see from Figure 5 that with increasing Mach number M the growth rate is reduced, hence the flow becomes less unstable. For even higher Mach numbers, the jet flow may become stable with respect to axisymmetric disturbances, but then additional disturbance modes due to the varying basic temperature may appear as it has been found in plane shear layers (Gropengießer 1969, Blumen et al. 1975). The stabilizing effect due to the Mach number M is stronger with respect to $m = 0$ than to $m = 1$. While, at least for $M = 0$, the growth rate for $m = 0$ is larger at higher frequencies, we have for $M = 1.2$ a higher growth rate for $m = 1$ in the whole unstable frequency range. The phase velocity of both disturbances is, however, only weakly dependent on the Mach number M.

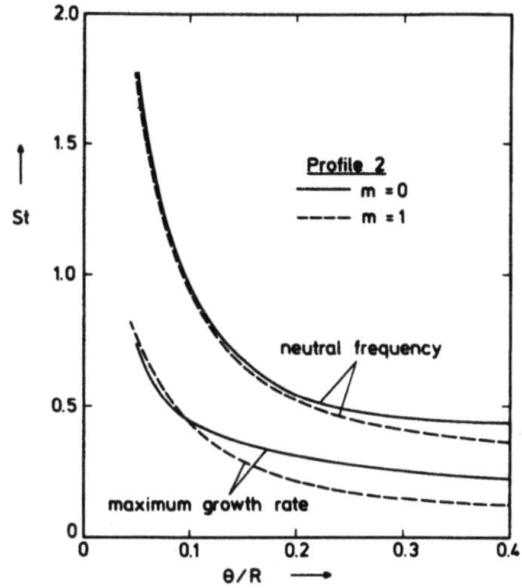

Figure 4. Strouhal number $St = \omega R/(U_0\pi)$ of maximum growth rate and of neutral frequency vs. shear layer thickness θ/R for axisymmetric ($m = 0$) and first azimuthal ($m = 1$) disturbances. Profile 2, isothermal, incompressible flow.

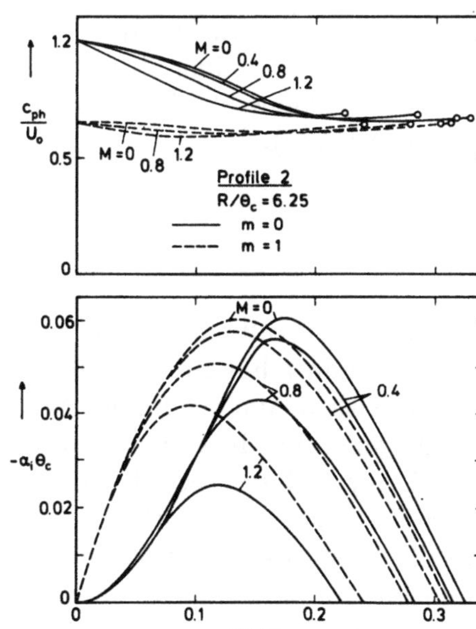

Figure 5. Phase velocity c_{ph} and spatial growth rate $-\alpha_i$ vs. frequency ω for various Mach numbers M. Axisymmetric ($m = 0$: —) and first azimuthal ($m = 1$: - - -) disturbance. Profile 2, $R/\theta_c = 6.25$, $T_0 = T_\infty$. Quantities are normalized by θ_c! (from [29])

For sufficiently high jet Mach numbers M the axial phase velocity c_{ph} can become supersonic with respect to the ambient fluid, i.e., the convection Mach number is $M_c \equiv c_{ph}/a_\infty > 1$. Then the question is whether the disturbances emit Mach waves. From the asymptotic behaviour of the pressure disturbances for large radial distance r, one can find for axisymmetric disturbances that in the ambient fluid oblique cylindrical wave fronts (surfaces of constant phase) are present. The wave normal direction is inclined with respect to the jet axis by an angle μ. The wave speed is equal to $c_w = c_{ph} \cos \mu$. It can be shown that for axisymmetric disturbances the wave speed is equal to the sound speed only for $M_c > 1$ and for the neutral frequency. The propagation angle μ then corresponds to the Mach angle. However for amplified disturbances the wave speed is lower, but can closely approach the sound speed, especially for low frequencies. For Mach numbers close to unity, the density fluctuations are of the same order of magnitude as the pressure fluctuations. Hence the wave fronts should become visible in Schlieren pictures, but they should not be denoted as sound waves, since their wave speed is lower than the sound speed.

Finally, it should be noted that for $M > 0$ multiple irregular modes exist, too. Upstream travelling disturbances for the plug flow have been discussed by Neuwerth (1974) and for a continuous velocity profile by Haertig (1979).

2.5 Temperature Effect

After having discussed the influence of the Mach number on the jet instability, we consider now a hot circular jet with a temperature ratio $T_0/T_\infty > 1$, but with a Mach number $M = 0$. Again the jet velocity profile will generally depend on the temperature ratio T_0/T_∞, too. The tendency to be expected from plane shear layers (cf. [14]) is a decrease of the shear layer thickness θ with increasing temperature ratio T_0/T_∞. But also here the same jet velocity profiles are compared accounting for any T_0/T_∞-dependence only in the relation between R/θ and x/D.

From the jet plug flow it is know (cf. [28]) that for increasing jet temperature $T_0 > T_\infty$ the flow becomes more unstable. This effect is in contrast to that obtained for temporally growing disturbances, as can be seen from the long-wave approximation by Gill and Drazin (1965). The results of the jet plug flow [28] show further that for a hot jet with $T_0/T_\infty > 1.6$ a clear distinction between "regular" and "irregular" modes becomes impossible, since both modes are mixed. The same phenomenon has also been found by Michalke (1971) for continuous jet velocity profiles. That data are replotted in Figure 6 using θ instead of θ_c. In that figure the reciprocal phase velocity U_0/c_{ph} and the growth rate $-\alpha_i\theta$ vs. frequency $\omega\theta/U_0$ are shown for the jet velocity profile 1 at large values of the jet parameter R/θ, for $T_0/T_\infty = 2$ and $m = 0$. For comparison, the isothermal data ($T_0 = T_\infty$) are included for the nearly plane shear layer ($R/\theta = 20000$). In this case the hot jet ($R/\theta = 27720$) has a slightly lower phase velocity, but a larger growth rate which, however, peaks at a slightly lower frequency as compared with the isothermal jet. Furthermore, the range of unstable frequencies is reduced by the heating. Similar results have been obtained for plane shear layers by Gropengießer (1969), Maslowe and Kelly (1971), and Kapur and Morris (1974).

For the jet parameter values $R/\theta = 69.3$ and 34.7 in Figure 6 the mode mixing occurs. The growth rate decays from its peak value to lower frequencies up till a minimum is reached. From there the growth rate increases to a finite value at zero frequency.

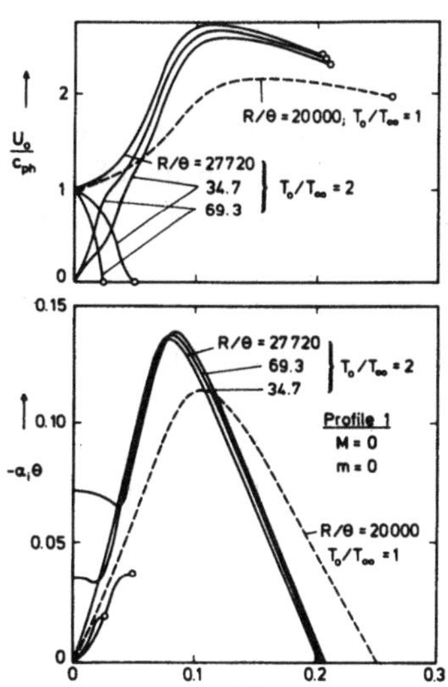

Figure 6. Reciprocal phase velocity U_0/c_{ph} and spatial growth rate $-\alpha_i$ of the axisymmetric disturbance ($m = 0$) vs. frequency ω for a hot jet ($T_0/T_\infty = 2$, $M = 0$), Profile 1, and various values R/θ. Isothermal jet ($T_0 = T_\infty$): - - - . (from [29])

Simultaneously, the phase speed tends to infinity for $\omega \to 0$. A second mode exists for low frequencies with vanishing growth rate at $\omega = 0$. Here with increasing frequency the growth rate increases, too, up to a critical frequency where the phase velocity becomes infinite, i.e., $\alpha_r = 0$. For this disturbance mode the phase velocity c_{ph} is always greater than the jet velocity U_0. The physical meaning of this irregular behaviour of both disturbance modes is quite unclear. Unfortunately, experimental data concerning the instability of hot jets are obviously not available.

2.6 Flight Effect

When a jet is produced by a jet engine in flight, the jet instability properties are changed by the flight velocity. In a coordinate system fixed to the nozzle exit the jet velocity decays radially from its maximum value U_0 to a value $U_\infty > 0$ which is equal to the flight velocity. Except for a small region close to the nozzle exit, the vorticity contained in the jet shear layer scales with the velocity differences $\Delta U = U_0 - U_\infty$. It is therefore taken for normalization.

The instability properties of the jet in flight depend additionally on the velocity ratio $U_\infty/\Delta U$. Some results for an isothermal ($T_0 = T_\infty$), incompressible ($M = 0$) jet with profile 2 have been derived in [32]. It is found that for both the axisymmetric and first azimuthal disturbances the peak value of $-\alpha_i\theta$ decreases with increasing external velocity U_∞, but the range of unstable frequencies is increased. The normalized phase velocity is always increased by the external flow velocity U_∞. Hence the phase velocity c_{ph} can well exceed the velocity difference ΔU. From theoretical considerations of equation (2.8) and (2.9) it has been found that the influence of the velocity ratio $U_\infty/\Delta U$ on the instability properties can be eliminated in a first approximation, if $-\alpha_i\theta\sigma_n$ and $(c_{ph} - U_\infty)/\Delta U$ are plotted vs. $\omega\theta/(\Delta U\sigma_n)$ where

$$\sigma_n = 1 + A_n U_\infty/\Delta U \tag{2.17}$$

is a stretching factor with the stretching parameter $A_n = U_0/c_{ph,n}$. The latter is the reciprocal neutral phase velocity of the static case $U_\infty = 0$. For jet parameter values R/θ not to large, the method is quite effective. For $R/\theta = 5$ the results are shown in Figure 7 for the growth rate and in Figure 8 for the phase velocity. It is obvious that the collapse of the curves for $U_\infty/\Delta U > 0$ with

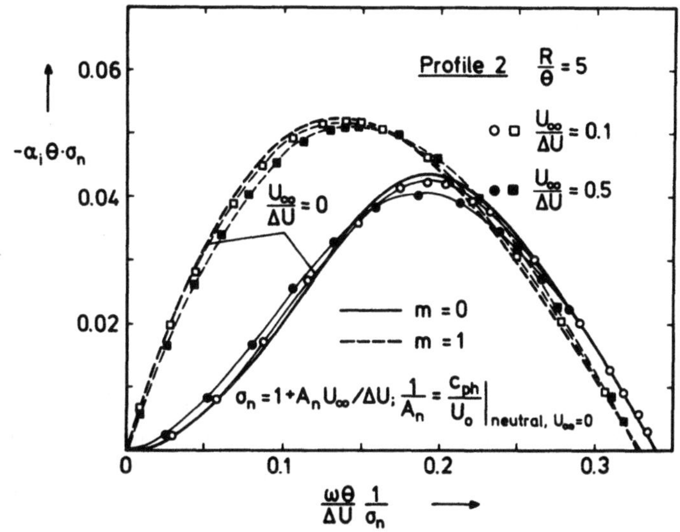

Figure 7. Reduced spatial growth rate $-\alpha_i\theta\sigma_n$ vs. reduced frequency $\omega\theta/(\Delta U\sigma_n)$ of axisymmetric ($m = 0$) and first azimuthal ($m = 1$) disturbances for Profile 2 ($R/\theta = 5$) and the velocity ratios $U_\infty/\Delta U = 0.1$ and 0.5. Stretching factor σ_n, isothermal, incompressible flow. (from [32]) Courtesy of Cambridge University Press.

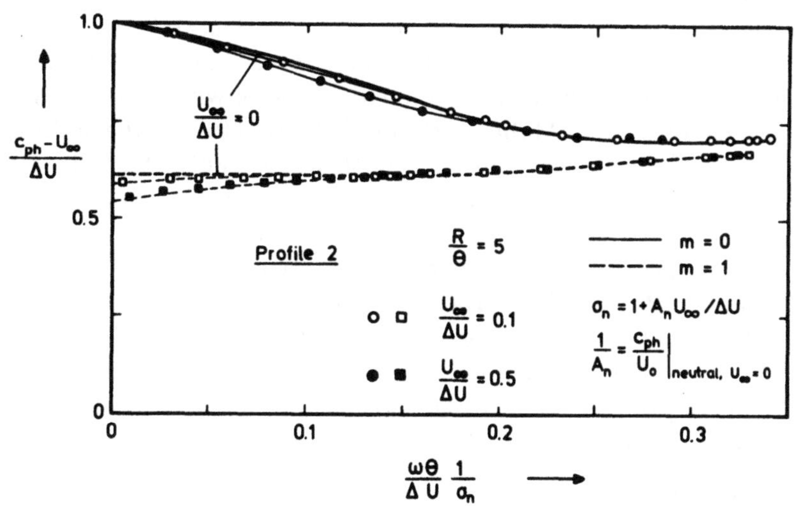

Figure 8. Reduced phase velocity $(c_{ph} - U_\infty)/\Delta U$ vs. reduced frequency $\omega\theta/(\Delta U\sigma_n)$ of axisymmetric ($m = 0$) and first azimuthal ($m = 1$) disturbances for Profile 2 ($R/\theta = 5$) and the velocity ratios $U_\infty/\Delta U =$ 0.1 and 0.5. Stretching factor σ_n, isothermal, incompressible flow. (from [32]) Courtesy of Cambridge University Press.

the static one is quite remarkable. For lower values of R/θ the agreement is even better. Physically, this result means that by the external velocity U_∞ the disturbed flow field is axially stretched and the frequency is increased by the factor σ_n compared with the static case. In the case discussed here, the stretching parameter A_n depends on the jet parameter R/θ and on the azimuthal order m. The variation of A_n with these parameters is, however, not very strong. Hence a universal stretching parameter $A = 1.4$ can be taken in a further approximation leading to a universal stretching factor σ. The results indicate that the universal stretching factor σ offers a suitable way to estimate the influence of flight on the instability. Measurements in a turbulent jet with external flow by Michel and Michalke (1981) have shown that the same similarity conditions are in a first approximation also applicable for the jet turbulence.

2.7 Reynolds Number Effect

So far the disturbance motion has been assumed to be unaffected by viscous diffusion and dissipation and by heat conduction. In actually operating jets the Reynolds number is generally large. The effect of the Reynolds number on the jet instability has mostly been taken into account only for incompressible, isothermal jets. Lessen and Singh (1973) did some calculations with spatially growing disturbances, and Mollendorf and Gebhart (1973) take additional bouyancy effects into account. In both papers only the fully developed jet has been considered, and the main attention has been paid to the critical Reynolds number where the jet becomes unstable at all. This aspect of laminar-turbulent transition is today of minor importance. Rather the variation of jet instability properties with Reynolds number are of more interest. Morris (1976) investigated the influence of the Reynolds number on the instability of various jet velocity profiles. The disturbance equations will not be given here, they can be found in the cited papers.

The main difference to the inviscid case is that due to the higher order of the differential equation there are additional asymptotic viscous solutions. If the kinematic viscosity ν is small or, for fixed other parameters, the Reynolds number is large, the viscous part of the solution oscillates very strongly with the radius r. As a consequence, very difficult problems arise for the numerical integration of the differential equation at large Reynolds numbers. Special numerical methods have to be applied. An additional problem is that for fixed Reynolds number and frequency an infinite set of eigenvalues exists. Of most physical interest is only that eigenvalue α which leads to the least stable disturbance. It is convenient to relate the Reynolds number Re to the (local) jet diameter $2R$ and the jet centerline velocity U_0. Then the Reynolds number is

$$Re = 2RU_0/\nu. \tag{2.18}$$

For jet velocity profiles with thinner shear layers as in the potential core of the jet, Morris (1976) found only a damping influence of viscosity with respect to the growth rate. For Profile 2 and a jet parameter value $R/\theta = 6.25$ his results are replotted in Figure 9 for $m = 0$ and 1. Above a Reynolds number $Re = 1000$ the influence of viscosity on the growth rate $-\alpha_i$ is evidently small. For lower values of Re, the axisymmetric disturbance is less unstable than the first azimuthal one.

Finally, the results of Morris (1976) indicate that the critical Reynolds number Re_c is for $m = 1$ in the order of $Re_c = 43$ to 49 nearly independent of the jet parameter ($2.2 \leq R/\theta \leq 50$), while for the axisymmetric disturbance the critical Reynolds number is $Re_c = 68.3$ for $R/\theta = 50$ and $Re_c = 110.6$ for $R/\theta = 6.25$ and becomes infinite towards the fully developped jet. Unfortunately Morris (1976) did not report anything about the existence of the irregular modes found in the inviscid case. Therefore it remains an open question whether these irregular modes are a purely inviscid phenomenon or not. Some results for the combined influence of Reynolds and Mach number on the jet instability have been given by Morris (1983).

2.8 Nonlinear Effects

In the framework of the linearized theory it is found that unstable disturbances grow exponentially in jet flow direction x. Hence it is clear that with increasing x the neglected nonlinear terms of the equations will become important and cannot be neglected further. These terms will modify the flow field calculated by means of the linearized theory.

In order to discuss the effects to be expected, let us consider the nonlinear vorticity equation for incompressible, inviscid flow and,

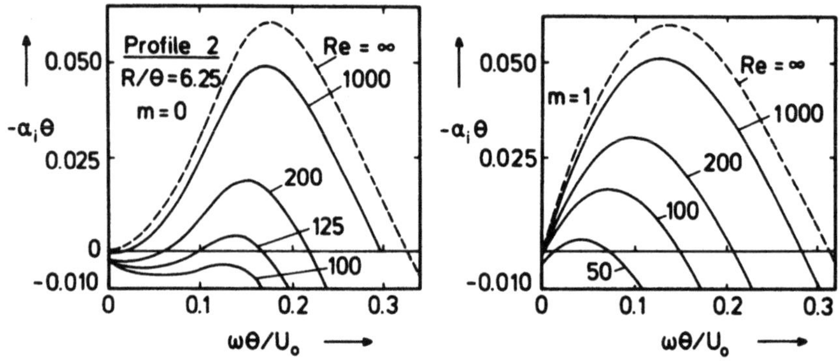

Figure 9. Spatial growth rate $-\alpha_i$ of the axisymmetric ($m = 0$) and first azimuthal ($m = 1$) disturbances vs. frequency ω for Profile 2 ($R/\theta = 6.25$) and various Reynolds numbers $\mathrm{Re} = 2RU_0/\nu$. Isothermal, incompressible flow. (from [36]) Courtesy of Cambridge Univ. Press.

for simplicity, only that for axisymmetric disturbances. Then the vorticity vector has only an azimuthal component $\Omega(x, r, t)$ which yields the inviscid vorticity equation

$$\frac{\partial}{\partial t}\left[\frac{\Omega}{r}\right] + c_x \frac{\partial}{\partial x}\left[\frac{\Omega}{r}\right] + c_r \frac{\partial}{\partial r}\left[\frac{\Omega}{r}\right] = 0. \tag{2.19}$$

The physical interpretation of this equation is that the expression Ω/r is constant along the particle path lines which are defined by

$$\frac{dx}{dt} = c_x(x, r, t); \quad \frac{dr}{dt} = c_r(x, r, t). \tag{2.20}$$

For a fixed time Ω/r is constant along the streak lines which can be visualized by smoke or by dye.

In the undisturbed circular jet the vorticity is concentrated in a sheet around the jet radius R. If the thickness of that shear layer is small compared with the jet radius R ($R/\theta \gg 1$), then Ω/r can approximately be replaced by Ω/R, and equation (2.19) corresponds to the vorticity equation of a plane shear layer. In this case, at a fixed time, the lines of constant vorticity are identical with the streak lines. When the sheet containing most of the vorticity is visualized by smoke, the disturbed jet shear layer looks like that shown in Figure 10, which has been taken from [31]. It is obvious that the shear layer rolls up and the vorticity contained in the shear layer is concentrated in vortices. Since the vorticity sheet is annular, ring votices are formed.

The opinion that the rolling-up process of the jet shear layer is a typical nonlinear effect, is, however, not quite correct. Rather the basic phenomenon is already contained in the linearized thoery. This will be demonstrated in two ways. Firstly, the streakline pattern can be calculated by means of the disturbed velocity field derived from the linearized theory. This has been made for the plane tanh-velocity profile in [31]. A result for the most strongly amplified disturbance is shown in Figure 11. We see a detail of the first stage of the rolling-up process which is in good agreement with the experimental results of Figure 10 although the local disturbance magnitude at these axial positions x is not all small. Secondly, it was shown by Michalke (1972) that for inviscid flow the vorticity in presence of an unstable disturbance should be of the type

$$\Omega(x,r,t) = \overline{\Omega}\,[r + \epsilon h(x,r,t)] \tag{2.21}$$

where $\overline{\Omega}$ is the undisturbed vorticity and ϵh is the particle displacement. For the most strongly amplified disturbance of the plane

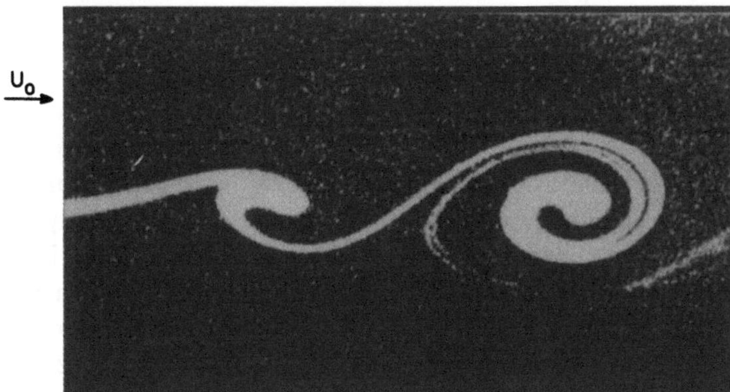

Figure 10. Streakline pattern of the disturbed shear layer of a circular jet visualized by smoke showing the formation of ring vortices. (from [31])

tanh-velocity profile and the same conditions as in Figure 11 the lines of constant vorticity have been calculatled by means of the displacement h derived from the linearized theory and are shown in Figure 12. Although only the linearized solution has been used, we clearly see again a rolling-up of the shear layer which is, however, slightly different in that detail shown in Figure 11. It should be no difference in both cases, if the full nonlinear solution would have been used. Hence we see that the nonlinear terms of the equation will modify the rolling-up process which is, however basically, already contained in the linearized theory.

Besides of the nonlinear development of a single disturbance, an important phenomenon of the instability process is the nonlinear interaction between disturbances of different frequencies. For axisymmetric disturbances the stream function ψ satisfying the continuity equation can be introduced in the vorticity equation (2.19). If two disturbances of frequency ω_1 and $\omega_2 < \omega_1$ are assumed in the flow, then the vorticity equation (2.19) yields that by the nonlinear interaction additional frequency components with frequencies $2\omega_1$, $2\omega_2$, $\omega_1 \pm \omega_2$ are generated. In order to satisfy equation (2.19), corresponding frequency components have to be added in the disturbance expansion.

Only in the case of subharmonic resonance $\omega_1 = 2\omega_2$ this is not necessary. The subharmonic frequency component (ω_2) has then to satisfy an equation

$$L(F_2, \omega_2) + \epsilon_1 \left[N(F_1, D^2\tilde{F}_2) + N(\tilde{F}_2, D^2 F_1) \right] / 2 = 0 \qquad (2.22)$$

where L and D^2 are linear operators and N is an operator which contains F_1 as coefficients. $F_1(x,r)$ and $F_2(x,r)$ are the amplitudes of both frequency components, ϵ_1 is the disturbance magnitude of the fundamental disturbance. The tilde ~ denotes the conjugated complex value. In the linearized approximation the function F_1 is of the type $F_1(x,r) = f_1(r)\exp(i\alpha_1 x)$. Hence equation (2.22) is of a type similar to the Mathieu equation with coefficients oscillating and exponentially growing with respect to x. For the plane shear layer and for temporally growing disturbances Kelly (1967) has analyzed the problem. For spatially growing disturbances the situation is something different. Formally, one can obtain a solution of equation (2.22) by expanding F_2 with respect to ϵ_1. To zero order one obtains

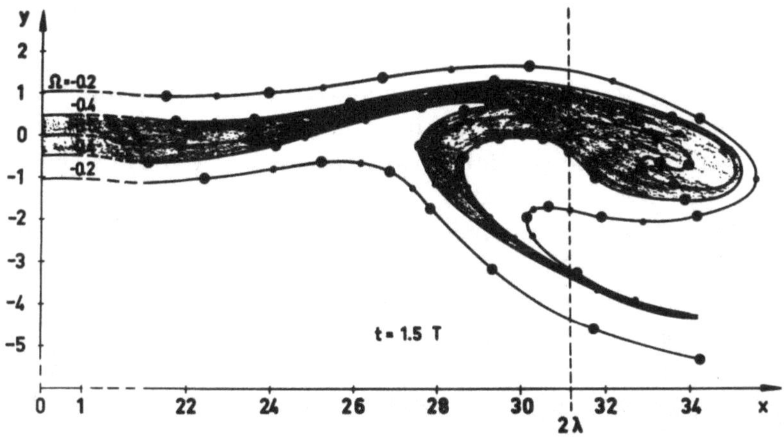

Figure 11. Streakline pattern in a plane shear layer calculated for the most strongly amplified disturbance of the tanh-velocity profile due to inviscid, linearized theory. (from [31])

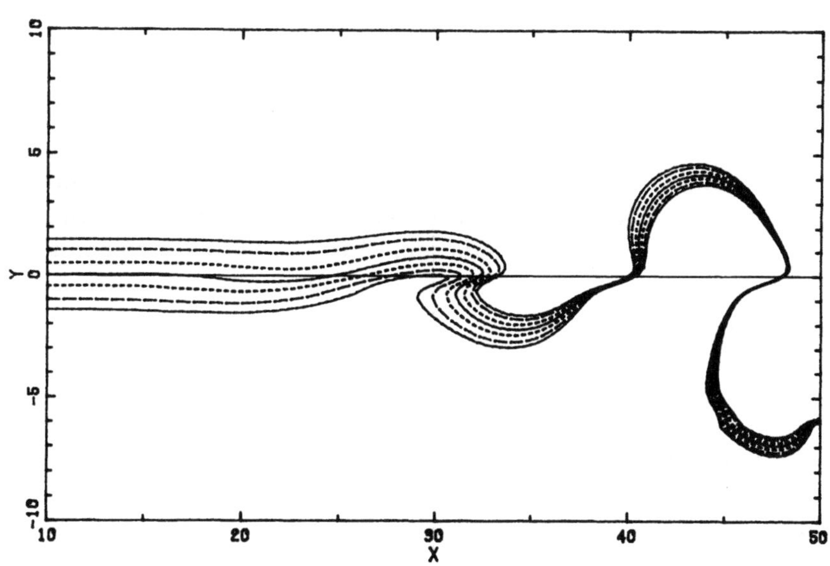

Figure 12. Lines of constant vorticity in a plane shear layer calculated by means of the displacement h for the most strongly amplified disturbance of the tanh-velocity profile due to inviscid, linearized theory. Same conditions as in Figure 11.

the usual disturbance equation for the frequency ω_2 which yields the eigenvalue α_2.

To $0(\epsilon_1)$ an inhomogeneous equation is obtained which has always a solution satisfying the boundary conditions provided $\alpha_1 - \tilde{\alpha}_2 \neq \alpha_2$. Since the spatial growth rate $-(\alpha_{i1} + \alpha_{i2})$ of the $0(\epsilon_1)$-term in the expansion is greater than that of the primary subharmonic term, $-\alpha_{i2}$, the expansion will be of restricted convergency with respect to x. Nevertheless the subharmonic disturbance with frequency ω_2 will be substantially changed by the resonance with the fundamental disturbance of frequency $\omega_1 = 2\omega_2$. The exceptional case $\alpha_1 - \tilde{\alpha}_2 = \alpha_2$ would require $\alpha_{r1} = 2\alpha_{r2}$, or equivalently $c_{ph1} = c_{ph2}$, and $\alpha_{i1} = 0$. For a parallel flow these conditions will be hardly satisfied. However, for a slowly diverging basic jet flow (Section 3) an unstable disturbance of high fixed frequency ω_1 travelling downstream in the jet can meet a condition where the local instability becomes neutral (Figure 4). The other condition $c_{ph1} = c_{ph2}$ is then surely not exactly satisfied. Nevertheless the vortex pairing can be explained by this condition as was mentioned by Petersen (1978) for a circular jet and by Ho and Huang (1982) in a plane shear layer.

A similar interaction can occur, if an axisymmetric and a first azimuthal disturbance are simultaneously present in the jet. If the axisymmetric disturbance is strong (i.e., a ring vortex) compared with the first azimuthal one, then the motion of the vortex ring can be estimated by looking at the velocity field generated by the first azimuthal disturbance and superposed to the vortex motion. It can be shown that for $m = 1$ the velocity component normal to the jet axis can be written as

$$\mathbf{c}_n \equiv c_y + ic_z = \mathbf{c}_1(x, r, t) + \mathbf{c}_2(x, r, t)\, e^{i2\phi} \tag{2.23}$$

in the linearized approximation. Both contributions are schematically shown in Figure 13 in the plane normal to the jet axis for fixed r and t. Under the influence of this velocity field, the ring vortex will be shifted out of the jet axis and deformed in his circular structure. Additionally the axial velocity component of the first azimuthal disturbance will tend to tilt the vortex ring axially. If there is no fixed phase relationship between both disturbances as is to be expected, the motion of consecutive ring vortices will show a certain jitter, as observed in experiments.

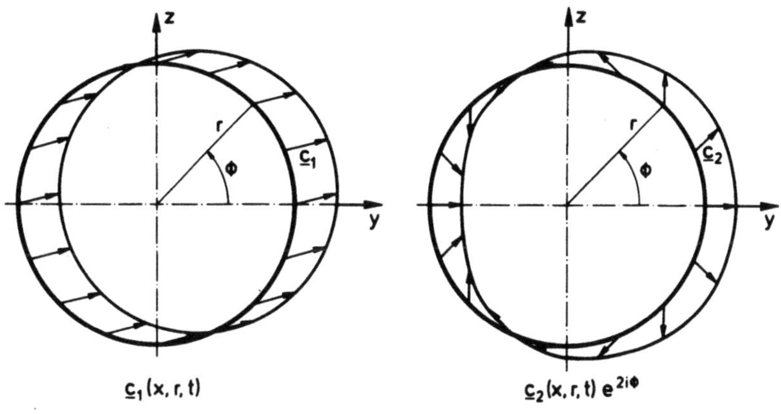

Figure 13. Velocity component $\mathbf{c}_n$ normal to the jet axis of the first azimuthal disturbance: $\mathbf{c}_n \equiv c_y + ic_z = \mathbf{c}_1(x, r, t) + \mathbf{c}_2(x, r, t)\exp(2i\phi)$.

3. Instability of Slowly Diverging Jet Flow

The assumption of an undisturbed parallel jet flow used so far is not strictly correct for an actual jet. Rather the shear layer thickness increases in downstream direction x. Hence the jet parameter R/θ decreases slowly with increasing x. Once a relation $\theta(x)$ like equation (2.6) is known, one could take the local parallel flow instability properties as a first approximation for the slowly diverging jet flow, as Chan (1974) did. However, this procedure is unsatisfactory, since it does not take into account that additional terms appear in the linearized disturbance equation. Since the axial basic velocity component is now $\overline{U}(r, s)$ where $s = \epsilon x$ is a slow variable for $\epsilon \ll 1$, the continuity equation requires a small radial velocity component $\epsilon\overline{V}(r, s)$. This term and $\partial\overline{U}/\partial x = \epsilon\partial\overline{U}/\partial s$ lead to additional terms of $0(\epsilon)$ in the linearized disturbance equation.

Woolley and Karamcheti (1973) discussed the problem with respect to free shear flows to $0(\epsilon^0)$ for finite Reynolds numbers and presented some results for a plane jet in connection with the edgetone generation [50]. However, problems arise, when terms of $0(\epsilon^1)$ are considered. For a circular jet Crighton and Gaster (1976) solved the inviscid problem by application of a multiple-scale expansion for axisymmetric disturbances. Plaschko (1979) extended the theory for non-axisymmetric disturbances. The basic principle of that theory is to expand the disturbances with respect to ϵ. To $0(\epsilon^0)$ the ex-

contained in $\overline{U}(r,s)$ and thus also in the eigenvalue $\alpha(s)$ acts as a flow parameter. To $0(\epsilon^1)$ the disturbance equation can be written as an inhomogeneous equation, which has only a solution, if the zero order term has a well-defined growth with respect to the slow variable s. This yields the pressure disturbance to $0(\epsilon^0)$:

$$p' = A_0 p_0(r, \epsilon x) \exp\left[i\left(\int_0^x \alpha(\epsilon x')\, dx' + m\phi - \omega t \right) - \int_0^{\epsilon x} (l(s)/k(s))\, ds \right] \tag{3.1}$$

where $l(s)$ and $k(s)$ are integrals over r, containing, among others, the adjoint eigenfunction of the homogeneous equation. $l(s) = 0$ for the purely parallel flow. A_0 is an initial disturbance magnitude and p_0 is the eigenfunction for the velocity profile $\overline{U}(r, \epsilon x)$. We see that the growth of the disturbance in downstream direction x is here much more complicated compared with the parallel flow instability. If a local eigenvalue $\overline{\alpha}$ by $-i\partial(\ln p')/\partial x$ is defined, one obtains

$$\overline{\alpha} = \alpha(\epsilon x) - i[\partial(\ln\, p_0(r, \epsilon x))/\partial x - \epsilon l(\epsilon x)/k(\epsilon x)]. \tag{3.2}$$

It is clear that the local phase velocity $\overline{c}_{ph} = \omega/\overline{\alpha}_r$ calculated with the real part of equation (3.2) depends not only on x, but, by the second term, also on the radial position r and is different for different flow quantities as pressure p', axial velocity u' or others.

Numerical calculations have been made by Crighton and Gaster (1976) and by Plaschko (1979) for the jet velocity Profile 2 with the jet parameter R/θ related to x by equation (2.6). In this case $\epsilon = d\theta/dx$ is sufficiently small. The corresponding radial velocity component $\overline{V}(r,s)$ has been calculated numerically from the continuity equation.

For the parallel flow the normalized eigenvalue $\alpha\theta$ is a function of $\omega\theta/U_0$ and R/θ as was shown above. Hence, for a disturbance with fixed frequency ω or Strouhal number $St = \omega R/(U_0\pi)$, the neutral case will be reached at a certain downstream distance x_n/R which is relatively small for high frequencies as can be derived from Figure 4 and equation (2.6) for $m = 0$ and 1. At this value x_n the adjoint eigenfunction becomes singular at the critical layer r_c where $\overline{U}(r_c, \epsilon x_n) = \omega/\alpha_r$. This leads to strong difficulties in the numerical evaluation of the integrals $k(s)$ and $l(s)$ of equation (3.1). This has been noted by Crighton and Gaster (1976) and by Plaschko (1979). In order to continue the computation beyond the neutral axial position x_n, it is necessary to circumvent the poles in the complex r-plane as it was already mentioned in connection with the

axial position x_n, it is necessary to circumvent the poles in the complex r-plane as it was already mentioned in connection with the damped inviscid disturbances. Plaschko (1983) has calculated the position of that poles for profile 2. For Strouhal numbers $St < 0.35$ all disturbances remain unstable (poles below the real r-axis) within the potential core region of the jet ($x/R \leq 12$) in agreement with the results of Figure 4.

By this method Plaschko (1983) was able to calculate the downstream development of the disturbances up to the end of the potential core in $0(\epsilon^0)$. The gain $G(p')$ of the pressure disturbances relative to $x = 0$ can be defined by $G = |p'(x, r, \phi, t)/p'(0, r, \phi, t)|$. It is shown in Figure 14 for $St = 0.5$, $r = 1.05\,R$ and $m = 0$, 1, and 2. It is obvious that the gain grows exponentially for small x, but then reaches a maximum between $x/R = 4$ and 6 depending on the azimuthal order m. Beyond the maximum the gain decreases downstream. Hence the growth of the disturbances is limited by the slow divergence of the jet flow, a fact that in the past has mostly been attributed to nonlinear effects. Crow and Champagne (1971) found experimentally this type of linear saturation in a turbulent jet for low excitation levels. Comparison with Figure 4 shows that the peak of the gain coincides with the local neutral position x_n, approximately, but it should be noted again that the gain depends on the radial position r and on the quantity considered, as was shown by Crighton and Gaster (1976).

The local phase velocity $\bar{c}_{ph}$ of the pressure fluctuations calculated by Plaschko (1983) at $r = 1.05\,R$ and for the Strouhal number $St = 0.5$ is shown in Figure 15 as function of the downstream distance x/R for various azimuthal order m. It is obvious that the phase velocity $\bar{c}_{ph}$ has a minimum close to $x/R = 2$ and decreases with increasing azimuthal order m. Concerning the local properties such as phase velocity c_{ph} of the locally parallel flow and $\bar{c}_{ph}$ of the slowly diverging flow, the opinions are controversial. While Crighton and Gaster (1976) emphasized that "any attempt to discuss the variation of $\bar{c}_{ph}$ with x purely from an examination of the local parallel flow stability problem is meaningless", Morris and Tam (1979) stated from their results for a supersonic jet that "all the calculations show that the gross features of the instability wave solution are given by the locally parallel flow approximation". Both conclusions may be more a question of the point of view, and they may be influenced by the great difference in the jet Mach numbers in both cases considered.

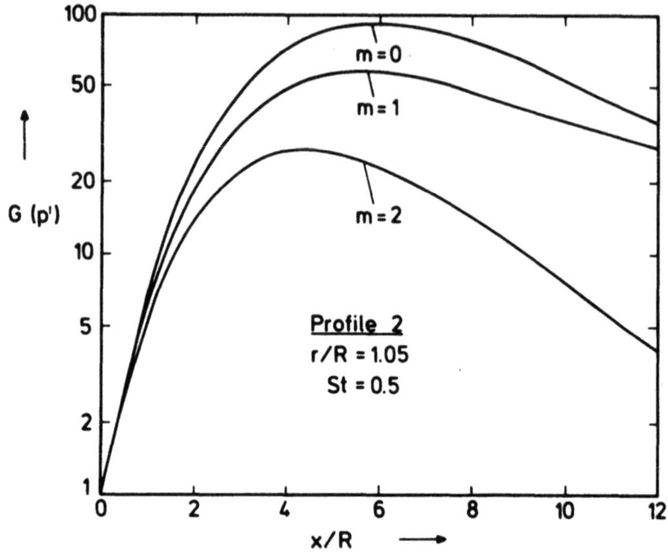

Figure 14. Gain $G(p')$ of pressure disturbances related to $x = 0$ at $r = 1.05\ R$ and a Strouhal number $St = 0.5$ vs. downstream distance x/R for Profile 2 and various azimuthal order m. Inviscid, isothermal, incompressible flow. (from [42])

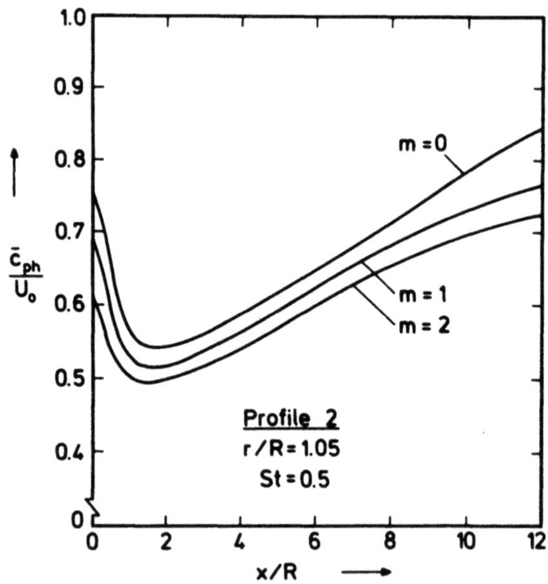

Figure 15. Local phase velocity $\overline{c}_{ph} = \omega/\overline{\alpha}_r$ of pressure disturbances at $r = 1.05\ R$ and a Strouhal number $St = 0.5$ vs. downstream distance x/R for Profile 2 and various azimuthal order m. Inviscid, isothermal, incompressible flow. (from [42])

4. Spatially Initial Value Problem

So far the jet flow has been considered as infinitely extended in the up- and downstream direction x. Hence the influence of the nozzle where the jet is generated was completely neglected. The initial conditions at the nozzle or pipe exit plane are shown schematically in Figure 16. First of all, the basic velocity profile $\overline{U}$ changes from a wall boundary layer velocity profile to a jet velocity profile with an inflexion point. Secondly, for a weakly disturbed flow, the particle displacement h of the separating streak line has to be zero at the nozzle exit plane. It is relatively little known about the influence of these initial conditions on the jet instability properties.

Concerning the behaviour of the displacement h close to the nozzle exit, the type of the Kutta condition to be applied has been controversial for long time. In a review paper, Crighton (1981) finally stated that the theoretical and experimental results must be seen as strong support for the application of the full Kutta condition in most high-Reynolds number problems. Then the displacement h leaves the nozzle lip tangentially.

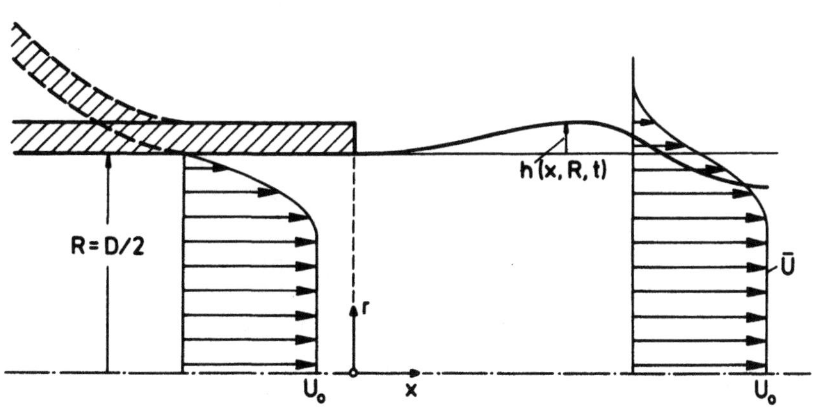

Figure 16. Spatially initial conditions of a jet at the nozzle or pipe exit.

In a plane case of a semi-infinite plate with uniform velocity on one side some results are known. The analytic solution of Bechert and Michel (1975) for this problem shown in Figure 17 offers the possibility to discuss the influence of the trailing edge-initial condition on the vortex sheed instability. The incompressible source imposes a transverse velocity $v_Q = \hat{v}_Q(x)\exp(-i\omega t)$ at the position of the undisturbed vortex sheet leaving the infinitely thin plate at $x = 0$. Here v_Q satisfies the boundary condition along the plate $(x < 0)$, i.e., v_Q is zero for $x < 0$, and is for $x > 0$ close to the trailing edge proportional to $x^{-1/2}$.

The main results, discussed in more detail and generalized by Bechert (1982, 1983) are as follows:

1. The full Kutta condition $h = 0(x^{3/2})$ is the only solution satisfying all boundary conditions.
2. The influence of the initial conditions (plate with trailing edge) on the vortex sheet instability vanishes at about one disturbance wavelength $\lambda = 2\pi/\alpha_r$ downstream of the trailing edge. Hence for larger distances the neglection of the initial conditions is admissible.
3. The far-downstream amplitude $(x > \lambda)$ of the exponentially growing disturbance is determined solely by the exciting source. If the distance r_0 of the source from the trailing edge is greater than a wavelength λ, only the excitation field immediately downstream of the trailing edge is important.

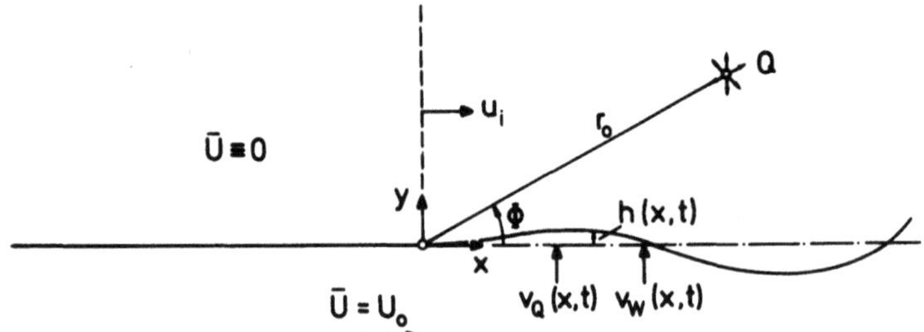

Figure 17. Vortex sheet leaving a semi-infinite plate of zero-thickness excited by a source.

4. The instability waves excited by the source are much stronger for upstream source position $\Phi > 90°$ than for downstream ones $\Phi < 90°$.

5. Without exciting source ($v_Q \equiv 0$) no instability waves exist.

The last result is most surprising. It is known that in actual free shear layers amplified disturbances are always present for high Reynolds numbers, even in absence of any recognizable, external excitation. Is it the upstream turbulence level of the flow which excites the disturbances? Or is it sufficient to excite the flow externally for a short time and then to switch off the source, in order to generate and maintain amplified disturbances in the shear layer? To answer these questions a generalisation of the problem to arbitrary time dependence of the source is necessary. It can be shown that for the semi-infinite plate problem the vortex sheet displacement $h(x,t)$ has to satisfy the following differential equation

$$\left[\left(\frac{\partial}{\partial t}+U_0\frac{\partial}{\partial x}\right)^2+\frac{\partial^2}{\partial t^2}\right]h(x,t)=2\frac{\partial v_Q(x,t)}{\partial t}. \tag{4.1}$$

This equation is similar to that describing the sound wave propagation in a one-dimensional channel with a supersonic mean flow. It is known from that case that waves can propagate only in the downstream direction. Furthermore, if the source will be switched off, the wave pattern generated up to this instant will propagate downstream with silence behind the wave tail. The same result is obtained from the solution of equation (4.1). The vortex sheet displacement is given by

$$h(x,t)=\frac{i}{U_0}\int_0^x dx'$$
$$[v_Q(x',t-(1+i)(x-x')/U_0)-v_Q(x',t-(1-i)(x-x')/U_0)] \tag{4.2}$$

The condition $h \equiv 0$ for $x < 0$ is satisfied provided $v_Q(x,t) \equiv 0$ for $x < 0$. It can be derived from this solution that after a switch-off of the source at time T_0 the displacement h of the vortex sheet is zero for $x/U_0 < t-T_0$. Hence there is no further oscillation of the vortex sheet after the excited wave has passed away, although axial velocity fluctuation u_i on both sides of the vortex sheet are still further existing up till the instability wave has decayed. Obviously no feed back mechanism between the downstream travelling instability wave and the trailing edge is active in this flow configuration.

The situation is, however, quite different, if the trailing edge of the plate consists of a corner with a wall along the positive y-axis, as shown by the dashed line in Figure 17. In order to satisfy the boundary condition along that wall, one has to compensate the axial velocity u_i normal to the wall by a source distribution q_w along with the y-axis. This source distribution q_w induces an additional normal velocity component v_w at the position of the undisturbed vortex sheet ($y = 0$). Hence the instability wave is determined not only by the source velocity v_Q, but also by this wall induced velocity v_w. Since the axial velocity component u_i at the wall is induced by the instability wave, it depends on v_Q and v_w, too. Therefore a feed-back loop is closed. If the source is switched off, neither u_i nor v_w will vanish immediately. As a consequence, the vortex sheet is further excited by itself. One can write down an integral equation for v_w, but its solution seems to be difficult. Hence it is not possible to decide whether the instability wave once excited is maintained by the self-excitation or decays gradually.

It is most likely that for a circular jet a similar feed-back mechanism exists. Numerical calculations by Maestrello and Bayliss (1982) support this idea. They calculated the response of an inviscid, compressible circular jet to an axisymmetric sound pulse by solving numerically the linearized Euler equations. The spreading mean flow was taken from experiments, and the pipe which generates the jet had a finite thick pipe wall. The numerical results show that a small sustained oscillation exists in the flow long after the original pulse has passed. It can be identified as a continual shedding of vortices from the nozzle lip. This result indicates a feedback mechanism, but may be taken again as a hint that some essential properties of jet instability are quite well modelled by the linearized equations.

References

[1] Batchelor, G. K. and Gill, E. A. "Analysis of the Stability of Axisymmetric Jets," *J. Fluid Mech. 14* (1962), 529–551.

[2] Bechert, D. "Excited Waves in Shear Layers." *Deutsche Luft- und Raumfahrt, DLR-FB 82–83,* 1982.

[3] ————. *A Model of the Excitation of Large Scale Fluctuations in a Shear Layer.* AIAA Paper 83-0724, 1983.

[4] Bechert, D. and Michel, U. "The Control of a Thin Free Shear Layer With and Without a Semi-Infinite Plate by a Pulsating Flow Field." *Acoustica 33* (1975), 287–307.

[5] Blumen, W., Drazin, P. G., and Billings, D. F. "Shear Layer Instability of an Inviscid Compressible Fluid, Part 2." *J. Fluid Mech. 71* (1975), 305–316.

[6] Chan, Y. Y. "Spatial Waves in Turbulent Jets." *Physics of Fluids, 17* (1974), 46–53.

[7] Chan, Y. Y. and Leong, R. K. "Discrete Acoustic Radiation Generated by Jet Instability." *Canadian Acoustics and Space Institute Trans. 6* (1973), 65–72.

[8] Crighton, D. G. "Acoustics as a Branch of Fluid Mechanics." *J. Fluid Mech. 106* (1981), 261–298.

[9] Crighton, D. G. and Gaster, M. "Stability of Slowly Diverging Jet Flow." *J. Fluid Mech. 77* (1976), 397–413.

[10] Crow, S. C. and Champagne, F. H. "Orderly Structure in Jet Turbulence." *J. Fluid Mech. 48* (1971), 547–591.

[11] Gaster, M. "A Note on the Relation Between Temporally-Increasing and Spatially-Increasing Disturbances in Hydrodynamic Stability." *J. Fluid Mech. 14* (1962), 222–24.

[12] Gill, A. E. "Instabilities of 'Top-Hat' Jets and Wakes in Compressible Fluid." *Physics of Fluid. 8* (1965), 1428–30.

[13] Gill, A. E. and Drazin, P. G. "Note on Instability of Compressible Jets and Wakes to Long-Wave Disturbances." *J. Fluid Mech. 22* (1965), 415.

[14] Gropengießer, H. "Beitrag zur Stabilität freier Grenzschichten in kompressiblen Medien." *Deutsche Luft- und Raumfahrt, DLR-FB 69-25,* 1969.

[15] Haertig, J. "Theoretical and Experimental Study of Wavelike Disturbances in a Round Jet With Emphasis Being Placed on Orderly Structures." In *Mechanics of Sound Generation in Flows.* Ed. E. A. Müller, New York: Springer Verlag, 1979, pp. 167–173.

[16] ————. *Une Solution Analytique du Probléme de l'instabilité d'un Jet Libre Rond Compressible.* Institut Franco-Allemand de Recherche de Saint-Louis, ISL-R 119/81, 1981.

[17] Ho, C. M. and Huang, L. S. "Subharmonics and Vortex Merging in Mixing Layers." *J. Fluid Mech. 119* (1982), 443–473.

[18] Kambe, T. "The Stability of an Axisymmetric Jet With Parabolic Profile. *J. Phys. Soc. Japan 26* (1969), 566–575.

[19] Kapur, S. S. and Morris, P. J. "A Model for the Orderly Structure of Turbulence in a Two-Dimensional Shear Layer." In *The Generation and Radiation of Supersonic Jet Noise.* Ed. H. E. Plumblee. Marietta, Georgia: Lockheed-Georgia, 1974, I-76–I-83.

[20] Kelly, R. E. "On the Instability of an Inviscid Shear Layer Which is Periodic in Space and Time." *J. Fluid Mech. 27* (1967), 657–689.

[21] Leconte, J. "On the Influence of Musical Sounds on the Flame of a Jet of Coal-Gas. *Phil. Mag. 15* (1858), 235–239.

[22] Lessen, M., Fox, J. A., and Zien, H. M. "The Instability of Inviscid Jets and Wakes in Compressible Fluid." *J. Fluid Mech. 21* (1965), 129–143.

[23] Lessen, M. and Singh, P. J. "The Stability of Axisymmetric Free Shear Layers." *J. Fluid Mech. 60* (1973), 443–457.

[24] Maestrello, L. and Bayliss, A. "Flowfield and Far Field Acoustic Amplification Properties of Heated and Unheated Jets." *AIAA Journal, 20* (1982), 1539–46.

[25] Maslowe, S. A. and Kelly, R. E. "Inviscid Instability of an Unbounded Heterogeneous Shear Layer." *J. Fluid Mech. 48* (1971), 405–415

[26] Mattingly, G. E. and Chang, C. C. "Unstable Wave on an Axisymmetric Jet Column." *J. Fluid Mech. 65* (1974), 541–560.

[27] Michalke, A. "On Spatially Growing Disturbances in an Inviscid Shear Layer." *J. Fluid Mech. 23* (1965), 521–544.

[28] ————. "A Note on the Spatial Jet-Instability of the Compressible Cylindrical Vortex Sheet." *Deutsche Luft- und Raumfahrt, DLR-FB 70-51,* 1970.

[29] ————. "Instabilität eines kompressiblen runden Freistrahls unter Berücksichtigung des Einflusses der Strahlgrenzschichtdicke." *Z. Flugwiss. 19* (1971), 319–328. English translation *NASA Tech. Memo. 75190,* 1977.

[30] ————. "The Instability of Free Shear Layers—A Survey on the State of the Art." *Progr. Aerospace Sci. 12,* 1972. Ed. D. Küchemann, New York: Pergamon Press, 213–239.

[31] Michalke, A. and Freymuth, P. "The Instability and the Formation

of Vortices in a Free Boundary Layer." *AGARD Conference Proc. No. 4, Separated Flows, Part II,* 1966, 575–595.

[32] Michalke, A. and Hermann, G. "On the Inviscid Instability of a Circular Jet With External Flow." *J. Fluid Mech. 114* (1982), 343–359.

[33] Michalke, A. and Schade, H. "Zur Stabilität von freien Grenzschichten." *Ing. Arch. 33* (1963), 1–23.

[34] Michel, U. and Michalke, A. "Prediction of Fly-Over Jet Noise Spectra." *AIAA Paper No. 81-2025,* 1981.

[35] Mollendorf, J. C. and Gebhart, B. "An Experimental and Numerical Study of the Viscous Stability of a Round Laminar Jet With and Without Thermal Bouyancy for Symmetric and Asymmetric Disturbances." *J. Fluid Mech. 61* (1973), 367–399.

[36] Morris, P. J. "The Spatial Viscous Instability of Axisymmetric Jets." *J. Fluid Mech. 77* (1976), 511–529.

[37] ———. "Viscous Stability of Compressible Axisymmetric Jets." *AIAA Journal 21* (1983), 481–482.

[38] Morris, P. J. and Tam, C. K. "On the Radiation of Sound by Instability Waves of a Compressible Axisymmetric Jet." In *Mechanics of Sound Generation in Flows.* Ed. E. A. Müller. New York: Springer Verlag, 1979, 55–61.

[39] Neuwerth, G. "Theorie der Ausbreitung wellenförmiger Störungen entgegen der Strömungsrichtung in einem Unterschall-Freistrahl." *Deutsche Luft und Raumfahrt, DLR-FB 74-20,* 1974.

[40] Petersen, R. A. "Influence of Wave Dispersion on Vortex Pairing in a Jet." *J. Fluid Mech. 89* (1978), 469–495.

[41] Plaschko, P. "Helical Instabilities of Slowly Diverging Jets." *J. Fluid Mech. 92* (1979), 209–215.

[42] ———. "Axial Coherence Functions of Circular Turbulent Jets Based on Inviscidly Calculated Damped Modes." Submitted to *Physics of Fluid,* 1983.

[43] Rayleigh, Lord. "On the Instability of Jets." *Proc. London Math. Soc. 10* (1879), 4–13.

[44] Schade, H. "Zur Stabilitätstheorie axialsymmetrischer Parallelströmungen." *Ing. Arch. 31* (1962), 301–316.

[45] Tam, C.K.W. "Supersonic Jet Noise Generated by Large Scale Disturbances." *J. Sound Vibr. 38* (1975), 51–79.

[46] Tyndall, J. "On the Action of Sonorous Vibrations on Gaseous and Liquid Jets." *Phil. Mag. 33* (1867), 375–391.

[47] Watson, J. "On Spatially-Growing Finite Disturbances in a Plane

Poiseuille Flow." *J. Fluid Mech. 14* (1962), 211–221.

[48] Wille, R. "Beiträge zur Phänomenologie der Freistrahlen." *Z. Flugwiss. 11* (1963), 222–233.

[49] Woolley, J. P. and Karamcheti, K. *A Study of Narrow Band Noise Generation by Flow Over Ventilated Walls in Transonic Wind Tunnels.* Wright-Patterson Air Force Base, Ohio: Air Force Office of Scientific Research, *AFOSR-TR-73-0503,* 1973.

[50] ——————————. "Role of Jet Stability in Edgetone Generation." *AIAA Journal 12* (1974), 1457–58.

Large-Scale Coherent Structures in Free Turbulent Flows and Their Aerodynamic Sound

J.T.C. Liu
Brown University
Providence, RI 02912

Abstract

Observed physical features of large-scale coherent structures in free shear flows are first interpreted from a point of view that involves the results from conservation principles, including, for instance, the role of fluctuating disturbances on the spreading rate, mode-mode interactions, the growth and decay of coherent structures and their negative production rate and energy exchanges between the various scales of motion. This sets the stage for discussing the large-scale coherent structures as a source of aerodynamic sound in turbulent jets. The aerodynamic sound is worked out from Lighthill's formulation and it is found that the lower frequency sound, which comes from lower frequency coherent structures that peak further downstream, radiates preferentially nearer the jet axis; the peak radiation moves away from the jet axis as the frequency increases with the higher frequency contributions coming from coherent structures that peak in regions nearer the nozzle lip; the peak frequency near the jet axis being independent of the jet velocity. These striking resemblances with observations are a direct consequence of Lighthill's formulation of the aerodynamic sound problem, being brought about by taking into account as sources the spectrally-dependent growing and decaying large-scale coherent structures in a real turbulent jet.

1. Introduction

Impetus towards the study of sound generation by coherent structures in shear flows is contributed in no small way by K. Karamcheti and his collaborators (see, for instance, Karamcheti (1956); Karamcheti et al. (1969); Krothapalli et al. (1982)). It is thus rather appropriate to discuss the subjects of coherent structures and their far sound field on the occasion of the tenth anniversary of the founding of the Joint Institute for Aeronautics and Acoustics at Stanford University, and of the honoring of one of its founders and director, K. Karamcheti. I am personally celebrating the fourteenth anniversary of meeting K. Karamcheti at the 1969 NASA Headquarters Meeting on Aerodynamic Sound.

While the existence of coherent structures in free laminar shear flows is attributable to dynamical instabilities, the recognition that similar structures in free turbulent shear flows are a manifestation of hydrodynamical instabilities is fairly recent (Liepmann, 1962). The scientific *discovery* of coherent structures in turbulent flows is traceable to Corrsin (1943) and Townsend (1947). They found that the flow near the outer edge of a turbulent jet or wake is only intermittently turbulent. In the flow between rotating cylinders, Pai (1939) and MacPhail (1941) both found that Taylor (1923) vortices persisted well after fine-grained turbulence came into being. In the wake behind a circular cylinder for Reynolds numbers above 150, Roshko (1952) found that far downstream nearly regular large-scale structures coexisted with turbulence of much finer scales; similar observations were also reported by Taneda (1959). Although the theoretical basis for the existence of such large-scale structures remained unclear at the early stages, their technological importance to mixing in combustion problems and to sound generation were recognized (Liepmann, 1952). The recent striking observational evidence of propagating wavy large-scale structures is given, for instance, by Brown and Roshko (1974) and Papailiou and Lykoudis (1974). We refer, for instance, to Cantwell (1981) and Hussain (1983) for surveys of more recent widespread observations of such structures. In this paper we shall present the use of a much broader view of hydrodynamical instabilities, based on conservation principles, towards the elucidation and connection of facts given to us by observations, thereby contributing to a firmer physical understanding of the mechanisms at play than would be possible via kinematics alone. The basic concept from hydrodynamic stability tells us that unstable wave disturbances in certain frequency ranges are more efficient in extracting energy from the mean flow than other ranges and, in the case of coexistence with turbulence of much finer scales, certain range of modes are more efficient in transferring energy to the fine-grained turbulence. Thus the imbalance between the rate of energy supply and loss, which has rather strong spectral dependence, determines the evolution of the coherent structures in a developing turbulent free shear flow. The streamwise lifespan, the peak amplification and its location are thus all functions of the frequency. We shall show that the coherent structures are responsible for the highly directional radiation pattern, which is frequency dependent, that is found in the far sound field of a turbulent jet thereby bringing intimately together the appropriate sound sources (Mankbadi and Liu, 1981) and the resulting aerodynamic sound field (Lighthill, 1952, 1962) in accordance with features of observations.

2. Mode-Mode Interactions and Mean Flow Spreading Rate

To begin the discussion of mode interactions, it would be most helpful to first illustrate the streakline patterns, obtained calculationally by Williams and Hama (1980) from the superposition of wavy disturbances of the fundamental mode and its subharmonic, upon a hyperbolic tangent mean velocity profile. This is reproduced in Figure 1. The top two patterns are, respectively, the fundamental and the subharmonic with constant amplitudes. The remaining three patterns are for the superposition of subharmonic to fundamental amplitude ratio of 1:1 with zero phase difference, 2:1 with zero phase, and 2:1 with $\pi/2$ phase. These bear striking resemblances with the visual observation of dye streak behavior in a mixing layer (e.g., Freymuth, 1966; Winant and Browand, 1974; Ho and Huang, 1982). However, the streakline calculations of Williams and Hama (1980) come from a linear superposition of two constant amplitude wave disturbances, the pairing and roll up are the consequence of the wave-interference. The simulated wave amplitudes of the fundamental and subharmonic are both constant and the switching of modal structure, as the visual appearance of streaklines would otherwise suggest, is absent. We are thus cautioned by this illustration, that dye streak behavior are not necessarily indicative of unique physical circumstances without simultaneous quantitative measurements.

We shall discuss quantitative measurements suggesting mode-mode interactions between the fundamental disturbance wave and its subharmonic in a shear layer, such as that reported by Ho and Huang (1982). Their shear layer is essentially one undergoing transition, and the presence of such distinct modes is brought about by forcing at the subharmonic frequency. The significance of Ho and Huang's (1982) work lies in the identification of the visually observed location of *pairing*, indicated by the accumulation of dye streaks, with the occurrence of the measured sectional-energy maximum of the subharmonic (actually, the kinetic energy associated with the streamwise velocity fluctuation, integrated across the shear layer); however, there was no abrupt switching from the fundamental frequency and wavelength to those of the subharmonic. Reproduced in Figure 2 (corresponding to Mode II of Ho and Huang (1982)) is the evolution of the measured sectional-energy associated with the streamwise velocity fluctuation. The 2.15 Hz curve corresponds to the forced, subharmonic component, the 4.30 Hz curve is the fundamental. Although the peak amplitudes of the two modes are distinct, the fading in of the subharmonic occurs in regions of active fundamental development and, in turn, the fading out of the fundamental takes place in regions where the subharmonic is active. The measurements most certainly suggest a natural occurrence of the switch-on and switch-off process compared to an abrupt switch in the modal content that might have been suggested by visual observations of dye streaks alone. (This has far-reaching contrasting consequences in the aerodynamic sound problem to be discussed subsequently.)

The theoretical understanding of mode-mode interactions in a developing shear layer is best undertaken for the case of a laminar shear flow without the involvement of the fine-grained turbulence at the outset (Liu and Nikitopoulos, 1982). The measureable sectional energy content of each mode is essentially the square of the amplitude of the coherent structure. For each frequency, it is a fixed streamwise envelope, under which the propagating wavy disturbance enters from its initiation upstream and exits downstream, due to the possible *dissipative* mechanisms or persists in some equilibrium balance. The study of mode-mode interactions has, as its aim, the understanding of the direction of energy transfer between the modes and its effect on establishing the spatial distribution of this *envelope* and hence the rate of spread of the shear flow (Nikitopoulos, 1982; Liu and Nikitopoulos, 1982).

If an essemble of disturbances exist in a shear flow, and we split the modes into *odd* (denoted by $(\tilde{\ })$) and *even* (denoted by $(\tilde{\tilde{\ }})$), then the energy transfer from the even to the odd modes is given by (Stuart, 1962):

$$\overline{\tilde{u}_i \tilde{u}_j \frac{\partial \tilde{\tilde{u}}_i}{\partial x_j}}$$

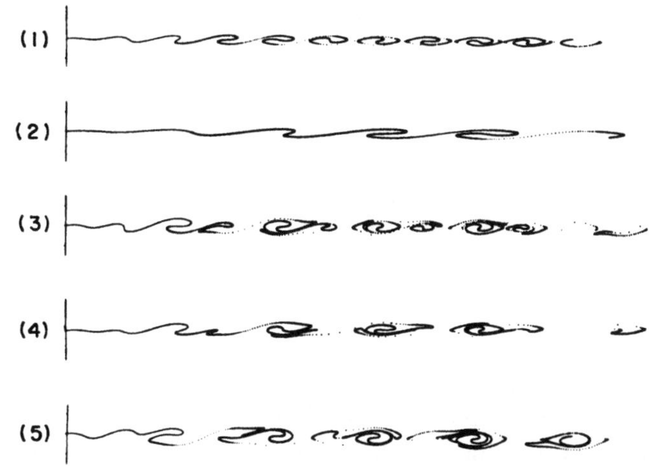

Figure 1. Calculated streaklines is a shear layer, from Williams and Hama (1980). (1) fundamental alone; (2) subharmonic alone; (3) fundamental/subharmonic amplitude ratio 1:1, zero phase; (4) 1:2, zero phase; (5) 1:2, $\pi/2$ phase.

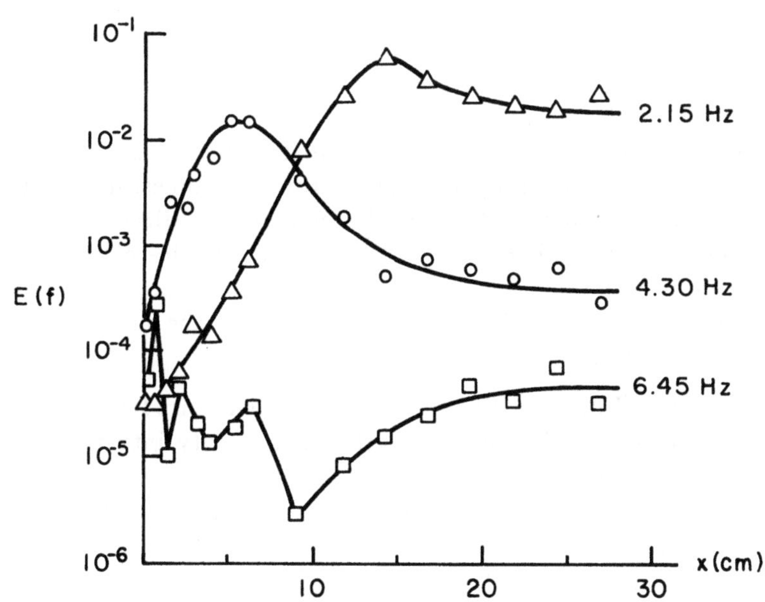

Figure 2. Measured streamwise development of the energy content across the shear layer associated with the axial velocity, $\int_{-\infty}^{\infty} \overline{\tilde{u}^2}\, dz/2\overline{U}^2\theta_0$. Forcing at subharmonic frequency 2.15 Hz at $\tilde{u}/\overline{U} = 0.10\%$, $R = 0.31$, phase relation uncontrolled. (Ho and Huang, 1982).
Courtesy of Cambridge University Press.

where the average is taken over the largest periodicity of the disturbances. The mechanism is the work done (by the stresses of the odd modes) against the appropriate rates of strain (of the even modes). It is clear that the phase relation between the stresses and the rates of strain determines the direction of energy transfer, and that the amplitudes determine the strength of this transfer (the "Kelly mechanism", Kelly, 1967; Liu, 1981).

For a spatially developing shear layer, Liu and Nikitopoulos (1982) considered the interaction between the subharmonic mode (a single *odd* mode) and its fundamental (a single *even* mode). If the energy content of the fundamental mode across the shear layer is denoted by E_2 and that of the subharmonic by E_1, then the overall energy transfer mechanism between the modes is proportional to $E_1 E_2^{1/2}$. In contrast, the respective energy production rate from the mean flow is proportional to E_1 and to E_2. This is also the case for the rate of viscous dissipation. The energy content across the shear layer, having been made dimensionless by that of the free stream, is much less than unity according to observations. In this case, the estimate here shows that the individual energy production from the mean motion to dominate over that of the mode-mode energy transfer. In other words, the mode-mode interactions are dominated by implicit nonlinear interactions via the mean motion rather than by the more explicit direct energy transfer mechanism. However, the latter is most certainly important towards affecting the details of the amplitude distribution in the streamwise direction. In the experiments of Ho and Huang (1982) there are modes other than the fundamental and the subharmonic present, including initially weak fine-grained turbulent disturbances, and these are not included in the analysis (Liu and Nikitopoulos, 1982).

To begin (Liu and Nikitopoulos, 1982), we start from the Navier-Stokes equations, and split the total flow quantity into that for the mean flow U_i and the overall disturbance $(\tilde{u}_i + \tilde{\tilde{u}}_i)$, with $(\bar{\ })$ as the usual Reynolds' (1895) time average. For purposes of obtaining the measurable *envelope* (or amplitude) equations, we obtain the kinetic energy equations for the different scales of motion:

Mean Motion:

$$\frac{\overline{D}}{Dt}\frac{U_i^2}{2} = -(\overline{\tilde{u}_i\tilde{u}_j} - \overline{\tilde{\tilde{u}}_i\tilde{\tilde{u}}_j})\frac{\partial U_i}{\partial x_j} - \nu\left(\frac{\partial U_i}{\partial x_j}\right)^2 + \overline{D}, \tag{2.1}$$

Odd Modes:

$$\frac{\overline{D}}{Dt}\frac{\overline{\tilde{u}_i^2}}{2} = -\overline{\tilde{u}_i\tilde{u}_j}\frac{\partial U_i}{\partial x_j} - \overline{\tilde{u}_i\tilde{u}_j\frac{\partial \tilde{\tilde{u}}_i}{\partial x_j}} - \nu\overline{\left(\frac{\partial \tilde{u}_i}{\partial x_j}\right)^2} + \overline{\tilde{d}}, \tag{2.2}$$

$\tilde{\text{p}}$roduction   mode interaction

Even Modes:

$$\frac{\overline{D}}{Dt}\frac{\overline{\tilde{\tilde{u}}_i^2}}{2} = -\overline{\tilde{\tilde{u}}_i\tilde{\tilde{u}}_j}\frac{\partial U_i}{\partial x_j} + \overline{\tilde{u}_i\tilde{u}_j\frac{\partial \tilde{\tilde{u}}_i}{\partial x_j}} - \nu\overline{\left(\frac{\partial \tilde{\tilde{u}}_i}{\partial x_j}\right)^2} + \overline{\tilde{\tilde{d}}}, \tag{2.3}$$

$\tilde{\tilde{\text{p}}}$roduction

where the overbar on the substantial derivative indicates that it is taken following the mean motion. In the disturbance energy equations, the first term on the right side is the *production* and these have counterparts in the mean flow energy equation; the second term is the mode-mode energy exchange; the third term is the rate of viscous dissipation. The so called *diffusion* effects are symbolically represented by $\overline{D}, \overline{\tilde{d}}$ and $\overline{\tilde{\tilde{d}}}$ for the various scales of motion. If we specialize the odd modes to a single subharmonic mode and the even modes to the fundamental, the kinetic energy equations (2.1)–(2.3) integrated across the two-dimensional shear layer then take the form:

$$\begin{aligned}\frac{1}{2}\frac{d}{dx}&\left[\int_{-\infty}^{0} U\,(U^2 - U_{-\infty}^2)\,dz + \int_{0}^{\infty} U\,(U^2 - U_{+\infty}^2)\,dz\right]\\ &= -\int_{-\infty}^{\infty}\left(-\overline{\tilde{u}\tilde{w}} - \overline{\tilde{\tilde{u}}\tilde{\tilde{w}}}\right)\frac{\partial U}{\partial z}\,dz - \overline{\Phi},\end{aligned} \tag{2.4}$$

$$\begin{aligned}\frac{1}{2}\frac{d}{dx}\int_{-\infty}^{\infty} U\,(\overline{\tilde{u}^2} + \overline{\tilde{w}^2})\,dz &= \int_{-\infty}^{\infty}(-\overline{\tilde{u}\tilde{w}})\frac{\partial U}{\partial z}\,dz\\ -\int_{-\infty}^{\infty}&\left[\overline{\tilde{u}^2\frac{\partial \tilde{\tilde{u}}}{\partial x}} + \overline{\tilde{u}\tilde{w}\left(\frac{\partial \tilde{\tilde{u}}}{\partial z} + \frac{\partial \tilde{\tilde{w}}}{\partial x}\right)} + \overline{\tilde{w}^2\frac{\partial \tilde{\tilde{w}}}{\partial z}}\right]dz - \overline{\tilde{\phi}},\end{aligned} \tag{2.5}$$

$$\frac{1}{2}\frac{d}{dx}\int_{-\infty}^{\infty} U\left(\overline{\tilde{\tilde{u}}^2}+\overline{\tilde{\tilde{w}}^2}\right)dz=\int_{-\infty}^{\infty}(-\overline{\tilde{\tilde{u}}\tilde{\tilde{w}}})\frac{\partial U}{\partial z}\,dz$$
$$+\int_{-\infty}^{\infty}\left[\overline{\tilde{u}^2\frac{\partial \tilde{\tilde{u}}}{\partial x}}+\overline{\tilde{u}\tilde{w}\left(\frac{\partial \tilde{\tilde{u}}}{\partial z}+\frac{\partial \tilde{\tilde{w}}}{\partial x}\right)}+\overline{\tilde{w}^2\frac{\partial \tilde{\tilde{w}}}{\partial z}}\right]dz-\bar{\tilde{\tilde{\phi}}}, \tag{2.6}$$

where x is the streamwise coordinate measured from the start of the mixing layer, z is the vertical coordinate measured from the center of the mixing layer, u, w are the x, z fluctuation velocities, U is the mean velocity with $\pm\infty$ denoting the upper and lower streams, respectively, $\overline{\Phi}$ is integral of mean flow viscous dissipation, the lower case ϕ represents the corresponding dissipation rates of the fluctuations. Equations (2.4)–(2.6), where two-dimensional wavy disturbances in a two-dimensional mean flow has been assumed, form the basis for obtaining the evolution equations for the sectional energies of the disturbances. Following earlier work (see, for instance, Liu, 1981), the disturbances are assumed to take the separable form of the product of an unknown amplitude $A_i(x)$ with a vertical distribution function given by the local linear stability theory,

$$\begin{bmatrix}\tilde{u}\\ \tilde{w}\end{bmatrix}=A_1(x)\left[\begin{pmatrix}\phi_1' e^{-i\beta t}\\ -i\alpha_1\phi_1 e^{-i\beta t}\end{pmatrix}+\begin{pmatrix}\text{c.c.}\\ \text{c.c.}\end{pmatrix}\right]$$
$$\begin{bmatrix}\tilde{\tilde{u}}\\ \tilde{\tilde{w}}\end{bmatrix}=A_2(x)\left[\begin{pmatrix}\phi_2' e^{-2i\beta t-i\theta}\\ -i\alpha_2\phi_2 e^{-2i\beta t-i\theta}\end{pmatrix}+\begin{pmatrix}\text{c.c.}\\ \text{c.c.}\end{pmatrix}\right],$$

where ϕ_i here denotes the eigenfunction of the local linear theory and is a function of the rescaled vertical variable $\eta = z/\delta(x)$, where $\delta(x)$ is a length scale of the mean flow yet to be identified and $(\ )'$ denotes differentiation with respect to η; $\beta = 2\pi f\delta(x)/\overline{U}$ is the dimensionless local frequency, f is the physical frequency and $\overline{U} = (U_\infty + U_{-\infty})/2$, the local wave numbers α are also scaled by $\delta(x)$; θ is the relative phase between the fundamental component (2β) and its subharmonic (β) and c.c. denotes the complex conjugate. The velocities and lengths are considered to be made dimensionless by $\overline{U}$ and δ_0 (so that $\delta(0) = 1$), and time by $\delta_0/\overline{U}$. The mean velocity profile is taken to be

$$U = 1 - R\tanh\eta,$$

where

$$R = (U_{-\infty} - U_\infty)/(U_{-\infty} + U_\infty).$$

(a)

(b)

(c)

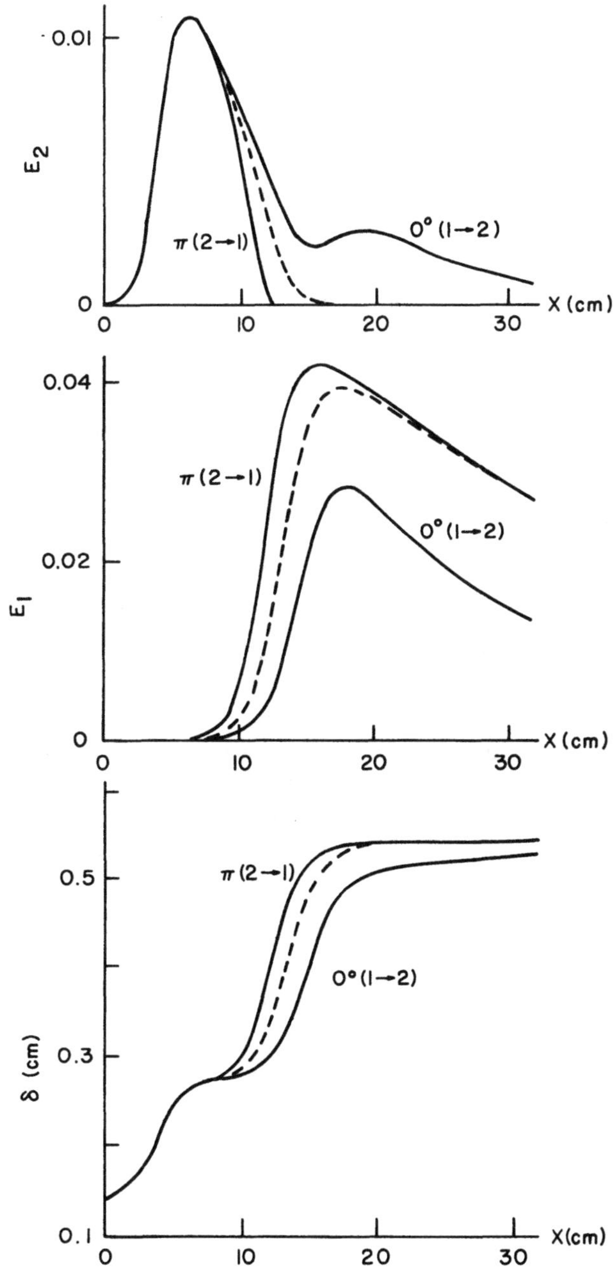

Figure 3. Streamwise development of calculated (a) fundamental energy content E_2; (b) subharmonic energy content E_1; (c) shear layer thickness for phase relation $0°$ ($E_1 \to E_2$), $\pi(E_2 \to E_1)$ and mode-mode energy transfer set equal to zero (in dashed lines). $R = 0.31$. (Liu and Nikitopoulos, 1982).

the subharmonic, and this is responsible for the earlier decay of the fundamental energy E_2 (Figure 3a) as well as the earlier peak, augmenting the direct energy supply from the mean motion, of the subharmonic energy E_1 (Figure 3b). The dashed line in Figure 3 indicates the result obtained from setting the energy transfer mechanism between the two modes to zero. The resulting growth of the mean flow is shown in Figure 3c, the first plateau is due to the peak in the fundamental, the second due to the peaking of the subharmonic according to (2.7). Because the interaction between the mean flow and the amplified disturbances is strong, the rapid spreading rate is a part of the nonlinear interaction process and ought not to be a presumed known input. The shear layer thickness due to the subharmonic is very nearly double that due to the fundamental in Figure 3c. That is, the ratio of the two plateau is nearly 2. However, this is somewhat dependent upon the initial conditions and ought not to be a general *rule of thumb*.

These calculations strongly resemble the quantitative measurements of Ho and Huang (1982), including the mean flow spreading rate reproduced in Figure 4. Their momentum thickness is half the value of the shear layer thickness of Figure 3c. There is also an approximate doubling of the thickness noticed earlier. The plateaus are clearly attributed to the net energy loss from the mean flow directly to the disturbances according to (2.7). The interaction between the disturbance modes has but an *indirect* effect. The continued subsequent spreading of the mean flow (Figure 4) in the experiments is attributable to other disturbances which are not accounted for here.

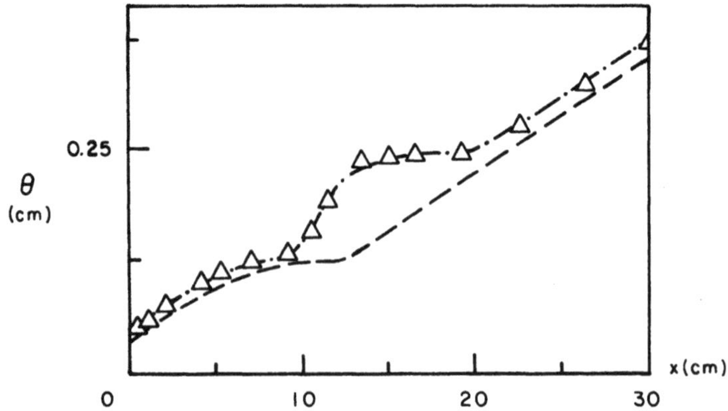

Figure 4. Measured development of momentum thickness, same conditions as Figure 2. (Ho and Huang, 1982). Courtesy of Cambridge University Press.

These simple ideas can be easily extended to include the presence of fine-grained turbulence, or isolated to the absence of any disturbances, in which case the shear flow spreads because of viscosity alone.

In Liu (1981), Kelly's mechanism has been discussed in a much broader context than the weakly nonlinear theory from which it was obtained. In the light of the calculations of Nikitopoulos (1982) given here for strongly amplified disturbances, we should mention that the details of Kelly's work (1967) for subharmonic growth are obtained for a neutral fundamental component, and this would correspond to a location in x in Figure 3 where $dE_2/dx = 0$. Nikitopoulos' (1982) results indicate that for $\theta = 0°$ and the fundamental component being neutral, energy flows from the fundamental to the subharmonic in support of Kelly (1967). However, this does not necessarily imply that the overall subharmonic growth rate is greater for $\theta = 0°$ than, say, for $\theta = \pi$ because the strength of the energy supply from the mean flow is dependent upon the initial amplitudes. In the context of strongly amplified disturbances in a developing mean shear flow, however, this issue is largely academic as the region in question appears quite obscure to the overall nonlinear development illustrated in Figure 3.

Liu and Nikitopoulos (1982) have also studied the three-mode interaction problem. This, and the two-mode problem briefly discussed here, shall appear elsewhere in greater detail.

To conclude this section, we want to emphasize that the spreading rate of the mean flow is proportional to the rate at which energy is removed from the mean flow. For a purely laminar viscous flow, only viscous dissipation contributes to the spreading $\overline{\Phi} \sim \nu U_\infty^2/\delta$ where ν is the kinematic viscosity; thus $\delta \sim \sqrt{x}$ as expected. For a laminar flow undergoing transition, the rate of energy transfer to originally small disturbances, reflected by the $-\overline{\tilde{u}\tilde{w}}$ Reynolds stress conversion mechanism (including, for simplicity in notation an *essemble* of coherent modes), now competes with the viscous dissipation. When the disturbances have become sufficiently finite, a marked deviation from the purely viscous spreading rate would be noticed (see, for instance, Sato and Kuriki, 1961; Ko, Kubota, and Lees, 1970). In the presence of both a fundamental disturbance and its subharmonic, such as in the experiments of Ho and Huang (1982), where the peak in the finite amplitudes are distinctively separated in space, the growth of the shear layer undergoes successive plateaus; the vigorous shear layer growth regions are associated with active energy extraction from the mean flow for the disturbance amplification, and the plateau regions associated with decaying disturbance amplitudes. In Ho and Huang's (1982) experiments, the shear layer continues to spread after the plateau regions; it is most likely that transition has taken place, in that the existing fine-grained turbulence having been sufficiently strained by the coherent structures, is now contributing towards the mean flow spreading rate via the fine-grained turbulence $-\overline{u'w'}$ Reynolds stress mechanism. For large-scale coherent structures in a turbulent shear flow both $-\overline{\tilde{u}\tilde{w}}$ and $-\overline{u'w'}$, but depending on their relative strength, contribute to the growth of the mean shear flow. In the downstream region where a particular mode of coherent structure has rearranged its velocity distribution; such that $-\overline{\tilde{u}\tilde{w}}$ is opposite the sign of $\partial U/\partial z$, then energy is returned to the mean motion from this particular mode and this contributes to the decrease of the spreading rate. The observed phenomenon of negative energy production associated with coherent structures (see, for instance, Fiedler et al., 1980) is not unusual and we shall discuss this further in connection with the growth and decay of coherent structures.

3. Mechanisms for Coherent Structure Growth, Decay, and Interactions With Fine-Grained Turbulence

The quantitative experimental study of coherent structures in natural turbulent shear flows is made difficult, not only by the *jittering* of their phases (Thomas and Brown, 1977) but also by the presence of an essemble of modes with frequency ratios not necessarily conducive to well-defined interactions of the subharmonic-fundamental-harmonic type. These, together with the frequency dependent growth and decay processes of the coherent structures, would most likely become obscured in correlation measurements. The study of well-controlled coherent structures in turbulent shear flows in the same spirit of Schubauer and Skramsted (1948) was suggested by Liepmann (1962), as was the splitting of the flow into the large-scale coherent structure, fine-grained turbulence and mean motion (see also Townsend, 1956). However, it was then not clear as to how the large-scale fluctuations could be sorted out from the overall fluctuations. The imposition of a fixed-phase disturbance on a turbulent shear flow and the use of the phase average enables one to sort out the coherent structure from the random fine-grained turbulence, augmenting the usual Reynolds (1895) averaging procedure (which is the time average for spatially developing flows). The experimental realization of the method was discussed by Karlsson (1959), Hussain and Reynolds (1970), Kendall (1970), Binder and Favre-Marinet (1973), Favre-Marinet and Binder (1979), Cantwell (1981) and recently reviewed by Hussain (1983).

In order to fix ideas, we refer to the experimental results of Favre-Marinet and Binder (1979), reproduced here in Figure 5. They imposed a rather large coherent disturbance in the nozzle amounting to about 30% of the jet exit velocity and at a Strouhal number (St) of 0.18. The hot wire measurements on the jet axis is shown in this figure in terms of the streamwise velocities normalized by the local jet centerline velocity. The coherent structure amplified and then decayed along the jet while the fine-grained turbulence level increased at a faster rate than it would in the absence of the imposed coherent structure. The spreading of the jet was also enhanced with a more rapid decay of the mean centerline velocity. In this case, it appears that a *burst* of fine-grained turbulence follows the demise of the coherent structure.

The mechanism of coherent structure growth can be understood through its interaction with the mean motion via the dominant rate of energy production $-\overline{\tilde{u}\tilde{w}}\partial U/\partial z$ in the case of the shear layer,

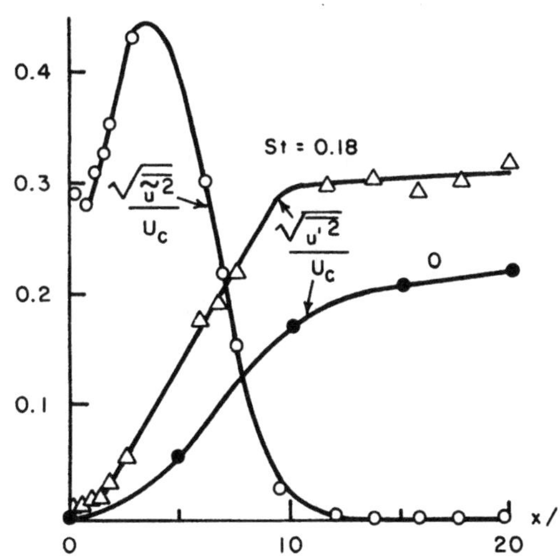

Figure 5. Measured coherent and turbulent axial velocities on the centerline of a forced turbulent jet. (Favre-Marinet and Binder, 1979).

which is somewhat simpler for illustrating the physical situation than the round jet problem (Mankbadi and Liu, 1981). In a controlled shear layer, Fiedler et al. (1980) measured this mechanism along the line of maximum shear (see Figure 6) and compared it to the total fluctuation production mechanism. While the random fluctuation production remained positive, that associated with the coherent structure increased, reflecting the amplification process, and then decreased to below the axis reflecting negative production or a return of kinetic energy to the mean shear flow further downstream. Rather than to describe this situation as a *negative viscosity phenomena*, this is best understood through the basic idea that the coherent structures are propagating instability waves, and as such are dynamically unstable in free turbulent flows. Their propagation into regions of spreading mean flow has allowed them to rearrange the phase shifts between the velocity components, on the average, so as to become *damped* disturbances in the *inviscid* sense, a situation which is well understood from studies in hydrodynamic stability. The typical development of the production mechanism of coherent structures in a turbulent shear flow is shown in Figure 7. The situation is the time-developing shear layer (Gatski and Liu, 1980) where the time t is qualitatively equivalent to the streamwise distance x of a spatially growing shear layer. (Although not shown

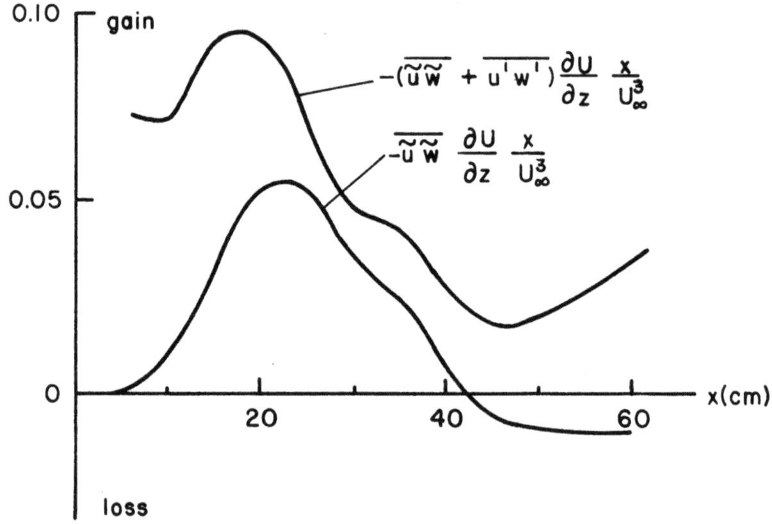

Figure 6. Total fluctuation and coherent energy production mechanisms, measured along the line of maximum mean shear of a forced turbulent shear layer. (Fiedler et al., 1981).

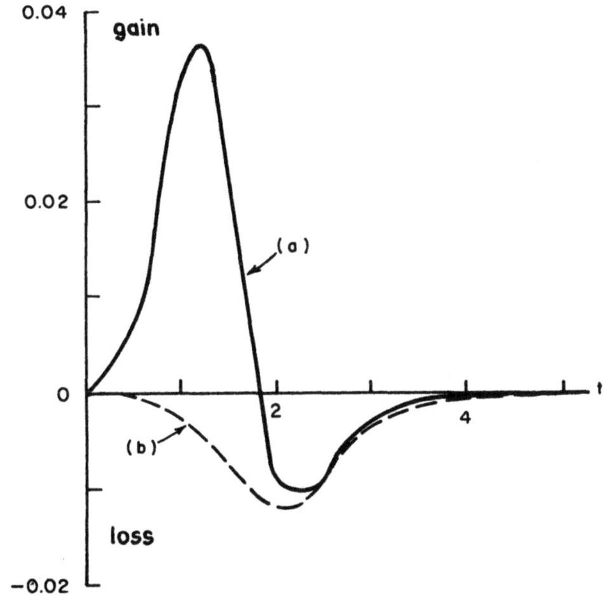

Figure 7. Development of large-scale structure energy transfer mechanisms: (a) energy transfer from the mean flow:

$$\int_{-\infty}^{\infty} -\overline{\tilde{u}\tilde{w}}\frac{\partial U}{\partial z} dz,$$

(b) energy transfer to the fine-grained turbulence:

$$\int_{-\infty}^{\infty} -\left[\overline{\tilde{r}_{xx}\frac{\partial \tilde{u}}{\partial x}} + \overline{\tilde{r}_{xz}\left(\frac{\partial \tilde{u}}{\partial z} + \frac{\partial \tilde{w}}{\partial x}\right)} + \overline{\tilde{r}_{zz}\frac{\partial \tilde{w}}{\partial z}}\right] dz$$

From Gatski and Liu (1980).

in detail in Section 2, the subharmonic and fundamental components undergo a similar development in their respective energy-exchange mechanisms with the mean flow.) The development of this *production* mechanism on part of the large-scale structure in a free turbulent shear flow shares the same essential physical mechanism as in a laminar flow except that the rate of this development is modified (Liu, 1971). The observations, such as those of Fiedler et al. (1980) is thus not entirely surprising from the perspective of a more broader use of ideas from hydrodynamic stability in a developing shear flow as we have done. However, the return of kinetic energy to the mean motion via the Reynolds stresses of the large scales is only partially responsible for the coherent mode decay.

We now must turn to the mechanism of interaction between the large-scale coherent structure and the fine-grained turbulence. Using the measurements of Favre-Marinet and Binder (1979) in Figure 5 as an illustration, it is rather obvious that there is a strong coupling between the large-scale coherent structures and the fine-grained turbulence. (The energy transfer mechanism between the various scales of motion for a round turbulent jet is thoroughly discussed in Mankbadi and Liu (1981).) Again, for simplicity, we shall discuss this physical mechanism in terms of the plane shear layer, which consists of the product of stresses and the appropriate rates of strain:

$$\overline{\tilde{r}_{ij}\frac{\partial \tilde{u}_i}{\partial x_j}},$$

where $\tilde{r}_{ij}$ are the modulated turbulent stresses, set up by the shaking up of the fine-grained turbulence by the large-scale coherent structure. They are obtained by subtracting the time-averaged from the phase-averaged turbulent stresses. The rates of strain $\partial\tilde{u}_i/\partial x_j$ are those of the coherent structure. (Of course, boundary layer approximations do not apply here!) We can see that the direction of energy transfer depends upon the relative phases between the modulated stresses and the appropriate rates of strain of the coherent structure. An eddy viscosity assumption would immediately make the fine-grained turbulence *dissipative* to the large-scale structure everywhere in the flow. We know from measurements (e.g., Hussain, 1983) that this is not so locally. In the study of coherent structures in turbulent shear flows the explicit, though approximate, calculation of the modulated stresses and the characteristics of the coherent structure show that, while locally energy transfer can take place from the fine-grained turbulence to the coherent structure as we see in Figure 8 (Gatski and Liu, 1980), the fine-grained turbulence is dissipative to the coherent structures in the global sense as shown in Figure 7 in this example (Gatski and Liu, 1980). The efficiency of the rate of energy transfer is strongly dependent upon the spectral characteristics of the coherent structure and this plays an essential role towards the decay of the large-scale structure in a real developing turbulent shear flow (Liu, 1981). The extraction of energy from the mean flow by the fine-grained turbulence causes an additional spreading of the mean flow and thereby enhances the early onset of energy return to the mean flow from the coherent structure. This is an implicit mechanism on the part of the fine-grained turbulence towards the eventual demise of the coherent structures. In general,

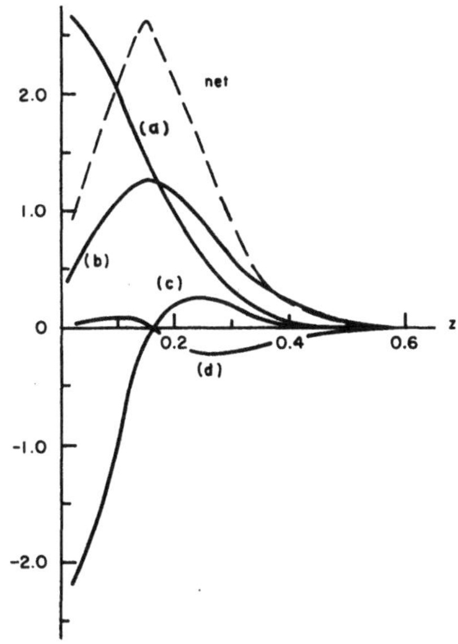

Figure 8. Vertical distributions of the individual energy transfer mechanisms between the large-scale structure and the fine-grained turbulence in a shear layer at $t = 1.50$ corresponding to the energy maximum of the large-scale structure, where

$$\text{(a)}\ -\overline{\tilde{r}_{xz}\frac{\partial \tilde{w}}{\partial x}},\quad \text{(b)}\ -\overline{\tilde{r}_{zz}\frac{\partial \tilde{w}}{\partial z}},\quad \text{(c)}\ -\overline{\tilde{r}_{xz}\frac{\partial \tilde{u}}{\partial z}},\quad \text{(d)}\ -\overline{\tilde{r}_{xx}\frac{\partial \tilde{u}}{\partial x}},$$

and net denotes the sum of (a)–(d). From Gatski and Liu (1980).

coherent structures in a turbulent jet or shear layer persist with a streamwise lifespan measured in terms of only several wavelengths, whereas they persist much further downstream in a laminar flow. In a flow undergoing transition, such structures are halted by almost abrupt bursts of fine-grained turbulence. The mechanism of interaction between the large-scale coherent structure and the fine-grained turbulence as we have discussed here is thus essential to the study of coherent structures in either turbulent or transitional shear flows.

The net effect of the global production and *dissipation* mechanisms give rise to the growth and decay of the large-scale structure energy as shown in Figure 9, again, taken from Gatski and Liu (1980); whereas the fine-grained turbulence energy achieves a *burst* from an initial nearly self-preserving development to a final one at a high level (Figure 10). These bear remarkable resemblance to the observations (Figure 5) reported by Favre-Marinet and Binder (1979) as well as those reported elsewhere.

The other *detailed* feature of observations that should be discussed here concerns the scale of fine-grained turbulence. Take, for instance, the observations of Brown and Roshko (1974) of the high Reynolds number mixing layer. The graininess of the fine-scaled turbulence becomes coarser as the shear flow develops and spreads

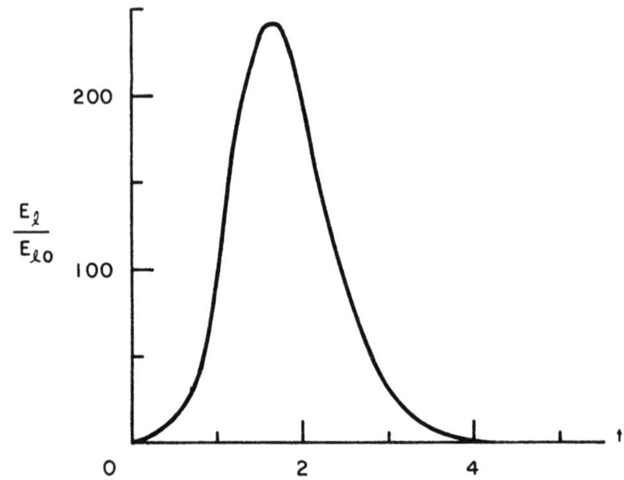

Figure 9. The development of large-scale structure global energy (Gatski and Liu, 1980).

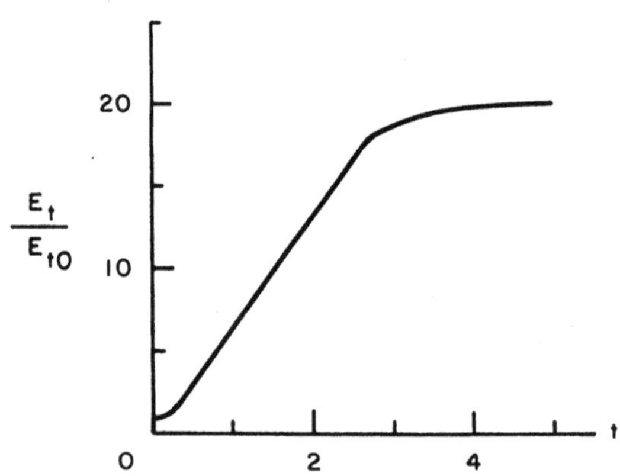

Figure 10. The development of fine-grained turbulence global energy (Gatski and Liu, 1980).

downstream, accompanied by the large-scale structure becoming very nearly extinct to visual observations. According to estimates of Gatski and Liu (1980), see Figure 11, the *global* scale of the fine-grained turbulence, referred to its initial value $L_\epsilon/L_{\epsilon 0}$, eventually develops very nearly as the shear layer thickness δ; whereas the aspect ratio of the large-scale structure referred to its initial value H/H_0 eventually collapses. The spreading of the shear layer in this case is sustained by coarsing of the scale of the fine-grained turbulence.

The *burst* in activities of both scales of motion contribute to the spreading of the shear layer. The active development of coherent structures in shear flows is essentially a nonequilibrium event. As such, their most fascinating behavior and practical exploitation for control purposes in technological problems lie in the region of shear flows before self-preservation, if at all, is reached. The many ways in which a turbulent jet can be controlled is fully discussed in Mankbadi and Liu (1981).

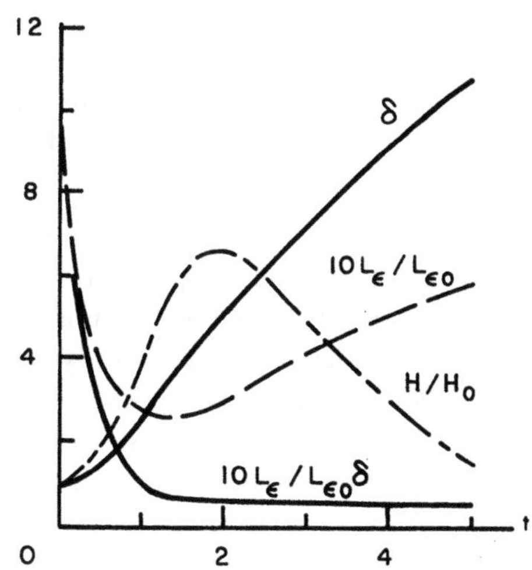

Figure 11. The development of shear layer thickness δ, large-scale structure aspect ratio H and the fine-grained turbulence length scale L_ϵ (Gatski and Liu, 1980).

We have used some of the results in Gatski and Liu (1980) to illustrate certain concepts and the theoretical understanding of the features of observations with the hope that this may stimulate further well conceived experiments. For further details of the theoretical problems we refer to Gatski and Liu (1980), Mankbadi and Liu (1981), and Liu (1981). At this stage it might be helpful to comment on the description of the large-scale structures from various points of view. The splitting of the total flow quantity Q into a Reynolds mean flow $\overline{Q}$ and *turbulence* q' requires the usual Reynolds average to obtain the conservation equations for $\overline{Q}$ as well as those of $\overline{q'q'}$. The Reynolds stress modeling, in spite of arguments at various levels of sophistication, is not universal and this is primarily attributable to the nonuniversality of instability mechanisms of the large-scale coherent structures included in $\overline{q'q'}$. The splitting of the total flow quantity into $\overline{Q}$ and the fluctuation quantity $\tilde{q}+q'$, where $\tilde{q}$ is the large-scale structure and q' the *turbulence*, provides the framework for the study of the interaction of the various scales of motion. This is effective only if $\tilde{q}$ is obtained via explicit calculation with its natural hydrodynamical instability properties unhindered by closure. However, the closure of the *background* fine-grained turbulence is more likely to be universal (if such a rationally satisfactory procedure could indeed be found). The details of the fine-grained turbulent fluctuations are not necessary but they act in a statistical sense (phase averaged) on the large-scale structures. Perhaps the large-scale computational efforts on fine-grained turbulence would eventually yield the results useful for the study of their interactions with the large-scales.

An equivalent way of looking at this problem is to lump the Reynolds mean motion and the coherent structure as a single coherent quantity $(\overline{Q}+\tilde{q})$. In obtaining the conservation equations for this quantity from the Navier-Stokes equations, only the phase average is required. The set of equations, including the phase-averaged fine-grained turbulent stresses $\langle q'q'\rangle$, are identical in form to the original Reynolds equations for $\overline{Q}$ and $\overline{q'q'}$ (see Gatski and Liu, 1980). The $(\overline{Q}+\tilde{q})$ and $\langle q'q'\rangle$ framework is very cumbersome to work with and useful only in the tackling of the simplest shear layer problem. However, interesting physical information is obtained by performing the usual Reynolds average on the result as was done by Gatski and Liu (1980). On the other hand, for the study of the more difficult problem of the round turbulent jet (e.g., Mankbadi and Liu, 1981), the $\overline{Q}$, $\tilde{q}$, $(\langle q'q'\rangle - \overline{q'q'})$ and $\overline{q'q'}$ framework is much more practical.

These apparently different points of views are equivalent, provided one does the *bookkeeping* correctly.

In closing this section, a comment about the *Taylor's hypothesis* applied to the large-scale coherent structures in free turbulent shear flows is warranted. We have taken the much broader view of hydrodynamic stability in showing that the coherent structures in turbulent shear flows are indeed a manifestation of *instabilities*. As such one can distinguish the laboratory situation (e.g., Brown and Roshko, 1974) in which the propagating large-scale structures grow and decay in the spatial (streamwise) direction, from a tilting tube experiment (e.g., Thorpe, 1971) in which the disturbances grow in time in a time-developing mean flow. There are no transformations possible between the two fully nonlinear problems. However, if the large-scale structures were weak, then the relation between temporal and spatial small disturbances follows from a transformation discussed by Gaster (1962). In the transformation between the growth rates, the group (rather than phase) velocity is the relevant velocity, but then the validity of such a transformation lies in the smallness of the growth rates. This condition is not always met in free shear flows, and most certainly not for the strongly amplified structures in regions of interest.

4. Sound Generated by Large-Scale Structures in a Turbulent Jet

In the last two sections we have dealt with certain concepts in large-scale coherent structures as they relate to observations, using only the simple aspects of the problem for the illustration of ideas. For the application of these ideas to the technologically important problem of a round turbulent jet we refer readers to Mankbadi and Liu (1981) and Liu (1981). The question as to the effect of these coherent, propagating wavy structures on the observed jet noise field is not just one of scientific curiosity but of enormous practical implications. We have seen that the large-scale structures within a jet can be controlled (Mankbadi and Liu, 1981). If they are of any importance to the jet noise field, the possibility of ultimately controlling the jet noise through the control of the large-scale structures (and the fine-grained turbulence) is most appealing.

We shall first state some of the results of low speed jet noise observations, as typified by those of Lush (1971), that might be explainable with the large-scale structures as a source of sound in a real turbulent jet. In the far jet noise field the observed lower frequency spectral peaks dominate at observation angles, θ, closer to the jet axis; the peaks shift to higher frequencies as θ increases towards 90°; the peak frequencies f_p at low θ do not scale as the Strouhal number ($St = fd/U_j$, where f is the frequency, d the jet nozzle diameter, and U_j the jet exit velocity), but remain constant over a wide range of U_j. Furthermore, the low frequency sources appear to be located downstream of the jet, whereas the high frequency sources appear to be located nearer the nozzle lip. These appear to be the *main issues* in the jet noise problem. The observed overall sound scaling as U_j^8 has long been reconciled with Lighthill's (1952) formulation of the aerodynamic sound problem. The *main issues* described above, which are some of the more detailed aspects of observations, still apparently remain if one considers an oversimplified interpretation of Lighthill's (1952) theory in terms of isotropically radiating, frequency-independent quadrupoles as sources. Then the convected quadrupoles subjected only to a Doppler correction, could give rise to higher frequency radiation nearer the jet axis. As we shall see, the large-scale structures radiate nonpreferentially in direction and this is highly dependent on their frequency content. These properties will be shown to have a far field contribution that bear striking resemblance to some of the details of observations.

We are aware, of course, that considerable work exists that is directed towards the understanding of the effect of flow on the sound generated within the jet (Phillips, 1960; Pao, 1973; Lilley, 1971; Ribner, 1964; Dowling et al., 1978), a problem that is not explicitly addressed by Lighthill's (1952) formulation. On the other hand, Lighthill's framework is one which is sufficiently simple, so as to enable one to work out the aerodynamic sound explicitly from the large-scale structures at the outset. This can be done without encountering the difficulties presented by other, alternative acoustic analogies in face of an actual calculation with a source distribution that is obtained from conservation principles (e.g., Mankbadi and Liu, 1981). In this section, the issue raised by Lush's (1971) experiments discussed earlier is addressed directly by the results from Lighthill's (1952) framework with the properly described sound sources in the form of a (Strouhal number) *spectrum* of large-scale structures of axisymmetric and spiral modes. The importance of large-scale coherent structures in sound generation was recognized much earlier by Mollo-Christensen (1960, 1967).

Lighthill's (1952) formulation of the aerodynamic sound problem is here taken as fundamental to the sound generated by real turbulent jets. The far-field pressure at location x and time t can be written as

$$p_s(\mathbf{x},t) = \frac{1}{4\pi R a_0^2}\int_V dV\left[\frac{\partial^2}{\partial t^2}(\rho c_r^2)\right], \tag{4.1}$$

where c_r is the full velocity component in the observer's direction along R, a_0 is the ambient sound speed, [ ] indicates quantities within are evaluated at the retarded time, ρ is the fluid density which will be taken as constant and V is the (finite) volume occupied by the sources. For convenience, equation (4.1) is recast, via Michalke and Fuchs (1975) into a form involving the (scalar) pressure fluctuations for $c_r/a_0 < 1$,

$$p_s(\mathbf{x},t) = \frac{1}{4\pi R a_0^2}\int_V dV\left[\frac{\partial}{\partial t}\left(f_1 p\frac{\partial c_r}{\partial y_r}\right) + \frac{1}{a_0}\frac{\partial^2}{\partial t^2}(f_2 p c_r)\right], \tag{4.2}$$

where $f_1 = 2/(1-M_r)^3$, $f_2 = (2-M_r)/(1-M_r)^2$, $\partial/\partial y_r$ is the derivative in the R-direction, $M_r = U_r/a_0$ and U_r is the mean velocity along R. We note here that equation (4.2) is convenient for the subsequent study of the so-called shear noise of coherent structures but rather cumbersome for the corresponding self-noise, which, fortunately, turns out to be much less important in regions of dominant radiation.

It is first conjectured that the large-scale coherent structures in the turbulent jet, whose existence is now well-recognized, would be responsible for the spectrally-dependent highly-oriented radiation patterns in the aerodynamic sound field. Accordingly, only contributions arising from the coherent structures are retained in the aerodynamic sound integral. The neglected fine-grained turbulence as far as the sound field is concerned is thought to otherwise contribute to the broadband, nearly isotropic radiation. The present source description follows Mankbadi and Liu (1981), but suitably modified to include an ensemble of $n = 0$ axisymmetric and $n = 1$ spiral modes in the relevant Strouhal number range. The coherent structures interact with the mean flow and the fine-grained turbulence as an ensemble through energy exchanges dictated by rates according to their individual spectral characteristics. Because such coherent structures are relatively *weak* in a real, developing turbulent jet, their mutual interactions are neglected as a first approximation (relative to energy production from the mean flow, as shown in Section 2). A typical streamwise distribution of their sectional energy content, corresponding to $E \sim |A|^2$ for the round jet problem (Mankbadi and Liu, 1981), slightly modified from our previous discussion of the plane shear layer, is shown in Figure 12. We can already note and emphasize that higher Strouhal number large-scale structures occur closer to the nozzle lip and lower Strouhal number structures further downstream.

The source distribution for each frequency is of the form (Mankbadi and Liu, 1981):

$$\tilde{q}\,(\mathbf{x},t) = A(x)\,\hat{q}(x,r)\,\exp\left(i\int_0^x \alpha_r(\xi)\,d\xi - i\omega t + in\phi\right) + c.c.,$$

where $\tilde{q}$ is either the velocity or pressure, $A(x)$ is the amplitude, $\hat{q}$ is the local shape distribution from the local linear theory, α is the wave number, ω the frequency, ϕ the azimuthal angle, n is the azimuthal wave number, and c.c. denotes the complex conjugate. The details (Mankbadi and Liu, 1984) can be more simply summarized. The sound sources, in a stationary coordinate system and evaluated at the appropriate retarded time, give rise to an equivalent streamwise distribution of line radiators after performance of the azimuthal and radial integrations in the aerodynamic sound integral. The far field sound intensity can now be written in the form for each frequency

$$I\,(\mathbf{x},St,n) = \frac{\rho_0 M^8 a_0^3 St^2}{4R^2}\left|\int_0^\infty A(x,St,n)\,e^{i\gamma(x,\theta,St,n,M)}F_{sh}(x,\theta,St,n,M)\,dx\right|, \tag{4.3}$$

where ρ_0 is the ambient density, the subscript sh denotes the *shear noise*, $M = U_j/a_0$, γ is an axial interference function and F_{sh} is a radial interference function (see Michalke and Fuchs, 1975). The streamwise oscillation of the equivalent sources is determined by an axial interference function strongly influenced by the wave number of each individual mode, whereas the streamwise growth and decay of the source envelope is determined primarily by the coherent structure amplitude (e.g., Figure 12) whose spectral dependence is also strong. The streamwise net imbalance of the source contribution, reflected by the axial integration in the aerodynamic sound integral, gives rise to the far sound field.

These are more vividly illustrated in Figure 13 where a typical integrand of the aerodynamic sound integral, equation (4.3), is shown.

With regard to the effective source distribution shown in Figure 13, we should emphasize that the growth and decay process of the source envelope is here physically obtained via the physical mechanisms that have been the central attention of a series of works (e.g., Liu (1981) as well as the previous sections of this paper). One can see that a source stretching from $x = \pm\infty$ at constant amplitude and at the same wavelength generates no sound. Here, for a given Strouhal number and azimuthal mode, the wavelength is slightly modulated by the advection of the wavy structure into regions of variable mean flow. From the discussion in Section 2, the fundamental-subharmonic interaction problem is seen as an overlapping, gradual switch-on and switch-off process. An over anxious model in the form of an abrupt switching of modal content, which is absent from the real physical problem, would generate artificial sound. The latter is also generated rather arbitrarily if arbitrary streamwise cutoffs were imposed upon an otherwise wavy structure of constant amplitude. The key to the aerodynamic sound problem here is seen as allowing the proper spectral-dependent physical mechanisms to contribute to the growth and decay of the sources.

The spectrally-dependent directional radiation obtained is shown in Figure 14. The spectra are taken to be the contributions from the $n = 0$ and 1 Strouhal modes corresponding to Figure 12. It is found that, in general, the radiation is primarily in the direction of the jet exhaust; the $n = 0$ modes' radiation patterns resembling those of longitudinal quadrupoles and that of the $n = 1$ modes resembling those of lateral quadrupoles. However, the $n = 0$ modes tend to peak at Strouhal numbers less than those of the $n = 1$ modes. The superposition gives a directional-spectal behavior that strikingly resembles that of observations: lower frequency sound radiates preferentially in the forward direction and as the frequency increases the peak radiation moves towards the lateral directions.

As observed from Figure 12, we emphasize again that contributions to the high frequency sound come from coherent structures that peak nearer the nozzle lip, whereas contributions to the low frequency sound come from such structures that peak further downstream in the jet. The calculated spectral shapes (Figure 14) are narrower than observations by typically a deficit of 4–7 dB per

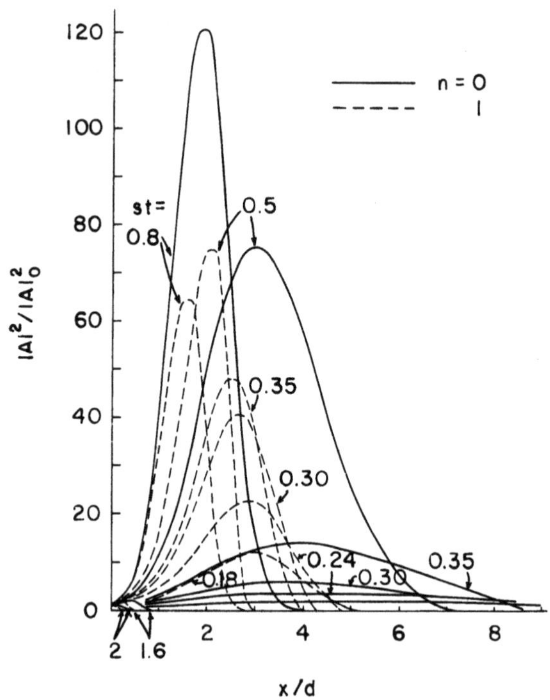

Figure 12. A Strouhal number (St) spectrum of large-scale structures in a turbulent jet for the $n = 0$ and 1 modes (Mankbadi and Liu, 1984).

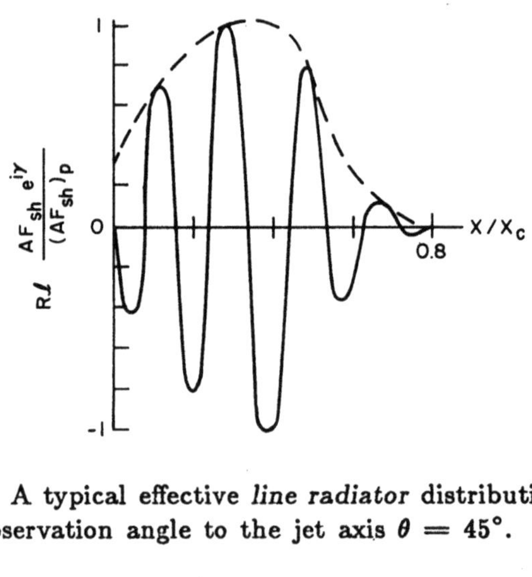

Figure 13. A typical effective *line radiator* distribution, St = 0.80, $n = 0$, and observation angle to the jet axis $\theta = 45°$. (Mankbadi and Liu, 1984).

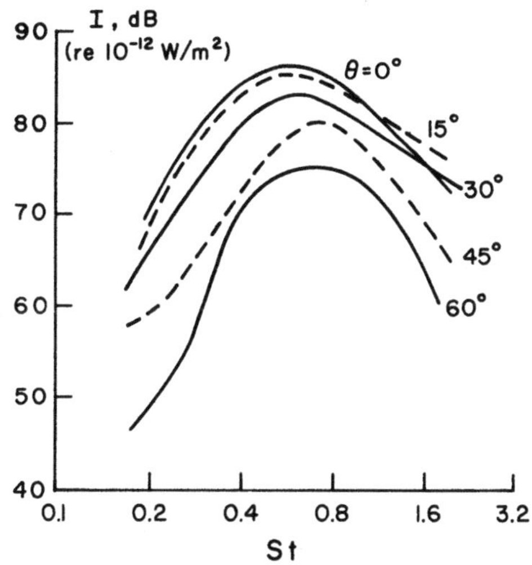

Figure 14. Calculated spectra of sound intensity due to the large-scale structures (shear noise) at various emission angles, $U_j = 195\ \mathrm{ms}^{-1}$ (Mankbadi and Liu, 1984).

octave on both the high and low frequency sides, and this is most likely attributable to the nearly isotropic radiation caused by the broadband fine-grained turbulence whose direct contribution to the sound field is not accounted for. For the same reason, the calculated aerodynamic sound field has a large deficit compared with observations in the vicinity of the 90-degree region. The dominant contributions to the radiation come from the so-called shear noise in the forward arc, whereas both the shear and self-noise of the coherent structures become equally insignificant to the same order in the 90-degree region.

Although the source distribution within the jet is calculated for an identically incompressible fluid, it is used in a limited sense to study the effect of jet exit velocity on the peak radiation frequency in the forward direction: it is found that the peak value of fd/a_0 takes on a value of about 0.30 independently of the jet exit velocity and this compares favorably with an observational value of about 0.20. In general, the angular distribution of the peak frequency due to coherent structures radiation compared favorably with observations. Compressibility effects which somewhat limit the amplification of coherent structures, as well as the effects of higher azimuthal modes whose radiation would peak at higher frequencies and larger lateral directions, warrant further study in the light of the present considerations. The present work, however, has already shown that the consequences of Lighthill's formulation of the aerodynamic sound problem agree with major features of observations, and that this is brought about by taking into account as sources the growing and decaying large-scale coherent eddies whose development within the turbulent jet and whose radiational properties are all strongly dependent upon their spectral contents. This work is seen as a supplement to all the earlier work on radiation from turbulent eddies of much smaller correlation volume (Mankbadi and Liu, 1984).

5. Further Discussions of the Jet Noise Problem

In the work just described, the dominant aerodynamic sound sources are considered to be the coherent large-scale structures that are inherently present in turbulent jets. The description of the sources is obtained from approximate but dynamically (or energetically) consistent conservation principles. As such, this is to be contrasted with empirical or analytically integrable but modelled source distributions (e.g., Crow, 1972). The source distribution here consists of the two lowest order azimuthal modes ($n = 0, 1$) over a Strouhal number spectrum relevant to jet noise. The question that naturally arises is what additional features would the inclusion of higher azimuthal modes present. We know, in general, that the $n = 0$ mode radiates effectively as if it were a longitudinal quadrupole and that the $n = 1$ mode a lateral quadrupole (but the details depend upon the frequency). Insight to the higher mode behavior can be gained by comparing approximately (Mankbadi and Liu, 1984) the relative features of the $n = 2$ and $n = 1$ modes. If we attribute the dominating effect on radiation to come from effects associated with $\partial U/\partial r$, then the $n = 1$ mode would peak at 45° and the $n = 2$ mode at about 55°. These crude estimates are, of course, modified by the spectrally-dependent radial and axial interference functions. However, it can be inferred from the present work and from Michalke and Fuchs (1975) that the peak radiation contributed by higher modes would occur at higher frequencies than the lower order azimuthal modes. This picture essentially reinforces the notion derived here that Lighthill's (1952) fundamental aerodynamic sound formulation gives rise to accordance with certain of the details of observations provided that the properly described structure of the sources is accounted for.

We have found that initial conditions are important to coherent structure development (e.g., Mankbadi and Liu, 1981). Because the downstream development of the coherent structures away from the nozzle lip is necessarily of a nonequilibrium nature and is therefore rather sensitive to the initial conditions; such as the initial levels of the coherent structure and of the fine-grained turbulence, the mean flow status at the nozzle exit as well as the azimuthal mode and Strouhal number concerned. The set of initial conditions used here for the numerical example is representative of that of natural or weakly forced jets and as such, sets a particular pattern of downstream distribution of, say, the value of the coherent structure amplitude peaks, the location of such peaks, the mean flow spreading rate, and the level of fine-grained turbulence. These are by no means universal because, in general, of the nonequilibrium nature of the interactions between the different scales of motion which are necessarily sensitive to initial as well as environmental conditions. From this, we can understand that the observed natural turbulent jets and their far sound field to be anything but universal in detail. For instance, the spreading rate of natural turbulent mixing layers is not unique (see the discussion in Alper and Liu, 1978) and that the spectrum shape of the sound from turbulent jets bears only qualitative resemblances to one another (Zaman and Yu, 1984). It is precisely because of the coherent structures' sensitivity to initial and environmental conditions that a variety of methods for their control is discussed by Mankbadi and Liu (1981). For instance, an initially thicker boundary layer weakens the downstream development of the coherent structures. The same purpose is also achieved by increasing the initial levels of the fine-grained turbulence, such as via a fine-meshed screen (Arndt et al., 1972; Wei and Niu, 1983). An implicit way of controlling the coherent structures comes from an understanding of the energy transfer mechanisms between such structures and the fine-grained turbulence (Mankbadi and Liu, 1981; Liu, 1981). Forcing a particular mode of the coherent structure that has the most efficient energy transfer rate to the fine-grained turbulence would increase the levels of the *dissipative* recipient, and this in turn would curtail the development of the other components of the large-scale structures, much in the same manner as the placing of a series of screens in the jet. The amplification of broadband jet noise by a pure tone excitation has already been demonstrated in the work of Bechert and Pfizenmaier (1975), the reduction of sound radiated by forcing was demonstrated by Moore (1977) and these are seen as the consequence of large-scale structures interacting with the fine-grained turbulence (Liu, 1981).

The radiation pattern in the far sound field derived here is entirely reconcilable with the location of sound sources in a manner that is fully consistent with observations. The theoretical picture here is in accordance with and essentially summarizes the earlier experimental efforts directed at jet noise source location (Lassiter and Hubbard, 1956; Howes et al., 1957; Potter and Jones, 1968). Though controversial at the time, the measured pressure fluctuations in the near jet noise field is now understood to be a direct consequence of the local activities of the coherent structures (Liu, 1974; Merkine and Liu, 1975), consistent with the spectrally dependent streamwise activity and life span discussed in the present work and in Mankbadi and Liu (1981). The contribution of the present work is seen, as that in the observations, to have brought more intimately together the mechanisms of the sources and the resulting aerodynamic sound field of real turbulent jets.

In certain instances when the fully turbulent jet conditions are not achieved or desired in the laboratory, the questions that need to be raised concern the mechanisms at work within the transitional jet structure and their effect on the far sound field (Huerre and Crighton, 1983; Laufer and Yen, 1983). We have strongly emphasized that the coherent structure source distribution, which can be effectively depicted as a line radiator as was done by Crow (1972), generates the far sound field through a net imbalance of quasi-periodic positive and negative contributions under a growing and decaying axial envelope determined through frequency-dependent nonlinear interactions. It has been known for some time (Michalke, 1971) that the coherent structures' shape distribution responds to the local profiles of the mean shear flow regardless of whether it is turbulent or laminar, however, the streamwise development of the *envelope* depends strongly on the nonlinear imbalances between the mechanisms of energy supply and *dissipation*. In a purely laminar jet the most unstable mode and its harmonics are likely to generate discrete sound in the absence of strong broadband disturbances. The coherent structures persist much further downstream owing to the much less efficient dissipative mechanism of a laminar flow. Inherent in real flows, however, is the possible presence of weak broadband fine-grained disturbances. They become strained and amplified by the developing coherent large-scale structure resulting in the latter's earlier demise (Liu and Merkine, 1976; Alper and Liu, 1978; Mankbadi and Liu, 1981). This process is necessarily Reynolds number dependent, since the rate of energy extracted by the fine-grained disturbances from the coherent structure (and from the mean flow) must overcome the rate of viscous dissipation. Thus, there is a Reynolds number, for the same initial and environmental conditions, below which the broadband fine-grained disturbances, even if artificially generated, would remain inactive. This is, of course, the physical interpretation of the so-called "*critical Reynolds number*" for the fine-grained turbulence to develop, which again is dependent upon the initial and environmental conditions and by no means necessarily universal. For sufficiently large Reynolds numbers for transition to take place in the noise producing region of the jet, then the early decay of the coherent structure would have a noticeable influence upon the far sound field. A *sudden* decay of the coherent structure, whether artificially placed in the computations (Michalke, 1969) or due to the transition process in a real jet, is to the far sound field effectively the placing of a dipole at the location of the *cutoff*. The flow structure in transitional jets is much more sensitive to initial and environmental conditions than a fully turbulent jet issuing from nozzles with turbulent boundary layers. As such, the careful study of the structural aspect of the transitional jet itself must necessarily be made an integral part of the quantitative study and understanding of its aerodynamic sound field. The present framework for the source description also provides a basis for the study of coherent structures in a laminar flow undergoing dissipation by weak fine-grained turbulence such as in transitional jet, thereby providing the proper *switch-on* and *cutoff* processes for the aerodynamic sound problem. The details, however, remain to be explored in the light of similar experiments.

This work is partially supported by the Fluid Mechanics Program, National Science Foundation through Grant NSF MEA78-22127 and the National Aeronautics and Space Administration, Langley Research Center through Grant NAG-1-379.

References

[1] Alper, A. and Liu, J.T.C. "On the Interactions Between Large-Scale Structure and Fine-Grained Turbulence in a Free Shear Flow, II. The Development of Spatial Interactions in the Mean," *Proc. R. Soc. Lond. A. 359* (1978), 497.

[2] Arndt, R.E.A., Tran, N., and Barefoot, G. "Turbulence and Acoustic Characteristics of Screen Perturbed Jets," *AIAA Paper No. 72-644*, 1972.

[3] Bechert, D. and Pfizenmaier, E. "On the Amplification of Broadband Jet Noise by Pure Tone Excitation," *J. Sound Vib., 43* (1975), 581.

[4] Binder, G. and Favre-Marinet, M. "Mixing Improvement in Pulsating Turbulent Jets," in *Fluid Mechanics of Mixing.* Ed. E. M. Uram and V. W. Goldschmidt, New York: Am. Soc. Mech. Engrs., 1973, 167–72.

[5] Brown, G. L. and Roshko, A. "On Density Effects and Large-Scale Structure in Turbulent Mixing Layers," *J. Fluid Mech. 64* (1974), 775.

[6] Cantwell, B. J. "Organized Motion in Turbulent Flow," *Ann. Rev. Fluid Mech., 13* (1981), 457.

[7] Corrsin, S. "Investigations of Flow in an Axially Symmetric Heated Jet of Air," *NACA Adv. Conf. Rep. No. 3123,* 1943; also *W-94.*

[8] Crow, S. C. "Acoustic Gain of a Turbulent Jet," Annual Meeting Div. Fluid Dyn., Am. Phy. Soc., Boulder, Colorado, *Paper IE.6,* 1972.

[9] Dowling, A. P., Ffowcs Williams, J. E., and Goldstein, M. E. "Sound Production in a Moving Stream," *Phil. Trans. R. Soc. Lond., A288* (1978), 321.

[10] Favre-Marinet, M. and Binder, G. "Structur des Jets Pulsants," *J. Méc., 18* (1979), 356.

[11] Fiedler, H. E., Dziomba, B., Mensing, P., and Rösgen, T. "Initiation, Evolution and Global Consequences of Coherent Structures in Turbulent Shear Flows," Int. Conf. on the Role of Coherent Structure in Mod. Turbulence and Mixing, June 25–27, 1980, Madrid.

[12] Freymuth, P. "On Transition in a Separated Boundary Layer," *J. Fluid Mech., 25* (1966), 683.

[13] Gaster, M. "A Note on the Relation Between Temporally-Increasing and Spatially-Increasing Disturbances in Hydrodynamic Stability," *J. Fluid Mech., 14* (1962), 222.

[14] Gatski, T. B. and Liu, J.T.C. "On the Interactions Between Large-Scale Structure and Fine-Grained Turbulence in a Free Shear Flow, III. A Numerical Solution," *Phil. Trans. R. Soc. Lond., A. 293* (1980), 473.

[15] Ho, C. M. and Huang, L. "Subharmonics and Vortex Merging in Mixing Layers," *J. Fluid Mech., 119* (1982), 443.

[16] Howes, W. L., Callaghan, E. E., Coles, W. D., and Mull, H. R. "Near Noise Field of a Jet Engine Exhaust," *NACA Rep. No. 1338,* 1957.

[17] Hussain, A.K.M.F. "Coherent Structures—Reality and Myth," Uni-

versity of Houston, *Report FM-17,* 1983; also *Phys. Fluids, 26* (1983), 2816.

[18] Huerre, P. and Crighton, D. G. "Sound Generation by Instability Waves in a Low Mach Number Jet," *AIAA Paper No. 83-0661,* 1983.

[19] Karamcheti, Krishnamurty. "Sound Radiation From Two-Dimensional Rectangular Cutouts in High Speed Flow," *Ph.D. thesis,* California Institute of Technology, 1956; also Stanford Univ. *JIAA TR-1.*

[20] Karamcheti, K., Bauer, A. B., Shields, W. L., Stegen, G. R., and Wolley, J. P. "Some Features of an Edge-Tone Flow Field," in *Basic Aerodynamic Noise Research.* Ed. I. R. Schwartz, *NASA SP-207* (1969), 275–304.

[21] Karlsson, S.K.F. "An Unsteady Turbulent Boundary Layer," *J. Fluid Mech., 5* (1959), 622.

[22] Kelly, R. E. "On the Stability of an Inviscid Shear Layer Which is Periodic in Space and Time," *J. Fluid Mech., 27* (1967), 657.

[23] Kendall, J. M. "The Turbulent Boundary Layer Over a Wall With Progressive Surface Waves," *J. Fluid Mech., 41* (1970), 259.

[24] Ko, D.R.S., Kubota, T., and Lees, L. "Finite Disturbance Effect in the Stability of Laminar Incompressible Wake Behind a Flat Plate," *J. Fluid Mech., 40* (1970), 315.

[25] Krothapalli, A., Hsia, Y., Banganoff, D., and Karamcheti, K. "On the Structure of an Underexpanded Rectangular Jet," Stanford Univ. *JIAA, TR-47,* 1982.

[26] Lassiter, L. W. and Hubbard, H. H. "The Near Noise Field of Static Jets and Some Model Studies of Devices for Noise Reduction," *NACA No. 1261,* 1956.

[27] Laufer, J. and Yen, T. C. "Noise Generation by a Low Mach Number Jet," *J. Fluid Mech., 134* 1983, 1.

[28] Liepmann, H. W. "Aspects of the Turbulence Problem, Second Part," *ZAMP 3* (1952), 407.

[29] ——————. "Free Turbulent Flows," in *Mécanique de la Turbulence (Coll. Intern. du CNRS à Marseille).* Ed. CNRS, Paris, 1962, 211–26.

[30] Lighthill, M. J. "On Sound Generated Aerodynamically: I. General Theory," *Proc. R. Soc. Lond., A. 211* (1952), 564.

[31] ——————. "Sound Generated Aerodynamically," (The Bakerian Lecture, 1961), *Proc. R. Soc. Lond., A. 267* (1962), 147.

[32] Lilley, G. M. "Sound Generation in Shear Flow Turbulence," *Fluid Dynamics Trans., (Poland) 6* (1971), 405.

[33] Liu, J.T.C. "Nonlinear Development of an Instability Wave in a Turbulent Wake," *Phys. Fluids 14* (1971), 2251.

[34] ————. "Developing Large-Scale Wavelike Eddies and the Near Jet Noise Field," *J. Fluid Mech. 62* (1974), 437.

[35] ————. "Interactions Between Large-Scale Coherent Structures and Fine-Grained Turbulence in Free Shear Flows," in *Transition and Turbulence*. Ed. R. E. Meyer, Academic Press, 1981, 167–213.

[36] Liu, J.T.C. and Merkine, L. "On the Interactions Between Large-Scale Structure and Fine-Grained Turbulence in a Free Shear Flow, I. The Development of Temporal Interactions in the Mean," *Proc. R. Soc. Lond., A 352* (1976), 213.

[37] Liu, J.T.C. and Nikitopoulos, D. E. "Mode Interactions in Developing Shear Flows," *Bull. Am. Phys. Soc., 27* (1982), 1192.

[38] Lush, P. A. "Measurements of Subsonic Jet Noise and Comparison With Theory," *J. Fluid Mech., 46* (1971), 477.

[39] MacPhail, D. C. "Turbulence in a Distorted Passage and Between Rotating Cylinders," *Ph.D. thesis*, Univ. of Cambridge, 1941; also in *Proc. 6th Int. Congr. Appl. Mech.*, Paris, 1946.

[40] Mankbadi, R. and Liu, J.T.C. "A Study of the Interactions Between Large-Scale Coherent Structures and Fine-Grained Turbulence in a Round Jet," *Phil. Trans. R. Soc. Lond., A 298* (1981), 541.

[41] ————————————. "Sound Generated Aerodynamically Revisited—Large Scale Structures in a Turbulent Jet as a Source of Sound," *Phil. Trans. R. Soc. Lond.*, 1984 (in press).

[42] Merkine, L. and Liu, J.T.C. "On the Development of Noise-Producing Large-Scale Wavelike Eddies in a Plane Turbulent Jet," *J. Fluid Mech., 70* (1975), 353.

[43] Michalke, A. "Sound Generation by Amplified Disturbances in Free Shear Layers," *Deutsche Luft-und Raumfahrt, Rep. No. 69-90,* 1969. Trans. E. Morse, Brown University.

[44] ————. "Instabilität eines Kompressiblen runden Freistrahls unter Berücksichitigung des Einflusses der Strahlgrenzschichtdicke," *Z. Flugwiss, 19* (1971), 319.

[45] Michalke, A. and Fuchs, H. V. "On Turbulence and Noise of an Axisymmetric Shear Flow," *J. Fluid Mech., 70* (1975), 179.

[46] Mollo-Christensen, E. "Some Aspects of Free-Shear-Layer Instability and Sound Emission," *NATO-AGARD Rep. No. 260,* 1960.

[47] ——————. "Jet Noise and Shear Flow Instability Seen From an Experimenter's Viewpoint," *Trans. ASME J. Appl. Mech.,*

E89 (1967), 1.

[48] Moore, C. J. "The Role of Shear-Layer Instability Waves in Jet Exhaust Noise," *J. Fluid Mech., 80* (1977), 321.

[49] Nikitopoulos, D. E. "Nonlinear Interactions Between two Instability Waves in a Developing Shear Layer," *Sc.M. thesis,* Brown University, 1982.

[50] Pai, S. I. "Turbulent Flow Between Rotating Cylinders," *Ph.D. thesis,* Calif. Institute of Technology, 1939; also *NACA Tech. Note No. 892,* 1943.

[51] Pao, S. P. "Aerodynamic Noise Emission From Turbulent Shear Layers," *J. Fluid Mech., 59* (1973), 451.

[52] Papailiou, D. D. and Lykoudis, P. S. "Turbulent Vortex Streets and the Entrainment Mechanism of the Wake," *J. Fluid Mech., 62* (1974), 11.

[53] Phillips, O. M. "On the Generation of Sound by Supersonic Turbulent Shear Layers," *J. Fluid Mech., 9* (1960), 1.

[54] Potter, R. C. and Jones, J. H. "An Experiment to Locate the Acoustic Sources in a High Speed Jet Exhaust Stream," *Wyle Lab. Res. Staff Rep. WR 68-4,* 1968.

[55] Reynolds, O. "On the Dynamical Theory of Incompressible Viscous Fluids and the Determination of the Criterion," *Phil. Trans. R. Soc. Lond., A 186* (1895), 123.

[56] Ribner, H. S. "The Generation of Sound by Turbulent Jets," *Adv. Appl. Mech., 8* (1964), 103.

[57] Roshko, A. "On the Development of Turbulent Wakes From Vortex Streets," *Ph.D. thesis,* Calif. Institute of Technology, 1952; also *NACA Rep. No. 1191,* 1954.

[58] Sato, H. and Kuriki, K. "The Mechanism of Transition in the Wake of a Thin Flat Plate Placed Parallel to a Uniform Flow," *J. Fluid Mech., 11* (1961), 321.

[59] Schubauer, G. B. and Skramstad, H. K. "Laminar Boundary Layer Oscillations and Transition on a Flat Plate," *NACA Rep. No. 909,* 1948.

[60] Stuart, J. T. "Nonlinear Effects in Hydrodynamic Stability," in *Proc. Xth Int. Congr. Appl. Mech.,* Stresa, Amsterdam: Elsevier, 1962.

[61] Taneda, S. "Downstream Development of Wakes Behind Cylinders," *J. Phys. Soc. Japan, 14* (1959), 843.

[62] Taylor, G. I. "Stability of a Viscous Liquid Contained Between two Rotating Cylinders," *Phil. Trans. R. Soc. Lond., A 223* (1923), 289.

[63] Thomas, A.S.W. and Brown, G. L. "Large Structure in a Turbulent Boundary Layer," in *Proc. 6th Australasian Hydraulics and Fluid Mechanics Conference,* Adelaide, Australia, 1977, 407–10.

[64] Thorpe, S. A. "Experiments on the Instability of Stratified Shear Flows-Miscible Fluids," *J. Fluid Mech., 46* (1971), 299.

[65] Townsend, A. A. "Measurements in the Turbulent Wake of a Cylinder," *Proc. Roy. Soc. Lond., A 190* (1947), 551.

[66] ——————. *The Structure of Turbulent Shear Flow,* Cambridge: Cambridge Univ. Press, 1956.

[67] Wei, Z. L. and Niu, Z. N. "The Disturbances Affect Brown-Roshko Structures in Plane Mixing Layer," in *Structure of Complex Turbulent Shear Flow-IUTAM Symposium Marseille 1982.* Ed. R. Dumas and L. Fulachier, New York: Springer-Verlag, 1983, 137–45.

[68] Williams, D. R. and Hama, F. R. "Streaklines in a Shear Layer Perturbed by Two Waves," *Phy. Fluids, 23* (1980), 442.

[69] Winant, C. D. and Browand, F. K. "Vortex Paring: The Mechanism of Turbulent Mixing-Layer Growth at Moderate Reynolds Number," *J. Fluid Mech., 63* (1974), 237.

[70] Zaman, K.B.M.Q. and Yu, J. C. "Power Spectral Density of Subsonic Jet Noise," *J. Sound and Vib.,* 1984 (to be published).

Part III
V/STOL Aerodynamics

The Induced Aerodynamics of Jet and Fan Powered V/STOL Aircraft

Richard E. Kuhn
V/STOL Consultant
111 Mistletoe Dr.
Newport News, VA 23606

Nomenclature

A	total jet exit area.
d	jet diameter.
d_e	diameter of equivalent single jet having area A.
D	planform diameter.
$\overline{D}$	angular mean diameter of planform (see Reference 37).
C_p	pressure coefficient, $C_p = \Delta P/q$.
h	height above ground.
h_v	maximum height at which ground vortex is felt.
L	lift.
ΔL	jet induced increment of lift.
M	moment.
ΔM	jet induced increment of moment.
ΔP	jet induced increment of pressure.
P_T	jet total pressure.
p	atmospheric pressure.
per	perimeter of jet.
q	free stream dynamic pressure.
q_n	jet dynamic pressure.
r	radius to measuring station.
R	radius of plate.
S	planform area.
T	total jet thrust.
V_e	effective velocity ratio, $V_e = \sqrt{q/q_n}$.
X	distance to leading edge of ground vortex flow.
α	angle of attack.

β	side slip angle.
ϵ	downwash angle.
δ	jet deflection angle.

Subscripts

∞	out of ground effect.
h	hover.
v	ground vortex.

1. Introduction

Various aspects of the aerodynamics of V/STOL and STO/VL aircraft have been the subject of many studies over the past 20 to 30 years. Most of these studies have been experimental investigations because at low speeds the flow fields around these aircraft involve large deflections of local flow relative to the free stream as well as significant amounts of viscous mixing. Some of these studies have been specialized investigations undertaken to study specific phenomena and others have been tests to gather design and performance data on proposed airplane configurations. Out of these studies a good understanding of most of the flow phenomena involved in hovering and low speed flight has evolved, and in some cases empirical methods for estimating the effects of these flows on the aerodynamic characteristics of these airplanes have been derived. More recently, theoretical studies have been undertaken and analytical tools for predicting the aerodynamic characteristics are being developed. These methods rely on the basic understanding of the flow fields that is in hand and most also rely at least partially on empirical correlations.

This article presents an overview of the flow phenomena involved in the hovering and low speed flight of jet and fan powered V/STOL and STO/VL aircraft and indicates some of the empirically based methods that are available for estimating the aerodynamic characteristics. Other papers presented at this symposium will discuss the analytical methods and techniques that are being developed.

2. Basic Flow Fields

The primary flow phenomena involved in the hovering and low speed flight of V/STOL and STO/VL aircraft that will be discussed in this article are shown schematically in Figure 1. In hovering out of ground effect the induced effects are small. Flow is entrained by the lifting jet inducing small suction pressure on the lower surface of the aircraft causing a small download.

Close to the ground, however, the download can be considerably larger. With a single jet configuration the impinging jet flows radially outward from the point on the ground at which it impinges and the entrainment area is greatly increased, causing an increase in download. With multiple jets on the other hand, and up flow or *fountain* is created where the wall jets flowing outward from the impingement points of adjacent jets meet. This fountain flow produces a lifting force when it impinges on the configuration, partially offsetting the download created by the entrainment action of the wall jet flow on the ground.

In transition between hover and conventional flight, out of ground effect, the lifting jet stream is swept rearward by the action of the free stream and the jet flow is rolled up into a pair of vortices. These vortices along with the entrainment action and blockage effect

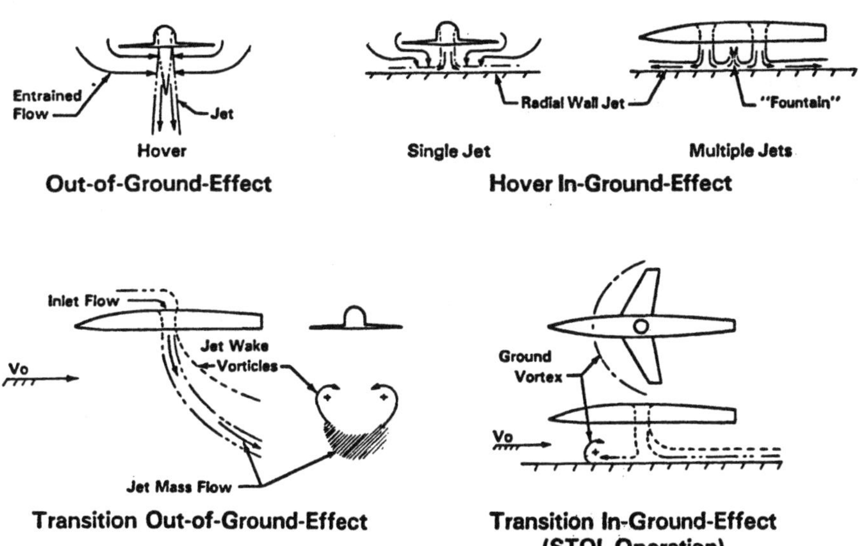

Figure 1. Basic flow fields.

of the jet induces suction pressures behind and beside the jet and positive, or lifting, pressures ahead of the jet. In most cases these induced pressures result in a net loss in lift and a nose up moment. However, jets or fans located at or near the trailing edge of a lifting surface can induce a favorable lift through their *jet flap* action. In addition the flow into the inlets to the lift jets or fans causes a ram drag and generally a nose up moment. Also in crosswinds or sideslipping flight these inlet and exit flows can induce significant rolling and yawing moments.

All the above phenomena are present but modified by the proximity of the ground during STOL operation. In addition the jet sheet flowing forward on the ground is opposed by the free stream and rolls up into a ground vortex. This ground vortex induces an additional download or lift loss on the configuration which is at least partially offset by the ground induced reduction in the download due to the wake vortex system. The position and strength of the ground vortex are also the primary factors in determining the extent of the hot gas, dust and debris, and spray that may be ingested into the fan or jet inlets in STOL operation.

All of these flow phenomena and empirical methods for estimating their effects are discussed more completely in the following sections.

3. Jet Induced Forces and Moments

3.1 Hovering Out of Ground Effect

A small lift loss is induced by the entrainment action of the jet or fan streams when the aircraft is hovering out of ground effect. As shown in Figure 2 (Gentry and Margason, 1966) the entrainment action of the jet issuing from the lower surface of a circular plate induces suction pressures on the lower surface of the plate. Although the pressures are small, their integration over the surface of the plate can produce a significant lift loss and this loss increases as the plate (or planform) area increases. The entrainment action occurs on the jet surface. The amount of ambient air entrained and the suction pressures and lift loss will increase if this entrainment area is increased; for instance if multiple jets or noncircular jets are used. These downloads have been the subject of several investigations (Gentry and Margason, 1966; McLemore, 1966; Shumpert and Tibbets, 1969; Henderson et al., 1980) which have shown that (Figures 3 and 4) they can be estimated with reasonable accuracy.

A more complete review of investigations of the lift losses induced out of ground effect and of methods for estimating these losses, including the effects of wing height, are presented in Henderson et al. (1980), Section 2.2.1.

3.2 Hovering in Ground Effect

3.2.1 Single Jet

The lift losses can be considerably greater in ground effect, particularly for the single jet case. When the jet impinges on the ground, as depicted in Figure 5, it is turned outward by the ground and generates a wall jet that flows radially outward from the impingement point. This radial wall jet becomes the primary entrainment mechanism and as shown on the right side of Figure 5 draws ambient air in through the gap between the edge of the planform and the wall jet. The suction pressures induced on the configuration are higher near the edge than toward the center of the planform but are fairly uniform. These pressures increase as the height, and therefore the gap, decrease causing an increase in the lift loss as the height is reduced.

At still lower heights a condition can occur where the edge of the plate or configuration reaches the upper edge of the wall jet, as depicted on the left in Figure 5. Below this height ambient air cannot be drawn in because the gap is closed and the air entrained by the inner part of the wall jet must come from the part of the wall jet nearing the edge of the plate. The result is a doughnut shaped trapped vortex and rapidly increasing pressures over the inner region (Bower, 1981; Spreeman and Sherman, 1958). Fortunately this condition is not encountered on practical aircraft configurations. The data shown in Figure 5 (Spreeman and Sherman, 1958) were obtained on a configuration that had a plate diameter to jet diameter ratio larger than would normally be encountered on practical V/STOL airplane configurations.

The first definitive work on the lift loss in ground effect (above the critical height) was done by Wyatt (1962). He showed that the

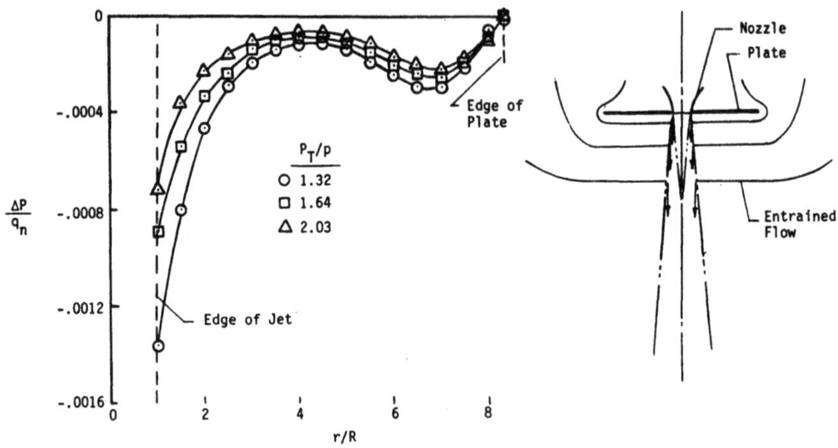

Figure 2. Flow field and pressure distribution induced on flat plate out of ground effect. (Gentry and Margason, 1966).

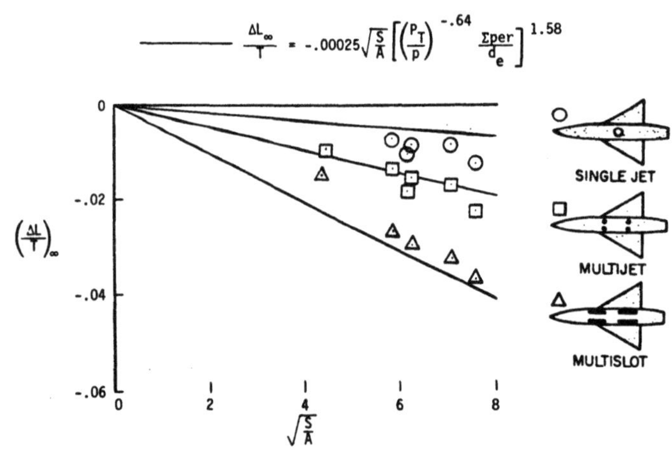

Figure 3. Out-of-ground-effect lift loss induced on three model configurations; $\frac{P_T}{p} = 1.89$. (Gentry and Margason, 1966).

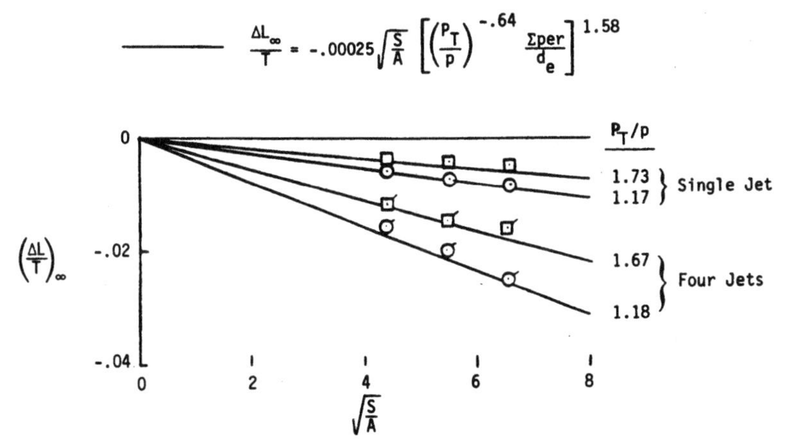

Figure 4. Out-of-ground-effect lift loss induced on flat plate by J-85 engine. (McLemore, 1966).

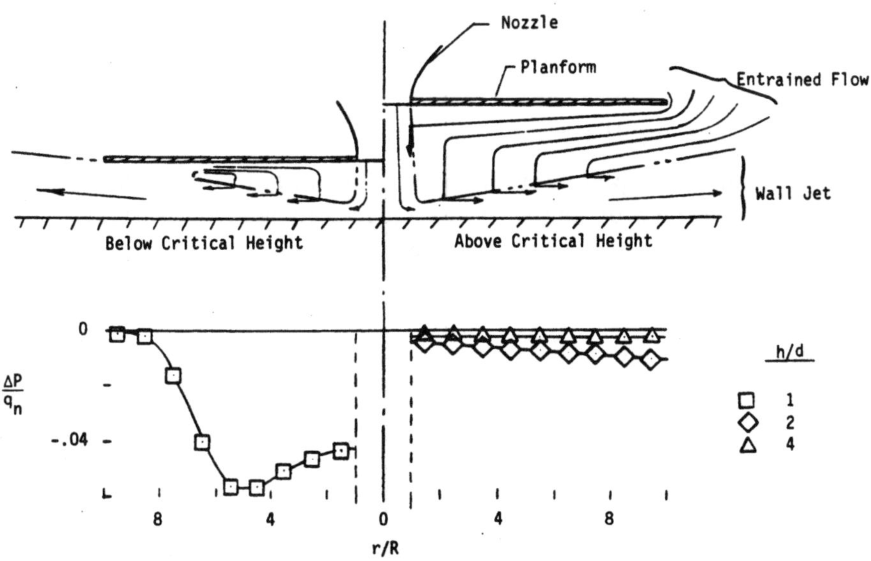

Figure 5. Pressure on plate in Hover. (Spreeman and Sherman, 1958).

suckdown data for a wide range of planform to jet diameter ratios could be correlated on the basis of $h/(D-d)$ as shown in Figure 6. He also showed that the lift loss on noncircular planforms could be estimated by using their angular mean diameter $\overline{D}$ which effectively weights the inner regions of the planform higher than the extremities.

Wyatt's experimental data were taken at low nozzle pressure ratios and his original expression for estimating the lift loss due to ground effect tended, at low heights, to overpredict the lift losses experimentally obtained by other investigators using higher nozzle pressure ratios. Wyatt's original expression was modified in Henderson, Clark, and Walters (1980), Section 2.2.1 to include the effects of pressure ratio. The revised expression for the ground effect on single jet configurations is shown in Figure 7. This expression fits Wyatt's original data reasonably well as shown in Figure 6. It also fits the limited large-scale hot-jet data fairly well as shown in Figure 7. (The estimates for the X-14 assume that the two jets are close enough together to be treated as a single jet.)

Figure 7 also illustrates the significance of using the angular mean diameter $\overline{D}$ of the planform in the estimates. The X-14 airplane has a larger total planform to jet area ratio than the J-85/plate configuration but experiences less suckdown which is adequately estimated by basing the estimate on $\overline{D}$.

A more complete review of the investigations of the ground induced suckdown of single jet configurations is given in Henderson, Clark, and Walters (1980), Section 2.2.1, along with methods for accounting for wing height.

3.2.2 Multiple Jets and Fountain Flows

When two or more jets are used, the wall jets flowing radially outward from the impingement point of adjacent jets meet on a stagnation line midway between the jets and are directed upward in a fan shaped *fountain flow* as shown in Figure 8. The impingement of this fountain flow on the bottom of the configuration creates a lift force that at least partially offsets the wall jet generated suckdown. In addition, when three or more jets are used, a *fountain core* is created at the centroid of the jet pattern where the fan shaped fountain from each pair of adjacent jets meet. This fountain core usually creates a significantly stronger lift increment than that created by the two jet fountain.

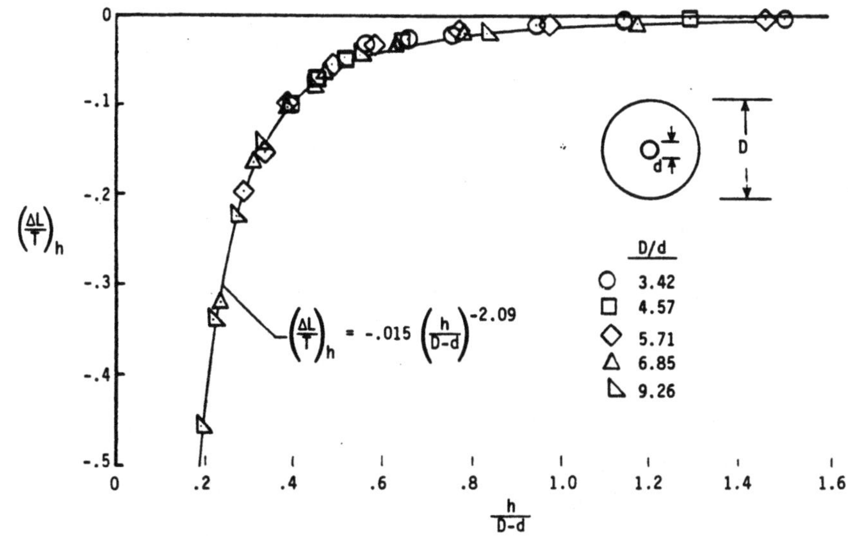

Figure 6. Lift loss in ground effect for a range of plate sizes. (Wyatt, 1962) $\frac{P_T}{p} < 1.5$.

Exp.	Est.	Conf.	Ref.	S/A	D̄/d	P_T/p
○	———	X-14 Airplane	25	150	10.8	~1.8
□	– – –	J-85 Eng./Plate	12	142	13.4	1.45

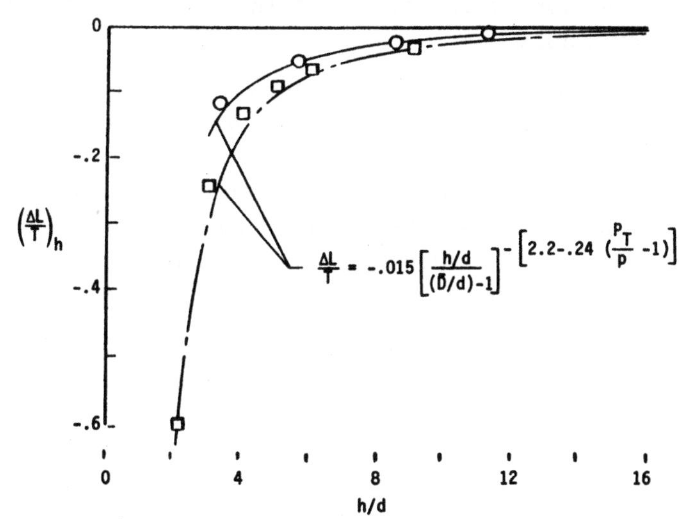

Figure 7. Comparison of estimates of hovering lift losses with large scale data.

As shown in Figure 9, however, the ground induced suckdown for some multiple jet configurations can be greater than the suckdown that would be expected for that configuration with a single centrally located jet of the same total jet area. The mechanism by which this can occur is illustrated in Figures 10 and 11 (Hall and Rogers, 1969). Flow visualization studies by Hall (1969) showed that the upward flow of the fountain and the downward flow of the jets generated a vortex like flow between the fountain and each jet (Figure 10). While the fountain flow generated lifting pressures in the impingement region (Figure 11), the vortex flows generated equally high suction pressures between the fountain and the jets. These suction pressures are about five times as large as the average suction pressure that would be generated by an equivalent centrally located single jet. These additional suction pressures produce a multiple jet suckdown force that must be accounted for before the lift produced by the fountain flow can be evaluated. In the case of the two jet configuration shown in Figure 9, these additional suction pressures generate an additional multiple jet suckdown that is larger than the fountain lift.

Unfortunately, these fountain-induced vortex flows have received little attention and there is no additional pressure distribution data of the type shown in Figure 11 to tell us anything about the effects of height, jet spacing, planform configuration and size, or the effects of using more than two jets on these vortex flows and the pressures they induce.

An empirical method for estimating both the multiple jet suckdown and the lift due to the fountain flow was developed from available force data in Kuhn (1981). The fountain lift part of the method is based on a modification of the fountain lift theory of Yen (1979). The additional multiple jet suckdown part was developed by subtracting the fountain lift increments estimated by the modified method of Yen (1979) from the experimental data.

The method of Kuhn (1981) also includes approximate expressions for estimating the effects of the lower surface contour of the fuselage and LID's (lift improvement devices) on the fountain increment. If the lower surface of the configuration is contoured, the fountain lift will be reduced because all the fountain flow will not have been stopped and turned to the horizontal by the bottom of the fuselage. Some of the flow will still have an upward component of momentum as it follows the contour of the body and flows upward

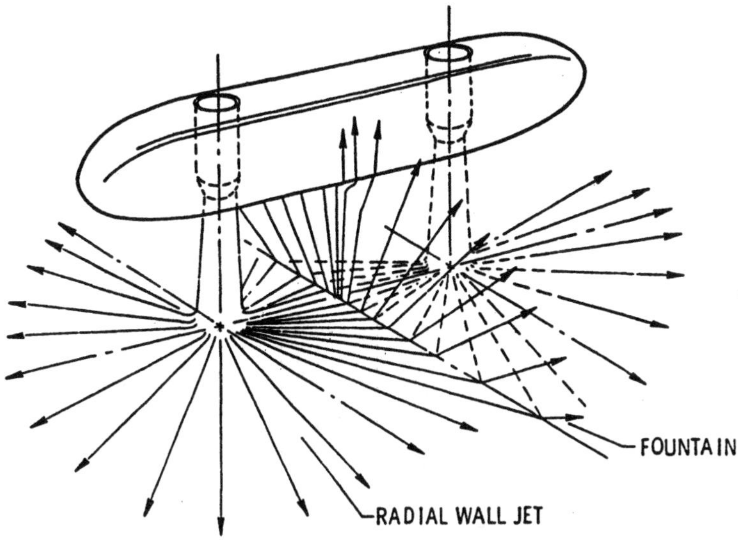

Figure 8. Fountain flow generated between a pair of jets.

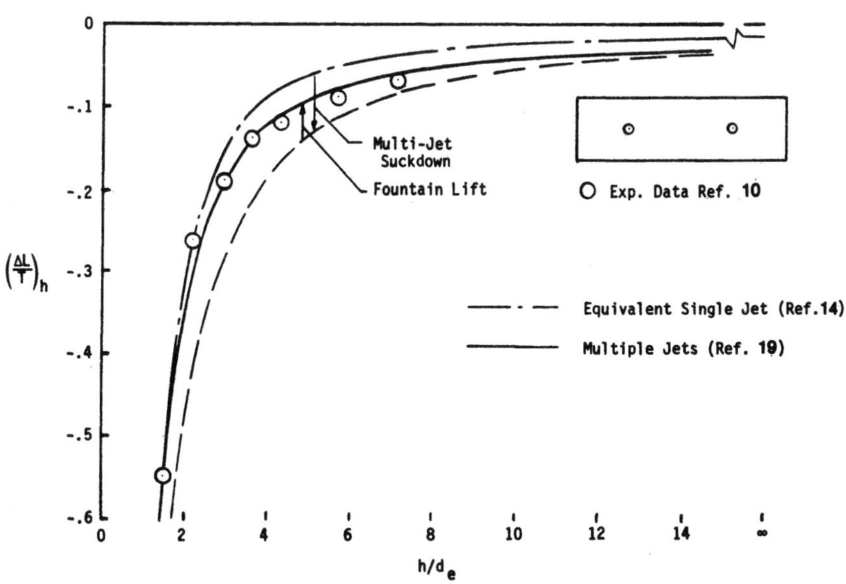

Figure 9. Fountain lift and the additional multi-jet suckdown generated on a simple 2-jet configuration.

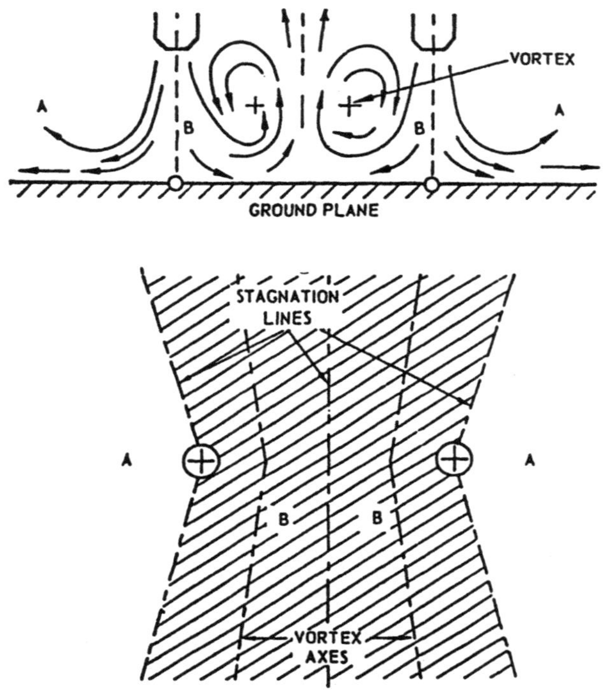

Figure 10. Flow field between two jets hovering in ground effect (Hall and Rogers, 1969).

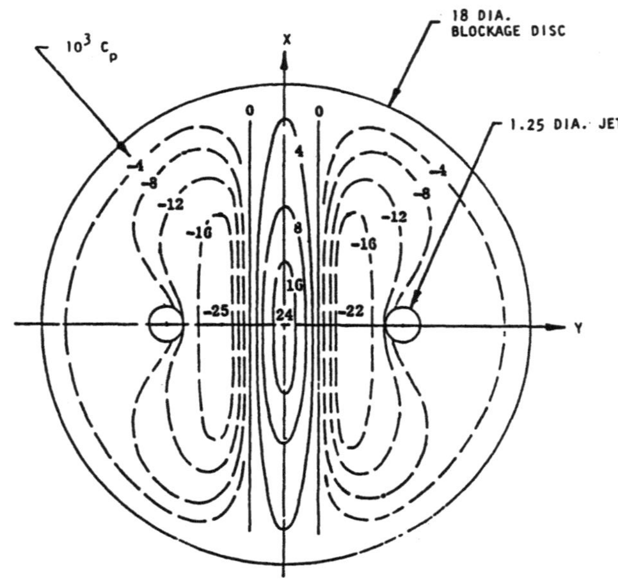

Figure 11. Pressure distribution due to fountain flow from two jets in ground effect (Hall and Rogers, 1969).

around the sides of the fuselage.

Frequently, strakes or LID's are installed on the lower surface of the body in an attempt to increase the fountain lift. In theory, if all the fountain flow could be contained and turned vertically downward, the fountain lift could be doubled. However, practical configurations fall short of this. Usually, there is insufficient space available for the LID's to enclose sufficient area, and it may be impractical to enclose the complete perimeter (or the rear portion may be left open to direct the hot gasses aft in order to minimize hot gas reingestion).

As illustrated in Figures 9 and 12, the method of Kuhn (1981) does a reasonably good job of reproducing the data from which it was derived. However, there is considerable need for additional work. As shown in Figure 13, the present method considerably overestimates the fountain lift on this simple two-jet configuration, probably because the body width relative to the jet diameter is over twice as wide as the few similar configurations used in developing the method.

Also, the method is limited to vertical, round jets of equal thrust and the data base on which the effects of body contour and LID's are based is very limited. Some additional data on the ground effects and flow fields of multiple jet configurations are contained in Bower (1981), Hall (1967), Foley and Sansone (1980), Johnson et al. (1979), Stewart and Kuhn (1983), Kotansky and Glase (1978), Sansone and Foley (1981), Sherrib (1978), Hill et al. (1982), and Williams and Wood (1966).

3.3 Transition Out of Ground Effect

3.3.1 Lift and Moments Induced

In the transition from hovering to conventional flight, the streams from the lifting jet or fan engines are swept back by the free stream (Figure 14) and are rolled up into pairs of vortices. These vortices, along with the entrainment action and blockage effect of the jets, induce suction pressure on the bottom of the configuration beside and behind the jets and a smaller region of positive pressures ahead of the jet (Figure 15). In most cases, these induced pressures result in a nose up moment and a net loss in lift on the configuration (Figure 16).

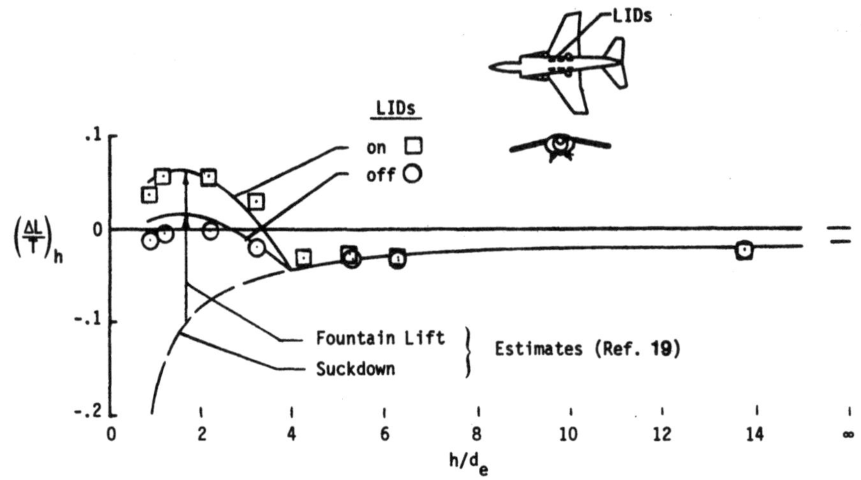

Figure 12. Lift induced on Harrier configuration hovering in ground effect (Johnson et al., 1979). "Copyright © the American Institute of Aeronautics and Astronautics; reprinted with permission of the AIAA."

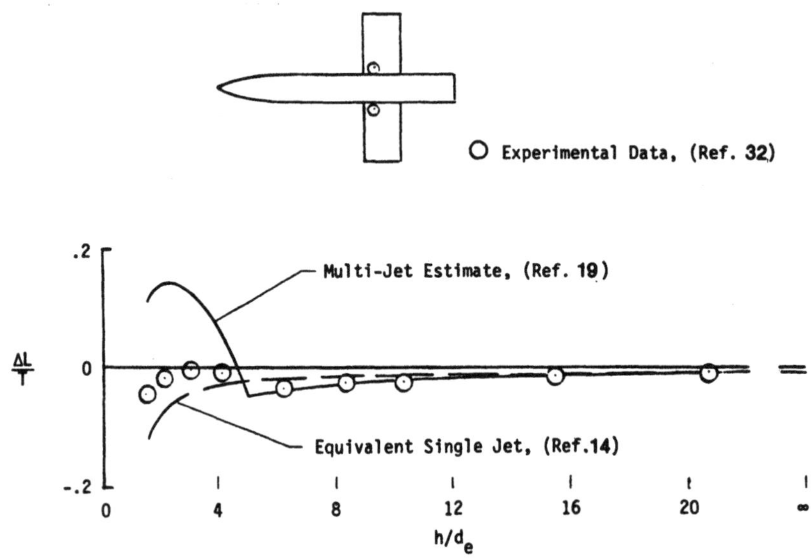

Figure 13. Lift induced on simple 2-jet high wing configuration hovering in ground effect.

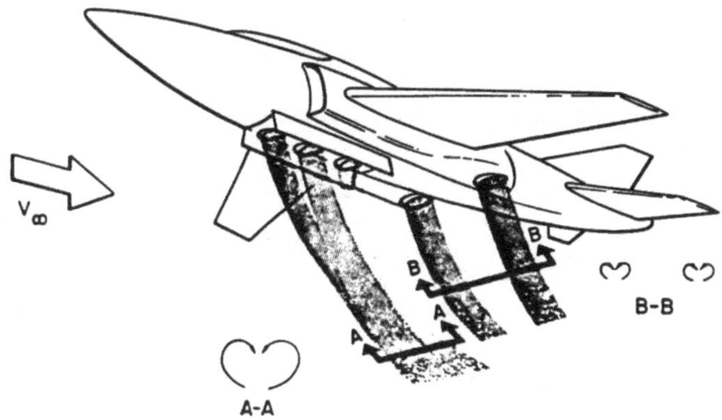

Figure 14. Jet/free stream interaction in transition.

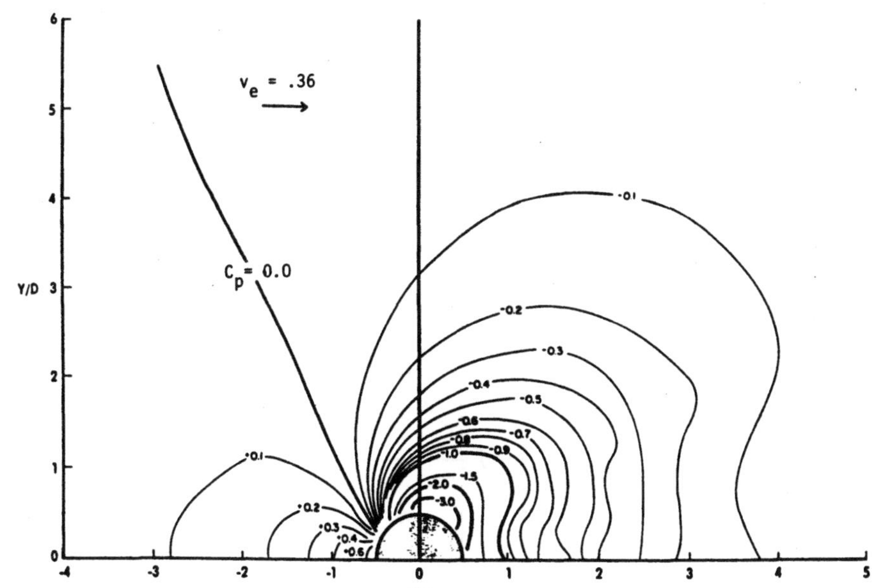

Figure 15. Contour lines of constant pressure coefficient induced on a flat plate by the jet/free stream interaction.

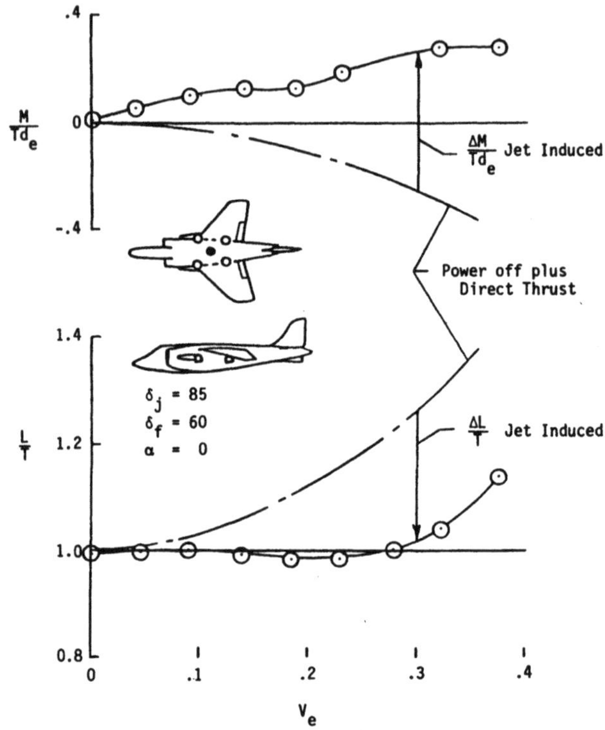

Figure 16. Jet induced effects on lift and pitching moment.

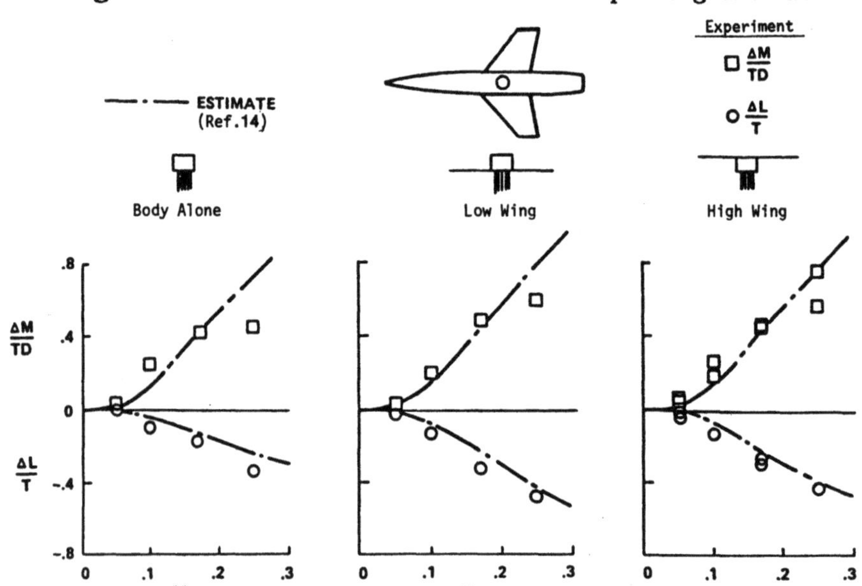

Figure 17. Jet wake effects on simple single-jet wing-body configuration.

The roll up of the jets into a vortex pair, the paths of the jet mass flow and the vortices and the pressures they induce have been the subject of many investigations (see, for instance, NASA SP-218, 1969; Aoyagi and Snyder, 1981; Beatty and Kress, 1979; Carter, 1969; Fern and Weston, 1975, 1979; Kuhlman et al., 1978; Kuhn, 1981; Margason et al., 1972; Mineck and Schwendemann, 1966; Perkins and Mendenhall, 1981; Spreeman, 1967; Volger, 1966; Williams and Wood, 1966; Wooler et al., 1972). The induced effects increase rapidly with aircraft velocity in transition. Williams (1966) showed that the appropriate parameter for correlating these effects of free stream velocity with jets of various pressure ratios and temperature is the effective velocity ratio, V_e, which is the square root of the free stream dynamic pressure to jet dynamic pressure ratio. (Many investigators prefer to use the parameter R which is simply the reciprocal of V_e.)

The induced effects also depend on the ratio of planform area to jet area, the number and location of the jets, wing height, etc. Wooler (1972) developed the first method for predicting these induced effects that was widely accepted. Other more sophisticated analytical methods (but still relying on some empirically derived constants) are currently being developed.

A purely empirical but simple to use method is presented in Henderson, Clark, and Walters (1980), Section 2.2.2.1. The basic effects of effective velocity ratio and planform area to jet area ratio were obtained by integration of the pressure distributions of Fearn and Weston (1975) and presented in chart form. Factors to adjust the basic lift loss for the effects of planform, position of the jet(s) in the planform, lateral and longitudinal spacing of the jets, jet aspect ratio, jet deflection and wing height were obtained from correlations of the experimental force data available. The method assumes the induced pitching moment increments can be estimated by multiplying the induced lift loss increment by an appropriate arm. The effective arm is estimated from charts and adjustment factors derived from the available moment data.

A comparison of the estimates with experimental data for a simple single jet wing/body configuration is presented in Figure 17. The data show that most of the induced effects are carried on the body and that there is some gain (less lift loss) in going to a high wing. The data for the high wing configuration also show the scatter in the data that makes the verification of prediction methods difficult. Figure 18 shows the comparison of the estimates with experimental data for a multiple jet configuration; in this case, the AV-6A Kestrel (predecessor to the Harrier) aircraft. Figure 18 also shows that the induced effects for a multiple jet configuration are much smaller than for a configuration with one jet of the same total area.

Although for most configurations the jet induced flows cause a lift loss, it has been found, as shown in Figure 19 (Carter, 1969), that if the jet is moved aft to a position near or aft of the wing trailing edge a favorable lift increment is induced. Apparently, the jet at or near the trailing edge is acting like a small span jet flap. In fact, the procedure for estimating these favorable induced lifts that is included in the method of Henderson, Clark, and Walters (1980), Section 2.2.2.1 for estimating jet induced effects is based on jet flap data. A comparison of the estimate by the method of Henderson, Clark, and Walters (1980) with data from Mineck and Schwendemann (1966) is presented in Figure 20. This model was tested with the jets in two locations: near the wing leading edge and at the wing trailing edge. Figure 21 shows the shift from adverse to favorably induced lift when the jets are moved aft but the method somewhat over-estimates the experimental data.

When McDonnell Douglas was starting the design of the AV-8B, the advanced version of the Harrier (Johnson et al., 1979), they were aware of the advantages of having at least part of the jet thrust near the wing trailing edge and of the data in Carter (1969). The design of the engine and balance considerations precluded moving the jets aft but, since the program was to include a new wing, they took advantage of the opportunity and designed a new large chord slotted flap that in effect moved the wing trailing edge (when the flap was deflected) closer to the rear nozzles. They could not eliminate the lift loss induced by the front jets but as shown in Figure 21 the favorable lift induced by the rear jets balances out the lift loss of the front jets and gives the configuration a net induced lift increment of about zero. This elimination of the induced losses plus the larger wing and more powerful flap gives the AV-8B a range/payload capability almost double that of the AV-8A (with the same engine) in the STOL overload operating mode.

3.3.2 Downwash at the Tail

The jet wake vortex system which causes the induced lift and moment on the configuration also induces an additional downwash

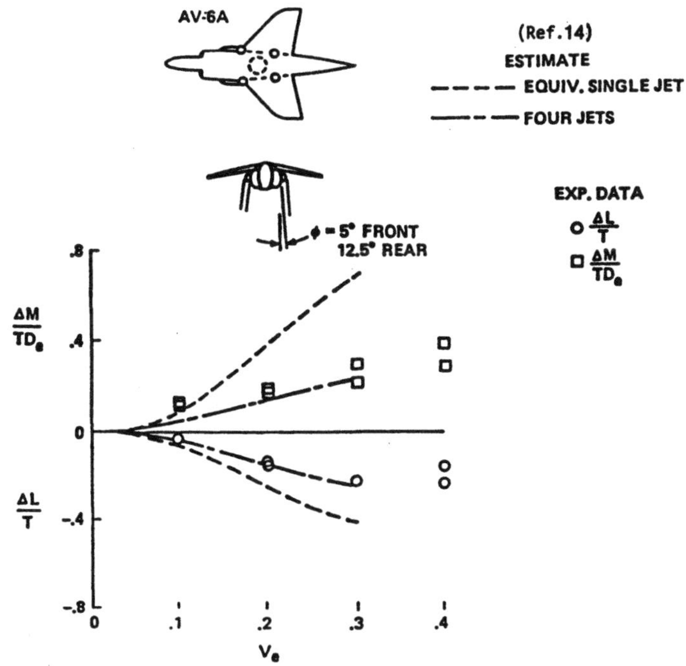

Figure 18. Jet wake system effects on the four jet configuration of Johnson et al. (1979). "Copyright © the American Institute of Aeronautics and Astronautics; reprinted with permission of the AIAA."

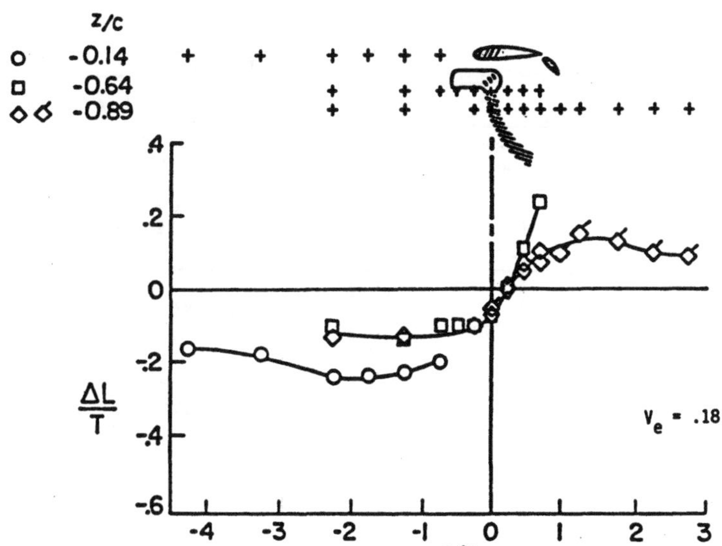

Figure 19. Effect of longitudinal position of jet on induced lift (Carter, 1969).

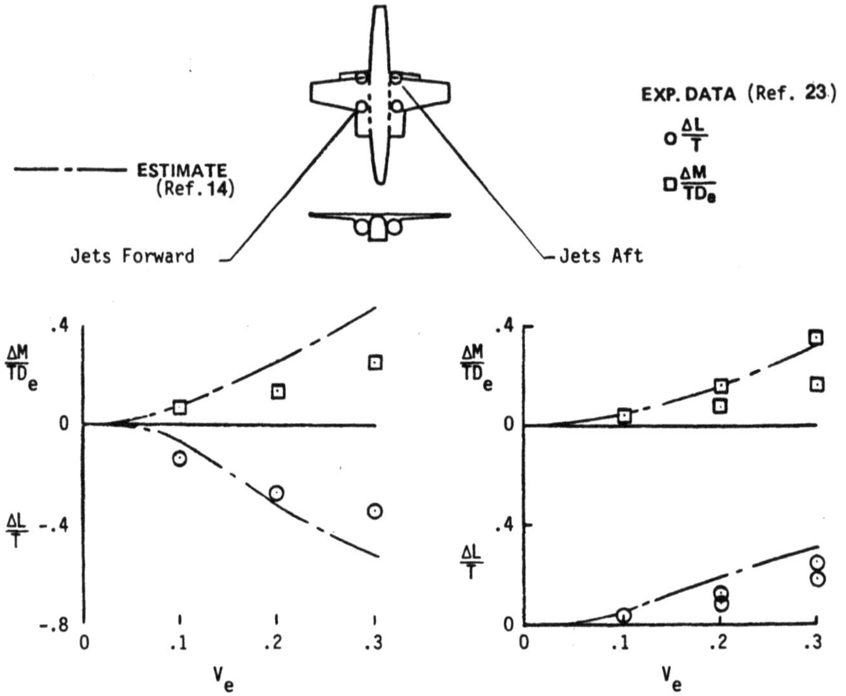

Figure 20. Comparison of estimate with experiment for two jet positions.

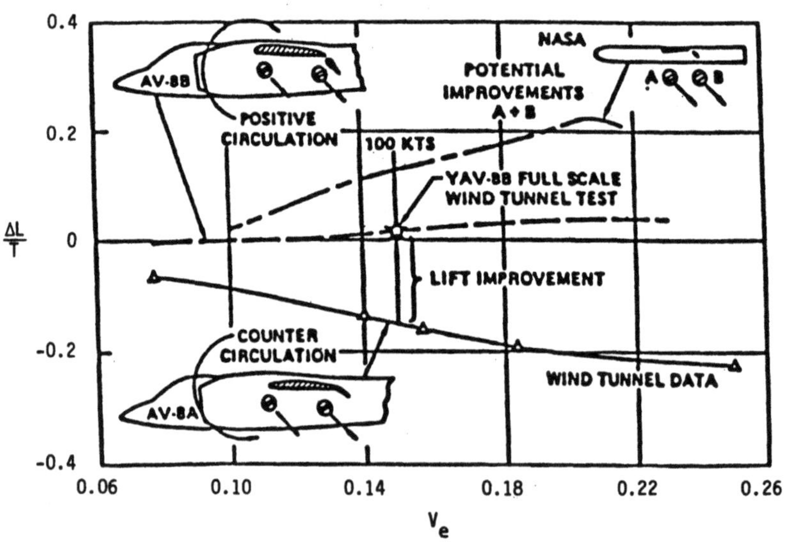

Figure 21. Effect of wing/flap design on reducing lift loss on AV-8B (Johnson et al., 1979). "Copyright © the American Institute of Aeronautics and Astronautics; reprinted with permission of the AIAA."

at the horizontal tail. There have been very few investigations of this jet induced downwash. One good study which included the effects of tail height and jet location is reported in Mineck and Schwendemann (1966). The effect of tail height from this study is presented in Figure 22. As would be expected, the downwash increases as hovering is approached (decreasing effective velocity ratio) because of the increasing jet strength relative to the free stream. The downwash also increases as the tail is lowered probably because the tail is getting closer to the vortices in the jet wake.

The effect of jet deflection is shown in Figure 23. The higher downwash for the lower jet deflection is probably due to the jet wake vortices moving up closer to the tail for the lower jet deflection. This configuration shows the surprising result of the downwash decreasing as the velocity ratio is reduced. In this case, the wing was carrying lift in the power off case as indicated by the level of power off downwash shown. The jet induced lift loss reduces the net wing lift and, because this induced lift loss tends to be concentrated inboard, probably also significantly alters the span load distribution. These factors may be decreasing the wing generated downwash faster than the increase due to the jet wake vortex system. These results suggest that the area of jet induced downwash needs considerable further study.

3.3.3 Lateral/Directional Characteristics

The same jet wake system that generates induced lift and pitching moment increments can affect the rolling moment, yawing moment and side-force characteristics of the configuration. As depicted schematically in Figure 24, at an angle of sideslip the jet wake system is displaced laterally with respect to the configuration and the pressure distribution that it generates on the body, and to a lesser extent on the wing, is shifted toward the downstream side of the configuration generating a jet induced rolling moment. An increment of side force and yawing moment are also induced on the body and although the jet wake system is usually far below the body, the force data available show that a significant sidewash is induced at the vertical tail.

A simple empirical method for estimating the jet induced increments of rolling moment, yawing moment, and side force was developed from the available data in Kuhn (1981). As shown in Figure 25, the method does a reasonable job of reproducing the data from which it was developed. Figure 25 also shows that at the higher

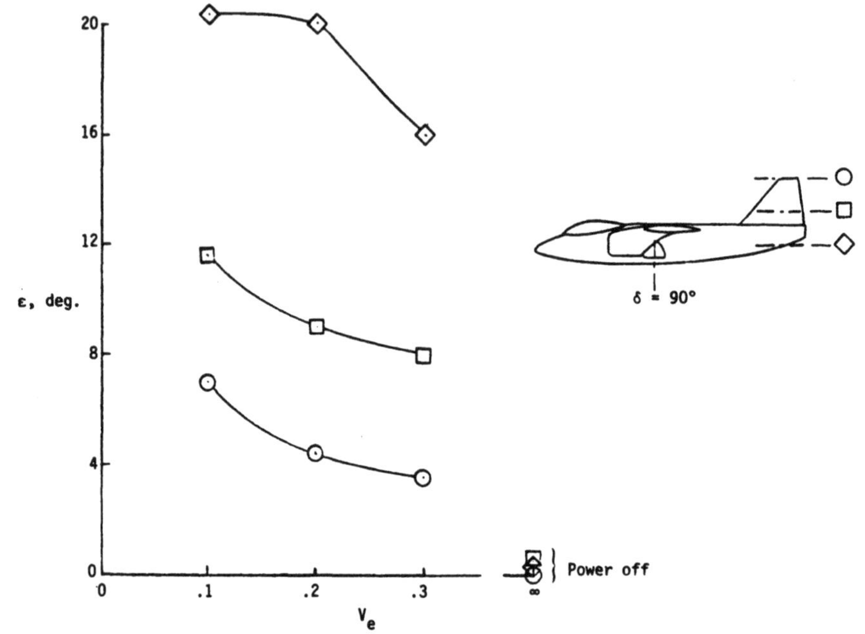

Figure 22. Jet induced downwash for configuration of Mineck and Schwendemann (1966).

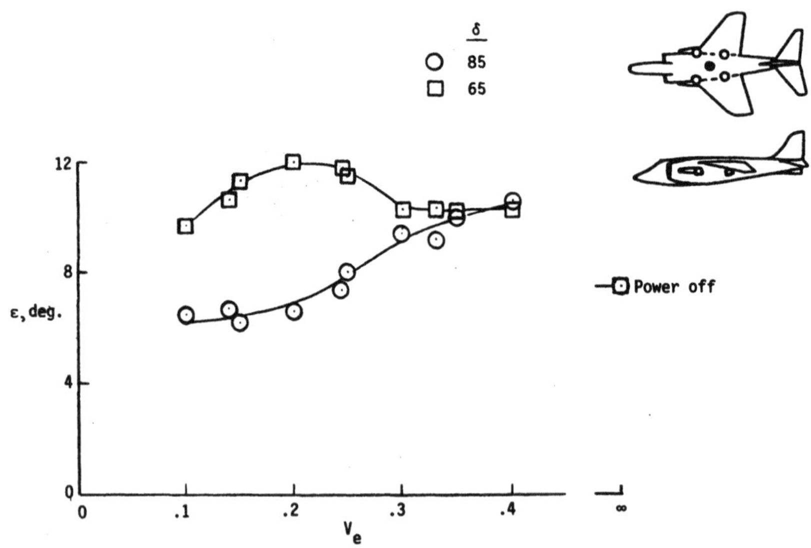

Figure 23. Jet induced downwash for configuration of Johnson et al. (1979). "Copyright © the American Institute of Aeronautics and Astronautics; reprinted with permission of the AIAA."

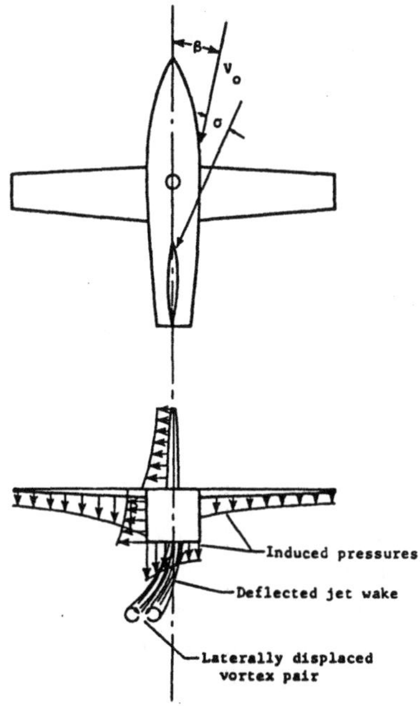

Figure 24. Schematic of effect of side slip on jet induced pressures.

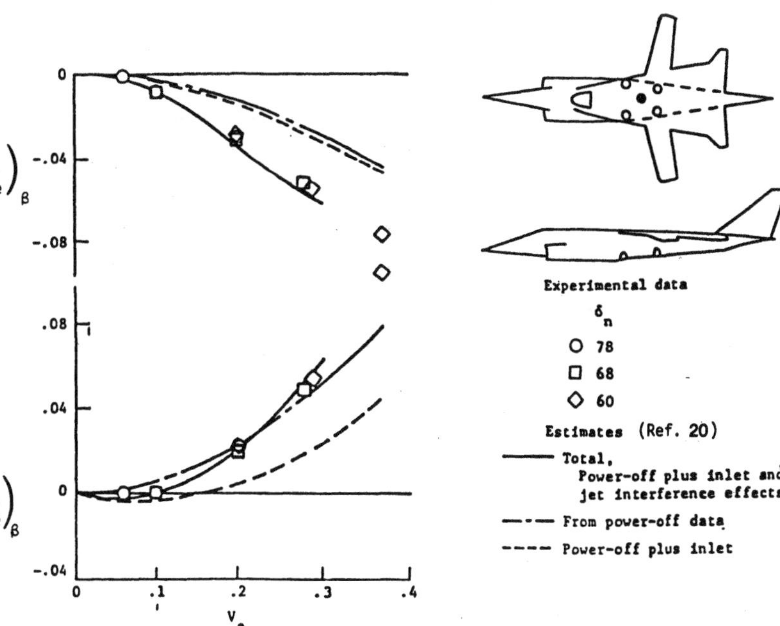

Figure 25. Comparison of estimates of lateral/directional characteristics with experimental data.

velocity ratios, the jet induced sidewash increases the vertical tail contribution to yawing moment enough to compensate for the destabilizing effect of the inlet flow. However, the inlet flow causes the configuration to be directionally unstable at low speeds.

The rolling moment due to sideslip however is approximately doubled by the jet effects. The combination of the inlet induced directional instability at low speeds plus the large jet induced rolling moment due to sideslip has been responsible for the loss of several of the early jet V/STOL research aircraft and may have been a factor in some of the Harrier accidents.

The problems created by the jet induced rolling moments can be aggravated by the jet induced effects on the effectiveness of the roll control jets as shown in Figure 26. Usually on jet V/STOL configurations, roll control in hovering and at low speeds is obtained from bleed air ducted to control jets at or near the wing tip. The interaction of these jets with the free stream induces suction pressures on the wing similar to those induced on the body by the main jets. These suction pressures can significantly reduce the control effectiveness particularly if the control jets are located forward on the wing. Control jets should be located as close to the tip and as far aft as possible to minimize this problem.

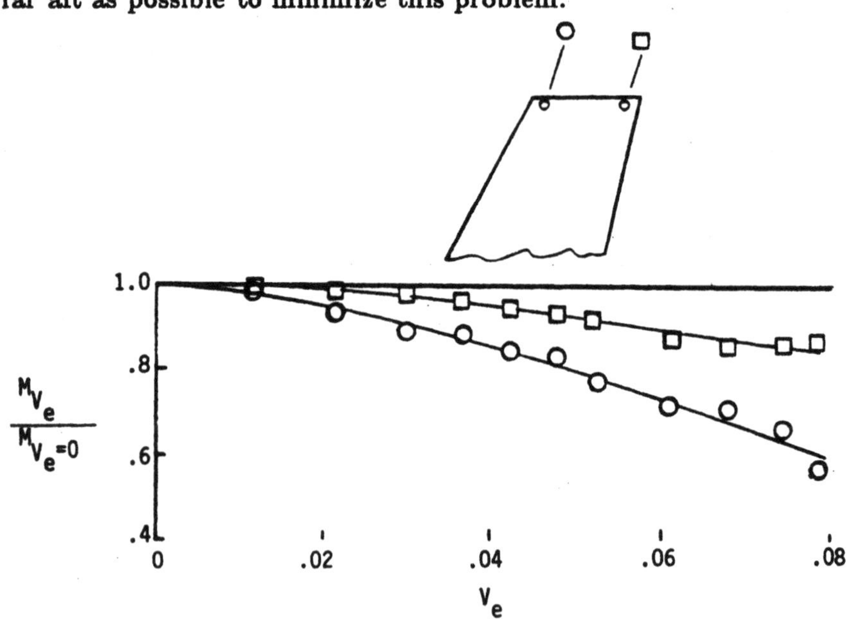

Figure 26. Effect of velocity ratio and control jet position on effectiveness of wing tip mounted roll control jets (Spreeman, 1967).

3.4 STOL Operation (Transition in Ground Effect)

3.4.1 Ground Vortex Formation

When the aircraft is operating at transition speeds in ground effect, all the induced effects discussed above are still present but modified by either the presence of the ground or the free stream. In addition, a ground vortex is formed (Figure 27) by the action of the free stream in opposing and rolling up the wall jet flowing forward from the impingement point of the lifting jets. The location of this ground vortex and its induced effects as well as the effects of the ground and free stream on the other induced effects discussed above were the subjects of the investigation reported in Stewart and Kuhn (1983).

The ground vortex induces suction pressures on the ground as well as on the lower surface of the configuration. In order to obtain a better understanding of the ground vortex and its effects, pressure distributions were measured on the ground board during the study of Stewart and Kuhn (1983). These pressure distributions give an indication of the location and strength of the ground vortex as well as the extent of the negative pressure region generated by the ground vortex. As expected, the ground vortex moves aft as the free stream velocity and the height are increased (Figure 28), and can be swept behind the projected centerline of the jet at high heights.

The strength of the ground vortex, as indicated by the maximum negative pressure coefficient induced on the ground board, decreases rapidly at the higher heights. The height, h_v, at which the pressure goes to zero is the height at which the jet is deflected aft by the free stream to the point that none of the impinging flow is projected forward and the ground vortex ceases to exist. As depicted in Figure 29, the lift loss for some configurations experiences a significant break at this height. The height h_v is the limit of the effects of the ground vortex but not the limit of ground effects. The variation in lift above this height is due to the effects of the ground on the lift loss due to the jet wake system.

At low velocity ratios the ground vortex effects can be felt to extreme heights as shown in Figure 30. Unfortunately, some of the early investigations of ground effects were carried to heights of only about 10 to 12 diameters and the data taken at the highest height may not have been truly out of ground effect.

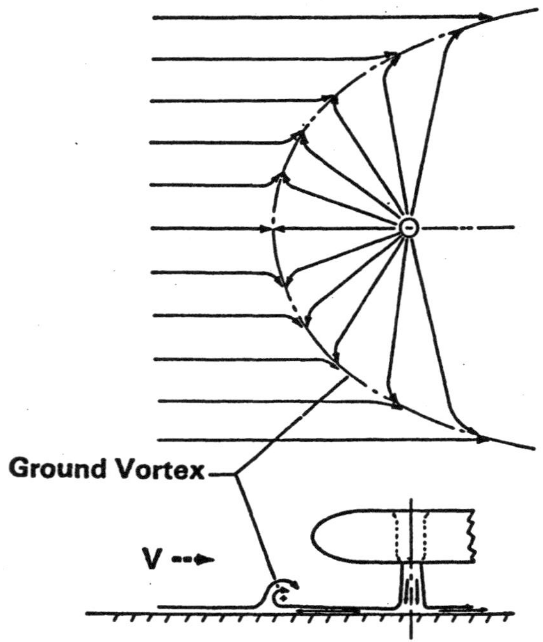

Figure 27. Formation of ground vortex.

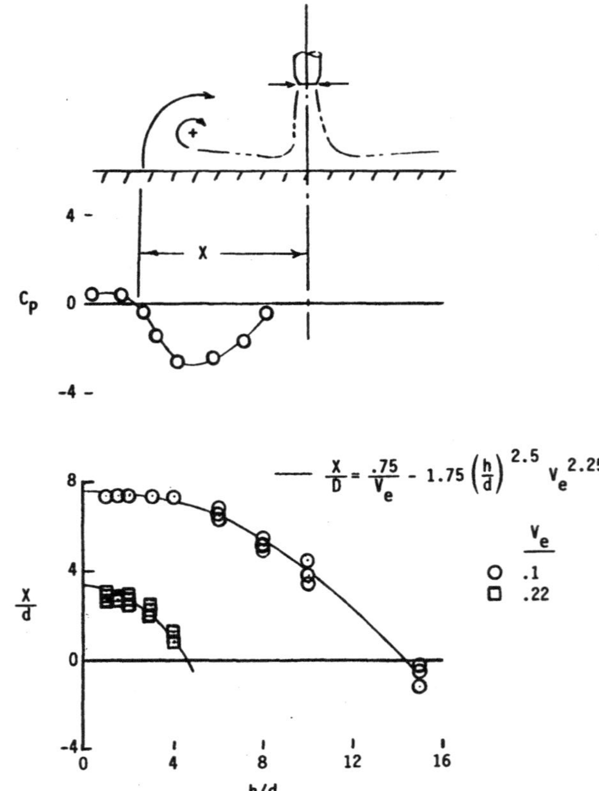

Figure 28. Maximum forward projection of wall jet and ground vortex flow.

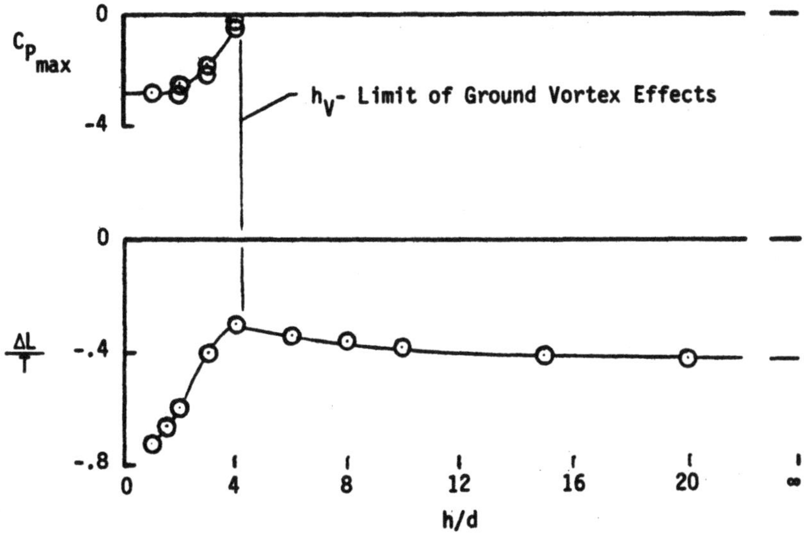

Figure 29. Typical variation of ground vortex induced pressure and lift loss with height for single jet configuration, $V_e = .2$.

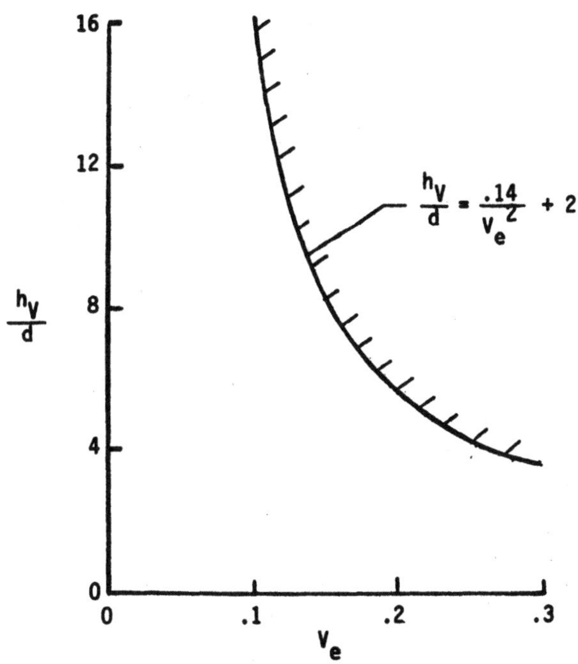

Figure 30. Maximum height to which ground vortex effects are felt for vertical jets.

3.4.2 Lift and Moments

An empirical method for estimating the effects of the ground on the jet induced effects in transition was developed in Stewart and Kuhn (1983). The method is designed so that it reduces to the methods for estimating ground effects on the hovering lift losses at zero forward speed and to the method for estimating the free air lift losses in transition out of ground effect. The method attempts to account for the total flow field experienced in ground effect as shown schematically in Figure 31. In addition to the ground vortex, the vortices from the jet wake system are present but deflected and probably altered in strength by the proximity of the ground.

At low heights and velocity ratios, the ground vortex is located well ahead of the configuration and the radial wall jet that causes the suckdown when hovering in ground effect is operating pretty much as it does in hover. However, as the height and/or the velocity ratio increase, the ground vortex moves back and the area under the configuration that can generate a suckdown is reduced. The method attempts to account for the resulting reduction in the *hover suckdown* lift loss.

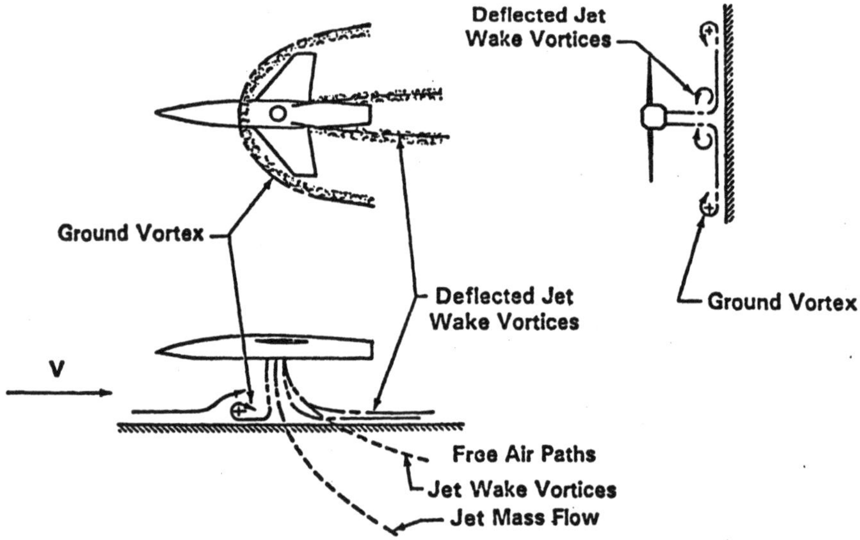

Figure 31. Flow field in ground effect.

The increments of lift and pitching moment induced on a simple single jet configuration in STOL operation as estimated by the method of Stewart and Kuhn (1983) are compared with experimental data in Figure 32. The increase in lift loss at very low heights is due to the ground vortex and to the *hover suckdown* terms.

Although not shown in Figure 31, the fountain flows produced by multiple jet configurations are also considered in the method of Stewart and Kuhn (1983). The magnitude of these fountain effects depends on the planform area enclosed within the jet pattern and on the arrangement of the jets. The positive lift increments induced can be very large at low heights but decrease rapidly with height; thus their effects are opposite to and tend to offset the lift losses induced by the ground vortex. The lift gain due to the fountain effects and the lift loss due to the ground vortex do not go to zero at the same height however, and at low velocity ratios the lift loss due to the ground vortex extends to considerably greater height than the lift gain due to the fountain. For the four jet configuration shown in Figure 33 on the other hand, the lift gain due to the fountain is greater than the ground effect loss resulting in a reduction in the lift loss at low heights for this velocity ratio.

An interesting part of the investigation reported in Stewart and Kuhn (1983) was the comparison of the ground effects on a direct thrust and a distributed jet configuration on the same simple wing/body model (Figure 34). The direct-thrust, circular-jet configuration, because the jets are at the wing trailing edge, experiences a favorable jet induced lift out of ground effect, and this lift increases rapidly as the ground is approached. These results are compared with data from the distributed jet configuration operating at close to the same level of induced lift out of ground effect. This configuration experiences a sharp decrease in induced lift close to the ground. The reasons for the dramatic difference in results close to the ground are not known but they are probably associated with the extent and strength of the ground vortices which pressures measured on the ground board show to be much greater for the slot jet configuration. These results suggest that the arrangement and spanwise extent of the jet of a jet flap configuration on the ground effects should be investigated further.

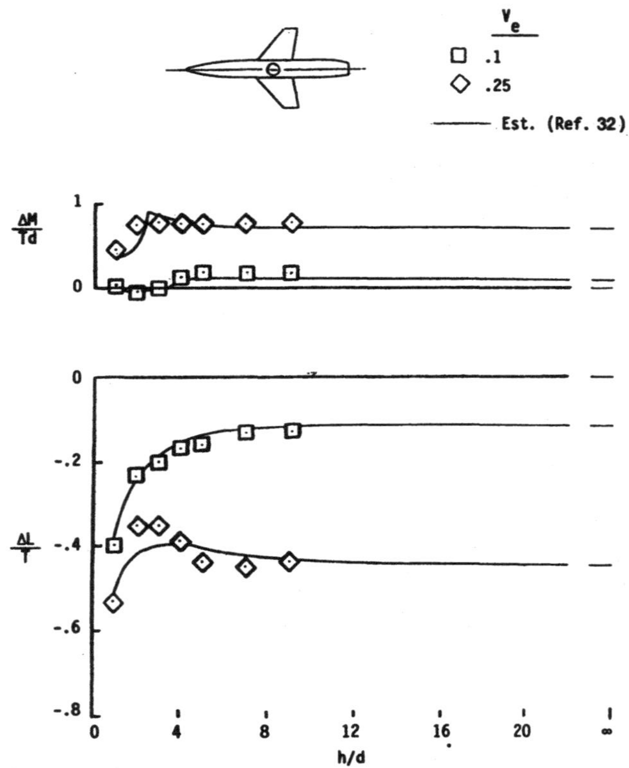

Figure 32. Comparison of estimates of effect of ground proximity with data for simple single jet configuration.

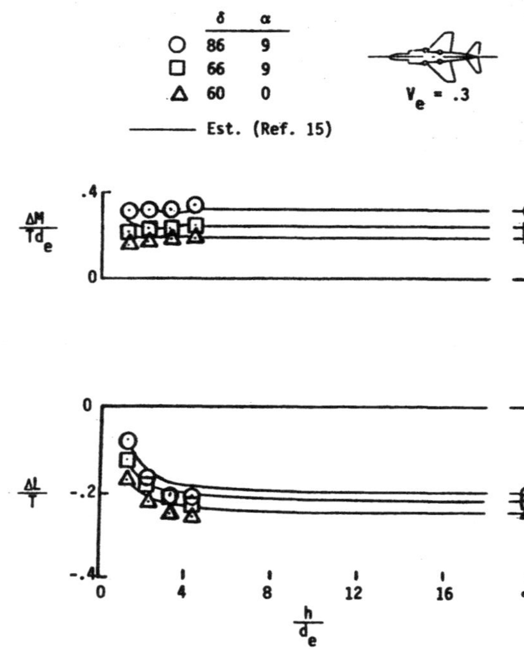

Figure 33. Comparison of estimate of effect of ground proximity with data for four jet configuration (Margason et al., 1972).

4. Loose Ends and Areas Needing Further Study

The basic phenomena that determine the aerodynamic characteristics of jet and fan powered V/STOL aircraft at low speeds are understood and in some areas our ability to predict these characteristics is adequate. The jet induced lift loss out of ground effect is small and can be adequately predicted with the methods available. Similarly, the ground effects in hover for single jet configurations have received a lot of attention and can be estimated with available methods.

On the other hand, the fountain effects and the related additional suckdown of multiple jet configurations needs more work. Studies of the effects of jet arrangement and spacing on both the positive and negative pressures induced between the jets are needed to provide a better basis for prediction methods for multiple jet configurations. This work should include, in addition to jet spacing and arrangement, the effects of jet deflection and differential jet thrust and size.

The data base for the effects of body contour and LID's is weak. Body contour effects depend on the form of the fountain. There are several sets of data for lengthwise fountains (from dual jet configurations with jets on either side of the fuselage) but none for crosswise fountains (dual jets fore and aft in the fuselage) and very little for fountains from multiple jet arrangements. These effects need to be systematically investigated.

The moments induced in hover have received little attention. There are some data from the tests of specific configurations in the literature but no systematic studies of the effects of the distribution of planform area and there have been no attempts to develop methods to predict the moments induced by asymmetric distribution of area.

The interaction of the jet flow and the free stream out of ground effect has received a lot of attention. The jet path, roll up into a vortex pair and the pressures induced on flat plates by this jet wake system for single jet configurations have been extensively studied, and methods for predicting the induced forces and moments are being developed. Some studies of multiple jet configurations have also been made and this work should be continued.

The downwash at the tail, on the other hand, has received very little attention. The limited data available indicate that large increases in downwash can be induced by the action of the jets and that there is a significant interaction with the effects of the lift change induced on the wing by the jet. Accurate methods for estimating the downwash throughout the transition speed range are needed for control system design.

The effects of ground proximity on the induced effects at transition speeds have only recently begun to receive attention. Most of the limited data available have been obtained over a fixed ground board, and although Vogler (1966) indicated that the belt had little effect on the force data, the correlation of the position of the ground vortex presented in Figure 35 appears to show a large effect of the boundary layer on the ground board.

Four of the investigations were conducted in a wind tunnel or with a blower to produce the free stream. The fifth (Abbott, 1967) used the moving model technique. In a wind tunnel or with a blower, a boundary layer is created between the free stream and the ground board over which it is flowing. There is no boundary layer with the moving model technique. The wall jet moving forward against the free stream is very thin (the maximum velocity in the wall jet occurs at a height of less than about 20% of the jet diameter above the ground). When this thin wall jet is opposed by the relatively lower velocity in the free stream boundary layer rather than by the full free stream velocity, the separation point will move forward. The investigation of Schwantes (1973) sets out to simulate atmospheric wind conditions and the boundary layer that would be present in a cross wind situation. The boundary layer was, therefore, made very thick and this investigation shows the maximum forward penetration (Figure 35). The moving model data (Abbott, 1967), which had no boundary layer, on the other hand, shows the smallest penetration. Nothing is known about the boundary layers for the other investigations but a photo in Weber and Gay (1975) indicates the ground board used was relatively short indicating that the boundary layer may have been relatively thin.

Also the effects of pressure ratio on the induced forces and moments have not been investigated but a comparison of the pressures induced on the ground by the ground vortex by the investigation of Stewart and Kuhn (1983) and the investigation of Colin and Olivari (1969) indicates that there may be a significant effect of pressure ratio on the strength of the ground vortex. An investigation of the effects of pressure ratio and of the need for the moving belt ground

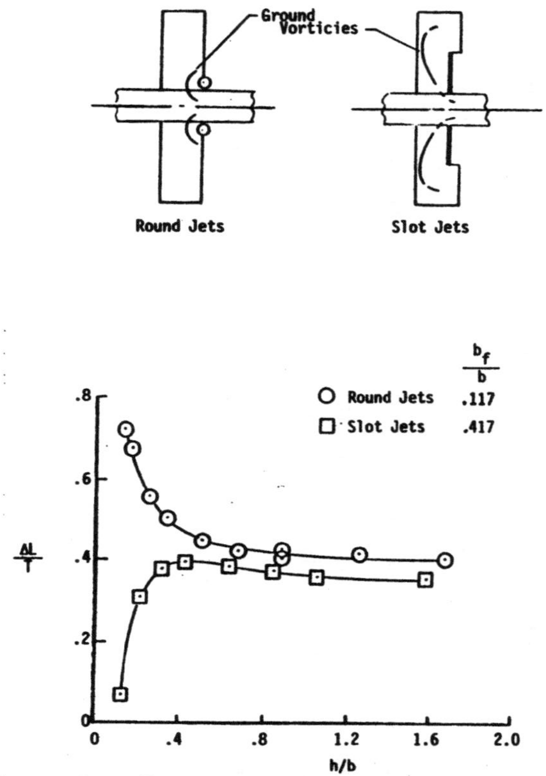

Figure 34. Comparison of ground effects on circular (direct thrust) and slot jet (jet flap) configuration (Stewart and Kuhn [20]).

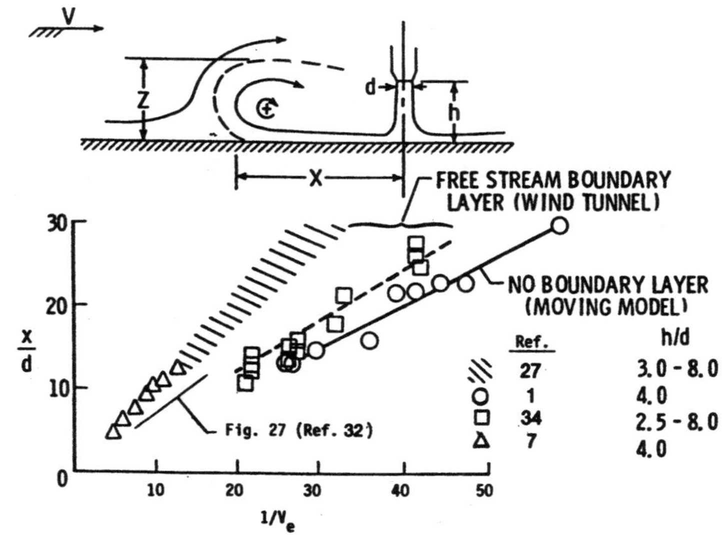

Figure 35. Effect of ground boundary layer on forward extent of ground vortex.

board on the position and strength of the ground vortex and its effect on the induced forces and moments should be the next order of business in this area.

There are data on only one configuration (Margason et al., 1972) from which the effects of the ground on the jet induced downwash at the tail can be determined. The investigation of the effects of configuration variables on the jet induced downwash at the tail recommended above should include the effects of ground proximity.

Finally, the data of Figure 34 indicate that there is a significant effect of the *span* of nozzles located near the wing trailing edge on the effects of ground proximity. These results suggest that there may be an optimum nozzle/wing configuration that will maximize the out of ground effect jet induced lift gains and minimize the ground induced lift losses experienced by jet flap configurations. This prospect appears worthy of further investigation.

5. Concluding Remarks

Most of the basic phenomena that determine the aerodynamic characteristics of jet and fan powered V/STOL aircraft in hover and transition flight, both in and out of ground effect, are understood, and the design principles to use to minimize adverse effects and take advantage of favorable effects are known. However, our ability to make accurate estimates of many of the induced effects is limited. We can get in the right *ball park* with the empirical and semi-empirical methods available, but an experimental program is required in the early stages of an aircraft development in order to obtain the accurate data needed for final design purposes.

The fact that this experimental program is needed creates a problem. Aircraft development programs are very expensive and there will be few new aircraft developed in the future. The more unknowns there are, the more difficult it is to justify a new program. A continuing program to develop and refine our tools for predicting the low speed aerodynamic characteristics of V/STOL aircraft is needed.

References

[1] Abbott, W. A. "Studies of Flow Fields Created by Vertical and Inclined Jet When Stationary or Moving Over a Horizontal Surface," *RAE ACR CP No. 911,* 1967.

[2] "Analysis of a Jet in a Subsonic Crosswind," *NASA SP-218,* September 1969.

[3] Aoyagi, K. and Snyder, P. K. "Experimental Investigation of a Jet Inclined to a Subsonic Crossflow," *AIAA 81-2610,* December 1981.

[4] Beatty, T. D. and Kress, S. S. "Prediction Methodology for Propulsive Induced Forces and Moments of V/STOL Aircraft in Transition/STOL Flight," *NAVAIRDEVCEN Report No. NADC-77119-30,* July 1979.

[5] Bower, W. W. "A Summary of Jet-Impingement Studies at McDonnell Douglas Research Laboratories," *AIAA 81-2613,* December 1981.

[6] Carter, A. W. "Effects of Jet-Exhaust Location on the Longitudinal Aerodynamic Characteristics of a Jet V/STOL Model," *NASA TN D-5333,* July 1969.

[7] Colin, P. E. and Olivari, D. "The Impingement of a Circular Jet Normal to a Flat Surface With and Without Cross Flow," *AD688953,* European Research Office, United States Army, January 1969.

[8] Fern, R. L. and Weston, R. P. "Induced Pressure Distribution of a Jet in a Crossflow," *NASA TN D-7816,* July 1975.

[9] ————————. "Velocity Field of a Round Jet in a Cross Flow for Various Jet Injection Angles and Velocity Ratios," *NASA Technical Paper 1506,* October 1979.

[10] Foley, W. H. and Sansone, J. A. "V/STOL Propulsion-Induced Aerodynamics Hover Calculation Method," *General Dynamics, NADC-78242-60,* February 1980.

[11] Gentry, G. L. and Margason, R. J. "Jet-Induced Lift Losses on VTOL Configurations Hovering in and Out of Ground Effect," *NASA TN D-3166,* Feb. 1966.

[12] Hall, G. R. "Scaling of VTOL Aerodynamic Suck-down Forces," *AIAA Journal of Aircraft,* July–August 1967, 393–94.

[13] Hall, G. R. and Rogers, K. H. "Recirculation Effects Produced by a Pair of Heated Jets Impinging on a Ground Plane," *NASA CR-1307,* May 1969.

[14] Henderson, C., Clark, J., and Walters, M. "V/STOL Aerodynamics and Stability and Control Manual," *NAVAIRDEVCEN-80017-60,* January 1980.

[15] Hill, W. G., Jr., et al. "An Investigation of Scale Model Testing of VTOL Aircraft in Hover," *ICAS-82-5.2.3,* August 1982.

[16] Johnson, D. B., et al. "Powered Wind Tunnel Testing of the AV-8B: A Straightforward Approach Pays Off," *AIAA Paper No. 79-0333,* January 1979.

[17] Kotansky, D. R. and Glase, L. W. "Investigation of Impingement Region and Wall Jets Formed by the Interaction of High Aspect Ratio Lift Jets and a Ground Plane," *NASA CR-152174,* September 1978.

[18] Kuhlman, J. M., et al. "Experimental Investigation of Effect of Jet Decay Rate on Jet-Induced Pressures on a Flat Plate," *NASA CR-2979,* April 1978.

[19] Kuhn, R. E. "An Engineering Method for Estimating the Induced Lift on V/STOL Aircraft Hovering In and Out of Ground Effect," *NADC-80246-60,* January 1981.

[20] ———. "An Engineering Method for Estimating the Lateral/Directional Characteristics of V/STOL Configurations in Transition," *NAVAIRDEVCEN Report No. NADC-81931-60,* Feb. 1981.

[21] McLemore, H. C. "Jet-Induced Lift Loss of Jet VTOL Configurations in Hovering Condition," *NASA TN D-3435,* June 1966.

[22] Margason, R. J., et al. "Wind-Tunnel Investigation at Low Speeds of a Model of the Kestrel (XV-6A) Vectored-Thrust V/STOL Airplane," *NASA TN D-6826,* July 1972.

[23] Mineck, R. E. and Schwendemann, M. F. "Aerodynamic Characteristics of a Vectored-Thrust V/STOL Fighter in the Transition-Speed Range," *NASA TN D-7191,* January 1966.

[24] Perkins, S. C., Jr. and Mendenhall, M. R. "A Study of Real Jet Effects on the Surface Pressure Distribution Induced by a Jet in a Crossflow," *NASA CR-166150,* March 1981.

[25] Rolls, S. L. "Jet VTOL Powerplant Experience During Flight Test of X-14A VTOL Research Vehicle," *AGARDOgraph 103,* October 1965, 543–57.

[26] Sansone, J. A. and Foley, W. H. "An Empirical Method for Estimating the Effect of Ground Proximity on the Jet-Induced Lift of V/STOL Aircraft Employing Rectangular Jets," *NADC-79298-60,* August 1981.

[27] Schwantes, E. "The Recirculation Flow Field of a VTOL Lifting Engine," *NASA TT F-14912,* June 1973.

[28] Sherrib, H. E. "Ground Effects Testing of Two, Three, and Four Jet Configurations," *AIAA 78-1510,* August 1978.

[29] Shumpert, P. K. and Tibbets, J. G. "Model Tests of Jet-Induced Lift Effects on VTOL Aircraft in Hover," *NASA CR-1297,* March 1969.

[30] Spreeman, K. P. "Free Stream Interference Effects on Effectiveness of Control Jets Near the Wing Tip of a VTOL Aircraft Model," *NASA TN D-4084,* 1967.

[31] Spreeman, K. P. and Sherman, I. R. "Effects of Ground Proximity on the Thrust of a Simple Downward-Directed Jet Beneath a Flat Surface," *NACA TN 4407,* September 1958.

[32] Stewart, V. R. and Kuhn, R. E. "A Method for Estimating the Propulsion Induced Aerodynamic Characteristics of STOL Aircraft in Ground Effect," *NADC 80226-60,* August 1983.

[33] Vogler, R. D. "Ground Effects on Single and Multiple-Jet VTOL Models at Transition Speeds Over Stationary and Moving Ground Planes," *NASA TN D-3213,* January 1966.

[34] Weber, H. A. and Gay, A. "VTOL Reingestion Model Testing of Fountain Control and Wind Effects," *Prediction Methods for V/STOL Propulsion Aerodynamics, 1, Naval Systems Command,* 1975, 304–33.

[35] Williams, J. and Wood, M. N. "Aerodynamic Interference Effects With Jet-Lift V/STOL Aircraft Under Static and Forward-Speed Conditions," *RAE Technical Report 66403,* December 1966.

[36] Wooler, P. T., et al. "V/STOL Aircraft Aerodynamic Prediction Methods Investigation," Air Force Flight Dynamics Lab., *AFFDL-TR-72-26,* January 1972.

[37] Wyatt, L. A. "Static Tests of Ground Effect on Planforms Fitted With a Centrally-Located Round Lifting Jet," *Ministry of Aviation CP749,* June 1962.

[38] Yen, K. T. "On the Vertical Momentum of the Fountain Produced by Multi-Jet Vertical Impingement on a Flat Ground Plane," *NADC-79273-60,* November 1979.

Advances in Ejector Thrust Augmentation

Paul M. Bevilaqua
Rockwell International Corporation
Columbus, OH 43216

Abstract

The additional thrust required to give an aircraft V/STOL capability may be obtained by diverting the engine exhaust flow through a thrust augmenting ejector. Entrainment by the primary jet induces a flow of air through the ejector duct, thereby increasing the thrust of the jet. Progress in developing a theory of ejector operation, and related efforts in modelling and prediction will be summarized. Studies of turbulent mixing and duct design which have resulted in improvements in ejector performance will also be described. Finally, research problems of current interest and the likely direction of future airplane programs will be discussed.

1. Introduction

An ejector is simply a duct in which entrainment by a jet of primary fluid is used to pump a secondary stream. By diverting the exhaust of a turbojet engine through an ejector pump, as shown schematically in Figure 1, significant increases in the jet thrust can

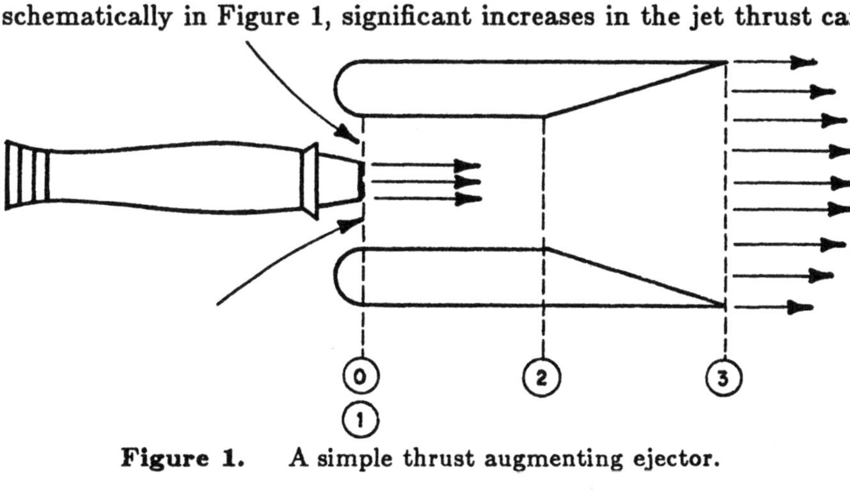

Figure 1. A simple thrust augmenting ejector.

be obtained. In this application, the ejector functions like the fan on a high bypass engine: thrust is increased by accelerating a large mass of air drawn from the atmosphere. Thus, an ejector can be used to give an aircraft having a basic thrust-to-weight ratio less than one the additional thrust necessary for vertical takeoffs and landings. When the ejector is integrated with the wing, the exhaust flow acts as a jet flap to increase wing lift and thereby provide good short takeoff and landing performance, and a smooth conversion to conventional flight. In addition, the mixing of the primary jet and entrained air within the ejector reduces the temperature and velocity of the lift jets, which reduces the hazard to ground personnel and the potential for damage to equipment.

Because ejector thrust augmentation provides a solution to many of the problems of achieving V/STOL capabilities, considerable research has been directed at understanding and improving ejector performance, and four ejector demonstrator aircraft have been built. Porter and Squyers (1981) have compiled a list of more than 1600 ejector research papers. It will not be possible to discuss all the important papers in this review. However, I will try to indicate the principle directions ejector research has taken, and summarize recent accomplishments and the current status. The proceedings of the recent AFOSR workshop (Braden et al., 1982) on ejector thrust augmentation provides a useful overview of current work, and some insight into the problems of ejector development.

In the next section of this paper, the current understanding of ejector operation will be described, and related progress in modelling and prediction will be outlined. Component development which has led to improvements in ejector performance will be summarized in the following section. The last section is a discussion of current research and the likely direction of future aircraft programs.

2. Modelling and Prediction

2.1 Principle of Thrust Augmentation

The force which augments the thrust of the primary jet is produced in essentially the same way as the lift on a wing. The air entrained by a free jet flows into it at an angle to the jet axis, as shown in Figure 2. However, passing the jet through a duct, as shown in Figure 3, redirects this flow into the ejector inlet, and along the jet axis. In the process, a circulation is generated around each of the duct sections. Thus, a force analogous to the lift on an airfoil in a moving stream is developed on the walls of the duct.

The thrust augmentation ratio, ϕ, can therefore be defined as the ratio of the primary jet thrust, T_0, plus the force on the duct, F, to a reference thrust. Although von Karman (1949) suggested using the thrust of the jet in the absence of the duct as the reference thrust, a better choice, suggested by Heiser (1967) is the isentropic thrust obtained by expanding the mass of the primary jet, $\dot{m}_0$, to atmospheric pressure. This choice makes it possible to compare the performance of ejectors with different nozzles, and to evaluate the net effect on performance of reducing the jet thrust in order to increase the force on the duct. The isentropic reference thrust, $\dot{m}_0 U_*$, will be used in this paper. Therefore, the augmentation ratio will be defined as

$$\phi = \frac{T_0 + F}{\dot{m}_0 U_*} \tag{2.1}$$

or, equivalently,

$$\phi = \frac{T_e}{\dot{m}_0 U_*} \tag{2.2}$$

in which T_e is the thrust of the mixed flow at the exit of the ejector.

The fundamental parameters on which the thrust augmentation depends may be identified by dimensional analysis. If it is assumed that the augmentation is a function of the jet thrust, the duct geometry, and the physical properties of the fluid, dimensional analysis yields

$$\phi = f(\mathrm{Re}, M, L/W, \delta) \tag{2.3}$$

in which Re and M are the jet Reynolds number and Mach number, and L/W and δ are the length to width ratio and divergence angle of the ejector duct. The augmentation also depends on parameters like the surface roughness and ambient turbulence level, but these should be controlled to minimize their effect. Temperature effects are usually assumed to be implicit in the variation of the Mach and Reynolds numbers; however, temperature has an independent effect on the turbulent mixing, which will be discussed in a later section.

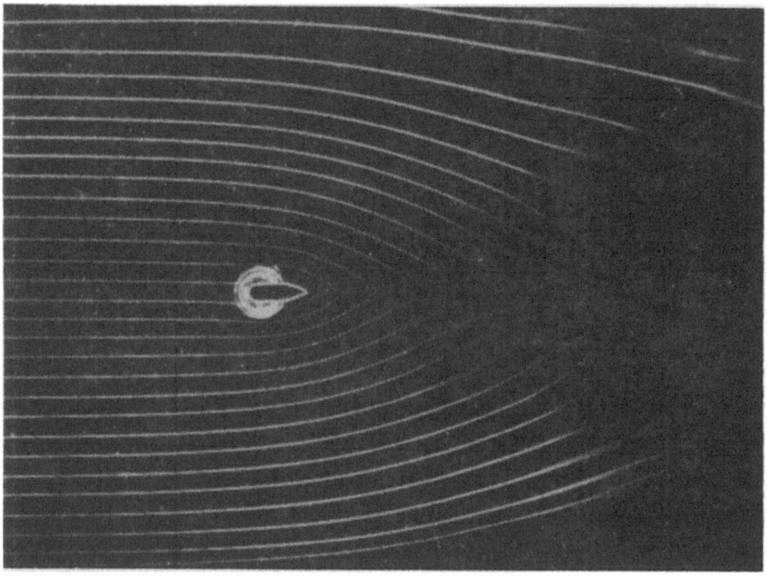

Figure 2. Jet streamlines.

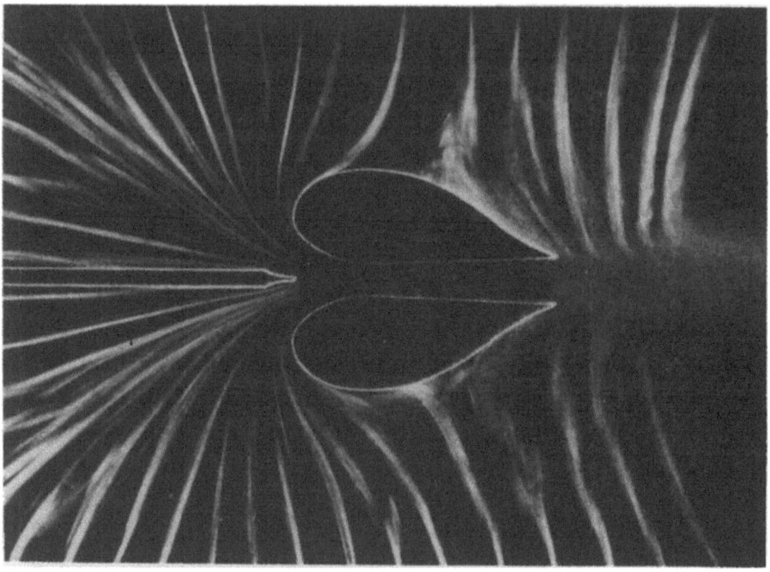

Figure 3. Ejector streamlines.

2.2 Analysis of Ejector Performance

Due to the complexity of the ejector flowfield, which includes interacting regions of turbulent and irrotational flow, the development of methods for analyzing ejector performance has been difficult. The most widely used methods are based on an approach suggested by von Karman (1949). The ejector geometry is described by its inlet and diffuser area ratios, and the jet mixing is characterized by mean values of the flow parameters at a few cross sections. As an example of this approach, consider the simple ejector shown in Figure 1. The nozzle exit area is denoted A_0, the cross-sectional area of the duct is A_2, and the diffuser exit area is A_3. The continuity equation for the control volume which coincides with the mixing section is

$$\rho U_0 A_0 + \rho U_1 (A_2 - A_0) = \rho U_2 A_2 \tag{2.4}$$

in which U_1 is the secondary velocity at the inlet plane. If it is assumed that the static pressure is constant in each cross section, then the momentum equation is

$$\rho U_0^2 A_0 + \rho U_1^2 A_1 + P_1 A_2 = \rho \beta U_2^2 A_2 + P_2 A_2 \tag{2.5}$$

in which $\beta = \int U_2^2 dA_2 / \overline{U}_2^2 A_2$ is a shape factor for the mean velocity profile; it is a simple way of specifying the degree of mixing at the ejector exit.

Bernoulli's equation is used to calculate the static pressure on each face of the control volume. For the flow through the inlet, we have

$$P_1 = P_a - \frac{1}{2}\rho U_1^2 \tag{2.6}$$

in which P_a is the atmospheric pressure. If it is assumed that the flow is diffused without further mixing, the pressure on the downstream face is

$$P_2 = P_a - \frac{1}{2}\rho U_2^2 / \beta [1 - (A_2/A_3)^2]. \tag{2.7}$$

The continuity equation for the diffuser

$$\rho U_2 A_2 = \rho U_3 A_3 \tag{2.8}$$

has been used to eliminate the exit velocity, U_3. Similarly, if η is the nozzle efficiency, the initial velocity of the jet becomes

$$U_0^2 = \eta (U_*^2 + U_1^2) \tag{2.9}$$

These equations can be solved for U_1, U_2, P_1, and P_2 in terms of the initial jet velocity and the shape factor for mixing. Skin friction and other real fluid effects can be included by the use of appropriate loss coefficients, as for the nozzle efficiency.

Using the continuity equation for the flow through the diffuser, the augmentation ratio can be written as

$$\phi = \beta(U_2/U_*)^2(A_2/A_0)(A_2/A_3) \tag{2.10}$$

A prediction of the thrust augmentation ratio may be obtained by substituting the solution for U_2/U_* into this relation. In Figure 4, the performance of an ideal, loss-free ejector has been shown as a function of the ejector geometry. The augmentation is seen to increase with the inlet and diffuser area ratios. In Figure 5, incomplete mixing ($\beta > 1$) is seen to cause a loss of performance.

This approach has been useful in understanding basic processes and in parametric studies of various factors which affect the level of augmentation (Kentfield, 1978; McCormick, 1967; Payne, 1966; Salter, 1975). However, the uniform flow assumptions are restrictive, and the predictions must be carefully interpreted. For example, the prediction that the augmentation monotonically increases with the inlet area ratio, as seen in Figure 4, is incorrect. As the inlet area ratio increases, the sections of the duct move away from the jet; thus, the augmentation ratio should actually reduce to one, the value for an isolated jet from an ideal nozzle. Another approach to the study of ejector performance has been to apply aero-thermodynamic cycle analysis (Minardi, 1981; Petty, 1979; Porter, 1981). However, as the authors of these studies point out, the isentropic flow assumptions which such analyses require may imply physically unrealistic flow processes. In particular, the jet mixing process is inherently nonisentropic. Thus, great care must be taken in interpreting the results of cycle analysis.

These parametric methods can be used as part of an experimental development program to estimate the effect of particular changes to a baseline ejector configuration. However, to be useful in predicting the performance of a new ejector design, an analytical method must be capable of predicting the turbulent mixing within the ejector. The development of these methods has been one of the principle advances in ejector technology during the past decade. Gilbert and Hill (1973) developed a finite difference scheme, which uses a mixing length model for the turbulence to analyze two-dimensional

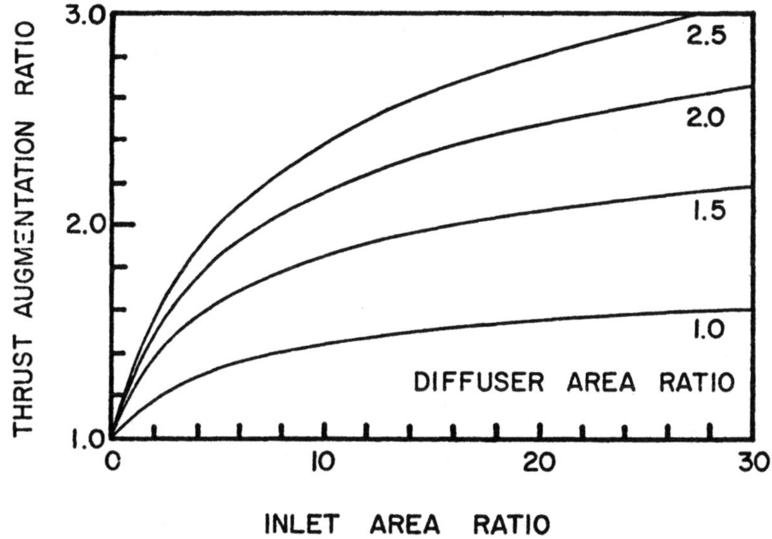

Figure 4. Thrust augmentation of an ideal ejector, according to momentum theory.

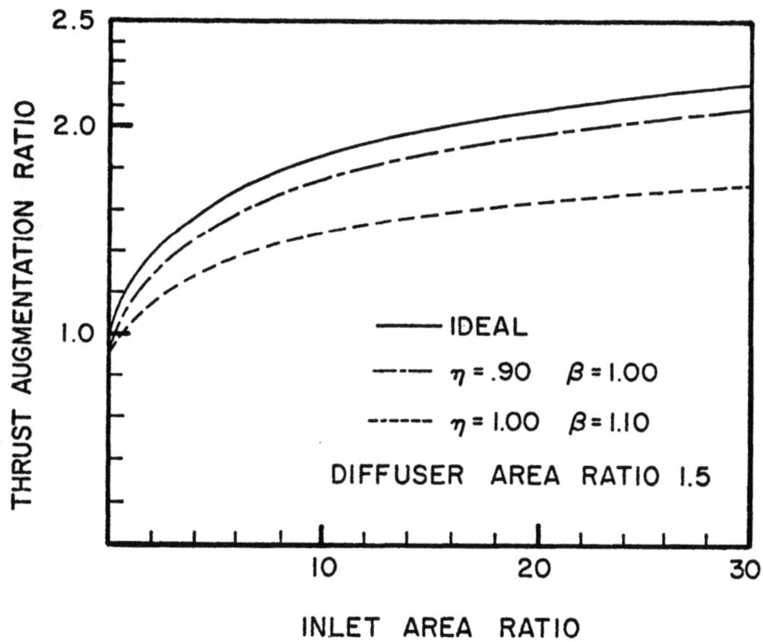

Figure 5. Effect of losses on ejector thrust augmentation.

ejectors. The thin shear layer approximation is applied to reduce the governing elliptic equations to a parabolic set, which can be solved by marching through the ejector in the streamwise direction. The secondary flow is determined as a function of the jet mixing rate by iterating on the inlet velocity until the computed exit pressure equals the atmospheric pressure outside the ejector.

In order to predict the entrainment of jets from multiple slot and lobe nozzles, DeJoode and Patankar (1978) developed a three-dimensional analysis, which uses the two-equation turbulence kinetic energy model of Launder and Spalding (1974). By assuming that the transverse velocity components would be smaller than the axial component of the flow through the ejector, they devised a partially parabolic scheme. Thus, a marching procedure was used for calculating the mean pressure gradient and streamwise velocity component, and an elliptic procedure was used for the secondary velocities and pressures in the transverse planes. With this scheme also, the solution is determined by iterating on the inlet velocities.

Such jet mixing analyses are useful for comparing the performance of alternate nozzle designs. However, the parabolic flow assumption introduces the same error at large inlet area ratios as the simpler methods do: the augmentation ratio does not reduce to unity in the limit of an isolated jet. Also, when the ejector is short or the diffuser angle is large, the exit pressure is less than atmospheric pressure. This pressure difference is supported by the momentum of the exhaust jet, while the jet momentum depends, in turn, on the pressure difference. Thus, the exhaust pressure becomes a floating boundary condition, and a unique solution for the flow through the ejector cannot be determined.

In order to overcome these limitations, a fully elliptic procedure is needed. Bevilaqua (1978) devised a technique for including elliptic effects by iterating between a viscous solution for the increase in jet thrust and an inviscid solution for the force on the duct. These forces are modelled by the interaction between a line of sinks, which represent the jet entrainment, and a pair of vortex sheets, which represent the vorticity bound in the sections of the duct. As an example of this approach, consider the flow field induced by an isolated turbulent jet, as shown in Figure 2. If the jet is assumed to be self-preserving, all velocities decay as $x^{-1/2}$, and so the entrainment can be represented by a line of sinks varying as

$$\sigma = \sigma_0(x/t)^{-1/2} \tag{2.11}$$

in which t is the nozzle gap, and σ_0 is a constant which depends on the rate of entrainment. The computed streamlines of this flow are shown in Figure 6.

The strength of the vortex sheets are determined using inviscid panel methods to satisfy the boundary condition that the duct must be a streamline of the flow induced by the jet. In a simple vortex lattice model, the continuous vorticity distribution is replaced by n discrete vortices of strength γ_j, as sketched in Figure 7. The boundary condition of no flow through the flaps is satisfied at n control points, midway between the vortices, by setting the normal components of the velocities induced by the vortex sheets equal but opposite to the normal velocities induced by the jet. This results in a system of n simultaneous algebraic equations for the unknown vortex strengths, which can be solved by a variety of familiar methods.

The interaction which produces the thrust augmentation may be understood by considering sink and vortex segments a distance r apart, as shown in Figure 7. The sink induces a velocity of magnitude $\sigma/2\pi r$ at the vortex. Therefore, the vortex experiences a force of magnitude $\rho\gamma\sigma/2\pi r$, perpendicular to r. Similarly, the vortex induces a velocity of magnitude $\gamma/2\pi r$ at the sink, which experiences the force, $\rho\sigma\gamma/2\pi r$, equal but opposite to the force on the vortex. The net effect of the interactions between all the sinks and vortices is a thrust on the duct and an equal but opposite reaction on the jet. The thrust augmentation ratio can be estimated by substituting either force into equation 2.1.

This analysis can be used to study the effect of varying the length and shape of the duct, as well as the inlet and diffuser area ratios. Results for the case in which the ejector length and inlet area ratio were varied are shown in Figure 8. For constant duct length, it can be seen that the augmentation decreases as the inlet area ratio increases. In the limit as the duct sections are removed to infinity, the augmentation approaches one, as it should. The results for the case in which the ratio of duct length to width is constant may be found by reading across the curves for different flap lengths. No dependence on inlet area ratio is found. This is consistent with the results of dimensional analysis, equation 2.3.

Because the influence of the duct on the rate of entrainment has been omitted, there is an unrealistic increase in the augmentation at small inlet area ratios. The augmentation actually reduces in this limit, as suggested by the classical result also shown in the figure.

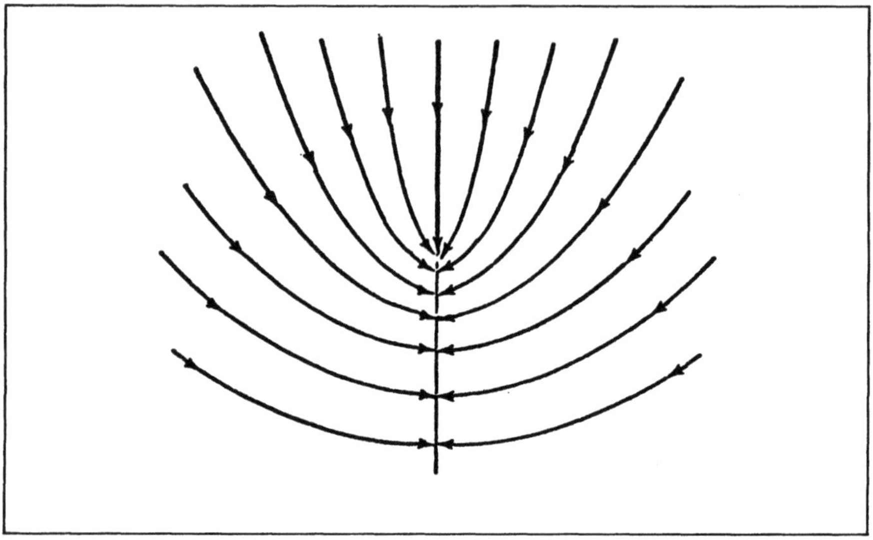

Figure 6. Computed streamlines of the flow entrained by a free jet.

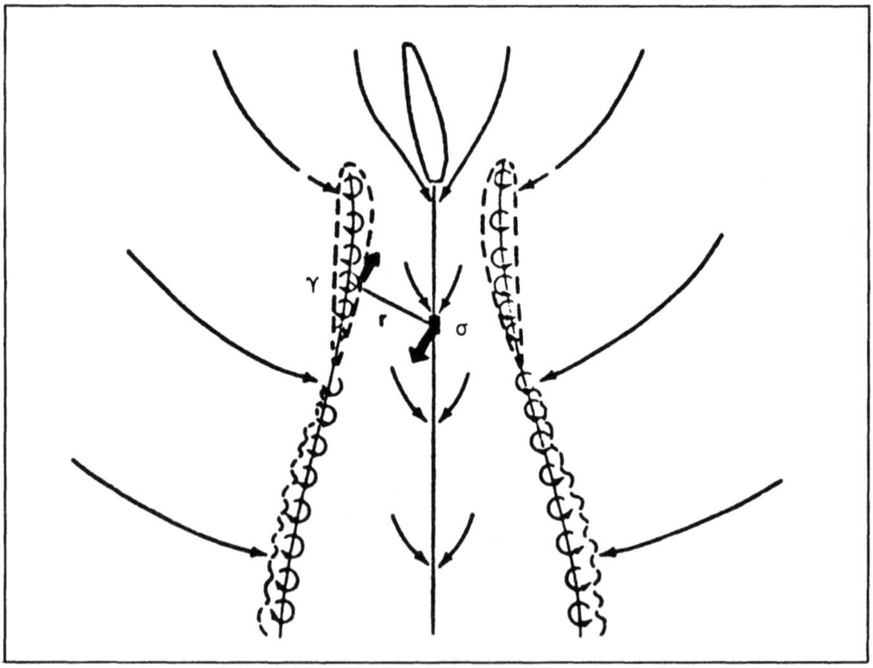

Figure 7. Vortex sheet model of a simple thrust augmenting ejector.

This effect can be predicted by using a parabolic solution for the turbulent mixing within the ejector to correct the sink strengths. Since the inviscid solution satisfies the Kutta condition, that the static pressure at the exit must be the same inside and outside the ejector, the exit pressure is determined for the parabolic analysis. A viscid/inviscid interaction analysis has been developed along these lines by Bevilaqua and DeJoode (1978). In Figure 9, the predicted variation of the thrust augmentation ratio with inlet area ratio is shown for a constant diffuser area ratio of 1.8. Because this solution merges the characteristics of the momentum and circulation theories of thrust augmentation, it has the correct behavior for both large and small inlet area ratios.

The most recent developments in ejector theory have been directed towards predicting the augmentation during forward flight. One-dimensional analyses, based on von Karman's approach, were developed by Hammond (1973) and Nagaraja et al (1973). These predicted a large loss in augmentation with flight speed, due to the rapid increase of ram drag. However, Alperin (1979) extended this analysis to cruising speeds, and found that ram compression of the secondary stream can lead to a supersonic mixed flow at the ejector exit. For this case, the augmentation ratio is predicted to be almost five times that predicted for the static case. This is an exciting prospect, however, because this performance is predicted for large inlet area ratios using the uniform flow assumption, it must be further studied. Currently, Alperin is performing a series of experiments aimed at testing these predictions.

Incorporating the ejector in the wing of a V/STOL aircraft, as shown in Figure 10 produces an effective lift/propulsion system. Such an aircraft could convert smoothly from jet borne hover to wing borne flight, because deflecting the ejector exhaust stream increases the wing lift by the jet flap effect faster than the jet lift is reduced. Linearized, thin airfoil methods for calculating the lift induced by an ejector of given thrust have been developed from Spence's (1957) jet flap theory by adding a sink to represent the flow into the ejector (Chan, 1970; Dillenius, 1979; Wilson et al., 1974; Woolard, 1975). Because the entrainment and thrust augmentation must be specified as input to these methods, they are limited to parametric studies. By extending the ideas of the viscid/inviscid interaction analysis, Bevilaqua, Woan, and Schum (1981) developed a method for predicting both the lift on the wing and the thrust augmentation ratio. The inviscid flow field consists of two separate regions with

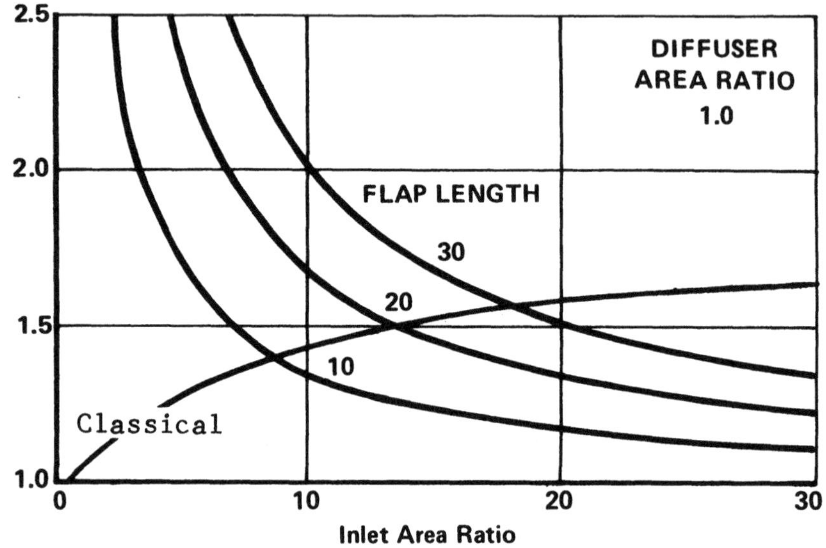

Figure 8. Effect of duct length on thrust augmentation, according to circulation theory.

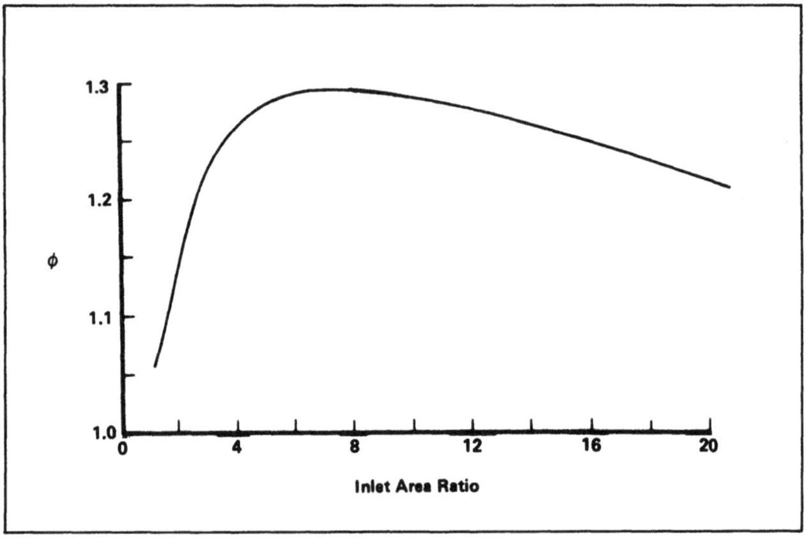

Figure 9. Predicted effect of inlet area ratio on thrust augmentation.

different total pressures: the jet wake and the surrounding free stream. A technique suggested by Küchemann and Weber (1953) was used to reduce this inhomogeneous flow to an equivalent homogeneous flow in which the total pressure is the same in both regions. Simply, the total pressure difference is related to the strength of a vortex sheet at the jet boundary

$$\rho U \gamma = \Delta H \tag{2.12}$$

The total pressure jump, ΔH, is considered to be known from the solution for the jet mixing. Vortex panel methods are then used to satisfy the boundary condition that the airfoil surface and jet boundary must be streamlines of the inviscid flow field. The calculated streamlines of the flow around a two-dimensional ejector wing is shown in Figure 11. The large circulation induced by the jet flap effect has moved the forward stagnation point near the trailing edge of the leading flap. This suggests that a leading edge device will be required to prevent flow separation and a resulting loss of lift and thrust. This approach can be extended to predict the performance of a complete aircraft incorporating ejectors in the wing or fuselage.

2.3 Ejector Scaling Laws

Although several sophisticated mathematical models of the ejector are now available, the complexity of the flow field nevertheless requires that testing be performed early in the ejector development process. There are no uncertainties regarding scale if testing is performed with full size ejectors and hot gas jets but, for reasons of cost and convenience, small scale models driven by cold air jets are preferred. Studies of ejector scale effects (Fought, 1960; Stewart, 1976) indicated that the thrust augmentation increases with ejector size. However, aircraft scale ejectors built more than a decade ago by the Lockheed, Boeing, and DeHavilland Aircraft Companies produced less augmentation than the laboratory models from which they were developed. Since the reasons for this are still not understood, considerable effort has gone into the study of ejector scale effects, and much testing is performed at full size to avoid scaling problems.

A major uncertainty is the effect of changes in the jet temperature on ejector performance. As previously noted, the first order effects of temperature on the physical properties of the jet (density,

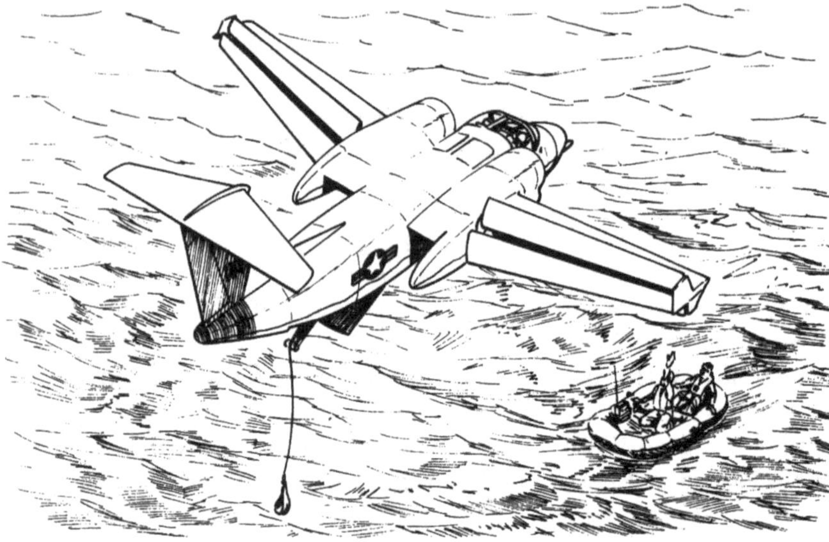

Figure 10. An ejector wing aircraft.

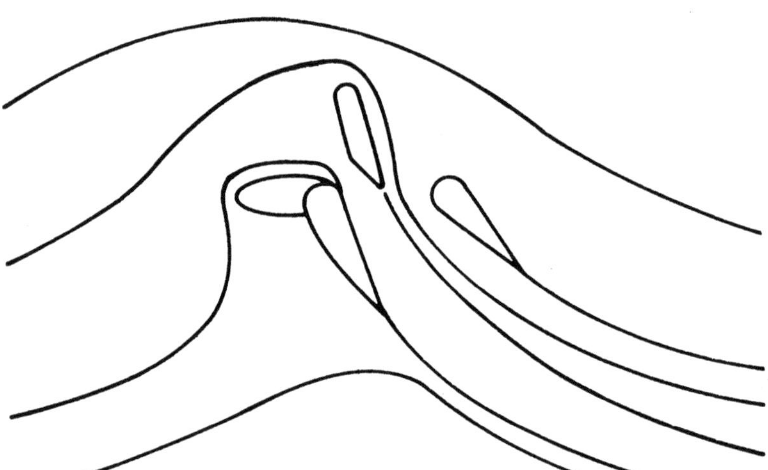

Figure 11. Computed streamlines of an ejector wing flow field.

viscosity, and compressibility) are implicit in scaling the Reynolds and Mach numbers. In particular, the Mach number is independent of temperature, since both the jet velocity and the speed of sound have the same dependence on temperature $V \sim T^{1/2}$. The jet thrust is also independent of temperature, because the change in density $\rho \sim T^{-1}$ is balanced by a change in velocity $V^2 \sim T$. The Reynolds number, on the other hand, is a function of temperature. However, changing the jet temperature 1000°R, which is typical of the exhaust of a jet engine, does not change the order of magnitude of the Reynolds number or the turbulent character of the flow.

The effect of temperature on the turbulent mixing is not as well understood. For small density differences, the mixing rate is proportional to the velocity difference between the two streams. Thus, a small temperature rise will probably increase the turbulent mixing due to the increase in the jet velocity $\Delta V \sim T^{1/2}$. If the density difference is large, the mixing rate is proportional to the momentum difference. Therefore, a large temperature rise will probably decrease mixing, $\Delta\rho V \sim T^{-1/2}$, although this has not been demonstrated.

The net effect of a 1000°R temperature rise, calculated using a finite difference solution for the mixing within an ejector (Bevilaqua and Combs, 1982) is a three percent drop in the thrust augmentation ratio. This is almost within the accuracy of the calculation procedure; however, the available data seem to support this result (Dejneka, 1980; Gates and Cochran, 1980; Phillips, 1975; Rabeneck et al., 1980). This is shown in Figure 12. Although all the data, except that of Lockheed, also included a scale or configuration change, the trend is consistent. Thus, a small increase in performance may be expected as a result of using cold air jets for model testing.

It is not possible to simultaneously match the Mach and Reynolds numbers of the prototype with a small scale model. However, because the Reynolds number of the prototype is large, Re $\sim 10^6$, changes due to the use of a quarter or tenth scale model only affect the smallest scales of the turbulence, which do not interact directly with the main flow. According to this principle of asymptotic invariance, the Reynolds number is not a significant parameter if its value is very large. Therefore, if the Reynolds number is large and the jet Mach numbers are matched, it may be concluded that scale models will give valid results. For one case in which the geometric scaling was nearly exact, Mefferd, Alden, and Bevilaqua (1978) obtained the same performance with a 0.2 scale cold flow model and a full size hot ejector. These ejectors are shown in Figure 13.

3. Component Development

3.1 Hypermixing Jet Nozzles

The process of ejector thrust augmentation is driven by the entrainment of the primary jets, which pumps the secondary stream through the ejector. Thus, increasing the jet entrainment rate is a principle means of increasing the thrust augmentation. Although the mechanism of entrainment is still not well understood, one hypothesis (Bevilaqua and Lykoudis, 1971) is that turbulent entrainment is primarily due to the action of large vortices. Significant increases in mixing and augmentation have been achieved with the so called hypermixing nozzles shown in Figure 14 (Bevilaqua, 1974; Quinn, 1973). The alternating deflection of the jet segments at the nozzle exit serve to introduce a row of streamwise vortices into the jet. These vortices enhance the turbulent mixing and thus increase the jet entrainment.

The increase in the rate of entrainment and the force on the duct depends on the size and deflection of the jet segments. But, this increase must be balanced against the tilt loss in the primary jet thrust due to deflecting the jet segments. Schum, DeHart, and Bevilaqua (1982) used a partially parabolic mixing program to study the effect of varying these parameters. In Figure 15 the effect of the initial jet deflection angle is shown. It can be seen that the mixing increases with the deflection angle. However, the tilt loss also increases. As shown in Figure 16, the deflection angle for which the tilt loss in primary thrust becomes greater than the gain due to mixing is about 20° in this case. A similar analysis of the segment aspect ratio indicated that this parameter has a smaller effect. The optimum combination of these jet parameters depends on the ejector inlet area ratio and length, so that these results cannot be generalized, but the performance increase due to increased mixing can be significant.

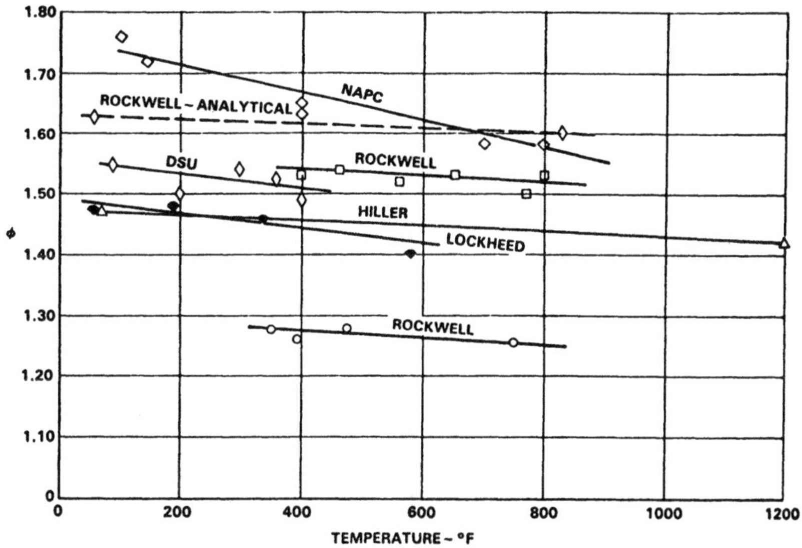

Figure 12. Effect of temperature on ejector thrust augmentation.

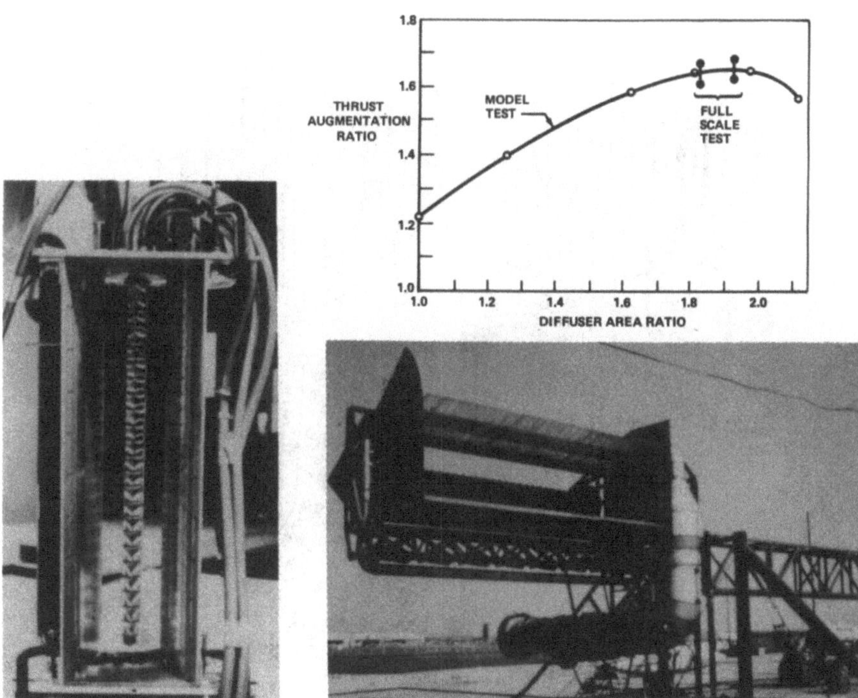

Figure 13. Comparison of prototype and 0.2 scale ejector wing models.

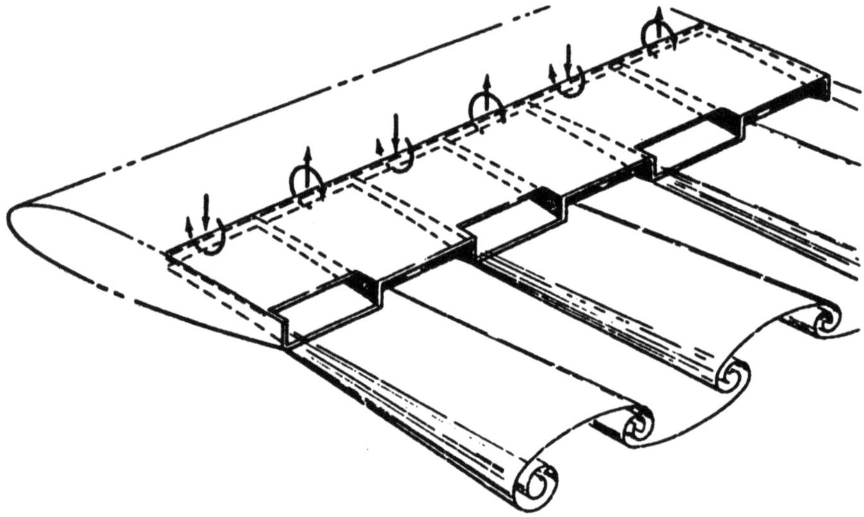

Figure 14. Streamwise vortices in a hypermixing nozzle.

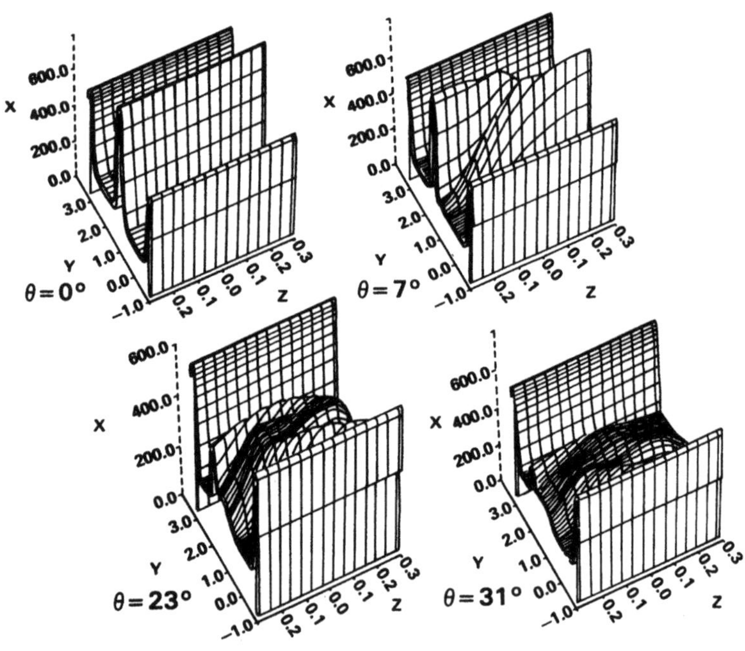

Figure 15. Predicted effect of hypermixing angle on jet mixing.

3.2 Section Geometry

The hypothesis that the augmenting thrust is related to the lift developed on a wing suggests the use of airfoil high lift technology to increase the augmentation. Although the ejector thrust corresponds more nearly to the leading edge suction than to the wing normal force, the magnitude of both forces depends on the circulation which satisfies the Kutta condition at the trailing edge. Thus, the thrust on the ejector duct may be expected to depend on the section geometry in the same way as the lift on an airfoil section.

The effect of adding camber and tabs to a straight duct was predicted using the inviscid vortex lattice analysis previously described (Bevilaqua, 1978). In Figure 17 it can be seen that the predicted effect of camber on the thrust augmentation is similar to its effect on airfoil lift: the slope of the thrust curve is unchanged, but the thrust at each diffuser angle is increased the same amount. The effect of deflecting a trailing edge tab was found to be similar to the effect of deflecting a wing flap: there is a gain in thrust at each diffuser angle in this case also.

These predictions were experimentally verified by Seiler and Schum (1979). The performance of a series of shaped and cambered duct sections, like those seen in Figure 18, were compared to the performance of a straight duct at the same inlet area ratio and diffuser area ratio. In Figure 19, the cambered section and both the long and short tabs are seen to generate greater thrust augmentation than the straight sections.

An experimental study of the effect of length on the thrust augmentation was reported by DeHart and Smrdel (1982). Length to width ratios of 1.5, 2.0, and 2.5 were compared. In Figure 20 it can be seen that increasing the ejector length increases the maximum augmentation by delaying diffuser stall to a higher diffuser area ratio. At smaller diffuser area ratios, there is no effect of length. This may be due to the fact that at a given diffuser area ratio the longer flap is at a smaller diffuser angle; that is, it is *flying* at a smaller angle of attack. However, this unexpected result needs additional study to be understood.

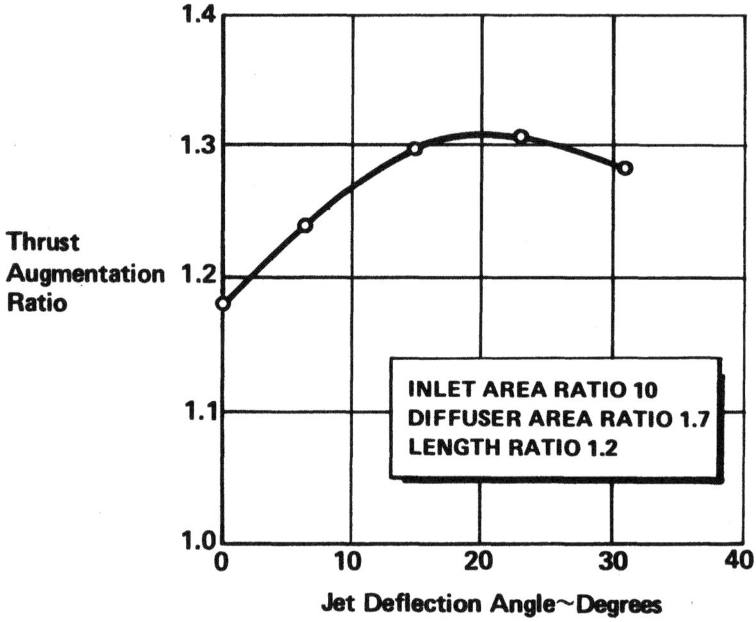

Figure 16. Predicted effect of hypermixing angle on thrust augmentation.

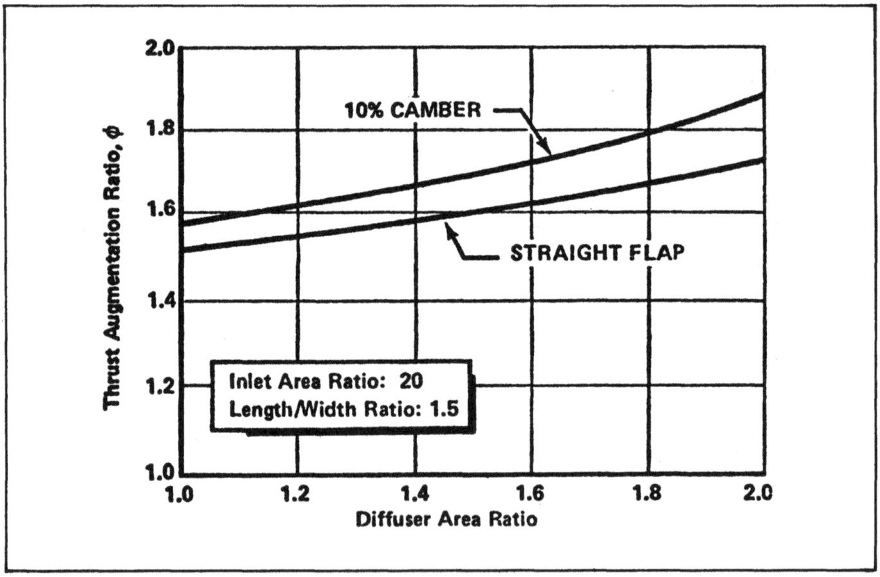

Figure 17. Predicted effect of camber on thrust augmentation.

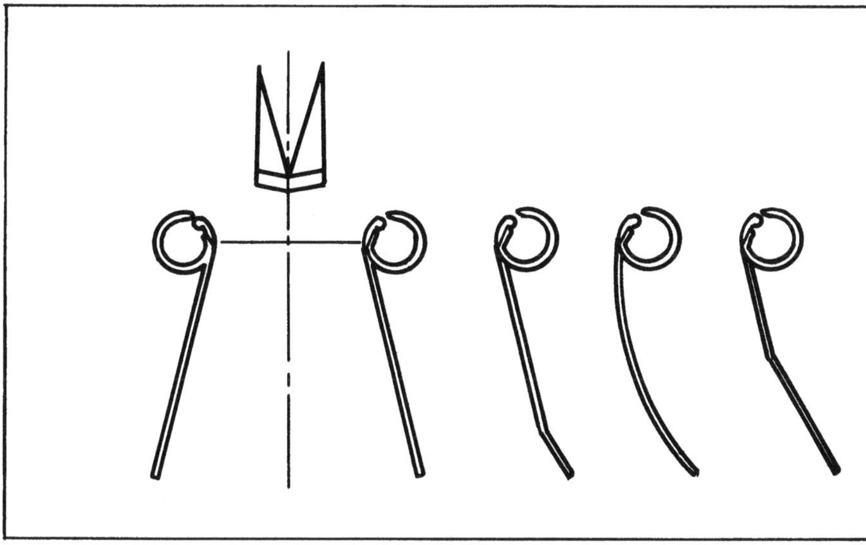

Figure 18. Typical cambered flaps.

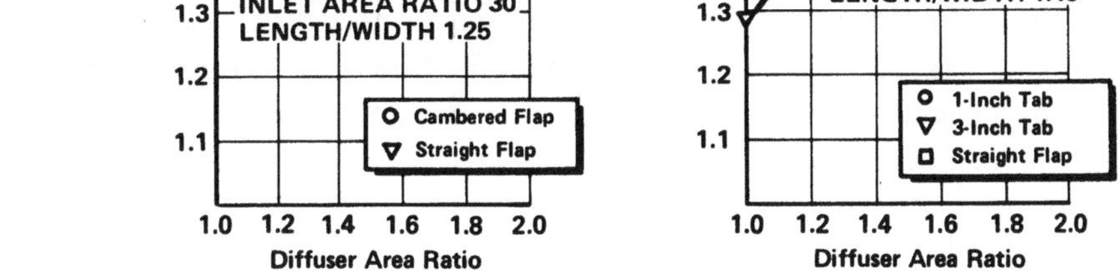

Figure 19. Measured effect of camber and tabs on thrust augmentation.

4. Demonstrator Aircraft

During the past 25 years, four demonstrator aircraft have been built to study the feasibility of using ejector thrust augmentation to develop V/STOL airplanes. Plan views of these four aircraft are compared in Figure 21. The Lockheed XV4-A was the earliest design (Nicholson and Lowry, 1966). It incorporated the ejectors in two longitudinal bays, one on each side of the fuselage. Although only 93 percent of the expected thrust augmentation was achieved in the demonstrator aircraft, $\phi = 1.3$, this provided sufficient lift for hover. During the flight test program, 81 hover tests were flown. These tests demonstrated that practical levels of augmentation could be achieved with aircraft hardware, and that operation of the ejector in gusts and in and out of ground effect did not stall the ejector or the engine. However, it was difficult to convert the aircraft from hover to wing borne flight. As the aircraft was accelerated, the large mass of air being pumped through the ejectors induced a ram drag and associated pitching moment on the aircraft. There was insufficient augmentation to both lift the aircraft and counter the pitching moment. The crash of the aircraft ended the flight test program before it could be determined if this was a fatal flaw in the design.

The XC8-A is a STOL research aircraft, which incorporates ejector flaps in two spanwise bays at the trailing edge of the wing (Ashleman and Skavdahl, 1972). The DeHavilland and Boeing Aircraft Companies modified a Buffalo transport aircraft for this research. The ejector is about the same size as that in the XV4-A, and achieved about the same thrust augmentation, $\phi = 1.26$. The addition of ejector flaps reduced the STOL takeoff distance to approximately three fourths that of the original aircraft, and likewise reduced the stall speed from about 65 knots to about 40 knots. Although it was not designed to hover, the good STOL performance of this aircraft demonstrated the performance gains that result from combining the ejector with the wing. The ten-year flight test program has been extended, and the aircraft is still being used for the study of noise abatement, and to establish STOL handling qualities.

The XFV-12A is a V/STOL aircraft which incorporates spanwise ejectors in the wing and canard (Bevilaqua and Combs, 1981). Since the thrust of each ejector can be modulated by varying the

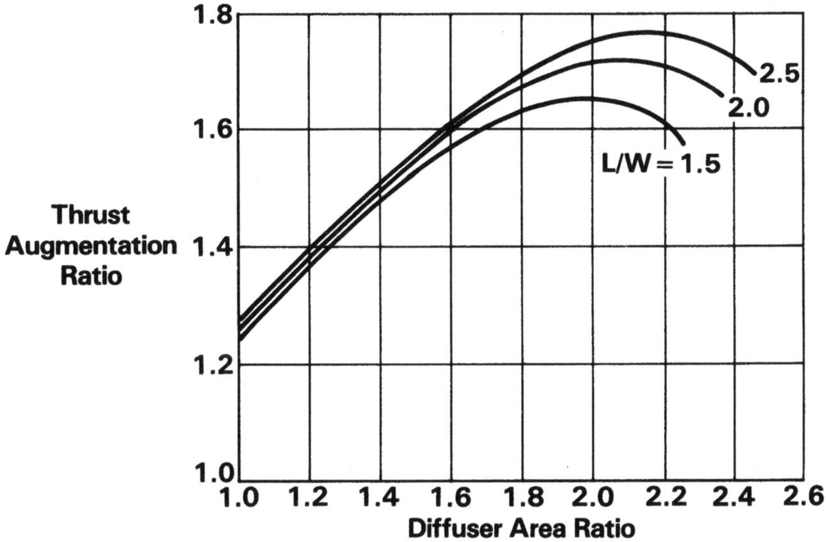

Figure 20. Measured effect of length on thrust augmentation.

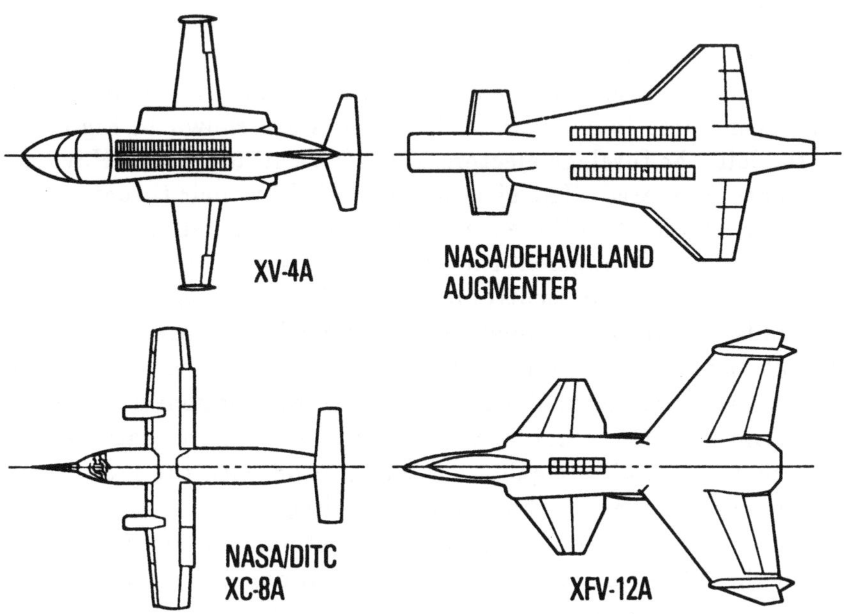

Figure 21. Demonstrator aircraft incorporating ejector thrust augmentation.

diffuser angle, separate reaction jets are not required for control during hover. This function is accomplished by differential modulation of the four ejectors to provide thrust vectoring, as well as thrust modulation. Tethered hover tests performed on the NASA Lunar Landing Gentry verified this control concept. However, although 96 percent of the expected augmentation was achieved in the demonstrator aircraft, $\phi = 1.4$, sufficient vertical lift for hover was not achieved due to poor deflection of the primary jets into the ejectors. The tethered testing was halted in order to develop a set of nozzle turning vanes, but the program was cancelled before this task was completed.

The fourth *aircraft* was not intended to fly, but was built by DeHavilland Aircraft as a very large model for testing in the NASA Ames 40×80 foot wind tunnel (Whittley, 1979). It has longitudinal ejectors on each side of the fuselage, similar to those of the XV4-A. The ejectors in the model develop sufficient augmentation to provide both lift and pitch control during takeoff. However, the large ram drag makes acceleration difficult, and current plans are to test primary nozzles which can be vectored aft to improve the acceleration.

Advances in the performance of aircraft ejectors, over the past decade were summarized by Murphy (1982) who compared the ejectors in these four aircraft. His results are shown in Figure 22. On the left, the fuselage ejectors in the XV4-A and the DeHavilland model are compared. The ejectors are scaled by their throat widths, since this shows the differences in the important length ratio. For these ejectors, the augmentation ratio is referenced to the measured nozzle thrust, not the isentropic thrust. However, since the nozzles are similar it can be assumed that their efficiencies are also similar, so that the comparison is meaningful. It can be seen that, even though the more recent design is shorter, the augmentation ratio is significantly higher. Similarly, the wing ejectors compared on the right show a significant gain in augmentation with the more recent design, even though it is shorter. The isentropic reference thrust has been used to compare these two ejectors. For both the XV4-A and the XFV-12A, the larger augmentation values represent the results of a lift improvement program completed after the flight test program, but never incorporated on the aircraft.

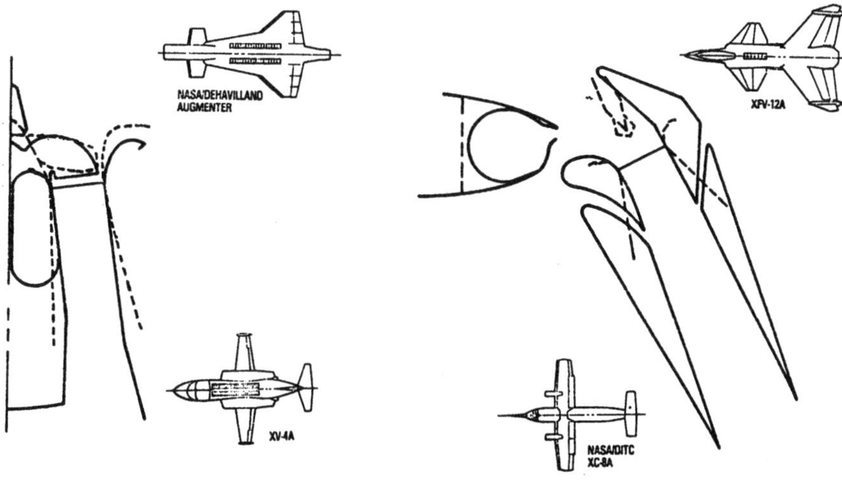

CONFIGURATION / CHARACTERISTIC	XV-4A	NASA/DHC AUGMENTER	NASA/DITC XC-8A	XFV-12A
1. A_2/A_0	10	20.5	9	20.1
2. L/D	4.1	3.0	5.1	1.5
3. Φ (RECTANGULAR)	*1.3-1.48	*1.74	1.26	1.62-1.64
4. SCALE	FULL	.7	.7	.2

*BASED ON MEASURED THRUST - NOT ISENTROPIC THRUST

Figure 22. Advances in aircraft ejector performance.

5. Current Trends

Ejector research was initially directed towards achieving high static thrust augmentation for vertical lift. More recently, the effort has focused on conversion aerodynamics and STOL performance. The importance of research in this area comes from the growing recognition that the range/payload performance of V/STOL aircraft can be significantly improved by operating with a short takeoff roll, while landing vertically (STOVL). Although some very sophisticated numerical methods have been developed to predict static augmentation, procedures being used to predict STOL performance are highly empirical and not easily generalized. This is due to the complexity of the interactions between the jet, wing, and free stream. The ejector increases the lift and thrust of the aircraft, both as a direct reaction to the thrust augmentation and by the jet-flap effect. Simultaneously, the forward motion generates a ram drag on the ejector and an induced drag on the aircraft. A systematic approach, which combines analysis and testing and proceeds from the study of isolated phenomena to the study of their interactions, will be necessary to develop methods for predicting the net force on the aircraft.

In order to justify the added weight of the ejector system, there is considerable interest in using the ejectors to improve the cruise or high speed performance of the aircraft. As previously noted, Alperin's (1979) supersonic solution is an interesting theoretical development which suggests that it may be possible to achieve significant augmentation at high speeds. Farbridge and Smith (1978) are investigating the possibility of using the ejector in cruise for boundary layer removal. In this application, the ejector flap is kept open during cruise. The inlet area ratio is relatively small, so that the thrust augmentation is slight; however, removal of the boundary layer delays the drag rise and extends the buffet boundaries. Testing to date has shown drag improvements comparable to those achieved with super-critical airfoils. Developments in both areas could be significant, and there is the interesting possibility of achieving both benefits on the same configuration.

Airframe development is currently directed towards the design of supersonic fighter/attack aircraft for production in the next decade. Rockwell International has developed a new configuration (Mark, 1982), which draws on their experience in testing the XFV-12A. In order to reduce the wave drag, the wing and canard have been combined into a single clipped delta wing planform, as shown in Figure 23. This configuration also permits the use of simple, rectangular ejectors at the trailing edge of the wing, although it was necessary to position the forward ejectors along the wing root to balance the aircraft in hover. In cruise, the mixed flow is diverted to a conventional exhaust nozzle. The engine inlets are located on top of the fuselage to minimize the loss in engine thrust which results from reingestion of the hot exhaust jet fountain during hover. A mixed flow turbofan engine operating at a nozzle pressure ratio of 3.5 and mixed flow temperature of 1200°F was selected to minimize duct losses without requiring a materials development effort.

General Dynamics and DeHavilland of Canada developed an F-16 based configuration (Flight International, 1983) which draws on DeHavilland's experience with the large wind tunnel model. This aircraft also has a delta wing, but the ejectors are located in the wing root only, as shown in Figure 24. The engine is a GE F110 turbofan. For vertical takeoff, the fan air is ducted to the ejectors, while the core air exhausts through a thrust vectoring nozzle.

Figure 23. Rockwell International supersonic V/STOL aircraft.

Figure 24. General Dynamics/DeHavilland supersonic V/STOL aircraft.

In cruise, the core flow is vectored aft, and the fan air is diverted through a separate exhaust nozzle. Current testing is aimed at solving the problems of ejector ram drag during conversion from hover. Although there are significant differences in these configurations, it can be seen that the mission requirements have produced many similarities. Testing to be performed during 1986 in the 80 × 120 V/STOL tunnel at NASA Ames will provide data on the performance of both configurations; however, selection of a concept will probably depend on required engine and materials developments, as well as operational requirements.

References

[1] Alperin, M. and Wu, J. "High Speed Ejectors," *AFFDL TR-79-3048,* May 1979.

[2] Ashleman, R. H. and Skavdahl, H. "Development of an Augmentor Wing Jet STOL Research Airplane Summary," *NASA CR-114503,* August 1972.

[3] Bevilaqua, P. M. "Evaluation of Hypermixing for Thrust Augmenting Ejectors," *Journal of Aircraft, 11, No. 6,* June 1974, 348–354.

[4] ———————. "Lifting Surface Theory for Thrust Augmenting Ejectors," *AIAA Journal, 16, No. 4,* April 1978.

[5] Bevilaqua, P. M. and Combs, C. P. "Theory and Practice of Ejector Scaling," paper presented at the Workshop for Aerospace Applications of Ejectors, *AFWAL TR 82-3059,* 849–864, 1982.

[6] ———————————————. "XFV-12A Development Program Summary Report," *AD-A108354,* November 1981.

[7] Bevilaqua, P. M. and DeJoode, A. D. "Viscid/Inviscid Interaction Analysis of Thrust Augmenting Ejectors," *Office of Naval Research Report ONR CR212-249-1,* Feb. 1978.

[8] Bevilaqua, P. M. and Lykoudis, P. S. "Mechanism of Entrainment in Turbulent Wakes," *AIAA Journal, 9, No. 8,* Aug. 1971, 1657–59.

[9] Bevilaqua, P. M., Woan, C. J., and Schum, E. F. "Viscid/ Inviscid Interaction Analysis of Ejector Wings," *NASA CR 166172,* April 1981.

[10] Braden, R. P., Nagaraja, K. S., and von Ohain, H. J. "Proceedings of the Workshop for Aerospace Applications of Ejectors," *AFWAL TR 82-3059, 1982.*

[11] Chan, Y. Y. "Lift Induced by Suction Flaps on Augmenter Wings," *Canadian Aeronautics and Space Institute Transactions, 3, No. 2,*

Sept. 1970, 107.

[12] DeHart, J. H. and Smrdel, S. J. "Ejector Shroud Aerodynamics," paper presented at the Workshop for Aerospace Applications of Ejectors, *AFWAL TR-82-3059,* 799–822, 1982.

[13] Dejneka, R. "Influence of Driving Air Temperature on Thrust Augmenting Ejector Performance," *NADC-PE-41,* August 1980.

[14] DeJoode, A. D. and Patankar, S. V. "Prediction of Three-Dimensional Turbulent Mixing in an Ejector," *AIAA Journal, 16, No. 2,* February 1978, 145–150.

[15] Farbridge, J. E. and Smith, R. C. "The Transonic Multi-foil Augmentor Wing," *Journal of Aircraft,* Nov. 1978, 755–761.

[16] Fought, D. E. "Test and Analysis of a Coanda Thrust Augmentation Nozzle," *MS thesis,* Pennsylvania State Univ., January 1960.

[17] Gates, M. F. and Cochran, C. L. "Evaluation of Annular Nozzle Ejector," *ARD-280,* November 1980.

[18] Gilbert, G. B. and Hill, P. C. "Analysis and Testing of Two-Dimensional Slot Nozzle Ejectors With Variable Area Mixing Sections," *NASA CR-2251,* 1973.

[19] Hammond, D. L. "Thrust Augmenting Ejector Analysis Utilizing Compressible Flow Theory," *AFFDL TM 73-111-PTB,* Aug. 1973.

[20] Heiser, W. H. "Thrust Augmentation," *Journal of Engineering for Power,* Jan. 1967, 75–82.

[21] von Karman, T. "Theoretical Remarks on Thrust Augmentation," in *Contributions to Applied Mechanics, Reissner Anniversary Volume,* Ann Arbor: J. W. Edwards, 1949, 461–468.

[22] Kentfield, J.A.C. "'Prediction of Performance of Low Pressure Ratio Thrust Augmentor Ejectors," *Journal of Aircraft, 15,* Dec. 1978, 849–856.

[23] Küchemann, D. and Weber, J. *Aerodynamics of Propulsion,* New York: McGraw-Hill, 1953, Chapter 3.

[24] Launder, B. E. and Spalding, D. B. "The Numerical Computation of Turbulent Flows," *Computer Methods in Appl. Mech. and Engrg., 3, No. 2,* March 1974, 269–289.

[25] Mark,L. "Study of Aerodynamic Technology for Single Cruise Engine V/STOL Fighter Attack Aircraft," *NASA CR-166270,* Feb. 1982.

[26] McCormick, B. W. *Aerodynamics of V/STOL Flight,* New York: Academic Press, 1967.

[27] Mefferd, L. A., Alden, R. E., and Bevilaqua, P. M. "Design and Test of a Prototype Scale Ejector Wing," *NASA CP 2093,* June 28–29, 1978, 437–448.

[28] Dillenius, M.F.E. and Mendenhall, M. R. "Theoretical Analysis of an Augmenter Wing for a VTOL Fighter," *Nielsen Engineering and Research TR 183*, May 1979.

[29] Minardi, J. E. "'Characteristics of High Performance Ejectors," *AFWAL TR-81-3170, 1981.*

[30] Murphy, R. D. and Farley, J. M. "Ejector Thrust Augmentation Lift Systems for Supersonic V/STOL Aircraft," paper presented at the Workshop for Aerospace Applications of Ejectors, *AFWAL TR 82-3059* (1982), 911-26.

[31] Nagaraja, K. S., Hammond, D. L., and Graetch, J. E. "One-Dimensional Compressible Ejector Flows," *AIAA Paper 73-1184*, Nov. 1973.

[32] Nicholson, R. and Lowry, R. B. "XV-4A VTOL Research Aircraft Program Summary Report," *USAAVLABS TR 66-45*, May 1966.

[33] Payne, P. R. "Steady State Thrust Augmentors and Jet Pumps," U. S. Army Aviation Material Laboratories, Ft. Eustis, VA, *AD 632-126, March 1966.*

[34] Petty, J. S. "Theoretical Performance Limits for Nonstatic Ejector Thrust Augmentors," *AFWAL-TR-79-120, 1979.*

[35] Phillips, J. D. "Temperature and Pressure Effects on Thrust Augmentation," *MS thesis*, the Ohio State Univ., 1975.

[36] Porter, J. L. and Squyers, R. A. "A Summary/Overview of Ejector Augmentor Theory and Performance," *Vought ATC TR R91100-9CR-47, I*, April 1981.

[37] ——————————. "A Summary/Overview of Ejector Augmentor Theory and Performance," *Vought ATC TR R91100-9CR-47, II*, April 1981.

[38] Quinn, B. P. "Compact Ejector Thrust Augmentation," *Journal of Aircraft, 10, no. 8*, Aug. 1973, 481–486.

[39] Rabeneck, G. L., Shumpert, P. K., and Sutton, J. F. "Steady Flow Ejector Research Program," Lockheed Aircraft Corporation, Georgia Division, *Final Contract Report NONC-3067(00)*, December 1980.

[40] Salter, G. E. "Method for Analysis of V/STOL Aircraft Ejectors," *Journal of Aircraft, 12*, Dec. 1975, 974–978.

[41] Schum, E. F., DeHart, J. H., and Bevilaqua, P. M. "Ejector Nozzle Development," *AIAA/ICAS Paper 82-4.3.4*, August 1982.

[42] Seiler, M. R. and Schum, E. F. "An Analytical and Experimental Investigation of Diffusers for V/STOL Thrust Augmenting Ejectors," *Journal of Aircraft, 16, no. 10*, Oct. 1979, 643–644.

[43] Spence, D. A. "The Lift Coefficient of a Thin, Jet-Flapped Wing," *Royal Society of London, Proceedings, 238,* Jan. 1957, 46.

[44] "Supersonic V/STOL Technology Shapes Up," *Flight International,* Jan. 15, 1983, 143–145.

[45] Stewart, V. R. "A Study of Scale Effects in Thrust Augmenting Ejectors," *Naval Air Development Center, TR NR 76H-2,* August 1976.

[46] Whittley, D. C. "The External Augmentor Concept for V/STOL Aircraft," Workshop on Thrust Augmenting Ejectors, *NASA CP 2093,* Sept. 1979, 449–471.

[47] Wilson, J. D., Loth, J. L., and Chandra, S. "Thrust Augmented Wing Sections in Potential Flow," West Virginia Univ., Aerospace Engineering, *TR-25,* Aug. 1974.

[48] Woolard, H. W. "Thin-Airfoil Theory of an Ejector-Flapped Wing Section," *AIAA Journal of Aircraft, 12, No. 1,* Jan. 1975, 26.

Progress Towards a Model to Describe Jet/Aerodynamic-Surface Interference Effects

*Richard L. Fearn**
University of Florida
Gainesville, FL 32611

Abstract

A first generation model is presented which relates the pressure distribution on an aerodynamic surface to properties of the jet plume. The characteristics of a jet in a crossflow that are of primary importance in determining the pressure distribution on the aerodynamic surface are assumed to be:

(1) a pair of contrarotating vortices associated with a jet in a crossflow,

(2) entrainment of crossflow fluid into the jet plume,

(3) a wake region near the aerodynamic surface and extending downstream from the jet orifice.

The model is applied to the configuration of a round jet exhausting perpendicularly through a flat plate into a uniform crossflow for a range of jet-to-crossflow velocity ratios from 3 to 10. It is demonstrated that the model is capable of describing the measured pressure distribution on the flat plate with model parameters that are compatable with the incomplete description of the vortex pair that is available. The force and moment on the plate are presented as functions of jet-to-crossflow velocity ratio.

Nomenclature

A	area of cross section to jet plume or model parameter.
a, b, c	model parameters.
C_p	pressure coefficient.
E, E_A, E_B	entrainment coefficients.

---

*Professor of engineering sciences.

E_{Ao}	entrainment parameter for jet in crossflow.
E_{so}	entrainment parameter for submerged jet.
F	blending function for wake correction.
g	Gaussian function used in wake correction.
M	pitching moment on reference circle of flat plate.
$\hat{n}$	unit vector normal to cross section of jet plume.
p, q	constants in a Padé approximation.
Q	mass flux.
R	effective jet-to-crossflow velocity ratio.
s	arc length.
T	jet thrust.
U_∞	crossflow speed.
x, y, z	coordinate axes
γ	dimensionless vortex strength.
ρ	density.

Subscripts

$1, 2, 3, \ldots$	distinguish between model, parameters.
c	refers to centerline.
m	locate or denote maximum value.
meas	measured value.
o	inital value.
pot	potential flow value.
v	refers to vortex properties.

1. Introduction

In the transition from hover to conventional flight, vertical or short take-off and landing (V/STOL) aircraft supplement wing-generated lift with direct thrust from lift jets, lift fans or tilt-proprotors. The proposed configurations all involve the injection of relatively high-velocity jets of air and/or exhaust gases into the crosswind caused by the forward motion of the aircraft. The interaction between these jets in a crossflow and the aerodynamic surface of the aircraft usually results in a loss of lift and an increment of nose-up pitching moment, both of which tend to increase with increasing forward velocity of the aircraft. This jet/aerodynamic-surface interference effect is one of the significant problems in V/STOL aerodynamics.

The need for a model to describe the jet/aerodynamic-surface interference effect has been recognized for well over a decade. For example, Rubbert et al. (1967) pointed out the need for a model that would be compatable with existing lifting-surface computer codes. In his summary remarks at a recent AGARD Symposium, Spree (1981) commented on the continued need for further development of potential-flow methods for this problem. Such a model should provide an adequate description of the pressure distribution on the aerodynamic surface of a given configuration for a range of jet-to-crossflow velocity ratios and for a range of jet injection angles. The logical framework for the model should be general enough to treat various jet and aerodynamic surface configurations.

One reasonable approach to such a complicated problem is to study the simplest configuration that retains the essential features of the jet/aerodynamic-surface interference during transition from hover to conventional flight. This approach has motivated numerous experimental studies of a subsonic round jet that exhausts through a flat plate into a subsonic crossflow including those of Jordinson (1958), Margason (1968), Wu et al. (1970), Soulier (1968), Fearn and Weston (1975, 1978, 1979), and Aoyagi and Snyder (1981). Attempts to model various properties of a jet in a crossflow have ranged from simple attempts to account for the trajectory of the jet plume (for example, Abramovich, 1963) to a discussion of solving numerically the time-averaged, three-dimensional Navier-Stokes equations (Baker et al., 1981).

The accumulation of information available for the simple configuration of a round jet, exhausting through a flat plate into a crossflow, has influenced the design procedures proposed to estimate the jet induced forces and moments on V/STOL aircraft configurations in the transition flight regime. Wooler's method (1972) is based largely on results of jet-in-crossflow experiments conducted prior to 1970. During the 1970's velocity measurements in the jet plume by Thompson (1971a), Kamotani and Greber (1972), Harms (1973), and Fearn and Weston (1978) established that a pair of diffuse contrarotating vortices constitutes a dominant and persistant feature of the jet plume. From velocity measurements in the jet plume Fearn and Weston (1974) estimated the physical properties of the vortex pair. In a master's thesis, Dietz (1975) incorporated this information into a prototype model to calculate the pressure distribution on a flat plate for a single case: perpendicular jet injection and a jet-to-crossflow velocity ratio of 8. The approach developed by Fearn and Weston (1974) to infer vortex properties of the contrarotating vortex pair from selected velocity measurements in cross sections to the jet plume has been applied to data for a rectangular jet of aspect ratio 4 by Thames and Weston (1978). Prototype jet interference models based on jet/flat plate studies are reviewed and evaluated in terms of their application to V/STOL design methodology by Beatty and Kress (1979). Another example of the utilization of results from the simple jet/flat plate configuration for estimating jet induced effects on V/STOL configurations is the classical engineering approach developed by Kuhn (1979).

Jet interference models can also be utilized to estimate wind tunnel wall effects when testing V/STOL configurations. Recent work on this topic has been done by Hackett (1982).

The purpose of this article is to describe the logical framework for a first generation model describing the jet/aerodynamic-surface interference effect. The model is based on pertinent physical characteristics of the flow-field of a jet in a crossflow, and it is applied to the case of a round jet exhausting perpendicularly from a large flat plate. A quantitative estimate of the properties of the diffuse vortex pair and simple assumptions about other characteristics of the flow are utilized to determine if the model is capable of describing the measured pressure distributions for a range of jet-to-crossflow velocity ratios. Although it has not been a primary goal of the study, the model should be applicable to the problem of estimating wind tunnel wall effects for V/STOL configurations.

2. Physical Description of a Jet in a Crossflow

Based on experimental investigations of a round subsonic jet injected from a flat plate into a subsonic crossflow, a realistic qualitative description of the flowfield can be constructed. Although there may be changes in quantitative detail, the flow features described below should be characteristic of a range of jet and aerodynamic surface configurations.

Extending a short distance from the jet exit plane is a core of jet fluid, that is characterized by relatively slow changes in flow properties; such as velocity and turbulent intensity. For no crossflow, this jet core is conical in shape and extends approximately six jet diameters from the jet exit plane (Albertson et al., 1950) before it is eroded and the flow becomes highly turbulent throughout the entire cross section of the jet. If a crossflow is introduced, the length of the jet core decreases with increasing crossflow/jet velocity ratio.

A shear layer exists at the boundary between the jet and crossflow fluids. The shear layer is thin near the jet orifice, and it is the rapid diffusion of this shear layer which erodes the jet core. The shear layer can be thought of as a region of concentrated vorticity, and for no crossflow the flowfield could be described in terms of the diffusion of the vortex layer representing the jet flow at the jet orifice. With the presence of the crossflow there is distortion of this vortex layer as well as diffusion. The importance to the development of V/STOL prediction methods, of adequately describing the distortion of the initial vorticity distribution into a contrarotating vortex pair, has been pointed out by Bradbury (1981). Some comments on this process have been made by Moussa, Trischka, and Eskinazi (1977) in connection with a study of the near-jet region for a jet issuing from a pipe into a crossflow (no flat plate). Also, Thompson (1971b) has looked at a vortex lattice model in studying the initial roll-up of the jet. There is available, however, no quantitative description of the formation of the contrarotating vortex pair observed in the far field from the distortion of the near-jet vortex structure. Both experimental studies utilizing laser instrumentation and an inviscid calculation of the roll-up of the initial vorticity distribution would provide much needed information for modeling the jet interference effect. Also complicating the flowfield near the jet exit plane is a wake region on the flat plate which extends downstream of the jet; this region is clearly visible from oil smear studies (Wu et al., 1970).

The location and growth of the jet plume, as it is deflected and swept downstream, can be observed by flow visualization. Deflection and decay of the initial jet of fluid can be detected by total pressure or velocity measurements in the jet plume. The curve tracing the locations of maximum jet speed from the jet core through subsequent cross sections of the jet plume may be called the jet centerline. This curve can be determined to a location where it is not possible to determine experimentally a local maximum in the axial velocity component in the jet plume. This occurs 15 to 20 jet diameters downstream of the jet orifice for the instrumentation used by Fearn and Weston (1978, 1979).

More than a few jet diameters downstream from the jet orifice, the dominant feature of the flowfield is a pair of diffuse contrarotating vortices. These vortices are swept along curved paths which lie on either side of the plane of flow symmetry and on the concave side of the jet centerline. As these vortices are swept downstream, diffusion of vorticity of opposite sign across the plane of flow symmetry causes a decrease in the strength (circulation) of each vortex. Rough estimates of the properties of this vortex pair are available for both round jets (Fearn and Weston (1974) and Krausche, Fearn, and Weston (1978)) and for rectangular jets of aspect ratio 4 (Thames and Weston, 1978).

Entrainment of the crossflow fluid into the jet plume probably occurs to a significant degree, but quantitative measurements of entrainment are not available.

Some of the features described above are illustrated in Figure 1 for a jet-to-crossflow velocity ratio of approximately 8. The stippled area represents the observed smoke plume within which the relative locations of the jet centerline and vortex penetration curves are shown. The diffuse contrarotating vortex pair is illustrated in a cross section to the jet plume and relative locations of the jet core and flat plate wake region are shown. No attempt has been made to depict the distortion of the intial vortex sheet into the vortex pair. Also shown in this figure are the coordinate axes used to describe the jet properties.

3. Model

According to the description of a jet in a crossflow presented in the previous section, a model of the jet interference effect should account for the following significant features of the flowfield: primary vortex system, entrainment, and a wake region.

3.1 Primary Vortex System

Given a quantitative description of the distortion of the vorticity distribution at the jet orifice into the contrarotating vortex pair and the diffusion of this vortex pair, one could calculate the vortex induced velocity field. For field points external to the jet

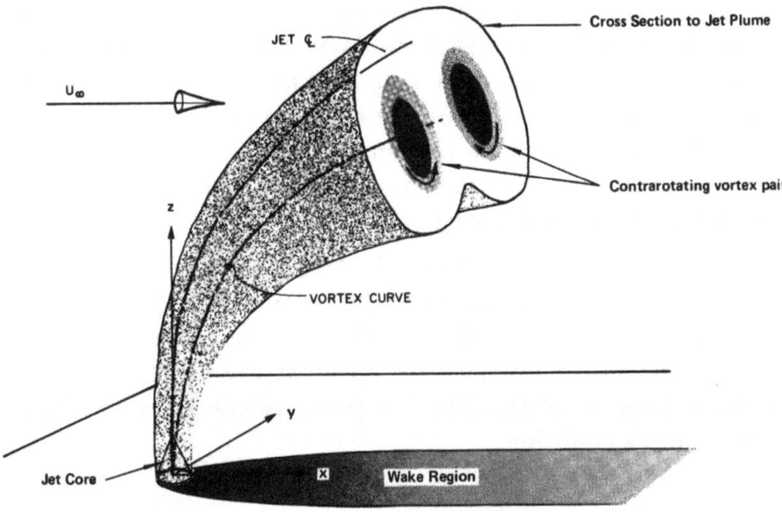

Figure 1. Sketch illustrating some features of a jet in a cross flow.

plume, it would probably be convenient to use a multipole expansion of the vorticity distribution.

3.2 Entrainment

Assuming that a boundary between the jet plume and the crossflow can be defined, the mass flux at a given cross section of the jet plume could be written $Q(s) = \int_{A(s)} \rho\vec{V} \cdot \hat{n} dA$, where s is arc length along the jet centerline from the location of the jet orifice to the cross section $A(s)$. The entrainment rate would thus be written as dQ/ds.

For a jet in still air (no crossflow), entrainment is well defined (Ricou and Spalding, 1961) and the effect of entrainment on the flow external to the jet plume can be represented by a line sink placed along the axis of the jet. The strength of this sink is determined from the entrainment rate. For a co-flowing jet (external flow parallel to the jet), there is a natural inflow from the external stream into the growing jet plume. The mass flux across each cross section increases more rapidly than the increase due to the growing jet boundary (Squire and Trouncer, 1944). The effect of entrainment on the external flow can thus be represented by a line sink along the jet axis where the sink strength is determined from the entrainment rate *and* the growth of the jet boundary. For a jet in a crossflow, it is assumed that entrainment is a significant feature of the flow, and that the effect of entrainment on the external flow can be represented by a distribution of sinks which represent the difference between the entrainment rate and the rate at which an undisturbed crossflow would add fluid to the growing jet plume.

A potential flow model to calculate the velocity and pressure fields can be constructed based on these two effects. For a field point external to the jet plume, the velocity can be calculated from the singularity distribution chosen to represent the vortex and entrainment characteristics of the jet. The pressure can then be calculated from Bernoulli's equation.

3.3 Wake Region

A characteristic of the jet in a crossflow that does not fit into a potential flow model is the wake region. It might be possible to describe the wake region by distributed vorticity or by a free vortex sheet, but it may be preferable to resort to an empirical correction to the pressure distribution on the aerodynamic surface calculated from potential flow theory.

The usefulness of the approach described above to model the jet interference effect for V/STOL configurations will depend on the successful completion of the several steps.

(1) An adequate quantitative description of the pertinent characteristics of the flowfield for selected simple configurations must be available as input to the model.

(2) The model must be verified by demonstrating that it is capable of describing the desired properties, for example, pressure distribution on the aerodynamic surface through which the jet exhausts or the velocity induced at wind tunnel walls.

(3) It must be possible to describe the change in the pertinent characteristics of the flow as a function of jet and aerodynamic surface characteristics without resorting to extensive experimental studies of the jet plume for each configuration of interest.

The example considered in the following section represents an attempt to establish a preliminary verification of the model utilizing the partial description currently available for the flowfield of a round jet/flat plate configuration.

4. Example

The most thoroughly studied configuration for a jet in a cross flow with application to V/STOL aerodynamics is a round jet exhausting perpendicularly through a flat plate. Even for this case, the significant features of the flow needed for a jet interference model have not been described adequately. The present paper utilizes information that is available to determine if the propsed jet interference model is compatible with measured pressure distributions for a range of jet-to-crossflow velocity ratios for this simple configuration. Input to the model is determined from information presented by Fearn and Weston (1974, 1978).

Currently, there is no quantitative description of the vortex system in the region near the jet orifice where distortion of the initial vortex distribution into the contrarotating vortex pair occurs. Once the vortex pair is established however, there is available a model to infer the properties of the diffuse vortex pair from selected velocity measurements in cross sections of the jet plume.* To calculate the pressure distribution on the surface through which the jet exhausts, only the velocity field external to the jet plume is needed. At a given cross section to the jet plume, the properties of each diffuse vortex are represented by the first term in a multipole expansion of the vorticity in the appropriate half-plane ($y = 0$ is assumed to be a plane of flow symmetry). In a region near the jet orifice, the strength of each vortex is expected to increase with downstream distance due to the vortex pair being fed by the distortion of the original vortex system. In a region sufficiently far downstream, the strength of each diffuse vortex is expected to decrease with downstream distance due to diffusion of vorticity of opposite sign across the plane of flow

*In this model, the distribution of vorticity at a given cross section to the jet plume is approximated by the superposition of two Lamb vortices. Although this description is not consistent with the basic equations of fluid dynamics, the velocity field induced by this two-dimensional diffuse vortex model is able to describe adequately the experimentally determined velocities in cross sections to the jet plume. This model for the diffuse vortex pair is compared with the results of a numerical solution of the Navier-Stokes equation for the diffusion of a contrarotating vortex pair by Weston et al. (1984).

symmetry. The representation of this diffuse vortex pair by the first term of a multipole expansion of the vorticity in each half-plane results in a pair of filament vortices of changing strength. In the absence of any quantitative description of the distortion of the original distribution of vorticity into the contrarotating vortex pair, the properties of the vortex pair are extrapolated to assumed initial values at the jet exit plane.

The effect of entrainment on the external flow is represented by a line sink placed along the jet centerline, and for the potential flow calculations, the flat plate is represented by the method of images. For the purpose of the present paper, the wake region downstream of the jet orifice is described by a simple empirical correction to the potential flow calculation.

The quantitative input for the jet interference model applied to perpendicular injection of a round jet through a flat plate represents a description of significant properties of a jet in a crossflow where they are available, and speculation based on available information where quantitative descriptions are not available. All variables are dimensionless. Distances are nondimensionalized by the jet diameter; velocities, by the crossflow speed; and circulation, by dividing by twice the product of the jet diameter and the crossflow speed.

The computer code written to calculate the jet/aerodynamic-surface interference effect (JASI), is written in portable Fortran and has a modular form so that updates and modifications should be relatively simple to implement. The current version of the code (JASIB) is available on request.

The quantitative input to the model for the example under consideration is given in the following subsections.

4.1 Primary Vortex System

The filament vortices representing the contrarotating vortex pair are specified by describing the vortex penetration curve (z_v), the vortex spacing ($2y_v$), and the vortex strength (γ) as functions of x and jet-to-crossflow velocity ratio, R.

Although semiempirical equations have been proposed for the vortex penetration curve, they are significantly more complicated than a power curve which provides an acceptable fit to experimental data: $z_v = A_1 x^{c_1}$ where A_1 and c_1 are parameters that may be functions of R. A power curve also appears to provide an adequate description of the vortex spacing: $y_v = A_2 x^{c_2}$.

Except in a region near the jet orifice, it appears that the jet trajectory can be described by the mutually induced drift velocity of the vortex pair superimposed with the crossflow velocity. As a first order approximation to the mutually induced drift velocity of the vortex pair at some location along the vortex curve, a two-dimensional model with point vortices is used. The mutually induced drift speed of the two-dimensional vortex pair is $\gamma/(2\pi y_v)$. The vortex drift velocity at any point on the vortex penetration curve is taken to be in the direction of the principal normal to the curve and to have the value given above, where γ and $2y_v$ are interpreted as local values for the dimensionless circulation and vortex spacing. The vector sum of the vortex drift velocity and the crossflow velocity is calculated and compared with the tangent to the vortex penetration curve at that point. The following table for $R = 8$ is constructed from the power curve representation for the vortex penetration curve given above and from the independent determinations of vortex properties presented in Table 1 of Fearn and Weston (1974).

x	γ	y_v	$\sin^{-1}\left(\frac{\gamma}{2\pi y_v}\right)$*	$\tan^{-1}(z'_v)$†
2.13	4.77	1.15	41°	45°
5.21	5.23	1.81	27°	31°
8.34	4.72	2.09	21°	25°
15.21	4.36	2.42	17°	18°

*Angle that vector sum of vortex drift velocity and freestream velocity makes with x-axis.

†Angle that tangent to empirical vortex penetration curve makes with x-axis, $z'_v = dz_v/dx$.

There are similar results for other velocity ratios. The relatively close agreement between these angles suggest that the trajectory of a jet in a crossflow might be described in terms of a free vortex condition. Imposition of this condition provides an equation relating the geometry of the vortex penetration curve to the properties of the vortex pair which may be written $z'_v/(1+z'^2_v)^{1/2} = \gamma/(2\pi y_v)$. If the vortex penetration and vortex spacing are known as functions of x, then this free vortex condition determines the vortex strength γ as a function of x. If this description for γ is used for all $x \geq 0$, then for realistic representations for y_v and z_v (e.g., power curves given above) one finds a maximum in the γ versus x curve. Denote the location of the maximum by x_m and the corresponding value of γ by $\gamma_m = \gamma(x_m)$. For the power curve representations of y_v and z_v, the maximum occurs at

$$x_m = \left(\frac{c_1^2 c_2 A_1^2}{1 - c_1 - c_2} \right)^{\frac{1}{2(1-c_1)}}$$

This equation provides a relationship between a property of the vortex system and model parameters. For example, A_1, c_1, or c_2 could be replaced by x_m as a model parameter if that were desirable.

There is insufficient knowledge of the flowfield in the region near the jet orifice to determine an adequate description of the near-jet primary vortex system. In order to complete the model, the distortion of the original vorticity distribution into the contrarotating vortex pair is ignored, and the properties of the vortex pair are extended to specified values at the jet exit plane. For convenience, the near-jet region is defined by $0 \leq x \leq x_m$. The power curve description for the vortex penetration and vortex spacing are utilized in this region, but the vortex strength is modified. The vortex strength in this region is described by a curve that increases from $\gamma = \gamma_o$ at the location $x = 0$, $y = 0$, $z = 0$ to join smoothly with the vortex strength described by the vortex drift model at (x_m, γ_m). To describe the vortex strength in this region of ignorance, a Padé approximate is used

$$\gamma = \frac{p_o + p_1 x + p_2 x^2}{1 + q_1 x + q_2 x^2} \qquad \text{for } 0 \leq x \leq x_m.$$

The parameters p_o, p_1, p_2, q_1, and q_2 are determined from the conditions

$$\begin{aligned} \gamma(0) &= \gamma_o, \\ \gamma'(0) &= \gamma'_o, \\ \gamma''(0) &= \gamma''_o, \\ \gamma(x_m) &= \gamma_m \end{aligned}$$

and

$$d\gamma/dx|_{x_m} = 0$$

4.2 Entrainment

The effect of entrainment on the external flow is modeled by placing a sink filament along the curve describing the jet centerline. The jet centerline curve is well documented, and a power curve representation is utilized for simplicity: $z_c = A_4 x^{c_4}$. By definition, the centerline curve lies in the plane of flow symmetry and intersects the flat plate at the center of the jet orifice.

The strength of the line sink placed along the centerline curve depends on both entrainment rate and the rate of growth of the jet boundary. An entrainment coefficient can be defined by $E = (dQ/dS)/Q_o$ where Q_o is the mass flux across the jet orifice. The entrainment coefficient is considered to be the sum of two parts: $E = E_A + E_B$, where E_B represents the entrainment that would occur for inflow across the growing jet boundary in an undisturbed cross flow, and E_A denotes additional entrainment. It is the entrainment component represented by E_A which has an effect on the flow external to the jet plume. If the entrainment rate and boundary geometry were known for a jet in a crossflow then the strength of the sink placed along the centerline curve could be determined from $E_A = E - E_B$. The author is aware of no direct experimental determination of entrainment properties for a jet in a crossflow. Since this paper is concerned only with the flow field external to the jet plume, only the entrainment component E_A is needed. It is admitted that not enough is known about entrainment for a jet in a crossflow to calculate this quantity. To complete the model it is speculated that the entrainment coefficient E_A is of the same mathematical form as the entrainment coefficient for a submerged jet (Alberston et al., 1950) $E_A = E_{Ao}[2(s/s_o) - (s/s_o)^2]$ for $s \leq s_o$, and $E_A = E_{Ao}$ for $s \geq s_o$, where E_{Ao} and s_o are parameters that may be functions of the velocity ratio and s denotes arc length along the jet centerline curve.

4.3 Wake Correction

The experimental pressure distribution along the downstream

ray from the jet orifice is used to estimate a correction factor applied to the pressure distribution calculated by the potential flow model in the wake region. The assumed form of the correction is $C_P(x,y) = C_{p,pot}(x,y) + \Delta C_p(x,y)$, where

$$\Delta C_p(x,y) = C_{p,\mathrm{meas}}(x\text{-axis},0) - F(x,y),$$

and where $F(x,y)$ is a blending function which is chosen to be

$$F(x,y) = g(x,y)C_{p,\mathrm{pot}}(x,y) + [1 - g(x,y)]\, C_{p,\mathrm{meas}}(x\text{-axis},0).$$

The term $g(x,y)$ is a function whose range is $0 \leq g \leq 1$ such that $g = 1$ along the downstream ray (x positive) and $g \to 0$ when the point (x,y) is sufficiently far from the downstream ray that wake effects are negligible. The function $g(x,y)$ is chosen to be a Gaussian function of y, with half-width denoted by $y_{1/2} = a_6\sqrt{x}$. The function $g(x,y)$ is defined as:

$$g(x,y) = 0 \qquad \text{for } x \leq 0 \qquad \text{(no upstream wake correction)}$$

$$g(x,y) = \exp\left\{-\left(\frac{y - y_w}{y_{1/2} - y_w}\right)^2 \ln 2\right\} \qquad \text{for } x > 0.$$

4.4 Parameter Values

All parameter values are determined for the range of effective velocity ratios, $3.2 \leq R \leq 10.0$. The highest confidence is in the parameters specifying the vortex penetration (A_1, c_1) and the jet centerline (A_4, c_4) curves. These parameter values are taken directly from Fearn and Weston (1978) and they are considered to have an uncertainty of less than 10%.

$$c_1 = 0.4293, \qquad c_4 = 0.3385,$$

$$A_1 = a_1 R^{b_1}, \qquad \text{and} \qquad A_4 = a_4 R^{b_4}$$

where

$$a_1 = 0.3515, \qquad a_4 = 0.9751,$$

$$b_1 = 1.122, \qquad \text{and} \qquad b_4 = 0.9085.$$

The vortex penetration and jet centerline curves are shown for jet to crossflow velocity ratios of 4.2 and 8.0 in Figures 2 and 3. Also shown in the figures are some of the experimental data from which the above parameters were determined.

Of significantly lesser confidence are the parameters determining the vortex spacing and vortex strength. Through the vortex drift model, the parameters A_2 and c_2 can be written as functions of x_m and γ_m. Thus, either set of parameters can be used to complete the description of the vortex pair in the region $x \geq x_m$. The model uses the following parameter specifications which are consistent with the vortex spacing and vortex strength reported in Fearn and Weston (1974): $\gamma_m = 0.5 + .525R$. Instead of specifying x_m explicitely, it was found to be more convenient to specify $z'_m = 1.0$ for $3.2 \leq R \leq 5$, $z'_m = 1.3 - 0.06R$ for $5 \leq R \leq 10$, where $z'_m = dz_v/dx|_{x=x_m}$. The power curve representation of the vortex penetration curve can differentiated and solved explicitely for x_m.

$$x_m = \left(\frac{c_1 a_1 R^{b_1}}{z'_m}\right)^{\frac{1}{1-c_1}}$$

Parameters for the vortex spacing curve can be obtained from x_m, γ_m by using the vortex drift model as described above. The explicit equations are

$$c_2 = \frac{1 - c_1}{1 + c_1^2 A_1^2 x_m^{2(c_1-1)}},$$

$$A_2 = \frac{\gamma_m(1 - {z'_m}^2)^{1/2}}{2\pi x_m^{c_2} z'_m}$$

For velocity ratios of 4.2 and 8.0, the input curves for vortex spacing are shown in Figures 2 and 3, and the input curves for vortex strength are shown in Figures 4 and 5. In each case, the symbols denote vortex properties inferred from selected velocity measurements in the jet plume using a diffuse vortex model.

The parameters specifying the initial vortex properties (γ_o, γ'_o, and γ''_o), the effect of entrainment on the external flow (s_o, E_{Ao}) and the wake effect (a_6) represent physical effects for which there are currently no descriptions. These are taken to be free parameters that are determined to provide a reasonable fit to the experimental pressure distributions.

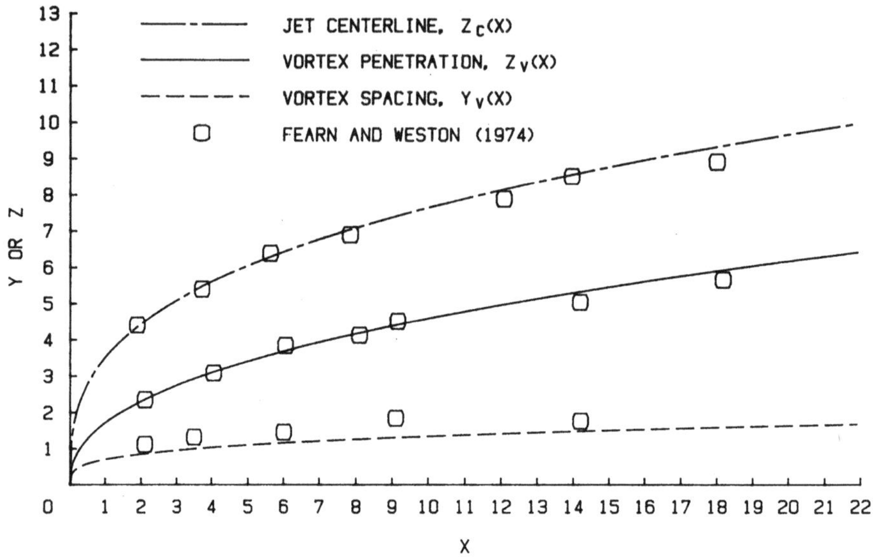

Figure 2. $R = 4.1$, geometry input for jet interference model.

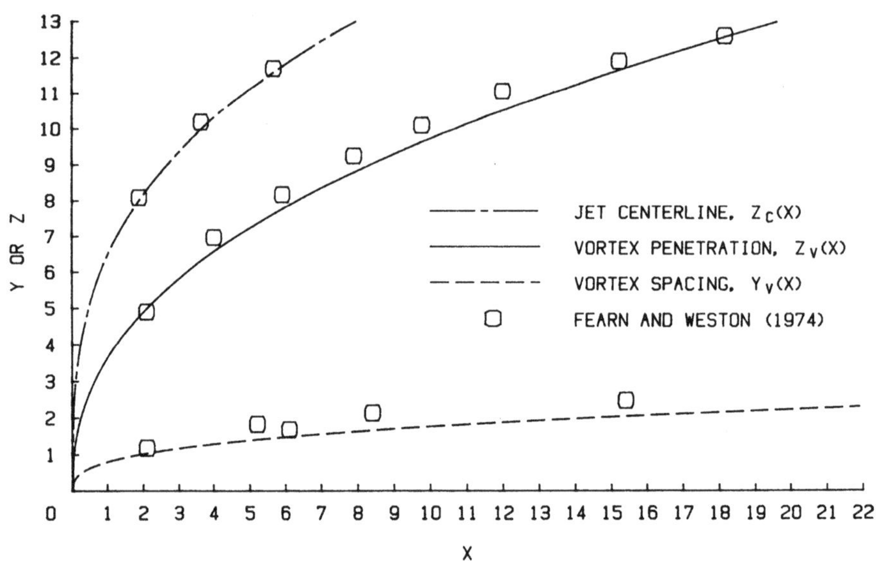

Figure 3. $R = 8.0$, geometry input for jet interference model.

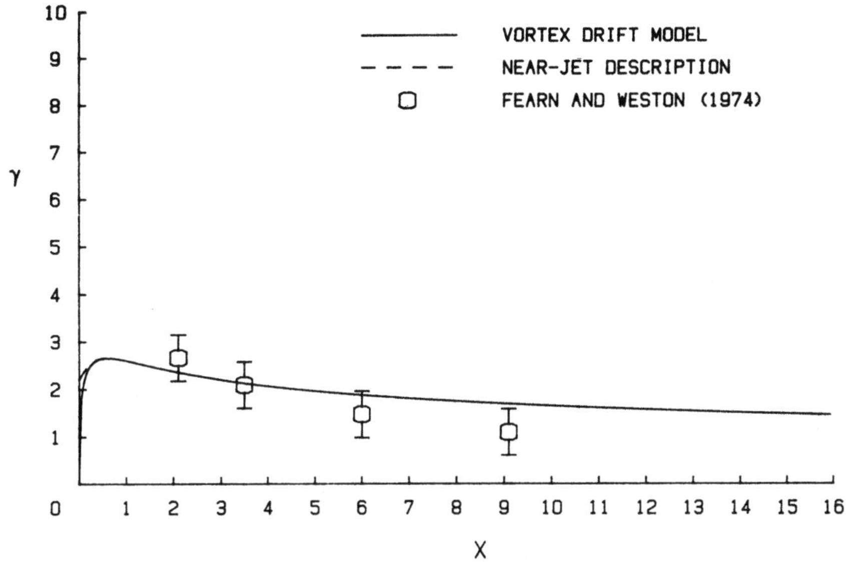

Figure 4. $R = 4.1$, vortex strength for jet interference model.

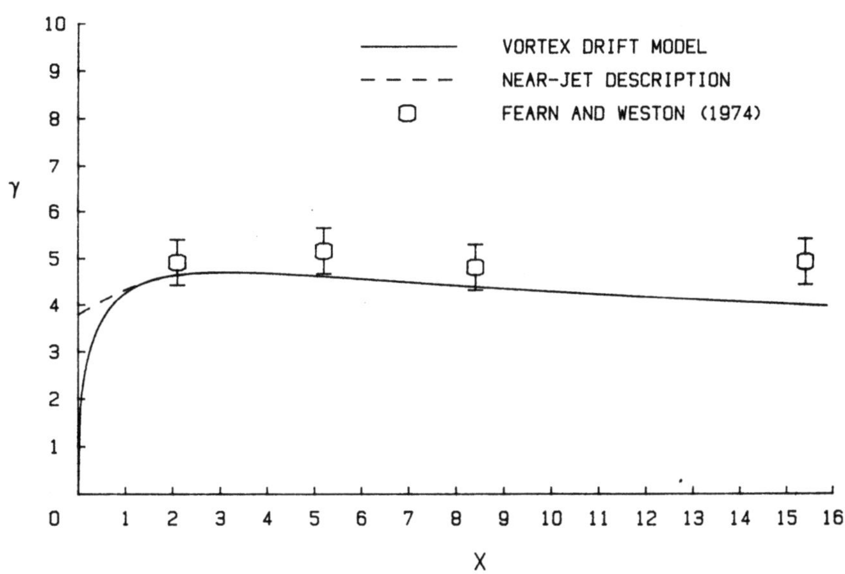

Figure 5. $R = 8.0$, vortex strength for jet interference model.

4.5 Numerical Considerations

In the computations, the filament singularities described above are approximated by straight line segments of constant strength. Near the jet orifice, the segments are of constant length, but for arc lengths along the appropriate curve (centerline or penetration) greater than a specified value, the length of each segment is taken to be proportional to the radius of curvature of the curve.

For the example considered, the pressure is calculated at each of 360 field points on the flat plate for each effective velocity ratio. Typically, 50 to 100 segments for each filament singularity are sufficient to obtain convergence of the model-calculated pressure distribution to three significant figures. Run time for a given velocity ratio is approximately ten seconds on an IBM 3033N.

A simple summation procedure is used to calculate the integrated effects of force and moment on the flat plate. For the purpose of comparison, the force and moment are calculated on a reference area consisting of the region between the circle of the jet orifice and a circle of diameter eleven times that of the jet orifice.

5. Results

Extensive parameter studies were conducted, and the ability of the model to fit the experimental pressure distribution (Fearn and Weston, 1975) for each effective velocity ratio was evaluated utilizing a nonlinear regression procedure (Helwig and Council, 1979). The following conclusions represent a compromise between quality of fit and simplicity of model.

(1) Allowing the vortex pair to intercept the flat plate at $(x, y) = (x_o, \pm y_o)$ instead of $(x, y) = (0, 0)$ results in no significant improvement in the ability of the model to fit the data. For simplicity of formulation, this observation is incorporated into the model description presented in the previous section.

(2) Model results are relatively insensitive to small variations of the initial slope and curvature of $\gamma(x)$. The following representations are chosen: $\gamma_o' = 2(\gamma_m - \gamma_o)/x_m$ and $\gamma_o'' = 0$.

(3) The arc length, s_o, at which the entrainment coefficient changes from a quadratic to a linear function can be written as:* $s_o = 0$ for $2.8 \leq R \leq 5$, $s_o \leq 6.2 - \frac{31}{R}$ for $5 \leq R \leq 10$.

(4) The wake parameter can be represented by: $a_6 = .45$ for $2.8 \leq R \leq 3.5$ and $a_6 = .1(1+R)$ for $3.5 \leq R \leq 10$.

(5) Values for the parameter representing initial vortex strength (γ_o) and entrainment coefficient (E_{Ao}) are the results of a two-parameter fit to the experimental pressure distribution for each effective velocity ratio studied. Parameter values versus effective velocity ratio are plotted in Figure 6.

Using the parameter set described above, the pressure distribution on the flat plate can be calculated for a specific value of jet to crossflow velocity ratio. The ability of the model to fit the experimental pressure distribution for several velocity ratios is shown in Figures 7 through 13. Contours of constant pressure coefficient are plotted for both model calculated and experimental pressure distributions. The equivalent doublet strength of the vortex pair

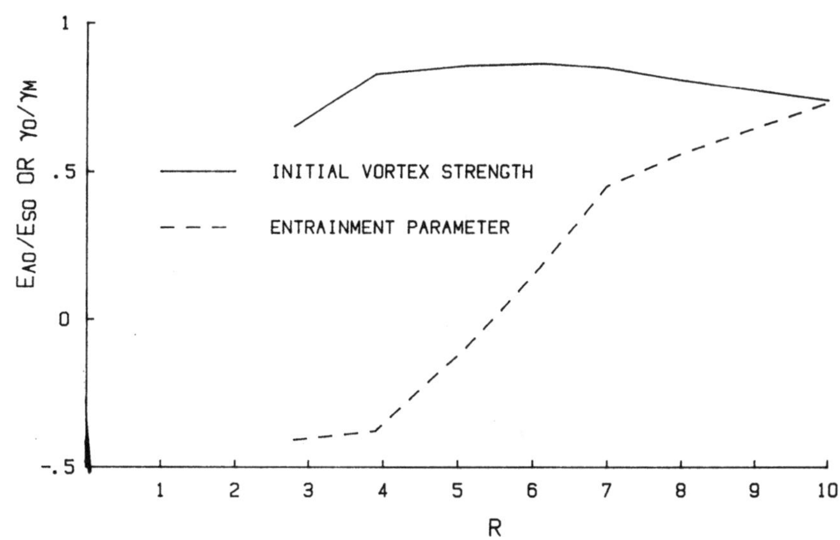

Figure 6. Input parameters for jet interference model.

*It should be pointed out that for a nominal effective velocity ratio of 3, the jet plume data was taken at $R = 3.2$ whereas the pressure distribution was taken at $R = 2.8$. Thus, some extrapolation of parameter values is necessary to utilize the model for the case of $R = 2.8$.

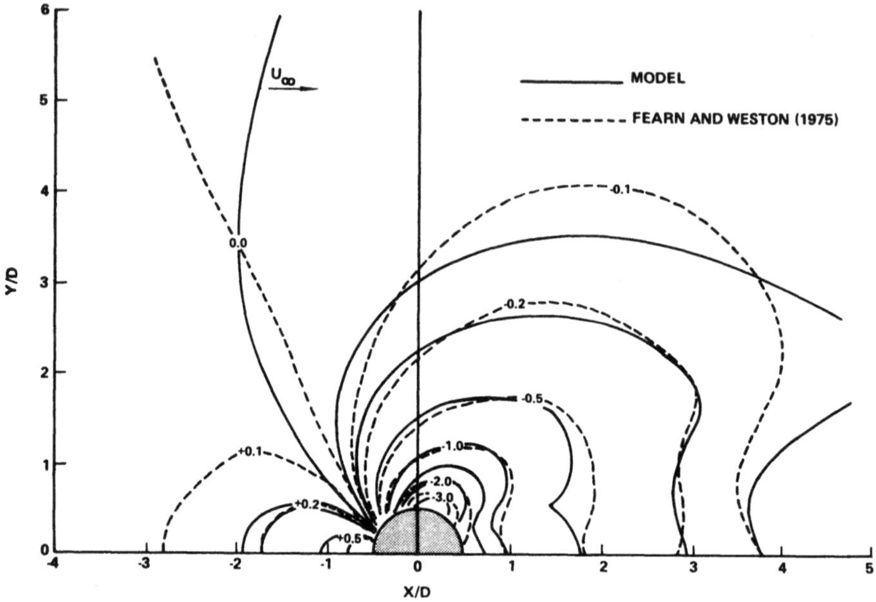

Figure 7. $R = 2.8$, contours of constant pressure coefficient.

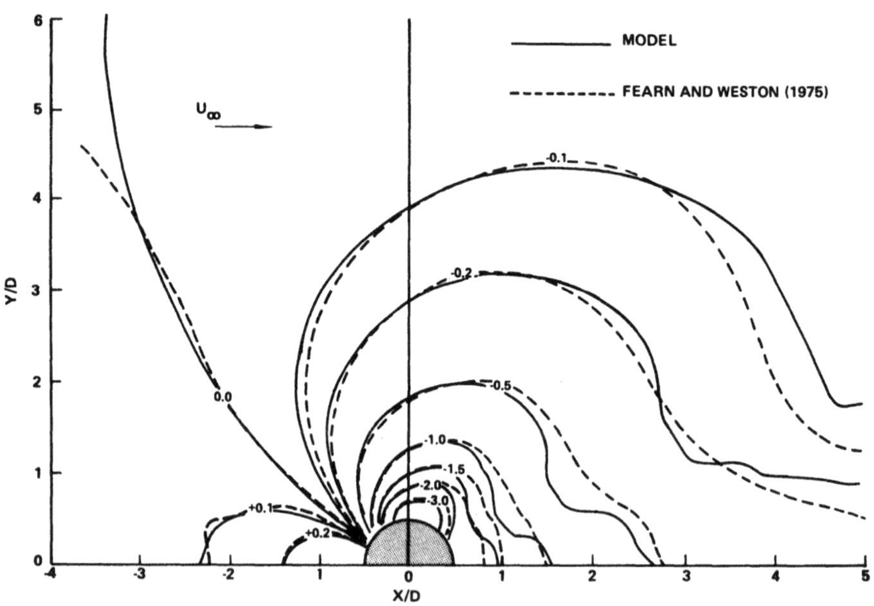

Figure 8. $R = 3.9$, contours of constant pressure coefficient.

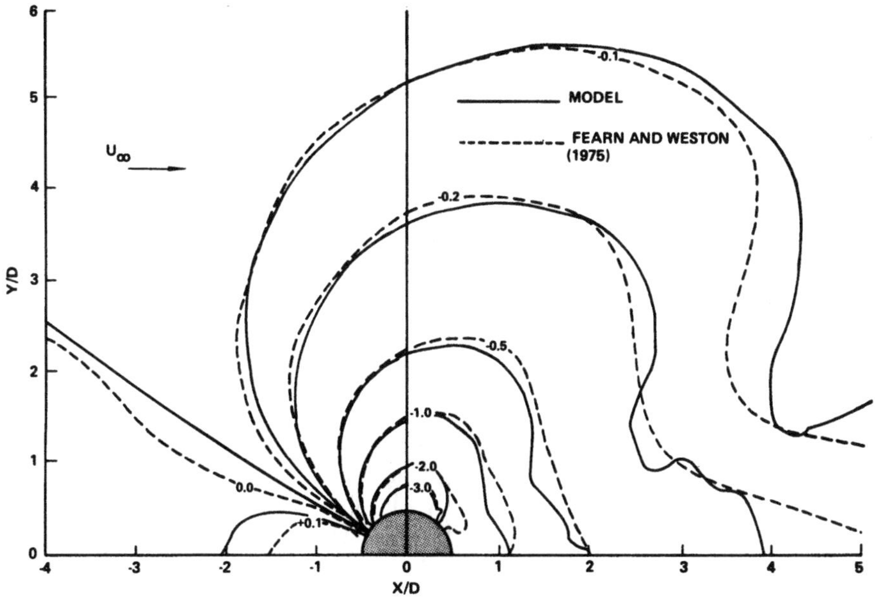

Figure 9. $R = 5.1$, contours of constant pressure coefficient.

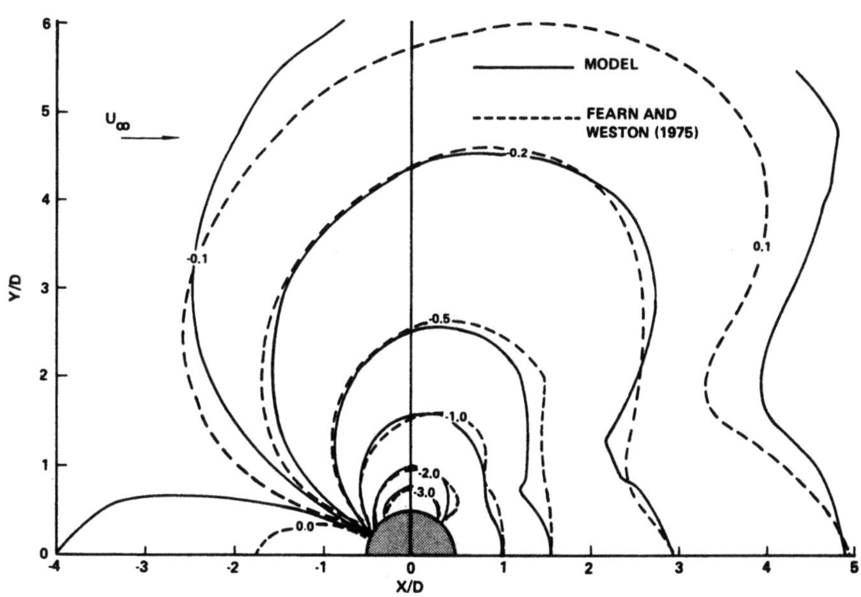

Figure 10. $R = 6.1$, contours of constant pressure coefficient.

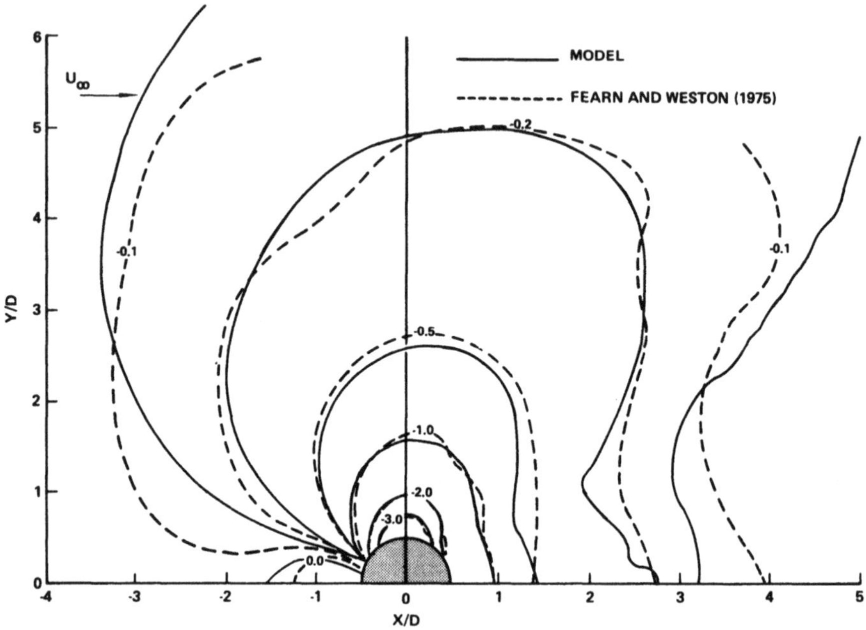

Figure 11. $R = 7.0$, contours of constant pressure coefficient.

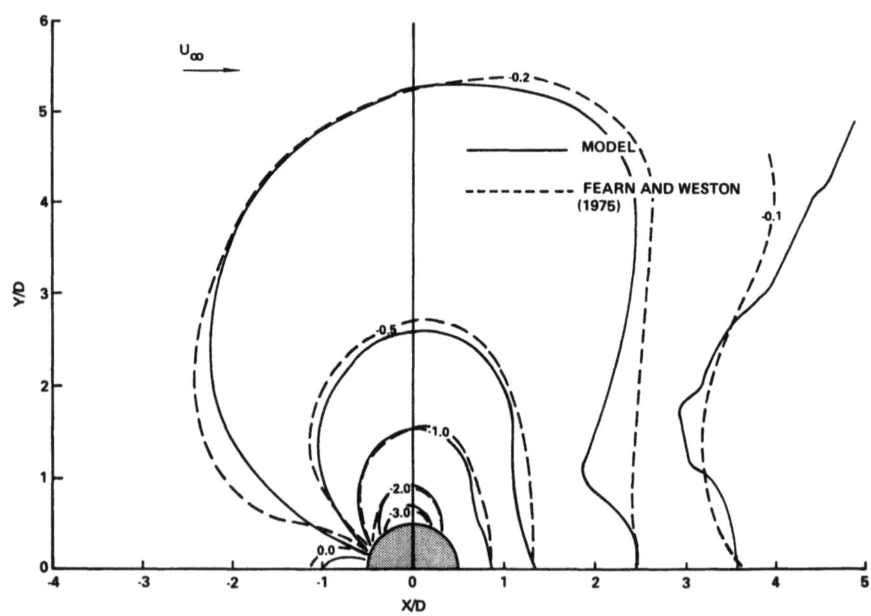

Figure 12. $R = 8.0$, contours of constant pressure coefficient.

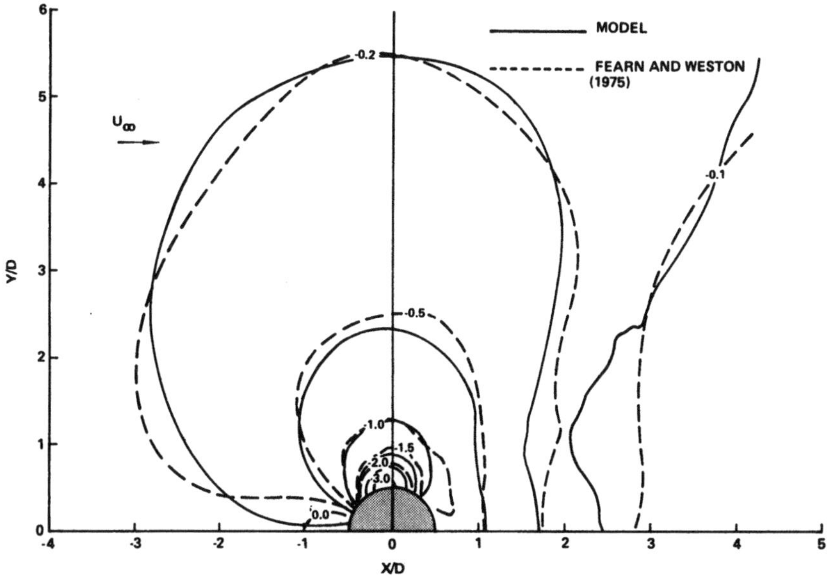

Figure 13. $R = 10.0$, contours of constant pressure coefficient.

models the blockage effect along the upstream ray. It appears that the potential flow model modified by a simple wake assumption is capable of providing a reasonable approximation to the experimentally determined pressure distribution. A quantitative measure of the ability of the model to fit the experimental pressure distribution is chosen to be an area weighted area normalized standard deviation of the pressure coefficient. Values of this standard deviation ranged from .034 to .053, being lowest for the midrange of velocity ratios.

Comparison of the force and moment calculations from the model generated and experimental pressure distributions are shown in Figures 14 and 15. The model appears to be capable of providing an adequate description of the force and moment on the reference circle of the flat plate for the range of velocity ratios considered.

The parameter values for E_{A_0} in Figure 6 display an interesting variation with jet-to-crossflow velocity ratio. For the largest velocity ratio considered, $R = 10$, the strength of the sink placed along the jet centerline is about 75% of that required to describe entrainment for a submerged jet (no crossflow). As the velocity ratio decreases, the sink strength decreases until it is negligible in the vicinity of $R = 5$. Further decrease in velocity ratio requires a source to be placed

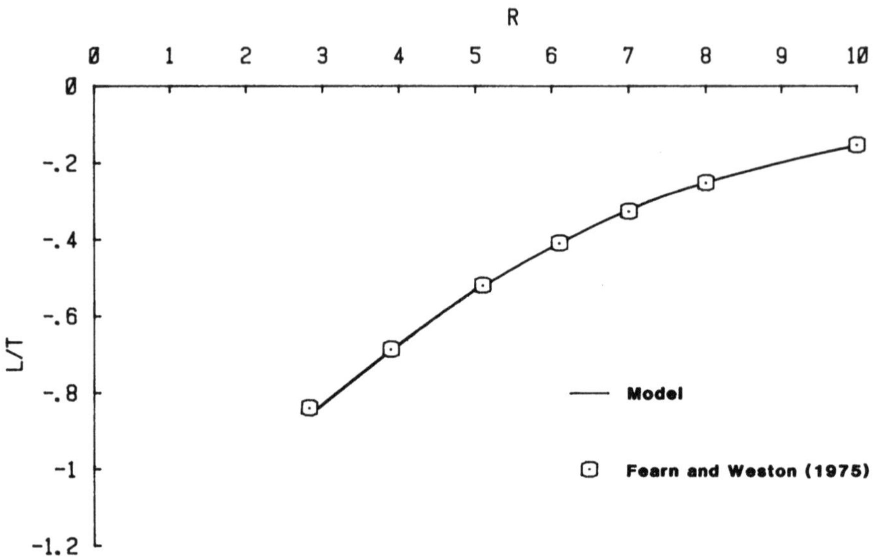

Figure 14. Lift loss on reference area.

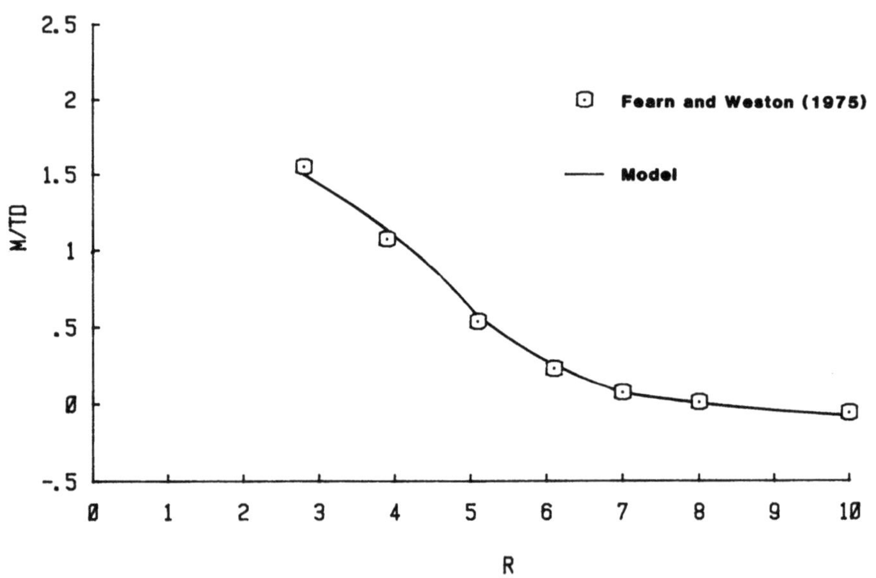

Figure 15. Pitching moment on reference area.

along the jet centerline. Two comments are offered concerning this behavior of the parameter E_{A_o}. Since there are several assumptions of a questionable nature made in the development of the model, there is no guarantee that the parameter E_{A_o} actually represents entrainment. If other parameters in the model are not representing the physical aspects of the flow that they are intended to represent, then it is certainly possible to obtain a reasonable pressure distribution with a physically unreasonable entrainment parameter. On the other hand, extensive parameter studies were made for velocity ratios of 2.8 and 3.9 with the constraint that E_{A_o} be positive (implying a sink). No combination of the remaining parameters could be found that provided a reasonable fit to the measured pressure distribution.

6. Conclusion

It appears that the jet/aerodynamic-surface interference model presented is capable of describing the surface pressure distribution for the simple example of a round jet exhausting perpendicularly through a flat plate into a crossflow. Evolution of the model can occur as more complete descriptions of the flow field become available. The usefulness of the model can be evaluated by considering additional simple examples and by patching the model to existing panel methods to study an extension of the model to calculate the jet interference effect for realistic V/STOL configurations.

The model presented in this paper is, of course, not unique and it certainly does not represent adequately all of the physics of the problem. It does, however, represent some of the physical phenomena in some regions of the flow field and it provides a framework for a model to calculate the flow external to the plume of a jet in a crossflow based on simple representations of vortex structure, entrainment, and an empirical wake correction. The very process of attempting to apply such a model to specific configuration leads one to identify pertinent questions for additional research. Probably, the most significant information needed for further evolution of the model is a quantitative description of the vortex structure in the region between the jet orifice and the location where the vortex pair is fully developed. An inviscid calculation of this rollup process would be most useful, especially if it were validated experimentally.

Acknowledgement

The work reported in this paper is supported by the NASA Ames Research Center under Grant NSG-2288 with Mr. K. Aoyagi as the technical monitor.

References

[1] Abramovich, G. N. *The Theory of Turbulent Jets,* Cambridge: MIT Press, 1963, 541–56.

[2] Albertson, L. M., et al. "Diffusion of Submerged Jets," *Trans. ASCE, Vol. 115* (1950), 639–64.

[3] Aoyagi, K. and Snyder, P. K. "Experimental Investigation of a Jet Inclined to a Subsonic Cross Flow," *AIAA Paper 81-2610,* 1981.

[4] Baker, A. J., Orzechowski, J. A., and Manhardt, P. D. "A Numerical Three-Dimensional Turbulent Simulation of a Subsonic V-STOL Jet in a Cross Flow Using a Finite Element Algorithm," *Report NADC-79021-60,* 1981.

[5] Beatty, T. D. and Kress, S. S. "Prediction Methodology for Propulsive Induced Forces and Moments of V/STOL Aircraft in Transition/STOL Flight," *Vol. 1, Final Report, NADC-77119-30,* 1979.

[6] Bradbury, L.J.S. "Some Aspects of Jet Dynamics and Their Applications for VTOL Research," *AGARD-CP-308,* 1981.

[7] Dietz, W. E., Jr. "A Method for Calculating the Induced Pressure Distribution Associated With a Jet in a Cross Flow," *M. S. thesis,* Univ. of Florida; also *NASA CR-146434,* 1975.

[8] Fearn, R. L. and Weston, R. P. "Vorticity Associated With a Jet in a Cross Flow," *AIAA J., Vol. 12, No. 12* (1974), 1666–71.

[9] ——————————. "Induced Pressure Distribution of a Jet in a Cross Flow," *NASA TN D-7916,* 1975.

[10] ——————————. "Induced Velocity Field of a Jet in a Cross Flow," *NASA TP-1087,* 1978.

[11] ——————————. "Velocity Field of a Round Jet in a Cross Flow for Various Jet Injection Angles and Velocity Ratios," *NASA TP-1506,* 1979.

[12] Hackett, J. E. "Living With Solid-Walled Wind Tunnels," *AIAA Paper 82-0583,* 1982.

[13] Harms, L. "Experimentelle Untersuchungen über das Strömungsfeld eines heissen turbulenten Strahles bei Queranströmung, Teil II.

Deutsche Forschungs—und Versuchsanstalt für Luft—und Raumfahrt E. V. ZNW, Goettingen, Germany, *Report 1-B 157-73 A 21*, 1973; Trans. as *NASA TTF-15706*, 1974.

[14] Helwig, J. T. and Council, K. A., eds. *SAS User's Guide*, Cary, NC: SAS Institute, 1979, 317–29.

[15] Jordinson, R. "Flow in a Jet Directed Normal to the Wind," *R&M No. 3074*, Brit. A.R.C., 1958.

[16] Kamotani, Y. and Greber, I. "Experiments on a Turbulent Jet in a Cross Flow," *AIAA J., Vol. 10, No. 11* (1972), 1425–29.

[17] Krausche, D., Fearn, R. L., and Weston, R. P. "Round Jet in a Cross Flow: Influence of Jet Injection Angle on Vortex Properties," *AIAA Journal, Vol. 16, No. 6* (1978), 636–37.

[18] Kuhn, R. E. "An Empirical Method for Estimating the Jet-Induced Effects on V/STOL Configuration in Transition," *Task Report, Contract No. N62269-79-C-0231*, 1979.

[19] Margason, R. J. "The Path of a Jet Directed at Large Angles to a Subsonic Free Stream," *NASA TN D-4919*, 1968.

[20] Moussa, Z. M., Trischka, J. W., and Eskinazi, S. "The Near Field in the Mixing of a Round Jet With a Cross-Stream," *J. Fluid Mech., Vol. 80, Part 1* (1977), 49–80.

[21] Ricou, F. P. and Spalding, D. B. "Measurements of Entrainment by Axisymmetric Turbulent Jets," *J. Fluid Mech., Vol. 2* (1961), 21–32.

[22] Rubbert, P. E., et al. "A General Method for Determining the Aerodynamic Characteristics of Fan-In-Wing Configurations; Vol. 1: Theory and Application," *Final Report, Contract DA 44-177-AMC-323(T)*, 1967.

[23] Soulier, A. "Essais à Sima pour Researcherches de base sur les Interactions de Jet—Répartition des Pressions de l'orifice du jet," Office National d'Etudes et de Recherches Aerospatiales, *Document no. 1/253 GY-Fascicule no. 1/5*, 1968; Trans. as *NASA TTF-14066*, 1972a.

[24] ————. "Essais à Sima pour Researcherches de base sur les Interactions de Jet—Répartition des Pressions et des Vitesses dans le Jet tuyere type idéale (àfroid)," Office National d'Etudes et de Recherches Aerospatiales, *Document no. 1/253 GY-Fascicule no. 2/5*, 1968; Trans. as *NASA TTF-14072*, 1972b.

[25] Spree, B. M. "Summary Remarks on the AGARD/FDP Symposium on the Fluid Dynamics of Jets With Application to V/STOL," *AGARD-CP-308*, 1981.

[26] Squire, H. B. and Trouncer, J. "Round Jets in a General Stream," *R.A.E. R&M, No. 1974*, 1944.

[27] Thames, F. C. and Weston, R. P. "Properties of Aspect-Ratios-4.0 Rectangular Jets in a Subsonic Cross Flow," *AIAA Paper 78-1508,* 1978.

[28] Thompson, A. M. "The Flow Induced by Jets Exhausting Normally From a Plane Wall Into an Airstream," *Ph.D. thesis,* Univ. of London, 1971a.

[29] Thompson, J. F. "Two Approaches to the Three-Dimensional Jet-In-Cross-Wind Problem: A Vortex Lattice Model and a Numerical Solution of the Navier-Stokes Equations," *Ph.D. thesis,* Georgia Institute of Technology, 1971b.

[30] Weston, R. P., Raj, P., and Fearn, R. L. "Implicit Finite-Difference Computation for the Diffusion of a Contrarotating Vortex Pair," *AIAA Journal,* accepted for publication.

[31] Wooler, P. T. "V/STOL Aircraft Aerodynamic Prediction Methods Investigation," *USAF AFFDL TR-72-26, Vols. I–IV,* 1972.

[32] Wu, J. C., et al. "Experimental and Analytical Investigations of Jets Exhausting Into a Deflecting Stream," *J. Aircraft, Vol. 7, No. 1* (1970), 44–51.

Multiple Jet Impingement Flowfields*

Donald R. Kotansky
McDonnell Aircraft Company
McDonnell Douglas Corporation
St. Louis, MO 63166

Summary

An engineering methodology based on an empirical data base and analytical fluid dynamic models has been developed for the prediction of propulsive lift system induced aerodynamic effects for multiple lift jet VTOL aircraft operating in the hover mode in and out of ground effect. The methodology takes into account the effects of aircraft geometry, aircraft orientation (pitch, roll) as well as height above ground. Lift jet vector and splay directions with respect to the airframe, lift jet exit flow conditions, and both axisymmetric and rectangular nozzle exit geometry are also accommodated. The methodology has been embodied in a computer code which accommodates configurations with up to six lift jets.

In ground effect, the prediction methodology proceeds logically from the aircraft lift jet exits through the free jets, jet impingement points, wall jets, fountain bases (stagnation lines) and fountain upwash flow and impact on the airframe undersurface. The induced suckdown flows are computed from the potential flowfield induced by the turbulent entrainment of both the free jets and wall jets in ground effect and from the free jets alone out of ground effect. Key elements of this methodology are emphasized in this paper including geometric considerations, computation of stagnation lines, fountain upwash formation and development, and fountain impingement on the airframe undersurface.

---

*Portions of this work were supported by the Naval Air Development Center (contracts N62269-76-C-0086 and N62269-81-C-0717), the NASA Ames Research Center (contracts NAS2-9646 and NAS2-10184), the Office of Naval Research (contract N00014-79-C-0130) and McDonnell Aircraft Company Independent Research and Development resources.

Nomenclature

A	area.
AR	aspect ratio.
D	jet exit diameter (circular nozzle), nozzle exit width (rectangular nozzle).
D_e	equivalent circular jet exit diameter.
F	force.
$f(\phi)$	radial momentum flux distribution about jet impingement point (normalized).
H	nozzle exit height above ground plane measured perpendicular to ground plane.
L	nozzle exit length (rectangular nozzle), lift.
LID	lift improvement device.
$\dot{m}$	mass flow.
$\dot{M}$	momentum flux.
N	normal distance above ground plane in wall jet, number of jets in idealized lift jet system.
NPR	nozzle pressure ratio.
R	radial distance.
S	nozzle exit centerline spacing.
S'	distance between jet impingement points on ground plane.
$R_{V/2}, S_{V/2}$	velocity profile width at point in profile where $V = 1/2\ V_{\max}$.
U	wall jet velocity.
V	jet, fountain velocity.
X, Y, Z	Cartesian coordinates with Z normal to the ground plane.
α	jet impingement angle measured from ground plane.
γ	momentum normalization correction.
θ	stagnation line slope in ground plane.
κ	fountain sidewash angle measured from the ground plane.
λ_{fI}	fountain upwash momentum flux transfer coefficient.
$\lambda_{\dot{M}}$	momentum flux magnitude recovery (conservation) factor.

ρ	density.
ϕ	azimuthal angle in ground plane; $\phi = 0°$ in direction of the horizontal component of free jet mean velocity in oblique impingement situations.
ϕ'	computational polar angle measured in the ground plane about a jet impingement point referenced to the line joining the jet impingement points.
ω	fountain upwash inclination measured from the ground plane.

Subscripts

$1,2$	jet designation.
F,f	fountain, final.
fI	fountain impingement.
j	jet.
je	jet exit.
max	maximum.
min	minimum.
N	normal.
o	initial.
R	radial.
S	suckdown.
TH	theoretical.
wj	wall jet.

1. Introduction

The design of successful high performance military VTOL aircraft requires a critical blend of new airframe and propulsive lift system technology with tried and proven CTOL high speed aircraft characteristics. The VTOL aircraft configuration analyst is beset with a myriad of performance requirements, airframe configuration variables and propulsive lift system options; yet there is a dearth of comprehensive vehicle performance prediction methods to aid in the identification of promising vehicle configurations. This, in combination with the many physical variables involved, can result in a largely subjective and empirical approach to vehicle configuration definition based on past experience. The unique operational requirements of VTOL aircraft that are responsible for this complex design dilemma are, of course, the vertical and transition flight requirements. These flight modes necessitate a knowledge of forces and moments on the vehicle which are unfamiliar to the CTOL aircraft designer. These flowfield effects are characteristically dependent on the selected propulsive lift system and its physical integration into the airframe due to the interactive nature of the lift jet induced aerodynamics with the airframe. These jet induced flowfield interactions occur in the transition and in the hover or vertical flight modes both in and out of ground effect. A purpose of this article is to provide insight and a quantitative basis for the modelling and prediction of multiple jet induced flowfields and the resulting aerodynamic forces and moments on the airframe in ground effect. A propulsion system designed without taking into account the induced forces may not provide sufficient thrust for an adequately controlled takeoff without a reduction in payload. In addition to these induced net loads, situations are encountered where unfavorable moments are produced on the airframe, resulting in significant stability and control problems. Until accurate, general purpose flowfield prediction techniques become available for these complicated viscous flowfields, dependence on wind tunnel testing for design and performance verification will be relatively high.

The induced forces (and moments) in and out of ground effect usually result from one of two reasonably well understood flow phenomena. These are jet entrainment and the formation of jet flow fountains. Jet entrainment causes otherwise static air to be set into motion, resulting in locally reduced static pressures on nearby airframe undersurfaces thus introducing negative aerodynamic loads. The jet entrainment effect occurs both in and out of regions influenced by the presence of the ground but is frequently accentuated as the distance between the nozzle exit and the ground is reduced. This is attributable to the proximity of the additional entrainment resulting from the ground wall jets. The flowfield about a VTOL aircraft hovering in ground effect is shown schematically in Figure 1.

The formation of jet flow fountains requires the impingement of the jets on the ground or shipboard landing pad and, therefore, is peculiar to operation close to the ground. The formation of fountains is configuration dependent in that multiple jets are required, and the jet impingement points, relative strengths of the jets, and

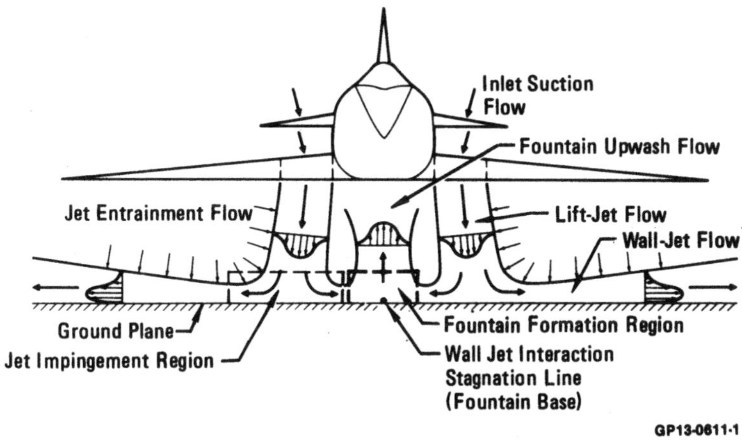

Figure 1. Flowfield about a VTOL aircraft hovering in ground effect.

jet impingement angles strongly influence the characteristics of the fountain. The upward convection of jet flow in the fountains usually results in a positive aerodynamic lift, caused by the positive pressurization of airframe undersurfaces containing and deflecting the fountain upwash flow. Because of the upward convection of the lift jet flow in the fountains, a degradation of propulsion system performance may result from exhaust gas ingestion into the aircraft inlets. In this respect, fountains can be detrimental to VTOL aircraft performance in proximity to the ground. Significant VTOL hover flowfield interactions and their resulting effects on vehicle performance in ground effect are summarized in Figure 2.

The following sections of this article will outline the empirically based methodology developed by the McDonnell Aircraft Company for the modelling and prediction of multiple jet VTOL aircraft flowfields in ground effect. This methodology was developed during the period from 1975 through 1982. The reader will be referred to appropriate published references for detailed mathematical developments and thorough documentation of experimental data. Key elements in the development of the methodology are emphasized in this article.

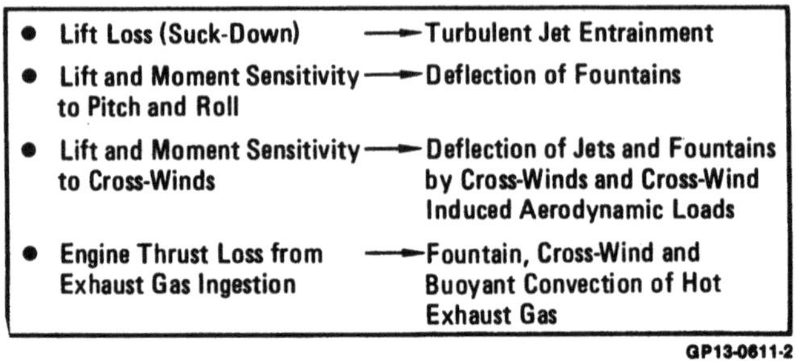

Figure 2. VTOL flowfield interactions in ground effect.

2. Geometric Considerations

The viscous (turbulent) flowfield between a multiple jet VTOL aircraft and the ground is strongly dependent on many geometric factors related to the geometry of the airframe, integration of the propulsive lift system including the geometry of the nozzle exits, and the spatial relationship of the airframe to the ground plane. The significant jet mean flow geometry and overall flowfield geometry include:

- lift jet system arrangement including the number of jets, jet exit spacings, and jet exit orientation with respect to the aircraft axes (including longitudinal *vector* and lateral *splay* angles),
- nozzle exit shapes and mean velocity distributions at the nozzle exits, and
- aircraft surface geometry, aircraft spatial orientation and height above ground.

Most of the above geometric variables were taken into account in the initial development of a ground flow field prediction methodology by the McDonnell Aircraft Company (MCAIR) (Kotansky et al., 1977). This work was based on the fundamental free jet flow development and impingement characteristics of round jets which, for the most part, were available in the published literature. Specifically, the free jet entrainment data of Kleis and Foss (1974) and the geometric and kinematic properties of free and impinging round jets established by Donaldson and Snedeker (1971) were used extensively. The data of Donaldson and Snedeker were also used to determine analytical expressions for wall jet entrainment.

A key element in the modelling of the ground surface flow field below a hovering VTOL aircraft is the azimuthal distribution of radial momentum flux about the individual jet impingement points on the ground plane. This distribution is strongly dependent on the exit shape of the jet nozzle (for low values of H/D) and the local impingement angle of the jet on the ground plane. The data of Donaldson and Snedeker shown in Figure 3 indicate that for a round turbulent jet this distribution is sensitive to the local jet impingement angle and relatively insensitive to the ratio of height above ground to nozzle exit diameter (H/D). These distributions of momentum flux together with the magnitudes of the total momentum fluxes emanating from the nozzle exits establish the location and (momentum) strength of the fountains in the flow field below the aircraft.

Another significant geometric consideration is the number and arrangement of the propulsive lift jets in the aircraft lift system. Figure 4 shows a generic idealized multiple lift jet arrangement spaced symmetrically about a hypothetical center of gravity. The octagonal central area circumscribed by the lines connecting the eight lift jet centerlines represents an idealized *inner region*. All area outside of this region extending to infinity constitutes the idealized *outer region*. The interaction lines of the wall jets resulting from the ground impingement of the lift jet pairs are shown as the dashed lines in Figure 4. These lines indicate the ideal locations of the wall jet stagnation lines which form in both the inner and outer regions. A stagnation line pattern as shown in Figure 4 would result from the ideal vertical impingement of round jets producing a $f(\phi) = 1$ distribution of wall jet radial momentum flux (see Figure 3). Figure 5 shows the maximum available fountain force as a function of the number of lift jets in this idealized multiple lift jet system. The lower curve in Figure 5 shows the maximum fountain force F_F available from the inner region only,

$$F_F \,(\text{inner region}) = \left(\frac{1}{2} - \frac{1}{N}\right) F_T \tag{2.1}$$

where F_T is the total lift system thrust and N is the number of jets in the idealized lift system.

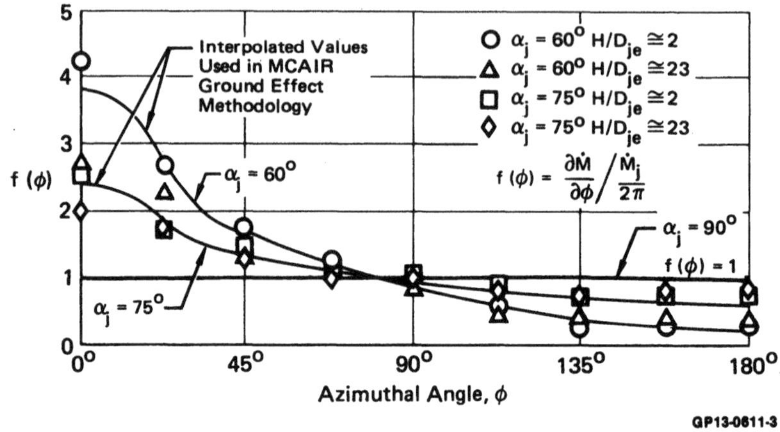

Figure 3. Azimuthal distribution of wall jet radial momentum flux for an impinging round jet.

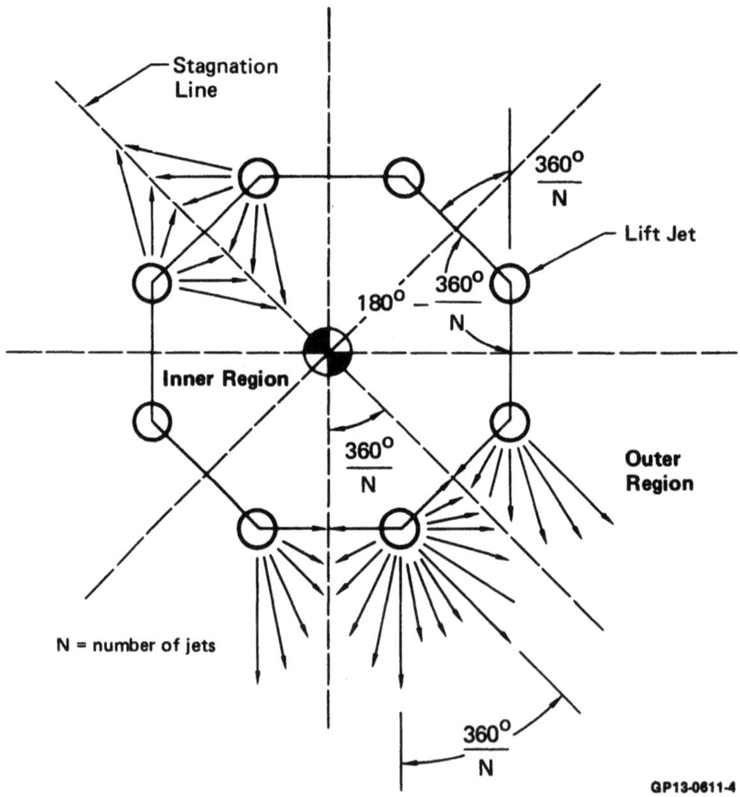

Figure 4. Idealized 8-jet lift system impinging on a ground plane.

It should be noted that a two jet lift system has no inner region, by definition, and therefore, all fountain force obtained from a two jet fountain is outer region force. The result shown in equation (2.1) is predicated on the assumption that all lift jet momentum flux that enters a volume below the aircraft whose planform is the inner region eventually is turned to vertical and imparts this vertical momentum to the airframe. The upper curve in Figure 5 indicates the total available ideal fountain force including contributions from both the inner and outer regions. The lift jet momentum flux entering the outer region volume is assumed to impinge on the airframe with an ideal sidewash angle of incidence resulting from the turning of the flow in an upward direction at the stagnation lines. Performing an integration of the vertical component of the outer region momentum flux and adding the result to the force from the inner region,

$$F_F\,(\text{outer region} + \text{inner region}) = \left(\frac{1}{2} - \frac{1}{N} + \frac{1}{\pi}\right) F_T \qquad (2.2)$$

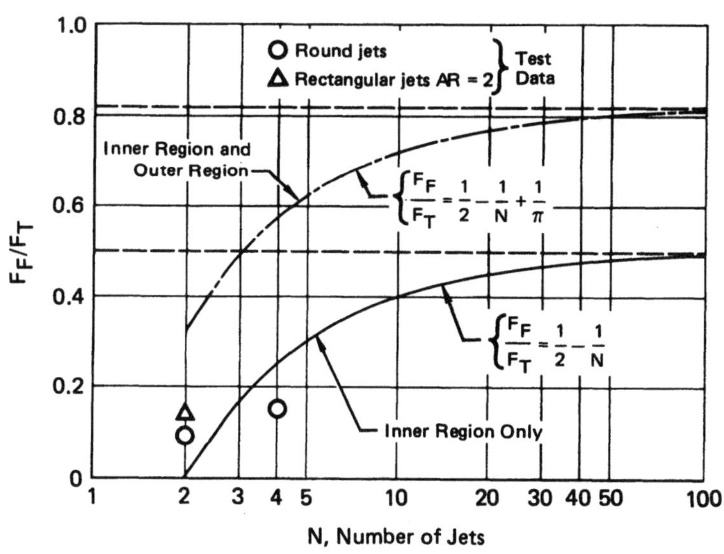

Figure 5. Variation of maximum ideal fountain force with number of round jets.

The potential benefits to be gained from additional fountain force resulting from the use of increasing numbers of jets up to six or even eight is obvious, but it must be realized that actual fountain forces produced on the airframe undersurface are much less than the ideal values shown in Figure 5 by a factor of one third to one quarter based on the ideal inner plus outer region values. (The reasons for this are discussed later in this article.) Limited data on experimentally measured values of fountain force are shown in Figure 5 for two and four jet configurations. These data were recorded on experimental configurations for which jet entrainment induced suckdown effects were minimal.

Recent interest in certain VTOL and STOL propulsive lift systems, such as thrust augmenting ejector systems and other concepts employing noncircular nozzles, has revealed a need for the understanding of the basic free jet and jet impingement characteristics of jets emanating from rectangular exit area nozzles. Consequently, two experimental programs (Kotansky and Glaze, 1978; 1980) were undertaken to investigate the wall jet characteristics produced by the impingement on a ground plane of jets emanating from low and high aspect ratio rectangular exit area nozzles. The primary purpose of these studies was to experimentally determine the azimuthal distributions of wall jet radial momentum flux about the impingement points of these jets for rectangular nozzles with exit area aspect ratios (L/D) of one, two, three, four, six, and eight for both vertical and oblique impingement.

The high aspect ratio rectangular nozzle data ($4 \leq L/D \leq 8$) obtained in the first investigation (for low nozzle pressure ratios only) (Kotansky et al., 1978) indicated that, unlike axisymmetric jets, the wall jet radial momentum flux distributions for vertical impingement are highly directional, even for the aspect ratio four ($L/D = 4$) nozzle at heights above ground as great as 16 nozzle widths. Consequently, the second study (Kotansky et al., 1980) was undertaken to extend the data base to include rectangular nozzles with exit area aspect ratios of one, two, and three. Wall jet radial momentum flux distributions were obtained for the three low aspect ratio nozzles as a function of jet impingement angle, nozzle exit height above ground and nozzle pressure ratio which included choked and under-expanded nozzle exit flow conditions. The ground flow field computer program originally developed for round jets was updated to include the impingement data obtained for both the high and low aspect ratio rectangular nozzles.

Figure 6 presents a qualitative view of the wall jet radial momentum flux distributions associated with impinging jets issuing from both axisymmetric and rectangular nozzles. The axisymmetric nozzle produces a uniform distribution in vertical impingement $f(\phi) = \text{const.}$ ($\alpha_j = 90°$). In oblique impingement, a peak occurs in the distribution in the $\phi = 0°$ direction with the relative magnitude of the peak increasing with decreasing impingement angle. The rectangular nozzles in vertical impingement, on the other hand, produce a prominent peak of the momentum flux distribution in a direction normal to the long sides of the nozzle exit, as indicated in Figure 6 by the $L/D = 4$ nozzle at $\alpha_j = 90°$. The primary effect of oblique impingement of the rectangular nozzle in pitch (rotation about the nozzle exit minor axis) is to shift the peak to an azimuthal location approximately coincident with the magnitude of the impingement angle. The $L/D = 1$ (square) nozzle at low pressure ratios exhibits characteristics similar to that of the axisymmetric nozzle. At high pressure ratios, however, the square nozzle exhibits characteristics common to both axisymmetric and rectangular nozzles with a primary peak occurring at $\phi = 0°$, as with axisymmetric nozzles, and a secondary peak occurring at an azimuthal angle approximately coincident with the jet impingement angle, as exhibited by rectangular nozzles.

The azimuthal distributions of wall jet radial momentum flux were determined by hot film anemometer surveys for each of the six single rectangular nozzle configurations and are presented in entirety in Kotansky et al. (1978) and Kotansky et al. (1980). This work is summarized in Kotansky et al. (1981a). Selected results are presented herein for an aspect ratio two nozzle for vertical impingement (Figure 7) and for oblique impingement in pitch (Figure 8) (rotation about the nozzle exit minor axis). The abscissa on each graph is the integral of the wall jet velocity squared, multiplied by the radial distance, R, from the jet impingement point to the point of measurement and is representative of the magnitude of the wall jet radial momentum flux. The radial distance, R, is included in order to account for the increase in area with radial distance. This allows comparison of data recorded at different radial stations. In vertical impingement, the wall jet radial momentum flux was measured over one quarter of the periphery of the impingement region from $\phi = 0°$, corresponding to a direction perpendicular to the nozzle exit minor axis, to $\phi = 90°$ corresponding to a direction perpendicular to the nozzle exit major axis. In oblique impingement, the wall jet radial

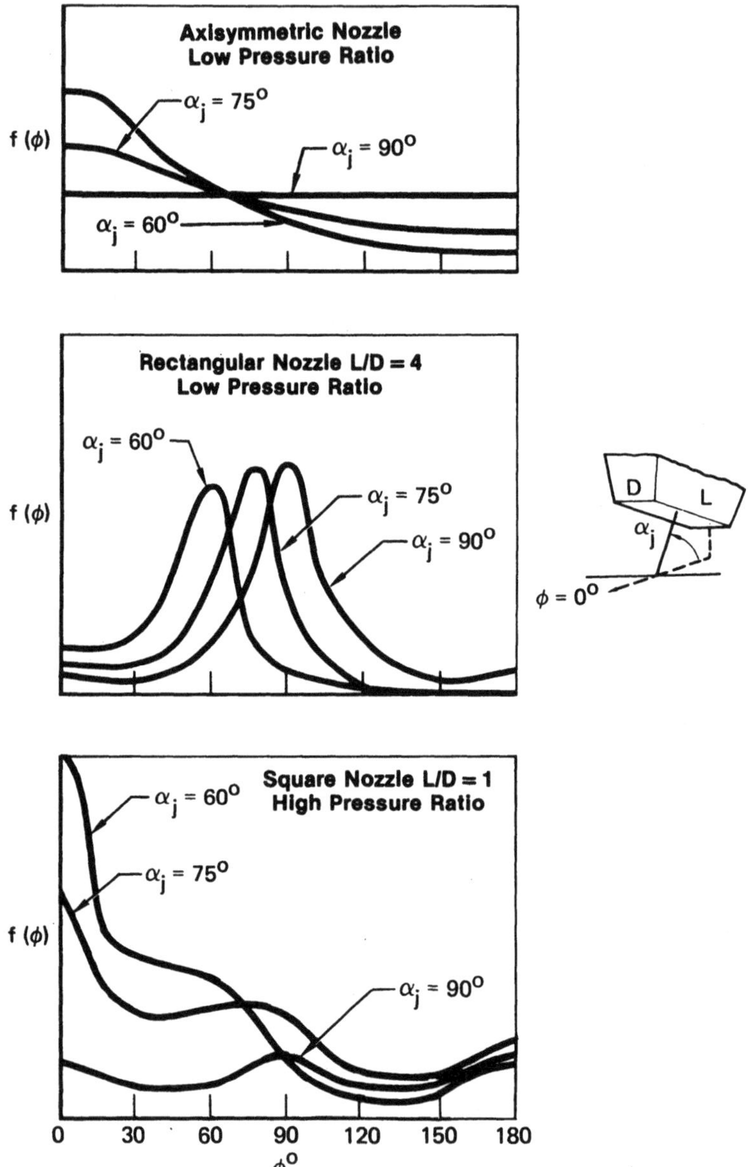

Figure 6. Jet impingement radial momentum flux distributions.

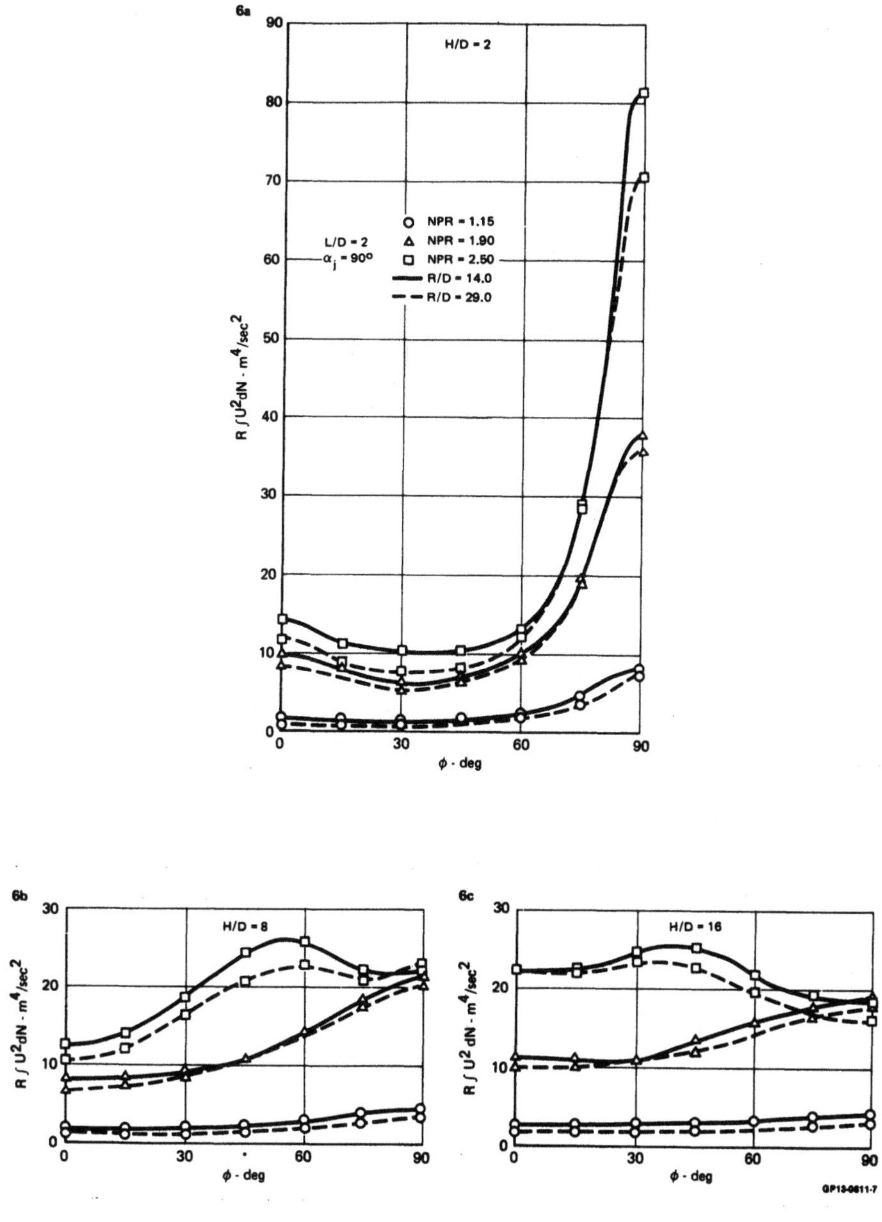

Figure 7. Azimuthal distribution of wall jet radial momentum flux; $L/D = 2$ nozzle, vertical impingement.

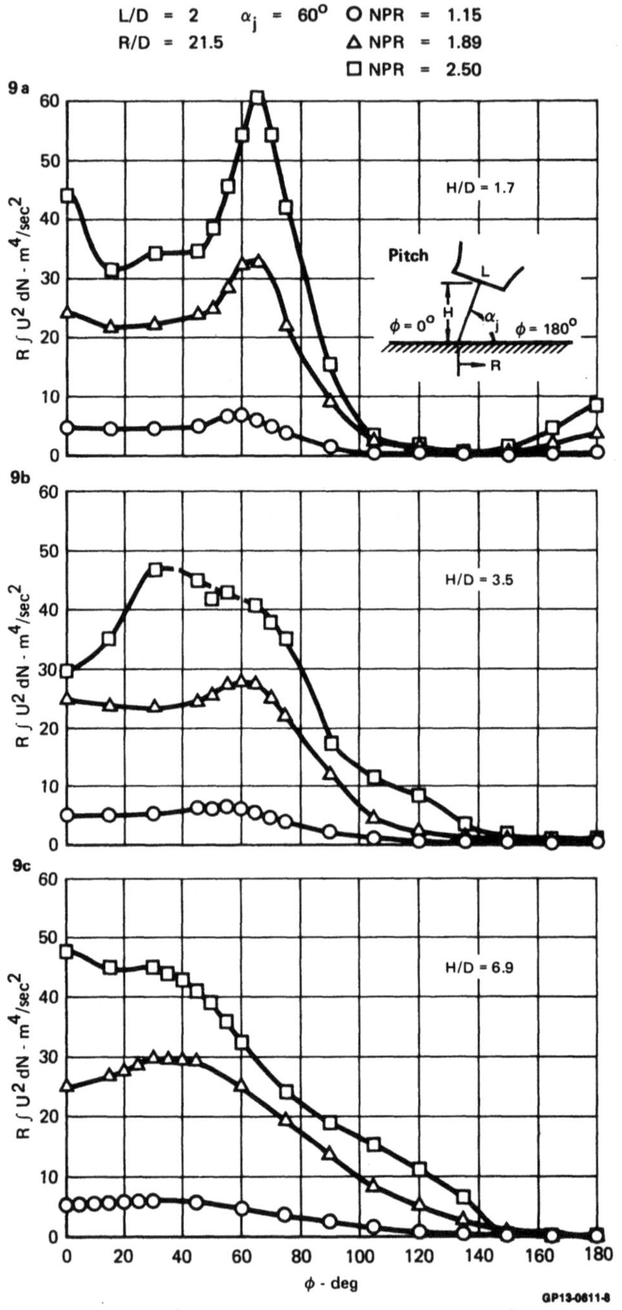

Figure 8. Azimuthal distribution of wall jet radial momentum flux; $L/D = 2$ nozzle, oblique impingement—pitch.

momentum flux was measured over one half of the periphery of the impingement region from $\phi = 0°$ (corresponding to the direction of the horizontal component of the free jet mean flow) to $\phi = 180°$.

In vertical impingement, the momentum flux distributions for the higher aspect ratio nozzles ($2 \leq L/D \leq 8$) all exhibited similar characteristics at low nozzle pressure ratios with a peak at $\phi = 90°$ (perpendicular to the nozzle exit major axis) as exemplified by the aspect ratio two nozzle results shown in Figure 7. However, at NPR = 2.50, although the aspect ratio two (and three) nozzles displayed the characteristic peak at $\phi = 90°$ for $H/D = 2$; at $H/D = 8$ the momentum flux distribution peak became broader and shifted to approximately $\phi = 60°$; and at $H/D = 16$, the peak shifted to approximately $\phi = 45°$. A minor peak is exhibited at $\phi = 0°$ (perpendicular to the nozzle exit minor axis) for the higher pressure ratios at $H/D = 2$, but vanishes at $H/D = 8$.

For oblique impingement in pitch, at low nozzle pressure ratios, the occurrence of a peak in the distribution of momentum flux at an azimuthal angle approximately coincident with the magnitude of the impingement angle is characteristic of rectangular nozzles ($L/D > 1$) and is exhibited by the $L/D = 2$ nozzle in Figure 8. The data in Figure 8 for the aspect ratio two nozzle indicate a broadening and a shift of the location of the momentum flux peak toward $\phi = 0°$ at the higher H/D's for a nozzle pressure ratio of 2.50. Similar behavior was observed for the aspect ratio three nozzle. These data indicate the sensitivity of the jet impingement flow field to nozzle exit area shape and, additionally, nozzle pressure ratio. Data for oblique impingement in roll (rotation about the nozzle exit major axis) for all nozzles may be found in the previously mentioned references.

3. Computation of Stagnation Lines and Fountain Upwash Inclination

In Kotansky et al. (1977), an analytical model was developed for the computation of wall jet stagnation lines formed between any pair of impinging jets for completely arbitrary conditions. Required input for these computations is simply the jet thrust scaled azimuthal distributions of radial momentum flux (as given in Figures 3, 7, and 8 for example). This model is termed the *Momentum Flux Method*.

The Momentum Flux Method established the location of the stagnation line between two jet impingement points by balancing the total momentum flux in the wall jets in a direction locally normal to the stagnation line in the ground plane. As a result of the requirement of a total momentum flux balance (normal to the stagnation line), the upward direction of the fountain in a vertical plane normal to stagnation line must itself be vertical, although a nonvertical sidewash component is allowed.

Green (1978) modified the criterion for determination of the location of the stagnation line by requiring a balance of wall jet momentum flux per unit area of the wall jet at the stagnation line. The balance of momentum flux per unit area is imposed in a direction locally normal to the stagnation line in the ground plane. Imposition of this criterion results in an imbalance of total wall jet momentum flux at the stagnation line in a direction normal to the stagnation line in the ground plane and, consequently, allows a nonvertical trajectory of the fountain upwash flow in a vertical plane normal to the stagnation line. This model for the determination of the stagnation line location is denoted as the *Momentum Flux Density Method.* The basic equations for the *Momentum Flux Method* (MFM) and the *Momentum Flux Density Method* (MFDM) models are compared in Table 1 (see Figure 9 for nomenclature definition).

Momentum Flux Method (MFM)	Momentum Flux Density Method (MFDM)
Stagnation Line Slope:	
$\tan\theta = \dfrac{\beta \sin\phi_2'}{1+\beta\cos\phi_2'}$	$\tan\theta = \dfrac{\beta \sin\phi_2'}{1+\beta\cos\phi_2'}$
where	where
$\beta = \dfrac{R_1}{S'}\sqrt{\dfrac{\gamma_2 f_2(\phi_2)\dot{M}_{je2}/\gamma_1 f_1(\phi_1)\dot{M}_{je1}}{R_2/R_1}} + \dfrac{R_2}{S'}$	$\beta = \dfrac{R_1}{S'}\left[\dfrac{\gamma_2 f_2(\phi_2)\dot{M}_{je2}/\gamma_1 f_1(\phi_1)\dot{M}_{je1}}{R_2^2/R_1^2}\right] + \dfrac{R_2}{S'}$
On the Line Joining the Jet Impingement Points:	
$\dfrac{R_2}{R_1} = \dfrac{\dot{M}_{je2} f_2(\phi_2)\gamma_2}{\dot{M}_{je1} f_1(\phi_1)\gamma_1}$	$\dfrac{R_2}{R_1} = \sqrt{\dfrac{\dot{M}_{je2} f_2(\phi_2)\gamma_2}{\dot{M}_{je1} f_1(\phi_1)\gamma_1}}$

GP13-0611-17

Table 1. Two methods for the computation of stagnation line location.

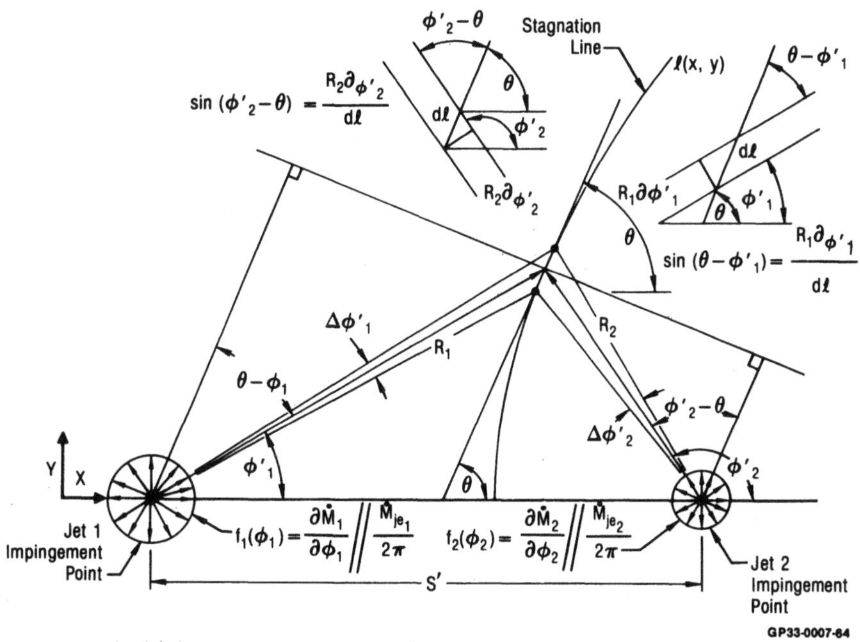

Figure 9. Wall jet interaction geometry.

Computations of wall jet stagnation lines using the MFM and the MFDM have been compared with actual stagnation line locations obtained experimentally through flow visualization techniques, and the results were presented, for example, in Kotansky et al. (1980). Based on these and other comparisons, the MFDM was found to give a more accurate prediction of wall jet stagnation line location. Utilizing the nomenclature as shown in Figure 9, the equation for the slope of the stagnation line in the ground plane for the MFDM is the following:

$$\tan\theta = \frac{\beta\sin\phi_2'}{1+\beta\cos\phi_2'} \tag{3.1}$$

where

$$\beta = \frac{R_1}{S'}\left[\frac{\gamma_2 f_2(\phi_2)\dot{M}_{je_2}/\gamma_1 f_1(\phi_1)\,\dot{M}_{je_1}}{R_2^2/R_1^2}\right] + \frac{R_2}{S'} \tag{3.2}$$

and, at the line joining the jet impingement points in the ground plane,

$$\frac{R_2}{R_1} = \sqrt{\frac{\dot{M}_{je_2} f_2(\phi_2)\gamma_2}{\dot{M}_{je_1} f_1(\phi_1)\gamma_1}} \tag{3.3}$$

Through the use of a control volume located on the stagnation line and a total momentum flux balance in a direction normal to the stagnation line in the ground plane, a relation between the momentum flux quantities in the two interacting wall jets and the horizontal component of momentum flux exiting the control volume in the fountain upwash is obtained (for negligible sidewash):

$$\begin{aligned}\frac{\gamma_1 f_1 \dot{M}_{je_1}}{2\pi}\sin(\theta-\phi_1')\partial\phi_1' - \frac{\gamma_2 f_2 \dot{M}_{je_2}}{2\pi}\sin(\phi_2'-\theta)\,\partial\phi_2' \\ + \rho V_f \cos\omega[U_{R_1}R_1\partial\phi_1' + U_{R_2}R_2\partial\phi_2']h = 0\end{aligned} \tag{3.4}$$

where V_f is the fountain upwash velocity, ω is the fountain inclination angle and h is the wall jet height. Assuming a total momentum flux magnitude conservation factor $\lambda_{\dot{M}}$ in the fountain formation region:

$$\begin{aligned}&\rho V_f[U_{R_1}R_1\partial\phi_1' + U_{R_2}R_2\partial\phi_2']\,h \\ &= \lambda_{\dot{M}}\left[\gamma_1 f_1 \dot{M}_{je_1}\frac{\partial\phi_1'}{2\pi} + \gamma_2 f_2 \dot{M}_{je_2}\frac{\partial\phi_2'}{2\pi}\right]\end{aligned} \tag{3.5}$$

Substituting equation (3.5) into equation (3.4), and then solving for $\lambda_{\dot{M}}\cos\omega$, the following general result is obtained for any location on the stagnation line:

$$\lambda_{\dot{M}}\cos\omega = \frac{\dfrac{\frac{\gamma_2 f_2 \dot{M}_{je_2}}{R_2}\sin^2(\phi_2'-\theta)}{\frac{\gamma_2 f_2 \dot{M}_{je_2}}{R_2}\sin(\phi_2'-\theta)}}{\dfrac{-\frac{\gamma_1 f_1 \dot{M}_{je_1}}{R_1}\sin^2(\theta-\phi_1')}{+\frac{\gamma_1 f_1 \dot{M}_{je_1}}{R_1}\sin(\theta-\phi_1')}} \tag{3.6}$$

For the special case of two round jets impinging vertically, on the line joining the jet impingement points,

$$\lambda_{\dot{M}} \cos\omega = \frac{1 - \sqrt{\dot{M}_{je_1}/\dot{M}_{je_2}}}{1 + \sqrt{\dot{M}_{je_1}/\dot{M}_{je_2}}} \tag{3.7}$$

Figure 10 demonstrates the relation between ω and the impinging jet pair momentum flux ratio as obtained from equation (3.7) with $\lambda_{\dot{M}}$ as a parameter. The validity of this model of fountain upwash flow inclination was investigated and verified with experimental data from the test program described in Kotansky et al. (1981b) and is summarized briefly in the paragraphs following.

An analytical model for the fountain sidewash inclination was presented in Kotansky et al. (1977). The result is

$$\tan\kappa = \frac{2}{\cot(\theta - \phi_1') + \cot(\phi_2' - \theta)} \tag{3.8}$$

where κ is the fountain sidewash inclination angle, and the other symbols are defined in Figure 9.

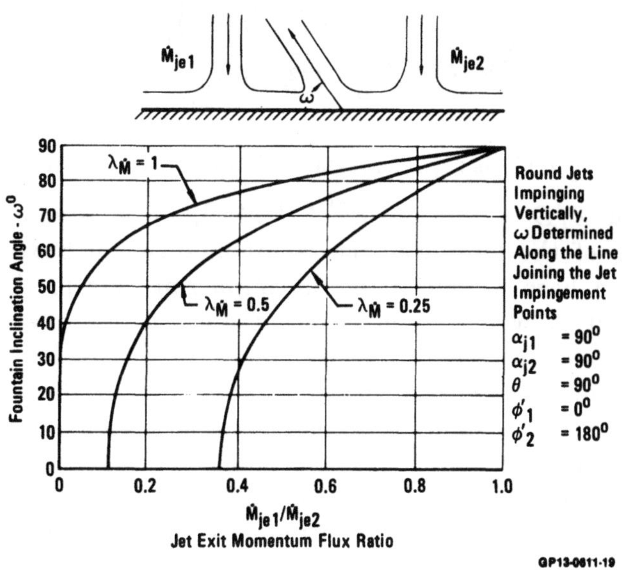

Figure 10. Fountain inclination angle.

4. Fountain Upwash Formation and Development

It has been observed consistently that fountain forces are substantially smaller than even conservatively estimated values based on lift jet thrust and momentum conservation. This is demonstrated quantitatively by the experimental data shown in Figure 5. For this reason, a thorough investigation of the fountain upwash formation and development process was conducted by MCAIR for fountains formed by two impinging round jets for a wide range of parametric test conditions. These results are reported in entirety in Kotansky et al. (1981b). An important result of this investigation was the quantification of a fountain upwash formation normal momentum flux recovery coefficient, $\lambda_{\dot{M}N}$.

The fountain upwash normal momentum flux recovery factor ($\lambda_{\dot{M}N}$) is defined as the ratio of the fountain normal momentum flux exiting the fountain formation region to the total wall jet radial momentum flux entering the fountain formation region (see Figure 1). In the subject investigation, the flow throughout the fountain region was incompressible ($\rho =$ constant), and the fountain formation region was assumed to be small so that radial area change effects were negligible ($dR \simeq 0$). Thus

$$\lambda_{\dot{M}N} = \frac{\left(\int_{X_0}^{X_f} V_N^2 \, dX\right)_f}{\left(\int_0^{N_1} U_R^2 \, dN\right)_{wj(1+2)}} \tag{4.1}$$

where $(\int_{X_0}^{X_f} V_N^2 dx)_f$ is representative of the fountain normal momentum flux at $Z/D = 1.0$ and $(\int_0^{N_1} U_R^2 dN)_{wj(1+2)}$ is representative of the sum of the local wall jet radial momentum fluxes entering the fountain base produced by the impinging jets in the system. A typical set of wall jet and fountain upwash flow mean velocity surveys for one test condition are shown in Figure 11. The surveys were obtained with a hot film anemometer. Since the local velocity ratios, $V_{\min}/V_{\max}$, for the fountain velocity profiles were considerably higher than those determined for the wall jet velocity profiles, the following procedure was established to define the limits of integration and to render consistency to the calculation process.

The upper limits of integration of the wall jet velocity profiles were established for each case by the normal distance, N_1, corresponding to the larger velocity ratio, $U_{\min}/U_{\max}$, of the two wall jet velocity profiles. The wall jet velocity and velocity squared profiles for jet 1 and jet 2 were then integrated and summed to yield $(\int_0^{N_1} U_R\, dN)_{wj}$ and $(\int_0^{N_1} U_R^2\, dN)_{wj}$. Now, since the fountain formation region is assumed to be small, with little exposed jet area available for mass entrainment, conservation of mass was assumed in the fountain formation region to yield

$$\left(\int_{X_0}^{X_f} V_N\, dX\right)_{f_{Z/D=1}} = \left(\int_0^{N_1} U_R\, dN\right)_{wj\,(1+2)} \tag{4.2}$$

Thus, the initial and final limits of integration (X_0, X_f) for the fountain velocity and velocity squared profiles were determined such that equation (4.2) was satisfied.

Figure 12 presents the fountain normal momentum flux recovery factor, $\lambda_{\dot{M}N}$ as a function of the jet exit momentum flux ratio, $\dot{M}_{je_{Low}}/\dot{M}_{je_{High}}$, for the data of Kotansky et al. (June 1980). The data indicate a general decrease in the fountain momentum flux recovery with a decrease in the nozzle thrust bias (increased $\dot{M}_{je_{Low}}/\dot{M}_{je_{High}}$). In addition, a slight increase in $\lambda_{\dot{M}N}$ is shown in the presence of a nozzle exit plane plate over that found without the plate. A polynomial curve-fit (also shown in Figure 12) was determined for the data and is given approximately by the following expression:

$$\lambda_{\dot{M}N} = 1.0 - 1.5\left(\frac{\dot{M}_{je_{Low}}}{\dot{M}_{je_{High}}}\right) + 1.5\left(\frac{\dot{M}_{je_{Low}}}{\dot{M}_{je_{High}}}\right)^2 - 0.5\left(\frac{\dot{M}_{je_{Low}}}{\dot{M}_{je_{High}}}\right)^3 \tag{4.3}$$

The behavior of the normal momentum flux recovery factor with the jet exit momentum flux ratio is not surprising. The strongest wall jet interaction and associated loss of mean flow energy or momentum occurs with two equal strength impinging jets. As one of the impinging jets becomes weaker (increased thrust bias), the wall jet interaction becomes weaker. For a high thrust bias, the weaker wall jet simply tends to deflect the stronger jet with an attendant reduction in mixing and in loss of mean flow energy.

Fountain geometric spreading characteristics and fountain mass flow were determined from the fountain upwash velocity profile data

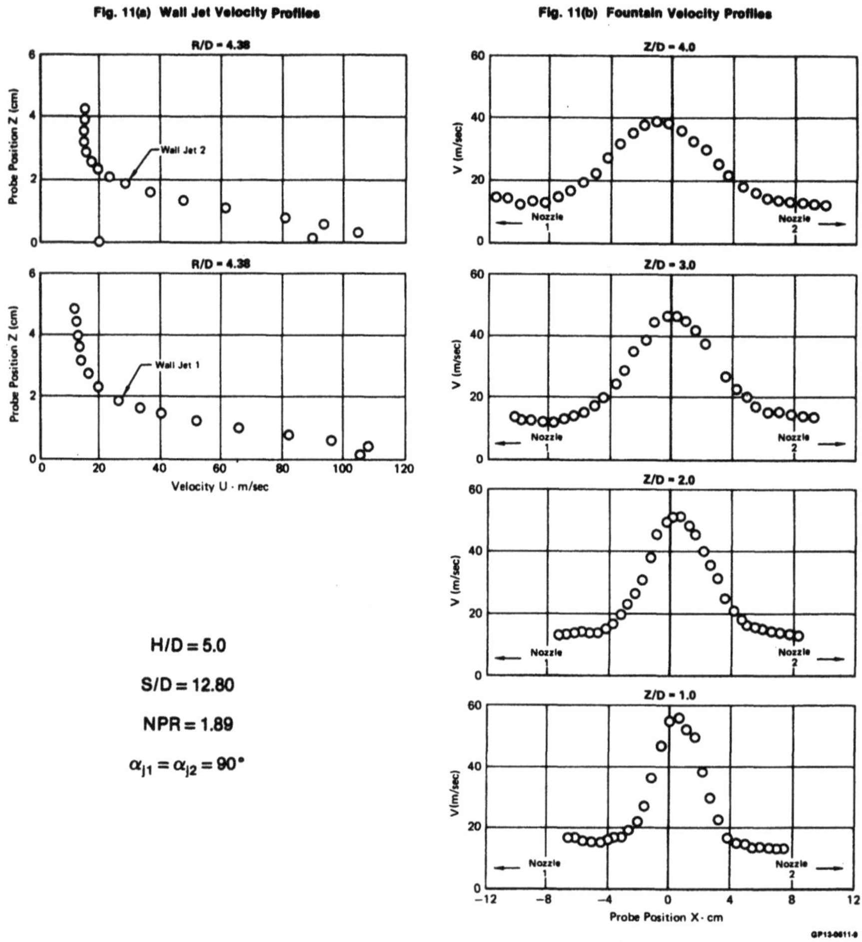

Figure 11. Symmetrical two-jet fountain.

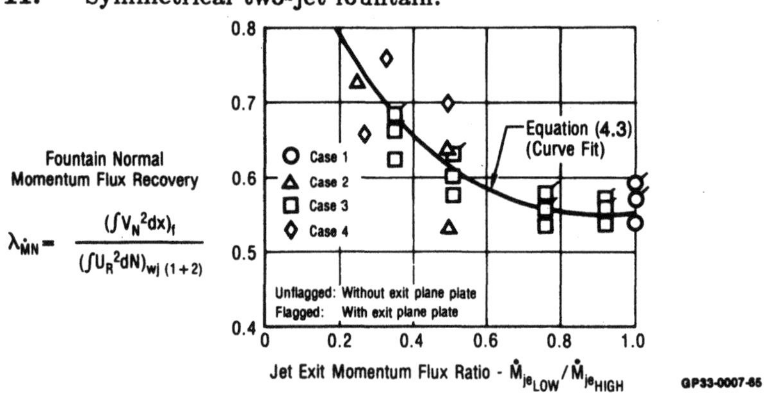

Figure 12. Fountain normal momentum flux recovery, based on conservation of mass flux through the fountain formation region.

for all of the cases investigated. For the data of Figure 11, the fountain spreading characteristics and mass flow are shown in Figure 13. Fountain spreading characteristics for four jet fountains formed from four round jets oriented in an approximately square array are shown in Figure 14. $R_{V/2}$ is an averaged half velocity radius obtained from hot film anemometer surveys of the fountain upwash at a number of azimuthal positions. The parametric data shown (Cases 10–13) represent variations in nozzle pressure ratio for pairs of jets in the array.

Fountain upwash trajectories were also determined for all of the cases investigated. The trajectories were defined as the loci of the maximum velocity points as determined from the fountain upwash velocity profiles. The fountain upwash inclination, ω, was determined as the angle between the ground plane and a straight line connecting the fountain base and the fountain upwash trajectory at $Z/D = 3$. These fountain upwash inclination data are plotted as a function of the jet exit momentum flux ratio $(\dot{M}_{je_{\text{Low}}}/\dot{M}_{je_{\text{High}}})$ in Figure 15. Also shown in the figure are the theoretical values of the fountain inclination, ω_{TH}, based on equation (3.7). Figure 15 indicates that for a fountain inclination determined at $Z/D = 3$, a momentum flux recovery factor, $\lambda_{\dot{M}}$, of approximately 0.65 results in a reasonable fit to the experimental data.

5. Fountain Impingement on the Airframe

The establishment of the above ground flow field modelling elements, together with recent improvements made to three-dimensional panel methods for the determination of jet entrainment induced suckdown, leaves only one empirical element remaining to complete the prediction methodology for a wide range of VTOL aircraft configurations. This remaining element is the quantification of the amount of momentum flux transferred to the airframe undersurface from the upwash momentum flux in the fountain. This quantity, λ_{fI}, is termed the fountain upwash momentum flux transfer coefficient. λ_{fI} is dependent on airframe undersurface shape and aircraft height above ground, H/D.

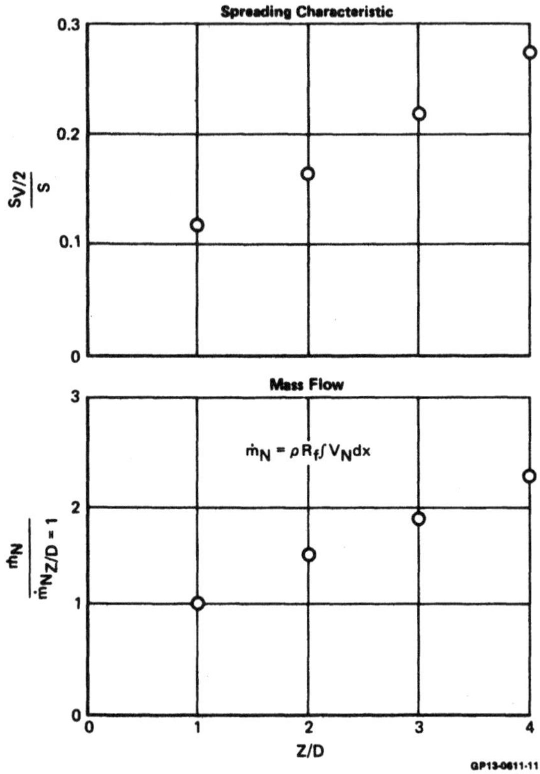

Figure 13. Fountain upwash characteristics symmetrical two-jet fountain.

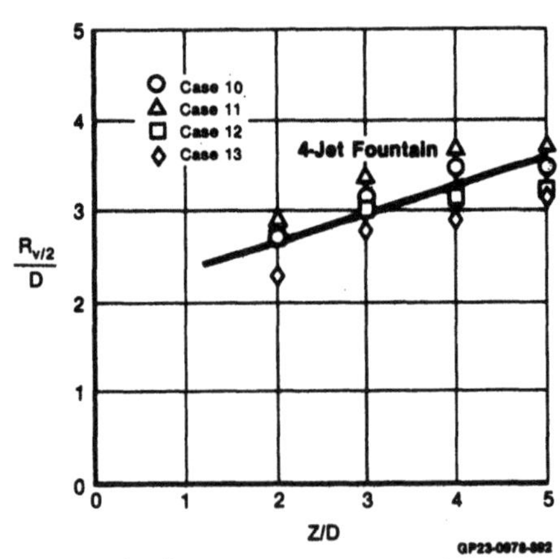

Figure 14. Fountain spreading characteristics four-jet fountains.

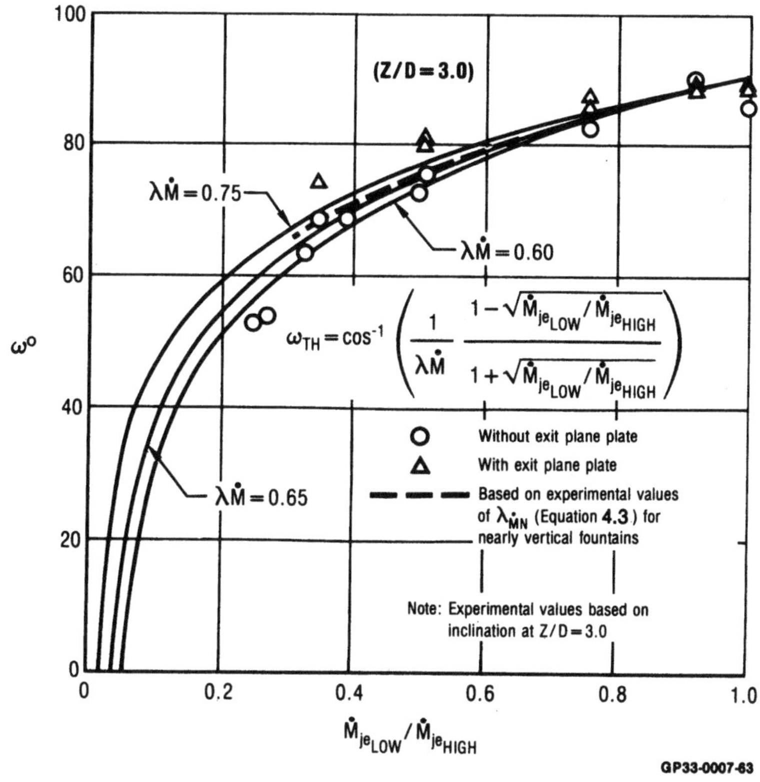

Figure 15. Two-jet fountain inclination—vertical impingement.

To demonstrate the usefulness of the existing methodology, fountain impingement forces were determined and compared with experimental data for simple bodies and planforms for which λ_{fI} was expected to be approximately unity and for which suckdown forces were expected to be insignificant. Results are shown in Figure 16 and were very encouraging. The deviations between the predictions and the data at H/Ds of 1 and under are, most likely, beyond the ability of the prediction methodology due to the basic changes in the flow structure beneath a body at these low H/Ds. That is, distinct regions of free jet flow, jet impingement, wall jets, fountain formation, and fountain upwash most likely do not exist at these conditions.

Fountain impingement forces measured on a swept low wing/ fuselage combination are shown in Figure 17. The data shown here were produced by the impingement of single and multiple radial jets

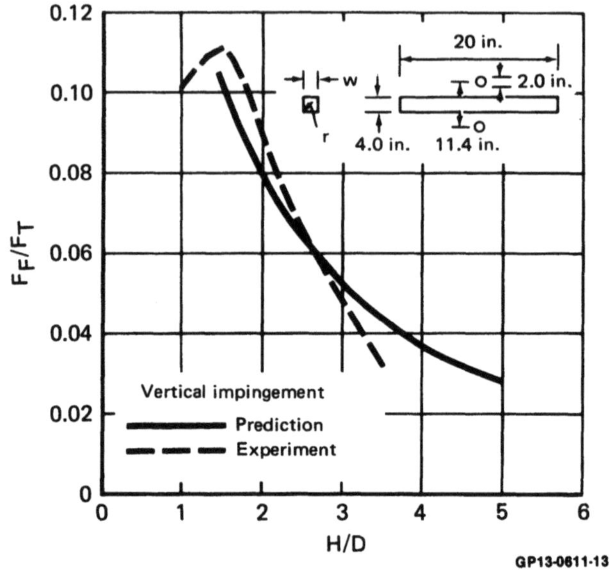

Figure 16. Accuracy of two-jet fountain impingement force prediction.

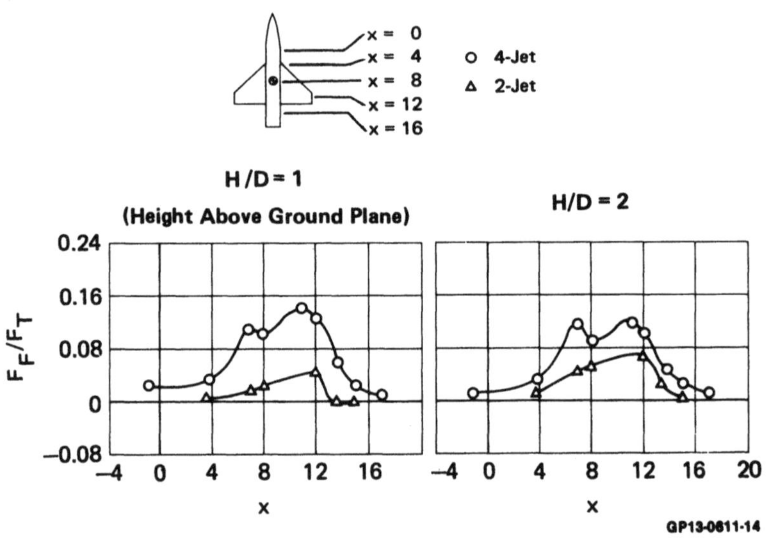

Figure 17. Fountain impingement test 2 and 4-jet fountain impingement on a fuselage/wing.

simulating fountain upwash flows produced by the vertical impingement on a ground plane of two and four round jets all with equal thrust. The normal force data shown have been nondimensionalized on the total thrust of the simulated jet lift system. In the two jet case, the plane of the radial jet was perpendicular to the aircraft longitudinal axis, and in the simulated four jet case, the radial jet planes were mutually orthogonal and aligned parallel to the aircraft longitudinal and lateral axes. The data in Figure 17 show the variation in net fountain force with longitudinal location of the center of the fountain as the fountain impingement region was moved from the forward fuselage to aft of the wing trailing edge. As would be expected, the maximum fountain force results when the lateral fountain legs impinge on the widest portion of the wing. A noticeable drop in fountain force magnitude occurs when the lateral fountain legs are moved aft of the wing trailing edge. These data also indicate the relative magnitudes of two versus four jet fountains. The maximum points on these curves have been plotted in Figure 5 for comparison with the theoretical maximum fountain force. The indicated increase in fountain force with increasing H/D for the two jet fountain is due to variation of fountain induced suckdown in this test installation which decreased as the aircraft model was raised above the ground. The comparative decrease in fountain force with H/D produced by the four jet fountain is characteristic of fountains with three or more jets which form a central core of upwash flow.

Fountain impingement momentum flux transfer coefficients, λ_{fI}, were determined for three realistic V/STOL aircraft configurations for which laboratory test data in ground effect were available. Fortunately, data were available for a two jet, a three jet, and a four jet configuration. λ_{fI} was determined as a function of height above ground by using the methodology presented herein to *back* λ_{fI} from the measured lift on the aircraft models in ground effect. This was accomplished via the following steps:

1. The suckdown force, F_s, due to the entrainment of both the free jets and the wall jets was calculated using an advanced panel method,

2. The fountain upwash momentum flux incident on the aircraft undersurface, $\dot{M}_{fI}$, was calculated using the ground flowfield methodology presented herein, and

3. λ_{fI} was calculated using the relation

$$\lambda_{fI} \dot{M}_{fI} = \Delta L - F_s \quad (5.1)$$

where ΔL was obtained from the laboratory test measurements on the three aircraft models.

The three aircraft configurations for which λ_{fI} was determined are shown in Figure 18. The computed values of λ_{fI} are shown in Figure 19. A detailed presentation of the application of the methodology to obtain these results may be found in Glaze et al. (1983). The results shown in Figure 19 clearly indicate the sensitivity of λ_{fI} to aircraft configuration geometry and aircraft height above ground. These results also clearly indicate the benefits of lift improvement devices.

As shown in Figure 19, additional lift may be obtained from fountain upwash impingement through the use of airframe undersurface protuberances in the form of fences or strakes which tend to capture the fountain upwash flow. The confining surfaces should extend through 360° of azimuth in an undersurface plane roughly parallel to the ground plane with the aircraft in the proper pitch and (zero) roll attitude for vertical translation. Permanently fixed surfaces suitable for high speed flight, such as longitudinal axis strakes and/or pods, can be utilized for this purpose, but lateral axis devices should be retractable. The confined undersurface area should be large enough to *capture* the central core of a multi-jet fountain including any fountain impact area translation on the airframe undersurface due to nominal aircraft pitch and roll excursions during hover. Sensitivities of lift improvement device related forces and moments to larger excursions in pitch and roll with and without cross winds should be investigated to assure adequate stability and control margins. This currently must be accomplished through wind tunnel testing of powered aircraft models in ground effect. Tests of this nature require accurate modelling of full scale airframe surface geometry including all undersurface contour details and protuberances. It is also desirable that full scale jet exit conditions in terms of nozzle exit geometry, nozzle exit velocity profiles and jet vector and splay angles be simulated carefully in the reduced scale tests.

The beneficial effect of passive lift improvement devices is demonstrated by the induced lift increment (ΔL) build-up data shown in Figure 20 which was measured in ground effect tests performed with a 15% scale powered model of the AV-8A. The data in Figure 20 indicate the positive lift increments obtained through the addition of gun pod strakes and a forward fence to the baseline vehicle which

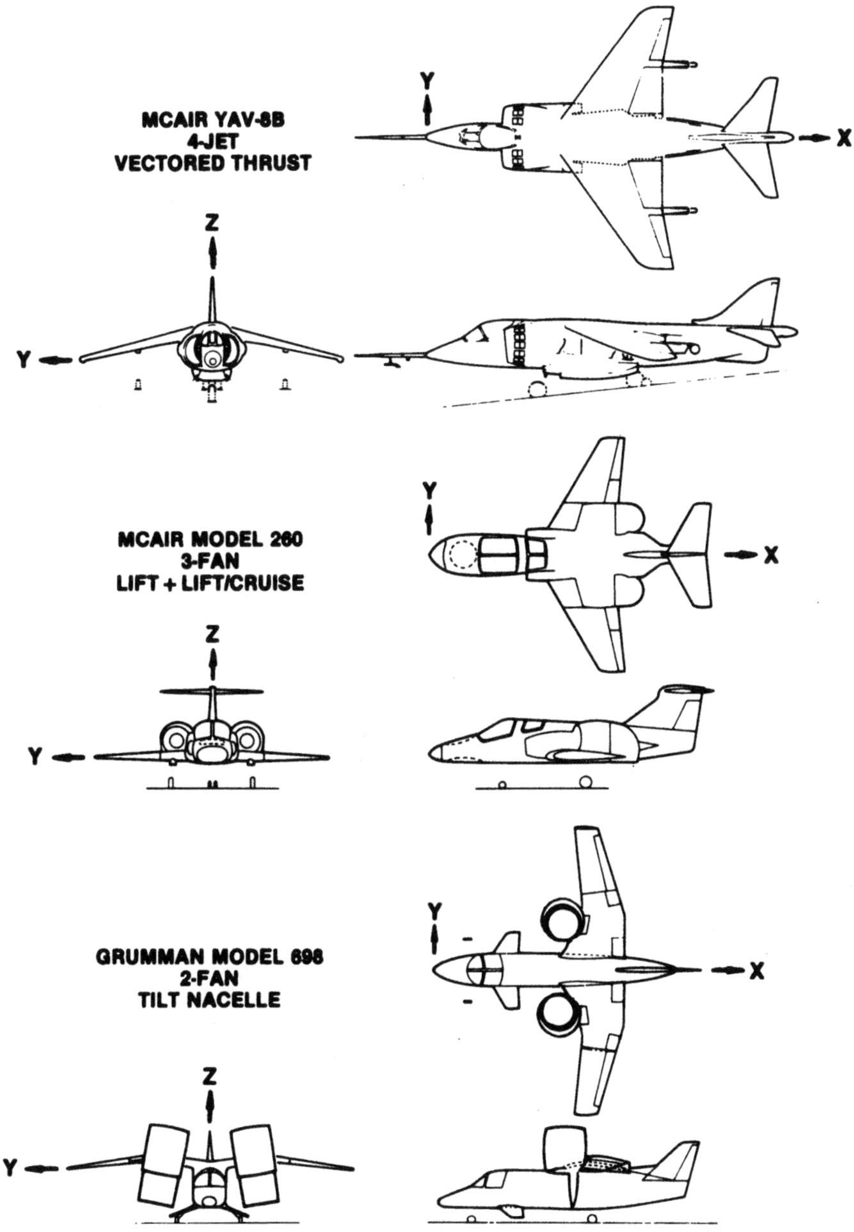

Figure 18. V/STOL aircraft selected for analysis.

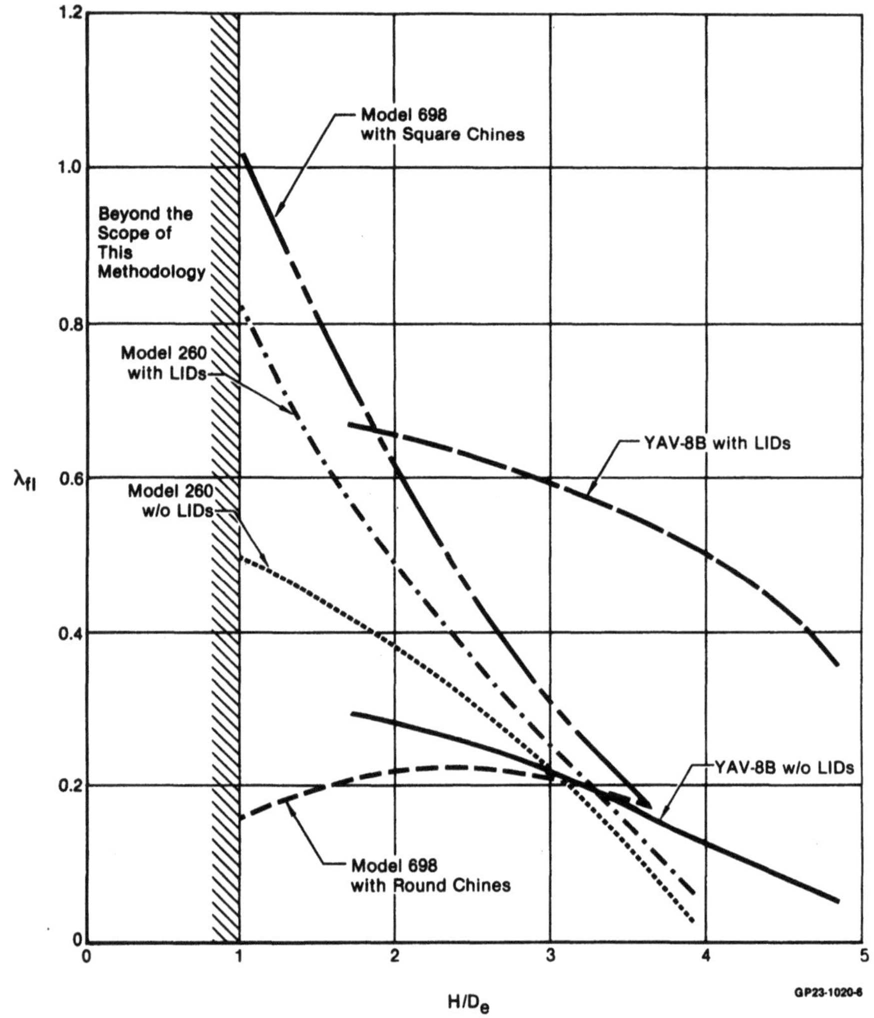

Figure 19. Fountain impingement momentum flux transfer coefficient summary.

included gun pods. The data shown were obtained with all four lift jets vectored perpendicularly to the ground plane, however, the nozzles were splayed outward in the nominal splay position (5% forward nozzles, 11.2° aft nozzles). The sensitivity of the fountain forces to changes in nozzle splay angle (or vector) is demonstrated by the data presented in Figure 21 which shows the additional positive lift increment gained by reducing the outward splay angle of all nozzles on the model to 0°. Similar positive lift increments have been measured on models of other aircraft configurations, and in some cases, positive lift improvement can be obtained by simply deflecting existing flaps on the baseline aircraft. Caution must be used to verify that local upwash flowfield changes caused by lift improvement devices do not result in lift engine exhaust gas ingestion.

The mechanism for the increase in fountain lift force due to the use of lift improvement devices (LIDs) appears to be a combination of two effects: (1) the confining surfaces form a concavity which is pressurized at some pressure increment less than or equal to the maximum dynamic pressure in the fountain upwash flow, and (2) the concavity or portions thereof act as turning vanes and increase the local turning of the impinged upwash flow beyond what would normally be obtained by undersurface impingement without the lift improvement devices. In the latter case, an ideally designed LID would double the beneficial force increment obtained by ideal fountain impingement on a flat airframe undersurface parallel to the ground. In the former case, the LID acts like the skirt on a typical ground effect machine. Both mechanisms are theoretically very sensitive to H/D in all multi-jet fountain situations, and the data in Figures 20 and 21 and other data tend to confirm this.

6. Closing Remarks

This article has emphasized empirical and analytical descriptions of certain key elements of multiple jet viscous flow fields below VTOL aircraft hovering in ground effect. In the actual flowfield below the VTOL vehicle, these flow phenomena occur elliptically, and interactions may occur which can alter the induced effects significantly from the physical descriptions presented herein. Two frequently occurring and strongly influential additional factors are strong cross-winds and aircraft transient motions whose frequency is high enough to preclude quasi-steady analysis based on steady state

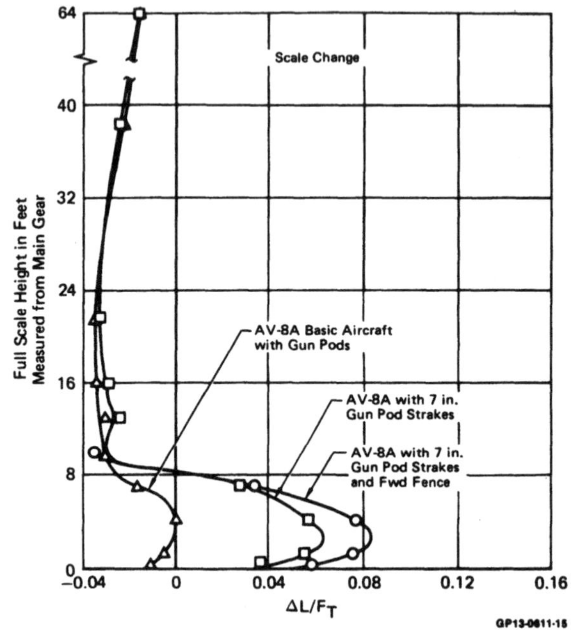

Figure 20. Fountain capture with lift improvement devices.

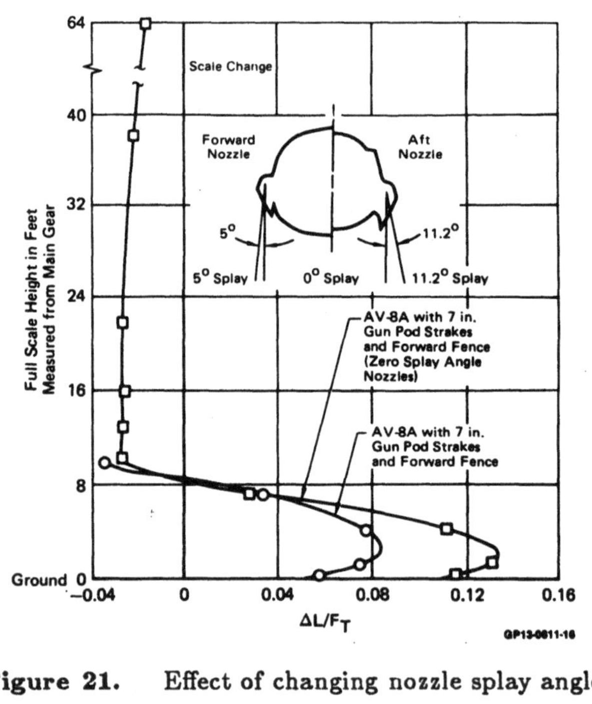

Figure 21. Effect of changing nozzle splay angle.

characteristics. The development of accurate prediction techniques for cross-wind effects on fountain upwash development is extremely complicated, although very important to the identification of both stability and control and exhaust gas ingestion problems. Critical areas requiring research for future V/STOL aircraft applications are summarized in Table 2.

Predictions of three-dimensional (and time dependent) viscous flows of this nature will not be available soon. It is likely that strong dependence on wind tunnel testing will be required for the foreseeable future in the vehicle design and development process. However, as discussed in the previous sections, the configuration analyst can utilize existing data and prediction techniques to guide the configuration synthesis process and to avoid serious problem areas. Kotansky et al. (1975) summarized alternate approaches to V/STOL aircraft flowfield prediction in ground effect and included projections on future prediction technique development.

In view of the frequent necessity for investigation of V/STOL aircraft jet induced effects through wind tunnel tests of small scale powered models in ground effect, it is productive to comment on the basic requirements of fluid dynamic similarity for the purpose

- SINGLE/MULTIPLE IMPINGING JETS WITH LOW SPEED FORWARD VELOCITY OR CROSS-FLOW
- JETS IN TRANSONIC/SUPERSONIC FLOWS INCLUDING JETS IN SUPERSONIC CROSS-FLOWS
- GRAZING IMPINGEMENT OF JETS ON AIRCRAFT FUSELAGE OR STABILITY AND CONTROL SURFACES
- INCLUSION OF ELEVATED JET TEMPERATURE EFFECTS IN ALL OF THE ABOVE
- BASIC STATIC IMPINGEMENT OF SUPERSONIC JETS ISSUING FROM CONVERGENT/DIVERGENT NOZZLES

GP23-0660-2

Table 2. Critical areas requiring research for V/STOL aircraft applications, jet flowfields.

of model testing. The first of these is simply exact, detailed scaling of the small scale model from the full-scale vehicle including all major protuberances, stores, and landing gear. Kinematic scaling of external and internal flows requires that primary velocity ratios (free stream to jet) be correctly simulated, and this would be accounted for automatically if ideal Reynolds number and Mach number simulation were achieved. However, in reality, Reynolds number and Mach number simulation are rarely achieved; in lieu of these, further application of kinematic scaling is desirable. For example, just as airframe undersurface geometry and nozzle exit geometry should be scaled directly, the nozzle exit flow velocity profiles should be representative of what might be expected from the full scale vehicle. This has a direct influence on jet entrainment induced suckdown and fountain formation and development. Kinematic scaling or simulation of jet exit velocity profiles in this manner may be thought of as a pseudo-Reynolds number and internal geometric simulation. Additionally, inlet suction flows should be included in the small scale model simulation in terms of inlet velocity ratio based on jet exit and/or free stream velocity. Nozzle exit Mach number simulation should be achieved if nozzle exit Mach numbers are supersonic or if nozzle exits are significantly under-expanded as shown dramatically by the data of Figures 7 and 8.

References

[1] Donaldson, C. du P. and Snedeker, R. S. "A Study of Free Jet Impingement, Part I—Mean Properties of Free and Impinging Jets," *Journal of Fluid Mechanics, 45, Part 2* (1971), 281–319.

[2] Glaze, L. W., Bristow, D. R., and Kotansky, D. R. "V/STOL Fountain Force Coefficient," *Naval Air Development Center Report No. NADC-81106-60,* January 1983.

[3] Green, K. A. Private correspondence, Naval Air Development Center to D. R. Kotansky, August 24, 1978.

[4] Kleis, S. J. and Foss, J. F. "The Effect of Exit Conditions on the Development of an Axisymmetric Turbulent Free Jet," *Third Year Technical Report, NASA Grant NGR 23-004-068,* Michigan State Univ., May 15, 1974.

[5] Kotansky, D. R. and Bower, W. W. "Forces and Moments Produced on a Two-Dimensional Body in a Strong Lift-Jet/Airframe/Ground Interaction," presented at the Workshop on Prediction Methods for

Jet V/STOL Propulsion Aerodynamics, Naval Air Systems Command, Arlington, VA., July 28–31, 1975, *1,* 288–303.

[6] Kotansky, D. R., et al. "Multi-Jet Induced Forces and Moments on VTOL Aircraft Hovering In and Out of Ground Effect," *Final Technical Report,* Naval Air Development Center, *NADC Report No. 77-229-30,* June 1977.

[7] Kotansky, D. R. and Glaze, L. W. "Investigation of Impingement Region and Wall Jets Formed by the Interaction of High Aspect Ratio Lift Jets and a Ground Plane," *NASA CR 152174,* Ames Research Center, Sept. 1978.

[8] ——————————. "Investigation of the Interaction of Lift Jets and a Ground Plane," *NASA CR 152343,* Ames Research Center, April 1980.

[9] ——————————. "Characteristics of Wall Jets Produced by the Impingement on a Ground Plane of Rectangular Jets of Aspect Ratio One Through Eight," *AIAA Paper No. 81-0012,* 19th Aerospace Sciences Meeting, St. Louis, Mo., January 12–15, 1981a.

[10] ——————————. "Investigation of the Effects of Ground Wall Jet Characteristics on Fountain Upwash Flow Formation and Development," *Report No. ONR-CR-212-261-IF,* Office of Naval Research, June 1980; also *AIAA Paper No. 81-1294,* 14th Fluid and Plasma Dynamics Conference, Palo Alto, CA., June 23–25, 1981b.

Recent Advances in Prediction Methods for Jet-Induced Effects on V/STOL Aircraft*

Ramesh K. Agarwal[†]
McDonnell Douglas Corporation
St. Louis, MO 63166

Abstract

This paper summarizes the currently used methodologies in aircraft industry for predicting forces and moments on a V/STOL aircraft in hover and transition modes of flight. These methodologies are based on a synthesis of various flow regions each of which accommodates a specific flow phenomena such as jet-ground-interactions, jet-in-crossflow, and fountain-airframe impingement. The progress made in recent years in theoretical modeling of the flowfield of representative jet-flow configurations: single-jet impingement, twin-jet impingement with fountain formation, and jet-in-crossflow is surveyed. The prediction methods ranging from semi-empirical approaches to the solution of Reynolds-averaged Navier-Stokes equations are discussed.

1. Introduction

Successful design and development of V/STOL aircraft requires accurate prediction of forces and moments caused by jet-induced effects during the take-off phase of flight. During hover and transition to/from conventional forward flight, extensive aerodynamic interactions occur between the lifting jets and the flow around the aircraft. Turbulent lifting jets mix with their surroundings leading to an induced downflow of air around the aircraft and a resulting suckdown force. At take-off or hovering in ground effect, the lifting jets impinge upon the ground and the resulting wall-jets collide to

---

*This research was conducted under the McDonnell Douglas Independent Research and Development program.

[†]Scientist, McDonnell Douglas Research Laboratories.

form a stagnation region from which a fountain emerges and impinges on the fuselage, causing significant changes in lift and pitching moment. In addition, possible recirculation of hot exhaust gases into the engine intakes could result in loss of thrust and engine performance. Figures 1 and 2 characterize the flowfield about a VTOL aircraft in hover and transition modes of the flight, respectively. Table 1 summarizes the influence of various jet-induced phenomena on aircraft forces and moments.

Table I. Influence of various jet-induced phenomena on aircraft performance.

Jet-flow phenomena	Induced effect on aircraft
1. Turbulent jet entrainment	Lift loss (suck-down)
2. Fountain impingement	Lift gain; fuselage heating
3. Deflection of fountains	Lift and moment sensitivity to Pitch and Roll
4. Deflection of jets and fountains by cross-winds	Lift and moment sensitivity to crossflow
5. Fountain, cross-wind, and buoyant convection of hot exhaust gas	Engine thrust loss from exhaust gas ingestion

Prediction of the turbulent flowfield around arbitrary VTOL aircraft-jet-configurations of the type shown in Figures 1 and 2, using the fundamental equations governing the viscous fluid motion (Navier-Stokes equations), is beyond the current state of the art in computational aerodynamics; it must await advances in three-dimensional grid-generation, robust and fast numerical algorithms, turbulence modeling and computer speed and memory (Kutler, 1983). After a considerable effort during the last decade, only recently has it been possible to calculate the inviscid flowfield around the complete aircraft using the full potential equation (Yu, 1981). Therefore, in VTOL design, jet-induced flowfields around the complete aircraft-jet configuration are predicted using the surface panel method; no other approach can solve the elliptic problem around the arbitrary complex boundaries of the airframe, the jets, and the ground.

In the surface panel method, the inviscid flow field around the aircraft is represented by distributions of elementary singularities (such as source, doublet, or vortex) on geometric panels (Bristow, 1980). The panel methods are widely used today in aircraft design

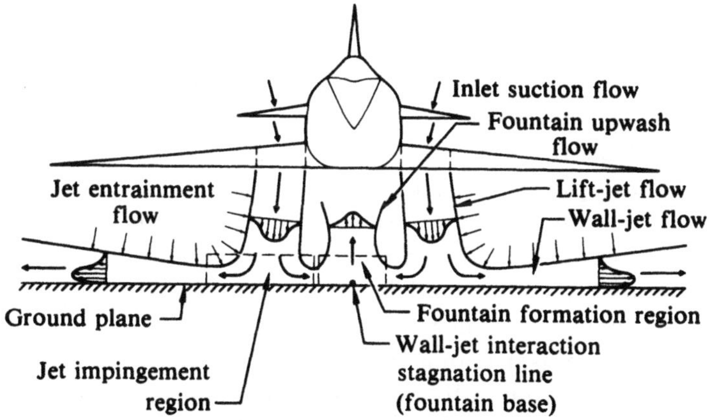

Figure 1. Flowfield about a VTOL aircraft hovering in ground effect.

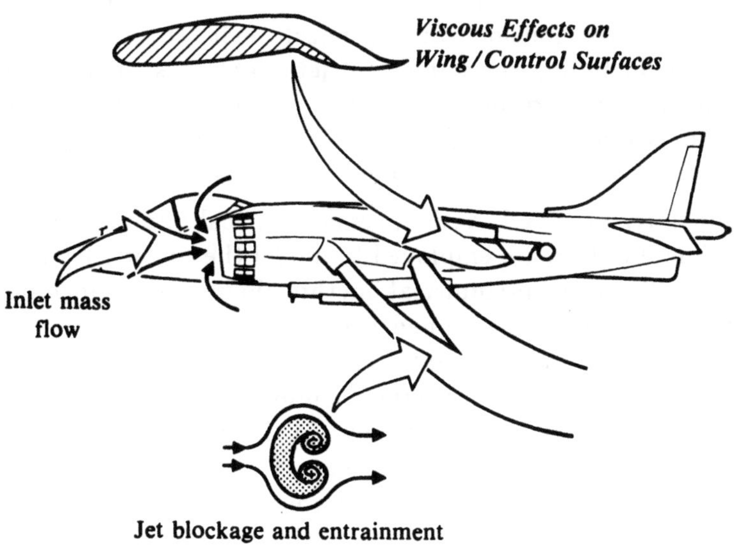

Figure 2. Flowfield about a V/STOL aircraft in the transition or STOL mode of flight.

for subsonic flow conditions. For a VTOL aircraft in hover (with or without ground effect) or transition, the aircraft, the multiple jets, and the ground plane are all subdivided into a large number of panels (for examples Figures 3 and 4) each containing a singularity (source/vortex) distribution. The usual zero normal velocity and Kutta condition for the aircraft surfaces are used, while the jets and the ground-plane wall-jet regions are treated as permeable surfaces through which mass influx is permitted. The prescription of the proper boundary condition on these latter surfaces requires data from individual jet-flow models which accommodate specific flow phenomena such as jet-merging, jet-in-crossflow, jet-ground interactions, or fountain-airframe impingement. Figure 5 shows the information needed from individual jet-flow models for calculating the suckdown force and fountain impingement force on a VTOL aircraft hovering in ground effect. Figure 6 shows the information needed from a jet in cross-flow model for calculating the induced lift on a VTOL aircraft in transition flight.

General modeling of a specific jet-flow phenomenon, such as jet-merging, jet-ground interactions, fountain upwash flow, or jet-in-crossflow, covering a wide range of geometric parameters and flow conditions, is a complex task. In the most general situation, the geometric considerations include (a) the nozzle exit shape and mean velocity distributions at the nozzle exit, (b) lift jet system arrangement including the number of jets, jet exit spacings, and jet axis orientation with respect to aircraft axes (including longitudinal *vector* and lateral *splay* angles), and (c) aircraft surface geometry, orientation, and height above ground.

Currently in VTOL design, semi-empirical approaches guided by experimental data are employed to model the specific jet-induced flowfields. These semi-empirical methodologies, that include the majority of the above-mentioned geometric variables, are updated and refined as more experimental data become available, especially on multi-jet merging and entrainment, multi-jet fountain formation, and multi-jet interference in crossflow. At the present time, however, most of the available experimental data from direct measurement and flow visualization are for a single round jet-normal (Bradbury, 1972; Bradshaw, 1959; Gardon and Akfirat, 1965; Giralt et al., 1977; Poreh et al., 1967; Schauer and Eustis, 1963; Scholtz and Trass, 1970; Tani and Komatsu, 1966) and inclined impingement (Donaldson and Snedeker, 1972; Snedeker and Donaldson, 1965) flowfield, twin-round jet impingement flow with fountain formation

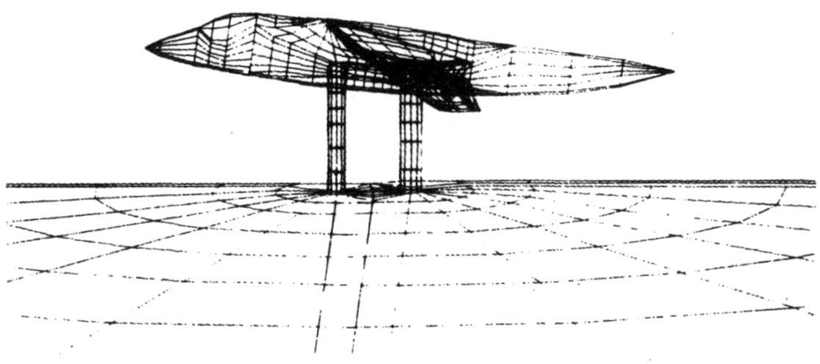

Figure 3. Paneled representation of YAV-8B in hover in-ground effect, $H/D = 2.76$.

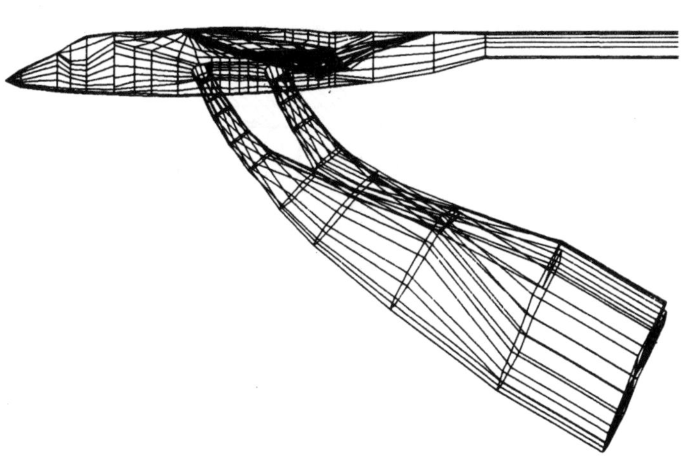

Figure 4. Paneled representation of YAV-8B (without tail) in transition, $V_j/V_\infty = 12.52$.

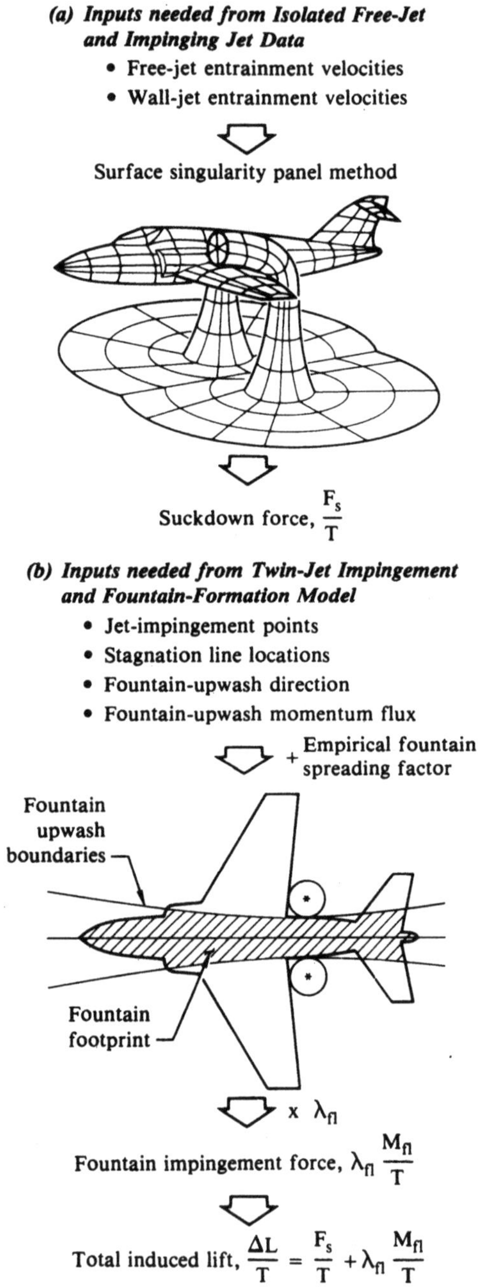

Figure 5. Inputs needed from individual jet-flow models for calculating the induced-lift on a V/STOL aircraft in hover.

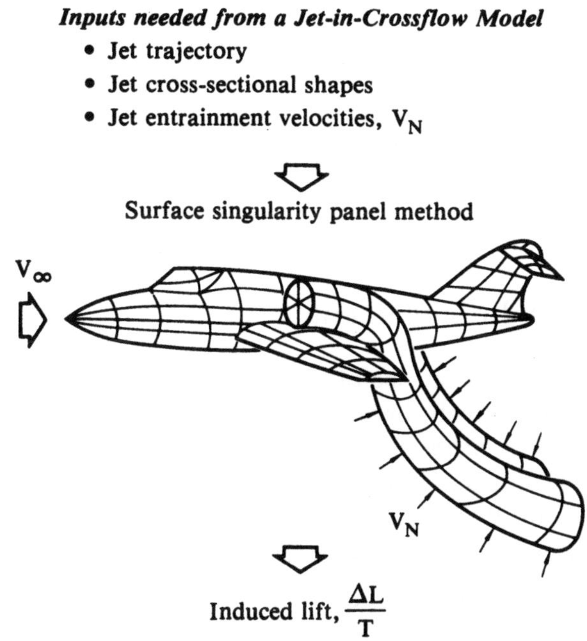

Figure 6. Inputs needed from a jet in crossflow model for calculating the induced-lift on a V/STOL aircraft in transition.

(Gilbert, 1983; Hill and Jenkins, 1979; Jenkins and Hill, 1977; Kotansky and Glaze, 1979, 1981; Saripalli, 1981) and the flowfield of a single round jet in crossflow (Aoyagi and Snyder, 1981; Antani, 1977; Bradbury and Wood, 1969; Callaghan and Ruggeri, 1948; Chassaing et al., 1974; Crabb et al., 1981; Fearn and Weston, 1974, 1975; Gordier, 1959; Jordinson, 1956; Kamotani and Greber, 1972; Keffer and Baines, 1963; Margason, 1968; Mosher, 1970; Ramsey and Goldstein, 1970; Taylor, 1977; Vogler, 1969; Wu et al., 1970). There are only a few experimental studies that deal with rectangular nozzles (Kotansky and Glaze, 1981, 1982) and three to four jet fountain flow (Kotansky, 1983). Therefore, most of the prediction methodologies ranging from semi-empirical approaches to the solution of the Reynolds-averaged Navier-Stokes equations have been developed for these idealized configurations which contain the essence of the jet-induced phenomena. The methodologies are general enough to be easily extended to multiple jet configurations with different geometric and flow parameters, as more experimental information and/or computational power becomes available.

This article surveys the progress made in recent years in modeling jet-flow phenomena relevant to V/STOL aircraft flowfields and summarizes the current status of numerical prediction methods and turbulence modeling for three-dimensional jet flows.

2. Single Jet Impingement

The flowfield of an isolated jet impinging on a flat plate has been studied extensively because of its application to V/STOL aircraft as well as to other areas such as jet cutting, paint spraying, arc welding with a shielding gas, and various heating and cooling applications. In VTOL aircraft design, the study of an isolated impinging jet is of importance because of the following two reasons:

1. For calculating the suckdown force on the aircraft in hover, the entrainment velocity distribution and jet-spreading parameter are needed for panelling the jet and specifying the boundary conditions on the jet-panels (Figure 5a). For a multiple jet configuration, in the current methodologies it is assumed that jets behave independently, at least for the purpose of calculating the suckdown force in hover out of ground effect.

2. For calculating the positive lift force on the aircraft due to fountain impingement, information is needed about the jet impingement points and stagnation line locations on the ground plane as well as momentum flux and direction of the fountain upwash (Figure 5b). Again, the single-jet impingement study provides the local distribution of radial momentum flux about the individual jet impingement points. This distribution is largely dependent on the exit shape of the jet nozzle (for low values of H (height of the jet above the ground plane)/D (jet diameter)) and the local impingement angle of the jet on the ground plane. These momentum flux distributions, together with the magnitudes of the total momentum fluxes emanating from the nozzle exits, establish the location and strength (momentum) of the fountains in the flowfield below the aircraft. This information is needed in the semi-empirical theories that predict stagnation line shapes and fountain upwash momentum flux and angle (see Section 3).

In recent years, theoretical models of varying complexity have been developed for calculating the flowfield of an isolated two-dimensional or axisymmetric jet impinging normal or at an angle to a flat plate. Rubel [56–58] has developed an inviscid rotational flow model which predicts the ground plane pressure, the stagnation point location, and the distribution of radial momentum flux in the azimuthal direction (for inclined impingement) quite well as shown in Figures 7–10. The model solves the Euler equations in velocity/vorticity form, however, it fails to provide information on entrainment and therefore cannot be used for predicting the suck-down forces on the airframe. Furthermore, it cannot include the effect of Reynolds number and, for hot jets, the heating of the impingement surface and the airframe. Nevertheless, the model provides information needed in developing semi-empirical theories of fountain formation (see Section 3). A few investigators have attempted to calculate the flowfield by dividing it into separate interacting regions (Figure 11) that are coupled later to provide the overall solution (Melnik and Rubel, 1983; Sparrow and Lee, 1975). However, this approach has achieved only limited success.

More recently, because of considerable advancement in computer speed and memory, numerical algorithms, and turbulence modeling for recirculating flows, attempts have been made to obtain the numerical solution of the Reynolds-averaged Navier-Stokes equations for bounded jets. Wolfshtein (1967) calculated the flowfield of a planar impinging jet using a one-equation (k) turbulence model while Bower et al. (1977) obtained the numerical solution for an axisymmetric impinging jet, also using a single-equation (k) turbulence model. These calculations did not give satisfactory agreement with the experimental data for small nozzle-to-plate distances. It was later demonstrated by Agarwal and Bower (1982) that the main cause of this disagreement between the numerical predictions and the experimental data could be attributed to the turbulence model employed in the calculations. The use of a single-equation turbulence model requires the specification of turbulence length scales in the entire computational domain. Wolfshtein (1967) specified a constant length scale over the entire flowfield except near the solid boundary, while Bower et al. (1977) arbitrarily specified a scale distribution. To avoid this difficulty of a priori specification of the turbulence length scale, Agarwal and Bower (1982) employed a two-equation ($k - \epsilon$) turbulence model in their study. They found that the turbulence length scales vary considerably in the flowfield

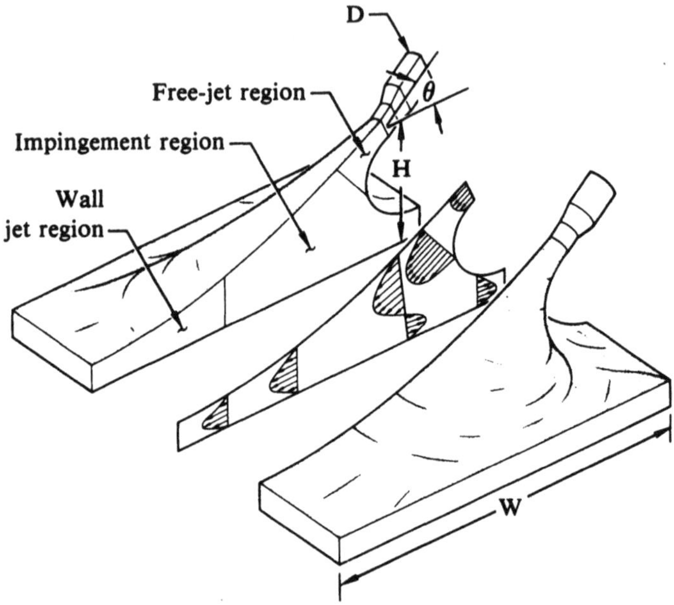

Figure 7. Schematic illustration of an obliquely impinging axisymmetric jet on a flat surface.

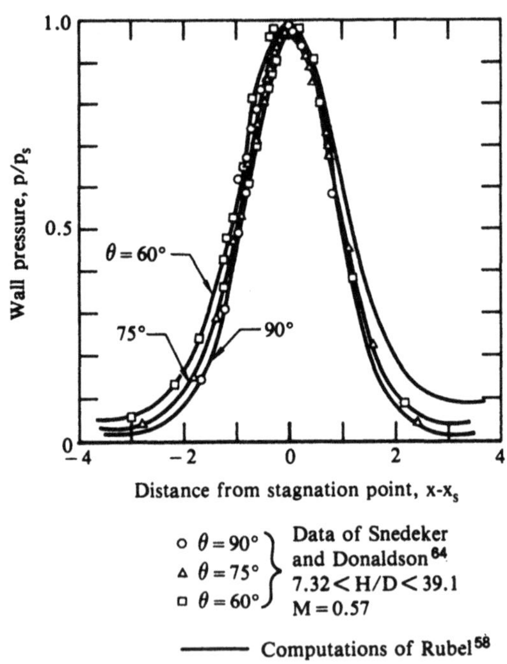

Figure 8. Symmetry plane pressure distribution at the wall for the oblique impingement of a round jet on a flat surface (see Figure 7).

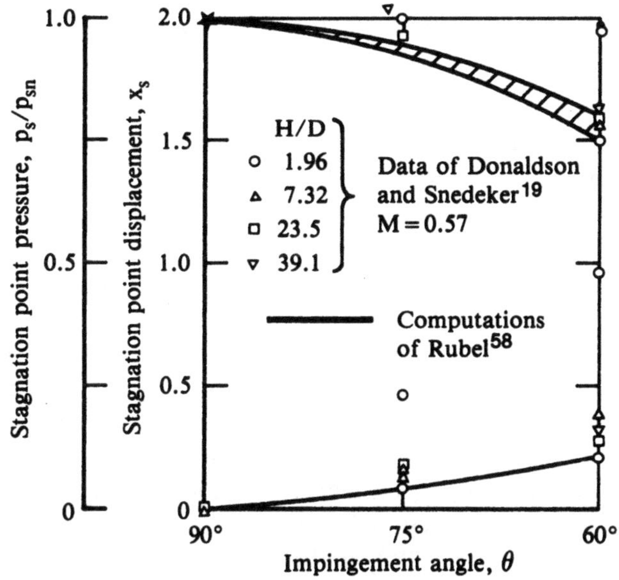

Figure 9. Effect of impingement angle on stagnation-point pressure and location for the case of oblique impingement of a round jet on a flat surface (see Figure 7).

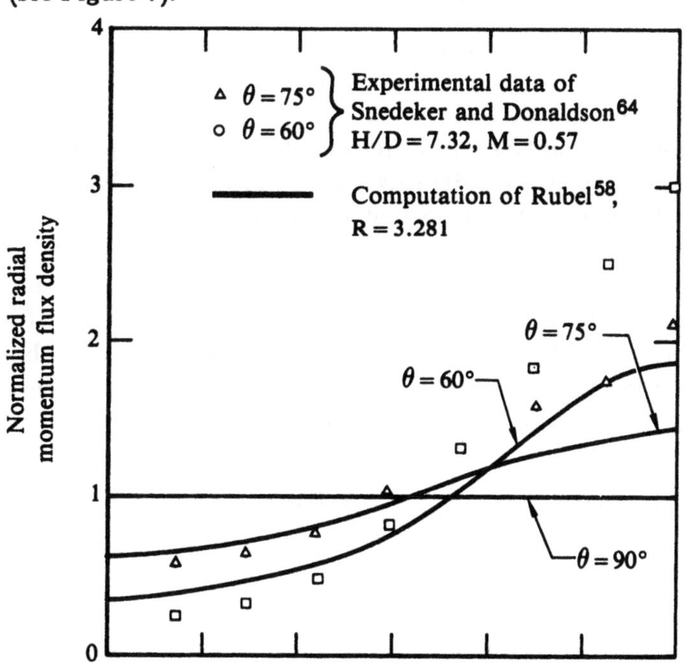

Figure 10. Azimuthal distribution of radial momentum efflux in case of oblique impingement of a round jet on a flat surface (see Figure 7).

because of its complexity characterized by a free jet transition and fully-developed region, stagnation region, wall-jet region and recirculating flow region (Figure 11).

Agarwal and Bower (1982) computed a number of planar and axisymmetric impinging-jet configurations, one simulating the undersurface of the fuselage (Figure 12). For the impinging jet configuration of Figure 12, comparison between the calculated and experimental ground plane and fuselage under-surface pressure distributions is shown in Figure 13, and Figure 14 shows the comparison of experimental and computed centerline velocity and turbulent kinetic energy distributions. The agreement between the experimental data and computations is good. Figure 15 shows the large variation in turbulence length scales in the flowfield as the Reynolds number of the jet changes; clearly, it is difficult to specify intuitively such a variation in turbulence length scales. Calculations performed by Agarwal and Bower (1982) included the effect of compressibility. Again, good agreement was obtained between the experimental data and the computations for the ground and the simulated fuselage under-surface pressure distributions (Figure 16).

In their study, Agarwal and Bower (1982), employed the low Reynolds number form of the two-equation $(k-\epsilon)$ turbulence model (1973). Although this form of the turbulence model allows integration of the flow equations and the turbulence model equations directly to the wall, it requires a densely packed grid to resolve the flowfield in the wall region; this aspect of the turbulence model restricts its use in three-dimensional applications. To circumvent this problem, the high-Reynolds number form of the $(k-\epsilon)$ model (Launder and Spalding, 1973) is often employed wherein wall functions (fully turbulent local equilibrium wall-law profiles) are used as the boundary conditions for the solution of the Reynolds-averaged Navier-Stokes and turbulence model equations, thus avoiding integration all the way to the wall.

Amano [3] and Amano and Brandt [4] have used the high Reynolds number form of the $(k-\epsilon)$ model in their study of the flowfield of a two-dimensional and an axisymmetric jet impinging normal to a flat plate. Figures 17–19 show the comparison between the calculations of Amano and Brandt [4] and experimental data of Tani and Komatsu (1966) for various flow properties at different jet heights above the ground plane. In Figure 17, comparison is also shown with the free-jet theory of Kleinstein (1964). Figure 20 shows

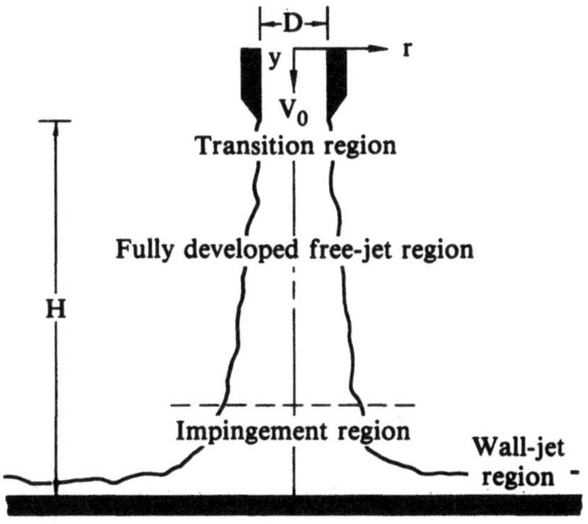

Figure 11. Schematic illustration of various flow regions due to normal impingement of an isolated jet.

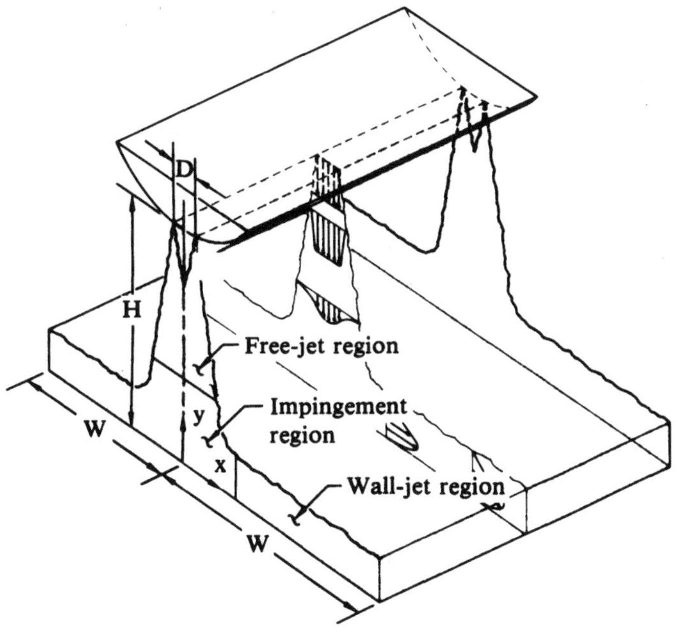

Figure 12. The planar impinging jet with a simulated fuselage undersurface.

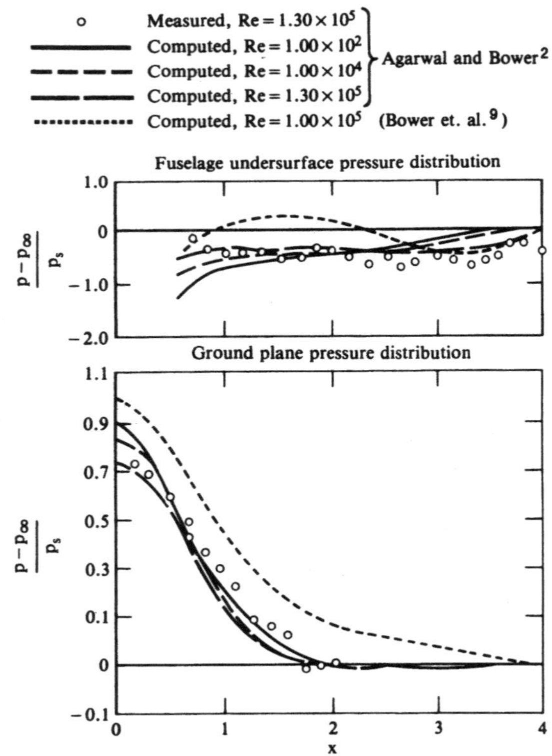

Figure 13. Comparison of experimental and theoretical pressure distributions for the incompressible planar jet in-ground effect (Figure 12), $H/D = 2$, $W/D = 3.68$.

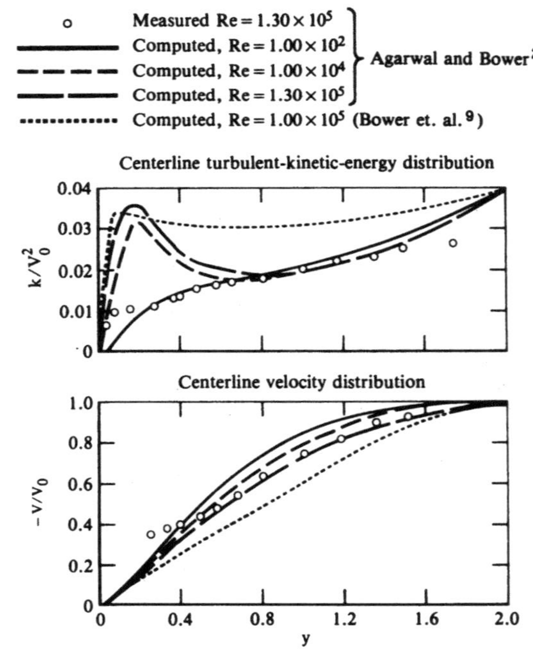

Figure 14. Comparison of experimental and theoretical centerline property distributions for the incompressible planar jet in-ground effect (Figure 12), $H/D = 2$, $W/D = 3.68$.

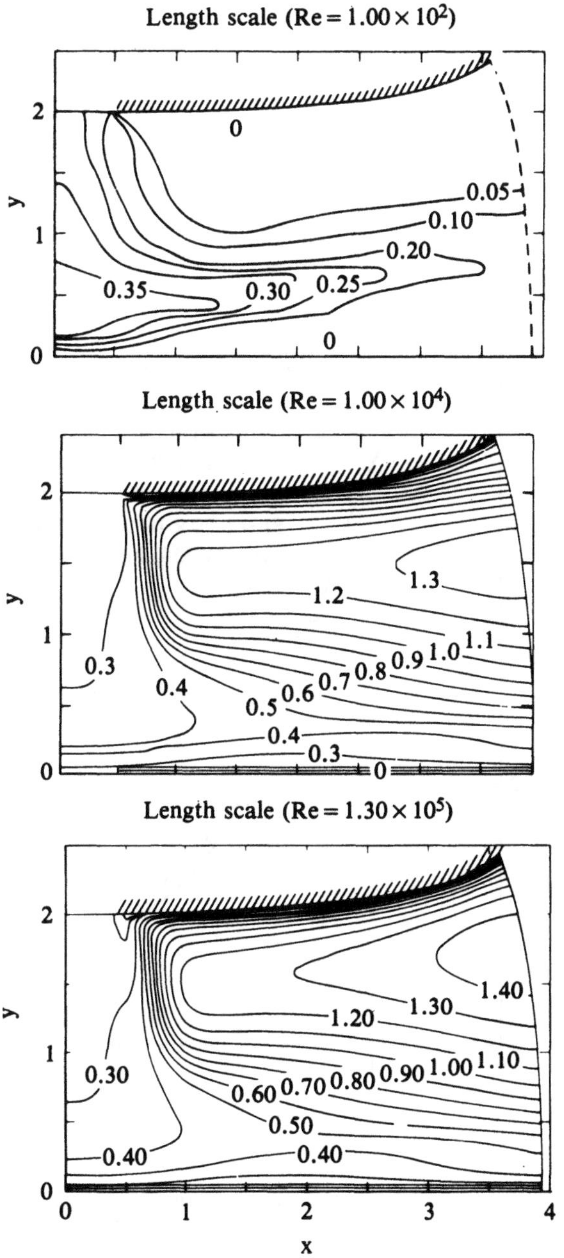

Figure 15. Turbulent length scale contours at different Reynolds numbers for the incompressible planar jet in-ground effect (Fig. 12), $H/D = 2$, $W/D = 3.68$, computed by Agarwal and Bower (1982). "Copyright © American Institute of Aeronautics and Astronautics; reprinted with permission of the AIAA."

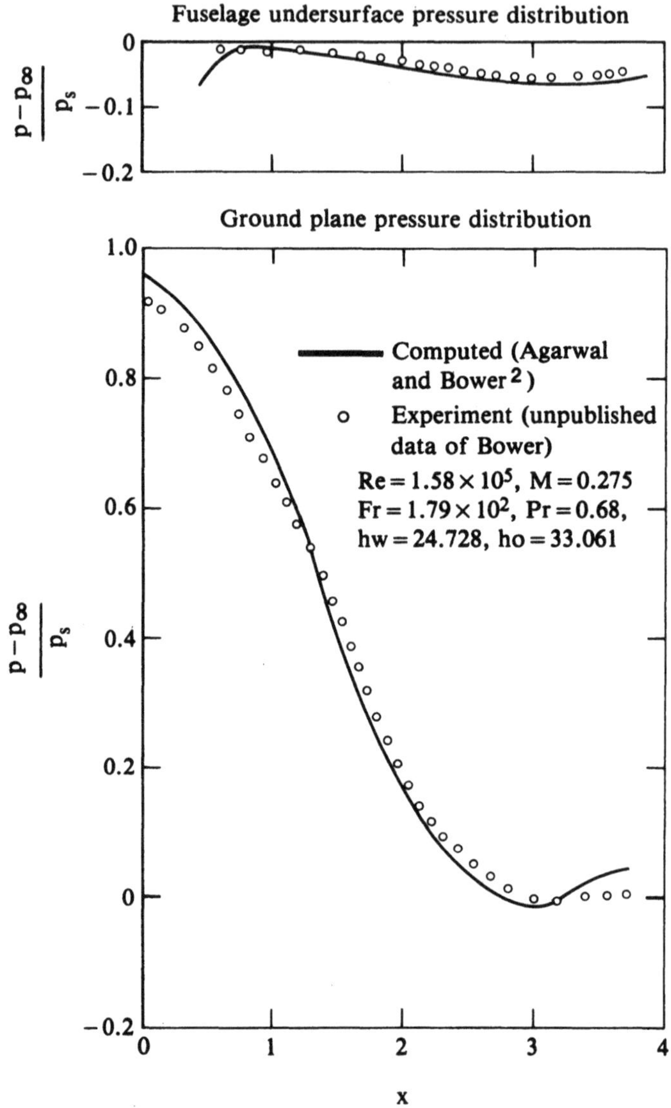

Figure 16. Comparison of experimental and theoretical pressure distributions for the compressible planar jet in-ground effect (Figure 12), $H/D = 2$, $W/D = 3.68$.

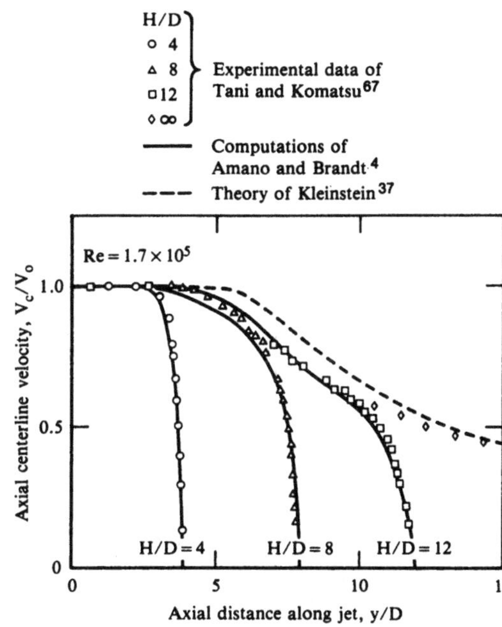

Figure 17. Decay of axial centerline velocity for an axisymmetric jet impinging on a flat plate.

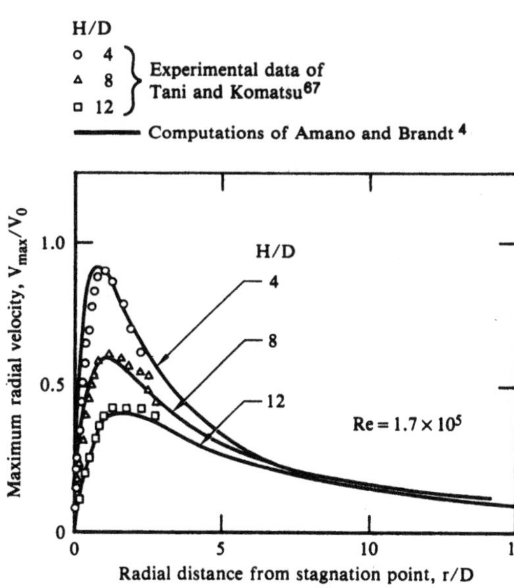

Figure 18. Maximum radial velocity along the wall for ar axisym-metric jet impinging on a flat plate.

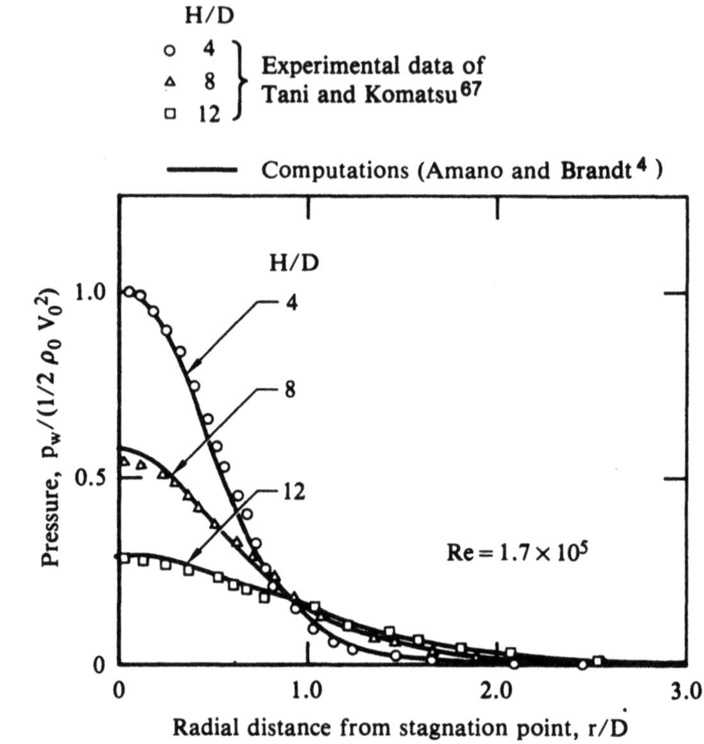

Figure 19. Pressure distribution along the wall for an axisymmetric jet impinging on a flat plate.

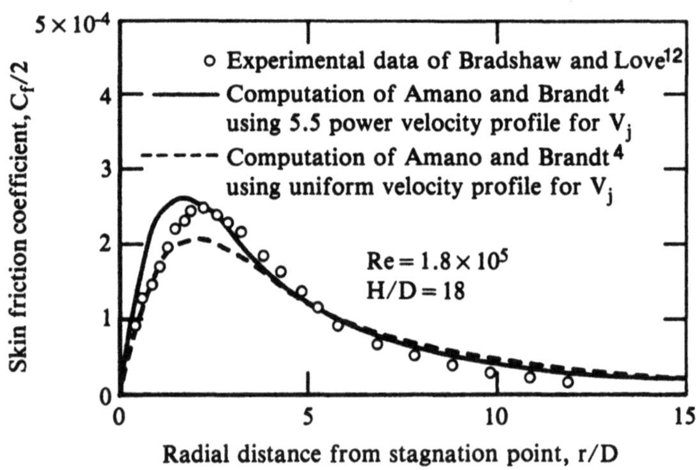

Figure 20. Skin friction along the wall for an axisymmetric jet impinging on a flat plate.

the comparison between experimental data of Bradshaw and Love (1959) and the calculations of Amano and Brandt [4] for skin-friction along the wall, and Figure 21 compares the universal skin-friction predicted by the $(k - \epsilon)$ model with the empirical relation of Poreh et al. (1967) derived from the experimental data. Figure 22 shows the influence of turbulence intensity in the nozzle and nozzle Reynolds number on computed skin-friction along the wall. For low jet heights above the ground plane and at high Reynolds number, a double peak is predicted in the skin-friction distribution. In all cases, agreement between the experimental data and computations is excellent. Thus the calculations of Agarwal and Bower (1982), and Amano and Brandt [4], demonstrate that it is now possible to predict 2-D and axisymmetric impinging jet flowfields reliably using the Reynolds-averaged Navier-Stokes equations and a two-equation turbulence model. Further research needs to be directed toward improving the efficiency of the calculations so that they become cost-effective for use in VTOL aircraft design practice.

For the case of oblique impingement of a round jet on a plane surface, the flowfield is three-dimensional and its calculation at high Reynolds numbers using the Reynolds-average Navier-Stokes equations with $(k - \epsilon)$ model requires a number of grid points that far exceeds the storage capacity of commonly available computers, such as CYBER 176. Furthermore, these calculations are presently quite expensive. Nevertheless, Bower (1982) attempted to calculate the flowfield due to oblique impingement of a round jet on a plane surface at very low Reynolds number using a fairly coarse grid so that the calculations could be performed on CYBER 176. The results of his calculations and their comparison with the experimental data of Snedeker and Donaldson (1965) is shown in Figure 23. Although the Reynolds number in the computations is many orders of magnitude smaller than that in the experiments, the agreement is satisfactory for the ground plane pressure and the stagnation point location. These flow properties are essentially governed by inviscid phenomena and the rotational potential flow model is adequate for their computation (Rubel, 1980). The calculations of Bower (1982) indicate, however, the possibility of computing the viscous flows in the near future on supercomputers, such as CRAY 1 and CYBER-205, using the Reynolds-averaged Navier-Stokes equations and more elaborate turbulence models.

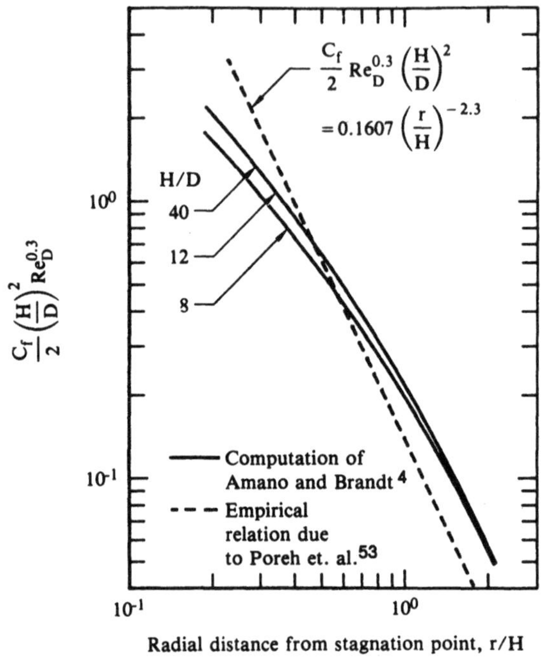

Figure 21. Universal skin friction along the wall for an axisymmetric jet impinging on a flat plate.

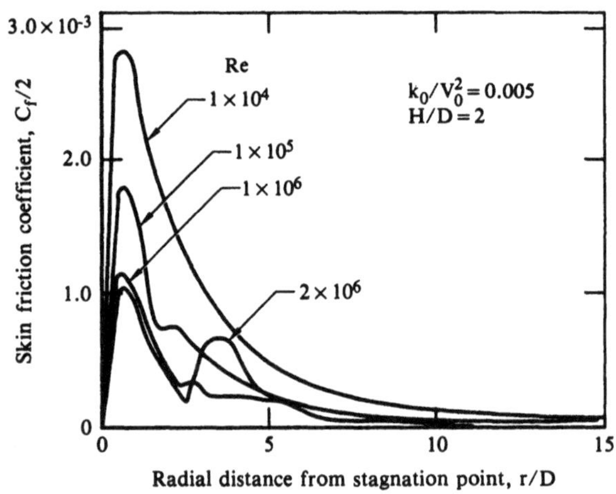

Figure 22. Influence of turbulence intensity in the nozzle and nozzle Reynolds number on computed skin friction along the wall for an axisymmetric jet impinging on a flat plate.

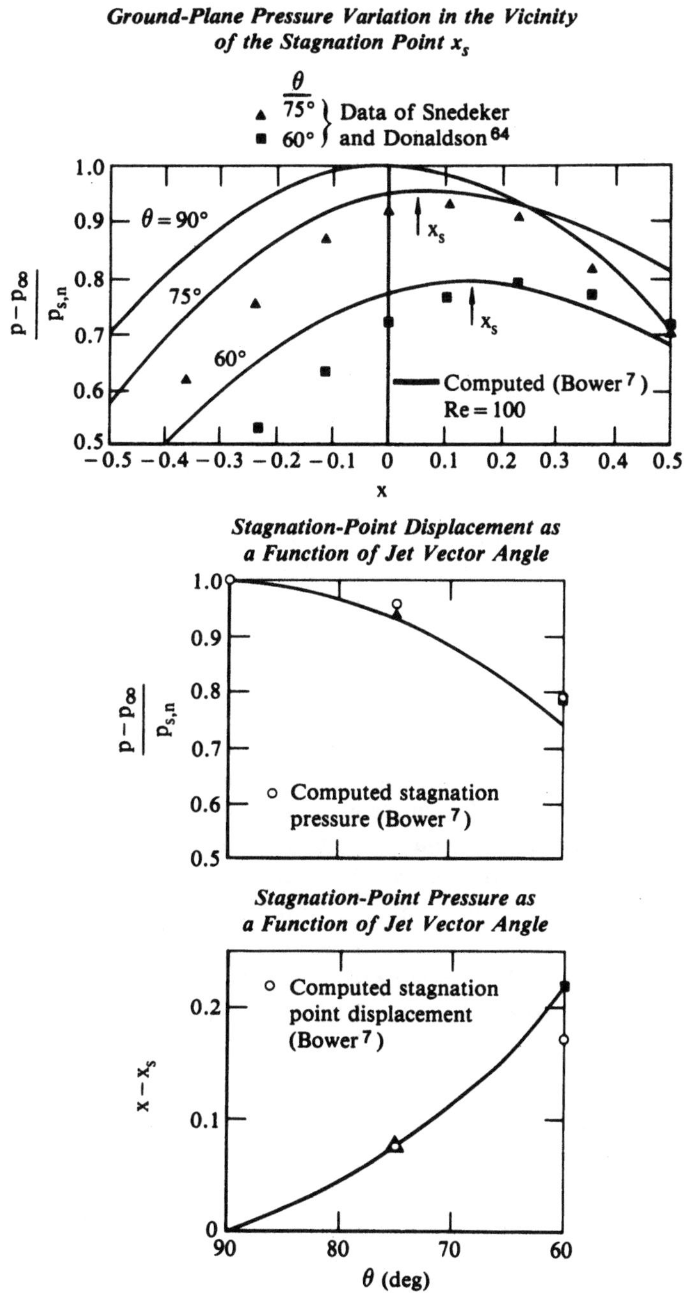

Figure 23. Flow properties for an isolated jet with inclined impingement.

3. Twin-Jet Impingement and Fountain Formation

On a VTOL aircraft hovering in ground effect, the multiple jet impingement on the ground plane generates a fountain which impinges on the fuselage undersurface of the aircraft resulting in a positive contribution to the lift force on the aircraft. However, the inclination of the fountain may be such that it produces adverse pitch and roll moments, and also may result in hot-gas ingestion by the inlets.

Because of the complex three-dimensional flowfield, the calculation of the lift force due to fountain impingement on the airframe, using the Reynolds-averaged Navier-Stokes equations at high Reynolds numbers, is currently not possible on present computers. For a twin-jet impingement configuration (Figure 24), Bower (1982) recently made an attempt to calculate the flowfield using the Reynolds-averaged Navier-Stokes equations in velocity/vorticity form in conjunction with $(k - \epsilon)$ turbulence model using a third-order accurate upwind scheme which is stable and accurate at high Reynolds numbers. Again because of the limitations on the number of grid points that could be used on CYBER 176, he could only calculate the flowfield at very low Reynolds numbers on a fairly coarse grid. Figure 25 shows the comparison between Bower's (1982) calculation at Re $= 200$ and the experimental data of Jenkins and Hill (1977) at Re $=$ $100,000$. The agreement is quite good in the impingement zones, but considerable difference remains in the fountain region. This result is expected since the diffusion effects are considerably larger at Re $= 200$ (used in the computations) compared to Re $= 100,000$ in the experiments. Figure 26 shows the comparison between the computations at Re $= 100$ and experimental data at Re $= 100,000$ for a larger spacing between the jets. The agreement is better in this case for the pressure distribution in the fountain region because the colliding wall jets have much smaller moments than for the closer spacing case of Figure 25. Figure 27 shows the comparison between the computations and the experimental data for the twin-jet impingement with unequal-strength jets. The calculations of Bower (1982) are the only one performed to date for the twin-jet configuration using the Reynolds-averaged Navier-Stokes equations with $(k - \epsilon)$ turbulence model. Because of the considerable computer resources required, it will be some time before this approach will be performed routinely in VTOL aircraft design. As indicated before, semi-empirical methods

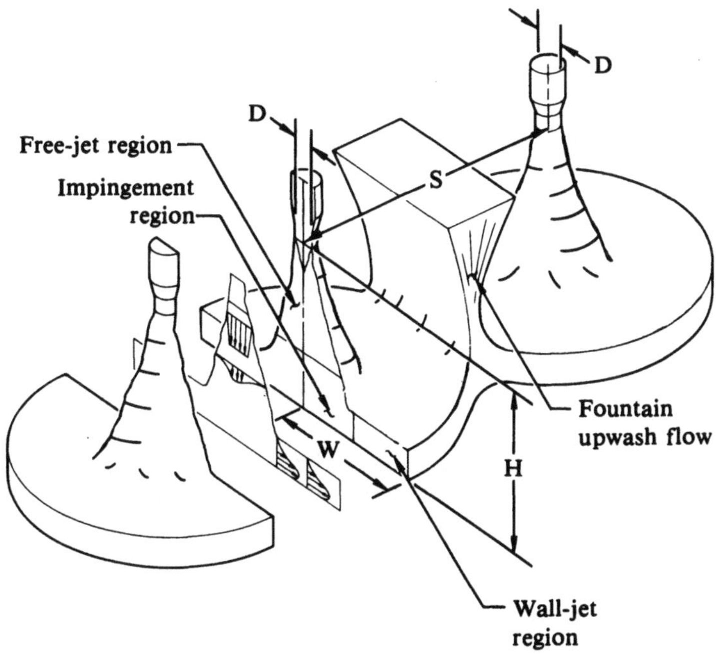

Figure 24. Schematic illustration of two interacting jets impinging perpendicularly on the ground with fountain formation.

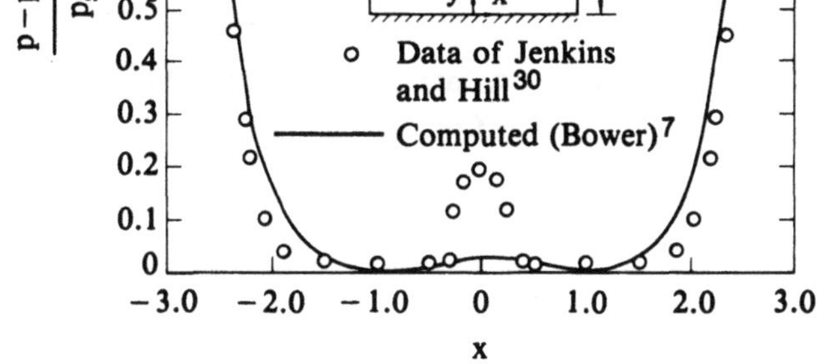

Figure 25. Ground-plane pressure variation for equal-strength jets with normal impingement ($S/D = 6$, $H/D = 3$, Re $= 200$).

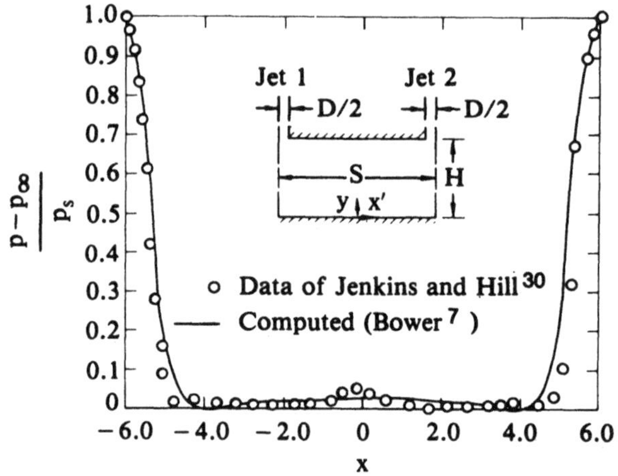

Figure 26. Ground-plane pressure variation for equal-strength jets with normal impingement ($S/D = 12$, $H/D = 4$, Re $= 100$).

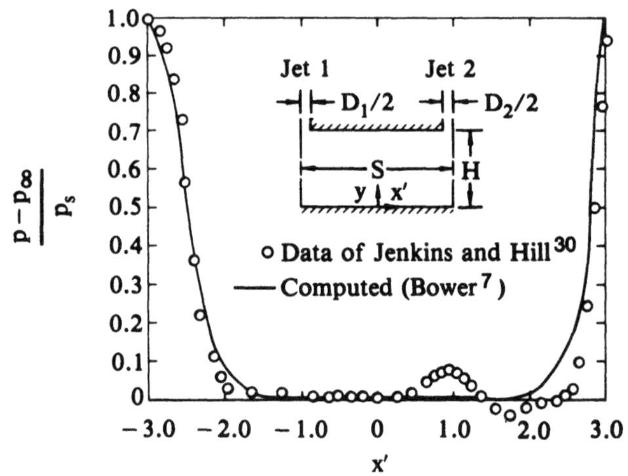

Figure 27. Ground-plane pressure variation for unequal-strength jets with normal impingement ($S/D_1 = 6$, $H/D_1 = 3$, $D_2/D_1 = 0.515$, Re $= 100$).

(Kotansky and Glaze, 1980; Kotansky et al., 1977; Siclari et al., 1980) are presently used to establish the location of stagnation lines, the fountain upwash angle and the momentum flux in the fountain.

The input required for these calculations is the azimuthal distribution of radial momentum flux which can be obtained from isolated jet impingement models or the experimental data (as given in Figure 10 for example). Kotansky et al. (1977) developed a method for the computation of stagnation lines and fountain upwash inclination which they call *the Momentum Flux Method*. The latter establishes the location of the stagnation line between the two jet impingement points by balancing the total momentum flux in the wall jets in a direction normal to the stagnation line in the ground plane. As a result of the requirement of a total momentum flux balance (normal to the stagnation line), the upward direction of the fountain in a vertical plane normal to stagnation line must be vertical, although a nonvertical sidewash component is allowed. In a later study, Kotansky and Glaze (1980) modified the criterion for determination of the location of the stagnation line by requiring a balance of wall jet momentum flux per unit area of the wall jet at the stagnation line. The balance of momentum flux per unit area is imposed in a direction normal to the stagnation line in the ground plane. This criterion permits an imbalance of total wall jet momentum flux at the stagnation line in a direction normal to the stagnation line in the ground plane and, correspondingly, allows a nonvertical trajectory of the fountain upwash flow in a vertical plane normal to the stagnation line. This model for determination of the stagnation line location is known as the *Momentum Flux Density Method*. The accuracy of the momentum-flux-density method in computing stagnation lines is demonstrated in Figure 28 which shows the comparison between the calculated stagnation line shapes with those obtained experimentally by flow-visualization techniques in water (Saripalli, 1981). Recently, Siclari et al. (1980) and Yen (1981) have proposed alternate methods for the calculation of stagnation lines and fountain inclination angle, which are based on the same physical concept but somewhat simplify the calculation procedure. The calculation of positive lift force due to fountain impingement on the aircraft requires knowledge of the stagnation line location, fountain inclination angle, and normal momentum flux in the fountain.

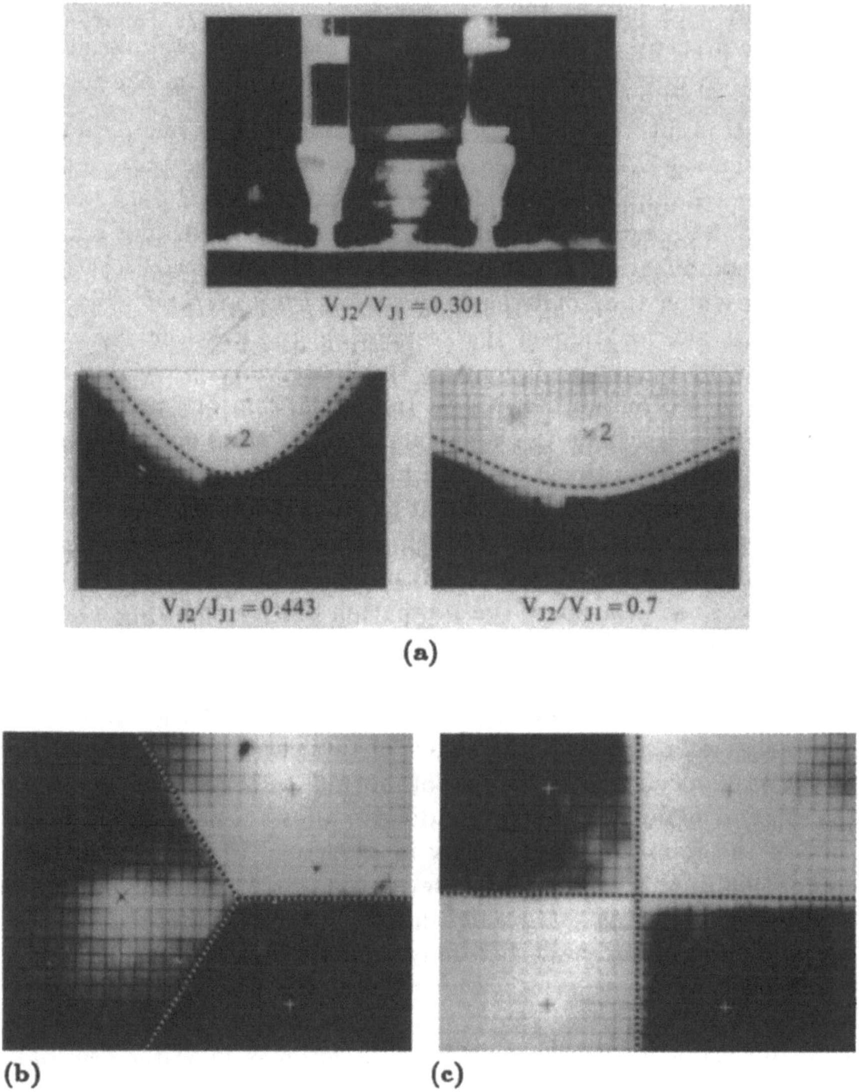

Figure 28. (a) Stagnation line patterns for twin-jet impingement flow visualization experiments of Saripalli: $S/d = 12$, $H/d = 4$, and $d_2/d_1 = 1$. The dotted line indicates the predictions using the momentum flux density method (Kotansky and Glaze). (b) Stagnation line pattern for three-jet impingement flow; $H/d = 4$, equal-strength jets. The dotted line indicates the calculated stagnation line pattern using the momentum flux density method. (c) Stagnation line pattern for four-jet impingement flow; $H/d = 4$, equal-strength jets. The dotted line indicates the expected stagnation line pattern using momentum flux density method.

4. Calculation of Jet-Induced Lift on a VTOL Aircraft Hovering in or out of Ground Effect

In the semi-empirical methodologies (Glaze et al., 1983; Siclari et al., 1976) currently used in the calculation of jet-induced lift on a VTOL aircraft hovering in or out of ground effect (IGE or OGE), the jet-induced lift $\Delta L/T$, where T is the total thrust of the jets, is defined as the sum of the jet-entrainment induced suck-down force (F_s/T) and the fountain impingement force (F_f/T). The fountain impingement force is defined as the product of the fountain impingement momentum flux transfer coefficient λ_{fI} and the fountain upwash momentum flux (M_{fI}/T) incident upon the aircraft; thus,

$$\frac{\Delta L}{T} = \frac{F_s}{T} + \lambda_{fI}\frac{M_{fI}}{T} \tag{4.1}$$

The value of λ_{fI} is determined empirically such that the total induced lift corresponds to the experimentally determined value.

For calculation of the suck-down force F_s, a surface singularity method is used in which the aircraft, lift-jet, and wall-jet surfaces are panelled. As an illustration, Figure 3 presents the panelled representation of YAV-8B in hover in ground effect. Zero flow normal to the panel surface is prescribed as a boundary condition on aircraft surfaces while nonzero normal velocity boundary conditions are prescribed on jet surface panels to represent jet-entrainment. For calculating the suck-down force in hover out of ground effect (OGE), a detailed jet model or a simple jet model can be used. Figure 29 shows the two jet models that have been employed at McDonnell Aircraft Company for calculating the suck-down force on the YAV-8B (Glaze et al., 1983). In the detailed jet model, the paneling represents the spreading jet geometry and in the simple jet model, equivalent cylinderical jet paneling has been used. For OGE calculations, currently free-jet data from the experiments (Hill, 1972; Kleis and Foss, 1974; Ricou and Spalding, 1961; Tentacoste and Sforza, 1967; Wygnanski, 1963) are used for specifying the entrainment velocities (Figure 30). For a multiple-jet configuration, it is assumed that the jets behave independently at least for the purpose of calculating the suck-down force. Influence of jet-merging and the fuselage undersurface geometry on the entrainment distribution is currently ignored because of lack of good experimental data and a theoretical prediction model. These effects are included simply by correcting the entrainment velocity distribution obtained from free-jet data or computational model by an empirical constant.

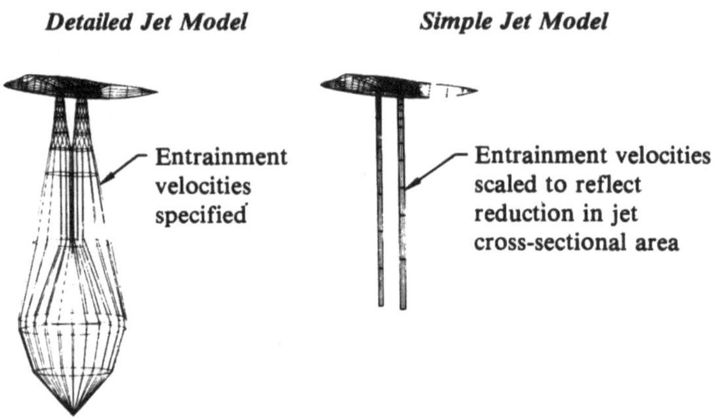

Figure 29. Alternate jet models for paneling the YAV-8B in hover out of ground effect.

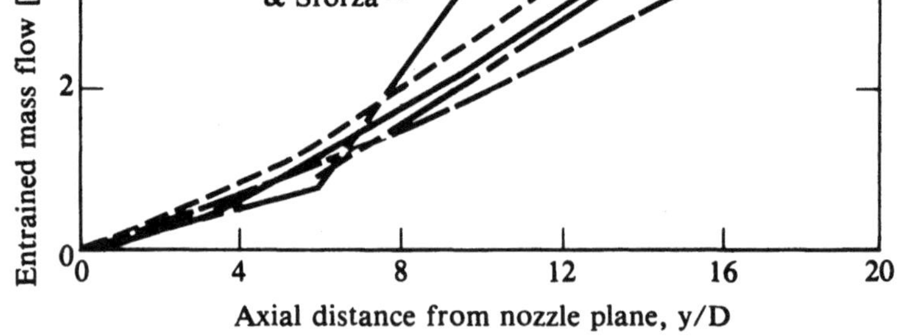

Figure 30. Experimentally obtained entrained mass flow distribution for circular free-jets.

For IGE suck-down force calculations, in addition to a free-jet entrainment velocity distribution, wall-jet entrainment velocity distributions are needed. The wall-jet data are again specified using the isolated jet-impingement data or a computational model.

For calculating the lift force due to multiple jet fountain up-wash momentum flux M_{fI} incident upon the aircraft, the following assumptions are made in the currently used methodologies:

1. The flowfield can be divided into an inner and an outer region. In the inner region, all local momentum contributions concentrate into a fountain. In the outer region, all local momentum contributions are coflowing along the stagnation lines (Figure 31).
2. The stagnation lines, to a first-order, are determined from the local momenta of a corresponding pair of jets, the contribution of other jets is assumed to be small.

With the above assumptions, equations can be established that are applicable to either the inner or outer region for a pair of jets. Any multi-jet configuration can be formulated from these equations by superposition and knowledge of the axis of symmetry.

The fountain upwash momentum flux contributions emanating from the inner region $-(M_{fI})_{\text{in}}$ and outer region $-(M_{fI})_{\text{out}}$ are determined graphically (Figure 32). To calculate $(M_{fI})_{\text{in}}$, the entire wall-jet momentumm flux within the inner region is assumed to be turned upward into a central fountain. Due to the lack of sufficient experimental data to define multiple-jet central fountain characteristics, the central fountain is assumed to be circular in cross section and to exhibit the same spreading characteristics as a twin-jet fountain. The total momentum flux contained within the central fountain upwash is determined as:

$$(M_f)_{\text{in}} = \lambda_M \sum_{i=1}^{n} \frac{\Delta\phi_i}{2\pi}(M_{je})_i \tag{4.2}$$

where M_{je} is the jet-exit momentum flux, λ_M is the fountain formation momentum flux recovery factor determined for two jet fountains (Figure 33 and Reference 38) and n is the number of lift jets. It is shown in (Kotansky, 1981) that λ_M is approximately 0.55 for two jets impinging vertically with equal thrusts and this value increases as the thrust differential between two impinging jets increases (Figure 33). Figure 34 shows the variation of fountain inclination

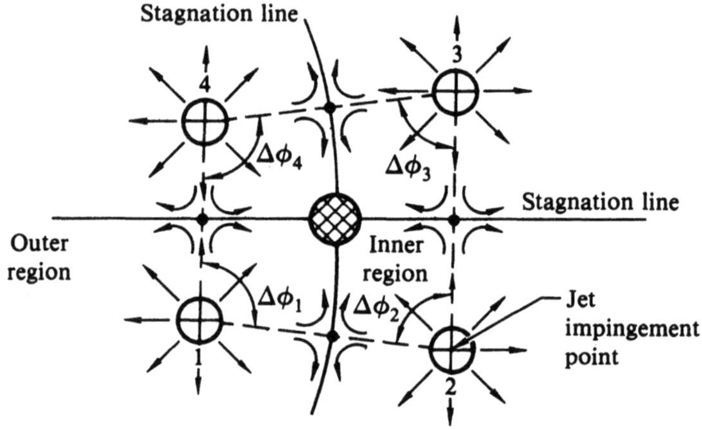

Figure 31. Inner and outer regions in a four-jet impingement configuration.

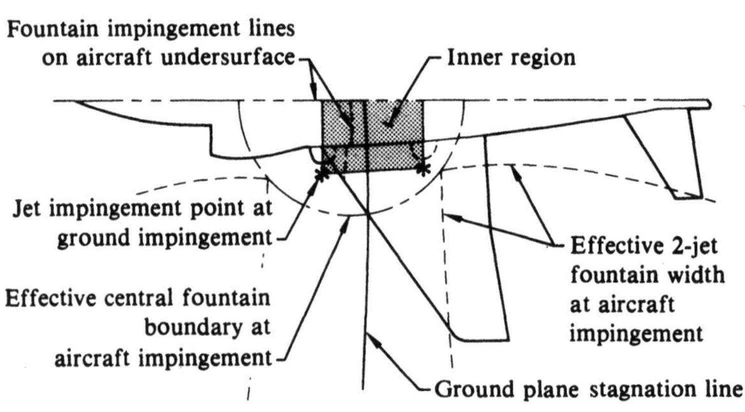

Figure 32. Fountain impingement force calculation model.

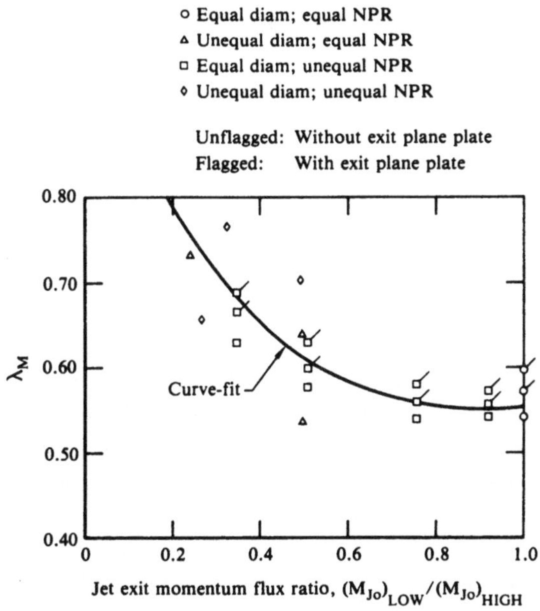

Figure 33. Fountain momentum flux recovery λ_M (= local momentum flux in the fountain upwash/total local momentum flux entering the fountain base from the ground wall jets), based on conservation of mass flux through the fountain formation region; NPR = nozzle pressure ratio.

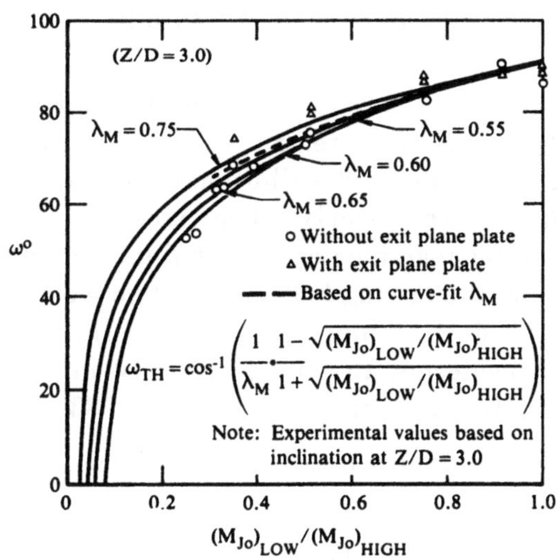

Figure 34. Two-jet fountain inclination (ω^o)—vertical impingement.

angle for twin-jet vertical impingement with jet exit momentum flux ratio for various values of λ_M (Kotansky, 1981). The inner region fountain impingement momentum flux is calculated based on the ratio of the cross-sectional area of the central fountain at impingement, A_{fc}, and the planform area intercepted by the central fountain, A_{ff} as (Figure 35):

$$(M_{fI})_{\text{in}} = \frac{A_{ff}}{A_{fc}}(M_f)_{\text{in}} \tag{4.3}$$

To calculate $(M_{fI})_{\text{out}}$, the fountain upwash momentum vectors are projected upward from the ground plane stagnation line to intersect the aircraft planform, yielding the fountain impingement centerline (Figure 32). Application of the empirical two-jet fountain spreading rate yields the effective fountain upwash boundaries and, consequently, the fountain *footprint* or that area of the aircraft planform intercepted by the fountain upwash. The two jet fountain spreading rate is determined empirically from the data (Kotansky and Glaze, 1980). Another way of computing the fountain impingement momentum flux in the outer region is to use the simple area ratio between the fountain *footprint* A_{ff}, and the fountain cross-sectional area at impingement, A_{fc}, (Figure 35) with:

$$(M_{fI})_{\text{out}} = \frac{A_{ff}}{A_{fc}}(M_{fn})_{\text{out}} \tag{4.4}$$

where $(M_{fn})_{\text{out}}$ represents the momentum flux contained within the fountain upwash in the outer region. This definition of $(M_{fI})_{\text{out}}$ inherently assumes a uniform distribution of momentum flux across the fountain. The apparent overlap of the cross-hatched areas of the inner region and outer region computational models shown in Figure 35 is consistent with the mixing between the two jet upwash flow and the central fountain flow. A more accurate description for $(M_{fI})_{\text{out}}$ is obtained by considering the nonuniformity in momentum flux distribution across the fountain (Glaze et al., 1983).

Now the total fountain impingement momentum flux is given by:

$$M_{fI} = (M_{fI})_{\text{in}} + (M_{fI})_{\text{out}} \tag{4.5}$$

As mentioned before, only a fraction of M_{fI} is converted into a force on the aircraft. This fraction is included by introducing an empirical coefficient λ_{fI} in equation (4.1) to determine the induced lift on the aircraft. λ_{fI} is determined correlating equation (4.1) with the experimental data for a number of flow conditions and aircraft configurations. An empirical correlation for λ_{fI} can then be used later for making a large number of computer runs for evaluating the performance of the VTOL aircraft for a wider range of parameters for which the experimental data do not exist. As an example, Figure 36 shows the suck-down and fountain-impingement forces on the YAV-8B aircraft in ground effect, and Figure 37 gives the empirical correlation curves for λ_{fI} developed at McDonnell Aircraft Company (Glaze et al., 1983) for AV-8A and YAV-8B configurations.

5. Jet in Crossflow

For calculating the effect of aerodynamic interference of a jet on the lift of a V/STOL aircraft in transition from hover to wing-borne flight, the current surface panel methods require a jet-in-cross flow model that can provide information about the jet-trajectory, jet cross-sectional shapes and jet-entrainment velocities (Figure 6), given the jet-to-free stream velocity ratio (V_j/V_∞), jet-injection angle (θ_j) and jet-exit diameter D. Again, because of the complex flowfield characterized by bending and spreading of the jet, the formation of contrarotating vortices, and the development of a wake region near the surface, the detailed prediction of the flowfield necessarily requires the solution of the three-dimensional Reynolds-averaged Navier-Stokes equations in conjunction with at least a two-equation turbulence model, such as $(k-\epsilon)$ model which does not require a priori specification of turbulence length scales in the flow domain. Because of the reasons outlined in the previous sections—lack of computer storage and cost effectiveness on most of the current main frame computers—routine calculations are not feasible yet. However, a number of noteworthy (Bower et al., 1983; Chien and Schetz, 1975; Claus, 1983; Jones and McGuirk, 1980; Patankar et al., 1977) attempts have been made in this direction, most of them for the flowfield of a round jet-in-crossflow issuing normal to a flat plate (Figure 38). All the investigators solve the time-averaged Navier-Stokes equations in conjunction with a high Reynolds number form of the $(k-\epsilon)$ model, the main difference being in the numerical finite-difference method employed and in the imposition of the boundary conditions on various surfaces of the computational domain. The calculations predict the jet trajectory,

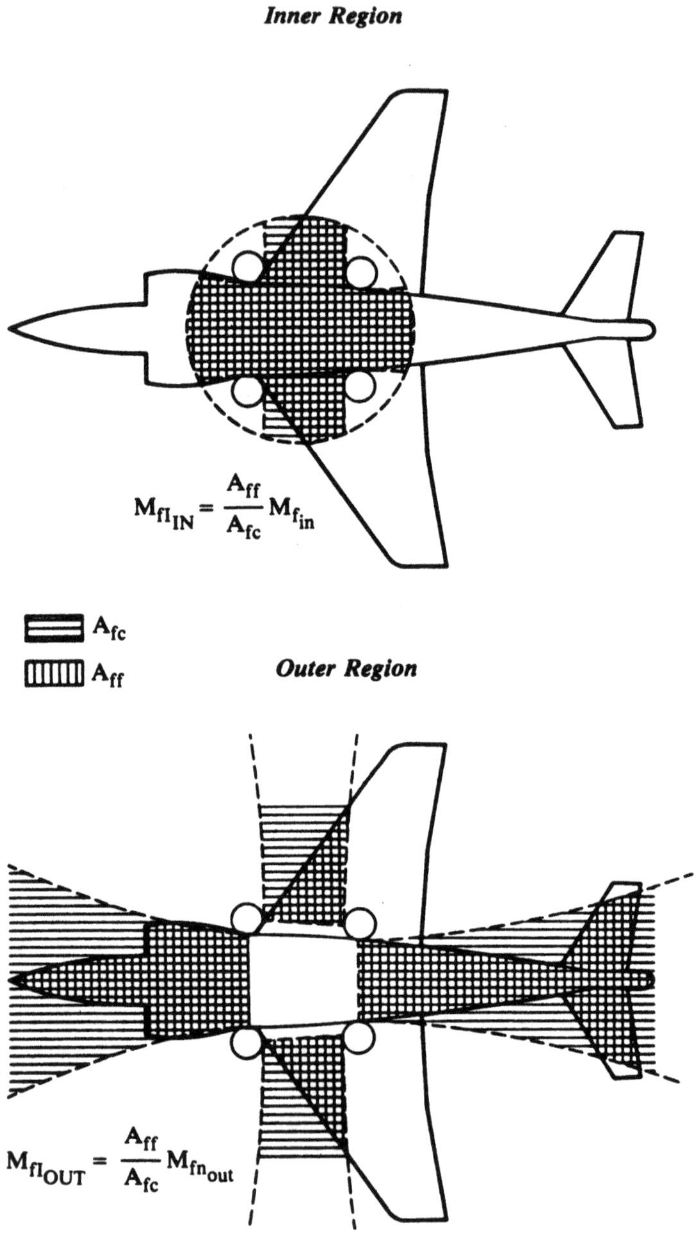

Figure 35. Fountain impingement forces (uniform momentum flux distribution).

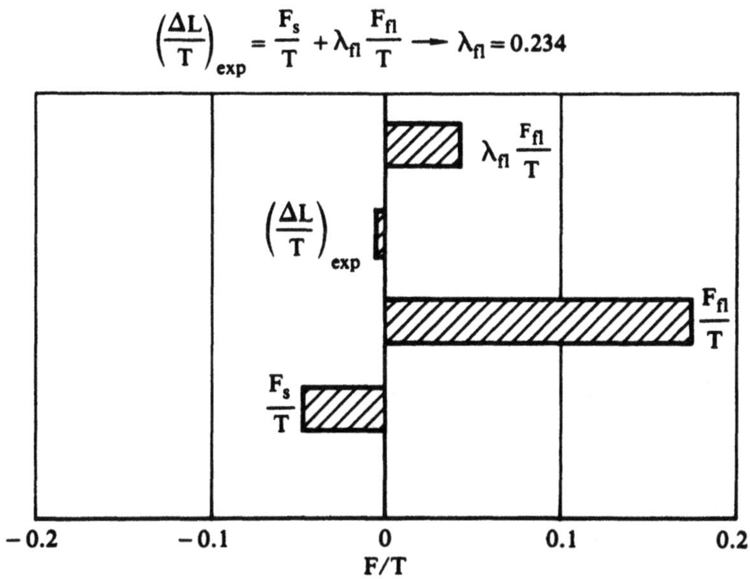

Figure 36. YAV-8B jet-induced forces in ground effect, $H/D = 3.1$.

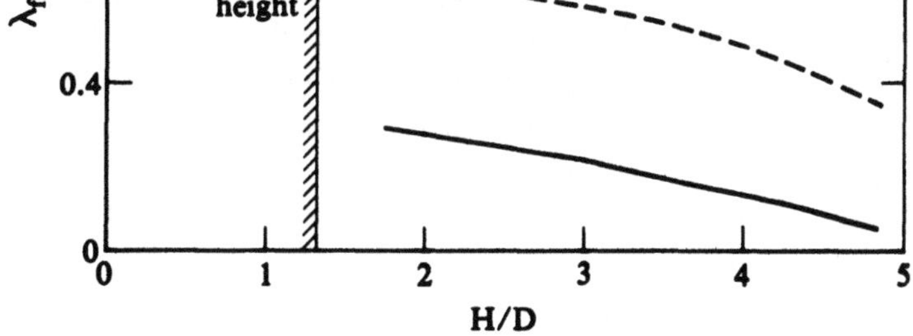

Figure 37. Fountain impingement momentum flux transfer coefficient for YAV-8B.

the decay of maximum velocity in the jet, and the velocity profile downstream of the wake region satisfactorily well (Figures 39–41). Calculations also show the formation of the contrarotating vortices and the pattern of entrainment in the jet. However, because of coarseness of the computational mesh, the resolution of the wake region and the detailed knowledge of the structure of the flow in the contrarotating vortex region must await further advances in computing power. Therefore, for providing inputs to the design codes, almost all the current prediction methods in transition aerodynamics are semi-empirical. A good review of these methods is given by Yen (1978). These semi-empirical methods are basically integral methods; best known among these are those of Wooler (1972), Snel (1974), Fearn and Weston (1975), and Adler and Baron (1979).

In Wooler's (1972) method, the jet is defined by three quantities: the centerline shape, the centerline velocity, and the jet-width normal to the freestream direction. The three differential equations for these quantities are the continuity equation and the momentum equations in the tangential and transverse directions relative to the jet centerline. Sufficiently far downstream, the jet cross section is assumed to be an ellipse with a major-to-minor axis ratio equal to 4. In addition, an empirical relation based on the experimental data is used for the entrainment velocity distribution. The blockage effect is simulated by a distribution of doublets along the jet centerline. The strength of the doublet along the centerline is obtained from the complex potential of a two-dimensional flow past the ellipse of the jet cross section. The jet entrainment is simulated by a distribution of constant-strength sinks over the surface formed by the centerline and major axis of the jet. The approximation adopted in the approach of determining the strength of doublets and sinks is the same as the small perturbation theory in aerodynamics. The induced velocity field is calculated by integrating the contributions over the extent of the jet. An image singularity system is used to satisfy the boundary condition at the surface. Extensive computations for the surface pressure distributions have been carried out by Wooler (1972) and compared with various experiments. Generally, the agreement between the computed results and measured data is good for the regions ahead of the jet. There are significant differences, however, between theory and data in the wake regions.

Snel's (1974) method is similar to Wooler's (1972), except that the jet flow is divided into two regions: a development region and a

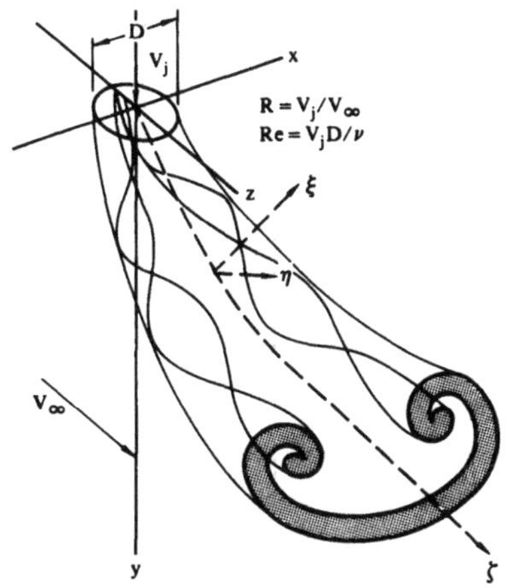

Figure 38. Schematic illustration of a jet in crossflow.

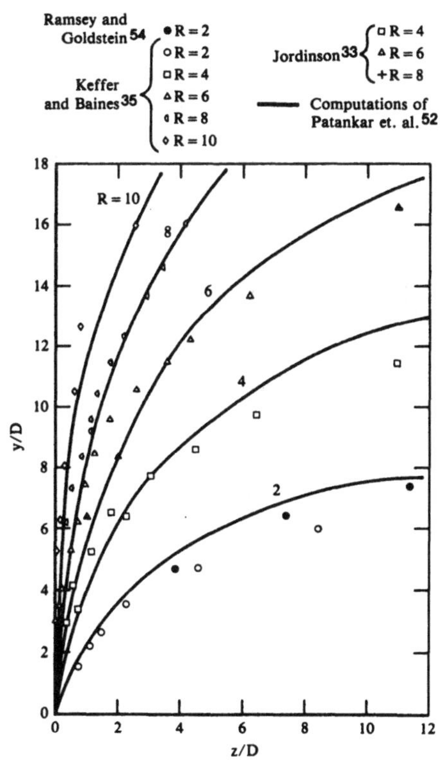

Figure 39. The jet centerline for different jet-to-mainstream velocity ratios for a jet in crossflow.

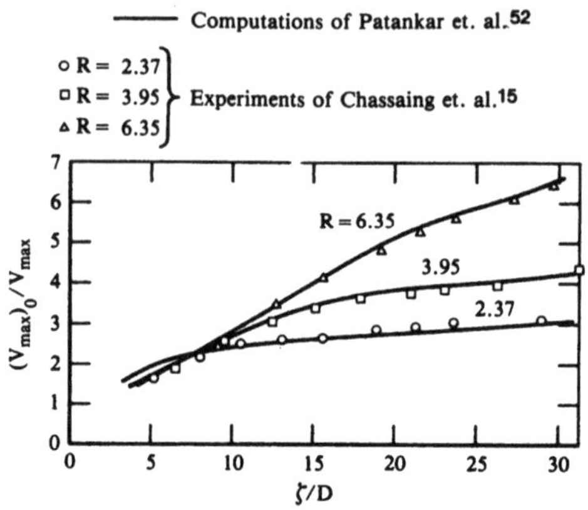

Figure 40. The decay of the maximum velocity in a jet in crossflow.

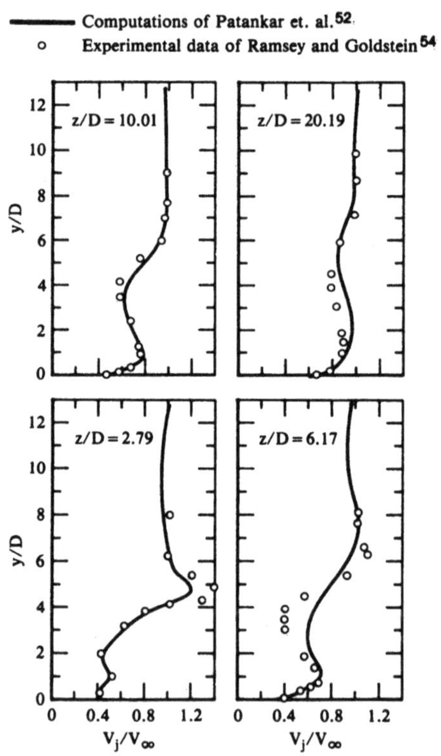

Figure 41. Comparison of computed and experimental velocity profiles for a jet-in-crossflow downstream of the wake.

fully-developed region. In the development region, a mass entrainment relation similar to Wooler's (1972) is assumed, but for the fully-developed region, a semi-empirical axial velocity decay relation is derived. With these modifications, Snel (1974) obtained results for the jet centerline generally in good agreement with the measurements. To calculate the induced flow-field, a panel method is used. The normal velocity at the jet surface is determined to satisfy the mass flux along the jet centerline. The calculated pressure distribution again had poor agreement in the wake.

In contrast to the methods of Wooler et al. (1972) and Snel (1974), Fearn and Weston (1975) simulate the jet flow by a diffuse-vortex model. The trajectories and strength of the vortices are prescribed by empirical relations, chosen to fit the experimental data. In the numerical evaluation of the induced velocity field, a series of finite strength straight-line filament vortices is placed along the vortex trajectory, with varying strength (determined experimentally). Image vortices are used below the plate to represent the boundary conditions on the plate. A series of sink elements also is placed along the jet centerline to simulate entrainment. The induced velocity at the plate surface is the sum of the velocity induced by the vortex and sink elements. In addition, a wake region is defined by a line through the origin at an angle of 45° to the crossflow direction as its boundary. In the wake region, the potential flow calculations are replaced by straight-line extensions at constant streamwise locations of the potential flow results at the boundary of the wake. This approach gives reasonable pressure predictions for velocity ratios greater than 5. At lower velocity ratios, the wake model is again inadequate. Furthermore, a severe limitation of the method is the requirement that the flow characteristics of the vortices must be known.

More recently, Adler and Baron (1979) proposed an integral method which predicts the inner structure of the deflected jet more accurately than the previously outlined methods. In their model, an improved jet growth function based on straight jet growth and vortex pair concept is employed to describe the curved jet; the cross section shape is not of an assumed simple geometry, but is approximately calculated resulting in a more realistic horseshoe shape, and quasi-three-dimensional nonsimilar velocity profiles (based on a realistic jet cross section shape) are used. Adler-Baron integral method gives good agreement with the experimental data for a wide range of jet to free stream velocity ratios and jet-injection angles.

Figure 42–45 show the comparison between the theoretical calculations and the experimental data for various flow properties.

At McDonnell Aircraft Company (Gilmer et al., 1983), Adler-Baron jet-in-crossflow model is employed in a surface-singularity panel method for calculating the induced lift due to jet-interference on a VTOL aircraft in transition mode of flight. Jet properties such as momentum and velocity centerline trajectories, entrainment rate, and the cross-sectional properties—such as area, shape, velocity profile, and mass-flux—are calculated using the Adler-Baron computer code. From these properties, a paneled representation of the jet is obtained (Figure 46). Panel normal velocities are also calculated to represent entrainment into the jet. The aircraft and jet paneling are then used in combination with a surface singularity technique to calculate the pressures, forces, and moments acting on the aircraft. For a velocity ratio $V_j/V_\infty = 7$, Figure 47 illustrates that the induced pressure distribution on the flat plate surface calculated by the panel method agrees closely with experimentally measured pressures (Fearn and Weston, 1975). The only significant difference is in the wake of the jet, where it is necessary to impose an empirical viscous correction to the potential flow pressure distribution. As mentioned before, none of the empirical theories model the wake accurately.

6. Calculation of Jet-Induced Lift on a V/STOL Aircraft in Transition Flight

The methodology employed for calculating the jet-induced lift on V/STOL aircraft in transition mode of flight is similar to that described for calculating the suck-down force in hover out of ground effect described in Section 4. As an illustration, we present here the calculations performed at McDonnell Aircraft Company (MCAIR) for a YAV-8B configuration (Gilmer et al., 1983). Figure 4 shows the aircraft-jet paneling for YAV-8B for a velocity ratio $V_j/V_\infty = 12.52$. The geometry and entrainment velocities for the paneled jets were calculated by the Adler-Baron computer program. An empirical correction to crossflow dynamic pressure was used to account for the effect of forward jet wake on the aft-jet properties. Figure 48 compares the predicted lift loss with experimental data obtained in MCAIR wind tunnel test programs. This figure shows good agreement between the experimental and calculated results for

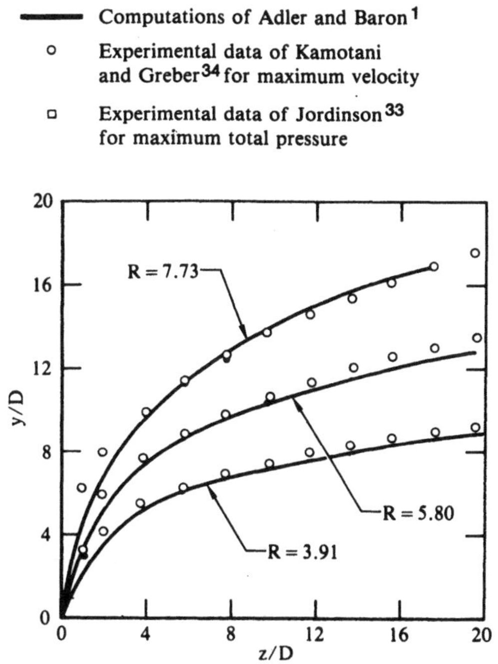

Figure 42. Comparison of computed and experimental centerlines for maximum velocity (or maximum total pressure) for a jet in crossflow, $\theta_j = 90°$.

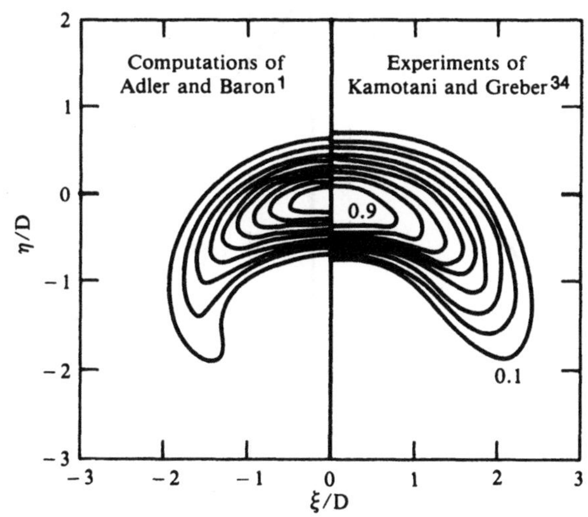

Figure 43. Comparison of cross sections and V = constant lines for $V_j/V_\infty = 3.91$, $z/D = 7$, and $\theta_j = 90°$.

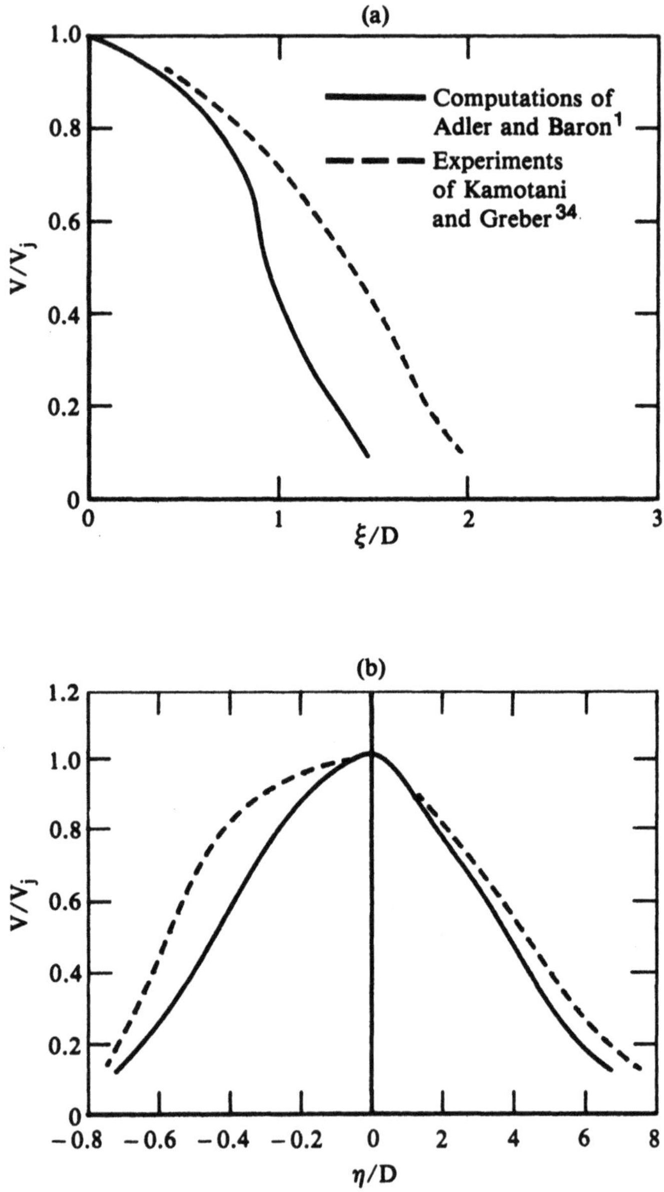

Figure 44. Comparison of velocity profiles for a jet in crossflow; (a) along $\eta/D = 0$, and (b) along $\xi/D = 0$.

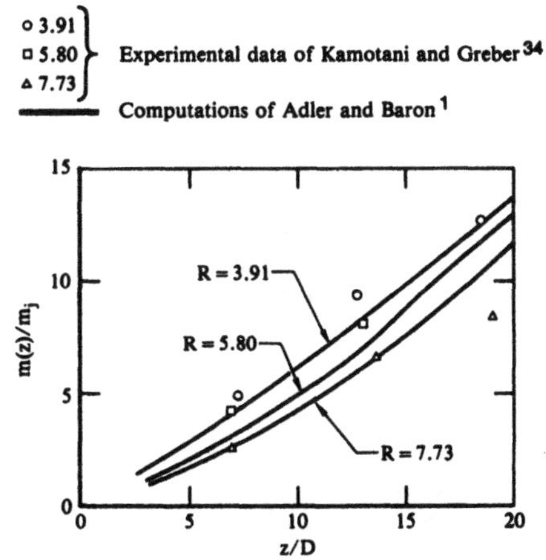

Figure 45. Comparison of the computed and experimental relative mass flux for a jet in crossflow, $\theta_j = 90°$.

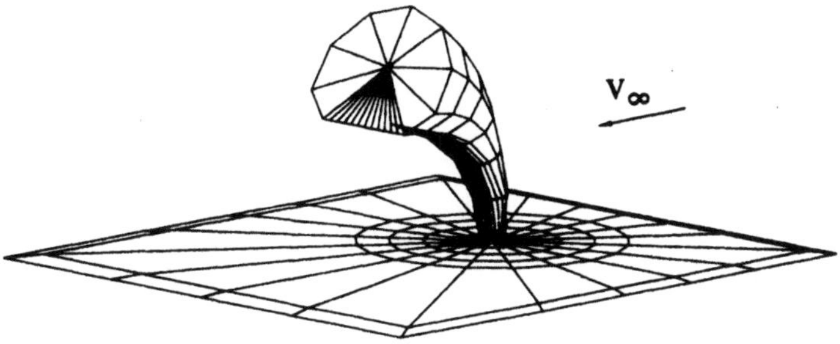

Figure 46. Paneling for a flat plate and circular jet in crossflow, $V_j/V_\infty = 7$, $\theta_j = 90°$.

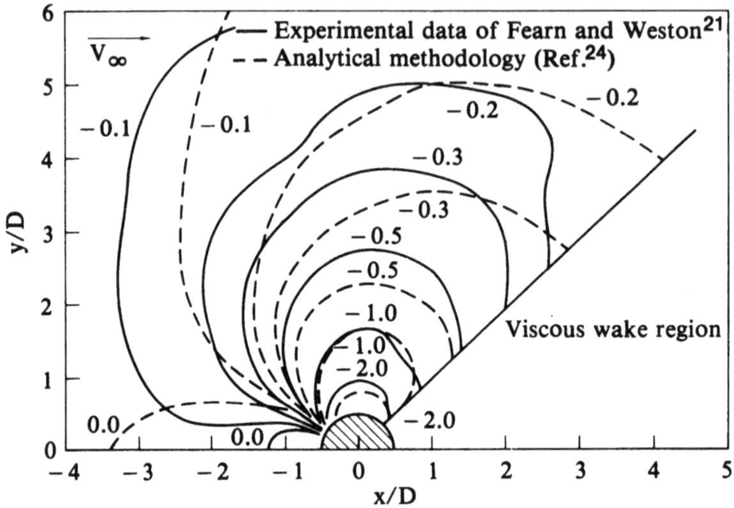

Figure 47. Comparison of induced pressure distribution for a jet in crossflow, $V_j/V_\infty = 7$, $\theta_j = 90°$.

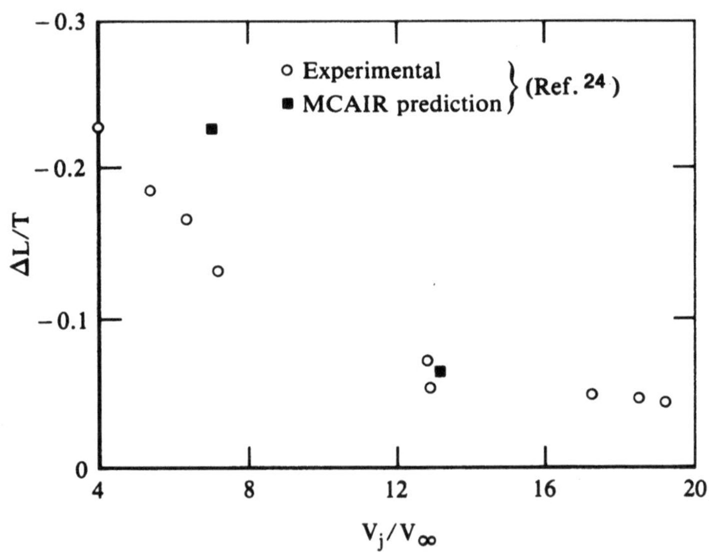

Figure 48. YAV-8B lift-loss in transition, $\theta_j = 60°$, $\delta_f = 50°$, $\alpha = 5°$.

$V_j/V_\infty = 12.52$, but a slight overprediction of the calculated results at $V_j/V_\infty = 7.09$. It should be noted that the viscous effects due to wing trailing edge flow separation have not been included for in these calculations.

7. Summary

This paper summarizes the currently used methodologies in aircraft industry for predicting forces and moments on a V/STOL aircraft in hover (in and out of ground effect) and transition modes of flight. These methodologies are based on a synthesis of various flow regions, each of which accommodates a specific flow phenomena, such as jet-ground interactions, jet-in-crossflow, and fountain-airframe impingement. The progress made in recent years in theoretical modeling of the flowfield of representative jet-flow configurations—single-jet impingement, twin-jet impingement with fountain formation, and jet-in-crossflow—is surveyed. The prediction methods ranging from semi-empirical approaches to the solution of Reynolds-averaged Navier-Stokes equations are discussed.

For two-dimensional and axisymmetric impinging-jet configurations, the flowfield calculations employing the Reynolds-averaged Navier-Stokes equations in conjunction with a two-equation turbulence model give excellent agreement with the data. Recent developments in numerical algorithms for calculating flows at high Reynolds numbers and turbulence modeling for recirculating flows have made possible the accurate computation of these flows. Three-dimensional flowfield calculations, however, require further advances in computing power. On the main-frame computers—commonly used today such as the CYBER 176, there are not enough grid points available to resolve the high-gradient regions at high Reynolds numbers.

In design practice, the prediction codes based on the Reynolds-averaged Navier-Stokes equations are not likely to be used routinely in the near future because of cost considerations; however, while semi-empirical prediction methods are orders of magnitude cheaper, their development has to be guided by experimental data. It is clear, by examining the current state of our understanding and ability to predict V/STOL related jet flow phenomena, that additional high-quality measurements and flow visualization data are needed for making further advances and refinements in semi-empirical prediction methods and turbulence-modeling for these flows. While the considerable experimental information available for free-jet flows, not only about the mean-flow properties but also about various turbulence quantities, has resulted in considerable development of computational methods and turbulence modeling for these flows, limited data are available for impinging jets.

Acknowledgement

The author wishes to acknowledge his gratitude to Dr. W. W. Bower of McDonnell Douglas Research Laboratories for introducing this subject to him and for providing suggestions that have improved the manuscript.

References

[1] Adler, D. and Baron, A. "Prediction of a Three-Dimensional Circular Turbulent Jet in Crossflow," *AIAA J., 17* (1979), 168.

[2] Agarwal, R. K. and Bower, W. W. "Navier-Stokes Computations of Turbulent Compressible Two-Dimensional Impinging Jet Flowfields," *AIAA J., 20* (1982), 577.

[3] Amano, R. S. "A Numerical Study of the Turbulent Impinging Jet," *Proc. of the ASME Symposium on Computers in Flow Predictions and Fluid Dynamics Experiments,* ASME Winter Annual Meeting, Washington, D. C., 1981.

[4] Amano, R. S. and Brandt, H. "Numerical Study of Turbulent Axisymmetric Jets Impinging on a Flat Plate and Flowing Into an Axisymmetric Cavity," unpublished.

[5] Antani, D. "An Experimental Investigation of the Vortices and Wake Associated With a Jet in Crossflow," *Ph.D. thesis,* Georgia Institute of Technology, 1977.

[6] Aoyagi, K. and Snyder, P. K. "Experimental Investigation of a Jet Inclined to a Subsonic Crossflow," *AIAA paper No. 81-2610,* 1981.

[7] Bower, W. W. "Computations of Three-Dimensional Impinging Jets Based on the Reynolds Equations," *AIAA Paper No. 82-1024,* 1982.

[8] Bower, W. W., Agarwal, R. K., and Peters, G. R. "Application of a New Navier-Stokes Solver to Three-Dimensional Jet Flows," *Proc. of the 18th Midwestern Mechanics Conference,* Univ. of Iowa, Iowa City, IA, 1983.

[9] Bower, W. W., Kotansky, D. R., and Hoffman, G. H. "Computations and Measurements of Two-Dimensional Turbulent Jet Impingement Flowfields," *Proc. of First Symposium on Turbulent Shear Flows*, Penn. State Univ., 1977.

[10] Bradbury, L.J.S. "The Impact of an Axisymmetric Jet Onto a Normal Ground," *Aeronautical Quart., 23* (1972), 141.

[11] Bradbury, L.J.S. and Wood, M. N. "The Static Pressure Around a Circular Jet Exhausting Normally From a Plane Wall Into an Air Stream," *Technical Note Aero 2978*, Royal Aircraft Establishment, Great Britain, 1969.

[12] Bradshaw, P. and Love, E. M. "The Normal Impingement of a Circular Air Jet on a Flat Surface," *ARC R&M, No. 3205*, 1959.

[13] Bristow, D. R. "Development of Panel Methods for Subsonic Analysis and Design," *NASA CR 3234*, 1980.

[14] Callaghan, E. E. and Ruggeri, R. S. "Investigation of the Penetration of an Air Jet Directed Perpendicularly to an Air Stream," *NASA Tech. Note 1615*, 1948.

[15] Chassaing, P., et al. "Physical Characteristics of Subsonic Jets in a Cross Stream," *J. Fluid Mech., 62* (1974), 41.

[16] Chien, J. C. and Schetz, J. A. "Numerical Solution of the Three-Dimensional Navier-Stokes Equations With Applications to Channel Flows and a Buoyant Jet in a Cross Flow," *J. Applied Mech., 62* (1975), 575.

[17] Claus, R. W. "Analytical Calculation of a Single Jet in Cross-flow and Comparison With Expepriment," *AIAA paper No. 83-0238*, 1983.

[18] Crabb, D., Durao, D.F.G., and Whitelaw, J. H. "Round Jet Normal to a Crossflow," *ASME J. Fluids Engg., 103* (1981), 142.

[19] Donaldson, C. du P. and Snedeker, R. S. "A Study of Free Jet Impingement; Part I, Mean Properties of Free and Impinging Jets," *J. Fluid Mech., 45* (1972), 281.

[20] Fearn, R. and Weston, R. P. "Vorticity Associated With a Jet in a Crossflow," *AIAA J., 12* (1974), 1666.

[21] ——————————. "Induced Pressure Distribution of a Jet in a Crossflow," *NASA TN D-7916*, 1975.

[22] Gardon, R. and Akfirat, J. C. "The Role of Turbulence in Determining the Heat-Transfer Characteristics of Impinging Jets," *Int. J. Heat and Mass Transfer, 8* (1965), 1261.

[23] Gilbert, B. L. "Detailed Turbulence Measurements in a Two-Dimensional Upwash," *AIAA paper No. 83-1687*, 1983.

[24] Gilmer, B. R., Miner, G. A., and Bristow, D. R. "Aircraft Aerodynamic Prediction Method for V/STOL Transition Including Flow Separation," *NASA CR-166467,* 1983.

[25] Giralt, F., Chia, C. J., and Trass, O. "Characterization of Impingement Region in an Axisymmetric Turbulent Jet," *Ind. Eng. Chem. Fundam, 16* (1977), 21.

[26] Glaze, L. W., Bristow, D. R., and Kotansky, D. R. "V/STOL Fountain Force Coefficient," *NAVAL Air and Development Center Report No. NADC-81106-60,* 1983.

[27] Gordier, R. L. "Studies on Fluid Jets Discharging Normally Into Moving Liquid," *Tech. Paper No. 28, Series B.,* St. Anthony Falls Hydraulics Lab., Univ. of Minnesota, 1959.

[28] Hill, B. J. "Measurement of Local Entrainment Rate in the Initial Region of Axisymmetric Turbulent Air Jets," *J. Fluid Mech., 51* (1972), 773.

[29] Hill, W. G. Jr. and Jenkins, R. C. "Effect of Nozzle Spacing on Ground Interference Forces for a Two-Jet V/STOL Aircraft," *J. Aircraft, 17* (1979), 684.

[30] Jenkins, R. C. and Hill, W. G. Jr. "Investigation of VTOL Upwash Flows Formed by Two Impinging Jets," *Grumman Research Dept. Report RE-548,* 1977.

[31] Jones, W. P. and Launder, B. E. "The Calculation of Low-Reynolds Number Phenomena With a Two-Equation Model of Turbulence," *Int. J. Heat Mass Transfer, 16* (1973), 1119.

[32] Jones, W. P. and McGuirk, J. J. "Computations of a Round Turbulent Jet Discharging Into a Confined Crossflow," *Turbulent Shear Flows, Vol. 2,* Springer-Verlag (1980), 233.

[33] Jordinson, R. "Flow in a Jet Directed Normal to the Wind," *R&M 3074,* Aeronautical Research Council, Great Britain, 1956.

[34] Kamotani, Y. and Greber, I. G. "Experiments in a Turbulent Jet in a Crossflow," *AIAA J., 10* (1972), 1925.

[35] Keffer, J. F. and Baines, W. D. "The Round Turbulent Jet in a Crosswind," *J. Fluid Mech., 15* (1963), 481.

[36] Kleis, S. J. and Foss, J. F. "The Effect of Exit Conditions on the Development of an Axisymmetric Turbulent Free Jet," *Third Year Technical Report, NASA Grant NCR 23-004-068,* Michigan State Univ., 1974.

[37] Kleinstein, G. "Mixing in Turbulent Axially Symmetric Free Jet," *J. of Spacecraft, 1* (1964), 403.

[38] Kotansky, D. R. "The Modeling and Prediction of Multiple Jet

V/STOL Aircraft Flowfields in Ground Effect," *AGARD-CP-308* (1981), 1000.

[39] ———————. "Multiple Jet Impingement Flowfields," *Proc. of the International Symposium on Recent Advances in Aeronautics and Acoustics,* Stanford Univ., 1983.

[40] Kotansky, D. R. and Giaze, L. W. "Development of an Empirical Data Base and Analytical Modeling of Multi-Jet V/STOL Flowfields in Ground Effect," *Proc. V/STOL Aircraft Aerodynamics, Vol. I,* Monterey: Naval Postgraduate School, 1979.

[41] ———————. Investigation of the Interaction of Lift Jets and a Ground Plane," *NASA CR 152343,* 1980.

[42] ———————. "Investigation of the Effects of Ground Wall Jet Characteristics on Fountain Upwash Flow Formation and Development," *Office of Naval Research Report ONR-CR-212-261-IF,* 1980.

[43] ———————. "Characteristics of Wall Jets Produced by the Impingement on a Ground Plane of Rectangular Jets of Aspect Ratio One Through Eight," *AIAA paper No. 81-0012,* 1981.

[44] ———————. "The effects of Ground Wall-Jet Characteristics on Fountain Upwash Flow Formation and Development," *AIAA paper No. 81-1294,* 1981.

[45] ———————. "Impingement of Rectangular Jets on a Ground Plane," *AIAA J., 20* (1982), 585.

[46] Kotansky, D. R., et al. "Multi-Jet Induced Forces and Moments on VTOL Aircraft Hovering in and out of Ground Effect," *NAVAL Air and Development Center Report NADC 77-229-30,* 1977.

[47] Kutler, P. "A Perspective on Theoretical and Applied Computational Fluid Dynamics," *AIAA Paper No. 83-0037,* 1983.

[48] Launder, B. E. and Spalding, D. B. "The Numerical Computation of Turbulent Flows," *Computer Methods in Applied Mechanics and Engg., 3* (1973), 269.

[49] Margason, R. J. "The Path of a Jet Directed at Large Angles to a Subsonic Free Stream," NASA TN D-4919, 1968.

[50] Melnik, R. E. and Rubel, A. "'Asymptotic Theory of Turbulent Wall Jets," *Proc. of the Second Symposium on Numerical and Physical Aspects of Aerodynamic Flows,* Cal. State Univ., Long Beach, CA, 1983.

[51] Mosher, D. K. "An Experimental Investigation of Turbulent Jet in a Crossflow," *Ph.D. thesis,* Georgia Institute of Technology, 1970.

[52] Patankar, S. V., Basu, D. K., and Alpay, S. A. "Prediction of the Three-Dimensional Velocity Field of a Deflected Turbulent Jet," *J. Fluid Engg., 99* (1977), 758.

[53] Poreh, M., Tsuei, Y. G., and Cermak, J. E. "Investigation of a Turbulent Radial Wall Jet," *Trans. ASME J. of Applied Mech., 89* (1967), 457.

[54] Ramsey, J. W. and Goldstein, R. J. "Interaction of a Heated Jet With Deflecting Stream," *NASA CR-72613,* 1970.

[55] Ricou, F. P. and Spalding, D. B. "Measurements of Entrainment by Axisymmetric Turbulent Jets," *J. Fluid Mech., 11* (1961), 21.

[56] Rubel, A. "Computations of Inviscid Rotational Jet Impingement Regions," *Proc. V/STOL Aircraft Aerodynamics, Vol. I,* Monterey: Naval Postgraduate School, 1979.

[57] ———. "Computations of Jet Impingement on a Flat Surface," *AIAA J., 18* (1980), 168.

[58] ———. "Computations of the Oblique Impingement of Inviscid, Rotational Round Jets Upon a Plane Wall," *AIAA paper No. 80-1331,* 1980.

[59] Saripalli, K. R. "Visualization of Multi-Jet Impingement Flow," *AIAA paper No. 81-1364,* 1981.

[60] Schauer, J. J. and Eustis, R. H. "The Flow Development and Heat Transfer Characteristics of Plane Turbulent Impinging Jets," *Report TR3,* Dept. of Mech. Engg., Stanford Univ., 1963.

[61] Scholtz, M. T. and Trass, O. "Mass Transfer in a Nonuniform Impinging Jet," *AICHE J., 16* (1970), 82.

[62] Siclari, M. J., Migdal, D., and Palcza, J. L. "The Development of Theoretical Models for Jet Induced Effects on V/STOL Aircraft," *J. Aircraft, 13* (1976), 938.

[63] Siclari, M. J., et al. "VTOL In-Ground Effect Flows for Closely Spaced Jet," *AIAA paper No. 80-1880,* 1980.

[64] Snedeker, R. S. and Donaldson, C. du P. "Experiments on Oblique Impingement of Underexpanded Jets on a Flat Plate," *ARAP Report No. 64,* Princeton, 1965.

[65] Snel, H. "A Model for the Calculation of the Properties of a Jet in a Crossflow," *NCR-TR-74080 U.,* 1974.

[66] Sparrow, E. M. and Lee, L. "Analysis of Flowfield and Impingement Heat/Mass Transfer Due to a Nonuniform Slot Jet," *ASME J. Heat Transfer, 97C* (1975), 191.

[67] Tani, I. and Komatsu, Y. "Impingement of a Round Jet on a Flat Surface," *International Congress of Applied Mechanics, Munich,*

Berlin: Springer-Verlag (1966), 672.

[68] Taylor, P. "An Investigation of an Inclined Jet in a Crosswind," *Aeronautical Quart.*, *28* (1977), 51.

[69] Tentacoste, N. and Sforza, P. "Further Experimental Results for Three-Dimensional Free Jets," *AIAA J.*, *5* (1967), 885.

[70] Vogler, R. D. "Interference Effects on Single and Multiple Round or Slotted Jets on a VTOL Configuration in Transition," *NASA TN D-2380*, 1969.

[71] Wolfshtein, M. "Convection Processes in Turbulent Impinging Jets," *Report SF/R/2*, Dept. of Mech. Eng., Imperial College of Science and Technology, 1967.

[72] Wooler, P. T., et al. "V/STOL Aircraft Aerodynamic Prediction Methods Investigation," *AFFDL-TR-72-26*, 1972.

[73] Wu, J. C., et al. "Experimental and Analytical Investigation of Jets Exhausting Into a Deflecting Stream," *J. Aircraft*, *7* (1970), 44.

[74] Wygnanski, I. "The Flow Induced by Two-Dimensional and Axisymmetric Turbulent Jets Issuing Normally to an Infinite Plane Surface," *Report No. 63-12*, Mech. Eng. Research Labs., McGill Univ., Montreal, Canada, 1963.

[75] Yen, K. T. "The Aerodynamics of a Jet in Crossflow," *Naval Air Development Center Report No. NADC-78291-60*, 1978.

[76] ————. "Vertical Momentum of the Fountain Produced by Multi-Jet Vertical Impingement on Flat Ground Plane," *J. Aircraft*, *18* (1981), 650.

[77] Yu, N. J. "Transonic Flow Simulations for Complex Configurations With Surface-Fitted Grids," *AIAA Paper No. 81-1258*, 1981.

Part IV
Experimental Techniques

A Wind-Tunnel Method for V/STOL Testing

W. R. Sears
University of Arizona
Tucson, AZ 85721

Nomenclature

A	matrix relating f and g in exterior flow 3)(equation 6.3).
B	matrix relating Δg and Δf in interior flow (equation 6.4).
C	jet-momentum coefficient $= (1/2)C_J U^2 c$ where C_J is the usual dimensionless jet-momentum coefficient, defined as (jet momentum flux)/(dynamic pressure of flight $\times$ wing area).
f, g	flow variables (functions) used to establish unconfined flow in the adaptable-wall scheme.
$g[f]$	the function g according to the computed outer flow determined by boundary values f and the far-field boundary conditions.
k	*relaxation factor* defined in Table 1, steps 4 and 6.
S	the interface between the inner and outer flow regions, defined by the instrumentation.
q	strength of source panel on Σ.
u, v, w	Cartesian components, in the x, y, z coordinate system, of the perturbation velocity vector; viz. $(u, v, w) =$ (fluid velocity) $-$ $\mathbf{U}$.
$\mathbf{U}$	simulated free-stream vector.
U	magnitude of $\mathbf{U}$.
v_t	tangential (to S) component of (u, v, w).
v_n	normal (to S) component of (u, v, w).
${V_n^{(p)}}_{ij}$	matrix giving v_n at point i due to unit increase of strength of jth panel, including indirect effect due to test model.
${V_t^{(p)}}_{ij}$	same for v_t.

x, y, z	Cartesian coordinate system (Figure 2); the x direction is parallel to the tunnel axis.
α	angle of attack of test model.
β	angle of test model relative to tunnel axis.
Γ	circulation; thus Γ_r = strength of rth dipole panel of the jet-flap model.
$\delta f, \delta v_n$	mismatch function defined in Table 1, step 4.
$\Delta f, \Delta g$	increment functions defined in equation (6.2).
θ_r	inclination of the rth wake panel relative to the tunnel axis.
Σ	surface formed by wall-control panels—*the tunnel wall.*

Subscripts

i	field-point number (point on S).
j	panel number (panel on Σ).
n	normal to S.
t	tangential to S.
r	panel number (panel of wing/wake model).

Superscripts

(p)	denotes quantities pertaining to the pth iteration.

1. Introduction

A persistent problem in wind-tunnel testing of V/STOL configurations is to simulate accurately their low-speed, high-lift regime of flight. This flight regime is, of course, characterized by very large lift coefficients produced by powered high-lift devices. This implies that the stream flow in the wind tunnel is grossly distorted; the effects of tunnel boundaries are large and the determination of the simulated flight-velocity vector is difficult. Since the direction of this vector is in doubt, the decomposition of force into lift and drag may be very inaccurate.

This paper is a progress report on a program of research intended to solve this problem. We begin with a brief review of the so-called *adaptive wall* concept, upon which our research is based, pointing out that its most important feature may be that it eliminates the time-honored, but questionable process of *wind-tunnel calibration*. We then proceed to consideration of a new idea in wind-tunnel design, which exploits this feature in a radical way. We present results of a numerical study in which the new type of tunnel was modelled, with a powered-lift airplane model installed, and the process of iterative tunnel modification simulated in some detail. The report closes with some results concerning convergence of the iterative scheme and discussion of a non-iterative method.

2. The Adaptive-Wall Concept

The adaptive-wall scheme has been with us for a decade (Ferri and Baronti, 1973; Sears, 1974); it has received attention in several countries (Binyon, 1982), and a number of adaptable-wall tunnels have been constructed for the transonic speed regime (Binyon, 1982; Sears et al., 1977; Satyanarayana et al., 1981). It seems likely, to the present author, that the concept is already familiar to his audience and that only the briefest outline is needed here, for emphasis.

The basic principle involved is that the proper simulation in a wind tunnel is assured when measured flow quantities at an interface S surrounding the test model are matched to a calculated, updated, outer-flow field that satisfies far-field boundary conditions. These far-field conditions specify the free-stream vector $\mathbf{U}$ in direction and magnitude.

The matching at the interface S must be done iteratively (or the equivalent), since the details of the outer field are not known a priori. Distributions of two flow variables must be measured at S; they are either consistent with the far-field boundary conditions, in which case the simulation is correct, or they are not, in which case adjustment of the wind-tunnel flow must be carried out to achieve proper simulation.

The practical procedure is to use one of the measured distributions, say f, as boundary values on S, carry out the calculation of the flow exterior to S using the far-field boundary conditions, and compare the result with the other measured data, say g, at S. If agreement is not found, the discrepancy constitutes a measure of the departure from correct simulation and can be used to set up a pattern of tunnel speed-control and wall-control settings to improve the simulation. This procedure is outlined in Table 1, where the notation $g[f^{(i)}]$ denotes the distribution of g calculated by using the values $f^{(i)}$ as inner boundary values and satisfying the far-field conditions. Here, also, $f^{(i)}$ and $g^{(i)}$ denote distributions measured in the ith iteration.

Table 1. Outline of the adaptable-wall iteration.

1. Set model configuration (α, flap angles, power, etc.) and initial tunnel configuration (arbitrary).
2. Measure $f^{(1)}$ and $g^{(1)}$ on interface S.
3. Calculate outer flow-field with $f^{(1)}$ as boundary values and read out $g[f^{(1)}]$.
4. Form $\delta^{(1)}g = g[f^{(1)}] - g^{(1)}$ and $g^{(2)} = g^{(1)} + k\delta^{(1)}g$.
5. Adjust wall-control organs to obtain $g = g^{(2)}$ on S.
6. Repeat from step 2: ($g^{(n+1)} = g^{(n)} + k\delta^{(n)}g$ etc.) until $\delta^{(n)}g$ is as small as required.

Thus, the adaptive-wall scheme requires that the wind tunnel possess three special features:

(1) **Instrumentation** capable of measuring two flow distributions, say f and g, on an interface S within the working section, surrounding the test model.

(2) **Computing equipment** to calculate the distribution $g = g[f]$ in the flow external to S.

(3) **Control organs** capable of modifying the tunnel walls and flow speed in such a manner as to change the values of f and g measured at S.

In existing embodiments of the scheme, the wall-control organs (3) control either the distribution of wall porosity and/or local exterior plenum pressure, or they control the actual geometrical shapes of solid tunnel walls. In the first case, f and g are chosen to be velocity components; in the second case, f has been taken to be the local static pressure at the wall and g the local slope of the wall. Considerable success has been achieved in both cases in the transonic regime.

The importance of the adaptive-wall principle in replacing the *tunnel calibration*—i.e., in fixing the simulated stream vector—has not always been fully appreciated, but it is probably more important than traditional *wind-tunnel wall interference* in many cases of testing at high Mach numbers, with shock waves, large blockage, or as in the present case, large flow deflection. This matter was investigated at some length by numerical simulation at the University of Arizona and reported in Sears (1983). According to these simulations, it was possible to begin iterations with gross errors in both magnitude and direction of the actual stream vector **U** and to arrive rapidly at the prescribed values by means of the iteration process of Table 1.

3. Application to V/STOL Testing

The suggestion that the adaptive-wall principle might be useful in V/STOL testing was made early in the history of this subject (Sears, 1974). There is, however, a complication in this application, namely that the vortical, powered wake that is characteristic of high-lift configurations must emerge from the closed interface S and must, therefore, be accounted for in the modelling of the exterior flow. Although this complication does not seem unsurmountable, it means that this wake, in some detail, must be identified in the process of measuring f and g and its effects on the whole flow field exterior to S must be included in the modelling process; the modelling process must be capable of predicting, from measurements at S, the geometry and the aerodynamics of the wake.

There seems to be only one way to circumvent this; namely, it might be possible to employ a very long interface, shaped and positioned so that the wake does not emerge from it. The wake would then not be involved in the calculations in any explicit way, although its effects would be present and correctly accounted for.

It was this reasoning that led to the idea of the *Arizona High-Lift Wind Tunnel*. In this tunnel the *downstream* direction—assumed to be fixed by tunnel architecture and by the placement of measuring instrument—is always reserved for the wake. (See Figure 1.) Thus the stream direction, angle of attack, etc., are always made such that the wake runs harmlessly downstream without impinging on instrument positions or, of course, on tunnel floor or ceiling. The simulated stream direction must therefore be variable, under control

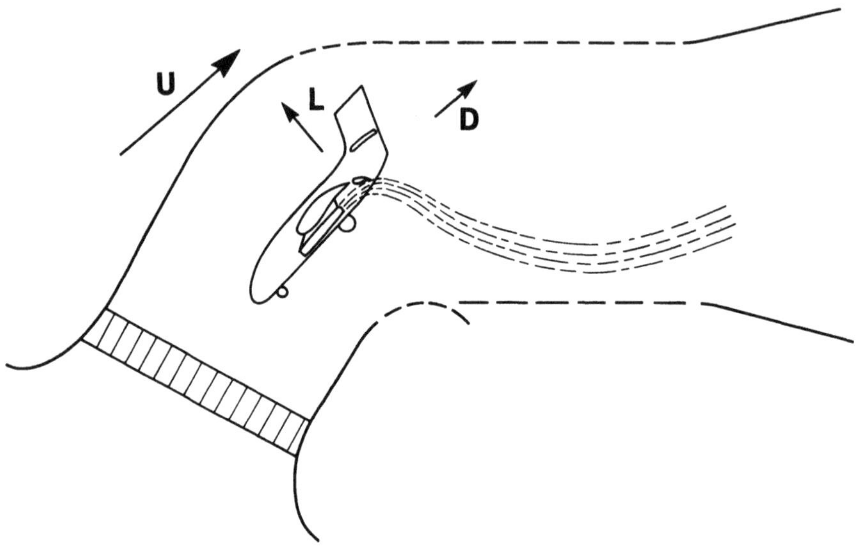

Figure 1. Sketch of a proposed adaptable wind tunnel for testing at high lift coefficients.

of the operator; this can be done in an adaptive tunnel as explained above.

In other words, *the adaptive-wall strategy permits the simulated stream direction to be divorced from the tunnel architecture, whereupon the fixed,* **downstream** *direction can be reserved for the troublesome wake.* No modelling of the vortical, powered wake is required.

In a practical embodiment, of course, the interface cannot be infinitely long, and neither can the working section with its array of wall controls. What is required is that this array and the array of instrumentation defining S be sufficient to define adequately the exterior flow field and to control adequately the flow field within the wind tunnel. Numerical simulations based on the panel methods that are described below have confirmed that very modest, practical dimensions are sufficient.

Table 2 is an outline of the test procedure to be followed in testing a given model configuration in a tunnel of the Arizona type. The model is placed in the tunnel at a first-guess geometric angle β, and it is ascertained (by readings from a simple, fixed rake of total-head tubes, for example) that the wake lies in a benign position. The desired angle of attack α then determines the required stream direction, arg $\mathbf{U}$, which is put into the computer as a far-field boundary condition. If the subsequent iteration to unconfined-flow conditions does not move the wake to an undesirable position, all is well. If it does, equal increments in β and arg $\mathbf{U}$ must be introduced to reposition the wake, and the iteration is continued. (Incidentally, our simulations suggest that the wake movements that occur during iteration are small, so that repositioning is not likely to be necessary.)

Table 2. Outline of test procedure.

1. Set model configuration: flap angles, power, etc.
2. Set initial tunnel configuration and approximate speed.
3. Rotate model so that wake trails downstream without impinging on instrumentation defining S.
4. Choose stream vector $\mathbf{U}$ to give desired angle of attack, and begin iteration to unconfined flow, as outlined in Table 1.
5. If wake changes position sufficiently to interfere with S, rotate model and stream vector equal amounts and proceed with iteration.

4. Simulations of Tests of a Jet-Flap Wing

A small demonstration version of the wind tunnel described above is under construction at the University of Arizona. In the process of design of this laboratory model, we have carried out a series of numerical simulations to demonstrate that a tunnel of this type, operated in the iterative mode described, can successfully test a powered-lift configuration.

To carry out this simulation, the controllable tunnel walls were modelled by source/sink panels and the test model—a jet-flap wing—by dipole panels. No attempt was made to model in detail either the flow through controllable wall panels or the flow through a variable-angle nozzle. Instead, the familiar uniform-strength panels were used, including four source panels on the front face of the rectangular box representing the working section—i.e., the five-sided box labelled Σ in Figure 2.

Thus, although we have in mind a tunnel like the one sketched in Figure 1, for example, we do not attempt to model its upstream geometry in detail. Rather, we assume that the rectangular panel arrays, Σ and S of Figure 2, represent a category of tunnels having instrumentation to measure distributions f and g on the open-ended interface S and having wall-control organs capable of performing the equivalent of the functions of the source panels of Σ. The latter is, of course, a strong assumption; it means that we are testing here the principles of the Arizona tunnel but not, so far as concerns wall-control organs, its detailed features.

The modelling of the test model (wing) and its wake is described later in this paper.

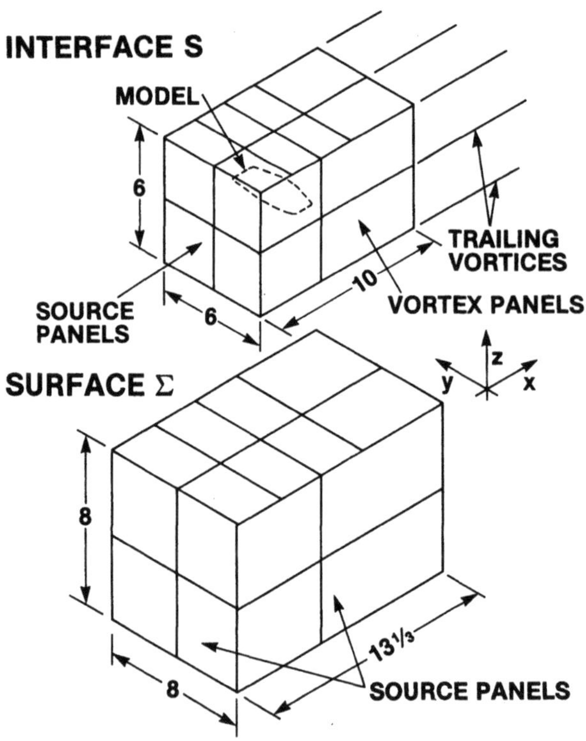

Figure 2. Panel arrays (interface S and tunnel walls Σ) and coordinate system.

4.1 Modelling of the Exterior Flow

It has been emphasized above that numerical calculation of the flow exterior to interface S is an essential feature of the adaptable-wall scheme. Thus, this part of our procedure constitutes both a simulation and a feature of actual operation of the tunnel. It is accomplished by means of an array of uniform-strength distributed-vortex panels on four sides of S and four source panels on the front face of S, as in Figure 2. The interface is a five-sided rectangular box, geometrically similar to Σ and symmetrically placed within Σ. Each vortex panel of S is actually infinitely long, since its longitudinal, concentrated, edge vortices extend back indefinitely in the plane of the panel.

4.2 Modelling of a Jet-Flap Wing

For our simulation of the testing of a high-lift wing model, it was necessary to have a tractable numerical model of the test article itself. For this purpose, the jet-flap wing is attractive, since it can be modelled rather simply. Its theory has been given in Spence (1956), Maskell and Spence (1959), Kerney (1967), and Tokuda (1971). The details of our present modelling, a panel method, are given in Lee (1981).

We use the array of dipole panels (rectangular vortex rings) sketched in Figure 3. There are 16 panels of uniform dipole strength; eight represent the wing and its mechanical flap and therefore have specified angles of incidence; the remaining eight represent the momentum-wake, whose shape is unknown and must be determined.

Two wake conditions determine this shape:

1. The wake is impermeable; thus the total normal velocity component—here evaluated at a single selected field-point on each panel—must vanish.

2. The normal force on the wake is proportional to the jet momentum flux and the wake curvature.

In the present approximation, wake circulation is concentrated in vortex filaments at the lines where the panels join. The force is therefore concentrated there too, and produces finite angular deflections there. We neglect the lateral velocity components at the wake panels; they also affect the wake shape, but the vertical deflection of

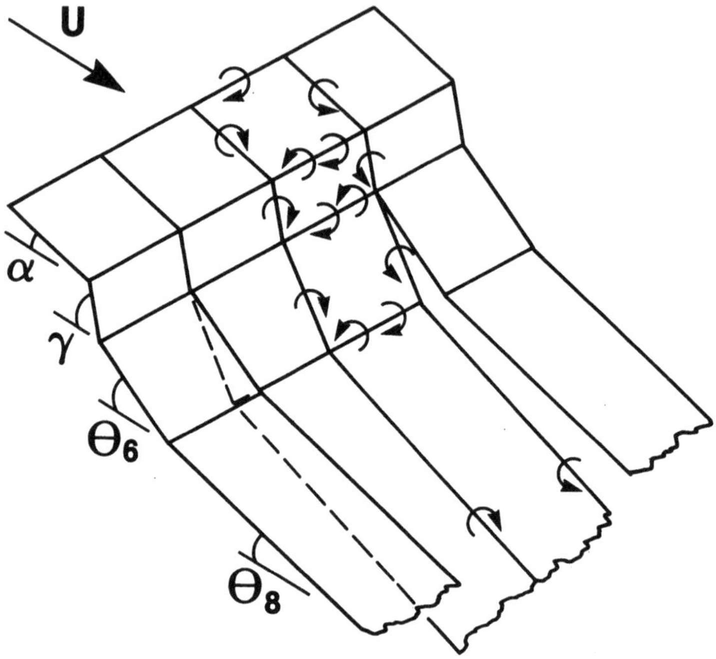

Figure 3. Sketch showing panel model of jet-flap wing and wake.

the wake is much larger near the wing, and therefore more important.

It will be recognized that the unknowns here are the 16 panel strengths Γ_r and eight wake-panel angles θ_r. These are determined by 16 kinematic conditions ((1), above) and eight force conditions (2). For large deflections (high lift), these relations are nonlinear. We obtain a solution, for a given wing and given jet-momentum flux, by assuming that the changes of wake deflections from one iteration to the next are small angles; this permits linearization of the trigonometric functions involved. Other quantities involved in conditions 1 and 2 are flow-field quantities and are evaluated from the preceding iteration. In other words, the solution is obtained by local linearization at each step, assuming small changes of wake deflection in the next step. This technique requires that a guess be made for the θ_rs to get the process started. The calculation at each step is then the solution of a set of linear algebraic equations, for which rapid matrix techniques are available.

Before this model was used in our wind-tunnel studies, it was exercised, by the author of Lee (1981), in cases of unconfined flow at various angles of attack and jet-momentum coefficients. The Cyber 175 computer at the University of Arizona was used. The effects of increasing the number of panels, both laterally and longitudinally, were also investigated. It was found that 10 to 15 iterations were required to reach convergence in unconfined flow. In the interests of brevity, further details concerning the jet-flap-wing model will not be presented here; interested readers are referred to Lee (1981).

4.3 Simulation of the Iteration

It is assumed next that this jet-flap wing is mounted inside the arrays of Figure 2 and that its angle of attack is very different from the value desired. Since the starting configuration of flow is arbitrary, we begin the process by assigning to the wing the circulation values Γ_r and wake angles θ_r corresponding to unconfined flow at the initial (wrong) angle of attack. The geometrical angle β of the wing is chosen to place the wake within S at the downstream end of the test section. The induced velocities at the field points of S then constitute the initial measured distributions $f^{(1)}$, $g^{(1)}$ in the tunnel; we are at step 4 of Table 2.

To be specific, we choose for the distribution f the velocity component tangent to S, viz. v_t, and for g the normal component v_n. (It will be seen that the kinds of panels chosen above for S and Σ, respectively, were dictated by these choices of f and g.)

Proceeding with the adaptive-wall strategy, we now activate the wall panels–the source panels of Σ—in accordance with the algorithm (Table 1). Our jet-flap model is now exercised again, this time in the presence of perturbations from the wall panels, leading to a new wing/wake configuration, and so on. If the procedure converges ($\delta v_n \rightarrow 0$), the total perturbations due to the activated wall panels are just those that cancel the errors of the initial situation, causing inner and outer fields to match at S. The outer field, it will be recalled, always satisfies the right conditions at infinity—viz. undisturbed flow at velocity U. The wing/wake geometry and circulation distribution should then agree with unconfined flow results for the same wing at the same values of α, U, and the jet-momentum coefficient C.

In each iteration it is necessary to adjust the wall controls (i.e., here, to set the strengths of the Σ-panels) so as to change the normal-velocity distribution from $v_n^{(p)}$ to $v_n^{(p+1)}$. In a real experiment, the way to do this would be to measure the influence matrix, say $V_n^{(p)}{}_{ij}$, giving the effects on v_n at i due to unit changes of the jth control organs. This could be done by making a small change in the setting of each control organ and measuring the resulting changes of v_{n_i}.

Our procedure, for this numerical demonstration, is to simulate exactly this process: we evaluate the combined matrix $V_n^{(p)}{}_{ij}$. This requires that the strength q_j of each source-panel of Σ be increased slightly, one at a time, the jet-flap wing be iterated to convergence with consideration of the resulting velocity increments at the wing-and-wake panels, and the resulting *combined* increments in v_{n_i} on S be calculated.

We can now invert $V_n^{(p)}{}_{ij}$ and calculate the total changes of source strengths q_j to produce the increments $k\delta^{(p)}v_n$ on S. In view of the nonlinearity, the calculated matrix pertains only to *wing/wake configuration p*. Moreover, we do not assume that our new q_is will produce exactly the desired distribution $v_n^{(p+1)}$ when all panels are simultaneously adjusted; hence we actually re-iterate the jet-flap wing in the presence of the velocity field of all our Σ-panels, together. The result is *wing/wake configuration* $p+1$. Its geometry and circulation distribution now provide the values $v_n^{(p+1)}{}_i$ and $v_t^{(p+1)}{}_i$. We are now ready for the next iteration.

It will be clear to the reader that each step of the iteration of the wind-tunnel configuration involves $N+1$ iterative calculations of jet-flap wing aerodynamics, where N is the number of wall controls. The numerical simulation we are describing, namely the complete simulation of a test of a high-lift model, is therefore a rather complex one, involving jet-flap-wing calculations, each of them iterative, within iterations of the wind-tunnel configuration toward unconfined flow. Nevertheless, the computer time needed for a full simulation turned out to be very little, as will be mentioned below.

5. Results of Simulations

The goal of our simulations has been to demonstrate that a tunnel of the Arizona category can be used to produce unconfined conditions around a jet-flap wing at large lift coefficients and large wake deflections. The panel arrays used are sketched in Figures 2 and 3; the tunnel has 28 control panels on Σ. For calculation of the outer flow, the interface S is made up of 24 vortex panels and four source panels, each carrying a field point.

The following are data taken from a typical numerical case carried through in our studies; this case constitutes the numerical demonstration for the present paper: (see Figure 4)

wing span	4.0
wing chord including flap	1.2
flap angle relative to wing	60 deg.
jet-momentum coefficient	$C = 0.5$
desired angle of attack	15 deg.
desired stream speed	$U = 1.0$
wing angle relative to tunnel axis (β)	−45 deg.

Thus, the simulated stream vector U should make an angle of 60 deg. to the tunnel axis.

Again the Cyber 175 of the University of Arizona was used. Each "experiment" in this case required about 20 iterations in the adaptive-wall procedure. Since there were 28 Σ-panels and lateral symmetry was assumed, the process of "experimentally" determining each member of the combined v matrix described above had to be carried out 14 times in each iteration step. Each of these determinations required at least seven iterations of the wing geometry. Nevertheless, a typical "experiment" was carried out, from arbitrary starting situation to unconfined flow, in about a minute of computer time.

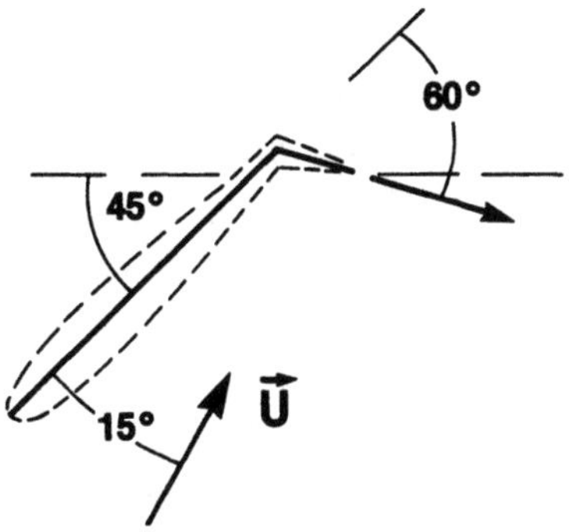

Figure 4.

Results of this calculation are presented in Figure 5 and Table 3. In Figure 5 are plotted (1) the mean absolute error in $\Gamma_r(p)$, (2) the maximum absolute value of the *mismatch-signal* $\delta^{(p)}v_n$ on S, and (3) the average absolute error in $v_n^{(p)}$ on S. The abscissa is the iteration-number, p. In this test case the properties and flow-field of the jet-flap wing in unconfined flow are known, so that the errors in (1) and (3) can be evaluated at any step of the iteration.

These results are also presented in Sears (1983). In Table 4 we present the results of some additional simulations that we carried out to investigate some practical matters:

A. Mean-square Fitting

When the number of wall controls on Σ is smaller than the number of field points on S, the determination of control settings q_i for given $k\delta v_n$ is made such as to obtain a least-mean-square-error fit. This technique was used in four simulations reported in Table 4, in which there were 44 field points but only 28 control panels. We conclude that larger values of k can be used when there are more field points and that this accelerates convergence.

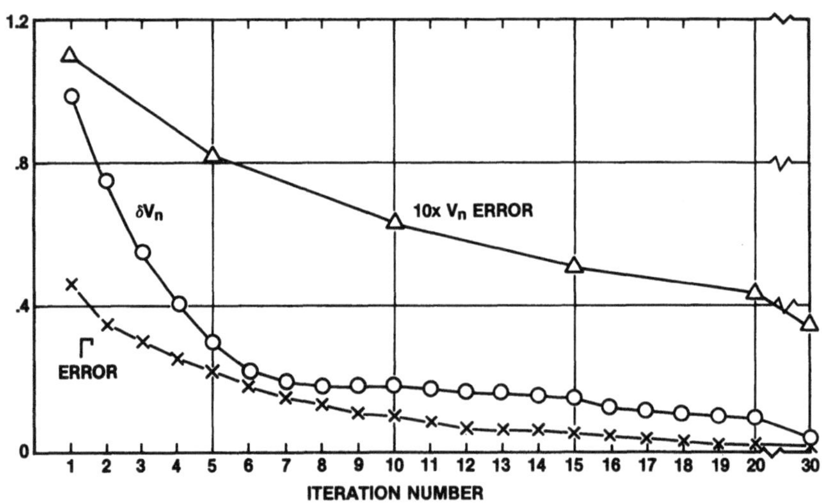

Figure 5. Iteration of jet-flap wing to unconfined-flow conditions in wind tunnel. $k = 0.05$. 28 Σ-panels, 28 points on S. Initial conditions: $\mathbf{U} = 1.0$ at 45° ($\alpha = 0°$). Final condition: $\mathbf{U} = 1.0$ at 60° ($\alpha = 15°$). $\times$ average absolute error in Γ_r. o maximum $|\delta v_n|$ on S. Δ average absolute true error in v_n on S. $C = 0.5$. The points are connected only for clarity.

Table 3. Results of iteration of jet-flap wing to unconfined flow in wind tunnel.

$k = 0.05$. Initial condition: $\mathbf{U} = 1.0$ at 45° ($\alpha = 0°$). Final condition: $\mathbf{U} = 1.0$ at 60° ($\alpha = 15°$). 28 Σ-panels, 28 S-points, wing centered in tunnel. $C = 0.5$.

Iteration Number	Γ_1	Γ_2	Γ_3	Γ_4	Γ_5	Γ_6	Γ_7	Γ_8	θ_5	θ_6	θ_7	θ_8
1	1.26	0.96	2.15	1.75	2.31	1.88	2.38	1.94	0.92°	3.97°	−5.93°	0°
3	1.44	1.11	2.33	1.90	2.49	2.03	2.57	2.08	1.00°	4.26°	−5.71°	0.73°
6	1.60	1.25	2.50	2.03	2.66	2.17	2.74	2.22	1.07°	4.57°	−5.48°	1.48°
10	1.70	1.33	2.60	2.12	2.77	2.26	2.85	2.30	1.19°	4.84°	−5.20°	2.09°
15	1.77	1.39	2.66	2.16	2.83	2.30	2.92	2.35	1.30°	5.04°	−4.97°	2.52°
20	1.80	1.41	2.69	2.18	2.86	2.33	2.95	2.37	1.36°	5.16°	−4.84°	2.74°
30	1.82	1.43	2.72	2.20	2.89	2.34	2.97	2.39	1.40°	5.24°	−4.73°	2.92°
unconfined	1.86	1.46	2.73	2.20	2.90	2.34	2.98	2.37	2.18°	6.02°	−3.31°	4.59°

Table 4. Results of additional simulations: jet-flap wing in wind tunnel.

Initial condition: **U** $= 1.0$ ($\alpha = 0°$). Final condition: **U** $= 1.0$ ($\alpha = 15°$). Number of Σ-panels $= 28$. $C = 0.5$.

Stream inclination Initial	Final	Number of field points	k	$\lvert\delta v_n\rvert$ max	Average abs. error in Γ_r	Average abs. error in v_n on S	Number of iterations
Model: wing only. Position: centered.							
45°	60°	28	0.05	0.039	0.0055	0.185	30
45°	60°	44	0.15	0.113	0.0055	0.272	25
45°	60°	44	0.20	0.133	0.0033	0.249	15
45°	60°	44	0.30	0.115	0.0035	0.284	15
45°	60°	44	varied: .1 to .2	0.115	0.0053	0.269	20
Model: wing and tail. Position: centered.							
45°	60°	28	0.05	0.145	0.0251	0.303	15
30°	45°	28	0.05	0.134	0.0119	0.318	15
Model: wing and tail. Position: raised 1/2 span.							
30°	45°	28	0.05	0.110	0.0090	0.291	18
15°	30°	28	0.05	0.114	0.0081	0.221	18

B. Variable Relaxation Factor

Convergence of the iteration scheme also seems to be accelerated if k is increased as the iteration proceeds, as in the run shown in Table 4 where k was gradually increased from .10 to .20 between the tenth and twentieth steps. (The subject of convergence-acceleration is mentioned again below.)

C. Effect of Horizontal Tail Surface

In Table 4 are also presented the results of simulations that included a modelled horizontal tail surface, placed and proportioned approximately to provide longitudinal moment-equilibrium. The tail was modelled in a manner similar to the wing, its circulation constituting a thirteenth unknown in the jet-flap wing model.

The data presented suggest that convergence toward unconfined conditions is impaired a bit when the tail is added, especially at large model inclinations, at which the tail is close to the interface S.

D. Changes of Wing Location and Stream Inclination

So long as the wake does not impinge on the instrumentation, there is a range of undisturbed-stream inclinations arg U, that can be used, and converged results should, ideally, be the same within this range. The range depends on the model's vertical location in the tunnel.

Table 4 shows two runs wherein the model was raised a half-span in the tunnel, permitting the model and stream to be inclined at smaller angles without having the wake impinge on the instrumentation. Improvement in results is shown.

6. Non-iterative Methods

Dowell (1981) has pointed out that for linear systems it is not necessary to iterate; the unconfined-flow solution can be calculated from the initial mismatch function $\delta^{(1)}$ by use of two influence matrices, one of which characterizes the outer flow and the other the inside flow including effects of the test model. His argument may be paraphrased in our terminology and notation as follows:

Suppose $f^{(1)}$ and $g^{(1)}$ are measured distributions as defined above and that $\delta^{(1)}g$, as before, represents the associated mismatch function:

$$\delta^{(1)}g = g\left[f^{(1)}\right] - g^{(1)} \tag{6.1}$$

Suppose also that f^* and g^* are the corresponding distributions for unconfined flow and that Δf and Δg are the increment distributions required; i.e.,

$$f^* = f^{(1)} + \Delta f \qquad \text{and} \qquad g^* = g^{(1)} + \Delta g \tag{6.2}$$

If the system is linear—e.g., it is adequately described by small-disturbance approximations—we can write, for the external flow,

$$g\left[f^{(1)}\right] = Af^{(1)} \tag{6.3}$$

and, for the internal flow,

$$\Delta f = B\,\Delta g \tag{6.4}$$

where A is the matrix relating g to f in the outer flow—easily constructed by the panel method used here, for example,—and B is the matrix relating f to g in the inside flow. B is, for example, the product of an influence matrix analogous to ${V_n^{(1)}}_{ij}$ above, namely ${V_t^{(1)}}_{ij}$, into the inverse of ${V_n^{(1)}}_{ij}$.

The first of equation (6.2) states that the external-flow calculation is linear, which is surely true for the inviscid, low-speed flow exterior to the fixed interface S. The second of equation (6.2) states the same for the internal flow—but only for the increments Δf and Δg. Although the internal flow for a high-lift case is neither linear nor inviscid, this may be a suitable approximation for small enough increments, viz. a *local linearization* around the initial situation (1).

The statement that f^*, g^* represent unconfined flow means that

$$g^* = g[f^*] \tag{6.5}$$

or

$$g^{(1)} + \Delta g = g\left[f^{(1)} + \Delta f\right] \tag{6.6}$$

$$= g\left[f^{(1)}\right] + g[\Delta f] \tag{6.7}$$

by virtue of the linearity of the operator $f[\ ]$. Combining equations (6.1) and (6.7), we have

$$\Delta g = \delta^{(1)} g + AB\, \Delta g \tag{6.8}$$

and therefore

$$\Delta g = (1 - AB)^{-1}\, \delta^{(1)} g \tag{6.9}$$

in familiar matrix notation.

This would mean that iteration could be dispensed with and unconfined flow attained in a single step if the matrices, say $V_{n\ ij}^{(1)}$ and $V_{t\ ij}^{(1)}$ were measured and used as in equation (6.9). Obviously, the required wall-control settings would be calculated easily by use of the appropriate one of these matrices.

Our experience to date with this technique, which is quite limited, had led us to the following tentative conclusions:

1. Use of equation (6.9) reduces the mismatch function δg much more in a single step than can be done with several iterations using uniform k.
2. When *mean-square-error* fitting is employed—i.e., when the number of controls is smaller than the number of field points, as mentioned above—the residual errors remaining after use of equation (6.9) may be considerably larger than can be achieved by iteration with uniform, constant k (as exemplified by the results in Table 3).
3. The Dowell procedure can be used in an iterative fashion; i.e., the process represented by equations (6.1) and (6.9) can be repeated. The residual errors mentioned in (2) can be rapidly reduced by this iterative use of Dowell's idea.
4. Dowell's procedure requires that both interior-flow influence matrices, the control matrices for both f and g, be measured. Since the process of measuring influence matrices is likely to be tedious, a question arises regarding the relative economies of this procedure and simple-minded iteration with constant k.

7. Conclusions

The adaptive-wall scheme, since it permits the operator to choose the simulated undisturbed-stream vector in a wind-tunnel test, offers the possibility of eliminating the problem of the wake in high-lift testing. Numerical simulations of the test of a jet-flap wing, including the iterative process of tunnel adaptation, appear to confirm this claim.

In the process of these simulations, a panel method for jet-flap wing and wake has been developed, and a number of detailed matters concerning convergence of the iterations have been explored.

Acknowledgement

This work has been carried out at the University of Arizona, Department of Aerospace and Mechanical Engineering, with support from the Office of Naval Research (Contract N 00014-79-C-0010) and the Air Force Office of Scientific Research (Grant AFOSR-82-0185). The author also gratefully acknowledges the able assistance of Daniel C. L. Lee and Karl Allmendinger—graduate assistants.

References

[1] Binyon, T. W., Jr. "Wall Interference in Wind Tunnels," *Technical Evaluation Rept., AGARD Advisory Report, No. 190,* AGARD/NATO, Sept. 1982.

[2] Dowell, E. H. "Control Laws for Adaptive Wind Tunnels," *AIAA J., 19* (1981), 1486–88.

[3] Ferri, A. and Baronti, P. "A Method for Transonic Wind Tunnel Corrections," *AIAA J., 11* (1973), 63–66.

[4] Kerney, K. P. "An Asymptotic Theory of the High-Aspect-Ratio Jet Flap," *Ph.D. thesis,* Graduate School of Aeronautical Engineering, Cornell University, Ithaca, NY, 1967.

[5] Lee, Daniel C. L. "The Modelling of a High-Lift Jet-Flapped Wing," *Master of Science Report,* Department of Aerospace and Mechanical Engineering, Univ. of Arizona, Tucson, AZ, 1981.

[6] Maskell, E. C. and Spence, D. A. "A Theory of the Jet Flap in Three Dimensions," *Proceedings of the Royal Society, A251,* June 1959, 407–25.

[7] Satyanarayana, B., Schairer, E., and Davis, S. "Adaptive-Wall Wind-Tunnel Development for Transonic Testing," *J. Aircraft, 18, No. 4* (1981), 273–79.

[8] Sears, W. R. "Self-Correcting Wind Tunnels," (The Sixteenth Lanchester Memorial Lecture), *Aeronautical J., 78* (1974), 80–89.

[9] ————. "Adaptable Wind Tunnel for Testing V/STOL Configurations at High Lift," *J. Aircraft, 20* (1983), 968–74.

[10] Sears, W. R., et al. "Interference-Free Wind-Tunnel Flows by Adaptive-Wall Technology," *J. Aircraft, 14, No. 11* (1977), 1042–50.

[11] Spence, D. A. "The Lift Coefficient of a Thin, Jet-Flapped Wing," *Proceedings of the Royal Society, A238* (1956), 46–68.

[12] Tokuda, N. "An Asymptotic Theory of the Jet Flap in Three Dimensions," *J. Fluid Mechnics, 46, Part 4,* 1971.

The Evolution of Adaptive-Wall Wind Tunnels

Sanford S. Davis
NASA Ames Research Center
Moffett Field, CA 94035

Summary

Adaptive-wall wind tunnel technology is examined as a natural extension of the continuing effort to mitigate wall-induced wind tunnel interference. The state of the art in interference assessment is briefly surveyed starting from its inception in the 1920s to present day practice. It is concluded that adaptive-wall wind tunnels will play a major role in future aerodynamic research, but that they will probably be used in a different manner than current wind tunnels. It is also concluded that more effort is needed to sort out the complex hardware/software/sensor relationships that will be required to support a large-scale, high-Reynolds-number, adaptive-wall wind tunnel.

1. Introduction

Wind tunnel practice now spans approximately three generations and, in many ways, may be considered a *mature technology*. In the development of new vehicles in the 1970s, \$15–30 million were spent on testing advanced aircraft, such as the C5A cargo carrier and the B-747 transport, in large wind tunnels. Such large investments were deemed prudent to minimize development risks, but there are always underlying concerns over the overall quality of the wind tunnel data.

There are two major limiting factors* in the application of wind tunne data to full-size vehicles: Reynolds-number simulation and wind tunnel wall interference. The *Reynolds-number gap* will soon be closed with the introduction of new cryogenic wind tunnels, and there is great promise that wall interference can be reduced

*Other factors may include support interference, model deformation, stream nonuniformities, and propulsive effects.

through the use of adaptive-wall technology. If both these limitations can be somehow eliminated, aerodynamicists will be able to validate new analysis and design tools by separating both viscous and wall effects for the first time.

The purpose of this paper is to examine the evolution of adaptive-wall technology in reducing the effects of wind tunnel wall interference. The historical record shows quite clearly that such advances as rational correction schemes and slotted-wall test sections were quickly adopted, and that they contributed greatly to aerodynamic science. However, at first sight, adaptive-wall technology is not so easy to justify in existing large wind tunnels. Its physical complexity mitigates against its incorporation as a simple add-on (as a slotted wall) or as a post-test correction (as in the classic approach). Adaptive-wall techniques will have to be proved to be both technically and economically justifiable before a large-scale commitment is made.

There are valid reasons for considering adaptive-wall wind tunnels on economical grounds. Typical model restrictions in transonic wind tunnels would be 1% in blockage and 2.5% in wing area/tunnel area. If these restrictions could be relaxed by a factor of two with adaptive walls, the same model could be tested in a smaller wind tunnel. Both capital and recurrent costs could probably be reduced by the same factor. Of course, an alternative approach would be to test a model twice the size in the large wind tunnel, but the gain in Reynolds number would probably not be worth the effort.

There are many formidable technical challenges in developing a feasible adaptive-wall for large three-dimensional wind tunnels. Choosing the appropriate mechanical arrangement, sensor system, and computational tools is a difficult task; especially when considered in light of the myriad other restrictions such as optical access, model support, and personnel access. The main point of this paper will be to argue that 60 years of research in wind-tunnel interference should not be ignored in the development of large, three-dimensional adaptive-wall wind tunnels. Full consideration should be given to the procedures that were developed in the past, and the salient features of the entire technology base should be considered in the design of three-dimensional adaptive-wall test sections.

The development of wall-interference theory in the 1920s is presented first, followed by a description of the effect of high-speed wind tunnels, in the 1930s and 1940s, which led to the invention of the slotted-wall wind tunnel. A modern wall-interference assessment method is discussed and the two-dimensional adaptive-wall wind tunnel that was developed at Ames Research Center (in a joint effort with the NASA/Stanford Joint Institute for Aeronautics and Acoustics) is described. In the final section, some approaches to the development of large adaptive-wall wind tunnels are suggested.

2. Early Investigations: 1920–1940

The classic approach to wall interference was developed in the 1920s along with the development of airfoil and wing theory. The correct boundary-value problem for the interference flow was first posed by Prandtl in the following manner: consider a small-disturbance velocity field whose effect at the wall is just sufficient to cancel a selected component of the free-air velocity field at the same location. The normal component must be extinguished at a solid wall and the axial component at a free jet. (The boundary condition is exact for the closed-wall case, but is only first-order accurate for the free jet.) Determining the effect of this interference velocity field on the forces and moments experienced by the wind-tunnel model is the objective of all correction schemes.

Prandtl appreciated the difficulties involved in solving the complete boundary-value problem in three dimensions. He subsequently introduced the simplifying method of solving an equivalent two-dimensional problem in the farwake (i.e., the Trefftz plane). In an early monograph, Prandtl (1921) applied these methods to determine the drag correction in open and closed wind tunnels with circular cross sections. This correction was confirmed in a wind tunnel test using the NACA 5-Foot Variable Density Wind Tunnel at Langley Field during the period 1925–27 (Higgens, 1927). These tests showed that Prandtl's correction method gave much better agreement with experiment than the empirical methods that were used previously. Figure 1 shows some data from this early test that validated the Prandtl theory in the U. S. for the first time.

A comprehensive survey of the subject of wall interference was presented in a monograph by Glauert (1933). In the analysis of three-dimensional wings, Prandtl's original method was extended in one direction to a variety of wind tunnel geometries and boundary conditions, and in the other to more precise wing characterizations.

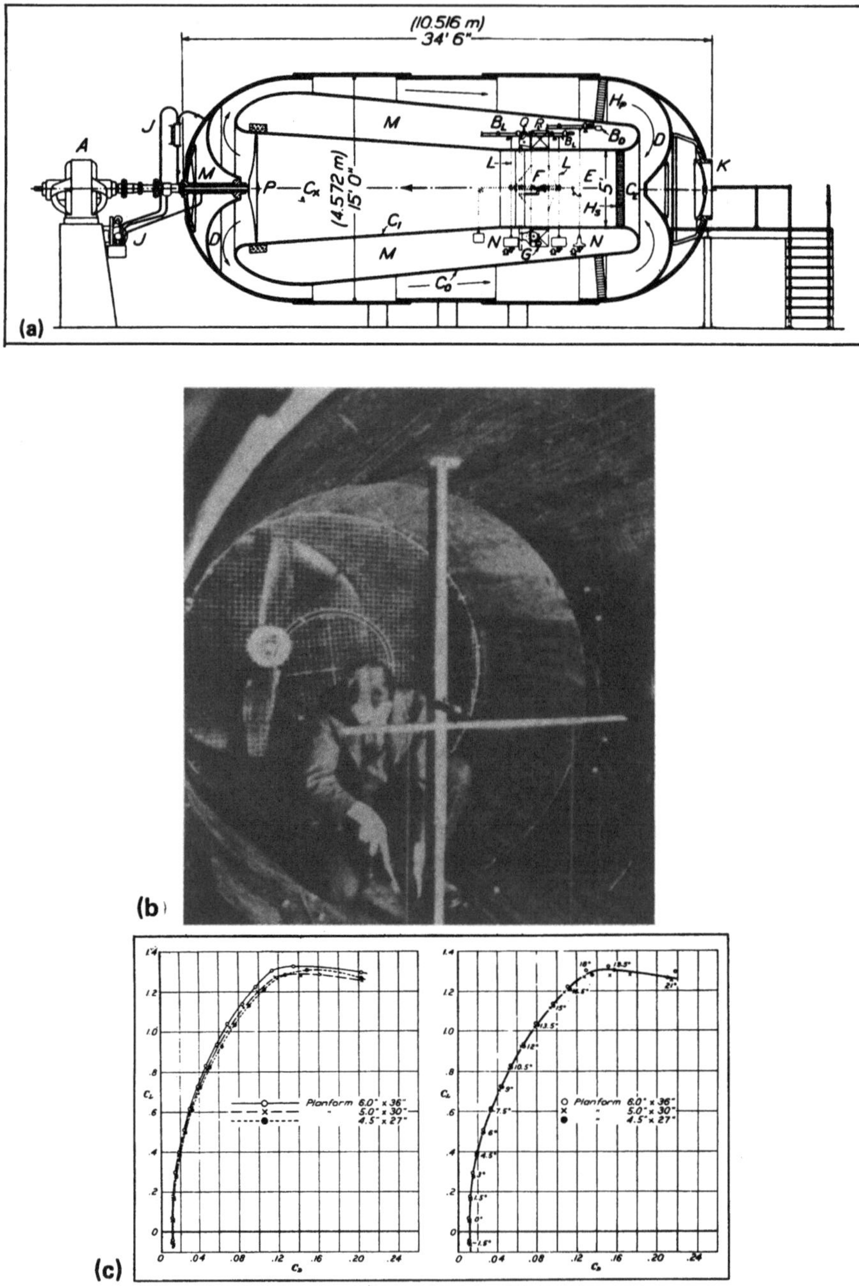

Figure 1. Verification of Prandtl's interference theory (adapted from Higgens, 1927). (a) NACA 5-foot variable density wind tunnel. (b) Test model mounted in test section. (c) Lift-drag polars comparing corrected (right) and uncorrected (left) data.

In two-dimensional airfoil analysis, Lock's (1929) early blockage corrections were extended to lifting airfoils and to streamwise curvature corrections using the newly developed thin-airfoil theory. In three dimensions, the powerful *method of images* was applied to wind tunnels with both circular and rectangular boundaries. Both solid and open boundaries were considered, and a useful set of theorems was described to relate interference among differing tunnels.

Most of these early classical results were obtained for small-span wings (wings whose ratios of span to tunnel width are very small); that is, the spanwise loading was ignored and the wing was replaced by a point vortex-doublet. A simple application of wing theory in the Trefftz plane reduced the interference effect to an induced upwash at the position of the vortex-doublet singularity:

$$v/V = \delta \frac{S}{C} kl$$

where v/V is the ratio of the induced upwash to the free-stream velocity, δ is the interference factor, S/C is the ratio of the wing area to the tunnel cross-sectional area, and kl is the lift coefficient defined as $L/\rho V^2 S$. The interference factor varies with the tunnel aspect ratio in a manner depicted in Figure 2. (This figure was adapted from Glauert's Figure 9.) It was certainly appreciated by these early workers that the opposing effect of open and closed tunnels could be used to propose low-interference semi-open configurations. Unfortunately, the semi-open wind tunnel shown in curve (2) in Figure 2 was given incorrectly in the original work. The discrepancy was not fully explained until Garner's (1963) analysis, the corrected curve (2) being shown in Figure 2. As will be discussed later, the concept of a mixed wind tunnel may provide a useful basis for the development of adaptive-wall wind tunnels.

Some researchers attempted to refine the wing representation by assuming uniform or elliptic loading, but the analysis was so complex that not many solutions were developed. During the 1930s, the method of vortex images was highly developed, and methods were introduced for correctly analyzing finite-span wings. Toussaint (1935) presented a detailed discussion of the vortex method in rectangular wind tunnels. The interference factor so obtained now depends on both the wind-tunnel aspect ratio and the wing semi-span to tunnel semi-width ratio (denoted by σ). Interference curves for two representative values of σ (0 and 0.4) are shown in Figure 3. It seems that tall, narrow tunnels are less sensitive to finite-span effects, but

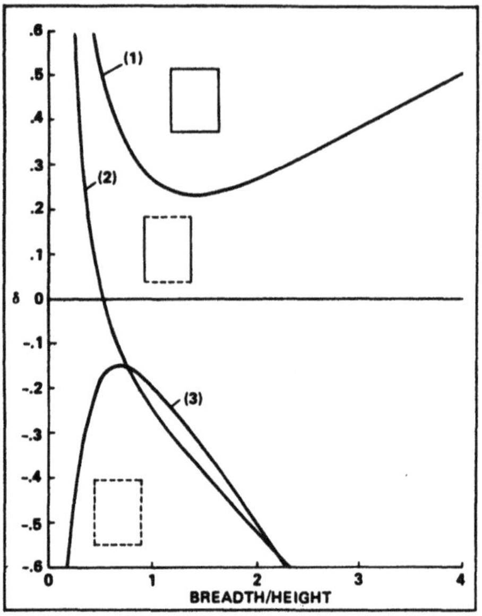

Figure 2. Wall interference theory for a small wing (vortex-doublet model) in a rectangular wind tunnel; curve (1)—closed tunnel; curve (2)—semiopen tunnel; curve (3)—open tunnel.

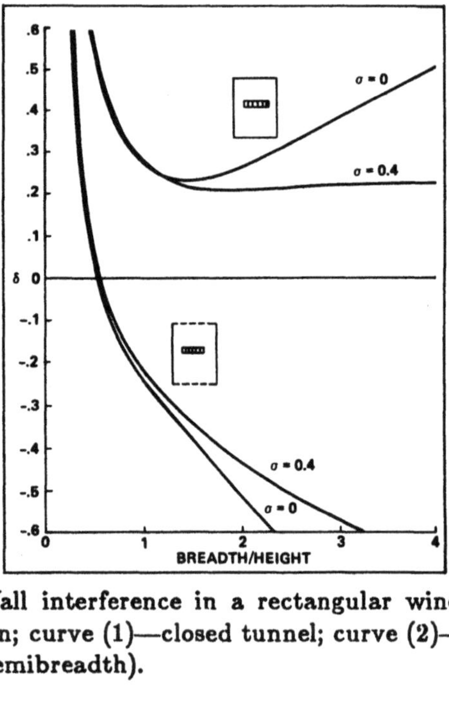

Figure 3. Wall interference in a rectangular wind tunnel showing effect of wing span; curve (1)—closed tunnel; curve (2)—semiopen tunnel (σ = semispan/semibreadth).

gradients in the lift interference factor are more severe. (The curves in Figure 3 were computed from formulas given in the analysis of Kraft (1973).)

Heretofore, all solutions were obtained for symmetrically located wings. In an important analysis, Silverstein and White (1935) expanded the theory in two important directions: to off-center wing locations and to three-dimensional streamwise interference effects. The latter effect is extremely important, for it includes the elusive streamwise curvature correction along with an estimate of empennage interference. For these problems, the entire horseshoe vortex must be considered, and practical correction methods were not achieved until the 1960s when high-speed computers became readily available.

To summarize the 1940 state of the art in wind tunnel interference assessment, the following assumptions and conditions were generally imposed: (1) an incompressible flow of an ideal fluid; (2) the interference was computed as an equivalent two-dimensional problem in the Trefftz plane; (3) the distribution of circulation across the span followed simple rules such as uniform or elliptic; (4) the wind tunnel boundary was replaced by a doubly infinite system of vortex images; (5) a characteristic upwash (either at a point or an average along the span) was used to compute angle-of-attack corrections; and (6) an infinitely long test section.

3. The High-Speed Era: 1940–1970

In the early 1940s, wall-interference technology developed in two major directions. In one direction, the classic theory was systematically expanded to include more realistic aircraft and wind tunnel configurations using sophisticated mathematical analyses. The second direction was related to a pressing practical problem: choking in high-speed wind tunnels. (In the early 1940s, high-speed meant velocities approaching the speed of sound.) Choking is the result of massive blockage-induced wall interference and was a real barrier to the understanding of high-speed flows.

A significant advantage of the first direction was its mathematical rigor. Possibly some of the most elegant solutions to the partial differential equations of mathematical physics were obtained in the continuing search for more exact interference potentials for complex geometries. Some of the significant classic work includes the papers of Pistolesi (1940) and Riegels (1939) on correction factors for circular and elliptical wind tunnels with partly open walls (these were early attempts to exploit the opposing effect of open and closed walls), and the work of Baranoff (1940) on the effect of tunnel walls in a compressible medium. In the treatment of unconventional geometries, mention is made of the paper by Batchelor (1944) on octagonal wind tunnels and the work of Kondo (1935) on wind tunnels consisting of circular arcs. More recent research was concentrated on more refined model representations such as variably loaded lifting lines (Acum, 1950), swept lifting lines (Eisenstadt, 1947), and lifting surface elements (Katzoff and Hannah, 1948). These solutions are complicated and were usually presented in the form of tables and charts tailored to specific wind tunnels. Even today, the tables prepared by Sivels and Salmi (1951) are routinely used to correct data in the Ames 12-Foot Pressure Wind Tunnel, a closed, 1942-vintage subsonic tunnel with a circular test section.

Despite many heroic attempts, a major defect of this approach was that only linearized solutions to the simpler class of boundary-value problems could be obtained. This effectively reduced the range of applicability to subsonic flows in closed or semi-open wind tunnels. In time, the disadvantages prevailed, and this research stream became less active.

The second direction has been by far the more successful, although its apparent success must be measured against the continuing lack of a rational interference assessment method. There are two major obstacles in this area: the nonlinear nature of the transonic flow equations and the complex wall geometries that have evolved as part of the solution to the choking problem.

The development of transonic wind tunnels in the 1940s and 1950s is an interesting story. Its development in the United States has been documented in some detail by Becker (1980). It had long been the practice in Europe to test high-speed models in small semi-open wind tunnels. These tunnels could be justified at low speeds from early analytical studies that showed that wall interference could be eliminated altogether (see Figure 2). However, larger semi-open tunnels could never be justified at high speeds because of large power losses. In 1946, the Langley 8-Foot Wind Tunnel was being repowered, and R. Wright, a young Langley researcher, was assigned the task of defining a new test section for the tunnel. Wright was aware of previous work on partly-open tunnels, but he also was wary of semi-open configurations. His compromise solution was a 10-slot arrangement that he analyzed with the tools then available. The resulting ventilated-wall geometry alleviated the choking problem without unacceptable power losses. The construction of a pilot-tunnel and the successful re-powering of the 8-Foot Tunnel paved the way for the current generation of transonic wind tunnels. Figure 4 shows the pilot tunnel along with a historically significant graph that indicated the first validation of the ventilated test section (Wright and Ward, 1955).

A new technology was developed, but at the expense of a complicated flow boundary. It appears that Busemann (Becker, 1980, p. 101) first broached the idea of an equivalent wall-boundary condition in discussions with Wright in 1947. This homogeneous wall-boundary condition remains unverified to this day, but is still used for lack of a better representation.

Once the ventilated wall was accepted for transonic testing, the search for elegant correction schemes based on first principles abated somewhat. Perhaps this was due to the complex geometries and the analytically difficult transonic flow fields. Perhaps it was also a result of the development of large, high-Reynolds-number wind tunnels that could accommodate reasonably sized models with acceptably small blockage and surface area ratios.

In the 1960s both research directions were consolidated in the publication of two significant works. The art and science of transonic wind-tunnel testing was described in a monograph by Goethert (1961) and the classically based research was codified in the encyclopedic work by Garner et al. (1966). These two works form a very complete history of wind-tunnel interference research from subsonic through low supersonic speeds.

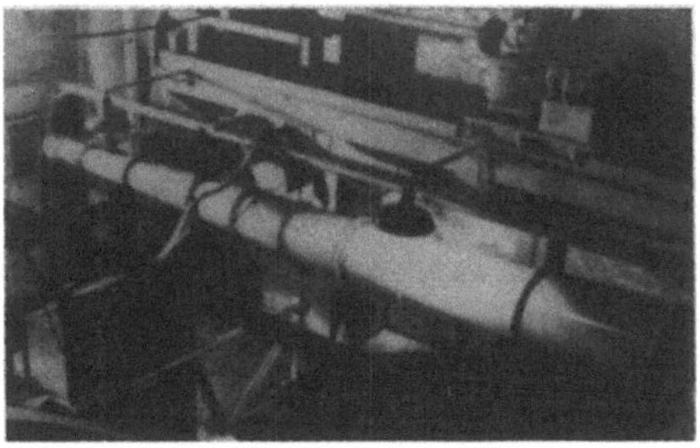

Figure 4a. (Caption on p. 556.)

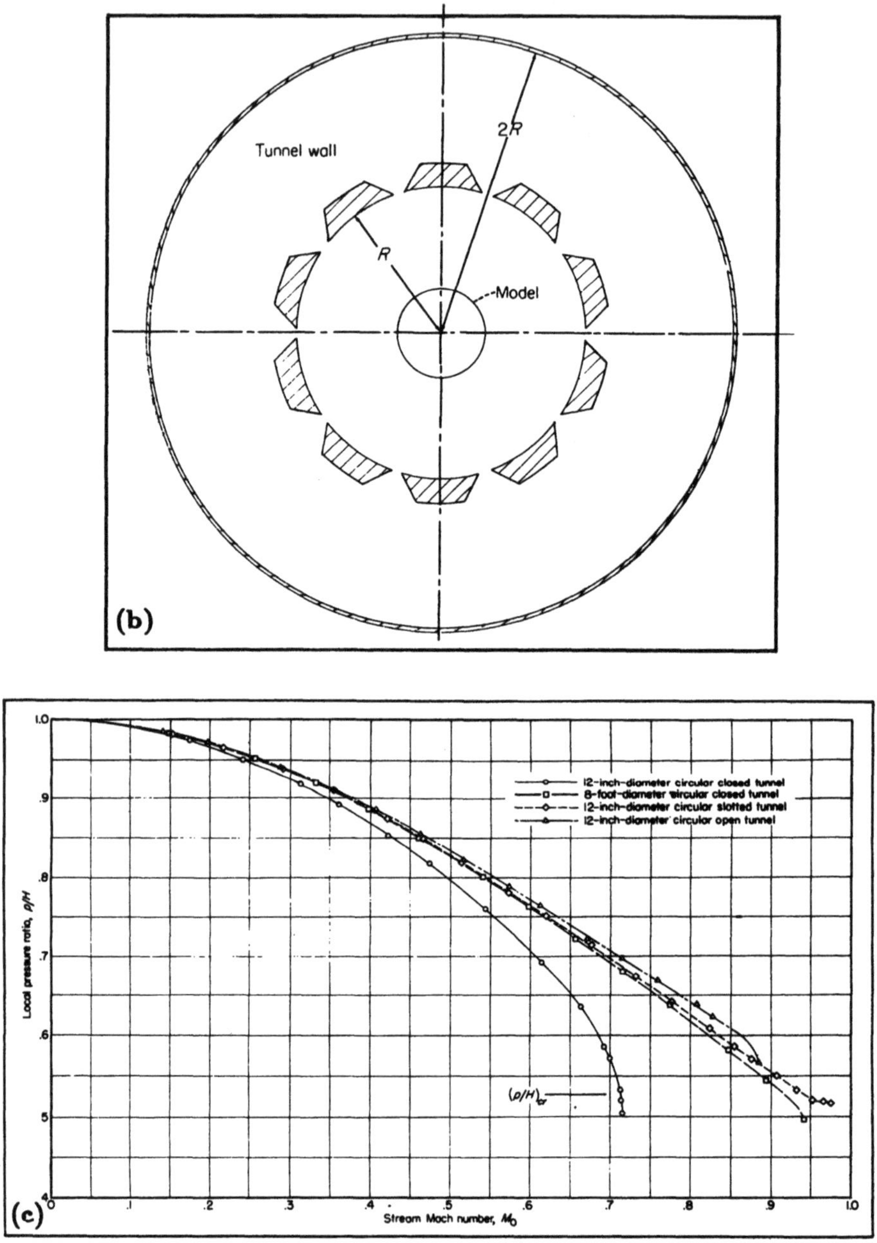

Figure 4. Verification of ventilated-wall wind tunnel for the alleviation of blockage interference (adapted from Wright and Ward, 1955). (a) 12-in. indraft-type pilot tunnel, (b) cross section of tunnel showing prolate spheroid model and 10-slot wall configuration. (c) Variation of midchord pressure with Mach number.

4. Post-1970 Period

In the early 1970s, technological advances included the development of highly complex maneuvering vehicles and large commercial aircraft that pushed existing test facilities to their operating limits. The effects of Reynolds number and wall interference were recognized as unknown quantities. In addition, the development of supercritical airfoils for transonic cruise caused a reassessment of standardized procedures for two-dimensional airfoil testing. Most wall interference techniques were based on subsonic correction methods overlaid on calculations that over-simplified the wall-boundary conditions. There was not a great deal of confidence in the resulting interference assessments, and corrections were often ignored altogether (AGARD, 1971).

With the successful computation of simple transonic flows in the early 1970s, there was much hope that the wall-interference problem could be solved once and for all. Wall-bounded flows were successfully computed (Murman, 1972), but a generally applicable correction method was not forthcoming, a result of the nonlinear nature of transonic flows. In attempting to compute transonic wall interference, even a linear approximation must include the basic unconfined flow (through variable coefficients in the linearized equation) at all distances from the body. This implies that the basic unconfined flow must be first computed, an ambitious task that is only partially successful with today's techniques. This major complicating factor highlights the great promise of adaptive-wall technology in which the flow near the model need not be computed, measured, or even considered.

Current developments in wind tunnel wall interference may be conveniently split into two groups: interference-assessment methods and adaptive walls. Progress in the interference-assessment methods has accelerated with the realization that measured boundary conditions are more reliable than those obtained from wall modeling. Modern assessment methods have been advanced in many countries, and are discussed in detail in recent AGARD proceedings (*AGARD,* 1982). One of the methods, that was developed in the United States by Kemp in 1975, is a useful case study and will be described briefly.

The objective of interference assessment is to compute an interference velocity field and to attempt to relate it to some free-air flow condition. The procedure is relatively straightforward for linear flows but not for nonlinear transonic flows. In the procedure advocated by Kemp (1979, 1980), three separate solutions to the transonic small-disturbance equation are required. The objective of the first solution is to use measured wall and airfoil pressure distributions to compute an equivalent airfoil shape. This equivalent airfoil is represented by a distribution of singularities that are used to compute the second free-air solution. This second solution is used to update the free-air Mach number to match velocities at a selected point on the airfoil. (In classic wall interference theory, this point is usually chosen at the midchord.) Any difference between these first two solutions is interpreted as residual interference. The third and final solution computes the angle-of-attack correction by predicting the angle of attack that satisfies the Kutta condition at the measured lift coefficient.

Some data from a series of tests in the 2- by 2-Foot Transonic Wind Tunnel at Ames Research Center, using floor and ceiling pressure rails (King and Johnson, 1980, 1983), were used to assess Kemp's method. These results are shown in Figure 5 where experimental data are compared with computations at both measured and corrected flow conditions. The solutions presented here were obtained from the Reynolds-averaged form of the thin-layer Navier-Stokes equations. Figure 5a shows results from a conventional NACA 64A010 airfoil at $M = 0.800$ at an angle of attack of 2°. A corrected Mach number (0.792) and angle of attack (1.39°) brings the wind-tunnel data into good agreement with free-air solutions. Figure 5b shows a similar comparison for a supercritical airfoil at cruise conditions. In this case the correction is in the right direction, but the predicted shock wave is slightly downstream of the measured position. Whether this defect is caused by inaccuracies in the interference assessment, by an imprecise model for the viscous flow, or by an *uncorrectable* interference cannot be answered at this time. However, it is clear that the availability of these assessment tools, when coupled with modern analytical methods, will certainly aid in our understanding of these complex flows.

Other methods for predicting wall interference, although based on linear theories, are being investigated. There is no doubt that they will be useful for correcting certain classes of wind-tunnel data, especially in large three-dimensional tunnels. In fact, these methods can be closely related to the emerging adaptive-wall technology since the sensor requirements are the same for both procedures, and can be used for both interference assessment and wall control.

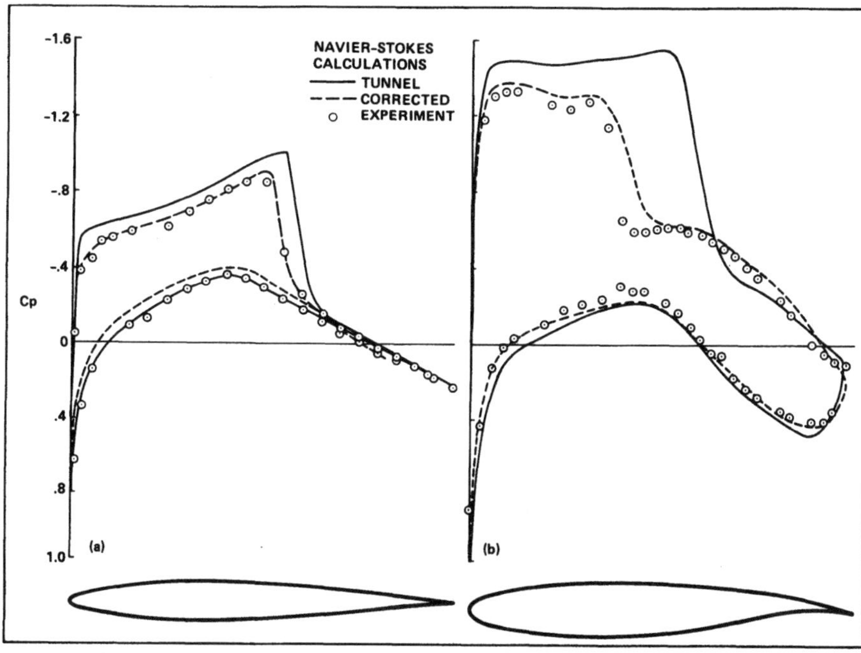

Figure 5. Wind-tunnel pressure data compared with Reynolds-averaged Navier-Stokes calculations using Kemp's interference assessment method (adapted from King and Johnson, 1980, 1983). (a) NACA 64A010 airfoil, Re = 2×10^6. Tunnel: $M = 0.800$, $\alpha = 2°$. Corrected: $M = 0.792$, $\alpha = 1.38°$. (b) Supercritical airfoil, DSMA 671, Re = 2.67×10^6. Tunnel: $M = 0.720$, $\alpha = 4.32°$. Corrected: $M = 0.696$, $\alpha = 3.38°$.

An adaptive wall is based on the premise that if streamlines above and below a test model (streamtubes in three dimensions) are deformed into their interference-free shape, then measurements on the model would yield interference-free data. The first analyses of adaptive walls were published independently by Ferri and Baronti (1973) and Sears (1975).

The work of Ferri and Baronti was aimed at developing a method for assessing wall interference by measuring two independent flow quantities near the tunnel boundaries. The flow variables chosen were angularity and wall static pressure. They pointed out the efficacy of developing a free-air solution in the process of interference assessment. They also considered the use of linearized theory for transonic wall-interference assessment and demonstrated their method with numerical computations.

In his paper, Sears advanced the main ideas behind the modern adaptive-wall wind tunnel. He introduced the idea of a control surface removed from the wall, where computed and measured flow properties are compared to determine their consistency with an unconfined flow. In Sears' method, one of the flow parameters is used as a boundary condition to compute an unconfined flow external to the chosen control surface. This calculation yields a prediction of the other flow property, which is in turn compared with the measured value to assess compatibility. The wall conditions (variable ventilation or geometry) are adjusted iteratively until the computed and measured values agree, which implies interference-free flow in the wind tunnel. The strongest feature of this approach is that the difficulty of modeling or calibrating the flow near the wall is avoided.

Since its inception, the adaptive wall has developed in three major directions. These are the streamlined-wall method, the variable porosity method, and the segmented-plenum arrangement. The streamlined-wall technique can be traced to early experiments at the National Physical Laboratory (NPL) in the 1940s and is being pursued by research groups in Berlin (Ganzer, 1982), Southampton (Wolf et al., 1982), and Toullouse (Archembaud and Chevallier, 1982). The variable-porosity method is being investigated at the AEDC in Tullahoma, Tennessee (Parker and Sickles, 1981; Parker and Erickson, 1982). Finally, the segmented-plenum arrangement is under consideration at Ames Research Center (Satyanarayana et al., 1981). All of these methods were demonstrated in two-dimensional transonic flow, and active research programs are now under way in three dimensions.

At Ames Research Center, a two-surface assessment method (Davis, 1981) was developed to exploit available laser anemometry systems. A typical result from the initial experiments is shown in Figure 6. Figure 6a shows a sketch of the Ames two-dimensional adaptive-wall test section. This 25cm by 13cm test section was fitted to a small-scale indraft wind tunnel. Figure 6b depicts how upwash measurements are used at $0.4C$ (lower row) to compute free-air flow at $0.67C$ (upper row). The predicted flow is compared with a second measurement to assess the compatibility of wind tunnel flow and the outer free-air flow. Finally, Figure 6c compares pressure distributions on the test airfoil before and after wall adjustment with other data obtained in a large wind tunnel. The data from this pilot test were used to develop the hardware and software to convert

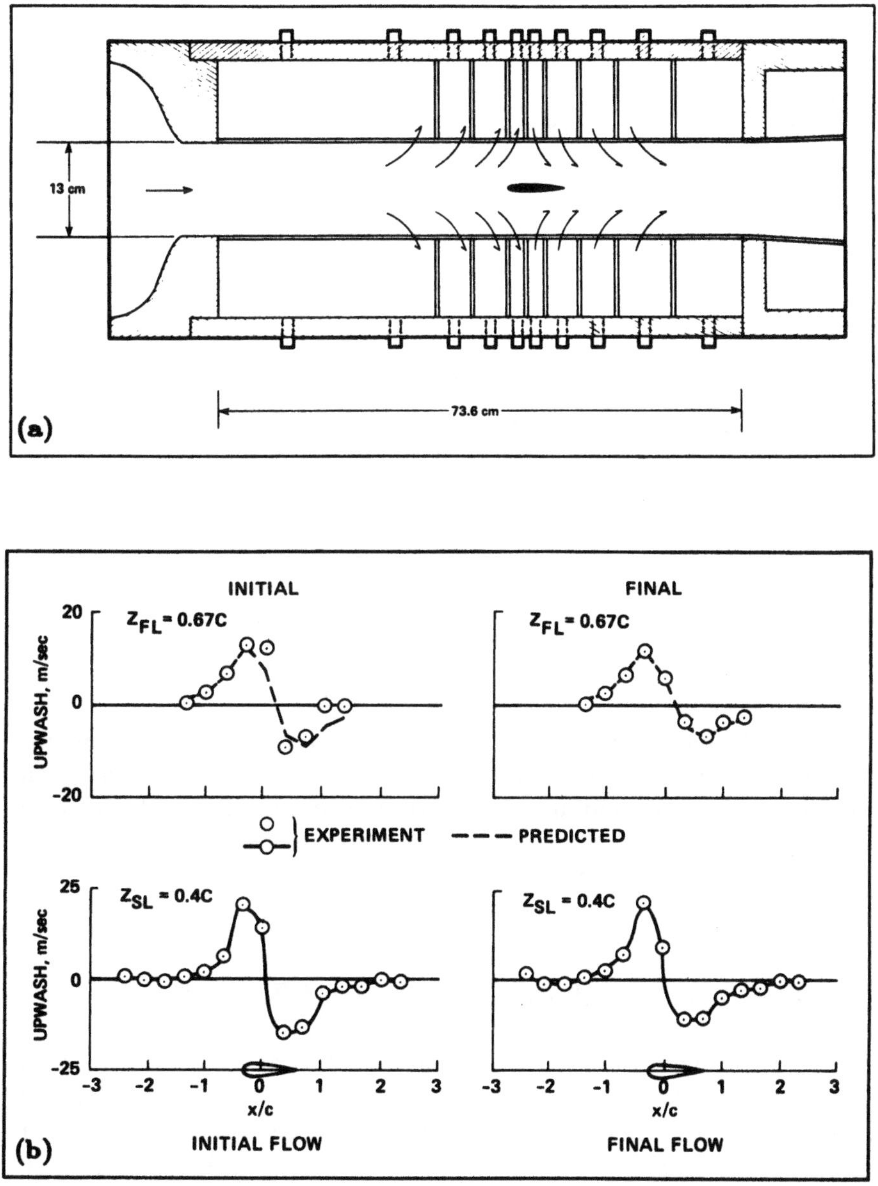

Figure 6. Ames two-dimensional adaptive-wall experiment, NACA 0012 airfoil. (a) Adaptive-wall test section. (b) Upwash measurements used for compatibility assessment.

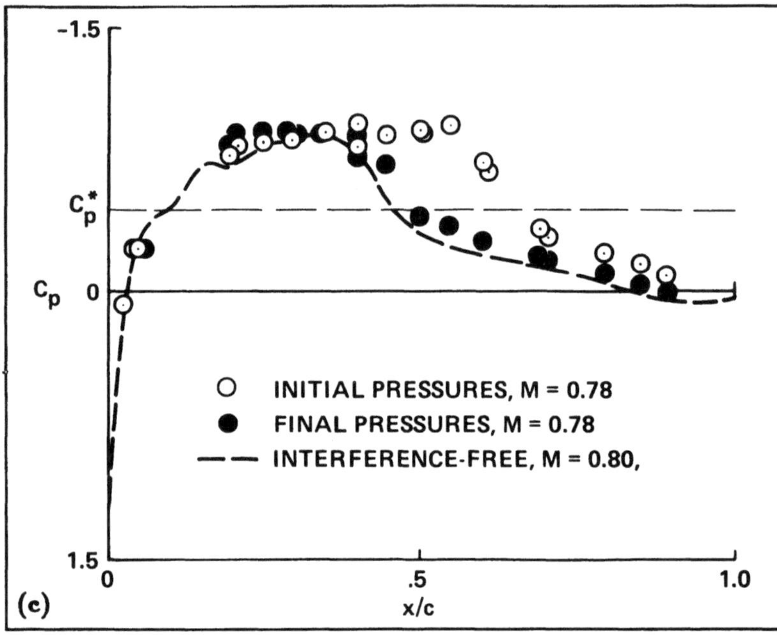

Figure 6c. Pressure distributions.

the Ames 2- by 2-Foot Transonic Wind Tunnel to a production-type, two-dimensional adaptive-wall facility. The two-level assessment method was also used in a three-dimensional pilot test (Schairer and Mendoza, 1982), and validated through the use of numerical simulations in subsonic flow (Mendoza, 1983).

5. The Role of Adaptive-Wall Wind Tunnels

Two-dimensional adaptive-wall wind tunnels are now operational at several research centers around the world, but direct extension of this technology to three dimensions is probably not feasible at this time. A major commitment to developing a large-scale high-Reynolds-number adaptive-wall wind tunnel must await the resolution of two problems: (1) finding a practical way to integrate the sophisticated hardware/software/sensor requirements in such a facility, and (2) demonstrating a need for this capability in such a large facility.

The successful resolution of these technical and economic problems may change the whole approach to wind tunnel utilization.

Current wind tunnel practice is to run through a wide operating envelope with an extensive and time-consuming testing program. With the availability of advanced computational procedures, in certain cases routine alpha or Mach sweeps may be treated numerically, while adaptive-wall wind tunnels can be used for point-design checks and for testing at flow conditions beyond the capability of current numerical modeling. In this sense, adaptive-wall technology becomes the perfect complement to computational fluid dynamics. It is recognized that model expense and setup time is a major cost factor in wind tunnel testing, but efficient testing techniques and the use of advanced computer-aided design and manufacture may reduce those large start-up costs. It is also foreseeable that alternatives to pressure orifices and force balances will be developed to monitor local and global flow conditions. (Plans are already under way at Ames Research Center to measure pressure distributions directly from holographic interferograms without the need for pressure models.)

With regard to the facility itself, historical and recent work suggests the possibility of a compromise type of tunnel for production testing. This would be a hybrid facility that would have both active and passive walls, with the passive walls used for optical access or other purposes. A combination of advanced computational tools, and the extensive analytical tools developed in the past could be used to compute residual interference on a point-by-point basis in the flow after the active walls were used to *preprocess* the flow into a form that is compatible with the computational and analytical envelope. For example, if a method could be developed to induce only a uniform distribution of interference velocity, classic corrections to Mach number and angle of attack could be directly applied with confidence.

In conclusion, it seems clear that the effect of wall interference is an important consideration in wind tunnel testing in general, and in transonic flows in particular. But isolating this effect from the other contaminants has proved to be a difficult task. With the adaptive-wall wind tunnel, it may now be possible to separate this effect in a practical manner. The ultimate justification will be when the cost-risk ratio (cost-of-testing/confidence-in-data) is shown to be significantly reduced with the use of adaptive-wall technology.

References

[1] Acum, W. "Corrections for Symmetrical Swept and Tapered Wings in Rectangular Wind Tunnels," *ARC R and M 2777,* 1950.

[2] AGARD. "Engine-Airplane Interference and Wall Corrections in Transonic Tests," *AGARD-AR-36-71,* 1971.

[3] ———. "Wall Interference in Wind Tunnels," *AGARD-CP-335,* Sept. 1982.

[4] Archembaud, J. and Chevallier, J. "Utilisation de Parois Adaptables pour les Essais en Courant Plan," *AGARD-CP-335,* Sept. 1982.

[5] Baranoff, A. "Tunnel Corrections for Compressible Subsonic Flow," *ZWB FB 1272,* 1940; *Trans. NACA TM 1162,* 1947.

[6] Batchelor, G. "Interference on Wings, Bodies, and Airscrews in a Closed Tunnel of Octagonal Section," *ACA 5, Australia,* 1944.

[7] Becker, J. "The High-Speed Frontier," *NASA SP-445,* 1980.

[8] Davis, S. "A Compatibility Assessment Method for Adaptive Wall Wind Tunnels," *AIAA J., 19,* Sept. 1981.

[9] Eisenstadt, B. "Boundary Induced Upwash for Yawed and Swept-Back Wings in Closed Circular Wind Tunnels," *NACA TN-1265,* 1947.

[10] Ferri, A. and Baronti, P. "A Method of Transonic Wind Tunnel Correction," *AIAA J., 11,* Jan. 1973.

[11] Ganzer, U. "On the Use of Adaptive Walls for Transonic Wind Tunnel," *AGARD-CP-335,* Sept. 1982.

[12] Garner, H. "An Anomaly in the Theory of Tunnel Wall Interference on a Lifting Wing," *Aero. Quart., 14* (1963), 31–40.

[13] Garner, H., et al. "Subsonic Wind Tunnel Wall Corrections," *AGARDograph 109,* Oct. 1966.

[14] Glauert, H. "Wind Tunnel Interference on Wings, Bodies, and Airscrews," *ARC R and M 1566,* 1933.

[15] Goethert, B. *Transonic Wind Tunnel Testing.* Pergamon, 1961.

[16] Higgens, G. "The Effect of the Walls in Closed-Type Wind Tunnels," *NACA Report 275,* 1927.

[17] Katzoff, S. and Hannah, M. "Calculation of Tunnel-Induced Upwash Velocities for Swept and Yawed Wings," *NACA TN-1748,* 1948.

[18] Kemp, W. B. "Transonic Assessment of Two-Dimensional Wind Tunnel Wall Interference Using Measured Wall Pressures, Part II," *NASA CP-2045* (1979), 473–486.

[19] ———. "TWINTAN: A Program for Transonic Wall Inter-

ference Assessment in Two-Dimensional Wind Tunnels," *NASA TM-81819,* 1980.

[20] King, L. and Johnson, D. "Calculations of Transonic Flow About an Airfoil in a Wind Tunnel," *AIAA Paper 80-1366,* 1980.

[21] ——————————. "Comparison of Supercritical Airfoil Flow Calculations With Wind Tunnel Results," *AIAA Paper 83-1688,* 1983.

[22] Kondo, K. "The Wall Interference of Wind Tunnels With Boundaries of Circular Arcs," *Report 126,* Aero Research Institute, Tokyo, 1935.

[23] Kraft, E. "Upwash Interference on a Symmetrical Wing in a Rectangular Ventilated Wind Tunnel, Part I: Development of Theory," *AEDC-TR-72-187,* March 1973.

[24] Lock, C. "The Interference of a Wind Tunnel on a Symmetrical Body," *ARC R and M 1275,* 1929.

[25] Mendoza, J. "A Numerical Simulation of Three-Dimensional Flow in an Adaptive Wall Wind Tunnel," *NASA TP,* 1984.

[26] Murman, E. "Computation of Wall Effects in Ventilated Transonic Wind Tunnels," *AIAA Paper 72-1007,* 1972.

[27] Parker, R. and Sickles, W. "Two-Dimensional Adaptive-Wall Experiment," *AEDC-TR-80-63,* Feb. 1981.

[28] Parker, R. and Erickson, J. "Development of a Three-Dimensional Adaptive Wall Test Section With Perforated Walls," *AGARD-CP-335,* Sept. 1982.

[29] Pistolesi, E. "On the Interference of a Wind Tunnel With a Mixed Boundary," *Comment. Ponotif. Acad. Sci., 4, No. 9,* 1949; *Trans. Cornell Aero. Lab.,* Dec. 1949.

[30] Prandtl, L. "Applications of Modern Hydrodynamics to Aeronautics," *NACA Report 116,* 1921.

[31] Riegels, F. "Correction Factors for Wind Tunnels of Elliptic Section With Partly Open and Partly Closed Test Sections," *Luft., 16,* 1939; *Trans. NACA TM 1310,* 1951.

[32] Satyanarayana, B., Schairer, E., and Davis, S. "Adaptive Wall Wind Tunnel Development for Transonic Testing," *J. Aircraft, 18,* April 1981.

[33] Schairer, E. and Mendoza, J. "Adaptive Wall Wind Tunnel Research at Ames Research Center," *AGARD-CP-335,* Sept. 1982.

[34] Sears, W. "Self-Correcting Wind Tunnels," *Aero. J., 78* (1975), 80–89; also *Calspan Report RK-5070-A-2,* July 1973.

[35] Silverstein, A. and White, J. "Wind Tunnel Interference With Particular Reference to Off-Center Positions of the Wing and the Down-

wash at the Tail," *NACA Report 547,* 1935.

[36] Sivels, J. and Salmi, R. "Jet Boundary Corrections for Complete and Semispan Swept Wings in Closed Circular Wind Tunnels," *NACA TN-2454,* 1951.

[37] Toussaint, A. *Experimental Methods—Wind Tunnels. Part I. Vol. III: Aerodynamic Theory.* Ed. W. F. Durand. Springer, 1935, p. 252.

[38] Wolf, S., Cook, I., and Goodyer, M. "The Status of Two- and Three-Dimensional Testing in the University of Southampton Transonic Self-Streamline Wind Tunnel," *AGARD-CP-335,* Sept. 1982.

[39] Wright, R. and Ward, V. "NACA Transonic Wind Tunnel Test Sections," *NACA Report 1231,* 1955.

Advances in Adaptive Wall Wind Tunnel Technique

Uwe Ganzer
Technische Universität Berlin
Germany

Summary

A brief outline is given of the adaptive wall concept. The research projects on this subject—carried out during the past ten years in the United States and Western Europe—are reviewed. The work done hereto at the Technical University of Berlin is discussed in some detail. It includes comprehensive tests of aerofoils in a tunnel with two flexible walls as well as tests of three-dimensional models in a tunnel with eight flexible walls. First complete wall adaptations for a lifting wing-body configuration were accomplished, giving evidence of the feasibility of the adaptive wall concept for 3D-model tests.

1. Introduction

The accuracy of wind tunnel testing is still far from being fully satisfactory. In particular for transonic flow conditions, uncertainties in the test results are great, amounting to an order equal to the order of the aerodynamic improvements an aircraft designer is trying to get. A recent example may illustrate this: Flight test results for the new Airbus A310 revealed a buffet boundary 10% better than estimated from the wind tunnel tests.

Quite a variety of reasons may be quoted for the uncertainties inherent in (transonic) wind tunnel testing. Usually the lack of Reynolds number similarity is named in the first place as the origin of the descrepancies between flight test and wind tunnel data. Besides this they may be attributed to the model deviating from a full scale configuration with the engine flow representation giving particular problems. Also, the various measurement techniques may include inaccuracies.

Above all there are uncertainties arising from the wind tunnel technique per se. The limited extension of the test section flow and the requirement of a model support system have, in general, an impact on the flow around the model: Tunnel wall interference and support interference deteriorate the measured aerodynamic phenomena.

There is also uncertainty arising from the fact, that wind tunnel calibration is usually done in the empty tunnel; with the model in the tunnel, the apparent main stream flow condition in terms of Mach number variation, flow angularity and turbulence level (and spectrum) may be different.

The problem is aggravated by a mutual interaction of the various phenomena. This may be illustrated by the example of a test in a wind tunnel at different Reynolds numbers. Such tests are supposed to show the impact of Reynolds number changes on the flow around the model only and allow extrapolation of model test data to full scale Reynolds number. But with the different Reynolds number, the flow conditions at the test section wall and around the model support also change.

These changes may be seen on one side as direct Reynolds number effects on the flow along the wall and the model support. But in addition to that secondary effects occur—the altered flow around the model results in a change of the oncoming flow for the model support. The different flow around model plus support will thereupon have different impact on the flow through a permeable test section wall (or along its solid wall). The great complexity of the situation leaves little hope for finding suitable correction methods to cope with all these problems yielding the required reliability of the test results.

The adaptive wall technique may, in principle, provide a solution to one of the probelms, i.e., the wind tunnel wall interference. In the following an account will be given of the advances made in adaptive wall wind tunnel technique during the past ten years, and it will be shown that there is good reason to be confident, that with this technique one of the large contributors to data uncertainty can be removed.

2. The Principle of the Adaptive Wall Technique

Explaining the principle of the adaptive wall technique is easiest by refering to the flow field around an aerofoil in a test section with flexible upper and lower wall, see Figure 1.

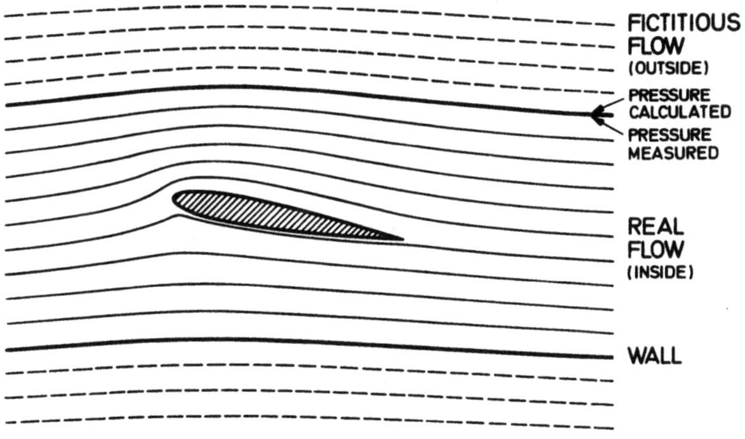

Figure 1. The principle of adaptive walls.

Neglecting the wall boundary layer in the first approach, the walls may be considered as substitute of streamlines: for interference free flow condition in the test section, the walls must have the shape of streamlines of an unconfined flow field. The test section flow is then the real part of such a flow field. A fictitious exterior part has to be supplemented to complete the picture.

The correct wall shape can be determined by a fairly simple iteraftive procedure. It is based on the assumption that the pressure on the two sides of a streamline must be the same. The pressure distribution along the inner side of the wall can be measured. Along the outer side of the wall, the pressure distribution can be calculated for the fictitious flow over the known wall contour. The wall shape is only correct if the calculated external and the measured internal pressure distribution are the same.

If there is a difference in the pressure distributions, this difference can be used for the iterative adaptation procedure. Something like a mean value between the two pressures is taken to calculate a new wall shape, which in the fictitious flow would produce this mean pressure distribution. (More precisely a relaxation factor is used so that not an exact mean value is taken but more weight is given to the calculated, external pressure distribution. This takes into account the different sensitivity of external and internal pressure distribution to changes in wall contour.)

The wall is then deformed, according to the calculated new shape, and again the pressure distribution is measured and compared with the one just prescribed for the fictitious exterior flow field. The procedure will be repeated until the differences are within a prescribed margin.

It has to be emphasized that a remarkable feature of this technique is that it refers solely to the flow condition along the wall. There is no model representation in this scheme and the model shape need not even be known! A more general explanation of the adaptive wall principle may now be deduced from the example discussed above. In fact the technique requires: (1) the measurement of two independent flow variables, and (2) a check, whether the two variables satisfy a functional relationship which is consistent with interference free flow conditions.

In the above discussed example, the two measured independent flow variables were the pressure distribution and the flow direction along the wall (given by the measured wall shape). In terms of the small disturbance theory, the pressure coefficient relates to the u-disturbance of the flow, i.e., the deviation of the local velocity component in main stream direction from its undisturbed value U_∞. The local flow direction on the other hand relates to the v-disturbance, i.e., the local velocity component normal to the main stream direction.

The relation between the two flow variables u and v under the assumption of an unconfined flow field is given by the solution of the Laplace equation, which reads for two-dimensional flow

$$u(x,h) = \frac{1}{\pi\sqrt{1-M_\infty^2}} \int_{-\infty}^{+\infty} \frac{v(\xi,h)}{x-\xi}\, d\xi.$$

The basic idea of using this relation for the check of interference-free flow condition was first published by Ferri and Baronti (1973) and Sears (1973).

A variation of this procedure was suggested later by Davis (1981). Instead of measuring two different flow quantities on one surface surrounding the model, he suggested to measure only one quantity but on two surfaces. For laser measurements, the v-component of the local velocity vector is the most convenient flow quantity. The decay of this quantity from one control surface to the other can again be described theoretically and is used to check the interference free flow condition.

3. A Survey on the Adaptive Wall Tunnel Projects

The adaptive wind tunnel wall technique requires three things: (1) a device to adjust locally the flow condition near the test section wall, (2) means to check how far the flow is free of wall interference, and (3) a strategy to arrive at wall interference-free flow condition.

The means for checking the flow condition have been discussed in the previous chapter and also the strategy to arrive at interference-free flow condition. The means are the two measured independent flow variables, and the strategy comprises the use of mean values between measured and calculated flow quantities and the calculation of an improved wall setting by the external flow field calculation.

In the following the various adaptive wall wind tunnel projects will be discussed. It will be shown that different constructional solutions have been chosen, and different techniques for measuring the two flow variables are being applied. A recent survey on the current projects was given during an AGARD Specialist's Meeting in London. The papers thereof were published by AGARD in 1982. A comprehensive literature review was made by Tuttle and Plentovich again in 1982.

Basically there are two different kinds of adaptive wall test sections: One type uses ventilated (perforated or slotted) walls with a local control of the flow through the wall. The other type employs flexible impermeable walls with the shape adjusted to the individual flow condition. Both types have been used for two-dimensional as well as for three-dimensional test sections.

3.1 Survey on the 2D Test Sections Design and Operation

A synopsis of the 2D test sections developed in the past ten years is given in Table 1.

2D-TEST SECTIONS

		TEST SECTION SIZE HxWxL [cm]	[model chord]	MODEL CHORD [cm]	WALLS	No. of Controls TOP+BOTTOM		STATUS
AEDC	1 foot	30.5x30.5x95	2 x 2 x6.3	15.2	PERFOR.	GLOBAL+LOCAL V.P. 2+2 SUBPLENA	●	Basic exploratory Research (finished)
CALSPAN	1 foot	30.5x25.4x142	2 x 1.7x9.3	15.2	PERFOR.	10+ 8 PCC	●●	Advanced explor. Research (finished)
NASA Ames I	13x25cm	13 x 25 x 74	1.7x3.2x9.4	7.8	SLOTTED	10+10 PCC	●	Basic exploratory Research
NASA Ames II	2 feet	61 x 61 x 153	2 x 2 x 5	30.5	SLOTTED	16+16 PCC	○○○	Production meas. starting operation
NASA Langley	0.3 m	36 x 36 x 144	1.8x1.8x7.1	20.3	FLEXIBLE	19+19 Jacks	○○○	Production meas. starting operation
ONERA S4 LCh	18 cm	18 x 18 x 75	2.3x2.3x9.4	8.0	FLEXIBLE	10+10 Jacks	●	Basic exploratory Research (finished)
CERT (ONERA) T2	40 cm	37 x 39 x 132	2 x 2 x6.6	20.0	FLEXIBLE	16+16 Jacks	●●●	Production meas. in operation
SOUTHAMPTON Univ. LS	6x12 in	15.2x30.5x107	1.1x2.2x7.8	13.7	FLEXIBLE	15+15 Jacks	●	Basic exploratory Research (finished)
SOUTHAMPTON Univ. TR	6 in	15.2x15.2x112	1.5x1.5x11	10.2	FLEXIBLE	19+19 Jacks	●●	Advanced explor. Research
TU BERLIN	15 cm	15 x 15 x69	1.5x1.5x7	10.0	FLEXIBLE	8+ 8 Jacks	●●	Advanced explor. Research

Table 1. Survey on the 2D test sections.

A. CALSPAN

The first project with adaptive *ventilated* walls was initiated by the work of Sears at CALSPAN Advanced Technology Center at the beginning of the 70s, see Figure 2. The plenum chambers behind the perforated walls are divided into ten segments on the top and eight segments on the bottom side, each connected to a pressure and a suction source through individual control valves. The tunnel stilling chamber was used as the pressure source while as a suction source an auxiliary compressor was employed.

The test section instrumentation consisted of two static pipes, each with 52 pressure orifices, and 18 flow-angle probes with each probe located above the center of the plenum segment. Further details of the test section, model, and instrumentation are given in the first reports by Vidal et al. (1975) and Sears et al. (1976).

Initial experiments were carried out using a 6 inch chord NACA 0012 aerofoil. The objectives of this work were, first of all, to demonstrate the feasibility of the adaptive wall concept and, second, to develop practical modes for operating such a tunnel.

These first experiments revealed a number of problems which then had to be solved: the existing auxiliary compressor provided insufficient control of the flow at high Mach numbers. The simples solution was the use of a smaller model of 4 inches chord.

Figure 2. The Calspan self-correcting wind tunnel.

The flow-angle probes for measuring the v-component on the control surface were found to be impractical by being very sensitive to contamination from oil present in the stream of the tunnel. A two-velocity-component static pipe was developed (Calspan pipe) and used for later tests.

The experiments were carried out all with manually operated controls. Transonic small pertubation theory was employed to compute the exterior flow field. In the final report of this research project Erickson et al. (1981) conclude that the adaptive wall facility did in principle allow a flow control toward interference free condition even when supersonic flow and shock waves extend to both tunnel walls. For an improved 2D test section design refined control near the model is recommended.

B. AEDC

At AEDC, two-dimensional adaptive walls were investigated during the period from 1976 to 1979. Different wall configurations were used, all ventilated. The main objective of the work was to find the most suitable wall configuration for an adaptive wall test section, in particular in view of an application to three-dimensional flow.

Among the wall configurations tested there was one with only uniformly variable porosity. The others had the capability of providing local control of the flow. A local variation of the angle of perforation was provided by holes in spheres which were sandwiched between two matched drilled plates. The spheres were attached to rods, which allowed the spheres to be rotated, and thus change the hole angle.

Another configuration consisted of matched perforated plate elements which could slide on a perforated ground plate. The amount of overlapping of the perforation then determines the local open area ratio.

A third configuration was a slotted wall with small vanes (baffles) in the slots. They could be adjusted individually and thus provide local control to the flow.

Finally a divided plenum chamber was used for the wall with uniformly variable porosity. It was simply made by using two localized subplena on each top and bottom wall inside the main plenum chamber.

In the manually controlled experiments with a 6-inch chord NACA 0012 model static pressure distribution along the control surface was measured with a static pipe. The flow angle was obtained with individual miniature aerodynamic probes. Transonic small pertubation theory was used to calculate the exterior flow field properties.

The results of the experiments were published by Kraft and Parker in 1979, and by Parker and Sickles in 1981. The control was found to give significant reduction in wall interference, although no complete unconfined flow condition was achieved. The various types of walls were found to perform slightly differently at different Mach numbers. For an application to 3D adaptive wall test sections the sliding plate segments were finally selected.

C. NASA Ames

At NASA Ames slotted top and bottom walls are used for the 2D tests. The small scale experiments were performed in the 25 by 13 cm tunnel with ten independently, manually controlled plenums behind each wall. Laser velocimetry was used to measure the v-component at two different levels both above and below the model. The velocimeter was traversed in the streamwise direction. It's motion and data aquisition was controlled by a dedicated minicomputer.

The measured velocities were used to compute from linear flow theory the required pressure adjustment in the plenum compartments. The computation relies upon influence coefficients which had been empirically determined in the empty test section before the experiments.

Convergence to interference-free flow condition was achieved, as long as the supersonic zone of the 3 inch NACA 0012 aerofoil remaind below the second control level. Satyanarayana, Schairer, and Davis reported more details of the tests in 1980.

Based on the experience with the small scale test section, a new adaptive-wall test section is being constructed at NASA Ames for the 2 by 2 foot transonic wind tunnel. The design incorporates several important improvements against the first test section. The flow through the slotted walls will be controlled by 64 slide valves, each driven by a stepping motor and all controlled by a small computer. The laser velocimetry will involve a very fast, computer-controlled traverse system of mirrors which will substantially reduce LV data aquisition time. Thus the provisions are made for a fully automatic adaptive wall control.

A nonlinear outer-flow solver is also being developed to cope with larger supersonic zones in the flow. A short note on the project was given in a paper by Schairer and Mendoza in 1982. At that time the operation of the new test section was scheduled for spring 1983.

D. ONERA

The first projects with adaptive *flexible* walls were started in the early 70s at ONERA and Southampton University. The NASA Langley test section with two flexible walls is a direct outcome of the work at Southampton University, which in fact was carried out under a NASA grant.

The test section is described by Ladson in 1979. It is very similar in design to the adaptive wall transonic test section of Southampton University. However, since the NASA Langley test section is supposed to be used for the 0.3 m cryogenic transonic tunnel, the flexible walls are made of steel instead of plastic material. They were milled with integrated joints for the jacks, and the wall thickness was varied between 1/8 and 1/16 inches. The test section was assembled and is expected to be installed into the wind tunnel some time next year (1984).

ONERA's first experiments with adaptive flexible walls were carried out in the S4LCh tunnel in Chalais-Meudon. It was considered basic exploratory research providing experience for the design of a larger test section for the CERT T2 tunnel. A compliant plastic material was used for the walls stiffened by longitudinal ribs. The ten jacks on each wall were operated manually.

Successful tests were made with a NACA 64A010 aerofoil at $M_\infty = 0.85$. Interference-free flow conditions were obtained after 3 to 5 iterations. It was found that departures from correct wall shapes by 5/10 mm (five times the wall setting tolerance) had only negligible effects on aerofoil pressure distribution. More details about this work were given by Chevallier in 1975.

The experience from this first facility was used by ONERA/ CERT to build the test section for the T2 wind tunnel in Toulouse. The size of the test section (37×39 cm) allows fairly large aerofoil models to be tested (20 cm chord). The tunnel can be pressurized up to five bars and since this summer (1983) it is operated at cryogenic conditions. The fully automated adaptive wall test section had been operated very successfully at normal temperatures for some time. With the use of cryogenic flow it will become the most advanced test facility in operation for aerofoil tests. This certainly will justify a more detailed description of the T2 test section design and its operation, see Figure 3.

The test section has two parallel side walls providing support for the airofoil model. Provisions are made for a quick model installation. The model will have to be cooled outside the tunnel for cryogenic tests. (Tunnel running time is only 30–60 s, with 10 s to establish steady flow condition.) The upper and lower wall of the test section are flexible steel plates 1.5 mm thick. Teflon is used to seal them against the side walls. Each wall is equipped with 91 pressure taps, and can be moved by 16 jacks. Maximum wall displacement is limited to 25 mm. Each jack is moved by a hydraulic actuator which is controlled by a step motor. One single step corresponds to a wall movement of 0.2 mm, thereby defining the wall setting tolerances. The wall position is controlled by a potentiometric displacement feeler with an accuracy of 5/100 mm, see Figure 4.

Figure 3. The T2 test section.

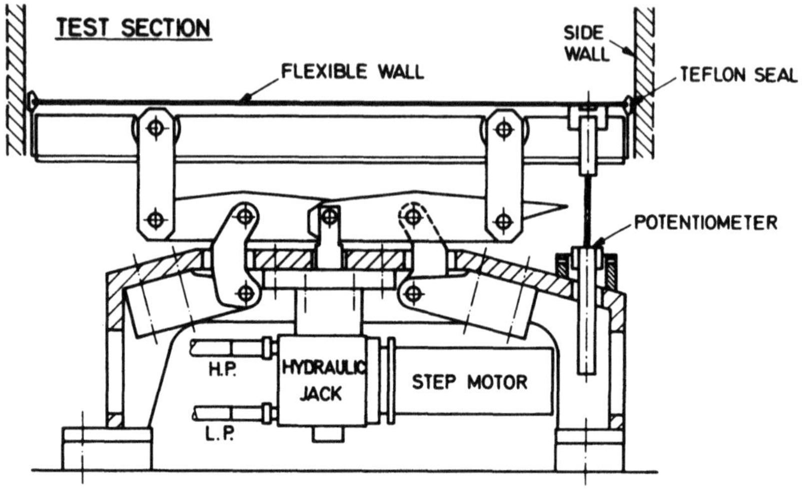

Figure 4. The jack system of the T2 test section.

Some sophisticated procedures are being used for the tunnel operation. At the start of a test series the initial wall shape is calculated by a flow field calculation using three singularities (source, doublet, and vortex) at the position of the model. The vortex strength is determined from an approximate estimation of lift, pitching moment and drag. For the following tests, then the last wall contour, obtained for a slightly different Mach number or angle of incidence, is used as a starting configuration.

The calculation of an improved wall configuration from the measured pressure distribution follows the classical line, except that for the choice of the relaxation factor, particular effort is made to ensure a fast convergence of the adaptation process. The estimated changes of the wall configurations are being split into four changes of the test section stream tube: a change of width, divergence, centerline camber, and centerline inclination. Different optimal relaxation factors have been found for these effects and are being used for the wall adaptation.

Finally, special procedures are used to calculate the correct main stream Mach number and to extrapolate the flow variables beyond the end of the test section, both based on the data measured along the walls. More details about the methods and the other operational procedures for the T2 wind tunnel are given by Archambaud and Chevallier (1982), and by Chevallier et al. (1983).

E. Southampton University

The work at Southampton University on adaptive wind tunnel walls started in a low speed tunnel. Goodyer reported 1975 on the first tests using a circular cylinder and a NACA 0012-64 aerofoil. The demonstration of the interference free flow condition achieved by contouring the walls was particularly impressive because of the large blockage of the models: a circular cylinder of 25% and 30% blockage and a NACA 0012-64 aerofoil of 10% blockage (tunnel height to aerofoil chord ratio 1.1).

Streamlining the walls was very time-consuming in these early experiments. Around eight iterations were required and the computation of one pair of wall shapes took typically one hour on a minicomputer.

The design of a new test section for a transonic tunnel of Southampton University was based on the experience with the first tunnel and detailed analytical work, as reported by Judd, Wolf, and

Goodyer (1976). A feature of the new test section are the relatively long walls with a corresponding large number of jacks (20 on each wall, with No. 20 used for Mach control). Again ribs are fitted to the flexible walls and attached to the jacks via metal flexures. The screw-jacks are driven by stepping motors and jack position is measured by linear potentiometers.

A PDP 11/34 computer is used for fully automatic control of the adaptive wall test section including model data aquisition. The time to complete a streamlining cycle varies from about $1\frac{1}{2}$ to 5 minutes.

The tunnel has been used extensively to test a 4 inch chord NACA 0012-64 model and, more recently, a supercritical NPL 9510 of 6 inch chord. The tests were carried to very high Mach numbers. At Mach numbers above 0.85 the streamlining process was found to become less stable with strong sensitivity of model pressure to changes in wall contour. More details are given by Goodyer and Wolf (1982) and Wolf, Cook, and Goodyer (1982).

F. Berlin

At the Technical University of Berlin, a 2D adaptive wall test section was developed with the explicit intention of applying this technique to a test section for 3D model tests. Thus, it was essential to bring the number of jacks per wall to a minimum and to provide a fast and very accurate automatic control for the wall position.

It was found that eight jacks are sufficient when fairly stiff wall material is used, such as fibreglass, giving smooth bending lines.

The control system relies upon DC electromotors to drive the jacks and high precision linear potentiometers touch on the walls to read the wall position. A fast control is achieved by supplying different voltage to each motor proportional to the amount of displacement required. Wall setting during one iteration typically takes about half a second.

A number of tests were made using a NACA 0012 and a supercritical CAST 7 aerofoil. The tests revealed that the supercritical aerofoil (in particular at its design condition) is much more sensitive to small changes in wall contour than the conventional aerofoil. Test results have been published by Ganzer in 1980 and 1982.

3.2 Representative Results Obtained in 2D Tunnels With Adaptive Walls

A few selected test results will be presented here in order to give an impression of the state of the art of 2D adaptive wall test sections. The first test result to be shown is a subcritical pressure distribution for the NACA 0012 aerofoil, see Figure 5. In this case, obviously, wall interference has been reduced to a negligible amount by adapting the walls.

At transonic speeds, in general, plain solid walls would not just lead to an overall change in pressure level, but to a severe distortion of the pressure distribution. Transonic blockage will occur, limiting the main stream Mach number, when sonic speed is reached in the smallest cross section between model and wall. Numerous test results obtained at transonic speed provide evidence that blockage is removed by wall adaptation and tests can be carried to fairly high Mach numbers. An example is taken from Goodyer and Wolf (1982), see Figure 6.

In this case the supersonic zone extended well up to the upper and lower wall (maximum local Mach number at the wall $M = 1.1$). Agreement with NASA data can still be improved by taking into account the local increase of boundary layer thickness due to the shock. If this is done, the data in the rear part of the aerofoil coincide.

The next diagram shows three sets of surface pressure distribution obtained for the CAST 7 aerofoil in the TU-Berlin tunnel. Comparison is made with data obtained for the same model in the DFVLR 1×1 m transonic wind tunnel, see Figure 7. The figures are selected as representative examples from a test series agreed to by the GARTEur AG 02. They include the most difficult flow condition, i.e., the sensitive condition near the design point and the highest Mach number. The discrepancies between TU Berlin and DFVLR TKG data are, at least to some extent, to be attributed to slightly different side wall flow conditions and a small streamwise Mach number gradient in the TKG tunnel.

The same model was also tested in the tunnel of Southampton University. For this test small plates were fixed on the aerofoil parallel to and in 1 cm from the side walls. These fences reduce the effective model span to 13 cm (1.3 chord) but shield off side wall interference effects which have been found to be very pronounced

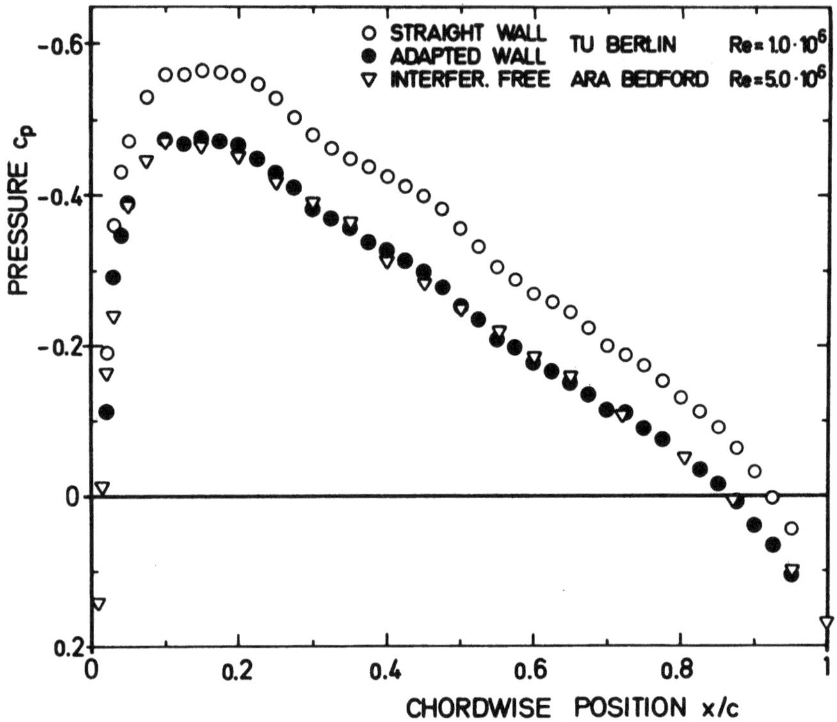

Figure 5. NACA 0012 pressure distribution at $M_\infty = 0.50$ and $\alpha = 0°$.

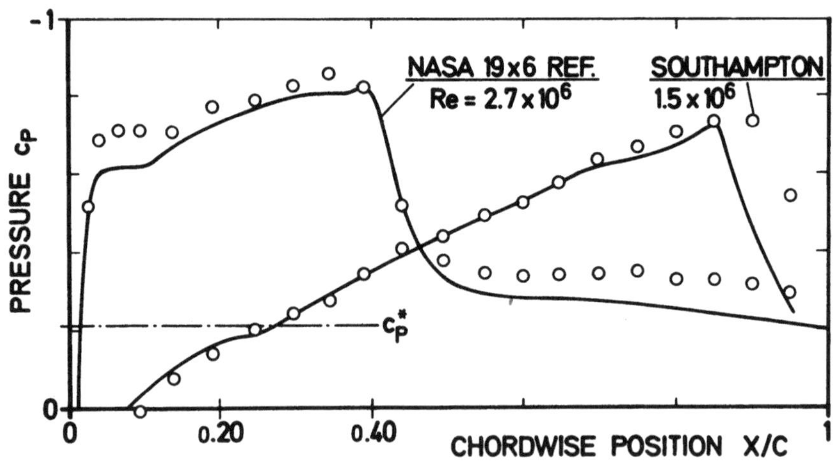

Figure 6. NACA 0012-64 pressure distribution at $M_\infty = 0.89$ and $\alpha = 4°$.

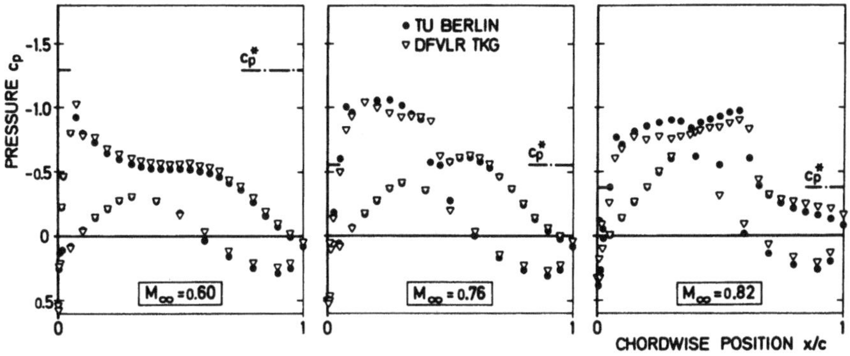

Figure 7. CAST 7 pressure distributions at $\alpha = 0.8°$ Re $= 1.2 \div 1.4 \cdot 10^6$.

without such fences, see Ganzer, Stanewsky, and Ziemann (1983). The results from the two tunnels, which have the same cross sectional size and the same Reynolds number of 1.4×10^6, agree fairly well as may be seen from the plot in Figure 8.

In this plot data are included for the T2 tunnel and other conventional tunnels as reported by GARTEur (1983). The data of TU-Berlin and University of Southampton show a different lift gradient and slightly different maximum lift. However, the correction of these data to the higher Reynolds number was just done by an empirically determined α-shift which can not be expected to lead to reliable results.

The good agreement of T2 data with other GARTEur results (ARA, NLR, ONERA S3Ma) on the other hand are indicative of the high standard reached with this facility. A final plot supports this conclusion, see Figure 9.

The T2 data have partly been checked by calculating residual interferences with a RAE method using measured boundary conditions, see GARTEur (1983). The residual interferences were typically of the order of $\Delta M = 0.002$ and $\Delta\alpha = 0.03°$. Such accuracy is about the best presently available in aerofoil testing and comes close to the requirements laid down in an *AGARD Conveners Report*, see Steinle and Stanewsky (1982).

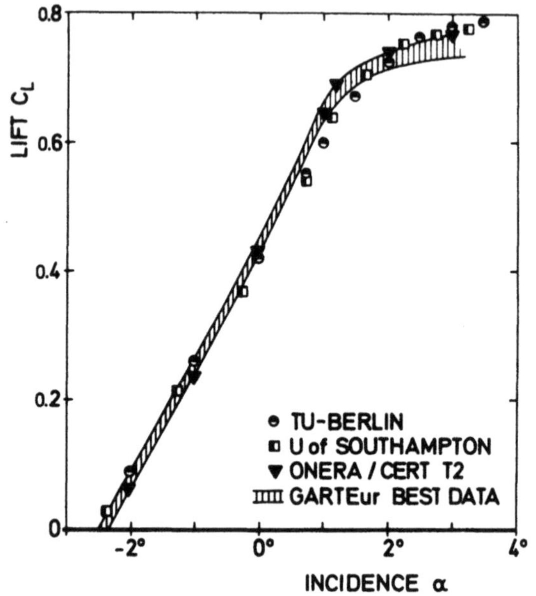

Figure 8. CAST 7 lift vs. incidence, data corrected to $\mathrm{Re} = 2.5 \cdot 10^6$, $M_\infty = 0.76$.

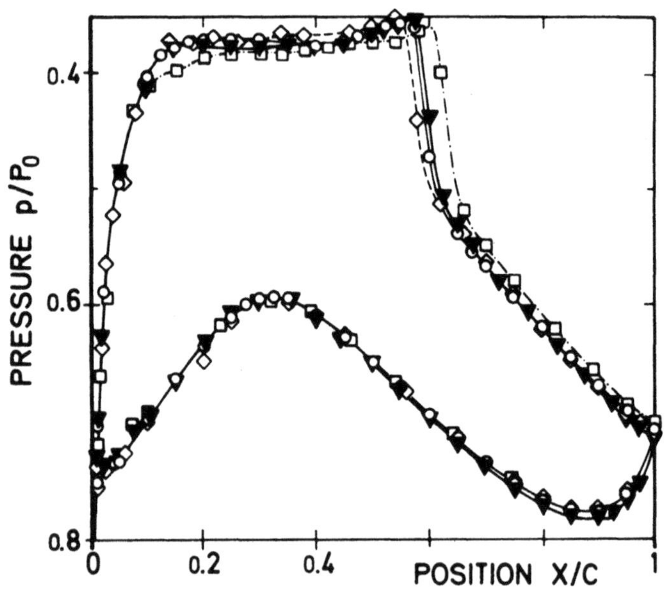

Figure 9. CAST 7 pressure distribution $M_\infty = 0.76$, $C_L = 0.73$, Re $= 6 \cdot 10^6$, CERT T2 compared with other GARTEur data.

It may be concluded, that the adaptive wall technique for 2D model testing is by now well established. The detrimental features of this technique are obvious: additional mechanical installations and additional wind tunnel time. In average two to three iterations are necessary. Each iteration requires at present 3 seconds (0.5 s for the wall pressure measurements using a PSI-System, 2 s for the external flow field calculation, and 0.5 s for the wall setting). But it appears that these disadvantages are well balanced by a reduced test section height and improved reliability of the test results.

3.3 Survey on the 3D Test Section Designs

The application of the adaptive wall technique to 3D model tests is seriously investigated only since the last three or four years. A survey on this work was given at an AGARD Specialist's Meeting, AGARD (1982). The current projects are listed in Table 2.

A. AEDC

At the AEDC the 1-Foot-tunnel (1T) is used for the project, equiped with a new test section, see Figure 10. The test section is featured by a segmented variable porosity wall configuration, which allows to change the local crossflow characteristic of the perforated wall. The top and bottom wall each have 24 individually adjustable segments, the side walls have eight. Stepping motors are driving the segments through a rack and pinion mechanism. Linear potentiometers are utilized as position sensing devices.

3D-TEST SECT.	TEST-SECTION		WALLS	No. of CONTROLS
	Cross-section	Length		
AEDC / CALSPAN	1 ft. square	3 1/8 ft	4 perforated	64 wall segm.
DFVLR Göttingen	80 cm diam.	240 cm	rubber tube	64 Jacks
NASA Ames	13x25 cm rect.	74 cm	2 slotted 2 solid	44 Plenum Ch. Comp.
Univ. of SOUTHAMPTON	6 in. square	44 in.	2 flexible 2 solid	40 Jacks
TU BERLIN	15 cm sqare	69 cm	2 flexible 2 solid	16 Jacks
	15x18 cm octagon	83 cm	8 flexible	78 Jacks

Table 2. Projects of 3D adaptive wall tunnels.

One of the major problems in using ventilated walls for an adaptive wall test section arises from requirement to measure two independent flow variables on a control surface surrounding the model. A two-velocity-component static pipe has been developed for this purpose, as shown in Figure 11.

Pressure orifices are located on the circular cross section pipe diametrically opposed along the direction of the normal to the interface. The pressure measured with one pair of orifices provides the local static pressure as the mean value and the local v-component of the flow from the difference of the two measured data. A detailed analysis of this technique is discussed by Nenni, Erickson, and Wittliff in 1982.

In the test section, two pipes are rotated along a circular surface about the centerline of the test section, automatically controlled by a computer. The v-component variation as determined from the measurement is used as a boundary condition for the calculation of the corresponding u-component. The difference between calculated and measured u-component then determines a merit function for the wall control. Since there is no direct relation between wall-control variable and the response of the flow variable, some elaborate procedure is necessary to tune the wall characteristic such that the desired u-distribution is achieved.

The AEDC adaptive wall test section and its control are described in more detail by Parker and Erickson (1982). At that time fully automated operation of the test section was scheduled for October 1982, but this has been somewhat delayed. No results have been published so far.

B. DFVLR

At the DFVLR Göttingen a 3D deformable rubber tube test section is being developed, see Figure 12. The rubber tube has a diameter of 80 cm and the wall thickness is 6 cm. Altogether 64 jacks are distributed around the deformable tube. The jacks will be driven by computer-controlled stepping motors. A one-step adaptation procedure has been developed which will allow to calculate the adapted wall configuration from only one set of wall pressure measurement.

The test section design was briefly described by Ganzer (1982). At present the test section is assembled and first tests are expected for the end of this year (1983).

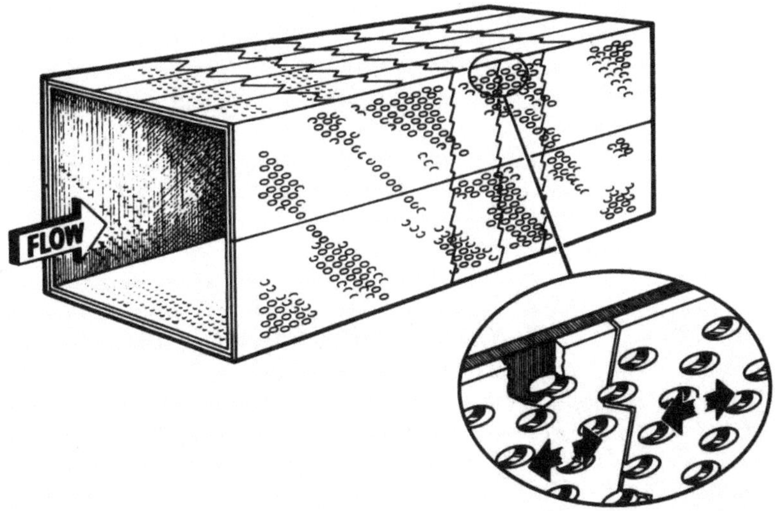

Figure 10. The AEDC/CALSPAN segmented, variable porosity wall test section.

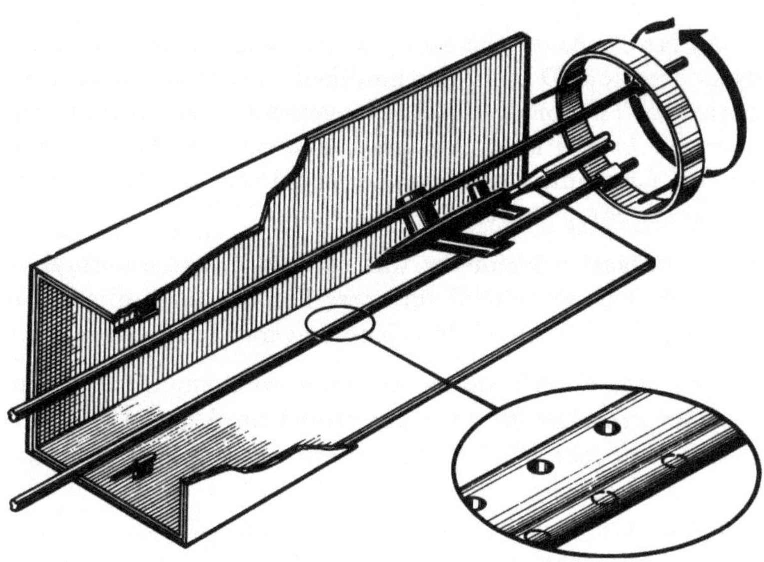

Figure 11. Interface flow variable measurement system.

Figure 12. The DFVLR deformable rubber tube test section.

C. NASA Ames

At NASA Ames the 25 by 13 cm wind tunnel with adaptive walls as used for 2D tests was modified to permit cross-stream wall adjustment. The longitudinal arrangement of the plenum chamber compartments was kept the same, but each of the six upper and lower compartments closest to the model was subdivided into three cross-stream compartments.

Again laser velocimetry was used to measure normal velocity components on two control surfaces. For the three-dimensional experiments, the velocimeter was also travered in spanwise direction.

The test model was a semispan wing mounted to one side wall. The experiments were performed at $M_\infty = 0.6$ at several angles of incidence. In all the experiments, wall interference was substantially reduced after the wall boundary conditions had been adjusted according to the adaptive wall procedure, although interference was not entirely eliminated. It was felt that the influence coefficient method used to adapt the walls was not fully adequate. The velocity changes required on the control surface could not be accurately produced. Schairer and Mendoza did report on the tests in 1982.

D. Southampton

At Southampton University, the 2D test section with two flexible walls is being used to investigate 3D configurations like a wing body combination. Wolf, Cook, and Goodyer reported in 1982 about the first experiments which demonstrated that model and support blockage can be relieved by wall contouring. Currently, work is in progress for developing a suitable algorithm to calculate the 3D interference effect and predicting the wall movements required to minimize these effects.

E. TU Berlin

At TU Berlin also the 2D test section with two flexible walls is being used for testing a body of revolution and a wing body combination. These tests are done in cooperation with Wedemeyer (DFVLR) and Lamarche (VKI) who did provide a computer program to calculate suitable wall shapes from measured wall pressure. The corresponding analytical work is described by Wedemeyer (1982). The program has been implemented at the TU Berlin computer, and first test's using a body of revolution at $M_\infty = 0.70$ and 0.84 yielded good agreement with interference-free reference data.

The development of a 3D test section with adaptive walls at the TU Berlin began in 1979. A configuration with eight flexible walls was chosen. All the walls are subject to a two-dimensional deformation similar to that arranged in the 2D test section. The eight walls from an octagon cross section of 18×15 cm in main dimensions. It is believed that this design is a reasonable compromise with respect to the desired three-dimensional wall configuration and a restricted mechanical complexity, see Figure 13. The flexible walls are made of steel 1.2 mm thick. Spring steel lamellas of 0.25 mm thickness are used to seal the corners between two walls. 73 lamellas are spot-welded on one wall and slide on the adjacent wall. The test section design is supposed to be applicable to cryogenic tunnels.

First test results have been published by Ganzer in 1982. Since it is the only 3D test section for which test results are available with fully adapted walls, its operational procedure and the test results will be discussed in some detail.

3.4 Operational Procedure of the TU-Berlin 3D Test Section

There are two dominant problems in operating the test section

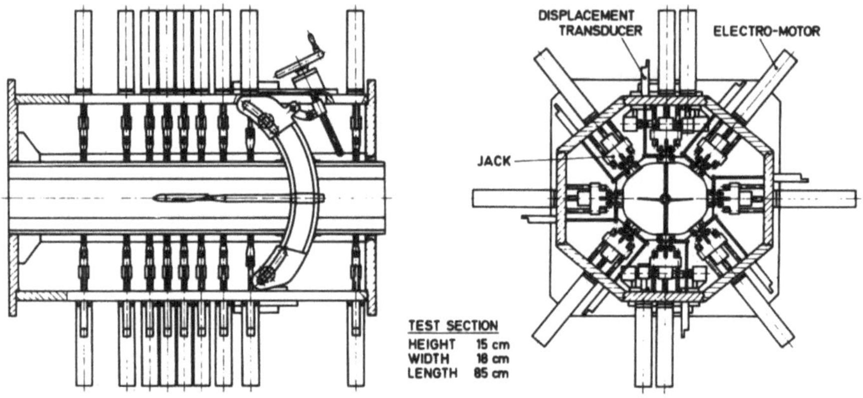

Figure 13. The TU Berlin 3D test section.

with adaptive flexible walls: a suitable algorithm is required to calculate the exterior three-dimensional flow field and a control system has to be provided which allows to adjust and control the wall shape.

The procedure of wall adaptation for three-dimensional flow condition was discussed recently by Ganzer (1983). In this case the external flow field calculation requires the solution of a three-dimensional differential equation. The use of linearized small pertubation theory again will be justified, as long as no severe transonic effects occur near the wall.

A boundary condition will be prescribed in terms of the measured pressure distribution. This in fact defines a u-disturbance on a control surface, which suggests to base the inverse flow field calculation on the Laplace equation for u:

$$\beta^2 u_{xx} + u_{yy} + u_{zz} = 0$$

The boundary condition is given by $u\,|_s = -c_p/2$ on the control surface and $u = 0$ at infinity. For ease of understanding, let's again consider the case of solid flexible walls which may be deformed into a streamline shape. The test section walls then can be chosen as the control surface.

For the solution of such a linear boundary value problem, the singularity method is well suited. In the present case source singularities are being distributed over the entire control surface. They produce a u-disturbance described by a Fredholm integral equation of the first kind in terms of the singularity strength σ.

This is solved by a first order panel method (constant singularity strength on each panel) which leads to a system of linear equations. The equations relate the source strength of the individual panels σ_j to the u-disturbance at the control points i as given by the boundary condition

$$u_i = \sum_{j=1}^{N} A_{ij}\sigma_j.$$

The matrix A results from the geometrical shape of the panelized control surface (test section geometry), and also depends on the main stream Mach number.

The singularity strength thus determined can be used to calculate any other flow variable. In order to calculate the corresponding wall shape, it is useful to determine the normal derivative of u-disturbance (normal to the control surface).

$$\left(\frac{\partial u}{\partial n}\right)_i = \sum_{j=1}^{N} B_{ij}\sigma_j$$

Again the matrix B depends only on the geometry of the control surface and the Mach number. In combining the two equations one obtains

$$\vec{u}_n = BA^{-1}\vec{u}$$

For a given Mach number BA^{-1} may be determined before a test, thus the calculation of u_n from a measured u requires only simple multiplication. The relation between the normal derivative u_n and the wall contour h can be explained by the use of the disturbance potential ϕ:

$$u = \partial\phi/\partial x$$

$$v = \partial\phi/\partial n$$

v is the disturbance velocity normal to the main stream and is equal to the streamwise gradient of the wall contour, $v = \partial h/\partial x$. This leads to

$$\begin{aligned} u_n &= \frac{\partial u}{\partial n} \\ &= \frac{\partial^2 \phi}{\partial x \partial n} \\ &= \frac{\partial}{\partial x}\left(\frac{\partial \phi}{\partial n}\right) \\ &= \frac{\partial^2 h}{\partial x^2} \end{aligned}$$

The second derivative of the wall contour may be written in difference form for the individual control points

$$\frac{d^2 h(x)}{dx^2} = \frac{h(x+\Delta x) - 2h(x) + h(x-\Delta x)}{(\Delta x)^2}$$

It is of great practical advantage that the conditions at the beginning and at the end of the test section can be introduced: $h(0) = h(L) = 0$. The resulting system of equations can be solved without problem yielding the required wall setting.

The wall shape as calculated from the exterior flow field calculation method is then set up and controlled by a computer-based control system. The control system used for the 3D test section employs a microprocessor as the key element in the control circuit. Every 100 ms it controls up to 80 jacks. It compares the actual wall positions as indicated by linear potentiometers with the nominal wall position. The nominal wall position is either calculated by the wind tunnel computer or prescribed by the wind tunnel user via the terminal.

Depending on the amount of difference between nominal and actual position, the microprocessor initiates either an action of the precontroller or the endcontroller. Both are software routines of the microprocessor. The precontroller fixes power values for the individual DC motors direct proportional to the required displacement of the jack. The one which has to produce the largest displacement will be provided with maximum power. Thus, a fast adjustment is ensured with minimal loads on the flexible walls. The endcontroller is of quasi-steady PD (proportional, differential) time-respond type. It becomes active only when the required wall displacement is very small (less than 0.3 mm).

The advantage of this control system is that by using the microprocessor, hardware requirements are reduced substantially and power requirements are kept small, which helps to avoid cooling problems.

The time required for a complete wall adaptation is first of all given by the number of iterations. The present operational experience with the 3D test section shows that in average 5 to 7 iterations are necessary. Each iteration then requires an external flow field calculation lasting 25 seconds on the HP1000F (including the interpolation of measured pressure data). Wall adjustment typically takes 0.5 seconds. Pressure measurements at the 192 wall pressure taps is done in less than 0.5 seconds, using PSI pressure measuring system.

3.5 First Results Obtained in the TU-Berlin 3D Test Section

The first tests in the 3D test section were made with a body of revolution, designed by ONERA as a calibration model. Pressure distributions at several transonic main stream Mach numbers are available for comparison, see Barche (1979).

Figure 14 shows the surface Mach number distribution as measured in the 3D test section with adapted and aerodynamically straight walls. The blockage of the model was 2%. This is a new, improved result as compared to the one presented by Ganzer in 1982. The improvement was achieved through the new exterior flow field calculation method. It is felt that agreement with the reference data is by now fully satisfactory.

The first measurements with a lifting configuration were made by using a DFVLR F4 wing-body-combination of 12 cm span. The blockage of the model was 1.6%.

A three-component balance of 8 mm diameter was especially build by the DFVLR for the TU-Berlin tests. The test results in terms of lift, drag, and pitching moment are shown in Figure 15. The data points obtained with adapted walls are compared with those measured with plane walls. Also, test results from the DFVLR 1×1 m Transonic Tunnel are included, where the same model was

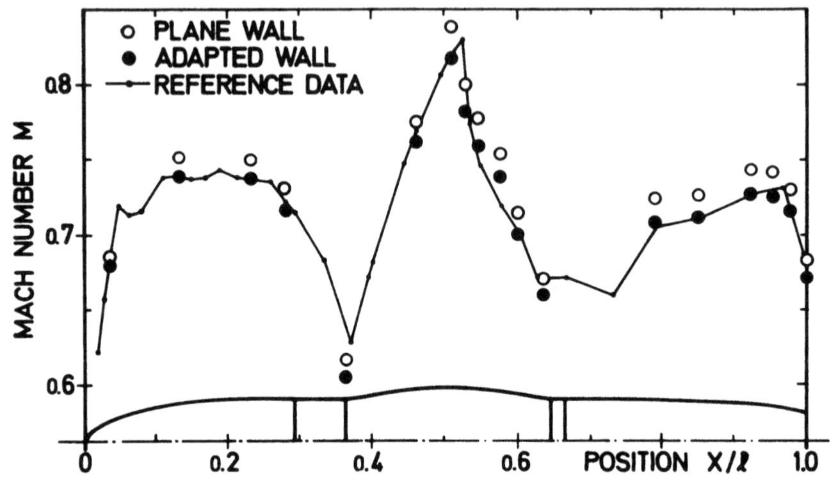

Figure 14. Mach number distribution for ONERA C body of revolution, $M_\infty = 0.70$, $\alpha = 0°$.

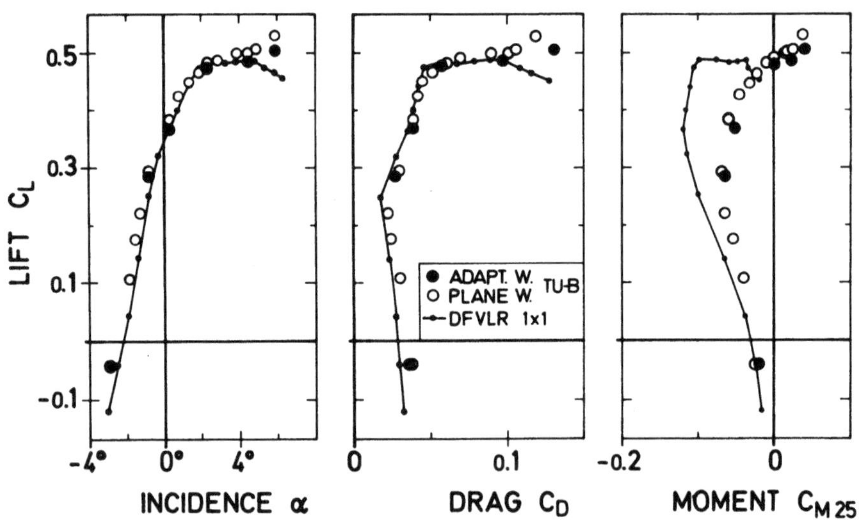

Figure 15. Results of force measurement for DFVLR F4 model $M_\infty = 0.70$.

used. Differences between adapted and straight wall data are not very pronounced because of the small model: the ratio between wing area and cross-sectional area of the test section was only 0.067. The results indicate, however, that wall adjustment has the expected effect in reducing the wall interference. Comparison with DFVLR data can, unfortunately, only be done with proviso. During the test in the 1×1 m tunnel, the model surface was slightly demolished by flow contamination concentrated on the centerline of the tunnel. So the model has now a different, somewhat rough surface.

As far as the adaptation procedure is concerned, the following diagrams are definitely more conclusive. In Figure 16 the changes of wall contours during the iterations are shown. The initial configuration was the aerodynamically straight wall. It can be seen that the adaptation procedure converges and that it yields plausible wall contours. A total of 13 iterations were required in this case. However, there are only 6 to 7 iterations necessary, when the initial wall contour is not the straight wall but a shape from another adaptation at similar flow condition is used.

The following Figure 17 gives an impression of the test section geometry with the F4 model. The required final wall displacements are shown at the individual jack positions. It has to be mentioned here that the reference wall configuration is the aerodynamically straight wall, which yields constant Mach number all along the test section walls when the model and sting are out but the quadrant is in.

The final proof of the reliability of the wall adaption procedure is given with the next figure—Figure 18. In this diagram two sets of wall pressure distributions are shown for the adapted and for the plane wall configuration. Measured pressure distributions are compared with the theoretical ones.

For the plane wall the theoretical pressure distribution is $c_p = 0 =$ constant. Thus the measured values give directly the discrepancy between calculated and measured data.

For the adapted case it is demonstrated that the measured pressure distribution compares very well with the one calculated for the exterior flow field. This agreement is considered as the decisive proof of the feasibility of the adaptive wall technique for 3D model tests.

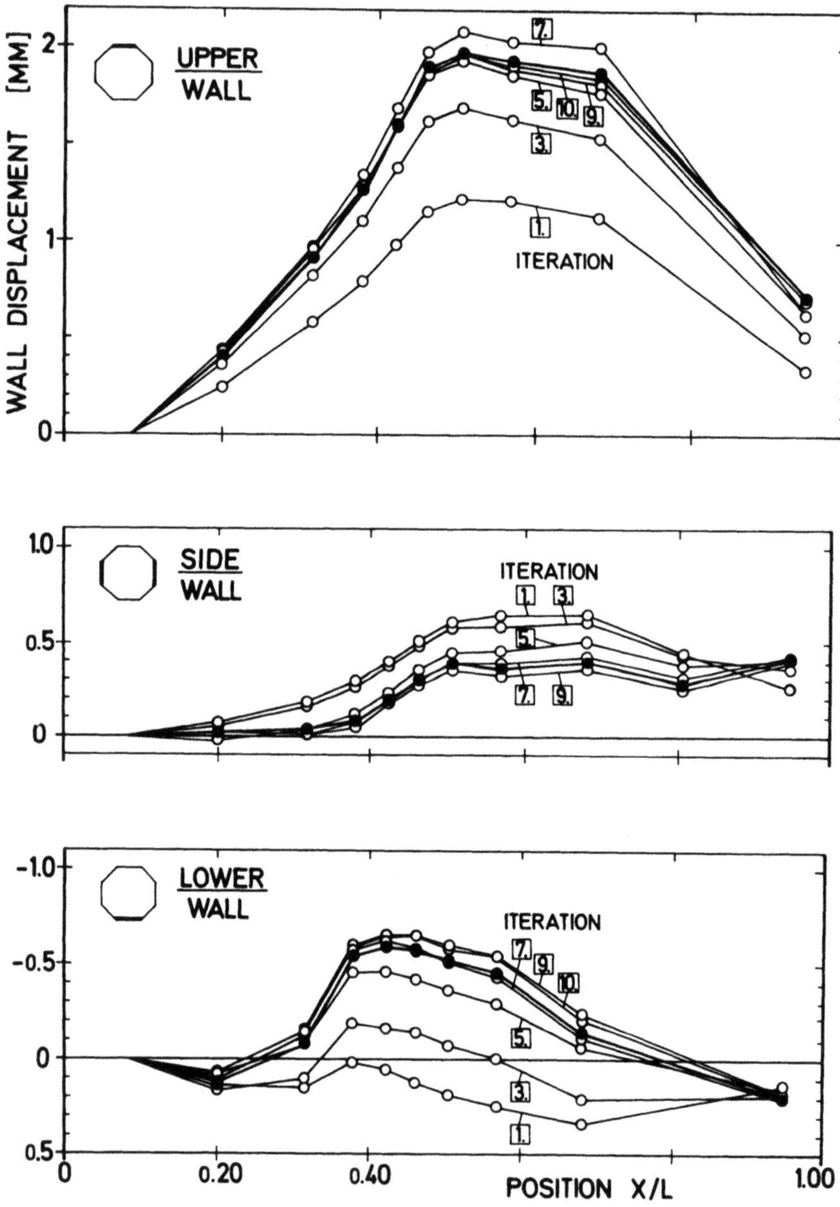

Figure 16. Wall contours during iteration. F4 model at $M_\infty = 0.70$, $\alpha = 4.4°$.

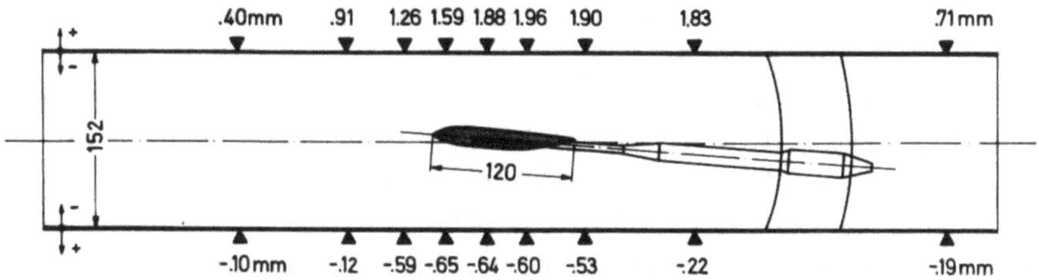

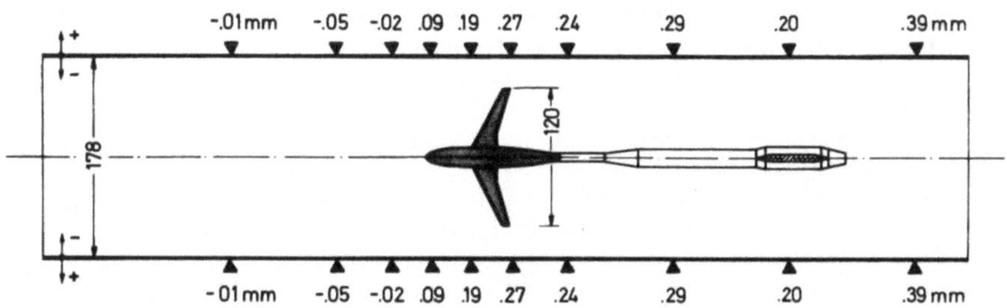

Figure 17. Adapted wall configuration for F4 model at $M_\infty = 0.70$, $\alpha = 4.4°$.

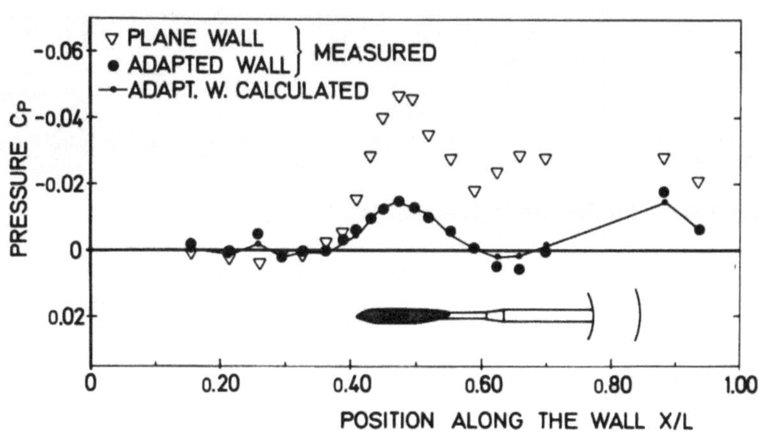

Figure 18. Pressure distribution along upper wall with F4 model at $M_\infty = 0.70$ and $\alpha = 0°$.

Further investigations are planned in this TU Berlin test facility with a new model. The model is shown in Figure 19. It has a span of 12 cm like the F4 model. The advantage of this configuration is that it allows to explore higher Mach number regions, possibly up to $M_\infty = 1.3$. On the other hand the model has a much larger wing area so that stronger forces occur on the balance and more pronounced pressure signals on the walls. This will ease the measurements. In addition to this, the body is thick enough to allow some surface pressure measurements. Ten pressure taps are distributed along the fuselage. Tests already have started with this model. Results will be published in near future.

Figure 19. Model for high speed tests.

4. Conclusions

The adaptive wall technique has been used quite extensively in testing 2D models, i.e., conventional aerofoils as well as supercritical aerofoils. The main purpose was first of all to demonstrate the feasibility of this technique and this was, in principle, achieved in all the projects.

Further developments then were aimed at a refinement of the measurement and the adaptation procedure, including fully automatic control and data aquisition. The refinement of measurements required first of all tremendous efforts when test sections with ventilated walls were used. The development of a two-velocity-component static pipe (CALSPAN-AEDC) and fast laser velocimetry are the outcome of this efforts. A microprocessor based automatic control of the adaptable walls is now standard for all test sections in operation.

Some improvement may still be necessary by employing transonic small pertubation theory instead of the linear theory, at least as a final check of the wall configuration. However, generally speaking the adaptive wall technique for 2D tests is well established by now and ready made for production test facilities.

The application of this technique to 3D flow conditions is still in the beginning and demonstrational phase. The first test results obtained in the TU-Berlin facility are considered as a general proof of the feasibility, at least for subsonic flow. But, in contrast to aerofoil testing, the tests of 3D configurations in many cases involve high transonic main stream Mach numbers up $M_\infty = 1.3$ and more. This requires first of all a transonic external flow field calculation, which is already used at the AEDC and is being developed at the TU Berlin. The high sensitivity of the flow at Mach numbers close to unity might result into difficulties for the adaptation procedure. Anyhow, the work up till now can be seen as fairly encouraging since there has been no evidence of unusual problems to be associated with the 3D flow.

Acknowledgement

This paper is based in part, on research carried out at the Technical University of Berlin sponsored by the German Minister of Research and Technology (BMFT) and the German Research Association (DFG).

References

[1] AGARD, "Wall Interference in Wind Tunnels," *AGARD-CP-335,* Sept. 1982.

[2] Archambaud, J. P. and Chevallier, J. P. "Utilisation de Parois Adaptables Pour les Essais in Courant Plan," *AGARD-CP-335*, Sept. 1982.

[3] Barche, J., ed. "Experimental Data Base for Computer Program Assessment," *AGARD-AR-138*, May 1979.

[4] Capellier, C., Chevallier, J. P., and Bouniol, F. "Nouvelle Méthode de Correction des Effectś de Parois en Courant Plane," *14e Colloque d'Aerodynamique Mémoire 2, Toulouse Nov. 1977 et La Recherche Aerospatiale no. 1*, 1978.

[5] Chevallier, J. P. "Soufflerie Transsonique a Parois Auto-Adaptable," *AGARD-CP-174*, Oct. 1975.

[6] Chevallier, J. P., et al. "Parois Adaptable à T2," *La Recherche Aerospatiale no. 4*, 1983.

[7] Davis, S. "A Compatibility Assessment Method for Adaptive Wall Wind Tunnels," *AIAA Journal, 19*, Sept. 1981.

[8] Erickson, J. C., et al. "Adaptive Wall Wind Tunnel Investigation," *Calspan Rep. RK-6040-A-2*, Febr. 1981.

[9] Ferri, A. and Baronti, P. "A Method for Transonic Wind Tunnel Corrections," *AIAA Journal, 11*, Jan. 1973.

[10] Ganzer, U. "Adaptable Wind Tunnel Walls for 2D and 3D Model Tests," *ICAS-80-23.3*, Oct. 1980.

[11] ————. "On the Use of Adaptive Walls for Transonic Wind Tunnel Testing," *AGARD-CP-335*, May 1982.

[12] ————. "Die Technik Adaptiver Windkanal-Wände," *Luftfahrtforschung und Luftfahrttechnologie, 3.* BMFT-Statusseminar, Mai 1983.

[13] Ganzer, U. and Igeta, Y. "Transonic Tests in a Wind Tunnel With Adapted Walls," *ICAS-82-5.4.5*, 1982.

[14] Ganzer, U., Stanewsky, E., and Ziemann, J. "Side-Wall Effects on Aerofoil Tests," to be published in *AIAA Journal*, 1983.

[15] GARTEur, "Report on the Work of the Action Group 02 on Two-Dimensional Transonic Testing Methods," to be published as *NLR-Report*, 1983.

[16] Goodyer, M. J. "A Low Speed Self Streamlining Wind Tunnel," *NASA TMX-72699*, August 1975; and *AGARD-CP-174*, Oct. 1975.

[17] Goodyer, M. J. and Wolf, S.W.D. "Development of a Self-Streamlining Flexible Walled Transonic Test Section," *AIAA Journal, 20, No. 2*, Febr. 1982.

[18] Judd, M., Wolf, S.W.D., and Goodyer, M. J. "Analytical Work in Support of the Design and Operation of Two-Dimensional Self-

Streamlining Test Sections," *NASA CR 145019,* July 1976.

[19] Kraft, E. M. and Parker, R. L. Jr. "Experiments for the Reduction of Wind Tunnel Wall Interference by Adaptive-Wall Technology," *AEDC TR-79-51,* Oct. 1979.

[20] Ladson, C. L. "A New Airfoil Research Capability," in *Advanced Technology Airfoil Research, 1, NASA-CP-2045, Part 1,* March 1979.

[21] Nenni, J. P., Erickson, J. C. Jr., and Wittliff, C. E. "Measurement of Small Normal Velocity Components in Subsonic Flow by Use of Static Pipe," *AIAA Journal, 20, No. 8,* Aug. 1982.

[22] Parker, R. L. Jr. and Sickles, W. L. "Two-Dimensional Adaptive Wall Experiments," *AEDC-TR-80-63,* Febr. 1981.

[23] Parker, R. L. Jr. and Erickson, J. C. Jr. "Development of a Three-Dimensional Adaptive Wall Test Section With Perforated Walls," *AGARD-CP-335,* May 1982.

[24] Satyanarayana, B., Schairer, E., and Davis, S. "Adaptive-Wall Wind-Tunnel Development for Transonic Testing," *J. Aircraft, 18, No. 4,* April 1981.

[25] Schairer, E. T. and Mendoza, J. P. "Adaptive Wall Wind Tunnel Research at Ames Research Center," *AGARD-CP-335,* May 1982.

[26] Sears, W. R. "Self Correcting Wind Tunnels," The Sixteenth Lanchester Memorial Lecture, London, May 1973, *Calspan Rep. RK-5070-A-2,* Juli 1973.

[27] Sears, W. R., et al. "Interference-Free Wind-Tunnel Flows by Adaptive-Wall Technology," *ICAS Paper 76-02,* Oct. 1976.

[28] Steinle, F. and Stanewsky, E. "Wind Tunnel Flow Quality and Data Accuracy Requirements," *AGARD-AR-184,* Nov. 1982.

[29] Tuttle, M. H. and Plentovich, E. B. "Adaptive Wall Wind Tunnels—a Selected, Annotated Bibliography," *NASA TM 84526,* Nov. 1982.

[30] Wedemeyer, E. "Wind Tunnel Testing of Three-Dimensional Models in Wind Tunnels With Two Adaptive Walls," *VKI Rep. 1982-36,* Sept. 1982.

[31] Wolf, S.W.D., Cook, I. D., and Goodyer, M. J. "The Status of Two- and Three-Dimensional Testing in the University of Southampton Self-Streamlining Wind Tunnel," *AGARD-CP-335,* May 1982.

[32] Vidal, R. J., Erickson, J. C. Jr., and Catlin, P. A. "Experiments With a Self-correcting Wind Tunnel," *AGARD-CP-174,* October 1975.

Prospects for Flow Measurements Based on Spectroscopic Methods

Donald Baganoff
Stanford University
Stanford, CA 94305

Abstract

The shadow, schlieren, and interferometric methods have been very useful in many areas of gas dynamics, particularly for the study of planar or axisymmetric compressible flows. With the development of the laser, the potential exists for the introduction of methods whereby the thermodynamic state of a general three-dimensional nonsteady compressible flow could be routinely measured at an arbitrary point in space. However, all known methods are presently limited by factors which have prevented their common use in general laboratory experiments and wind tunnels. As laser development continues and a wider choice of laser characteristics becomes available, it appears certain that continued research will lead to a routine application of one or more of these approaches. The known methods are reviewed in a general way by introducing a classification which places each into one of three categories. The grouping identifies the basic measurement concept employed, the relationship between the different approaches, their fundamental limitations, and reasons for choosing one approach over another. The motivation for employing laser-induced fluorescence as a diagnostic scheme for measuring density (as well as temperature and velocity) in certain flows is discussed, and the status of research in applying the method to various fluid mechanical problems is reviewed, together with a discussion of the outlook for continued development of the technique.

1. Introduction

Because many gas dynamic flows of technological interest are neither planar nor axisymmetric, and the usefulness of the shadow, schlieren, and interferometric methods in providing quantitative data related to the density distribution of these flows are greatly reduced, it has been a long standing goal of experimentalists to develop an equivalent method whereby an arbitrary two-dimensional slice of a flow could be studied, and data defining the state of the flow could be collected for any point in the slice. This need has been felt most strongly for measurements in cold low-density flows such as those found in small laboratory setups and wind tunnels where fluid mechanical studies are most easily carried out. The measurement of fluid velocity alone, at an arbitrary point in these flows, has been advanced significantly over the past decade because of the development of the laser and the introduction of laser doppler velocimetry (Stevenson, 1977; Yeh and Cummins, 1964). The laser has also been looked upon as providing the means for introducing an equivalent capability in measuring fluid density or temperature, but progress in this direction has been considerably slower for a number of reasons, the principal ones being weak signal strength and the need for sophisticated and expensive equipment.

A number of different approaches have been taken with regard to the general question of measuring the fluid state at an arbitrary point in a flow, and each approach has its advantages, disadvantages, and technical problems associated with its implementation. Some of these are of a fundamental nature, some require additional research, and others simply must await, for example, the further development of more rapid-pulsed high-powered lasers or tunable ultraviolet lasers. In order to understand some of the basic problems, the objective of research in the field, and some of the approaches taken, it is useful to attempt to arrange the various methods studied into three groups, where the grouping is based on the geometrical or physical arrangement associated with the basic concept used in making the measurement. In this way, attention can first be given to the fundamental problems faced in employing a particular approach, even if all other aspects of the method are handled in an ideal manner. The common element assumed in the grouping to be used is that the flow field is probed with a probe beam, and some effect arising from an interaction between the probe beam and matter is employed in making a measurement of the state of the flow.

The aim of this review is to describe recent work in employing laser-induced fluorescence as a tool in studying compressible flows, to identify present problems in implementing the technique and to discuss prospects for future development. The principal interest in laser-induced fluorescence stems from several important facts: the signal strength is quite strong when compared with other techniques;

the fluorescent signal is frequently in the visible part of the spectrum, which allows one to either view the flow or photograph it for later study; and sufficient signal strength is usually available to allow the use of continuous-wave lasers, as opposed to pulsed lasers, to view an entire transverse section of the flow in real time. This survey will be limited to fluorescence studies in gases. The method has been applied quite successfully to the special case of liquids, where Rhodamine 6G dye and argon-ion laser can be used to good advantage (Owen, 1976).

2. Classification of Methods

The geometry or physical arrangement employed in making use of a particular interaction between a probe beam and matter, that leads to the effect detected, can be used to conveniently classify many of the known methods into one of three categories. These three categories are represented by the diagrams in Figures 1, 2, and 3.

The scheme represented by Figure 1 is one where the interaction occurs over the entire length ℓ as the probe beam passes through the gas dynamic field, which in this case is schematically represented by a nozzle flow. The detector in this case is to be viewed as a generalized detector; i.e., one that responds to either beam deflection, intensity, polarization, phase with respect to a reference beam, or any other beam property. Examples which can be placed in this group are the shadow, schlieren, absorptive, interferometric, and holographic interferometric methods. In each case, the interaction length ℓ ultimately determines the sensitivity, or signal to noise ratio, for the method; and in each case gas-dynamic variations along the path ℓ cannot be resolved in a direct way.

Because the length ℓ is of the order of the dimension of the flow field studied, the interaction is generally quite strong and usually leads to a detected signal that can be clearly resolved. The long interaction length ℓ is the basic reason why photographs of gas-dynamic flows produced using the shadow, schlieren and interferometric methods frequently yield fairly distinct images. Furthermore, if a flow field happens to be a two-dimensional planar field, where gas dynamic variations do not occur along the interaction length ℓ, then the group as a whole produces rather remarkable records for study. In particular, the interferometric method permits a direct quantitative measure of the density at each point in a two-dimensional steady or nonsteady field, and the measurement can be made quite accurately.

On the other hand, for three-dimensional flows represented by the downstream flow in Figure 1, these methods do not allow for a point measurement of density without employing extraordinary means. For the case of an axisymmetric flow, the density field can be obtained from a single photographic record by use of one of several inversion methods. For a general flow, data from a number of photographs taken at different angles with respect to the flow can be used to reconstruct the density field using the principle of tomography. However, in each case a significant loss in spatial resolution over the original photograph is found, unless a truly large number of views and points are used in the reconstruction process. As a group, the photographic records themselves seem to imply that good spatial resolution can be obtained and that an alternate method should meet the same standard. However, once one allows for the loss in resolution which accompanies a reconstruction process, it is clear that an alternative measurement scheme has only to meet these reduced standards to become competitive.

Because of the highly successful application of tomographic reconstruction in the medical field, increasing interest is being given to its application to fluid mechanics. Nevertheless, because the technique requires a large number of views from a variety of different angles, only those flows for which free access is available from most every direction would allow the technique to be applied. Likewise, time-dependent flows would be difficult to study.

The category represented by Figure 2 makes use of an arrangement where the gas absorbs (or scatters) a small portion of the energy in the probe beam at each point along its path and then emits a certain fraction of the same energy in the form of photons, some of which are collected by a lens and imaged onto a detector. The principal feature of this arrangement is that the optical axis of the detector together with the axis of the probe beam define a single point in three-dimensional space, at which point the measurement is made. Examples falling in this category are: methods employing laser-induced fluorescence, electron-beam induced fluorescence, Mie scattering (light scattered from small particles in the flow), Rayleigh scattering (light scattered elastically from molecules), and Raman scattering (light scattered inelastically owing to energy exchange with internal energy states of the molecule). The main attraction of

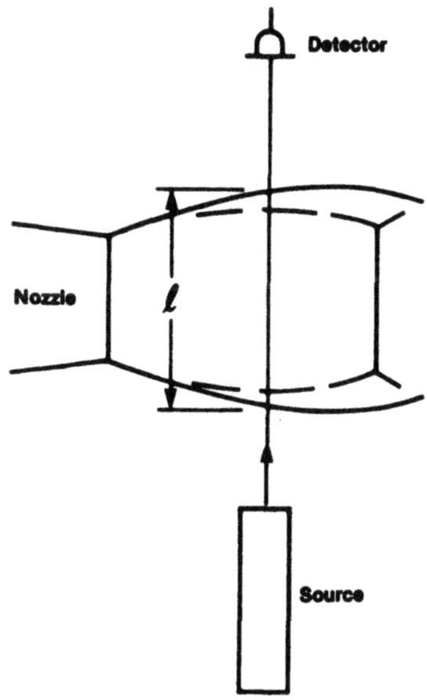

Figure 1. Schematic representation of the case where the measurement results from an interaction along the entire beam path in the fluid.

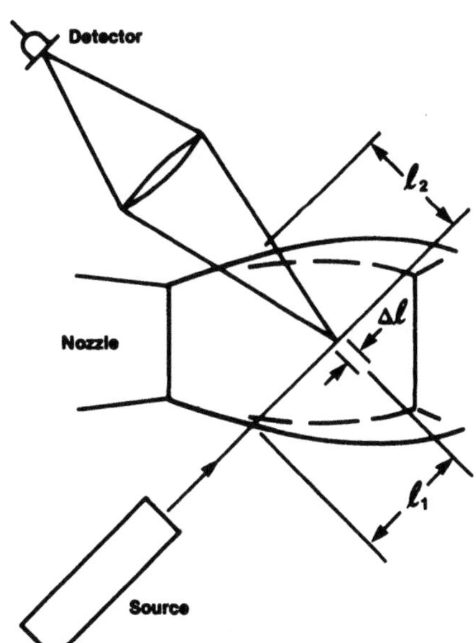

Figure 2. Schematic representation of the case where the measurement is based on photons collected from segment $\Delta\ell$, but the interaction occurs along the entire beam path.

this approach is that, in principle, it leads directly to a measurement at a point in space, as well as a function of time. The added attraction is that data can be obtained even if fairly limited access to the flow is encountered. In the case of laser-induced fluorescence, the beam can be spread into a sheet of laser light and the detector can be replaced by a photographic film and a record of fluorescence intensity associated with an entire transverse section of the flow would be obtained.

However, basic to the method depicted in Figure 2 is a fundamental problem which limits its utility even in its ideal form. In order to make an accurate measurement in the segment $\Delta\ell$, the amount of energy absorbed (or scattered) along the path ℓ_1 must be a very small fraction of the energy present in the probe beam, so that the beam intensity entering the segment $\Delta\ell$ is known, in general. The gas should therefore be a weak absorber (or scatterer). On the other hand, the strength of the emitted signal depends on the amount absorbed (or scattered) in the segment $\Delta\ell$ which means the gas should be a strong absorber (or scatterer). These conflicting requirements can only be satisfied by an intermediate value of the level of absorption (or scattering). Therefore, one should expect to find an upper limit on the distance ℓ_1 that can be used, as set by questions of overall detector sensitivity and the desired degree of decoupling between gas-dynamic processes occurring in ℓ_1 and those in $\Delta\ell$ where the measurement is made. In the case of Raman scattering, the scattering process is so weak that detector sensitivity and laser power become the limiting factors rather than the length ℓ_1, and in the case of laser-induced fluorescence with its strong scattering, the length ℓ_1 frequently becomes the limiting factor.

A similar limitation on the length ℓ_2 occurs if the radiation emitted from $\Delta\ell$ is also absorbed by the gas, called radiative trapping, which is assured to occur, for example, in the case of Mie scattering and in the case of laser-induced fluorescence if the fluorescence radiation happens to be at the same wavelength as the wavelength of the excitation beam. However, the overall attractiveness of a point measurement, or measurement over a plane section, can offset these limitations in many instances.

An excellent survey, from the point of view of the user, concerning the use of measurement techniques based on Raman scattering, Rayleigh scattering, and electron-beam fluorescence, together with a listing of the relevant literature on these topics, is found in a paper by Peterson (1979).

The basic limitation of the general scattering method represented by Figure 2 is, in principle, circumvented by the approach depicted in Figure 3. In this method two or more beams define a point in space at their common point of intersection, just as for the general scattering method. The principal idea employed here is that the effect sensed by the detector is only produced by the interaction of two (or more) beams at the intersection point and not at any other point. If this were fully realized in practice then the measurement would truly be a point measurement. Because two electromagnetic waves will not interact in free space or in a medium with constant properties (the superposition principle applies to Maxwell's equations when the coefficients in the equations are constants), the interaction is introduced by using a medium with properties that can be altered in some way by the electric field of one of the beams. For example, the absorption coefficient can be reduced in some cases by using a beam (write beam) of sufficient intensity, and this reduced absorption can affect the transmission of a second (read) beam which

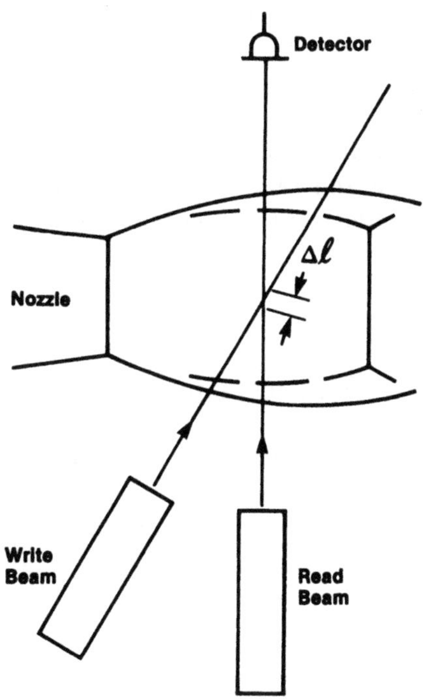

Figure 3. Schematic representation of the case where a nonlinear interaction in segment $\Delta\ell$ gives rise to the detected signal.

crosses it. This is the principle used in saturation spectroscopy (Schawlow, 1978) and in the use of the nonresonant ac Stark effect as discussed by Farrow (1982). Alternatively, the write beam may polarize the gas in its path and cause the plane of polarization of the read beam to rotate, as in polarization spectroscopy (Schawlow, 1978). The write beam intensity may be sufficient for the read beam to encounter a nonlinear susceptibility at the intersection and the combination produces a coherent scattered wave (in a new direction in general). This interaction represents the process occurring in coherent anti-Stokes Raman spectroscopy (Eckbreth et al., 1979). Each of these interactions may be considered as a special case of the general category of four-wave mixing (Yariv, 1978).

All of these effects depend in some way on a particular nonlinear property of gas molecules. These nonlinear effects, however, are not in general large effects and therefore the available signal strengths are quite small and difficult to detect. Because the effects are weak, implementation of the method generally requires high powered pulsed dye lasers which can deliver up to six orders of magnitude greater (pulse) power than continuous-wave lasers. Consequently, one must accept a low repetition rate on the measurement made or integration over time for faster repetition rates and continuous wave lasers. Therefore, if one is interested in a survey of a transverse slice of a flow, the data must be collected over an extended period of time and real time observation is not possible. In this regard, the method represented by Figure 2 provides a clear advantage.

3. Laser-Induced Fluorescence

The principal attractiveness of laser-induced fluorescence, for which the fluorescence wavelength is in the visible part of the spectrum, is that one can obtain a photographic record of data for an entire transverse section of a flow; and if the laser is a continuous wave laser, visually observe what is occurring in the section in real time, which is frequently much more instructive than having to study a process from a sequence of photographs. The principal drawback at present is that the candidate seed gases and lasers available are few in number, and these gases possess some undesirable properties. The eventual aim and hope is that tunable ultraviolet lasers will be developed to the present stage of dye lasers, then a much wider selection of candidate gases would become available for use. On this basis, all the techniques developed with the present limited combinations could be quickly applied to similar studies once the tunable ultraviolet laser is available.

A listing of data for three of the more actively studied seed gases used in various fluorescence experiments conducted in different laboratories is given in Table 1. Iodine is a solid at ambient conditions and has a vapor pressure of 0.2 Torr at room temperature. The absorption spectrum of molecular iodine vapor is quite complex but it has a strong line that nearly coincides with the gain profile of the 514.5 nm output of the argon-ion laser, making the argon-ion laser a very efficient excitation source for the iodine molecule. With the availability of several Watts of continuous power from the argon-ion laser, the iodine molecule can be quite energetically excited. Because of the strong fluorescence from iodine, the room temperature vapor pressure of iodine provides sufficient concentration for most purposes. Because the bulk of the fluorescence from molecular iodine (yellow-orange) is shifted some distance in wave-length from the 514.5 nm (green) excitation wavelength of the argon-ion laser, the fluorescence signal is not absorbed nearly so strongly as the excitation beam. Therefore, the distance ℓ_2 does not enter as a limiting factor in the use of iodine fluorescence. The use of iodine does however present some problems. Iodine vapor is a reactive as well as a mildly toxic substance. Aluminum must be anodized to protect its surface, low-carbon steel takes on a rusty look, ordinary vacuum grease decomposes, and some plastics and epoxies are degraded by exposure to iodine vapor. Although its level of toxicity is low, it is

Substance	State (20°C)	Vapor Press. (20°C)	Laser	Wave-length (nm)	Most Significant Problems
iodine	solid	0.2 Torr	Argon-Ion	514.5	Quenching, reactivity
biacetyl (butanedione)	liquid	40 Torr	Krypton-Ion or dye	413	Phosphorescence also present
sodium	solid	–	dye	590	Heating required, no water/oxygen

Table 1. Description of substances used in fluorescence experiments.

a substance which should not normally be allowed to escape into a laboratory environment because it seems to cause eye irritation in some people. Because of its relatively long radiative lifetime of approximately 1 μs, the conversion of excitation energy to thermal energy (quenching) becomes a problem and must be taken into account in the theory for the predicted fluorescence intensity. Because of this, the amplitude of the fluorescent signal is reduced by over two orders of magnitude as the pressure of the carrier gas (for example, nitrogen) is raised from 10 Torr to atmospheric pressure.

The organic molecule biacetyl (butanedione) is a liquid at ambient conditions and has a vapor pressure of 40 Torr at room temperature, which makes it easy to add as a seed gas to a room-temperature carrier gas. It has an absorption band in the near ultraviolet that conveniently spans the output wavelength of a continuous-wave krypton-ion laser at 413 nm and it fluoresces in the blue part of the spectrum. A power level of approximately 1 Watt is available from the krypton–ion laser, so adequate power is available for continuous excitation and real time observation. The radiative lifetime associated with its fluorescence is approximately 10 ns and consequently collisional quenching does not appear to be a problem at moderate pressures. Some of the problems associated with the use of biacetyl are that it also emits a phosphorescent (long duration) signal that is significantly stronger than the fluorescent (short duration) signal and this signal must be filtered; its vapor is combustible so it may only be used in high concentrations in an inert gas-like nitrogen; it has a very distinctive odor that penetrates an entire working area, even when very small quantities are released; and its useful life is affected by exposure to air, which causes the biacetyl liquid to form a viscous, sticky substance with loss of fluorescent properties.

Sodium is a solid at ambient conditions and has a melting point of 97.81°C. Its vapor has an absorption band in the orange-red part of the spectrum which lies nicely within the range of frequencies of the output of the Rhodamine 6G continuous wave dye laser when the laser is pumped by an argon-ion laser. Typical dye laser power available in this case is a fraction of a Watt, which is quite adequate because the sodium atom absorbs quite strongly in this region and converts the energy to fluorescence very efficiently. The radiative lifetime of the fluorescence signal is approximately 16 ns, so quenching is likewise a less serious problem in sodium. The principal problem associated with the use of sodium is the fact that it has a very low vapor pressure at room temperature, and therefore it must be heated before it can be used to seed a flow. If the flow itself is allowed to cool as in a supersonic flow, then serious problems may arise with condensation of sodium in the test section. However, sodium emits so strongly that only a very small amount is needed, which greatly reduces the condensation proble faced in the interior of the flow. Nevertheless, for solid boundaries like the walls of the test chamber on the surface of a model, the problem generally cannot be ignored. Also, sodium cannot be used in air containing water vapor because it decomposes water, forming hydrogen and NaOH, and the hydrogen then reacts with oxygen in air producing the flame which characterizes this well-known reaction. If it is used with a dry inert gas like helium or nitrogen then sodium presents a manageable problem. Additionally, the fluorescence wavelength is identical to the excitation wavelength and therefore absorption occurs along ℓ_2 as well (see Figure 2) and this length must be considered in the design of an experiment.

3.1 Iodine Studies

The suitability of using the 514.5 nm output of the argon-ion laser to excite iodine fluorescence probably became most readily known as a result of spectroscopic studies of the iodine molecule in a molecular beam and the more recent development of the technique of supersonic free jet spectroscopy to study the spectra of very cold gases (Levy et al., 1977). The absorption spectrum of iodine contains a very large number of lines and a number of these could be accessed by a continuous-wave dye laser if a lower level of laser power is acceptable. For high-power pulsed operation, the second harmonic output of three Q-switched, Nd:YAG laser lines at 530.6 nm, 530.2 nm, and 536.7 nm can be used (Byer et al., 1972).

All of the applications of laser-induced fluorescence in iodine to fluid mechanical problems have, essentially, made use of the 514.5 nm output of the argon-ion laser because of the convenience it offers. One of the first reported uses of iodine fluorescence in an experiment where gas-dynamic information was the principal aim, involved a study of aerosol dispersion in a low-density supersonic microjet (Schmidt-Ott et al., 1977). The technique made use of Mie scattering of the laser light to locate the aerosol particles, iodine seeding of the vacuum chamber to highlight the background, and no seeding of the jet core to locate the boundary between the core flow and the vacuum chamber gas.

The work led to a further study where emphasis was placed on flow visualization in a low-density supersonic jet (Hiller et al., 1977). The technique made use of the relatively long radiative lifetime for fluorescence in iodine, which is on the order of a microsecond, to cause a streak to appear downstream of a narrow laser beam directed perpendicular to the flow. With the use of some ten parallel beams positioned in a single plane, one is able to produce a rather interesting view of the velocity field in the plane. The method is most effective when nearly all fluid velocity vectors point in one direction and the magnitude of the velocity is greater than 1 km/s, for which the streak length produced is greater than 1 mm. However, the large variations of pressure which characterize supersonic nozzle flow make the streak lengths difficult to interpret in a quantitative way, because of the effect of quenching on the duration of fluorescence visibility and the length of the streaks.

A similar approach was used by Rapagnani and Davis (1979) to study the fluid mechanical mixing process introduced by small nozzle arrays used in chemical-laser cavities, on which the operation of the chemical laser is so dependent. The approach made use of iodine seeding in one flow (fuel) and no seeding in the second flow (oxidizer) and a single laser beam which was translated either parallel to the face of the nozzle array or downstream of the nozzle array, to visually map out the flowfield and study the mixing process in detail. The technique proved to be very useful in locating small defects in the nozzle array, which altered the flow from the desired pattern over the face of the array, and in demonstrating that the method gives very good spatial resolution on a small scale. This was contrasted with efforts to use chemiluminescence photographs which showed no nonuniformities because of the spatial integration they introduce.

A more extensive study of such a flow was carried out by Cenkner and Driscoll (1982), who used the same visualization scheme to study the usefulness of gas trips in enhancing the mixing process in supersonic mixing nozzles, and thus improve their performance. The composite nozzle flow consisted of a primary, secondary, and trip stream which could be individually seeded with iodine vapor for study. One of the more interesting observations made was that one has the opportunity to separate the effect of convective mixing from diffusive mixing by using gases of different molecular weights with respect to the molecular weight of iodine.

A more quantitative approach to the use of iodine fluorescence in experimental fluid mechanics was initiated by McDaniel (1981) in his work in further developing the method. His work led to a series of papers in which steps were taken to account for and make use of the special characteristics of the iodine molecule (McDaniel, 1981; McDaniel et al., 1982, 1983). Rather careful attention was given to developing a mathematical model for the iodine molecule and the use of its predictions in guiding experimental applications.

Figure 4 shows the relation between the 514.5 nm argon-ion laser gain profile and the absorption spectrum of iodine. The absorption line used is the one located 2 GHz from the center of the gain profile. The width and height of the line changes with gas pressure (pressure broadening), and in the pressure range from 50 Torr to atmospheric, the width of the line changes by over an order of magnitude, becoming wider and shorter with increasing pressure. When an etalon is placed in the laser cavity, the width of the laser line becomes a small fraction of the iodine line and the laser can be tuned across the iodine line to study its response. Typical data obtained for different laser detunings are shown in Figure 5 where the fluorescent intensity is displayed against nitrogen gas pressure. The data show that the relation between iodine fluorescence intensity and gas pressure is not a simple relation. For the high-pressure end of the curve, the intensity varies as p^{-2} where one power of p is due to collisional quenching and one power of p is due to pressure broadening of the absorption line.

An interesting and useful theoretical observation made by McDaniel (1981) is that, by sufficiently detuning the laser line, one is able to offset the decrease in fluorescent intensity resulting from collisional quenching by an increase in intensity in the wings of the line resulting from pressure broadening. This balance in opposing effects is the main reason why the curves in Figure 5, for the detuned cases, look so different from the tuned case. In the limit of large detuning, the fluorescence intensity would not depend on background gas pressure but only on iodine concentration, which would then give the desired linear relation between fluorescence intensity and flow-field density. If the line used in iodine were an isolated line, then the limiting case could be realized for a wide range of pressures. But

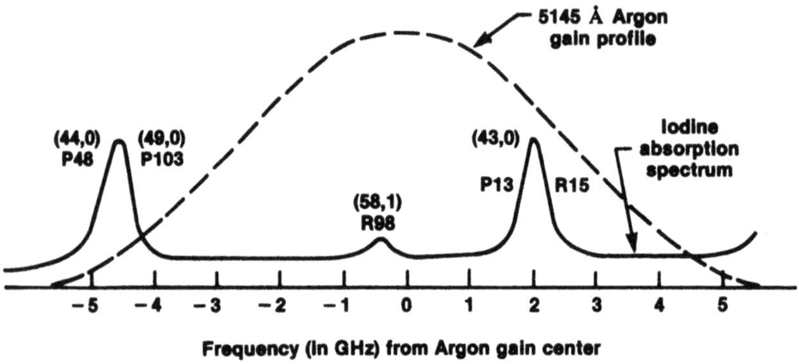

Figure 4. Relation between the iodine absorption spectrum and the 514.5 nm argon-ion laser gain profile.

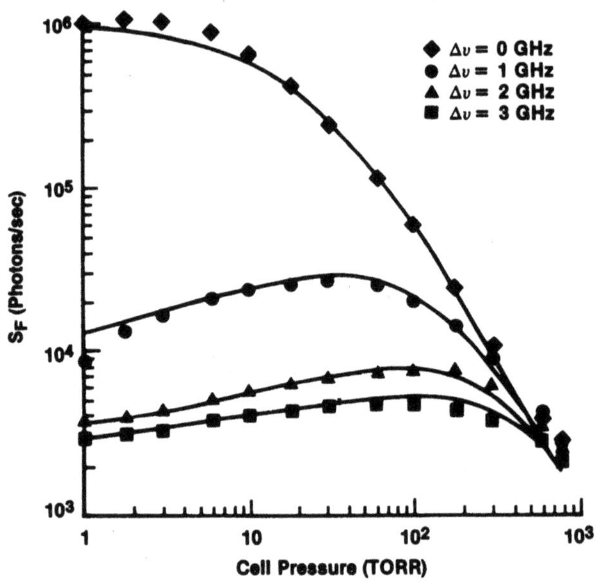

Figure 5. Iodine fluorescence intensity versus nitrogen pressure for various values of laser detuning, with narrowband excitation, and for the vapor pressure of iodine at room temperature.

Figure 4 shows that there are two neighboring lines within the laser gain profile and because the closest line is a hot-band transition, which grows rapidly in height with temperature, the effectiveness of detuning is lost beyond 3 GHz. However, if one of the more isolated lines in iodine and another laser wavelength were used, then larger detunings could be used and the method of measurement would be correspondingly improved. McDaniel showed in his work with an underexpanded supersonic nozzle that the fluorescent intensity distribution for the tuned case is not at all like the known density distribution in the nozzle, while for 3 GHz of detuning the intensity distribution closely matches the known density distribution. An example for 3 GHz detuning is shown in Figure 6.

A similar problem has been faced in the field of combustion research where one is interested in using fluorescence intensity to determine specie concentration in a flame. One of the approaches considered is to saturate the absorption line so as to cancel the effect of quenching (Daily, 1978). Although iodine appears to be an ideal candidate for such an approach, because of its relatively long radiative lifetime, McDaniel (1981) found that the method could not be used for a number of practical reasons.

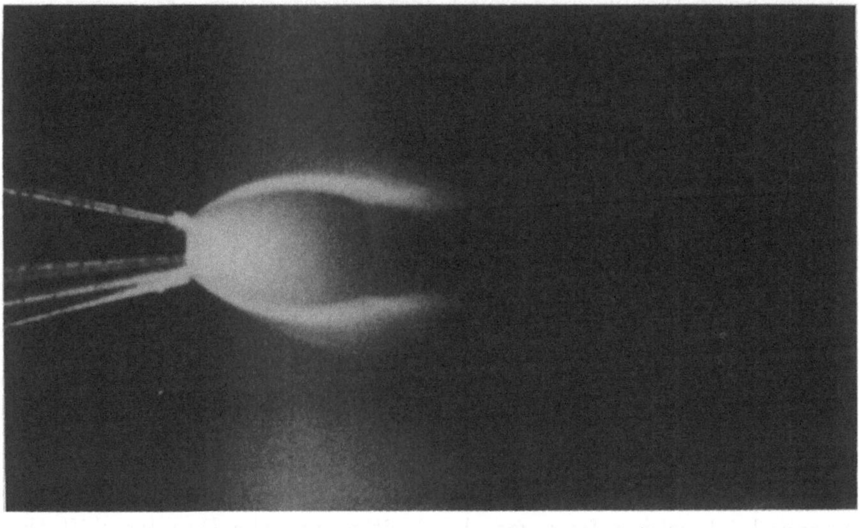

Figure 6. Iodine fluorescence intensity in an underexpanded jet for 3 GHz of laser detuning, showing the close relation between fluorescence intensity and gas density achieved by detuning.

When the direction of propagation of the laser beam is not perpendicular to the fluid velocity a Doppler shift of the absorbing transition occurs. For line center excitation and a fluid velocity of 300 m/s in the direction of propagation of the laser beam, the Doppler effect shifts an iodine transition by approximately half its width. Consequently, two fluid particles having velocity components along the path of the laser beam which differ by this amount will show a significant variation in fluorescent intensity. Figure 7 gives an example of the asymmetric intensity distribution one obtains for an underexpanded supersonic jet, owing to the finite radial component of velocities present, when the laser beam is perpendicular to the jet centerline and it is detuned by 500 MHz from the peak of the iodine absorption line for a static gas. McDaniel applied this technique to the measurement of velocity in a supersonic nozzle by placing the beam at an angle with respect to the jet centerline and showed that the measurements obtained were in very good agreement with predictions of ideal nozzle flow (McDaniel, 1983).

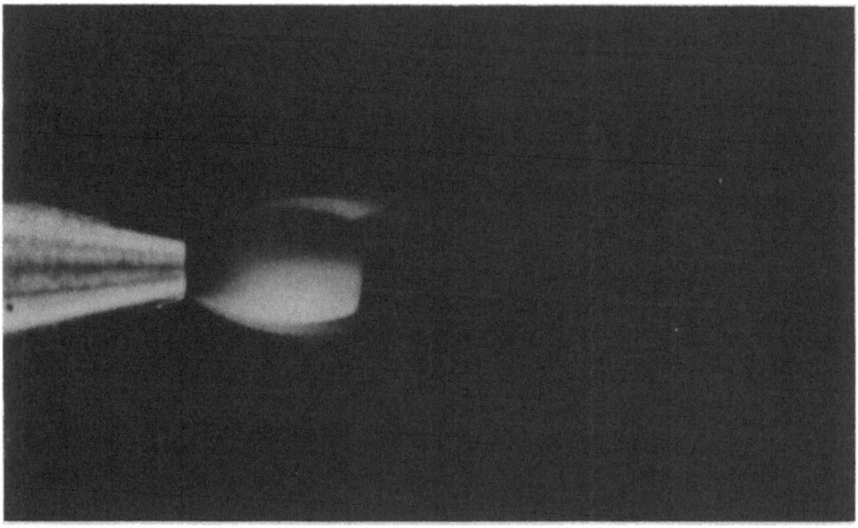

Figure 7. Asymmetry in the iodine fluorescence intensity in an underexpanded jet arising from radial velocities and the Doppler shift they introduce when the laser is detuned by 500 MHz from line center.

In certain applications one is more interested in measuring fluctuations in the state of a flow than in measuring mean values. Such a case arises when one considers an experiment where the far-field microphone signal is compared with the near-field density or pressure fluctuations in a cause-and-effect study of noise generation. A study by Ackermann et al. (1983) was carried out to determine whether iodine fluorescence could be used to measure fluctuations at an arbitrary point in a nozzle flow. When the theory was developed for small disturbances about a mean state, it was shown that the tuned condition eliminates the Doppler effect and the fractional change in fluorescence intensity is approximately equal to the fractional change in pressure, but with an opposite sign because of the negative slope in Figure 5. An experiment was conducted in an acoustic cell under conditions of resonance, to amplify the pressure perturbation generated, and the theoretical predictions were compared with experimental results. Figure 8 shows a comparison between the fluorescent signal and the microphone signal near the condition of acoustic resonance for both amplitude and phase. Some of the conclusions drawn from the work are: the best laser stability achieved was approximately one part in 10^3 and since a pressure perturbation of one part in 10^2 represents a very loud sound, laser stability is presently a limitting factor in making these measurements; because of the strong absorption at line center and the limited sensitivity set by laser stability, the length ℓ_1 in Figure 2 becomes a limiting factor and this length could not be increased beyond 10 cm without introducing significant errors; and some discrepancy exists between the theory used and the experiments because the theory does not predict all operating conditions equally well.

A problem area which seems ideally suited to the use of iodine fluorescence is being studied by Neal (1983). Figure 9 displays the vapor pressure for iodine and several alcohols as functions of temperature, and it is seen that octanol has a vapor pressure curve that falls quite close to that for iodine. Because iodine is soluble in octanol, one could consider forming a droplet from the solution and observe iodine fluorescence in the vapor cloud as the droplet evaporates; and with the close match in vapor pressure between the two substances, it is expected that the iodine concentration in the cloud would be high enough for direct observation. In addition, because evaporation can be studied as a constant pressure process, complications associated with quenching are not present.

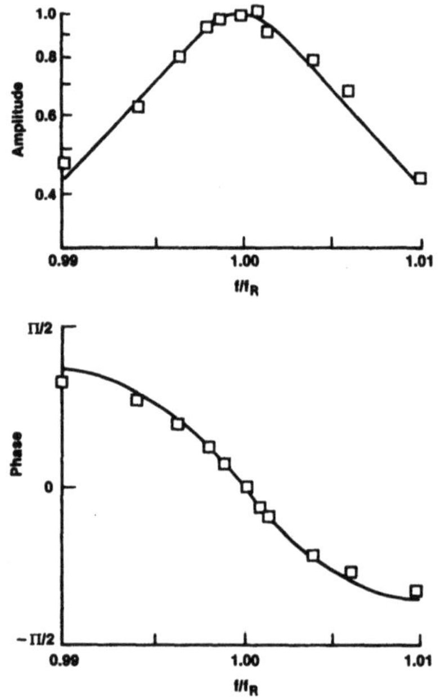

Figure 8. Comparison of amplitude and phase angle measurements for iodine fluorescence and microphone pressure near the resonant condition in an acoustic cell: □, fluorescence; —, microphone.

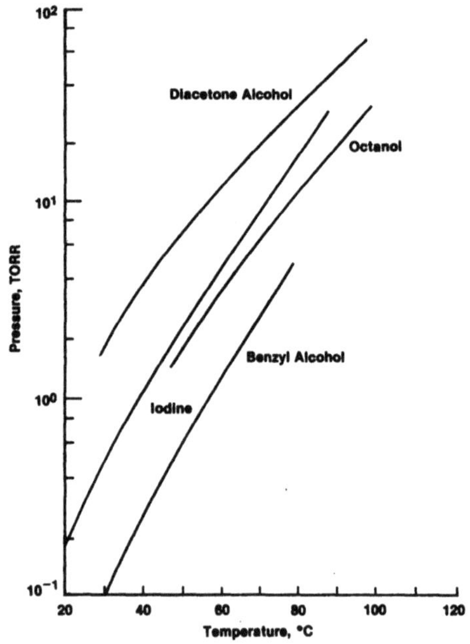

Figure 9. Comparison of vapor pressure curves for iodine and several alcohols.

The first step in judging the feasibility of the concept is to use a captured droplet to develop the technique and explore its potential. The diagram in Figure 10 shows an arrangement for such a study. The droplet is formed and suspended on a tube having a very small diameter and a uniform flow is introduced to carry away the vapor so iodine will not fill the chamber and present a background fluorescent signal. The laser beam is then formed into a sheet of light for visual observation or photography, and positioned at different locations to give sectional views of the vapor cloud. Figure 11 presents several photographs taken in this way. Small quartz tubes were drawn to a tip diameter as small as 100 μm, and a droplet was formed and allowed to grow near its tip. These droplets would grow to a size of approximately 1 mm in diameter and then fall off the tip. In the early stages of growth, surface tension would cause the droplet to climb the glass tube and settle at a position above the opening of the tube as seen in one of the photographs.

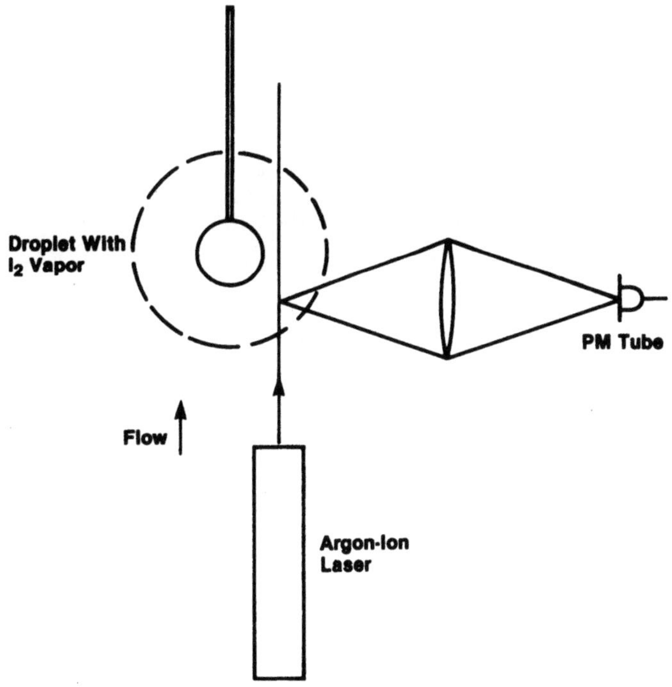

Figure 10. Schematic arrangement of experiment for the study of droplet evaporation with a captured droplet.

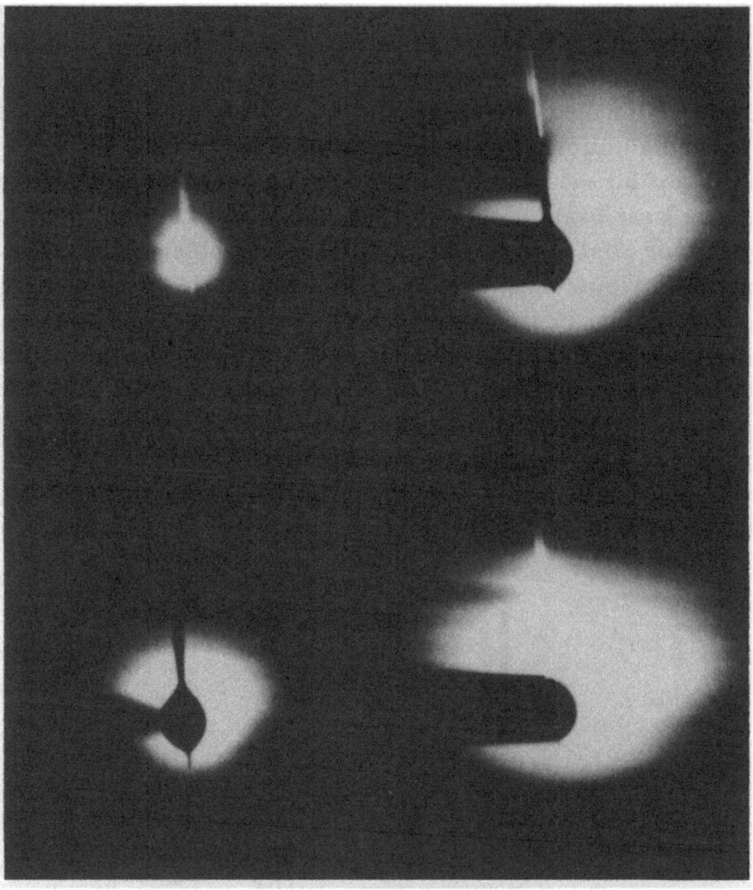

In his study, Neal was also able to image a small region of the vapor cloud onto a photomultiplier tube and obtain a trace as shown in Figure 12, as the droplet fell and new droplets formed on the tube tip. The trace is remarkable in that it shows the intensity falls nearly to zero and then recovers each time, indicating an effective cleansing of the iodine vapor by the mean flow and a high diffusion rate which reintroduces iodine vapor near each new droplet.

In a study by Melton (1983) which had a very similar objective, spectrally separated fluorescence emission from liquid and vapor diesel fuel was monitored by the addition of an exciplex-forming molecule to the fuel. Excitation in the ultraviolet was necessary and was achieved by frequency-doubling the output of a dye laser (to 313 nm), which was pumped by the second harmonic output of a Nd:YAG laser. The technique has the advantage of allowing one to work with diesel fuel but it appears to produce much less fluorescence intensity than the approach with iodine.

3.2 Biacetyl Studies

The use of biacetyl (butanedione) for flow visualization and a quantitative measure of den: was introduuced by Epstein (1977). He examined a wide range ol lecules, both organic and inorganic, in an earlier study (Epstein 72) before initiating his work with biacetyl. The method was ap d to the study of a laboratory model

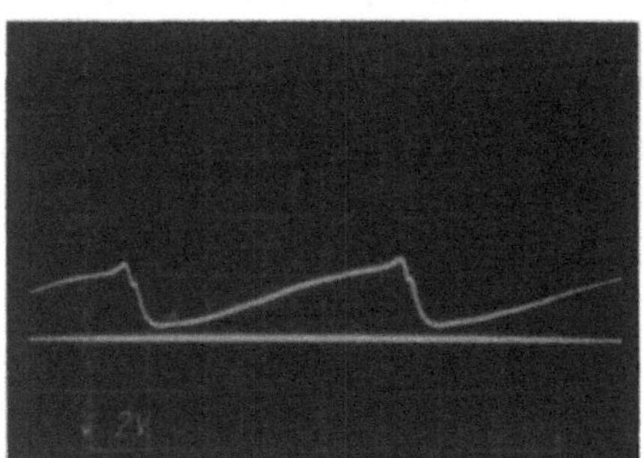

Figure 12. Iodine fluorescence intensity for a sequence of falling droplets past a fixed point of detection.

of a transonic compressor rotor for which the working fluid was a mixture of argon and freon-12. The facility operated in a blowdown mode and thus the biacetyl could be carefully premixed in the supply tank. The light source was a flashlamp-pumped dye laser which produced 0.3 μs pulses centered at a wavelength of 425 nm. The laser beam was expanded into a sheet of laser light and various transverse sections of the flow were illuminated for study. An image intensifier operating at gains ranging from 100 to 40,000 was used to amplify the light and contact photography at the tube's fiber optic output was used to obtain a picture. The light entering the image intensifier was filtered with a narrow band (10 nm) interference filter centered at 460 nm to block the phosphorescent signal.

A proof-of-principle study was conducted using a projectile launched into freon at a Mach number of 2.2. The freon was at standard temperature and pressure and the partial pressure of biacetyl was 30 Torr. The image showed the expected wave system about the projectile, and demonstrated that the capability for making quantitative measurements of density was clearly present. The test demonstrated that quenching is not an important problem in biacetyl under these conditions, and that the light amplification was sufficient to obtain an image of the transient flow.

The main study centered around the collection of photographs of slices of the flow in the transonic rotor for different test conditions and the conversion of the fluorescent intensity distributions to contour maps of fluid density. Care was taken to correct the density measurements for modifications introduced by the image intensifier, the optical system, and absorption in the gas. The attenuation through the rotor passage, a length of 10 cm, for a partial pressure of biacetyl of 10 Torr was reported to be about 50 percent. The collection of photographs analyzed and the overall quality of the results obtained by Epstein are quite impressive, and demonstrate clearly the value of the technique when the application is well suited. The study was very thoroughly done, and the only step which may have been appropriate to include is a calibration test to verify that the fluorescence intensity produced by the biacetyl is in fact proportional to the density of the gas for the range of conditions tested.

Work with biacetyl was also carried out by McKenzie et al. (1979), who were interested in developing a nonintrusive scheme for measuring density fluctuations in a turbulent flow. Calibration tests were conducted in a Mach 3 channel flow having cross-section dimensions of 2 cm high by 7.5 cm wide. The reservoir gas was at room temperature and consisted of a mixture of 520 Torr nitrogen and 20 Torr biacetyl. The channel was operated in a transient mode, like a blow-down wind tunnel, with a quasi-steady flow duration of 50 ms. The output beam from a 1.5 Watt continuous-wave krypton-ion laser operating at 413 nm was focused to a point in the flow and the fluorescence was collected and focused onto a photomultiplier tube. Care was taken to block the strong phosphorescent emission from biacetyl as well as other stray light in the system. During each run the fluorescent signal from biacetyl was compared with the output of a conventional hot-wire anemometer positioned very close to the focal point of the detection optics. In comparing the two traces, they appeared to be quite similar and the technique seemed very promising. The objective of the work was to develop the technique for the study of a turbulent boundary layer, where measurements were to be made close to a material surface. Biacetyl happens to be a substance for which its luminescent behavior in both the liquid and gas phase are quite similar. Therefore, any condensation present in a flow would add its own component to the signal detected, and would be difficult to remove. On further testing, it was found that a signal was also present when the laser wavelength was changed to a value where biacetyl should no longer fluoresce; consequently, it was concluded some condensation of biacetyl was present (McKenzie, 1981). The higher Mach number used and the much lower temperature encountered may have been the reason why problems with condensation became evident.

3.3 Sodium Studies

Sodium vapor is a very interesting substance to consider because it emits a very strong visible fluorescence and its excitation wavelength is accessible to a dye laser. The measurement of sodium-vapor densities—using the laser resonance fluorescence technique—is such a sensitive method that absolute atomic number densities, ranging over nine orders of magnitude with a low limit below 100 atoms/cm^3, have been reported for a laboratory environment (Fairbank et al., 1975). Of course, a much larger number density must be considered if sodium is to be used as a trace gas in making gas dynamic measurements in a very small volume. The number density of sodium atoms at room temperature is approximately 10^5 atoms/cm^3 (Fairbank et al., 1975) which is apparently too small a number for present applications, and most experiments making use of sodium fluorescence heat the sodium to raise the concentration to a much higher level.

Because of the need for heating, the ideal application is in the study of flames, where the environment supports a larger concentration of sodium, and a number of papers can be found dealing with the subject. Steps to use sodium at lower temperatures were taken by Blendstrup and Bershader (1978), who studied the enhanced refraction which develops near a resonance line. They were able to show that a significant improvement in the sensitivity of conventional optical methods for low gas densities becomes possible in this way.

A more ambitious step was taken by Miles and his co-workers (Cheng et al., 1983; Miles, 1975; Miles et al., 1978; Zimmermann and Miles, 1980), who seeded the flow in a hypersonic wind tunnel with sodium and made use of sodium fluorescence to determine the complete state of the flow, including velocity. Helium and nitrogen were used as working fluids to eliminate the troublesome reaction found with sodium and wet air. The heat required to vaporize the sodium was provided by a small oven, through which a purge gas flowed at a low rate and carried the sodium vapor through a needle (which was held at approximately 650°C) to a position a short distance upstream of the throat of the nozzle. The effect of the heating raised the effective reservoir temperature to 70°C. The test-section Mach number was 3.2, and the static pressure 12.7 Torr. The measurement yielded a rather detailed plot of the Doppler shifted D_2 spectrum of sodium, from which a computer generated fit provided the velocity and thermodynamic state of the gas. The results obtained were in reasonable agreement with other techniques.

Miles (1975) proposed the use of the sodium D_2 line for the measurement of fluid velocity by employing the principles of resonant Doppler velocimetry, and Miles et al. (1978) suggested the use of sodium fluorescence for quantitative flow visualization in a hypersonic flow. An interesting extension of their work on measuring both fluid velocity and the thermodynamic state of a flow was reported in a paper dealing with the use of laser-induced fluorescence to measure time-averaged turbulence components (Cheng et al., 1983). If the level of turbulence is low and small disturbance theory can be used, then cross correlations between the perturbed values of velocity, pressure, and temperature would be contained in the perturbed fluorescent signal when it is analyzed statistically in an appropriate manner. The reason the various cross-correlation terms would be present is related to the fact that the fluorescence intensity for narrowband excitation is, in general, a function of the fluid velocity, the thermodynamic state, and the laser wavelength; and for the conditions of small disturbance theory, the perturbed fluorescence intensity is a linear combination of all the perturbed quantities. Because the laser wavelength determines the relative importance of the fluid velocity, pressure, and temperature on the fluorescence intensity, the laser wavelength can be used as an independent variable (when using a dye laser) to separate some of the terms for individual evaluation.

4. Concluding Remarks

The method of laser-induced fluorescence is being explored and applied to a growing number of different gas-dynamic problems. As experience is being gained, it is becoming clear that selected problems can be handled very effectively with the method in its present state of development, and that the potential for further development is quite exciting. Applications, which have been studied, clearly demonstrate the value of being able to both observe and record quantitative data associated with an entire transverse section of a flow, frequently in real time.

Condensation is a significant problem with most of the substances that can be used with presently available laser wavelengths, and attention must be given in each case to applications that do not require operation in a critical regime. The problem becomes less severe as the concentration of seed gas is reduced, and if the flow of interest is not bounded by material surfaces, or if the primary interest is in evaporation itself—such as the study of droplet evaporation using iodine fluorescence. Otherwise, appropriate tests should be conducted in each case to determine whether condensation is present, and whether it affects the data collected.

As laser development leads to the availability of shorter and shorter wavelengths in practical applications, the list of gases, which can serve as seed gases in fluorescence experiments, will likewise grow. A very interesting listing is given by Fairbank et al. (1975), in which the ground-state transitions of the 87 elements, whose energy levels are known, and the kind of lasers either available or needed to excite these transitions, are tabulated. Of most interest to gas-dynamic applications is the column for future lasers having a wavelength of 230 nm or less, with which argon could be excited at 106.7 nm, krypton at 125.0 nm, xenon at 149.1 nm, and neon at 74.4 nm. The most interest here would be in argon, and clearly condensation would not be a problem in its use in a wide variety of exciting gas-dynamic applications. An equivalent listing for molecular substance would undoubtedly add a number of well-suited gases to the group.

Acknowledgement

This work was sponsored in part by the Air Force Office of Scientific Research under contract F49620-83-K-0004.

References

[1] Ackermann, U., Baganoff, D., and McDaniel, J. C. "Dependence of Laser-Induced Fluorescence on Gas-Dynamic Fluctuations With Application to Measurements in Unsteady Flows," to be published.

[2] Blendstrup, G. and Bershader, D. "Resonance Refractivity Studies of Sodium Vapor for Enhanced Flow Visualization," *AIAA Jour. 16* (1978), 1106.

[3] Byer, R. L., et al. "Optically Pumped Molecular Iodine Vapor-Phase Laser," *Appl. Phys. Lett. 20* (1972), 463.

[4] Cenkner, A. A. and Driscoll, R. J. "Laser-Induced Fluorescence Visualization on Supersonic Mixing Nozzles that Employ Gas-Trips," *AIAA Jour. 20* (1982), 812.

[5] Cheng, S., Zimmermann, M., and Miles, R. B. "Separation of Time-Averaged Turbulence Components by Laser-Induced Fluorescence," *Phys. Fluids, 26* (1983), 874.

[6] Daily, J. W. "Saturation of Fluorescence in Flames With a Gaussian Laser Beam," *Appl. Opt. 17* (1978), 225.

[7] Eckbreth, A. C., Hall, R. J., and Shirley, J. A. "Investigation of Coherent Anti-Stokes Raman Spectroscopy (CARS) for Combustion," *AIAA paper 79-0083,* New Orleans, January 1979.

[8] Epstein, A. H. "Fluorescent Gaseous Tracers for Three Dimensional Flow Visualization," *SM thesis,* MIT, 1972.

[9] ——————. "Quantitative Density Visualization in a Transonic Compressor Rotor," *Journal of Engineering for Power, Trans. ASME, 99* (1977), 460.

[10] Fairbank, W. M. Jr., Hänsch, T. W., and Schawlow, A. L. "Absolute Measurement of Very Low Sodium-Vapor Densities Using Laser Resonance Fluorescence," *J. Opt. Soc. Am., 65* (1975), 199.

[11] Farrow, R. L. "Spatially Resolved IR Absorption Spectroscopy by Optical Stark Modulation," *Appl. Opt. 21* (1982), 4183.

[12] Hiller, W. J. and Schmidt-Ott, W.-D. "Visualization of Low Density Gas-Jets by Laser Induced Fluorescence," in *International Congress on Instrumentation in Aerospace Simulation Facilities, 68,* 1977.

[13] Levy, D. H., Wharton, L., and Smalley, R. E. "Laser Spectroscopy in Supersonic Jets," in *Chemical and Biochemical Applications of Lasers."* Ed. C. B. Moore. New York: Academic Press, 1977, *2,* 1.

[14] McDaniel, J. C. "Investigation of Laser-Induced Iodine Fluorescence for the Measurement of Density in Compressible Flows," *Ph.D. thesis,* Dept. of Aero. and Astro., Stanford Univ., 1981.

[15] ——————. "Quantitative Measurement of Density and Velocity in Compressible Flows Using Laser-Induced Iodine Fluorescence," *AIAA paper 83-0049,* Reno, Nevada, January 1983.

[16] McDaniel, J. C., Baganoff, D., and Byer, R. L. "Density Measurement in Compressible Flows Using Off-Resonant Laser-Induced Fluorescence," *Phys. Fluids, 25* (1982), 1105.

[17] McDaniel, J. C., Hiller, B., and Hanson, R. K. "Simultaneous Multiple-Point Velocity Measurement Using Laser-Induced Iodine Fluorescence," *Opt. Lett., 8* (1983), 51.

[18] McKenzie, R. L. private communication.

[19] McKenzie, R. L., Monson, D. J., and Exberger, R. J. "Time-Dependent Local Density Measurements in Unsteady Flows," *AIAA paper 79-1088,* 14th Thermophysics Conference, June 1979.

[20] Melton, L. A. "Spectrally Separated Fluorescence Emissions for Diesel Fuel Droplets and Vapor," *Appl. Opt. 22* (1983), 2224.

[21] Miles, R. B. "Resonant Doppler Velocimeter," *Phys. Fluids, 18* (1975), 751.

[22] Miles, R. B., Udd, E., and Zimmermann, M. "Quantitative Flow Visualization in Sodium Vapor Seeded Hypersonic Helium," *Appl. Phys. Lett. 32* (1978), 317.

[23] Neal, D. R. "Application of Iodine Fluorescence to the Study of Droplet Evaporation," *Ph.D. thesis,* Dept. of Aero. and Astro., Stanford Univ., to be published.

[24] Owen, F. R. "Applications of Nonintrusive Instrumentation in Fluid Flow Research," *AGARD Conference Proceedings, No. 193,* London: Hanfor House, *paper 7-1,* 1976.

[25] Peterson, C. W. "A Survey of the Utilitarian Aspects of Advanced Flowfield Diagnostic Techniques," *AIAA Journal, 17* (1979), 1352.

[26] Rapagnani, N. L. and Davis, S. J. "Laser-Induced Iodine Fluorescence Measurements in a Chemical Laser Flowfield," *AIAA Jour., 17* (1979), 1402.

[27] Schawlow, A. L. "Laser Spectroscopy of Atoms and Molecules," *Science 202* (1978), 141.

[28] Schmidt-Ott, W.-D., von Dincklage, R.-D., and Hiller, W. J. "Production and Measurement of Dispersion Aerosols: Application to the Transport of Deuteron-Induced and ^{84}Kr-induced Reaction Recoils," *Nuclear Instruments and Methods, 144* (1977), 553.

[29] Stevenson, W. H. "Principles of Laser Velocimetry," in *Experimental Diagnostics in Gas Phase Combustion Systems.* Ed. B. T. Zinn. New York: AIAA, 1977, *53* of *Progress in Aeronautics and Astronautics,* 307.

[30] Yariv, A. "Phase Conjugate Optics and Real-Time Holography," *IEEE Jour. of Quantum Electr., QE-14* (1978), 650.

[31] Yeh, Y. and Cummins, H. Z. "Localized Fluid Flow Measurements With an He-Ne Laser Spectrometer," *Appl. Phys. Lett. 4* (1964), 176.

[32] Zimmermann, M. and Miles, R. B. "Hypersonic-Helium-Flow-Field Measurements With the Resonant Doppler Velocimeter," *Appl. Phys. Lett. 37* (1980), 885.

Laser Velocimetry for Transonic Aerodynamics

D. A. Johnson
NASA Ames Research Center
Moffett Field, CA 94035

Abstract

Applications of laser velocimetry to the measurement of turbulent flow properties of strong transonic viscous-inviscid interactions are reviewed. The data resulting from these studies are then discussed in relation to their importance in the development of improved viscous-flow calculation methods. Also considered are the current limitations of laser velocimetry, the need for further improvements in the method, and potential future applications.

Nomenclature

a	speed of sound.
c	chord length.
C_f	skin-friction coefficient.
H	boundary-layer shape factor, δ^*/θ.
l	Prandtl's mixing length.
M	Mach number.
P	static pressure.
P_t	total pressure.
u	streamwise velocity component.
v	normal velocity component.
w	cross-stream velocity component.
x	distance in streamwise direction.
$\tilde{X}$	reduced streamwise coordinate, $(x - x_o)/\delta_o{}^*$.
y	distance in normal direction.
δ	boundary-layer thickness.
δ^*	boundary-layer displacement thickness.
ϵ_t	turbulent eddy viscosity.
τ	Reynolds shear stress.

θ	boundary-layer momentum thickness.

Subscripts

e	boundary-layer-edge conditions.
max	maximum value of quantity.
o	origin of interaction.
s	surface.
∞	free-stream conditions.

Superscripts

$(')$	fluctuating quantity.
$(^{-})$	time-averaged quantity.
$\langle(\ )'\rangle$	rms value of fluctuating quantity.

1. Introduction

The strong viscous-inviscid interactions characteristic of transonic flow result in substantial penalties in performance for both commercial aircraft and rotorcraft. Usually, the loss in performance can be attributed to the interaction of an embedded shock wave with the turbulent boundary layer. For supercritical airfoil sections, this shock-wave/turbulent boundary-layer interaction can be relatively weak, but the penalty for this is a very strong viscous-inviscid interaction near the trailing edge.

Because of the need for improved aircraft performance at transonic speeds, considerable effort is being directed toward the development of numerical prediction methods for transonic aerodynamic flows. To date, the successful application of these methods has been limited to cases where the interaction between the inviscid flow and the viscous flow is relatively weak. For example, performance predictions of supercritical airfoil sections even for on-design conditions have proven to be a formidable challenge. Nobody claims to be able to predict performance at buffet or shock-induced stall conditions.

The major obstacle for these numerical methods is the accurate prediction of the turbulent-boundary-layer and wake response to the outer inviscid flow. For finite-difference, viscous-flow calculation methods, the closure model used to describe the turbulent Reynolds stresses is fundamental to the prediction of the boundary-layer and wake development. Although often overlooked, integral methods also require some form of turbulence model to describe the Reynolds shear stress (Cousteix, 1981). Examples of this are: (1) the popular lag-entrainment method of Green et al. (1977), which uses an empirical correlation between the maximum Reynolds shear stress and the entrainment coefficient, and (2) the recently developed method of Whitfield et al. (1980), which uses an empirical mean-velocity profile correlation and the Cebeci-Smith algebraic eddy-viscosity model to determine the Reynolds shear stress needed in the evaluation of the dissipation integral.

Although the modeling of the turbulent Reynolds stresses is essential, considerably less attention has been given to its measurement in transonic flows than has been given to the development of numerical prediction methods. This to a large extent is a result of the experimental difficulties associated with transonic flows. Some believe that the development of the large digital computer has diminished the need for experimental observation, but it is the author's opinion that just the opposite is true. With the rapid advances that are taking place in computer hardware and numerical solution methods, there is an ever increasing need for detailed turbulent flow measurements of complex flows (1) to aid in the evaluation of these numerical prediction techniques and (2) to provide new information which can be used to extend their applicability, whether it be a finite-difference or an integral method approach.

To best evaluate the accuracy of a prediction method for airfoil flows, measurements in the viscous flow regions of both the mean velocities and the turbulent Reynolds stresses are needed. There has been a tendency to discount the importance of the Reynolds stresses, since their measurement until recently was not feasible in many flows of engineering importance. With the development of the laser velocimeter technique, this is no longer true for transonic flows or separated flows. This powerful optical technique and its application to transonic airfoil flows are the subject of this paper.

The main body of the paper is broken into four sections. In the first section, general considerations regarding the application of laser velocimetry are discussed. This is followed by a brief history of the development of the technique for compressible flow measurements.

In the third section, some of the data that have been obtained in transonic flows using this technique are reviewed and discussed in relation to their significance to the field of transonic aerodynamics. Finally, the author gives his opinions as to where efforts in laser velocimetry should be directed in the futue.

2. Discussion

2.1 General Considerations

Potentially, the laser-velocimeter technique can provide local velocity measurements (mean and fluctuating) in flows that would be unobtainable by any other technique. It has the advantage of being noninvasive and, probably more important, for compressible flows, it is sensitive to velocity alone. Thus, the difficulties that compressible flows pose for hot-wire anemometry (Owen, 1983; Robinson et al., 1983) are not present for laser velocimetry. Also, flows with extremely high turbulence levels, such as separating or separated turbulent boundary layers, present no major difficulties for this technique. This, of course, is not the case for pitot pressure probes or hot-wire anemometry. Another important feature of laser velocimetry is that the normal velocity component is no more difficult to measure by this technique than is the streamwise velocity component, and with only a modest increase in effort the Reynolds normal and shear stresses, $\langle u' \rangle^2$, $\langle v' \rangle^2$, and $-\overline{u'v'}$, can be determined. (To be precise, the Reynolds stresses include a $\overline{\rho}$ term, but for transonic flows the variation of $\overline{\rho}$ is relatively small.) Thus, in theory it has the capability of measuring all the dependent velocity variables in the Reynolds-averaged x and y momentum equations, regardless of whether the flow is compressible and/or separated (for transonic flows, the terms involving density-velocity correlations are negligibly small).

The disadvantages of the technique are that it is difficult to apply in practice and requires very expensive instrumentation in comparison with other velocity measurement devices. A part of the difficulty is due to the complexity of the technique. Successful application demands an understanding of lasers, optics, photodetection, particle light scattering, particle dynamics, high-frequency electronics, digital data acquisition, and fluid mechanics. Also, compared with other velocity measurement methods, the signal-to-noise ratio of the signals to be processed is quite low. If care is not taken, it often can be too low for reliable measurements to be obtained. This is especially true when measurements are attempted near a solid surface. Like an artist or a craftsman, the experimentalist applying this technique must develop a trained eye, which comes from experience. At the 1980 AFSOR-HTTM Stanford Conference on Complex Turbulent Flows, Simpson (1980) made the following statement regarding even the application of the technique to low-speed diffuser flows:

> *"Unfortunately laser anemometry is expensive to use and requires some time to master. By the time most graduate students become competent to use it, they finish their studies, having produced only a small amount of useful data. There is the distinct need for future diffuser research to be conducted by organizations that use laser anemometry professionally with long-term personnel."*

Taking these factors into account, laser velocimetry must be classified as a technique that should be recommended primarily when more easily implemented techniques cannot do the job. The study of turbulent boundary-layer separation and transonic flows fall into this category. Another area in this category is the measurement of the Reynolds shear stress in the near-wall region of attached turbulent boundary layers. Laser velocimetry can provide measurements of this quantity much closer to a surface than can be achieved by hot-wire anemometry. This will be further discussed in a later section.

2.2 Brief History

A detailed description of the laser-velocimeter technique and the principles of operation will not be repeated here. There are several books and numerous technical papers which treat these subjects in detail. However, a brief summary of the history of the development of the technique seems in order.

It was not until the early seventies that laser velocimetry began to be applied to compressible flows. This was nearly ten years after Yeh and Cummins (1964) had demonstrated the feasibility of using optical heterodyning to detect the Doppler-shift of light scattered from moving particles. Essential to their pioneering work was the laser—a light source with a frequency bandwidth much smaller than the Doppler frequency shift to be measured.

In the early stages of development, most of the applications of this technique, referred to then as laser Doppler velocimetry (LDV), were confined to water flows. In airflows, it was difficult to achieve the particle concentrations required to ensure continuous signals as needed by the signal processors of that time (spectrum analyzers and frequency trackers). Also, the need for smaller particles to ensure that the particles would track the flow and the higher speeds associated with airflows both resulted in reduced signal-to-noise ratios as compared with water flows. In the late sixties, several developments took place which were to make measurements in airflows feasible even when the airspeed was supersonic. These were (1) the argon-ion laser, (2) the dual-scatter or fringe-mode method, and (3) the velocity detection of individual particles by burst-period counters. With the argon-ion laser, available laser power was increased a hundredfold from that available with the earlier He-Ne lasers. This power increase combined with the use of large light-collection solid angles as permitted by the fringe-mode approach resulted in substantial increases in measurement sensitivity; so much so, that velocity measurements from individual particles crossing the sensing volume became feasible even for supersonic flows if the light was collected at small scattering angles (i.e., forward-scatter).

The study of Johnson (1974) was the first demonstration that the laser-velocimeter technique (after the development of the fringe-mode method, it became popular to drop Doppler from the name) could be used to measure the turbulent Reynolds stresses of a compressible boundary layer. In this early compressible-flow application, a single-component system was used. The turbulent shear stress was obtained from measurements taken at two fringe orientations (+45 and −45 deg.) in a manner similar to that used with a single-slanted, hot-wire probe. Since that application, two further refinements in the technique have taken place which have significantly enhanced the measurement accuracy of the technique and its applicability to more complex flows. These are the use of frequency shifting by acousto-optical modulators and the simultaneous use of two spectral lines of the argon-ion laser. The former development made feasible the direct measurement of either the streamwise or normal velocity component independent of turbulence level by ensuring the minimum number of fringe crossings needed by the signal processing electronics. With this development, measurements even in separated turbulent flows became feasible. The combined use of frequency shifting and two laser wavelengths made possible the simultaneous measurement of the streamwise and normal velocity components from the same particle. This approach significantly improved the accuracy to which the Reynolds shear stress could be measured. In addition, with this measurement approach corrections of the results for sampling bias, which is inherent in single-particle laser velocimeter measurements (McLaughlin and Teiderman, 1973; Johnson et al., 1982b), could be made.

2.3 Transonic Airfoil Flows

For reasons discussed in the *Introduction*, the primary focus of laser velocimetry in transonic aerodynamics has been in the study of the turbulent boundary-layer and wake development of airfoils or of flow models that simulate certain features of airfoil flows. Two early studies (Johnson et al., 1976; East, 1976) clearly demonstrated the feasibility of using laser velocimetry in the study of transonic airfoil flows. One was the application of the technique to a supersonic shock-wave/turbulent boundary-layer interaction with an extensive separated flow region. This application demonstrated the potential of the technique for complex compressible boundary-layer flows. A schematic representation of the flow field investigated is shown in Figure 1. The incident shock wave was produced by a full span

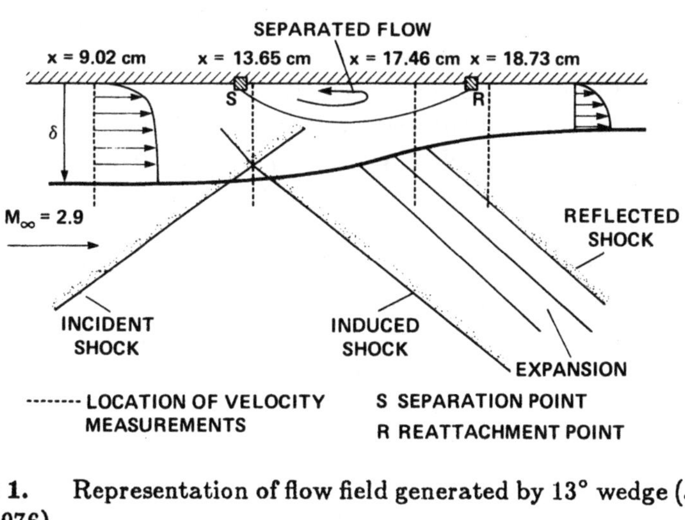

Figure 1. Representation of flow field generated by 13° wedge (Johnson et al., 1976).

wedge set at an angle of 13 deg to the approaching supersonic flow ($M_\infty = 2.9$). Indicated in this figure by dashed vertical lines are streamwise stations where laser velocimeter measurements and pitot-static probe measurements from an earlier study (Reda and Murphy, 1973) were obtained. Figure 2 compares the mean velocity results for these four streamwise stations; (a) upstream of the interaction, (b) near the separation point, (c) within the separated region, and (d) near the reattachment point. Included in this figure are the streamwise turbulence intensities measured with the laser velocimeter. A single-velocity component system with acousto-optical frequency shifting was used. Figure 2 clearly demonstrates the superiority of the laser velocimeter in measuring velocities near and within a turbulent separated flow region. The disagreement in edge velocities in Figure 2c and 2d was a result of errors in the pressure-probe measurements of static pressure resulting from large flow inclinations relative to the probe axis.

The other study (East, 1976) demonstrated the application of two-component laser velocimetry to transonic inviscid flow measurements. In this study, detailed measurements were obtained of the outer inviscid flow for a transonic shock-wave/boundary-layer interaction generated by a bump mounted on a wind-tunnel wall.

The studies of Seegmiller et al. (1978) and Johnson and Bachalo (1980) were the first applications of laser velocimetry to transonic airfoil models. A circular arc and an NACA 64A010 section were used, respectively. In both of these studies, two-color laser-velocimeter systems that incorporated frequency shifting were used, since the objective in both studies was to measure the Reynolds shear stress in addition to the mean velocity profiles. Similar measurements were obtained in the two studies, except that $\langle u' \rangle^2$ and $\langle v' \rangle^2$ were measured for the NACA 64A010 section, whereas only the sum of the two quantities was resolved in the circular-arc study. In both studies, the majority of the boundary-layer data obtained were restricted to the case of massive shock-induced separation. Except for measurements immediately downstream of the trailing edge, boundary-layer measurements were not reported for attached flow cases because the boundary layers developed under these conditions were too thin to be adequately resolved by the technique at that time. Also, the lack of any previous measurements of turbulent flow properties for shock-induced stall conditions gave impetus for these measurements. Shown in Figures 3 and 4 are mean velocity data obtained in these two studies. The mean velocity data of Figure 3 are

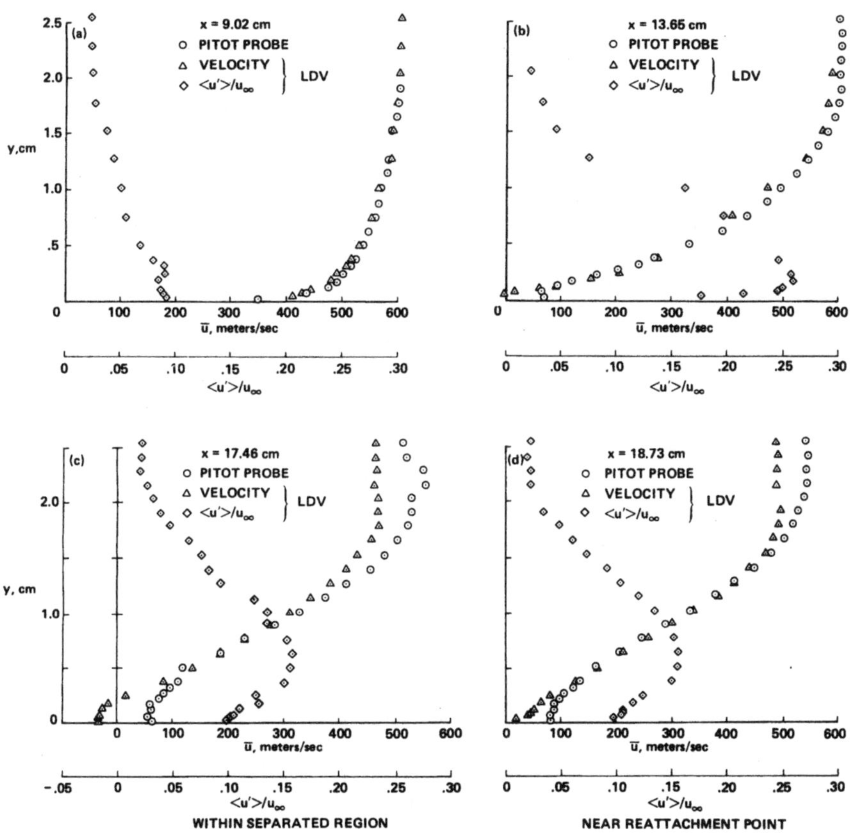

Figure 2. Mean velocity and turbulence-intensity profiles of supersonic shock-wave/turbulent boundary-layer interaction (Johnson et al., 1976). (a) Upstream of interaction. (b) Near separation point. (c) Within Separated region. (d) Near reattachment point.

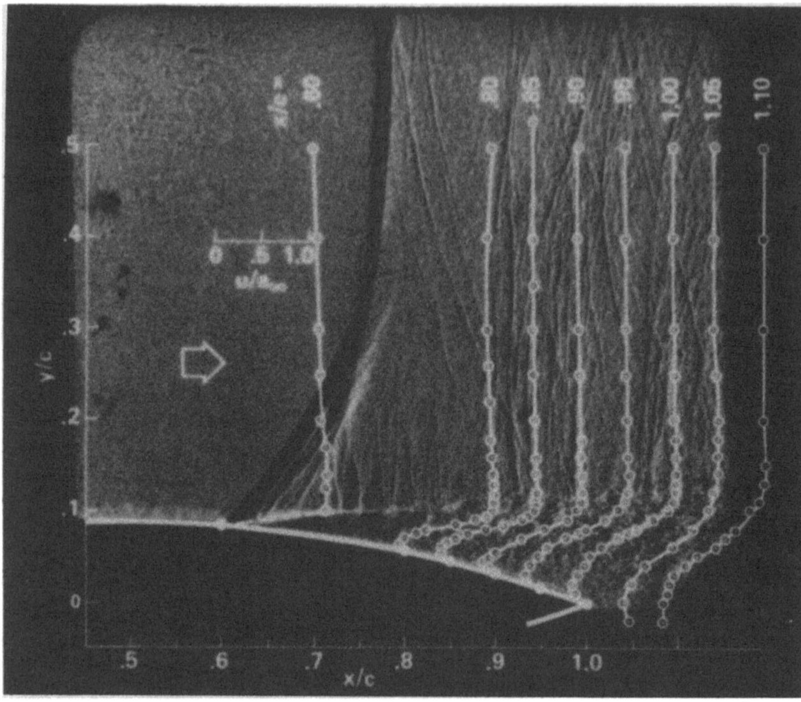

Figure 3. Mean velocity profiles of shock-induced separated flow on a circular-arc section (Seegmiller et al., 1978). "Copyright © the American Institute of Aeronautics and Astronautics; reprinted with permission of the AIAA."

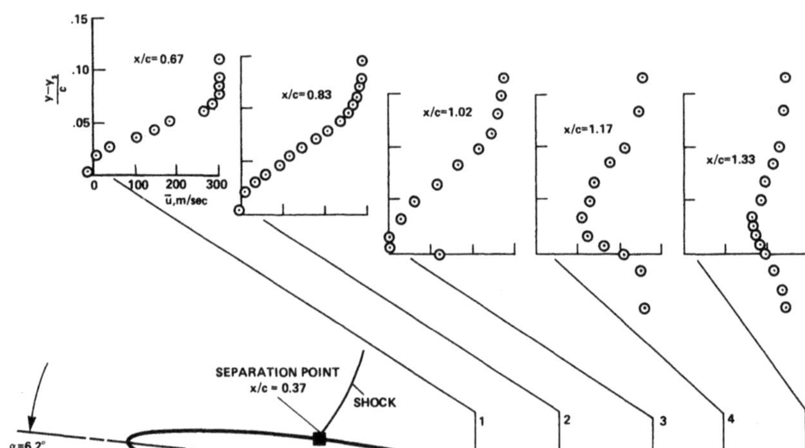

Figure 4. Mean velocity profiles of shock-induced separated flow on an NACA 64A010 airfoil section (Johnson and Bachalo, 1980). "Copyright © the American Institute of Aeronautics and Astronautics; reprinted with permission of the AIAA."

overlaid onto a shadowgraph visualization of the circular-arc flow; and an infinite-fringe interferogram, which gives a visual picture of the NACA 64A010 flow, is presented in Figure 5. The Reynolds shear stress results for the NACA 64A010 are shown in Figure 6. Similar results were obtained for the circular-arc section.

Other studies followed (e.g., Bachalo and Johnson, 1979; Delery, 1983; Viswanath, et al., 1980; Viswanath and Brown, 1983) in which flow models simulating certain features of airfoil flows were used. In Bachalo and Johnson (1979) and Delery (1983), flow models were chosen to study the boundary-layer development, whereas in Viswanath et al. (1980) and Viswanath and Brown (1983) the trailing-edge and near-wake flow development at high Reynolds number was emphasized. In these studies, a flow model was selected that produced

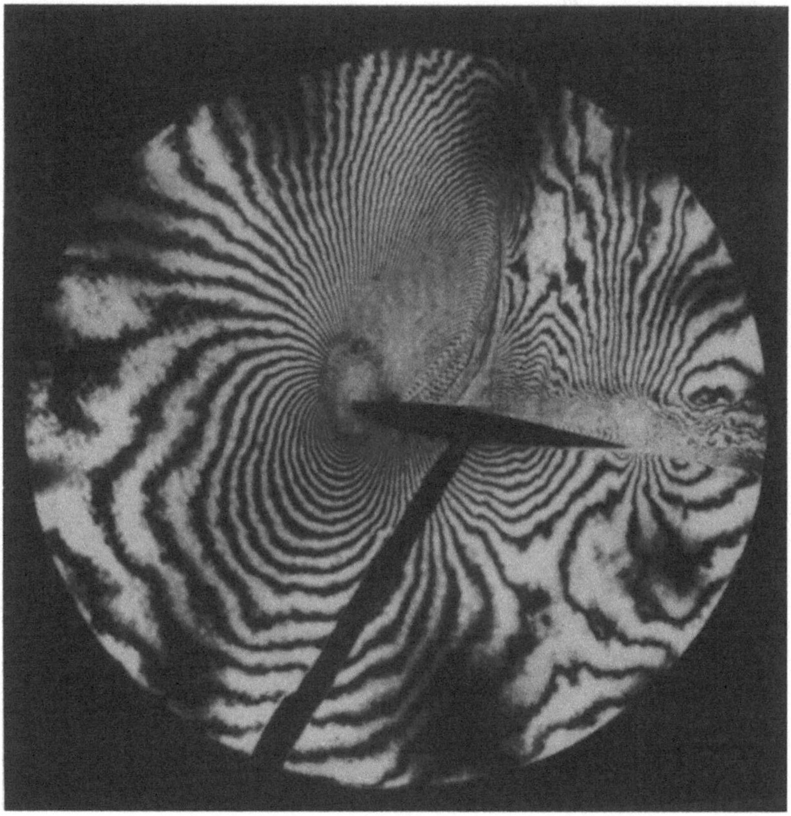

Figure 5. Infinite-fringe interferogram of shock-induced separated flow on an NACA 64A010 airfoil section (Johnson and Bachalo, 1980). "Copyright © the American Institue of Aeronautics and Astronautics; reprinted with permission of the AIAA."

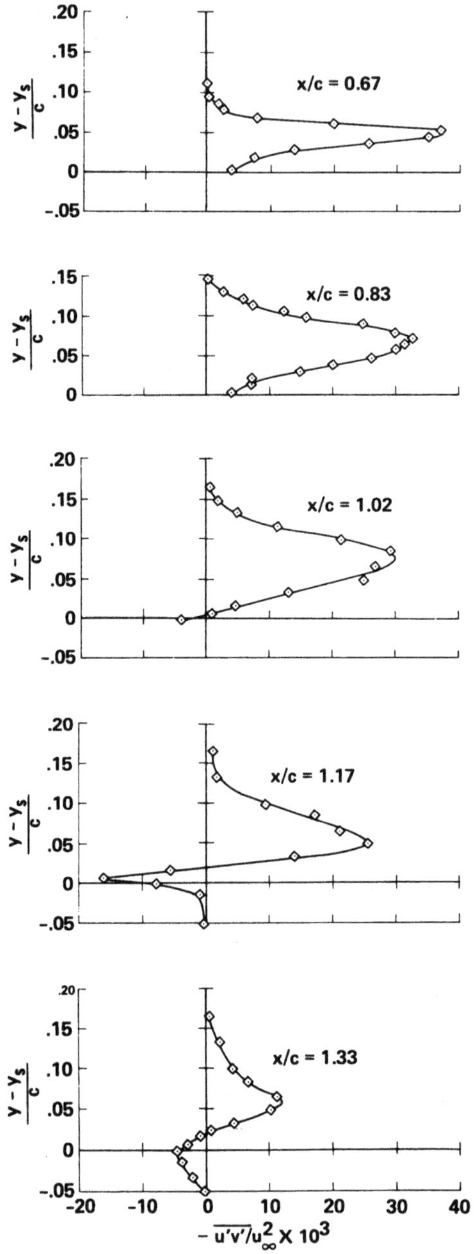

Figure 6. Reynolds shear-stress profiles of shock-induced separated flow on an NACA 64A010 airfoil section (Johnson and Bachalo, 1980).

a thick turbulent boundary layer in relation to the size of the test facility. As a result, considerably better resolution of the boundary-layer flow field was obtained in these studies. Recently (Johnson and Spaid, 1983), laser velocimetry was used for boundary-layer and wake measurements of a supercritical airfoil section at near on-design conditions. In that study, pitot probes were used to better resolve the near-wall mean velocities of the attached boundary layer. The laser velocimeter was used primarily to measure the turbulent Reynolds stresses within the boundary layer and the wake. However, it also provided an independent check of mean velocity in the wake and for a substantial part of the boundary layer. In addition, measurements of local flow angle within the viscous layers and static pressures at the edge of the viscous layers were realized with the laser velocimeter.

As is evident, the number of transonic aerodynamic studies in which laser velocimetry has been applied is rather limited. Besides the reasons stated earlier, limited tunnel availability for this type of research and the few researchers working in this area have also been contributing factors. However, because of these studies, a data base is now available that could never have been achieved by hot-wire anemometry.

In the following paragraphs, some of the features now known about turbulent boundary-layer behavior in transonic aerodynamics as a result of these studies will be discussed. First, we will consider the information that has been obtained about the streamwise mean velocity development in the vicinity of boundary-layer separation and in the fully separated flow region of transonic shock-wave/boundary-layer interactions. Presented in Figure 7 are streamwise mean velocity profiles measured at the point of separation and slightly downstream for the flow of Bachalo and Johnson (1979). The flow model consisted of an axisymmetric circular-arc section that was attached to a hollow cylinder. The axis of the cylinder was aligned with the oncoming flow ($M_\infty = 0.875$). A sketch of the model is included in Figure 7. From oil-flow visualizations, separation of the boundary layer was determined to occur at $x/c \simeq 0.7$, which was immediately downstream of the shock location. The streamwise distance is referenced from the leading edge of the circular-arc section. Although mean velocities were measured to within 0.25 mm of the surface, reverse flow was still not measured at $x/c = 0.75$ as seen from Figure 7 because of the extreme thinness of the separated zone in the early stages of separation. The character of the mean velocity profiles shown in Figure 7 are totally different from that measured by Simpson et al. (1981) in the vicinity of separation for a low-speed diffuser flow. The diffuser results, which were also obtained by laser velocimetry, are shown in Figure 8. The closest that measurements were obtained from the surface in this study was also 0.25 mm, but the boundary layer was considerably thicker. For this flow, the momentum defect in the inner part of the boundary layer ($y < \delta/3$) is quite large upstream of separation. This is the primary reason for the better definition of the time-averaged separation point in Figure 8, rather than the better near-wall resolution. At separation the shape factor for the transonic flow and the diffuser flow are $H \simeq 2.1$ and $H \simeq 4$, respectively. Clearly, there is not a one-to-one relationship between $C_f = 0$ and H, as is assumed in some boundary-layer integral schemes.

For a shock-induced separated flow, the character of the mean velocity profile at separation depends on the character of the boundary layer just ahead of the shock wave. This can be seen by comparing the streamwise mean profiles of Figures 2b and 7. In contrast, the diffuser flow slowly approaches separation and at separation the mean velocity profile shape appears nearly independent of its previous history. This is demonstrated by how well Coles' law-of-the-wall/law-of-the-wake representation fits the diffuser mean velocity profile at separation (see Figure 8). At separation ($C_f = 0$), Coles' expression reduces simply to $\overline{u}/u_e = \sin^2 \frac{\pi}{2}\frac{y}{\delta}$.

Within the fully developed separated flow region, a more universal character in the mean velocity profiles is observed as seen in Figure 9, where data are presented for the transonic shock-induced flow and the low-speed diffuser flow. For these two profiles, the shape factors are $H \simeq 7.5$ and $H \simeq 7.7$ for the transonic flow and the diffuser flow, respectively. Other measurements within the region of fully separated flow show a similar character to those of Figure 9 (see Figures 2c, 3, and 4).

Included in Figure 9 are mean velocity profiles predicted by an inverse boundary-layer method with the experimental displacement thicknesses specified. The popular Cebeci-Smith algebraic closure model (Cebeci and Smith, 1974) was used to describe the Reynolds shear stress. As evident, this closure model predicts a separated profile shape that is inconsistent with the experimental data. In addition, it predicts too large of a pressure recovery in the separated zone, as indicated by the edge velocities being lower than the

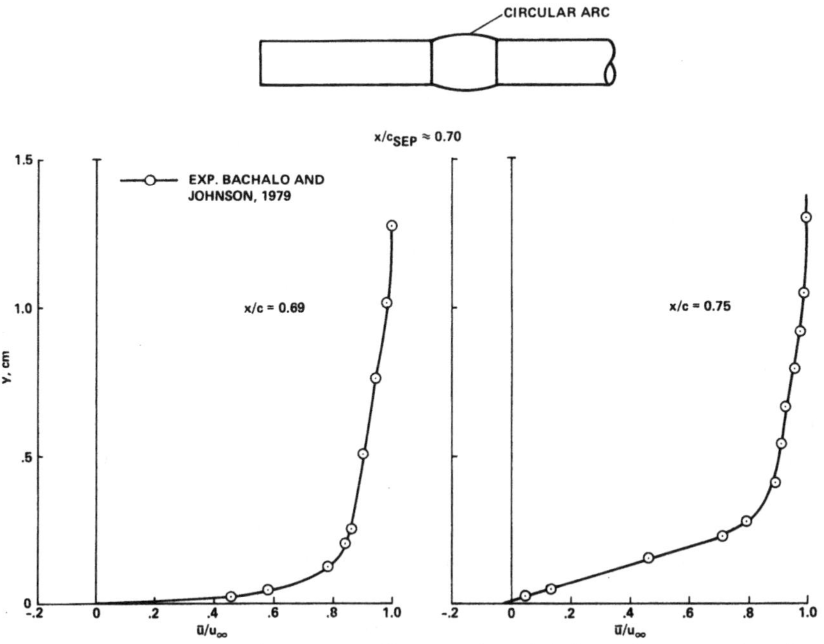

Figure 7. Mean velocity profiles in the vicinity of separation for a transonic shock-wave/turbulent boundary-layer interaction, $M_\infty = 0.875$ (Bachalo and Johnson, 1979). "Copyright © the American Institute of Aeronautics and Astronautics; reprinted with permission of the AIAA."

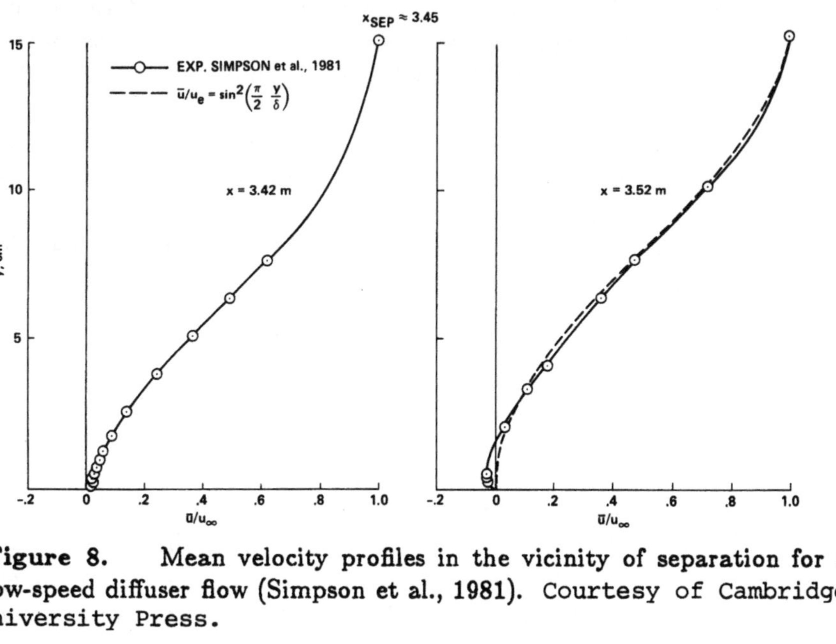

Figure 8. Mean velocity profiles in the vicinity of separation for a low-speed diffuser flow (Simpson et al., 1981). Courtesy of Cambridge University Press.

experimental values. Clearly, this simple algebraic model is inadequate in treating separated flows.

Of equal importance to the mean velocity in these flows is the information gained from laser velocimetry regarding the development of the turbulent Reynolds stresses. For brevity, only the Reynolds shear-stress development will be discussed. An inspection of the results that have been obtained for strong viscous-inviscid interactions reveals that the data all exhibit strong nonequilibrium effects. That is, the turbulent shear stress cannot be accurately described by using only local mean flow properties. This is well illustrated by the mixing-length distributions presented by Delery (1981) for two transonic shock-wave/turbulent boundary-layer interactions that he studied, one without separation and one with massive separation; these interactions were generated in a converging-diverging nozzle and a bump mounted on the tunnel wall, respectively. The mixing length distrubutions,

$$l = \frac{(-\overline{u'v'})^{1/2}}{\partial\overline{u}/\partial y}$$

determined by Delery, are shown in Figure 10. Included in the plots is the theoretical distribution suggested by Escudier (1965). This simple distribution is commonly used in finite-difference turbulent boundary-layer prediction methods, where the expression $\tau = \rho l^2(\frac{\partial\overline{u}}{\partial y})^2$ is used to represent the turbulent Reynolds shear stress. For boundary-layer flows in near-equilibrium, such as a zero-pressure-gradient boundary layer, this closure model works quite satisfactorily. However, this is not the case for boundary-layer flows that are subjected to rapidly changing pressure gradients as present in transonic aerodynamics. The comparisons shown in Figure 10 are indicative of the departures from this theory for shock-wave/turbulent boundary-layer interactions and trailing-edge flows (e.g., Viswanath and Brown, 1983).

Since τ cannot be simply related to $\rho l^2(\frac{\partial\overline{u}}{\partial y})^2$ in these flows, the experimental mixing lengths have no physical meaning. However, it is instructive to look at the development of this quantity, since it shows the rate at which τ changes in relation to the strain rate $\partial\overline{u}/\partial y$. In the wake region of the boundary layer (here the theory assumes that $l/\delta = 0.09$), the experimental results show that the experimental mixing lengths are less than the theoretical values ahead of where the maximum turbulent shear stress $\tau_{\max}$ occurs (this

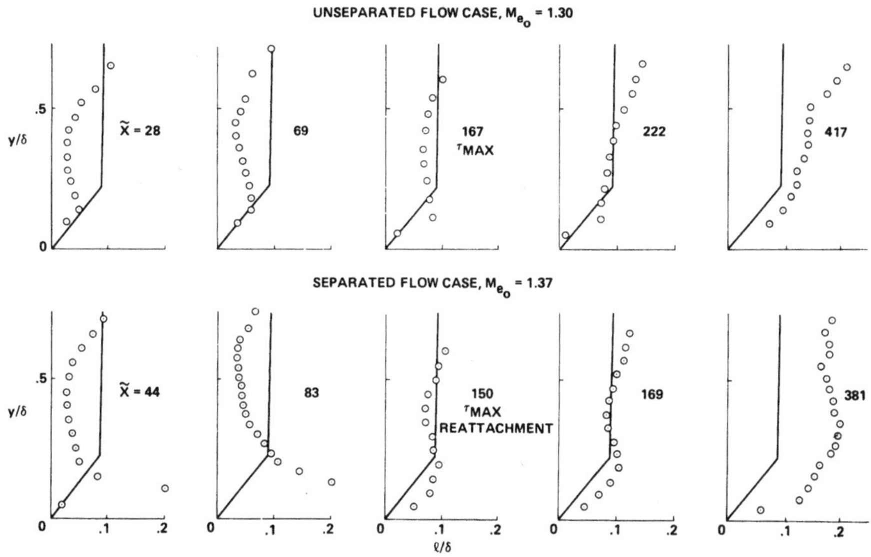

Figure 9. Mean velocity profiles in the fully developed separated region. (a) Transonic shock-wave/boundary-layer interaction (Bachalo and Johnson, 1979). (b) Low-speed diffuser flow (Simpson et al., 1981). Courtesy of Cambridge University Press.

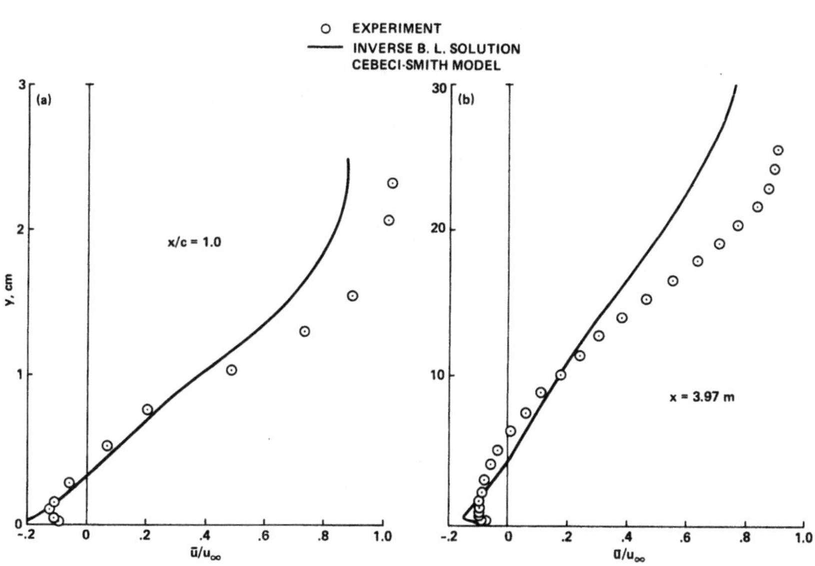

Figure 10. Mixing length distributions for unseparated and separated transonic shock-wave/boundary-layer interactions (Delery, 1981). "Copyright © the American Institute of Aeronautics and Astronautics; reprinted with permission of the AIAA."

streamwise location is indicated in Figure 10) and greater than the theoretical values downstream of this station. There is good agreement between the theory and experiment only where the turbulent shear stress attains its maximum value. This was similarly observed in the experiments of Seegmiller et al. (1978), Johnson and Bachalo (1980), and Bachalo and Johnson (1979). These trends are a result of the slow response of the turbulent shear stress to changes in the mean strain rate $\partial\overline{u}/\partial y$. In the outer or wake region of the boundary layer, the strain rate increases during the early part of the interaction zone. The turbulent shear stress also increases, but at a slower rate, as indicated by the smaller mixing lengths. Downstream, where boundary-layer recovery occurs, the strain rate in the outer part of the boundary layer decreases at a much more rapid rate than the turbulent shear stress, which accounts for the increasing mixing lengths in this region. This behavior is caused by convection effects, or what is commonly referred to as history effects. For the case where massive separation occurs, the effects of convection at stations $\tilde{X} = 44$ and $\tilde{X} = 83$ are even more dramatic in the inner part of the boundary layer. In this region of the flow $\partial\overline{u}/\partial y$ has undergone a rapid decrease, whereas the turbulent shear stress remains nearly unchanged from what it was farther upstream. Thus, large apparent mixing lengths result. In the axisymmetric bump study (Bachalo and Johnson, 1979) similar behavior was observed of the mixing length in the developing region of the separated layer: In the inner part of the separated region where $\partial\overline{u}/\partial y$ changes sign, $-\overline{u'v'}$ remained positive. This was also observed in the diffuser experiment of Simpson et al. (1981), but to a lesser extent. Thus, in this region, the Boussinesq relationship $\tau = \rho\epsilon_t(\frac{\partial\overline{u}}{\partial y})$, upon which all the turbulence closure models for airfoil flows are based, is not valid. Fortunately, the turbulent shear stress is very small in this region and may very likely not have to be predicted accurately.

Inherent in closure models like the Cebeci-Smith model is the assumption that the convection of turbulence can be neglected. Local equilibrium of the turbulence is assumed by equating the local production of turbulence to the local dissipation. This approach is clearly not valid for strong transonic viscous-inviscid interactions. More recently developed two-equation closure models, which use partial-differential equations to describe the development of the turbulence kinetic energy, and a dissipation length scale attempt to describe the nonequilibrium effects of flows subjected to rapidly changing pressure gradients. Modeling constants required by these formulations have been obtained from incompressible experimental data.

Solutions to the mass-averaged Navier-Stokes equations have been obtained (Johnson et al., 1982) for the axisymmetric bump flow of Bachalo and Johnson (1979), using both the Cebeci-Smith model and the two-equation Wilcox-Rubesin model (Wilcox and Rubesin, 1980). A comparison of the predicted and measured development of τ_{max} as a function of streamwise distance is presented in Figure 11a. Although the more sophisticated two-equation model does considerably better in predicting the slow decay in turbulent shear stress in the downstream region of the interaction, it does little better than the Cebeci-Smith model in predicting the rate at which the turbulent shear stress increases in the vicinity of the shock wave. As a result, the shock location is mispredicted almost as badly as it is with the Cebeci-Smith model, as seen in Figure 11b.

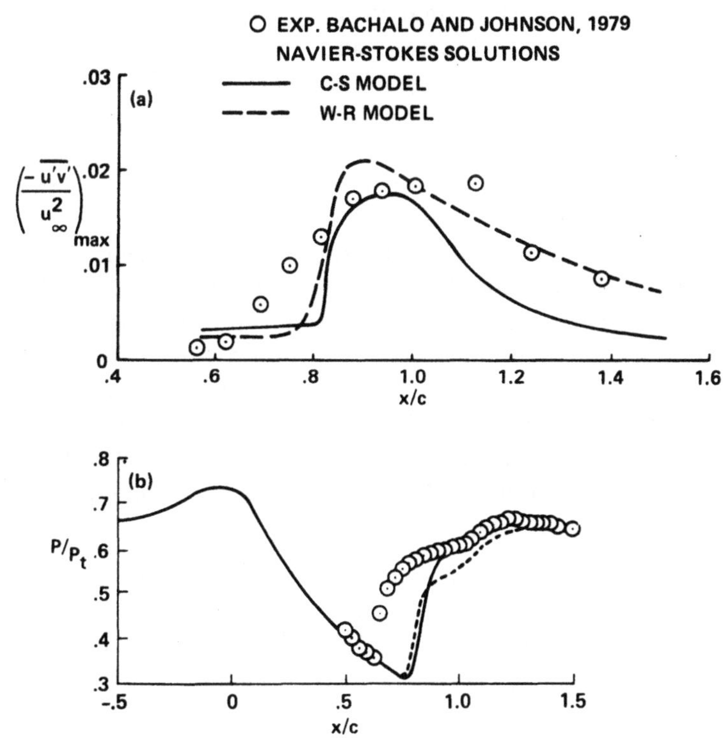

Figure 11. Navier-Stokes solutions of axisymmetric bump shock-wave/boundary-layer interaction (Johnson et al., 1982). (a) Streamwise variation in maximum turbulent shear stress. (b) Surface pressure distribution. "Copyright © the American Institute of Aeronautics and Astronautics; reprinted with permission of the AIAA."

This comparison and that of Figure 7 demonstrate the need for better closure models for strong transonic viscous-inviscid interactions. So far, laser-velocimeter data on transonic viscous-inviscid interactions have only been used to evaluate turbulence closure model formulations. The data will have its greatest impact when it begins to be used in developing new, improved turbulence models. Whether a finite-difference or integral approach is used, the strong convection effects present in these flows will have to be accurately taken into account.

2.4 Future Directions

The laser velocimeter technique is unique in its ability to measure the streamwise and normal velocity components, and the turbulent Reynolds stresses (normal and shear) of turbulent boundary layers that are on the verge of separating or that have undergone massive separation. In low-speed and transonic aerodynamics, these conditions are often present. In the author's opinion, this measurement capability has not been capitalized on in the field of boundary-layer research, as it should have been. For example, there are more compressible data for pressure-driven, separating boundary-layer flows than there are incompressible data. The low-speed diffuser experiment of Simpson et al. represents to the author's knowledge the only incompressible experiment where laser velocimetry has been used to measure the Reynolds shear stress for such a flow. This situation exists despite the well-known fact that current prediction methods are incapable of accurately describing the development of turbulent boundary layers that are subjected to strong adverse pressure gradients.

Currently, little is known about the Reynolds shear-stress behavior in the near-wall region of adverse pressure gradient boundary layers, although it is critical that it be accurately modeled in finite-difference calculation methods (e.g., Chapman and Kuhn, 1981). Better near-wall modeling is especially important in the prediction of skin friction. This is an area of research where laser velocimetry could make a significant contribution. In all the experiments cited in this paper, Reynolds shear-stress measurements were obtained within at least 1 mm of the surface. Compared with hot-wire anemometry, this is impressive. Even in low-speed flows, the closest that reliable Reynolds shear stress measurements can be obtained with hot-wire anemometry is about 2 mm from the surface. In the axisymmetric bump study of Bachalo and Johnson (1979), Reynolds shear stresses were consistently measured to within 0.25 mm of the surface. Based on this performance, measurements even closer to the surface should be possible in incompressible flows.

Unfortunately, the near-wall performance in the measurement of the streamwise velocity component alone is generally not as good as can be obtained by either pitot pressure probes or hot-wire anemometry. When extreme care is taken, these techniques can measure the streamwise velocity to within about 0.1 mm of a solid surface in low-speed flows. Similar performance can currently be obtained by laser velocimetry, but only under very restricted circumstances. For example, in the study of Johnson and Spaid (1983) reliable pitot pressure measurements were obtained to within 0.3 mm of the airfoil's surface, whereas, 0.7 mm was the best achieved with the laser velocimeter. The major obstacle in obtaining laser-velocimeter measurements close to a solid surface is the deterioration of the signal-to-noise ratio caused by laser light scattering from the surface (often referred to as flare). For the relatively small facilities used in the studies described in this paper, 1 mm was nominally the distance at which the noise from this source reached the level where measurements were no longer possible. With the axisymmetric model of Bachalo and Johnson (1979) the effects of flare were considerably reduced, which accounted for the much better near-wall performance. This example illustrates how model geometry can significantly affect the measurement capabilities in laser-velocimetry applications. In the author's opinion, more effort needs to be directed toward the development of schemes for rejecting or reducing this flare.

Another serious limitation of laser velocimetry is that the near-wall measurement capabilities tend to deteriorate with increased size of the test facility, whereas they do not for probe-type devices. In theory, the same sensing volume dimensions can be maintained as the facility size increases, but this is difficult to do in practice. More importantly, though, flare rejection becomes extremely difficult as the collecting optics are placed farther and farther from the measurement point. Thus, the approach of generating a thicker boundary layer in a larger facility to better resolve the boundary layer is likely to fail when the approach is taken to maintain all the laser velocimeter components outside of the test section.

With the difficulty solid surfaces pose, the temptation may be to apply laser velocimetry mainly to the wake regions of airfoils, wings, and other aerodynamic components. However, the accurate modeling of the wake region is predicated with knowledge of the turbulent boundary-layer properties that define the initial conditions. Thus, wake data alone are very restrictive in their potential to improve theoretical prediction methods.

A major challenge of future experiments will be to exceed what previous researchers have been able to accomplish. To do so, ingenuity in both the selection of the optical configuration and the flow model will be required. For example, the recent developments in fiber optics technology show great promise for laser velocimetry (Knuhsten et al., 1982). This technology should lead to the development of miniature laser-velocimeter systems, which could be used in a manner similar to probe-type devices. A sacrifice in the noninvasiveness of the technique would result but the ability to better resolve thick viscous layers in large facilities would be attained. Even though some form of probe or probes would have to be placed in the flow, the measurement point would still be remote. Thus, the interference effects would be less than with hot-wire or pitot probes.

So far no mention has been made regarding three-dimensional flows. This is an area of research where the capabilities of laser velocimetry have hardly begun to be exploited (e.g., Yanta andAusherman, 1983). Potentially, the other important Reynolds shear-stress component, $-\overline{w'v'}$, could be resolved much better using laser velocimetry than using triple-wire hot-wire anemometry. Measurements of this quantity to within the same proximity of solid surfaces as $-\overline{u'v'}$ should be possible. These measurements could greatly improve our understanding of three-dimensional boundary-layer development.

3. Concluding Remarks

The main purpose of this article was to demonstrate the importance of laser velocimetry in the field of transonic aerodynamics. With this optical technique, measurements have been obtained for strong transonic viscous-inviscid interactions that could not have been obtained by any other means. Although there have been only a small number of transonic experiments in which this technique has been applied, these experiments have begun to generate the type

methods. Especially important in these studies was the measurement of the turbulent Reynolds stresses that must be modeled in either a finite-difference or integral viscous flow calculation method. Nearly all of the approaches now used to describe the turbulent Reynolds stresses in the most advanced airfoil prediction methods predate the data discussed in this paper. These data will have their greatest impact when information available from them is used to develop better models for describing the turbulent Reynolds stresses. The data show the strong influence of convection on the Reynolds shear stress, which will have to be accurately modeled by the viscous flow prediction method.

As discussed in this paper, laser velocimetry, like any other measurement technique, has advantages and disadvantages. These should be kept well in mind in the design of future experiments to insure that the measurement capabilities required by the problem can be met. In cases where measurements are desired very close to a model's surface, care will be required in the selection of the optical arrangement and the flow model. Through intelligent applications of the technique, further understandings should result of transonic viscous flows.

References

[1] Bachalo, W. D. and Johnson, D. A. "An Investigation of Transonic Turbulent Boundary Layer Separation Generated on an Axisymmetric Flow Model," *AIAA Paper 79-1479,* 1979.

[2] Cebeci, T. and Smith, A.M.O. "Analysis of Turbulent Boundary Layers," New York: Academic Press, 1974.

[3] Chapman, D. R. and Kuhn, G. D. "Two-Component Navier-Stokes Computational Model of Viscous Sublayer Turbulence," *AIAA Paper 81-1024,* 1981.

[4] Cousteix, J. "Integral Techniques," AFSOR-HTTM-Stanford Conference on Complex Turbulent Flows: Comparison of Computation and Experiment. *Vol. II: Taxonomics, Reporters' Summaries, Evaluation and Conclusions.* Stanford, CA: Stanford University, 1981, 650–71.

[5] Delery, J. M. "Investigation of Strong Shock-Turbulent Boundary Layer Interaction in 2D-Transonic Flows With Emphasis on Turbulence Phenomena," *AIAA Paper 81-1245,* 1981.

[6] ————. "Experimental Investigation of Turbulence Properties in Transonic Shock/Boundary-Layer Interactions," *AIAA Journal,*

21 (1983), 180–85.

[7] East, L. F. "The Application of a Laser Anemometer to the Investigation of Shock-Wave Boundary-Layer Interactions," *AGARD Conference Proceedings No. 193, Applications of Noninstrusive Instrumentation in Fluid Flow Research,* Saint-Louis, France, *Paper No. 5,* 1976.

[8] Escudier, M. P. "The Distribution of the Mixing Length in Turbulent Flows Near Walls," *Report TWF/TN/1,* Imperial College, London, 1965.

[9] Green, J. E., Weeks, D. J., and Brooman, J.W.F. "Prediction of Turbulent Boundary Layers and Wakes in Compressible Flow by a Lag-Entrainment Method," *Aeronautical Research Council Reports and Memoranda No. 3791,* 1977.

[10] Johnson, D. A. "Turbulence Measurements in a Mach 2.9 Boundary Layer Using Laser Velocimetry," *AIAA Journal, 12, No. 5* (1974), 711–14.

[11] Johnson, D. A. and Bachalo, W. D. "Transonic Flow Past a Symmetrical Airfoil—Inviscid and Turbulent Flow Properties," *AIAA Journal, 18* (1980), 16–24.

[12] Johnson, D. A. and Spaid, F. W. "Supercritical Airfoil Boundary-Layer and Near-Wake Measurements," *Journal of Aircraft, 20* (1983), 298–305.

[13] Johnson, D. A., Bachalo, W. D., and Modarress, D. "Laser Velocimetry Applied to Transonic and Supersonic Aerodynamics," *AGARD Conference Proceedings No. 193, Applications of Noninstrusive Instrumentation in Fluid Flow Research,* Saint-Louis, France, *Paper No. 3,* 1976.

[14] Johnson, D. A., Horstman, C. C., and Bachalo, W. D. "Comparison Between Experiment and Prediction for a Transonic Turbulent, Separated Flow," *AIAA Journal, 20* (1982a), 737–44.

[15] Johnson, D. A., Modarress, D., and Owen, F. K. "An Experimental Verification of Laser-Velocimeter Sampling Bias and Its Correction," in *Engineering Applications of Laser Velocimetry,* ASME Winter Meeting, eds. H. W. Coleman and P. A. Pfund, 1982b, 153–62.

[16] Knuhsten, J., Olldag, E., and Buchhave, P. "Fibre-Optic Laser Doppler Anemometer With Bragg Frequency Shift Utilizing Polarisation-Preserving Single-Mode Fibre," *J. Phys. E: Sci. Instrum., 15, No. 11* (1982), 1188–91.

[17] McLaughlin, D. K. and Teiderman, W. G. "Biasing Corrections for Individual Realization Laser Anemometer Measurements in Turbulent Flows," *Physics of Fluids, 16, No. 12,* 1973.

[18] Owen, F. K. "An Assessment of Flow-Field Simulation and Measurement," *AIAA Paper 83-1721,* 1983.

[19] Reda, D. C. and Murphy, J. D. "Shock Wave-Turbulent Boundary-Layer Interactions in Rectangular Channels," *AIAA Journal, 11* (1973), 1367–68.

[20] Robinson, S. K., Seegmiller, H. L., and Kussoy, M. I. "Hot-Wire and Laser Doppler Anemometer Measurements in a Supersonic Boundary Layer," *AIAA Paper 83-1723,* 1983.

[21] Seegmiller, H. L., Marvin, J. G., and Levy, L. L., Jr. "Steady and Unsteady Transonic Flows," *AIAA Journal, 16, No. 12* (1978), 1262–70.

[22] Simpson, R. L. AFSOR-HTTM-Stanford Conference on Complex Turbulent Flows: Comparison of Computation and Experiment. *Vol. I: Objectives, Evaluation of Data, Specifications of Test Cases, Discussion, and Position Papers.* Stanford, CA: Stanford University, 1980, 253–73.

[23] Simpson, R. L., Chew, Y. T., and Shivaprasad, B. G. "The Structure of a Separating Turbulent Boundary Layer, Part I: Mean Flow and Reynolds Stresses," *Journal of Fluid Mechanics, 113* (1981), 23–51.

[24] Viswanath, P. R. and Brown, J. L. "Separated Trailing-Edge Flow at a Transonic Mach Number," *AIAA Journal, 21* (1983), 801–07.

[25] Viswanath, P. R., Cleary, J. W., Seegmiller, H. L., and Horstman, C. C. "Trailing-Edge Flows at High Reynolds Number," *AIAA Journal, 18* (1980), 1059–65.

[26] Whitfield, D. L., Swafford, T. W., and Jacocks, J. L. "Calculation of Turbulent Boundary Layers With Separation, Reattachment, and Viscous-Inviscid Interaction," *AIAA Paper 80-1439,* 1980.

[27] Wilcox, D. C. and Rubesin, M. W. "Progress in Turbulence Modeling for Complex Flow Fields, Including Effects of Compressibility," *NASA TP-1517,* 1980.

[28] Yanta, W. J. and Ausherman, D. W. "The Turbulent Transport Properties of a Supersonic Boundary Layer on a Sharp Cone at Angle-of-Attack," *AIAA Paper 83-0456,* 1983.

[29] Yeh, Y. and Cummins, H. Z. "Localized Fluid Flow Measurements With a He-Ne Laser Spectrometer," *Applied Physics Letters, 4, No. 10* (1964), 176–78.

Part V
Rotor Dynamics and Aerodynamics

The Aerodynamics and Dynamics of Rotors—Problems and Perspectives

Rene H. Miller
Massachusetts Institute of Technology
Cambridge, MA 02139

1. Introduction

The aerodynamic and dynamic characteristics of rotary wing devices such as helicopters, other VTOL aircraft and wind turbines differ markedly from those of fixed wings. Major differences are due to:

1. the high concentration of bound circulation over the outer portion of the blade resulting in an intense vortex trailed from the tip,
2. a curved spiralling wake remaining initially close to the rotor causing strong blade/vortex interactions,
3. the high centrifugal force field in which the blades operate,
4. the relatively large steady state displacements of the blades out of the plane of rotation.

This paper will explore some of the aerodynamic and aeroelastic problems associated with these characteristics and will try to place them in perspective as regards their operational importance. Simple analogs will be used, so far as possible, to help interpret the phenomena involved since the mathematical complexity usually required for their precise definition tends to mask the physical understanding so essential for design solutions. Recent advances have greatly increased our understanding of these problems, however a heavy reliance on empiricism is still required during the design process. Much remains to be done before theoretical treatments can be used with complete confidence in optimizing, or even predicting, rotor operational characteristics.

Rotor blades are long flexible beams maintained in equilibrium primarily by centrifugal forces. Since their ability to transmit bending moments is therefore limited, a fairly high coning angle of the order of 5° is required to transmit the lift forces to the hub of the vehicle through blade tension. These lifting forces must be at least equal to the gross weight of the aircraft. They are produced by blades weighing less than 5% of the gross weight and carrying a centrifugal force which is several times the gross weight, specifically, the gross weight divided by the equilibrium coning angle. When not rotating such blades are weak and limber devices in bending, as evidenced by the high damage rate when nonrotating blades are positioned or folded by hand, and yet when rotating such blades can produce blockage equivalent to the flat plate disk area, as evidenced by the deployment of rotor chutes in high speed wind tunnel tests.

In the presence of a centrifugal force field rotor blades are therefore highly efficient structural devices for transmitting lift loads to the fuselage. However, the out of plane displacement of the blade necessary to transmit these loads through the high centrifugal tension forces introduces the possibility of several types of dynamic instability. Since the coning angle, although small, is finite, normally second order terms must be treated as first order in the stability analyses. This phenomenon has been well recognized during the past 40 years of intensive helicopter development. In particular, coning can aggravate the serious case of classical blade bending-torsion flutter and increase the possibility of instability even with the elastic axis, center of gravity and aerodynamic centers of the blade coincident.

Additional instabilities arise from small inplane forces due to components of induced drag and coriolis acceleration and the low inplane damping of the semi-rigid rotors which have recently become of interest. These instabilities appear to be mild with a degree of negative damping of the same order as the positive damping due to structural hysterisis. More serious problems occur as a result of support flexibility which gives rise to ground and air resonance, and rotor whirl instabilities.

Precise aeroelastic analysis of these problems is complicated by the lack of an aerodynamic theory for rotors in hover and forward flight. Fortunately simple modification in the classical Rankine-Froude momentum theories originally developed for propellers over 100 years ago appears to be adequate for most aeroelastic analyses requiring unsteady aerodynamics in hover. In the case of steady flows, however, when it is desired to determine the loading on the blade either for performance optimization, noise reduction or structural analysis, a deeper understanding of rotor aerodynamics is required. Under all flight regimes, but particularly in hovering flight, the rotor interacts with its own wake causing large local variations of bound circulation along the blade. This as yet not well-understood phenomenon involves the interaction of a blade with the intense vorticity in the wake represented by the strong tip vortices generated by the rapid drop off of bound circulation near the tips of the blades.

After having insured aeroelastic stability, reasonable performance and vibration levels and low acoustic signatures, the designer is left with the need to provide satisfactory handling qualities and dynamic stability for the vehicle as a whole. If VTOL aircraft are to achieve the dispatch reliability required of a practical short-haul transportation system, they must be capable of complete IFR operations in hovering flight. This will require a degree of inherent stability and, more important, controllability that has not as yet been achieved. Fortunately, in this area the unknowns are fewer and the flight characteristics, at least for the conventional helicopter, reasonably well understood, as are the steps necessary to ensure operational suitability under all conditions of visibility down to zero/zero.

We thus have three problem areas which would be of interest to explore:

1. the aerodynamics of the rotor,
2. the dynamics of the rotor, and
3. the dynamics and control characteristics of the vehicle.

Of these, the least well-understood and probably the most complex is the first, the aerodynamics of the rotor. The first section will discuss this topic, concentrating on the problems of determining blade airloads in hover and forward flight. Following sections will examine the problems of vibration caused by these airloads, blade stability analysis, and, finally, vehicle dynamics.

In discussing all these topics major concentration will be placed on those potential areas of investigation which are the most promising for providing the design tools required to ensure a satisfactory vehicle. No attempt will be made to review the vast amount of literature now becoming available since several highly informative papers have appeared recently summarizing the state of the art. A more useful contribution would probably be to concentrate on the physics of the problem and to sort out the driving design parameters from the complex mathematical analyses necessary to treat these problems.

In the final analysis it will never be possible to completely model the dynamics and aerodynamics of a system as complex as a rotary wing vehicle. Theoretical analyses can contribute mainly signposts indicating the directions to try when problems occur in flight evaluation. Hopefully, such analyses should also provide design tools for minimizing the probability of such problems occurring. To this end the simplest forms of analysis which retain the essence of the problem are of greatest value in assisting the designer during the crucial stages of design and flight evaluation.

As mentioned above, the field is fortunate in having several excellent treatment of the problems of rotor aerodynamics and dynamics now available. A comprehensive text by Johnson (1980), includes a bibliography with over 2000 citations. The aerodynamics of the rotor has been well summarized in References [6] and [27]. Reference 28 contains a useful description of the physics of the flow. References 9, 18, and 31 summarize in clear fashion the dynamic and aeroelastic problems of VTOL aircraft. Reference 9 also contains a very useful review of more recent publications in this area. Among the major contributions are those of References 10 and 16. Many more have made valuable contributions to this complex area, but it is not the intent of this paper to review them in detail. Rather, this is left to the above cited publications.

2. Rotor Blade Airloads

2.1 Aerodynamics of the Rotor in Hover and Forward Flight

No closed-form solution exists, even in the idealized case, for the aerodynamics of a rotor in forward flight. An empirical solution based on momentum theory for an actuator disk which satisfies the boundary conditions in hover and at high forward speed (when the rotor may be approximated by a circular wing) was proposed by Glauert over 50 years ago and is in current use for performance estimation. In hovering flight the Rankine-Froude momentum theory first proposed in the 1860's is still the standard method for performance analysis. The Prandtl-Glauert vortex theory developed in the early 1920's for propellers, in which the vortex spiral is represented by a semi-infinite cylinder, is sometimes used. However, this theory is not applicable to the case of a hovering rotor since its basic assumption of induced velocities small compared to the inflow velocity is obviously not satisfied for the hovering case where the induced flow is the only axial flow through the rotor. Consequently, certain simplifying assumptions, such as the neglect of wake contraction in computing the velocities induced by the vortex cylinder, are inadmissible. Also, although the momentum theories for hovering and forward flight have proven to be simple and satisfactory methods for performance estimation, they are based on an assumed infinite number of blades and can not therefore be used for predicting blade loads, these loads being critically dependent on individual blade vortex interaction. Optimization of rotor performance and prediction of the vibratory loads must therefore eventually be based on vortex theories and, in particular, free wake numerical methods.

The problem is further complicated by what are believed to be real fluids effects, not as yet well understood. The importance of these effects became apparent when attempts were made to predict blade airloads in forward flight using a complete wake representation. Initial results obtained using a rigid wake for comparison with the experimental airloads of Reference 45 (one of the first comprehensive flight test programs in which blade pressure distributions were measured) showed reasonable agreement at the blade tip but deteriorating as the radius decreased. For example, Figure 1, taken from Reference 35, shows a rapid variation of load in the computed results at the 85% span location near the 90 degree azimuth which did not appear in the test data. It was suspected that this and other discrepancies were due to the rigid wake assumptions. Consequently, the free wake analysis technique of Reference 46 was developed. At the same time the wake geometry was determined experimentally from smoke tests on a wind tunnel model. Figure 2, taken from Reference 46 shows that at least qualitatively the theory correctly predicted the observed geometry. However, when this theory was used to predict airloads, agreement was much worse than in the case of the rigid wake, as shown in Figure 3.

Vortex bursting had occasionally been noticed during the wind tunnel tests and reported by other investigators; therefore a rapid increase in core size after first encounter was added to the analysis. Some improvement was realized, but it was only when vortex breakdown well ahead of first encounter was assumed did reasonable agreement result, as shown in Figure 4. More recently, laser velocimeter

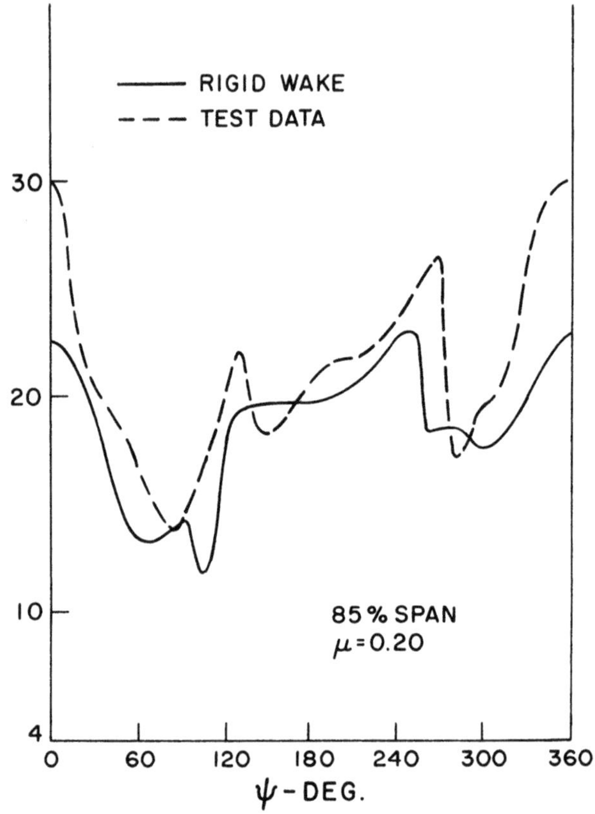

Figure 1. Comparison of computed and measured flight airloads for four bladed rotor of Reference 35 in forward flight. Rigid wake assumption.

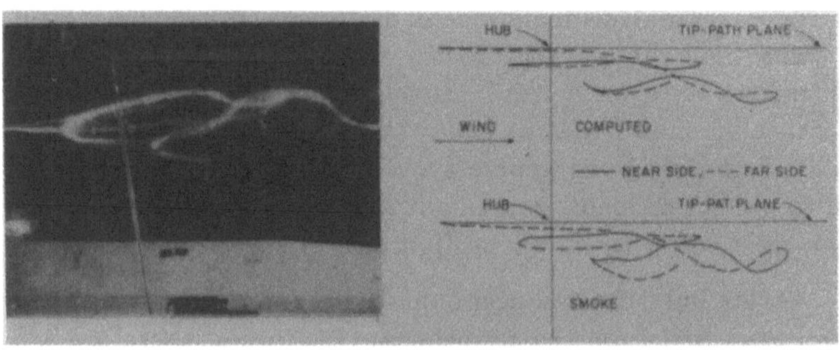

Figure 2. Predicted and experimental free wake geometry—μ = .10 [46].

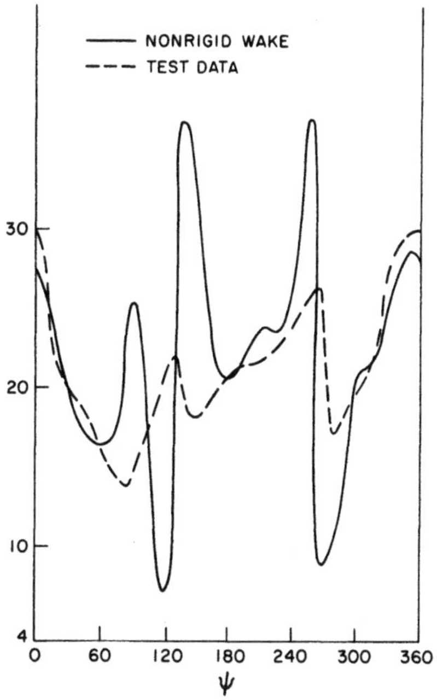

Figure 3. Comparison of computed and measured flight loads—initial free wake analysis of [46].

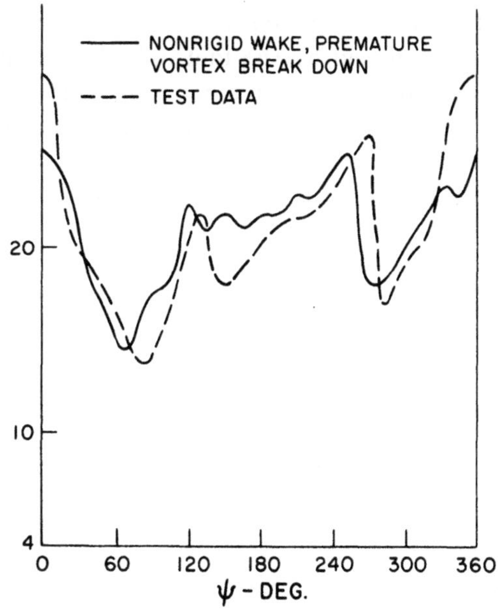

Figure 4. Same as Figure 3, but with vortex breakdown [46].

tests have indicated that the premature bursting postulated in Reference 46 may indeed occur, as shown in Figure 5 from Reference 4. The smoke studies of Reference 49 using stroboscopic light to stop the flow instantaneously also indicated the possibility of vortex expansion in hovering flight as shown in Figure 6.

Another phenomenon involving separation at the leading or trailing edges of the blade appears to occur at close blade vortex encounters, possibly due to the resulting rapid spanwise variations in flow along the blade. Reference 13 demonstrated the occurrence of such premature separations experimentally and showed that the maximum incremental lift coefficient which could be generated by a close blade/vortex encounter was limited to from about 0.2 to 0.3. Figure 7, taken from Reference 13, illustrates this phenomenon.

Tip losses are expected, particularly in the case of blades with rapid changes in geometry towards the tip. Rotor blades, unlike fixed wings, carry a high level of circulation near the tip. The bound circulation therefore must leave the blade in a strong tip vortex. Part of the tip losses must be attributed to the as yet not fully understood flow characteristics at the tip as this strong tip vortex is formed. In addition, separation could well occur as the flow proceeds around the tip from the lower to the upper surface.

Evidently, the vortices in the wake must contain viscous cores whose size may be estimated by the methods discussed in Reference 25. In addition, the roll up of the strong tip vortex results in distributed vorticity outside the core, as predicted by the Betz criteria of conservation of linear and angular momentum. These effects are small for a two-bladed rotor in hover but may be appreciable for four-bladed rotors.

2.2 Free Wake Modelling in Hover

In order to explore these problems more fully and to provide a better understanding of the physics of the problem, it is desirable to identify simplifications which would reduce the computational effort required to determine the wake geometries. One such simplification was used for the forward flight case in Reference 35 where the wake at encounter with a blade was replaced by an infinite vortex line and its effects otherwise ignored. It was shown that this approximation gave excellent agreement with the more complete solutions in which

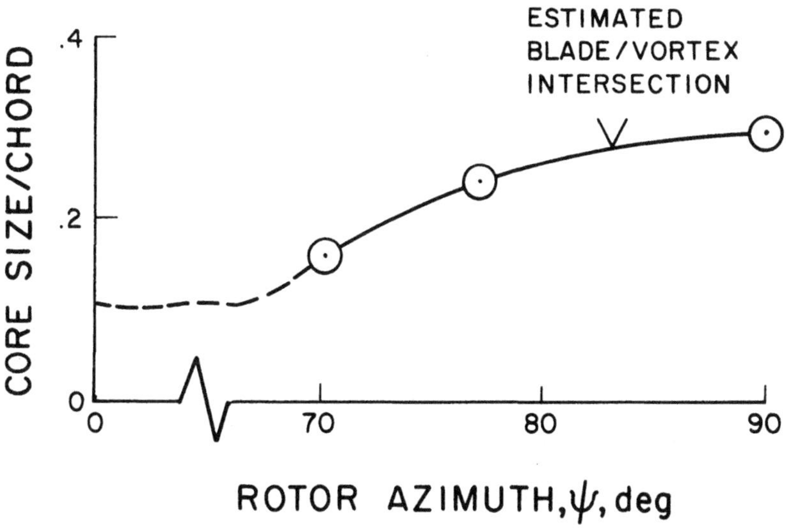

Figure 5. Experimental evidence of vortex breakdown [4].

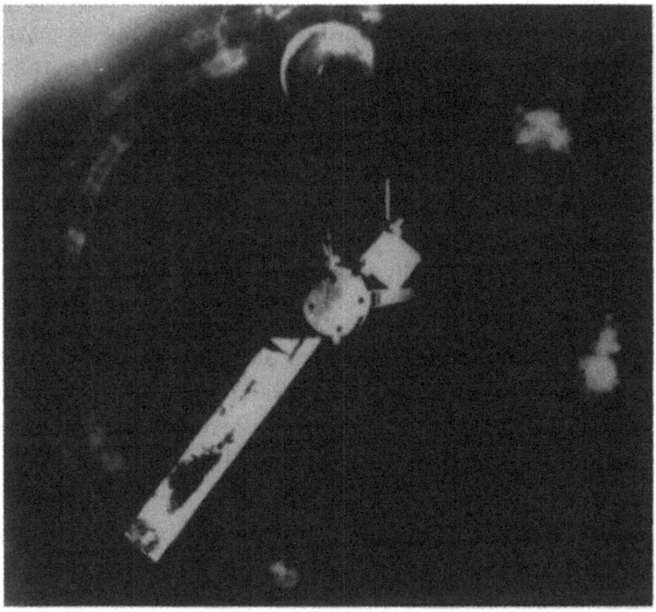

Figure 6. Expansion of tip vortex shown by stroboscopic flash in hover [49].

the entire wake was modeled, while reducing computational time to about 2% of the original. This method is now being extended to include a free wake in forward flight.

In the meantime, the same approach was used in Reference 39 to treat the more complex case of hovering flight. It was found that very simple models are also adequate for this case, for example, a quasi-two-dimensional one. The rotor wake is replaced by

1. a near wake consisting of semi-infinite vortex filaments attached to the blade in the plane of the rotor with the blade represented by a lifting line,
2. pairs of infinite line vortices below the rotor forming an intermediate wake, and
3. a far wake consisting of a pair of vortex sheets.

This geometry is possible because of the symmetry of the rotor in hovering flight. Viewed from the side, this wake model appears as a two-dimensional one (Figure 8). Wake displacements between any two vortices are determined from the vorticity transport law (vortex moves with the fluid) using the average of the velocities at the two vortices acting over a time increment corresponding to the individual blade passages. A similar model, but using vortex rings for the intermediate wake and vortex cylinders for the far wake, is shown in Figure 9. For simplicity, only one of the series of rings and one cylinder formed by the tip vortex is shown, although others will exist inboard depending on the roll-up schedule assumed. The two-dimensional model results from replacing the rings by pairs of infinite line vortices, one of which is located below the blade in question. Both models are thus applicable to a rotor with any number of blades.

Figure 10 taken from Reference 38, shows results obtained using these models compared to the experimental data of Reference 4 and the estimated wake positions of Reference 20. It was assumed that the near wake rolled up into vortex filaments, conserving linear momentum in the process, according to various schedules determined by the bound circulation distribution. In this case a tip, root, and center vortex were postulated. It should be noted that although there is clear experimental evidence for the existence of a strong tip vortex, there is as yet none for the inner wake vortices. Until more experimental data becomes available, any assumed roll-up schedule for the center portion of the blade should be considered more as a

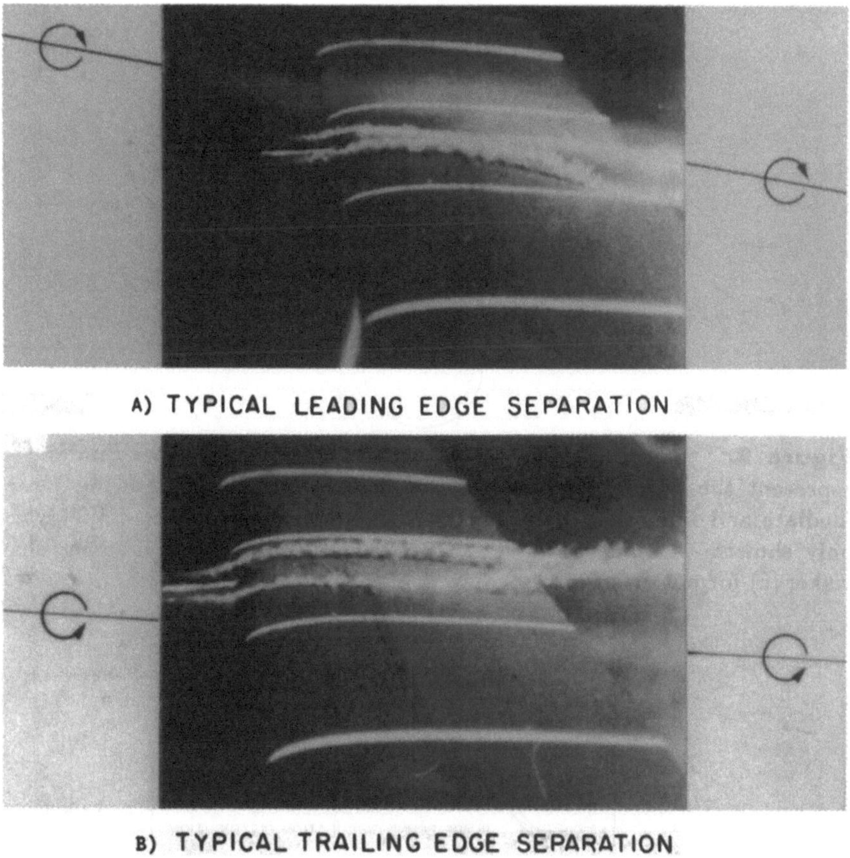

Figure 7. Typical leading (a) and trailing (b) edge separation occurring as a result of close blade vortex interaction [13].

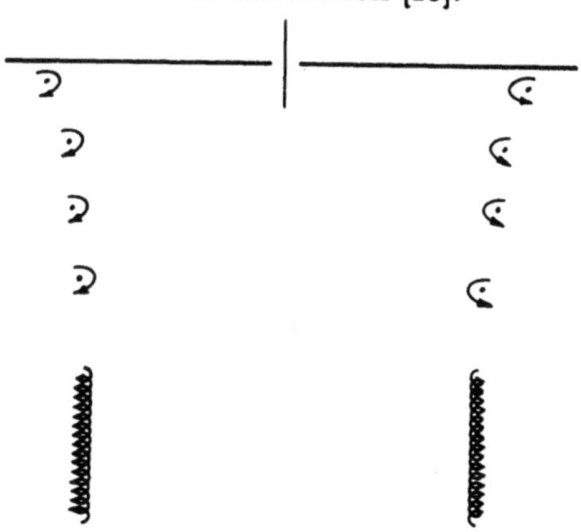

Figure 8. Geometry of model using line vortices and vortex sheets to represent the wake.

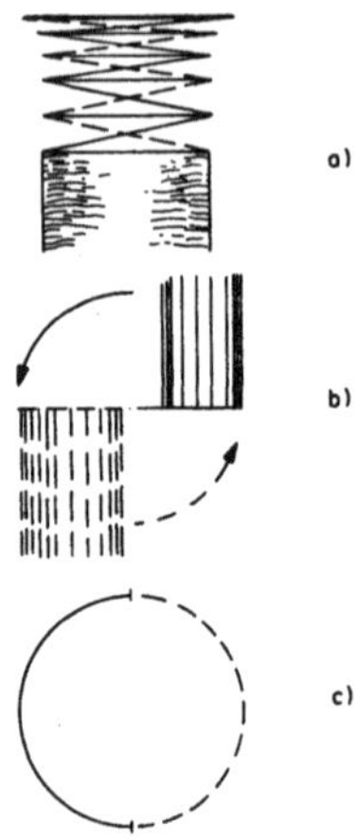

Figure 9. Geometry of model using vortex rings and cylinders to represent the wake. (a) side view of rotor wake model showing intermediate and far wakes formed from vortex spiral—2 blades. Tip vortex only shown: — Blade one, - - - Blade two; (b) plan view showing near wake; (c) formation of intermediate wake.

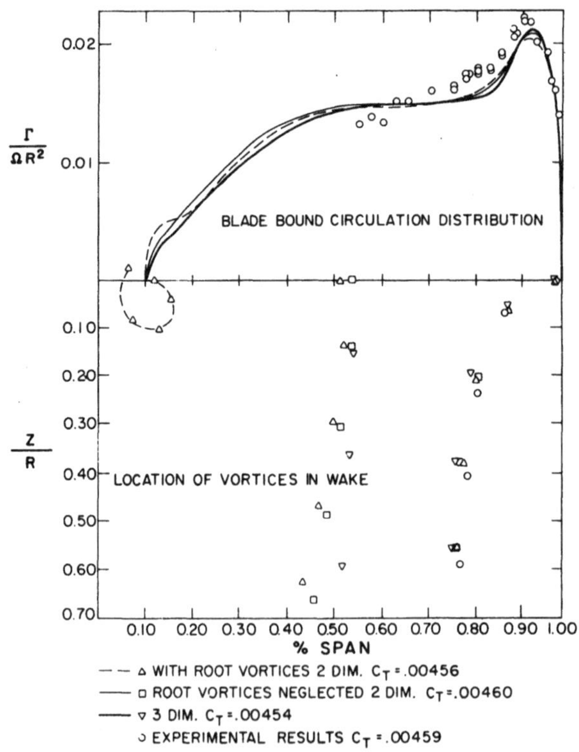

Figure 10. Blade bound circulation distribution and location of vortices in wake for two bladed rotor of [38].

convenient computational technique to account for the effects of the inboard trailing vorticity rather than as an exact prediction of the wake structure.

As expected, the center vortex descends at a more rapid rate than the tip vortex. Of particular interest is the root vortex, whose position is strongly influenced by the upflow at the center of the rotor frequently observed on hovering helicopters. The migration of this root vortex (with its attendant far wake) through the blade, causes computational difficulties resulting in slow and uncertain convergence. And yet its influence upon either the blade loading or rotor performance is negligible, as shown in Figure 10. Furthermore, it is probable that this vortex does not exist in practice in the form postulated, since it trails into a region where the induced velocities, and hence the flow, are highly disturbed by the hub and root fittings. The resultant mixing and diffusion of the root vortices from all blades make their true contributions to the wake velocities uncertain. Since the effects of the root vortex on the solutions are in any case negligible, its strength has been set equal to zero for the other solutions discussed here.

Evidently the bound circulation on the blade is heavily influenced by wake contraction. Neither the inflow, nor the circulation distributions are close to the uniform values assumed in classical vortex theories for the *ideally twisted* rotor blade. The theoretical model described here should be sufficiently simple computationally to permit heuristic and possibly formal techniques to be used for rotor performance optimization. Reference 38 contains a more detailed discussion of the theoretical development as well as the program description and listing for both wake models. Additional results are also given for a blade with a very highly tapered tip section (the ogee blade).

When this method is extended to four-bladed rotors, problems similar to those encountered in forward flight as discussed above occur. Unfortunately very little information is available on the airload distributions in hovering flight for four-bladed rotors and none with simultaneous measurement of wake geometries similar to the two-bladed results of Reference 4. Figure 11 shows a comparison between the analytical and flight test results for the four-bladed rotor of Reference 45. Best agreement was obtained when vortex bursting was assumed. The close proximity of the vortex to the blade at first encounter suggests that such breakdown may well occur. In order to allow for vortex bursting, an approximation suggested in Reference 46 was used where the velocity induced by the vortex is reduced by a factor $1/[1 + (\rho/a)^2]$, where a is the distance of the vortex from the blade and ρ is a factor representing vortex core growth. From References 4 and 25, vortex core sizes before breakdown of the order of 0.06 of the blade chord may be expected. Figure 11 indicates that reasonable agreement with the test data resulted only after the core size was increased by a factor of 10, to 0.6 of the blade chord. However, Figure 5 indicates core growths at encounter of the order of only three times the initial value, which suggests that some other phenomenon in addition to the vortex breakdown may be occurring, for example, the flow separation at close encounter discussed above (Figure 7). Also the more complete lifting surface representation for the blade in place of a lifting line with local lifting surface correction may be required. Finally, the assumptions implicit in the velocity averaging procedure discussed above for determining the wake displacements require further investigation, particularly in the near wake where the velocity variations depend on the nature of the roll-up process. These points are discussed further below.

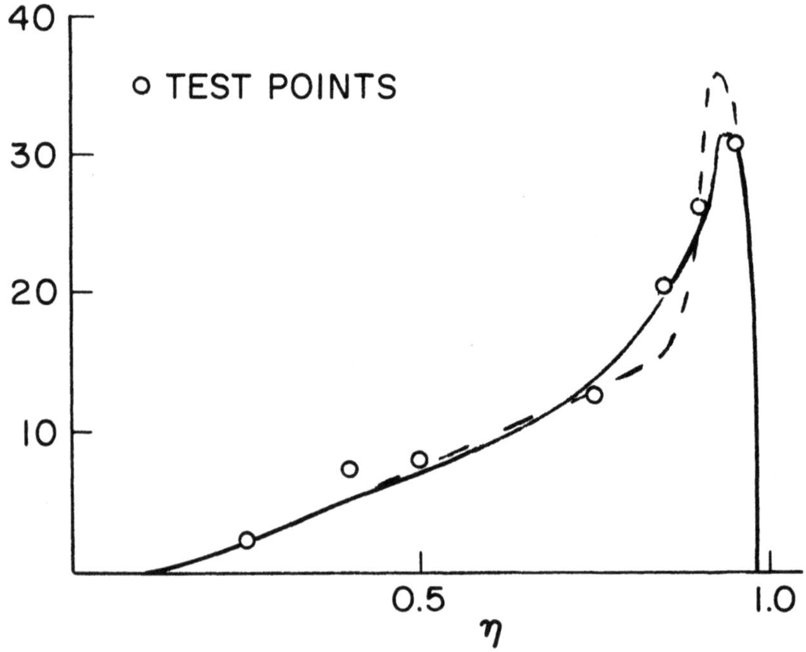

Figure 11. Comparison of computed and measured airloads for four bladed rotor of [45]. — free wake lifting surface solution with vortex breakdown; - - - without vortex breakdown.

2.3 Modelling Vortex Roll-Up in the Near Wake

As mentioned above the wake displacements in the near wake are determined by taking the average of the velocities at the vortices, specifically

1. the velocities in the wake immediately behind the blade in question at the computed spanwise location of its rolled-up trailing vortices, and
2. the velocities in the wake immediately behind the following blade (thus including the contribution from its trailed rolled up vortices) at the locations of the displaced rolled-up vortices from the first blade.

While this approach has the merit of consistency and results in good agreement with test data, both as regards geometry and bound circulation distribution, there is some question as to how well it models the near wake displacements during the roll-up process and before encounter with the following blade. For example, an alternative approach would be to compute the velocities in the wake immediately ahead of, rather than behind, the following blade neglecting the contribution in that region of the vortices trailed by the following blade. Figure 12 shows geometries and circulation computed using such an assumption for velocity averaging. The thrust coefficient is lower and both the circulation distribution and wake displacement at first encounter do not agree with the observed results. Evidently the original assumption of velocity averaging after blade passage results in a better representation of the complex roll-up process occurring in the near wake, but has the disadvantage of introducing a degree of empiricism in an otherwise rigorous analytical model.

Recent interest in the formation of vortices from fixed wing aircraft has stimulated the development of computation techniques for predicting the geometries of rolled-up three dimensional vortex sheets. In these techniques the sheet is treated as a collection of line or point vortices and their motions tracked as they distort under their mutual interference velocities using various computational methods. Some of the earliest work (Westwater, 1935) is quoted in Reference 3, pp. 589–590. Subsequent efforts to duplicate these results with a finer grid resulted in chaotic motions, particularly in the tighter portions of the spiral. Introduction of artificial viscosity resulted in more ordered solutions, but unfortunately, dependent on the degree of viscosity introduced.

Most recent efforts have concentrated on the Euler-Lagrange solutions in which a series of point vortices are tracked in a Lagrangian frame of reference and referred back to an Eulerian frame for solution of the equations of flow. This *cloud in cell* technique was applied to the roll-up of vortex sheets from a wing in Reference 2, and more recently in Reference 44. In Reference 48 the two dimensional approximation to the rotor wake developed in Reference 39 was applied to the computation of the wake roll-up in hovering flight. A typical solution is shown in Figure 13 which clearly indicates the roll-up process. Time did not permit completing the study to a converged solution. This work was continued using the three-dimensional model of Reference 39 which allows for curvature in the wake. Preliminary work concentrated on determining the effects of cell size as one measure of the degree of artificial viscosity introduced in the solution. Dependence of the solution on the assumed cell size remains a matter of concern.

In Reference 40 a further approach to the problem of a curved vortex sheet roll-up was attempted in which the wake was modeled as a curved series of vortex filaments and their roll-up predicted using the Biot-Savart relationship. Referring to Figure 14, it is necessary to compute interference velocities between any two vortex filaments everywhere in the near wake and integrate these velocities to obtain displacement. If it is desired to compute the velocity induced at any point A on one vortex filament by another vortex filament B, then it is clear that the velocity at A due to B would be induced primarily by that portion of B closest to A. The rest of the spiral vortex filament B may therefore be approximated by a vortex ring and the velocities computed readily by using the Biot-Savart relationships and logarithmic series solutions for the resulting elliptic integrals (Miller, 1981). The displacement of any vortex may be computed by integrating the total induced velocity on a vortex due to contributions from all other vortices in the near wake over an increment of time represented by a small change in azimuth, starting from the blade in question, and using standard techniques of integration such as fourth order Runge/Kutta formulas. With this simplified model it is possible to set up a much finer wake structure, thereby presumably achieving a more realistic wake roll-up.

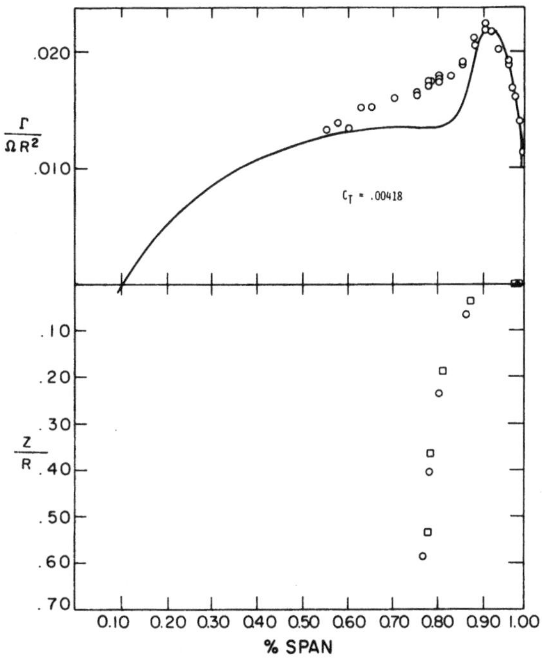

Figure 12. Blade bound circulation distribution and location of vortices in wake for rotor of Figure 10, but with effect of following blade trailed vortex neglected.

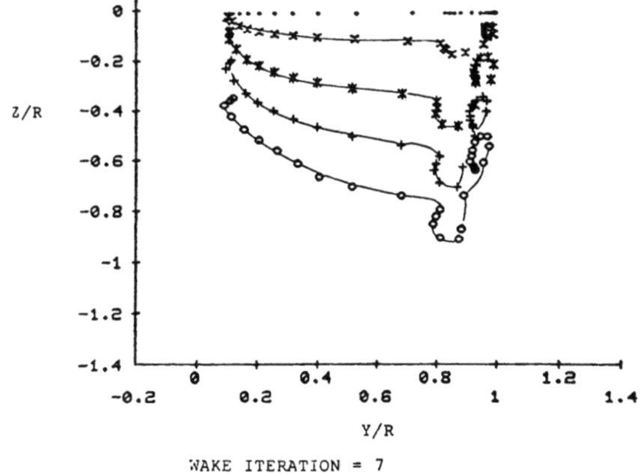

Figure 13. Wake displacements obtained from Euler/Lagrange solutions of [48].

This technique was first used to examine the roll-up of the curved vortex generated from the tip of the blade with the bound circulation distribution of Figure 10. A typical result is shown in Figure 15. Nineteen filaments were generated from the outer six percent of the blade with an assumed core size of 1% of blade span. Most of the vorticity is contained in the numbered vortex filaments and in particular the first six, as is evident from the table showing the strength of each vortex filament, G, where G is equal to $(1/\Omega^2 R)(d\Gamma/dr)(dr)$. The bound circulation, $\Gamma/\Omega^2 R$, has a maximum value of .02 at 94% span. It may be deduced from Figure 15 that the vortex rolls up very rapidly, first rising and then descending. Roll-up apparently occurs a few chord lengths behind the rotor blade, as may be expected from the experimental evidence.

The vertical displacement of what is apparently the vortex core is characteristic of a curved vortex sheet and not a straight sheet such as would be generated by one half span of a wing. In the latter case, it is well known that no vertical displacement of the centroid of vorticity will occur if the effects of blade-bound vorticity and of the opposite half wing are neglected. The difference between the roll-up of a straight and a curved series of vortex filaments will be evident from the following simple analysis.

Consider two infinite vortex filaments of strength Γ_1 and Γ_2 located a distance Δr from each other as shown in Figure 16a and with mutually induced velocities v_1 and v_2 given by

$$v_1 = \frac{\Gamma_2}{2\pi\Delta r} \qquad v_2 = -\frac{\Gamma_1}{2\pi\Delta r}$$

The vertical displacement of each vortex filament over a small increment of time Δt will be

$$x_1 = v_1\Delta t \qquad x_2 = v_2\Delta t$$

and the vertical centroid of the displaced vortices after time Δt will be

$$\begin{aligned}\bar{x} &= \frac{\Gamma_1 x_1 + \Gamma_2 x_2}{\Gamma_1 + \Gamma_2}\Delta t \\ &= 0.\end{aligned}$$

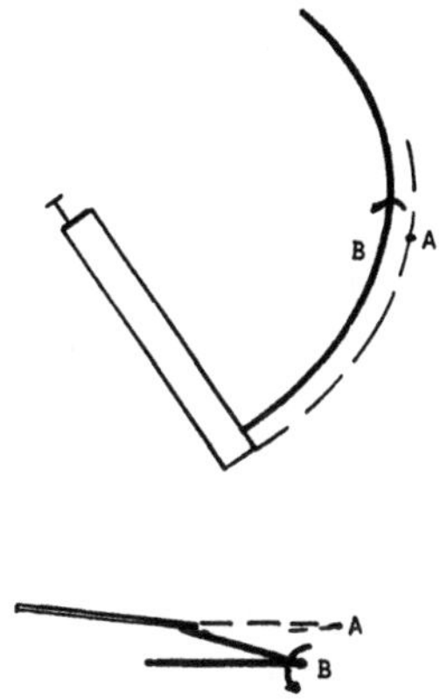

Figure 14. Geometry of near wake roll up model.

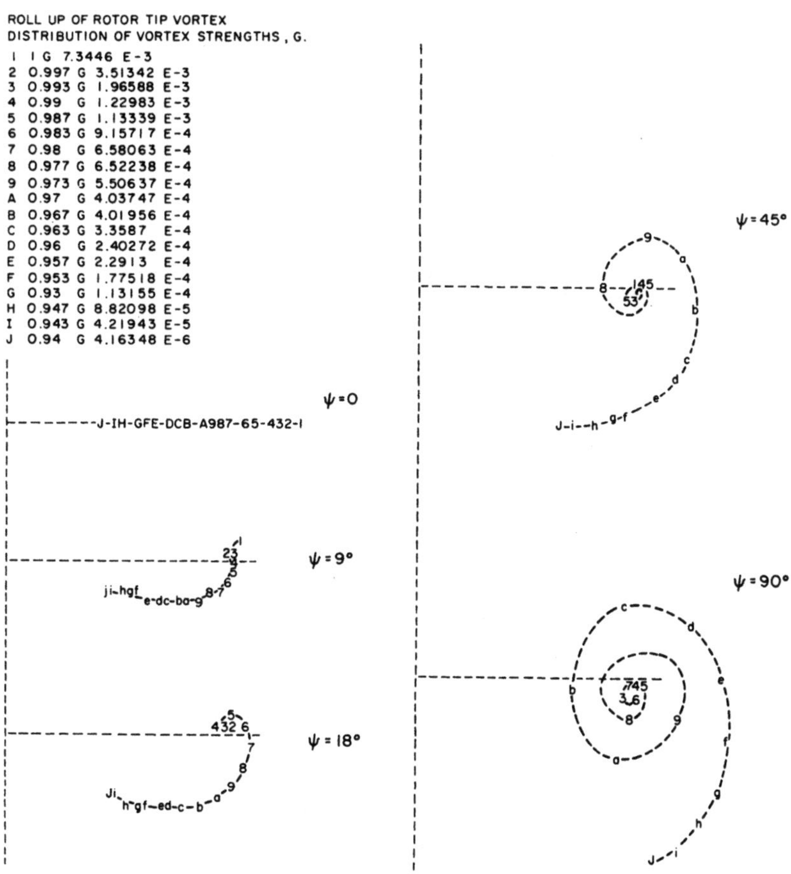

Figure 15. Predicted roll-up of tip vortex in near wake.

Evidently if the vorticity postulated above is generated by one wing halfspan then there will be a sheet of opposite vorticity from the other half span. For an elliptic load distribution the centroid of vorticity occurs at a distance $\pi/4$ of the half span from the plane of symmetry and hence the eventual steady state velocity of descent of each rolled up vortex will be $\Gamma/\pi^2 R$ where R is the semispan.

We are concerned here, however, with the initial roll-up mechanism and neither the effects of bound circulation, nor the effects of the far field are pertinent to the present discussion. Consider two curved vortices whose curvature is such that they may be approximated by two ring vortices a distance $\Delta r = r - \eta$ apart as shown in Figure 16b. The velocity induced at a point on a vortex ring of radius η by a vortex ring of radius r is

$$v = \frac{\Gamma}{4\pi R}\int_0^{2\pi} \frac{r(r-\eta\cos\phi)\,d\phi}{(r^2+\eta^2-2r\eta\cos\phi)^{3/2}}$$

Evaluating the integrals (Reference 38) gives

$$v = \frac{\Gamma}{4\pi R}\sqrt{\frac{k^2}{r\eta}}\,\{K - E[1-.5k^2(1+r/\eta)]/(1-k^2)\}$$

where

$$k^2 = \frac{4R\eta}{(r+\eta)^2}$$

and

$$1-k^2 = \left[\frac{r-\eta}{r+\eta}\right]^2$$

K and E are elliptic integrals of the first and second kind.

Substituting for k^2

$$v = \frac{\Gamma}{2\pi R}\cdot\frac{1}{r+\eta}\left[K - E\left(\frac{r+\eta}{r-\eta}\right)\right]$$

For the two vortex rings, $r_1 = \eta_2 = r$ and $\eta_1 = r_2 = \eta$. Since k^2 is thus the same for both rings, their vertical centroid after time

$\Delta t = \Delta\psi/\Omega$ will be

$$\bar{x} = \frac{\Gamma_1 v_1 + \Gamma_2 v_2}{\Gamma_1 + \Gamma_2}\Delta t$$
$$= \frac{1}{2\pi R} \cdot \frac{\Gamma_1 \Gamma_2}{\Gamma_1 + \Gamma_2} \cdot \frac{2K}{r+\eta} \cdot \Delta t$$

If Δr is small so that $r \to \eta$ then $K \to \ln[4(\frac{r+\eta}{r-\eta})]$, $r > \eta$. For a rotor turning at angular velocity Ω, $\Delta t = \Delta\psi/\Omega$ and the vertical displacement of the centroid of vorticity at $\Delta\psi = \pi$ will then be

$$\frac{\bar{x}}{R} = \frac{1}{2}\frac{\Gamma}{\Omega R^2} \cdot \frac{1}{r+\eta} \ln\left[4\left(\frac{r+\eta}{r-\eta}\right)\right]$$

Typically the peak value of circulation on the rotor is of the order $\frac{\Gamma}{\Omega R^2} = .02$ and if this circulation is assumed to drop off over the outer 6 percent of the blade in two vortex filaments at $r = 1$ and $r = .97$, each of strength $\Gamma/\Omega R^2 = .01$, it follows that $x/R = .014$. For the complete wing discussed above the equivalent value would be

$$\frac{\bar{x}}{R} = \frac{\Gamma}{\pi^2 R} \cdot \frac{\pi}{\Omega R} = .006.$$

This displacement, of which only the initial motion has been described above, will continue with contraction and expansion of the ring radii due to horizonal components of the induced velocities, eventually resulting in the classical mutual threading of vortex rings as each passes through the other in succession (Lamb, 1945).

The effect of core size on the computed displacement of the centroid of vorticity for the case of Figure 15 is shown in Figure 17. Evidently the solution is sensitive to the assumed core size. When the core size approaches zero, the results tend to become chaotic near the centroid of the spiral, as previous investigators have found, although the centroid of vorticity appears to descend in an ordered fashion.

The wake displacements due to the bound vorticity of one blade will, to first order, be cancelled by that of the following blade, resulting in no net displacement of the vortex filaments, whether curved or straight, due to bound circulation. The second order effects due to the variation of bound circulation strength along the blade have been shown in Reference 40 to be negligible. Consequently neither the effects of the bound circulation nor of the self-induced

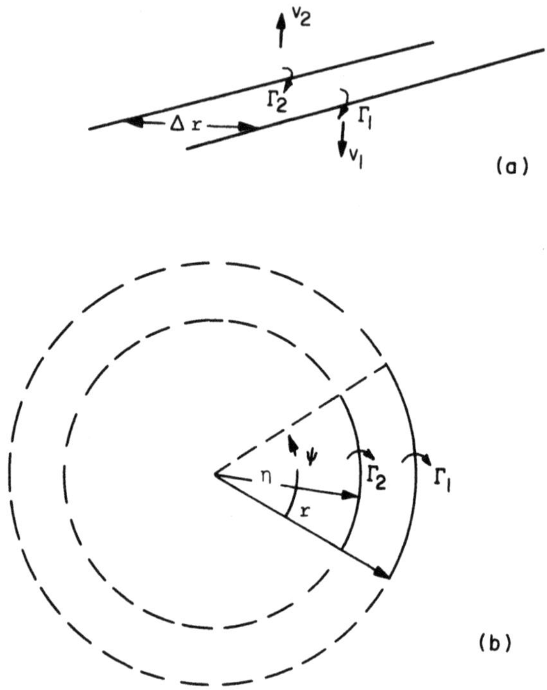

Figure 16. (a) Geometry for a pair of infinite vortex filaments; (b) Geometry for a pair of ring vortices.

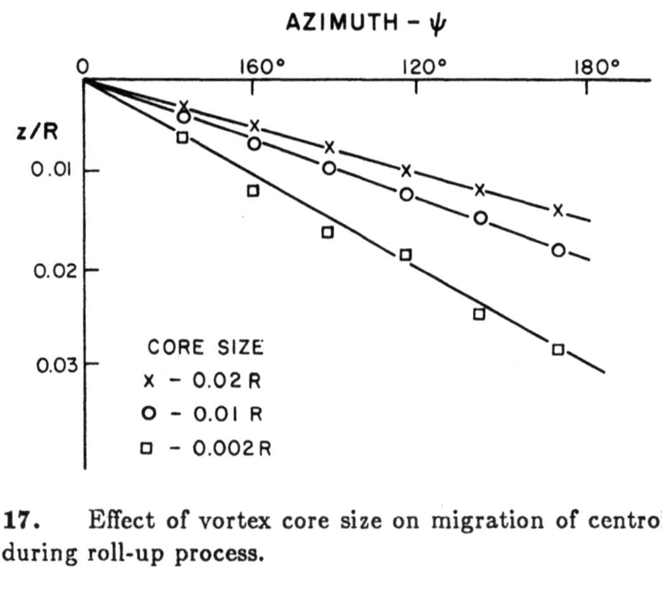

Figure 17. Effect of vortex core size on migration of centroid of tip vortices during roll-up process.

velocities (Lamb, 1945) of the vortex ring have been included in the results of Figure 17, in order to clarify the more important effects due to the mutually induced velocities of the free vorticities.

For a two bladed rotor Figure 10 indicates that the tip vortex at first encounter is located approximately 5% to 6% of the span below the following blade, as shown both analytically (Miller, 1981) and experimentally (Johnson, 1980). Consequently, the differences in the vertical displacement of the centroid of the rolled-up vortex with various assumptions as to core size, of the order of 2% of the blade span at first encounter, may not appear to be crucial. However blade airloads are sensitive to the location of the tip vortex at this first encounter, and the dependency of the solution on assumptions as to core size remains of some concern. Experimental evidence (Figure 5) indicates vortex core sizes of the order of 1% of the span, but these measurements are presumably of the rolled-up vortex core, which may not necessarily be the core size required to model the sheet as it leaves the blade. A great deal more analytical and experimental investigation is necessary before this problem may be completely resolved. It is possible that simplified forms of the Navier-Stokes equations for spiraling vortices will have to be developed in order to obtain a better understanding of the phenomenon of roll-up and migration of the tip vortex generated by a rotating blade as, for example, in Reference 29.

In a first attempt to extend this investigation to the complete wake, a solution was obtained using 24 curved vortex filaments for the near wake. A converged solution was first obtained using the fast free wake techniques of Reference 39, where the near wake is assumed to roll up almost instantly. The resulting velocities at the blade induced by this intermediate and far wake were then added to those computed from the 29 filament representation of the curved near wake. The results, shown in Figure 18, agree well with the basic characteristics evident from the test data of References 12 and 26. An intense tip vortex rolls up containing most of the bound circulation from its peak to the tip. This tip vortex is followed immediately inboard by what appears to be a quiescent area. A small secondary vortex of opposite sign then rolls up from the vortex filaments trailed as the bound circulation starts decreasing inboard from its peak value. This secondary vortex is well below the tip vortex and displaced inboard of its originating trailed vorticity. The next step in this investigation will involve recomputing the bound circulation with this new more detailed near wake geometry and reiterating to a converged solution.

2.4 Lifting Surface Solutions

Because of the proximity of the blade to the vortex at first encounter, it is logical to consider the use of a lifting surface rather than a lifting line representation of the blade. In Reference 5 it is shown that, for the case of a perpendicular intersection between vortex and blade, the Weissinger approximation to a lifting surface solution gives almost exact agreement with a more complex five panel solution. In this approach the control point is placed at the 3/4 chord location, and the velocity is determined from the vortex system consisting of a bound vortex at the 1/4 chord position and associated trailing vortices. Since the simplified free wake analysis described above uses an iterative technique, it is readily adaptable to the Weissinger approximation without appreciable increase in computer time. The results are compared in Figure 19 with the lifting line solution. It is evident that the circulation peaks are somewhat smoothed out as expected (Van Dyke, 1964) and C_T slightly lower, although the wake displacement at first encounter remains essentially the same.

To explore this point further, solutions were obtained for the multi-bladed rotors for which experimental results were presented in Reference 26. Figure 20 shows comparisons for these rotors between test data and both the lifting surface and lifting line solutions. The same core size of .03 of the blade radius as postulated above was used.

2.5 Effect of Stall

Rotors and wind turbines frequently operate with extensive areas of stall over the rotor disc, but little is known about the characteristics of the blade in the stalled regions when operating in a strong centrifugal force field. For the rotor the most serious problem is stall flutter on the retreating side which arises from the characteristic hysterisis loop in section lift as the blade moves in and out of stall (McCroskey et al., 1981). This loop is due to the tendency of the flow to separate at a higher angle with increasing angle of attack than for reattachment as the angle of attack decreases. Energy is

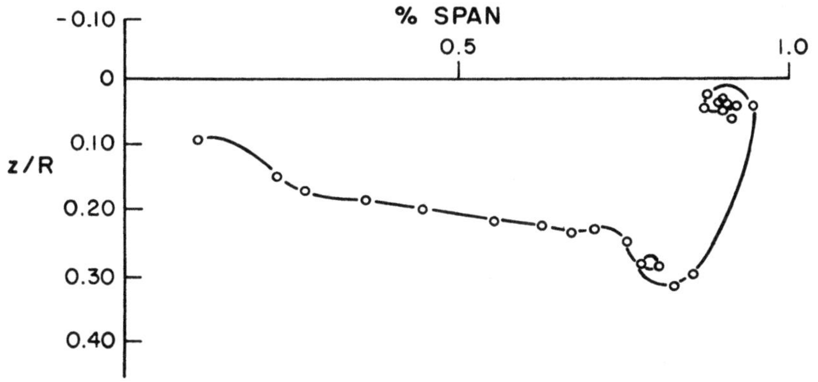

Figure 18. Roll-up of complete wake at $\psi = 180°$. Assumed core size of .01R.

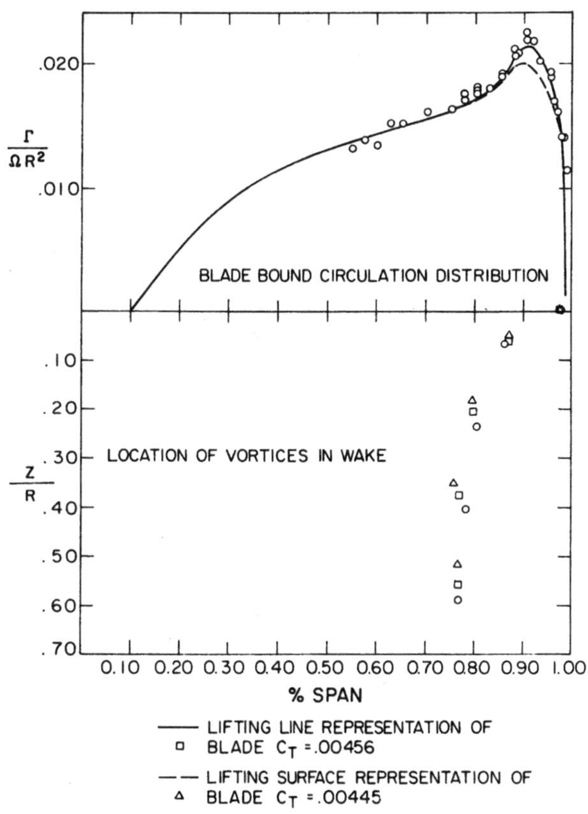

Figure 19. Blade bound circulation distribution and location of vortices in wake for two bladed rotor of [20].

thus fed into the blade, with the blade responding at its torsional frequency with increasing, and then decreasing, amplitude on the retreating side but remaining quiescent on the advancing side. The effect is aggravated by the tendency towards negative pitch damping in the reverse flow region since the aerodynamic center is then at 75% chord (rear neutral point).

The effects of stall flutter are proportional in severity to the amount of the rotor disc area effected by stall. Stall flutter is believed to be a source of high pitch link loads and vibration which frequently limit high speed performance. Increased torsional stiffness simply increases the frequency of the phenomenon, however pitch damping is highly effective in controlling the oscillation, as is individual blade control (Ham, 1983).

Stall obviously also affects performance directly due to increased drag. However the dynamic pressures in the stalled areas are low, and the penalty is not severe for helicopter rotors. Of more interest is the effect of stall on the performance of wind turbines, since these appear to operate unexpectedly well in regimes where extensive stall might be expected, and which could represent an appreciable portion of their duty cycle.

Figure 21 shows the measured performance of the Mod-O wind turbine (Viterna and Janetzke, 1981) compared to computed results using the free wake analytical technique (Miller, 1981) and the assumed post stall variations in lift and drag which best fit the observed results. Data on the behavior of airfoils well into stall and including rotational effects is not available. Experimental results for a 0012 airfoil (without rotational effects) are shown for comparison. Evidently, although generally similar, there are appreciable differences in the estimated behavior of the airfoil in the post stall region, which could be attributed either to the effects of the centrifugal force field on the separated flow or to the time varying angle of attack due to the earth's boundary layer in which the wind turbine operates. This could result in dynamic stall effects causing separation to occur at a higher angle of attack than in the static case.

2.6 Modelling the Turbulent Wake/Vortex Ring Condition

Also of interest in Figure 21 is the ability of the free wake analysis to predict performance in the vortex ring condition. This

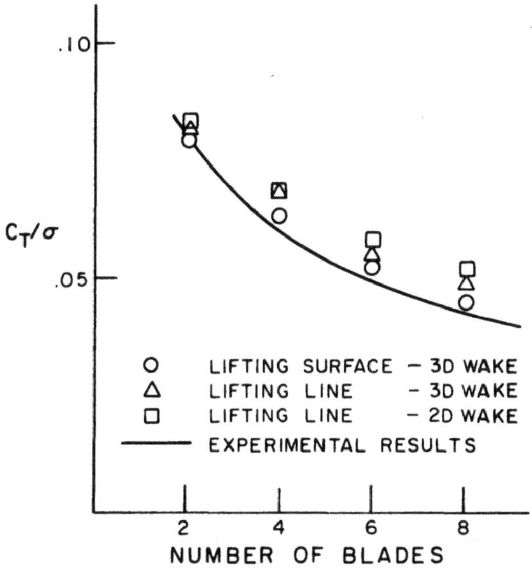

Figure 20. Effect of blade and wake modelling with increasing number of blades and comparison with experimental results of [26]. ($\theta_{75} = 8°$)

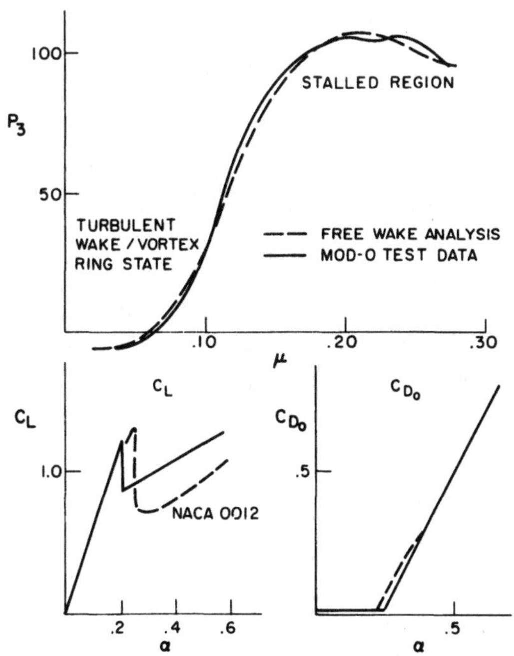

Figure 21. Comparison between measured performance of Mod 0 Wind Turbine and free wake analysis with lift and drag schedules as shown. α measured from zero lift angle. P_3 is electric power output in *KW* [52].

condition occurs when the velocity in the fully developed wake exceeds the free stream velocity, presumably causing a back flow. In helicopter operations, this flight regime is usually avoided because of the resulting severe degradation in handling qualities. The reasons for this degradation are not clear, but have been attributed to the formation of a standing vortex ring in the wake which breaks away periodically from the rotor. However it has been observed that the problem appears to be less severe with configurations in which the fuselage is more fully immersed in the rotor wake (tandem, coaxial, synchropter) indicating the possibility of interaction between the wake and the fuselage as the cause of the poor handling qualities in this flight regime.

In the case of the wind turbine however, experimental evidence indicates that steady operation in the vortex ring condition is possible, an unexpected result. This condition, indicated on Figure 21, was examined using the fast free wake techniques of Reference 38. Evidently operation in the turbulent wake/vortex ring condition is well predicted. Figure 22 shows the corresponding wake geometries and blade load distributions.

Thrust measurements would be a better test of the adequacy of the free wake analytical technique in the turbulent wake/vortex ring condition since uncertainties as regards the drag coefficient would thereby be eliminated. Unfortunately thrust data on the Mod-O wind turbine were not available. However, in Reference 7, test results were obtained for a model wind turbine rotor operating under various conditions, including well into the turbulent wake/vortex ring state. Figure 23 compares the analytical results with the experimental data. It is also interesting to observe the close agreement of the momentum theory when this theory is suitably modified by the experimentally determined corrections suggested in Reference 11.

To summarize the above discussions, the following lists some of the extensions to existing theory and additional experimental data required as an aid to further development of rotor aerodynamic design techniques:

- Definition of vortex core size at first encounter.
- Time history of vortex roll-up.
- Vorticity distribution outside vortex core.
- Number of vortex formations in inboard wake.

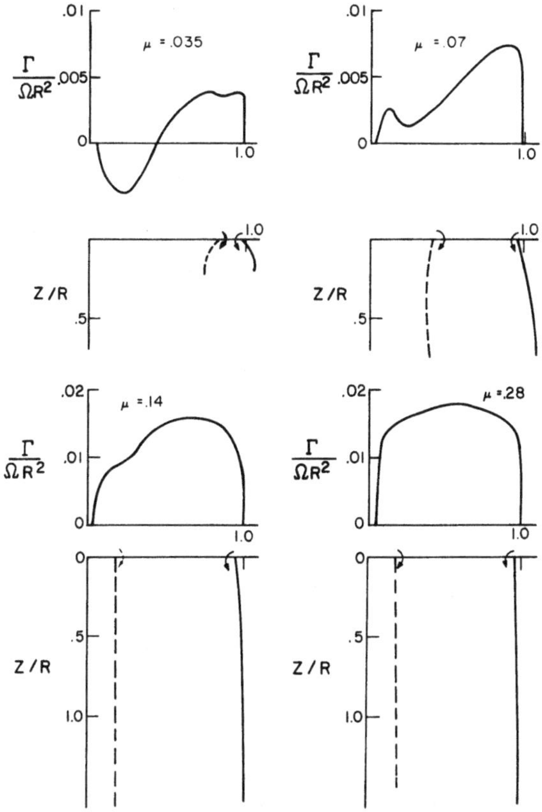

Figure 22. Selected wake geometries and blade loadings from free wake analysis of Mod 0 Wind Turbine.

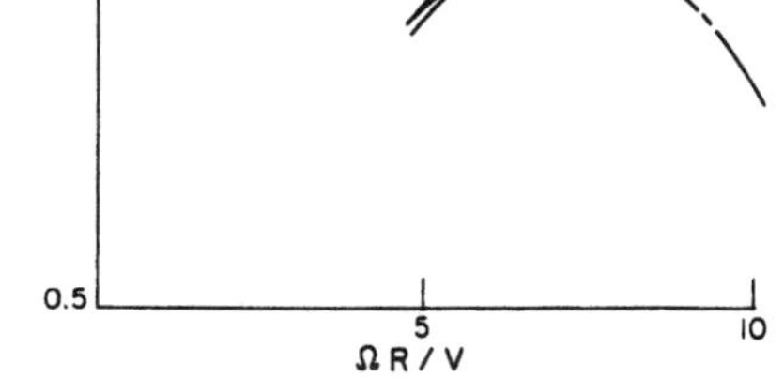

Figure 23. Wind turbine performance predictions in turbulent wake/ vortex ring state and comparison with experimental results of [7]. Courtesy U.S. Dept. of Energy, Office of Scientific and Technical Information, Oak Ridge, TN.

- Formation of tip vortex at blade (tip losses).
- Viscous flow effects on bound circulation due to vortex encounters.
- Experimental data on blade circulation for four and more bladed rotors.
- Experimental data on post stall behavior of airfoils operating in a centrifugal force field.

3. Unsteady Aerodynamic Effects and Vibration

3.1 Source of Vibration

In forward flight the time varying, essentially harmonic, airloads are one of the primary causes of vibration. Discussion of this problem may be simplified by considering the three basic factors which contribute to this vibration as separable, although heavily interacting, elements. These three elements are: (1) rotor aerodynamic loading, (2) blade and rotor dynamics, and (3) fuselage dynamics.

It is possible to show these elements and their interaction schematically, as in Figure 24. The interaction between the elements determines the degree of complexity required for numerical solutions. Interactions shown by broken lines are not believed to be of primary importance. At the lower advance ratios considered above (Figures 1–4) blade flapping and the first harmonic variation in speed at the rotor disc have little direct influence on the higher harmonic airloading. The effect of these conclusions is largely to uncouple the elements in the schematic of dynamics and hence to simplify the analysis, since each element can now be investigated separately.

The primary element is the steady rotor lift and bound circulation on the rotor blades. This bound circulation, as it leaves the blades, generates a spiral vortex system in the wake of constant strength dependent only on mean rotor thrust. The vertical component of induced velocity generated by this vortex system at a point on the blade, when combined with the horizontal velocity at the blade due to the blade rotation and forward speed, determines the induced angle. In forward flight the wake spiral is not symmetrically located below the rotor, but is distorted by the forward velocity as shown in Figure 25. The induced velocity due to this distorted wake may be computed in terms of the wake geometry by the expression given in the Reference 36. As might be expected, it is far from uniform over the disc. Consequently the blade is subjected to a constantly varying induced angle as it rotates and this is the primary source of the higher harmonic blade loading.

3.2 Modelling Unsteady Aerodynamic Effects

Since the blade is subjected to time varying air loads caused by the wake generated by the steady state lift, additional elements of vorticity must exist in the wake besides those due to the steady state lift. In particular, for example, the air load varying as the nth harmonic of rotor speed must add a variable strength to the trailing wake system which in turn will induce all harmonics of downwash at the rotor disc. Since the blade circulation is changing there must also be a vortex system shed from the trailing edge of the blade at any instant equal and opposite in magnitude to the change in bound circulation and this shed wake will also induce all harmonics of downwash at the rotor disc (Figure 25). Consequently interharmonic coupling is potentially important. For example, if the nth harmonic induced large steady state components of downwash, then the steady state lift would be appreciably changed by the harmonic airload.

The numerical solutions of the forward flight case reported in (Miller, 1964) were undertaken to explore the importance of this interharmonic coupling. It was found that the results of the analysis could be interpreted more easily by considering the wake in two parts, the *near* wake representing that portion in the immediate vicinity of the blade in question and the *far* wake consisting of that portion from a quarter of the disc away from the blade and extending to infinity down the spiral. The exact boundaries of the near and far wake are not of primary importance.

The near wake is relatively undistorted since it includes only a small portion of the spiral and could be closely approximated by a straight wake extending aft of the blade to infinity. It must therefore induce primarily the frequency of the bound circulation by which it was generated. The far wake induces all harmonics at the rotor disc due to its distorted spiral form, but it is swept further downstream and hence becomes of decreasing importance as the advance ratio increases. Consequently the contribution of a particular harmonic

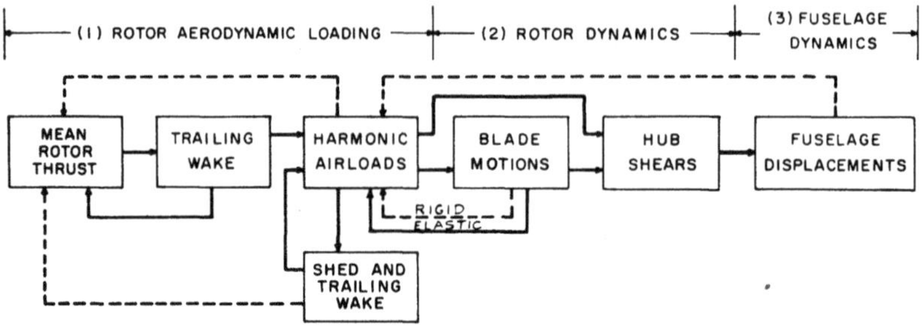

Figure 24. Block diagram of elements contributing to helicopter vibration showing interactions.

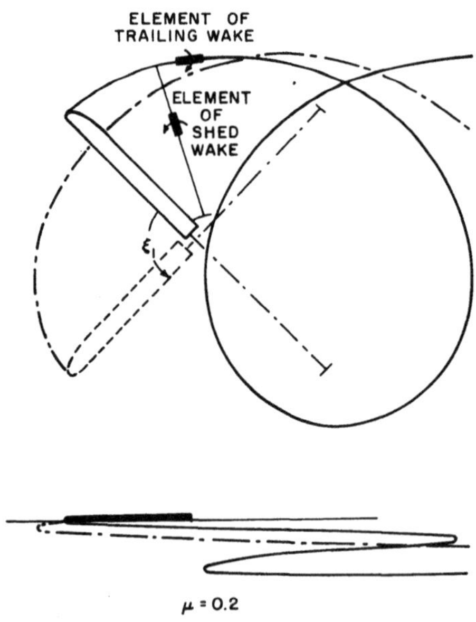

Figure 25. Wake geometry showing trailing tip vortex and element of shed wake.

of circulation, say the nth, to another harmonic, say the mth, of downwash at the blade usually is of an order of magnitude less than its contribution to the nth harmonic of downwash. Also the steady state or 0th harmonic of circulation which determines the rotor thrust is an order of magnitude greater than any other harmonic of air load and so is the downwash which its wake induces at the rotor disc. At the lower advance ratios, the nth harmonic of downwash at the blade thus consists primarily of two components:

1. The nth harmonic of downwash induced at the blade by the vorticity in the far spiralling wake generated by the steady state blade lift of 0th harmonic and

2. the nth harmonic of downwash induced by the vorticity in the wake generated by the nth harmonic variation in circulation.

The lift changes due to (1) alone may be considered as a quasi-static lift change arising from a harmonic variation in angle of attack due to the nonuniform wake. This harmonic variation may be computed once the rotor and wake geometry are known, and to first order is unaffected by any other blade motions or harmonic lift variations.

The downwash components (2) is an unsteady aerodynamic effect which results in a reduction and a phase shift of the quasi-static lift by an amount which again depends only on the rotor and wake geometry. This effect can therefore be tabulated as a generalized lift deficiency function $C(k)$ as has been done for a two-dimensional airfoil in, for example, Reference 50. In Theodorsen's nomenclature

$$C(k) = F + iG$$

where F represents the reduction in lift and $\tan^{-1} G/F$ represents the phase shift.

Once the change in blade circulation has been defined as being oscillatory in nature, the strength of the shed vortex and its position at any instant relative to the blade can be defined for any three-dimensional system such as a rotor in forward flight. Consequently the instantaneous velocity which the wake induces at any point on a blade can be computed. Similarly the strength of the trailing vortex at any point in the wake is defined and its induced velocity field established. The expressions for these induced velocities are given in Reference 36.

Computation of the air loads is complicated by the existence of singularities in the solution. These occur as the shed wake approaches the trailing edge of the rotor and whenever the blade passes through a trailing vortex line generated by itself or another blade. The treatment of the singularities and of the nonuniform flow field requires the use of lifting surface theory and hence the numerical evaluation of the downwash at several chordwise as well as spanwise stations. Approximate methods have therefore been developed to evaluate the unsteady aerodynamic effects.

3.3 Unsteady Aerodynamics in Hovering Flight

Before considering these treatments it is interesting to examine the effects of unsteady aerodynamics for the hover case discussed in the previous section. If it is assumed that the rotor has an infinite number of blades and if wake contraction is neglected, the vortex spiral may be replaced by a column of vorticity and the vortex strength at any point in the column defined by the nature of the assumed change in bound circulation. This circulation change could, for example, come from periodic change in pitch of the blade. For oscillations at harmonics of the rotor speed, the unsteady aerodynamic effects become of paramount importance and could result in values of the lift deficiency function, F, close to zero. More usual values for conventionally loaded rotors in hover are of the order of 0.5 which in effect means that the slope of the lift curve of the blade for oscillatory loads is reduced by 50%, evidently not a negligible effect.

It was shown in Reference 36 that the lift deficiency function $C(k) = F + iG$ can then be obtained in closed form for a three-dimensional model as

$$F = \frac{1}{1 + (\sigma\pi/4\lambda_0)}, \qquad G = 0$$

where σ is the blade solidity and λ_0 the mean inflow through the rotor. This result agrees with the two-dimensional results of (Loewy, 1957) and almost exactly with the numerical solution using a finite number of blades (Figure 27), except that G then does have a small value.

The interesting conclusion from this analysis is the fact that the lift deficiency function is independent of frequency and can be given by a simple expression depending only on blade solidity and downwash.

As shown in Reference 36 the same results may be derived from consideration of simple momentum and blade element theory. From momentum theory, for uniform inflow and a superimposed periodic thrust change,

$$T_n e^{in\Omega t} = 2 \times \text{ mass flow through rotor } \times w_n e^{in\Omega t}$$

where w_n is the velocity change through the rotor disc due to T_n.

Considering mean velocity only, $w_0 = \lambda_0 \Omega R$, in determining the mass flow,

$$T_n = 2 \times \rho\pi R^2 \times \lambda_0 \Omega R \times \lambda_n \Omega R$$

or

$$C_{T_n} = 2\lambda_0\lambda_n$$

From blade element theory for uniform flow through the rotor and periodic change in pitch, $\theta_n = e^{in\Omega t}$,

$$C_{T_n} e^{in\Omega t} = \frac{\sigma\pi}{2}(\theta_n - \lambda_n) e^{in\Omega t}$$

and with $C_{T_{qn}} = (\sigma\pi/2)\theta_n$ where $C_{T_{qn}}$ is the quasi-static lift change due to θ_n, it follows that

$$\frac{C_{T_n}}{C_{T_{qn}}} = \frac{1}{(1 + \sigma\pi/4\lambda_0)}$$

as in the vortex theory.

3.4 Unsteady Aerodynamics in Forward Flight

The singularities normally encountered in the solution of the downwash integrals have been cancelled, for the hovering case, by the assumption of an infinite number of blades which permits the solution in closed form given above. In the forward flight case, as in hover, the existence of a finite number of blades is the dominant factor in determining rotor loads and it is therefore necessary to devise techniques capable of handling these singularities. The simplest way to do this computationally is to follow the procedure suggested in (Miller, 1964) which shows that a reasonable approximation is achieved by stopping integrations at the rear neutral point in a manner similar to the modified lifting line solution suggested by Weissinger for the steady case as discussed in Section 2.

Typical results are presented in Figures 26 and 27 for a shed wake of constant radial strength. In Figure 26 the contribution of all components of the wake to the lift deficiency is plotted as a function of advance ratio. The increasing importance of the near wake as the advance ratio increases is evident, as well as the relative unimportance of the trailing wake. In fact at an advance ratio of 0.2 the reduction in lift, F, is substantially that which would be predicted by the two-dimensional theory. Since the hovering flight case, $\mu = 0$, is subject to the simple closed form solution discussed above, it is possible that for many engineering applications a reasonable approximation to this very complex analysis for F and G could be obtained by fairing a simple curve through the two known points at $\mu = 0.2$ and $\mu = 0$.

The convergence of $C(k)$ for all harmonics towards the classical two-dimensional solution for the higher advance ratios is also shown in Figure 27, together with the insensitivity of this function to frequency at hovering, which is also a conclusion of the closed form solution given above.

A further unsteady aerodynamic effect arises in forward flight due to the variations in velocity over the airfoil as the blade rotates. The velocity perpendicular to the blade in the plane of rotation at a radial distance r is

$$U = \Omega r + \mu \Omega R \sin \psi$$

where μ is the advance ratio, R the blade radius, and ψ the azimuth angle. Such a pulsating flow will change the values of $C(k)$ as the blade rotates, increasing it at the advancing side ($\psi = 0$ to π) as the shed wake is stretched and the reduced frequency is reduced, and decreasing it on the retreating size as the shed wake piles up behind the blade. It has been shown in Reference 19 that a reasonable approximation for typical advance ratios is to compute the value of $C(k)$ based on the local value of U.

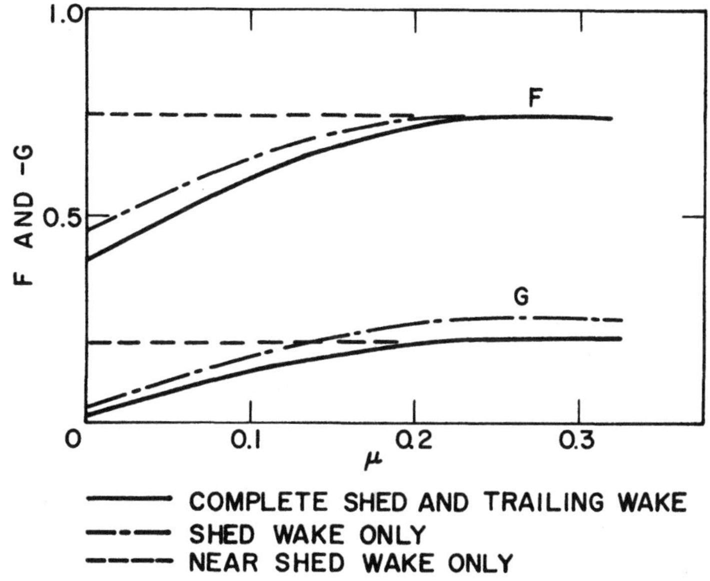

Figure 26. Effect of advance ratio, μ, on reduction in lift, F, and phase shift $\tan^{-1} G/F$, due to three-dimensional unsteady aerodynamic effects. Three blades: $\eta = 0.8$, $\lambda_0 = 0.05$. Third harmonic: $n = 3$.

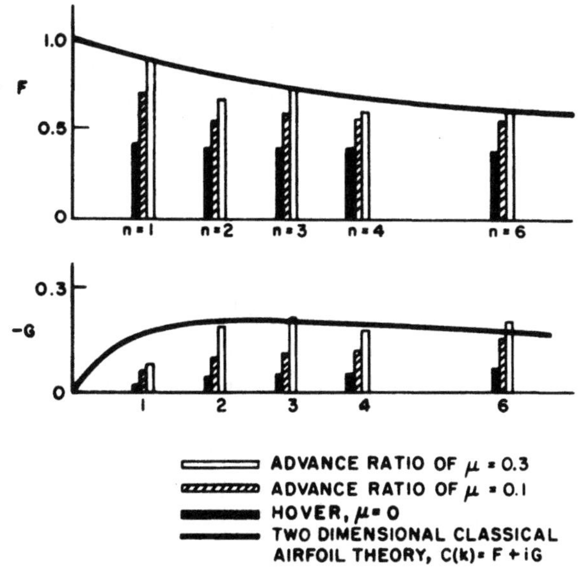

Figure 27. Three-dimensional lift deficiency F, and phase shift $\tan^{-1} G/F$ as a function of harmonic n. Conditions the same as in Figure 26.

3.5 Experimental Verification of Unsteady Aerodynamic Effects

It is of interest to compare the various solutions for the unsteady case with experimental results where available. Figure 28 shows the case of a wind turbine (Dugundji et al., 1978) operating in a variable inflow corresponding to the effects of the earth's boundary layer. This is a phenomenon occurring at rotor frequency, hence a low frequency phenomenon. For this case it appears that the simple momentum approach suggested above as expanded in Reference 43 gives results in reasonable agreement with a more complex solution in which the wake is represented by a series of vortex filaments and unsteady effects included as discussed above. In the case of a higher frequency phenomenon as, for example, a rotor passing through the shadow of a tower, the simple momentum theory is not adequate to describe the airload variations, as shown in Figure 29. Figure 30 more graphically displays the difference between the low and high frequency phenomena.

One of the earliest experimental evidences of the reduction in lift curve slope, F, and hence of the low damping associated with unsteady aerodynamic effects in hover was that reported in Reference 33 in which bending moments were measured on blades of varying stiffness. When the rotor was operated such that the blade first bending frequency was an integer of rotor speed, large increases in vibratory bending moments occurred. Typical results are shown in Figure 31. Similar results were recorded with stiffer blades up to eight per rev.

3.6 Vibration Control

Evidently the time varying airloads discussed above in forward flight will cause undesirable vibratory loads which may be amplified by blade and fuselage resonances. Much effort is spent in the design phases in attempting to place blade and fuselage frequencies such that the vibratory response will be minimum. However in the absence of an adequate definition of the airload distribution along the blade in forward flight, and the difficulty of determining the natural frequencies of the complex airframe structures, final vibration reduction at the present time can only be accomplished through extensive flight testing.

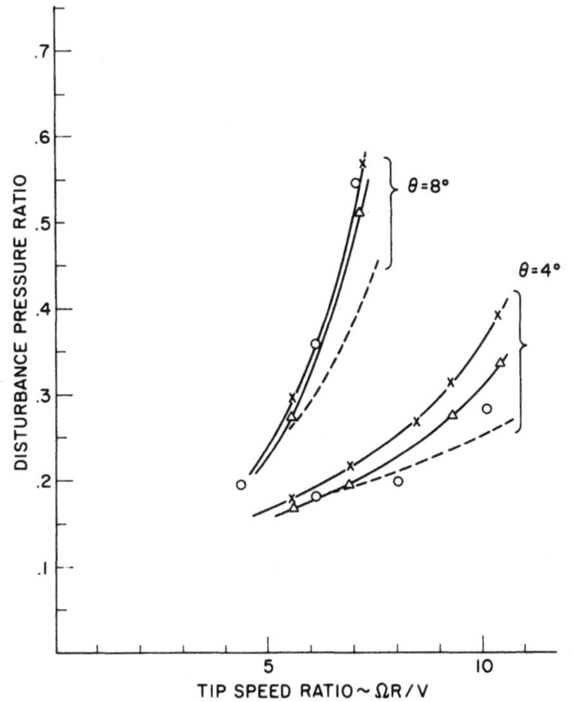

Figure 28. Unsteady aerodynamic effects—comparison between theory and experiment for wind shear effects. o: experimental results; ×: vortex theory—shed wake neglected; Δ: vortex theory with shed wake; - - - : modified momentum theory. Disturbance pressure ratio defined as periodic pressure change divided by steady state value.

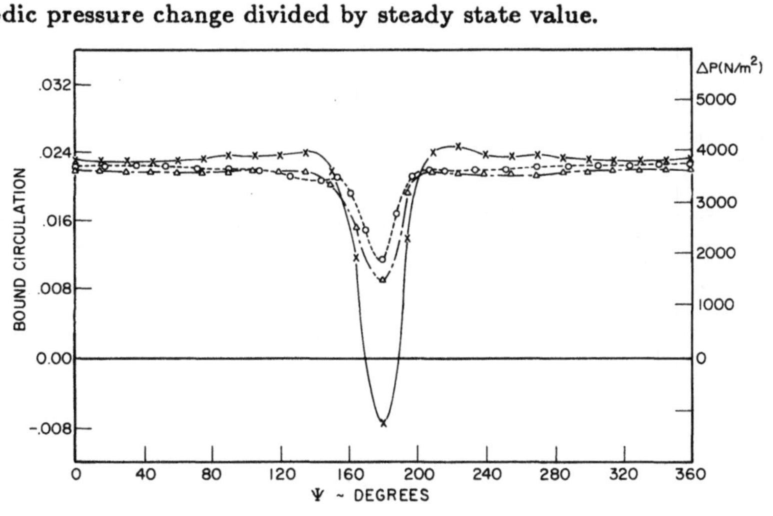

Figure 29. Unsteady aerodynamic effects—comparison between theory and experiment for tower shadow effects. o: experimental results; ×: vortex theory—shed wake neglected; Δ: vortex theory with shed wake.

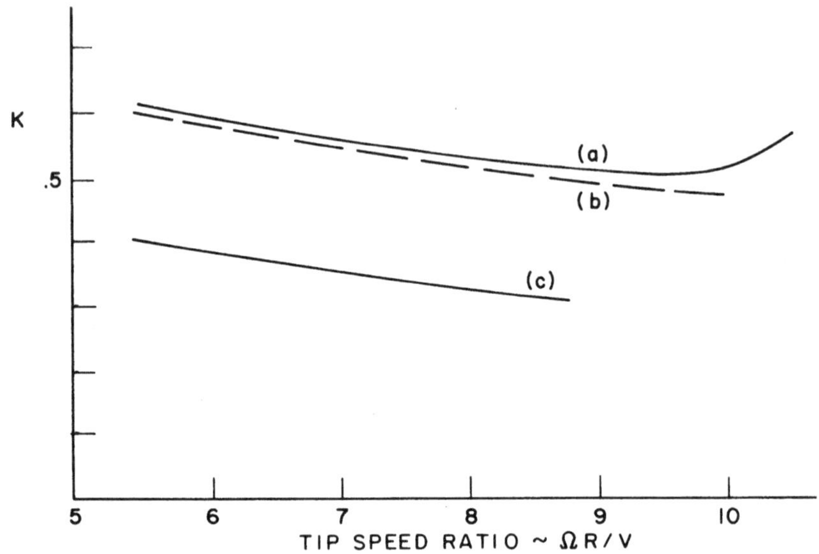

Figure 30. Lift deficiency factor, $K = L/L_q$ for the case of: (a) wind shear frequencies—vortex theory; (b) wind shear—modified momentum theory; (c) tower shadow frequencies—vortex theory.

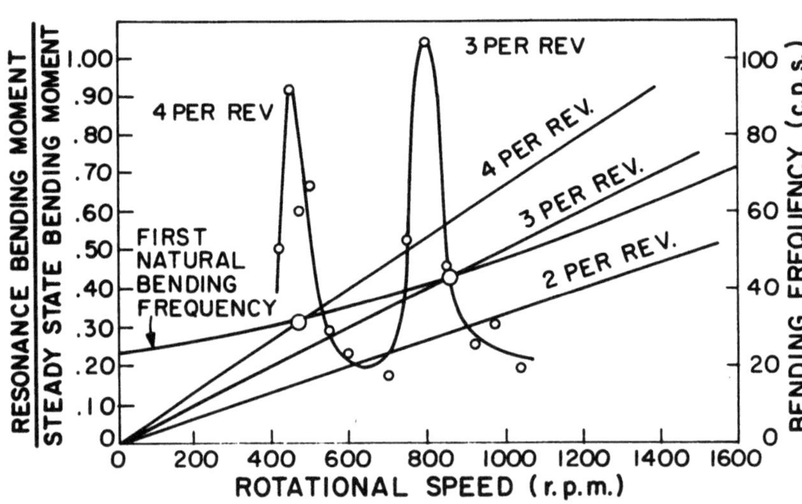

Figure 31. Rotor blade resonance study. Hovering, hinged at root condition. $\theta = 12°$, $EI = 3,000\ lb.\ in^2$.

It should be emphasized that the vibratory airloads come primarily from wake interaction. As pointed out above, the loads arising from the time varying velocity that a blade passes through in forward flight are otherwise small, of the order of μ^n where n is the harmonic of concern and μ is the advance ratio. A better knowledge of vortex interaction effects is therefore essential if minimization of vibration through aerodynamic modelling of the blade is to be achieved. In light of our present knowledge of rotor aerodynamics, artificial means of reducing vibration, such as vibration absorbers and isolators, appears to be a logical, and successful, design approach.

However a better approach would be to eliminate the vibratory input at its source by reducing the aerodynamic excitation coming from the rotor. Since first proposed many years ago, higher harmonic control has always been an option for doing this, but it is one which is fraught with many pitfalls and for this reason has been slow to gain acceptance. These pitfalls come from the need to have a fairly accurate definition of the blade airloads in order to predict the spatial and temporal distributions of the required harmonic input. The alternative is a search and rescue operation which may or may not be successful for all flight conditions and could result in higher blade stresses.

One of the earliest attempts at higher harmonic control (Miller and Ellis, 1956) used pitch/flap coupling to provide inherent vibration reduction. The blade center of gravity was placed ahead of the aerodynamic center and the torsional frequency, ω_0, tuned to the frequency of the disturbance, in this case two per rev. The blade flutter characteristics were analyzed in some detail, including the effects of coning, in order to ensure that the stabilizing effect due to the forward movement of the C. G. was sufficient to ensure freedom from flutter or divergence at the lowered torsional frequency. The computed reduction in vibratory hub shear, P_n, as the C. G. was moved forward is shown in Figure 32 for an arbitrary input consisting of a uniform pitch change along the blade. P_{n_R} is the response without pitch/flap coupling. Flight demonstration confirmed the favorable reduction in vertical vibration level of the rotor resulting from pitch/flap coupling. Unfortunately it did little to attenuate the primary source of vibration which in this case was the inplane component.

Most of the vertical vibratory input was probably generated

by the basic $2/rev$ aerodynamic loading evident in the test data of Reference 45, as reproduced in Figure 1, amounting to some 20% of the steady component at $\mu = .2$ and increasing approximately as μ^2. The contribution of the elastic mode in bending to the vibratory hub shears was small for the assumed disturbance and blade bending frequency, ν_2. This however will not generally be the case particularly for rotors with more than two blades. As shown in Reference 47, the spanwise and azimuthwise distribution of the forcing function must in such cases be carefully selected and based on the expected forcing airload distribution. With the present state of knowledge of rotor airloads the prediction of the required compensatory pitch inputs is clearly heavily dependent on flight experimentation. However, the recent successful flight tests (Wood, 1983) of a higher harmonic control system are an encouraging example of the potential effectiveness of this vibration control technology. The successful extension of this concept to individual blade control leads to the possibility of reducing not only vibration but also improving stall characteristics and reducing fatigue loads (Ham, 1983).

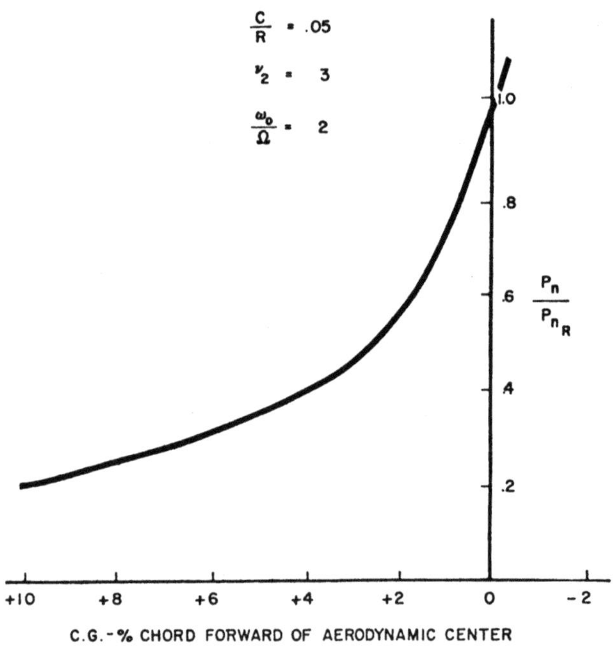

Figure 32. Vibratory load ratio for blades flexible in torsion as a function of C. G. location relative to A. C. $-n = 2$.

4. Dynamics of Rotors and Blades

4.1 Sources of Instability

Next to the problem of vibration control, ensuring individual blade stability has proven to be one of the most vexing design problems in rotor development. The recent advent of rotors without a lag hinge and associated dampers has introduced additional complexities in the blade stability equations because of the introduction of the lowly damped inplane mode, a potential problem for wind turbines and semi-rigid rotors. Fortunately lead-lag instabilities appear to be mild in nature introducing negative damping of the same order as the expected positive damping due to structural hysteresis and, as shown in Reference 17, may also be stabilized by elastic coupling between flap and lag, thus increasing the participation of the heavily damped flapping mode. Of more concern are variations in stability limits of the potentially more violent classical torsion-flexure flutter which is not so easily damped, although here again active control offers a means for stabilizing the motion.

When the rotor is mounted on a flexible support, such as a wing in the case of a tilt rotor aircraft, or the tower in the case of a wind turbine, another serious instability, in this case primarily mechanical, can occur known as ground resonance since it was first experienced with helicopters on the ground when the rotor inplane displacements coupled with the rocking mode of the helicopter on its wheels. This instability, in its simplest form, involves no aerodynamics and results from an unsymmetrical mode of blade motion about the lag hinges in which, in the case of a three bladed rotor, two blades move in opposite directions about these lag hinges thus displacing the collective blade C. G. off the center of rotation. In view of the large centrifugal forces acting on the blades, this displacement, if it occurs at the same frequency as the support system frequency, ω, results in a very serious instability capable of destroying a helicopter in a few second. In the case of rotors lacking polar symmetry, for example a two bladed rotor, Floquet or harmonics balance technique must be used to identify the regions of instability since time varying coefficients appear in the equations of motion (see for example, Reference 8).

The natural frequency of a blade about its lag hinges in the rotating system in terms of Ω is $\nu \sim \sqrt{e/d}$ where e is the lag hinge offset and d the distance to the blade mass centroid. When viewed in the fixed system the frequencies are evidently $\nu \pm \Omega$. When the regressive mode, $\nu - \Omega$, is equal to ω ground resonance can occur unless damping is provided both to ground and about the lag hinges. Evidently similar problems can occur with soft inplane rotors mounted on a wing both on the ground and in flight if coalescence of frequencies should occur, a problem compounded by the difficulties of introducing sufficient damping. Introducing coupling such that the aerodynamically heavily damped flapping mode is involved in the resultant motion, and careful placement of the natural frequencies ensures freedom from this type of instability for both tilt rotors and wind turbines (Hodges, 1979).

Another problem of concern for tilt rotor aircraft is the potential of an unstable whirling divergence or oscillation. With an articulated blade, forces perpendicular to the plane of rotation are generated as the rotor responds, for example, to the Coriolis accelerations engendered by a rate of tilt of the hub due to wing torsional displacements or to engine mount flexibility. These forces must be reacted by aerodynamic forces with components perpendicular to the plane of rotation. Because of the high inflow angles, inplane aerodynamic forces (induced drag) are also introduced acting in the same direction as the shaft tilt and hence increasing the initial tilt, resulting in an instability. On a rigid rotor cross flow resulting from hub tilt produces similar instabilities. Again, careful treatment of the blade and wing natural frequencies will alleviate this problem. Reference 21 contains an informative and comprehensive discussion of both these phenomena.

Evidently the problem of blade and rotor stability is a formidable one involving the dynamics of a rotor blade when the blade has relatively large steady state displacements out of the plane of rotation due to coning, and also requiring the careful computation of normally small inplane aerodynamic forces. And yet, as pointed out in Section 2, our state of knowledge of rotor airloads at the present time is insufficient to allow precise definition of either the steady state blade displacements or inplane airloads. For this reason simplified analyses which clarify the physics of the problem are of greatest help as a design guide. In the final analysis, it is unlikely that all the parameters of the problem will be sufficiently well defined to give complete confidence in a purely analytical prediction.

As an illustration of the sensitivity of the analyses to the assumed parameters, the case of blade divergence in flap/torsion will be examined in a simplified form which yet contains all the elements of significance in determining the flap/torsion blade stability margins.

4.2 Flap/Torsion Divergence

The two degree of freedom case of torsion/flexure flutter, including the effects of a finite coning angle, was considered in Reference 41, as discussed above. Review of the large number of second order terms, due to the finite displacement of the blade out of the plane of rotation, β_0 indicated that only one or two would have any appreciable effect on the blade stability. The most important was that associated with the tendency of the blade to pitch into the plane of rotation, as indicated by the diagram in Figure 33, a tendency resisted by the lift force acting on the blade. This lifting force is the primary reason for the existence of a coning angle.

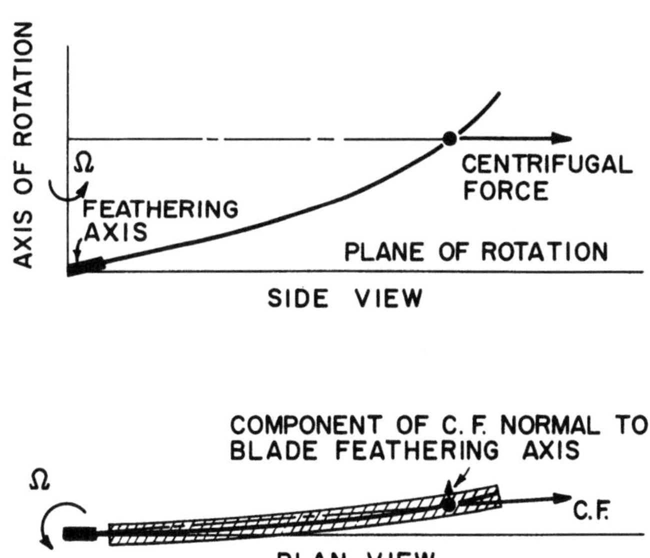

Figure 33. Side view and plan view of blade bending out of plane of rotation and twisting about a feathering axis located near center of rotation. Component of centrifugal force normal to feathering axis causes additional twisting.

Consider a blade displaced out of the plane of rotation through an angle β_0 with the feathering hinge remaining in the plane of rotation. This would be the case for a teetering rotor with the gimbal axis in the plane of rotation. Alternatively, β_0 could be the elastic deflection of the blade about an *equivalent* flapping hinge located near the root with a stiffness, k, equal to the bending stiffness of the blade in the first mode, a reasonable approximation for the very limber blades typical of rotary wing aircraft.

The centrifugal force $dm\Omega^2 r$ acting on an element of mass dm will produce a moment about the feathering hinge

$$-\int_0^R \Omega^2(r\beta\theta + \chi)\, r\beta\, dm = -I_b\Omega^2\beta^2\theta - I_R\Omega^2\beta$$

where χ is the C. G. offset from the feathering hinge positive aft and

$$I_b = \int_0^R r^2\, dm$$

$$I_R = \int_0^R \chi r\, dm.$$

There will be an additional moment, $I_f\Omega^2\theta$ (the *tennis racket* effect), where I_f is the moment of inertia of the blade about the feathering hinge. The negative signs indicate destabilizing moments. There is also, however, a stabilizing lift force acting on the blade:

$$dL = \frac{1}{2}\rho a b\Omega^2 r^2 \left(\theta_0 + \theta - \lambda\frac{R}{r}\right) dr$$

where θ_0 is the blade pitch setting. For an *ideally* twisted blade, $\theta_0 = \theta_T(R/r)$ where θ_T is the pitch setting at the blade tip. According to simple momentum theory, the inflow ratio λ ($\lambda = u/\Omega R$ where u is the inflow) is then constant with r. b is the blade chord and a the local slope of the lift curve.

However, there is also a horizontal destabilizing force due to drag given by

$$dH = \frac{1}{2}\rho\Omega^2 r^2 b C_{D_0} dr + \lambda\frac{R}{r} dL$$

The moment about the feathering hinge due to these forces is

$$M = \int (r\beta\theta + \chi_A)\, dL - \int r\beta dH - I_b\Omega^2\beta^2\theta - I_R\Omega^2\beta + I_f\Omega^2\beta + K\theta$$

where K is the torsional stiffness at the blade root in feathering and χ_A is the location of the aerodynamic center, positive aft.

Integrating over the blade radius and dividing by $I_b\Omega^2$ results in the derivatives

$$M_\theta = \frac{I_f}{I_b}(1 + \nu_\theta^2) - \beta_0^2 + \frac{\gamma}{2}\left[\frac{\beta_0}{3}(\theta_T - \lambda) - \frac{\beta_0\lambda}{3} + \frac{1}{3}\frac{\chi_A}{R}\right]$$

$$M_\beta = \frac{-I_R}{I_b} - \frac{\gamma}{2}\left[\frac{C_{D_0}}{4a} + \frac{\lambda}{2}(\theta_T - \lambda)\right]$$

where γ is the Lock number $= \rho abR^4/I_b$ and ν_θ the torsional frequency of the blade divided by Ω.

Similarly the moment about the flapping hinge is

$$m = I_b\Omega^2\beta - \int_0^R r dL + \nu_\beta^2 I_b\Omega^2\beta.$$

The steady state equilibrium of the blade requires that

$$\beta_0(1 + \nu_\beta^2) = \frac{\gamma}{6}(\theta_T - \lambda)$$

where ν_β is the nonrotating flapping frequency.

The flapping derivatives are

$$m_\theta = -\frac{\gamma}{6}$$

$$m_\beta = 1 + \nu_\beta^2.$$

The divergence limit occurs when

$$M_\theta m_\beta - M_\beta m_\theta = 0.$$

It is of interest to consider the case of a blade free to flap so that $\nu_\beta = 0$. Then β_0 becomes

$$\beta_0 = \frac{\gamma}{6}(\theta_T - \lambda),$$

and it is at once evident from the expression for M_θ that for this case the destabilizing inertia moment is cancelled by the aerodynamic moment, leaving only a residual due to the change in induced drag associated with the change in angle of attack, θ. M_θ then reduces to

$$M_\theta = \frac{I_f}{I_b}(1 + \nu_\theta^2) - \frac{\gamma}{2}\left(\frac{\lambda\beta_0}{3} - \frac{\chi_A}{3R}\right).$$

Divergence will occur, even with $\chi_A = \chi_I = 0$, when

$$\frac{I_f}{I_b}(1 + \nu_\theta^2) - \frac{\gamma}{6}\beta_0\lambda - \frac{\gamma^2}{16}\left[\frac{C_{D_0}}{4a} + \frac{\gamma}{2}(\theta_T - \lambda)\right] = 0.$$

Consider a helicopter rotor with

$$\theta_T = .10$$
$$\lambda = .05$$
$$\gamma = 12$$
$$\beta_0 = .1$$
$$\frac{C_{D_0}}{a} = .002$$

and

$$\frac{I_f}{I_b} = .001.$$

Then ν_θ must be greater than 5.0 for stability. Alternatively a shift in AC or C. G. of the order of 10% of the chord (for a blade of aspect ratio 10) will avoid divergence even with $\nu_\theta = 0$.

On the other hand consider a small wind turbine with a flapping hinge and

$$\theta_T = 0$$
$$\lambda = .125$$
$$\gamma = 3.35$$
$$\beta_0 = -.07$$
$$\frac{C_{D_0}}{a} = .0014$$
$$\frac{I_R}{I_b} = .003.$$

and

$$\frac{\chi_A}{R} = -.006.$$

Despite the negative χ_A (-10% chord), divergence is evidently not critical. Figure 34, however, shows the stability limits for the complete case of flap/lag/torsion flutter for this rotor. Reference 42 contains a more general discussion of the stability limits for this and similar cases.

4.3 The Effects of Aerodynamic Modelling on Dynamics

The effect of including unsteady aerodynamics in the case of flap/lag flutter, including the effects of the returning wake, are shown in Figure 35 (Miller and Ellis, 1956) compared to the quasistatic solution customarily used in blade stability analyses. Evidently the quasi-static approximation is conservative for this case. However when the source of instability is primarily mechanical, as in the case of ground resonance, and aerodynamic damping is introduced to insure stability the unsteady aerodynamic effects, as might be expected, tend to reduce the stability margins (Johnson, 1981).

The cancellation of the large aerodynamic and inertial reaction moments discussed above for the equivalent hinge model serves to emphasize the importance of using the correct displacement of the blade in the steady state condition, compatible with the assumed model displacements used for the dynamic analysis. For an elastic mode shape $g(r)$, $g(r) \neq r$, complete cancellation would evidently not occur. For this reason, and in view of the difficulties of determining the steady state displacements of the blade to any high degree of accuracy because of the limitation in existing aerodynamic theory as discussed in Section 2, any blade configuration which cannot be closely modelled by the equivalent hinge concept will give numerical results which must be considered as close approximations only.

Finally it is of interest to examine a beneficial aerodynamic effect of coning on the complete rotor system for the case of a wind turbine mounted on a tower. Reference 37 showed that a nonarticulated, rigid rotor with coning would be statically stable in yaw both upwind and downwind. This unexpected result is due to the *dihedral* effect of coning which produces a positive weathervaning tendency in the presence of the cross flow due to yaw in both cases.

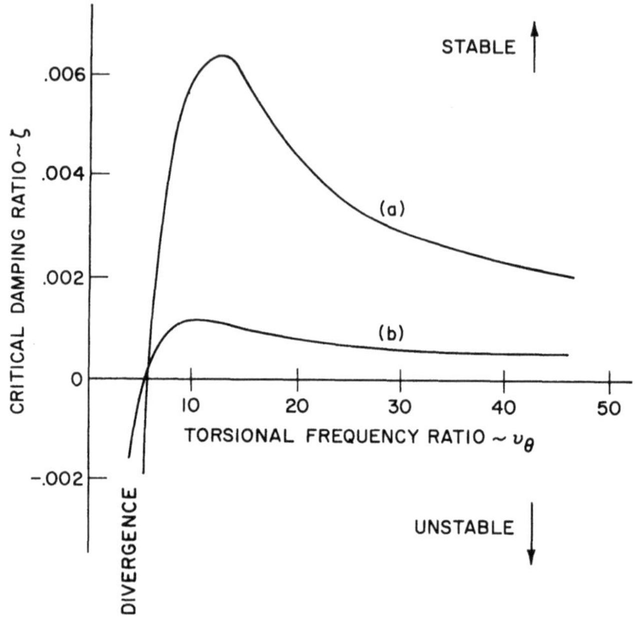

Figure 34. Effect of torsional frequency ratio, ν_θ (non-dimensionalized by Ω) on stability margins. (a) flap/torsion mode, (b) lag/torsion mode.

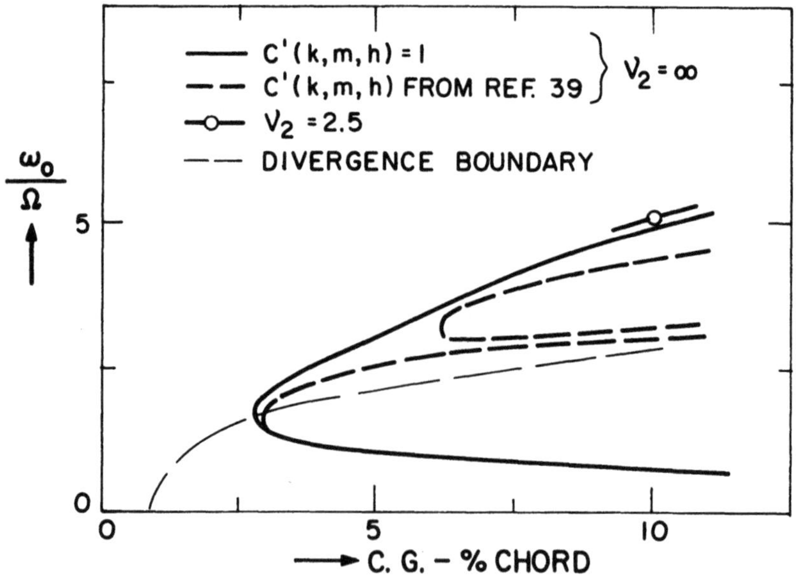

Figure 35. Fluttter boundaries in hovering showing effects of unsteady aerodynamics. ν_2 = frequency of first elastic mode in bending divided by Ω. Flat rotor ($\beta_0 = 0$).

Since wind turbine rotors operate under very high thrusts, and hence high elastic or built in coning angles, it may be expected that thrust changes with changing wind conditions could result in varying degrees of weathervaning stability or instability.

In considering the general problem of VTOL aircraft stability and control, the small effects discussed in this section, which so dominate the problem of blade and rotor stability, continue to be of significance in determining the aircraft aerodynamic stability derivatives. The more important problem however is the fundamental one of position control of a hovering vehicle in which inertial, rather than aerodynamic effects, predominate, as discussed in the next section.

5. Aircraft Stability and Control

5.1 Aircraft Dynamics in Hovering Flight

A VTOL aircraft comes closer to having a true all weather operating capability than any other transportation mode. Even the heliport need not be cleared of snow cover prior to landing, and the vehicle could operate under conditions when all ground travel has stopped. Furthermore, by virtue of its vertical flight capability and, equally important, the ability to move in any direction (including backwards), a VTOL aircraft such as a helicopter can descend under zero/zero conditions and land in an area not much greater than its rotors. Unfortunately, current VTOL handling qualities preclude complete exploitation of this unique capability.

It may be worth reviewing briefly the stability and control characteristics of VTOL aircraft in hovering flight in order to clarify the reasons for their inherently poor handling qualities. Horizontal positioning is the primary control function in low speed approach and landing with attitude being important only as it affects position. The great advantage of VTOL is the independent control available of vertical and horizontal velocity, in that forward speed is not required to produce lift, and thus the pilot can take whatever time is needed to position himself relative to his landing point, either visually or by following a microwave beam before ground contact. However, such horizontal positioning is a difficult task, particularly under IFR conditions.

The nature of the problem may be readily understood if it is realized that in order to position a hovering aircraft, say to move a horizontal distance x, (Figure 36) the pilot must control two accelerations resulting from the initial control deflection before the horizontal displacement is achieved. In order to provide a horizontal force to initiate the displacement, the pilot must first tilt the aircraft by application of a control moment either by a slight tilt of the tip path plane, as shown for the case of the rotor-supported aircraft in Figure 36, or by direct application of moments through auxiliary jets in the case of a jet lift vehicle. In either case the moment of inertia in pitch or roll of the vehicle must first be overcome by these control moments in order to tilt the lift vector so as to provide the force required to cause a horizontal motion. Since there is almost no inherent damping or static stability in pitch (or roll) in hovering flight, the application of the moment produces an angular acceleration and not the desired angular displacement. The pilot thus has the difficult problem of defining the angle of tilt, α, when the application of his control produces an angular acceleration, $\ddot{\alpha}$. Furthermore, this angle of tilt produces, at least initially, a linear horizontal acceleration $\ddot{x}$ when what is desired is a change in horizontal position, x. Thus, corrections in horizontal displacement must be achieved by controlling two accelerations, hence through two second order time lags, or four orders of integration, and this is a formidable task.

The elementary concepts of Figure 37 will serve to illustrate this point. It is simple to position an essentially massless object (Figure 37a) between two points, A and B, by the application of a regulated force, P, when the object is held by spring restraints. If the object is held by viscous dampers instead of springs (Figure 37b), moving it from one point to another precisely is more difficult, since the force applied produces a velocity rather than a displacement, and velocity can not be perceived directly by the operator as can displacement. Regulation of the applied force in order to obtain a desired displacement x must therefore be based on mental integration relating velocity to displacement, a more difficult task than when x is related directly to the applied force. If the object has no restraints, either spring or viscous, but has mass (Figure 37c), positioning it between two points, A and B, becomes quite difficult. The applied force now produces an acceleration and two orders of mental integration are required to judge displacement. Furthermore, upon removal of the force a velocity is left which requires opposite force to stop the motion, and hence a high degree of anticipatory

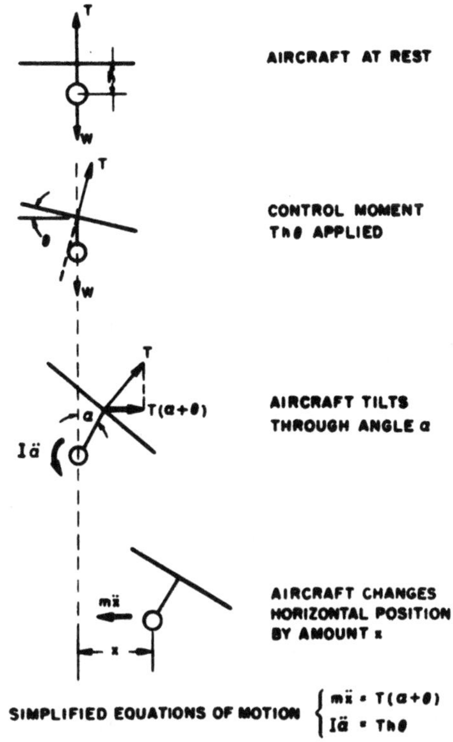

Figure 36. Simplified representation of VTOL control dynamics.

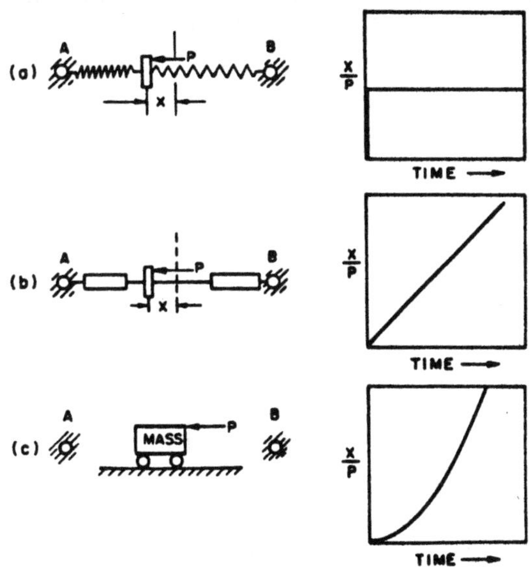

Figure 37. Illustrative control responses.

skill. It is evident that, when the pilot must control through two such acceleration responses and hence four orders of integration, the task is almost beyond human capability. For this reason an experienced helicopter pilot uses aircraft tilt (horizon reference) for his primary visual cue when hovering, thus eliminating the need for continually controlling through four orders of integration, possibly simultaneously in two directions, and only monitors ground position. The subjective effect of the horizontal inertia lag due to mass is thereby eliminated, and only two orders of continual mental integration are required, not easy but possible with sufficient training. Experience and skill allow him to relate subconsciously attitude changes to position changes. Under IFR conditions (no horizon reference) the problem is evidently aggravated.

The physics of the problem may be visualized by considering the positioning of a ball of mass m rolling on a plate pivoted about its center with moment of inertia I (Figure 38), with the plate tilt controlled by regulated air jets producing pulsed forces of $\pm P$. The equations of motion of this system are similar to those of the hovering helicopter shown in Figure 36 and are identical if the mass of the ball is small compared to that of the plate and if the control input θ in Figure 36 is small compared to α. Attempting to position the ball along the plate using the airjets illustrates the difficulties inherent in the helicopter control problem. A truer simulation would add the additional complication of pivoting the plate at its center, simulating combined pitch and roll control rather than resting it on a knife edge. Four control jets rather than two would then be required in order to control pitch and roll simultaneously.

5.2 Stability Augmentation

In order to improve helicopter control characteristics, stability augmentation systems have been introduced in VTOL aircraft using attitude sensing devices which, with suitable feedback loops, change the angular acceleration response in tilt to either an angular velocity or an angular displacement response, thus eliminating one or two orders of integration, and thereby considerably simplifying the control problem. However it is evident from the above discussion that the ultimate control system must also include feedback loops involving sensing of horizontal acceleration, velocity or position. In practice, a control system which provides a horizontal velocity vector proportional to control input rather than a horizontal displacement would appear to be the best solution, since position control will then not be limited and, upon neutralization of control, the aircraft will still hold position relative to inertial space. The concept may be demonstrated, using the simple model of Figure 38, by replacing the ball with a flat oiled puck (velocity feedback) and controlling plate tilt directly by hand (attitude stabilized). Figure 39 shows the basic block diagram of such a velocity vector control system suitable for hover and low speed flight.

The capability of such a system was demonstrated in the flight tests with the TAGS system reported in Reference 1. Descents were made under simulated zero/zero flight conditions (pilot under the hood) up to 90 degrees from the horizontal, using a velocity vector control system based on an inertial platform. Unfortunately the system was designed around navigation quality equipment and, being fly by wire, required triplication of this high quality equipment in order to ensure flight safety. This resulted in a high initial cost which could not be justified for the applications for which it was originally developed. It is well to remember, however, that the first application of the concept was as a drift eliminator for landing the Kaman drone helicopter (Kendall, 1954). The original design consisted of a pendulum back to back with an artificial horizon (to eliminate gravitational accelerations) and with a simple modifying network to integrate acceleration and convert it into a velocity feedback. An improved version of such a rudimentary system, designed as a control and not a navigation aid, could result in a major improvement in VTOL handling qualities for the critical approach system, with long term drift corrected by pilot input in response to external navigational aids.

5.3 The Effects of Aerodynamic Derivatives on Aircraft Stability

In the above discussion the effects of the aerodynamic stability derivatives have been ignored. Figure 40 shows the response in hover with and without aerodynamic derivatives. Evidently their effect in hovering flight is small, certainly for the short period response which is of primary concern to the pilot.

In forward flight the rotor adds stability derivatives in addition to those associated with wings, fuselage or tail surfaces. The most important is the instability with respect to angle of attack which can

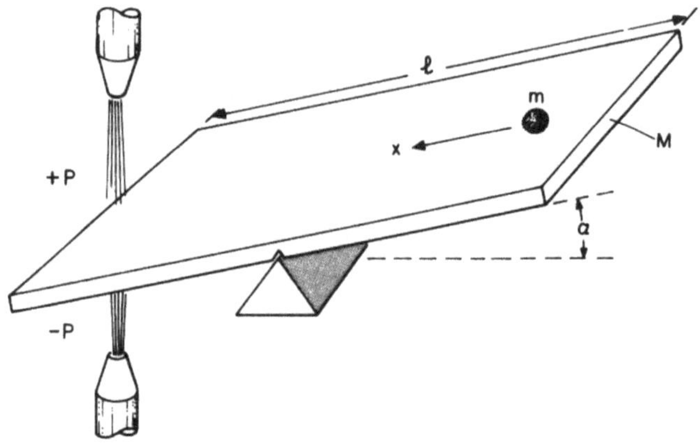

$$\ddot{x} = \frac{5}{7} g\alpha$$

$$\ddot{\alpha} = \frac{P\ell}{I} \qquad I = \frac{M\ell^2}{12} \qquad M >> m$$

$$\text{FOR MODEL}: x \simeq .030 \left[\frac{P\ell}{I}\right] t^4$$

$$\text{FOR HELICOPTER}: x \simeq .042 \left[\frac{T\theta h}{I}\right] t^4 \qquad \alpha >> \theta$$

Figure 38. Physical analog of helicopter stability characteristics in pitch or roll.

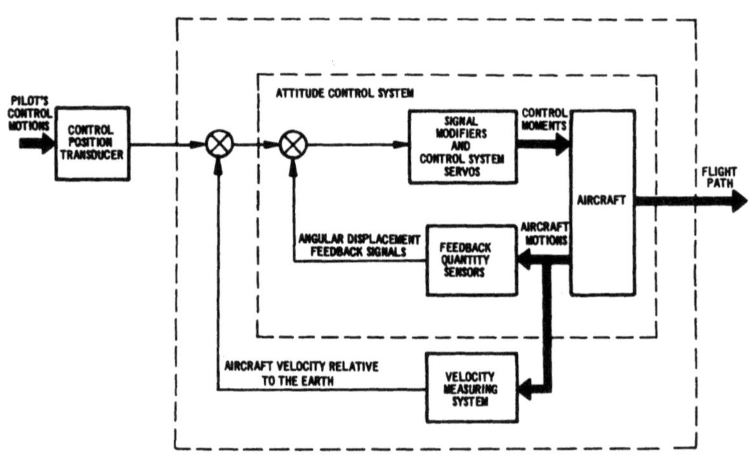

Figure 39. Mechanization of position referenced helicopter with velocity feedback loop.

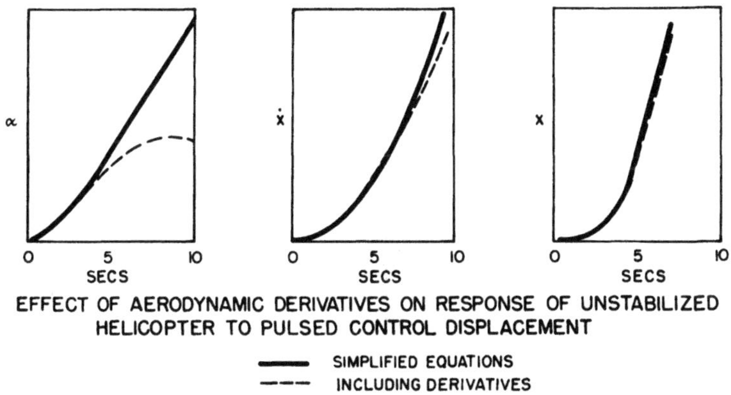

Figure 40. Response of hovering helicopter with and without aerodynamic derivatives.

become serious at higher forward speeds. This instability arises from the change in tip path plane inclination due to the change in inflow as the rotor angle of attack changes, which produce destabilizing moments.

Consider the well known expression for change in rotor tilt angle with forward speed for a flapping rotor as it accommodates to the different velocities on the advancing and retreating sides of the rotor disc by flapping (see Reference 19, p. 189),

$$\Delta\beta = \frac{-\frac{8}{3}\mu(\theta_0 - \frac{3}{4}\lambda)}{1 + \frac{3}{2}\mu^2}$$

where the negative sign indicates a rearward tilt of the tip path plane. μ is the advance ratio and λ the inflow velocity, both non-dimensionalized by the tip speed, ΩR. θ_0 is the blade pitch. An increase in angle of attack, $\Delta\alpha$ of the helicopter shaft will result in an instantaneous reduction in inflow, $-\mu\Delta\alpha$, so that

$$\frac{\partial\Delta\beta}{\partial\alpha} = \frac{-2\mu^2}{1 + \frac{3}{2}\mu^2}$$

indicating an increased rearward tilt of the tip path plane, producing a destabilizing moment on the aircraft growing as μ^2. The thrust coefficient is to first order in μ and for an untwisted blade,

$$C_T = \frac{\sigma a}{4}\left(\frac{2\theta_0}{3} - \lambda\right)$$

where a is the lift-curve slope. Then the nondimensionalized stability derivative is, according to this simple derivation,

$$m_\alpha \sim -2\mu^2\left(\frac{2}{3}\theta_0 - \lambda\right)$$

When, however, the complete expression is developed including the effects of horizontal forces in the plane of rotation, this stability derivative becomes

$$m'_\alpha \sim -2\mu^2\left(\frac{5}{6}\theta_0 - 2\lambda\right)$$

indicating a favorable effect due to inplane forces compared to the solution neglecting horizontal force effects. This is opposite from the effects of the horizontal forces on rigid rotor whirl stability, as discussed above where the inplane forces are due mainly to changes in cross flow as opposed to changes in inflow.

Typical values are, at $\mu = .2$, $\theta_0 = .12$, $\lambda_0 = .03$, giving

$$m_\alpha \sim -.0040,$$

compared to

$$m'_\alpha \sim -.0032.$$

For rotors which can transmit hub moments, such as those with large hinge offsets, or for hingless rotors, this beneficial effect tends to be masked by the large destabilizing moments resulting from inertia moments due to $\Delta\beta$.

Another important derivative existing both in hovering and forward flight comes from the damping due to the change in $\Delta\beta$ as the tip path plane responds to an aft tilting velocity, $\dot{\alpha}$, of the shaft. This effect is similar to that associated with whirl flutter discussed previously and arises from the need to maintain equilibrium about the flapping hinge in the presence of Coriolis accelerations resulting from $\dot{\alpha}$. It may be readily shown that

$$\frac{\partial \Delta\beta}{\partial \dot{\alpha}} = \frac{16}{\gamma\Omega}$$

and hence, in its simplest form

$$m_{\dot{\alpha}} \sim \frac{16}{\gamma}\left(\frac{2\theta_0}{3} - \lambda\right)$$

independent of μ to first order of μ. However, as pointed out in Reference 34, when the effects of horizontal forces are included the expression becomes

$$m'_{\dot{\alpha}} \sim \frac{16}{\gamma}\left(\frac{2\theta_0}{3} - \frac{3}{2}\lambda\right)$$

indicating an appreciable reduction in damping which could become negative for sufficiently high inflows such as on tilt rotors. For the same case as above and $\gamma = 8$ the derivatives are

$$m_{\dot{\alpha}} \sim .10$$

and

$$m'_{\dot{\alpha}} \sim .07$$

Although the error may be as much as 50% due to the neglect of inplane forces, the overwhelming effects of onboard stabilization equipment would ensure satisfactory handling qualities regardless of the existence of such destabilizing effects or potential reductions in damping. The problem of VTOL aircraft stability and control therefore reduces to one of providing, in hovering flight, artificial stabilization about, and equally important, along all axes and, in forward flight, providing in addition fixed control surfaces of adequate size, or resorting to the more versatile capabilities of artificial stabilization.

6. Concluding Remarks

The three major problems of rotary wing aircraft: vibration reduction, blade stability, and control in hovering flight, are all dependent on a precise definition of the aerodynamics of the rotor. Much progress has been made in understanding the complex fluid mechanics of a lifting surface interacting with its own wake, but much remains to be done before a complete definition of the spatial and temporal loading on a rotor blade is possible.

In the meantime the dynamics of the rotor system have been formulated to a high degree of precision and the important aerodynamic interactions identified. Such analyses provide valuable guidelines for design and for flight development. However, in the absence of a precise definition of the rotor aerodynamics, final design optimization must continue to be based on empiricism, hopefully over a continually narrowing range.

For many of these problems, modern control techniques offer a versatile and effective way of providing satisfactory performance in the presence of uncertainties arising from the not fully known and continually varying aerodynamics inputs. Better identification of the state variables as our knowledge of rotor aerodynamics grows and increased confidence in the required electronic equipment will probably make automation the preferred approach for handling the complex problems of rotor stability, vibration control, noise reduction, and vehicle handling qualities.

References

[1] Anonymous. *VTOL Advanced Flight Control System Studies for All-Weather Flight, Vol. 1, Task 1 Report,* Charles Stark Draper Laboratory, *USAAMRDL TR-75-13A,* July 1975.

[2] Baker, G. R. "The 'Cloud in Cell' Technique Applied to the Roll-Up of Vortex Sheets," *Journal of Computational Physics, 31* (1979), 76–95.

[3] Batchelor, G. K. *An Introduction to Fluid Dynamics,* Cambridge University Press, 1967.

[4] Biggers, T. C. et al. "Measurements of Helicopter Rotor Tip Vortices," *AHS Forum Proceedings,* May 1977.

[5] Brower, M. "Lifting Surface and Lifting Line Solutions for Rotor Blade Interaction With Curved and Straight Vortex Lines," *MIT ASRL TR 194-5,* November 1981.

[6] Cheeseman, I. C. "Developments in Rotary Wing Aircraft Aerodynamics," *VERTICA, 6, No. 3,* 1982.

[7] Dugundji, J., Larrabee, E. E., and Bauer, P. H. *Wind Energy Conversion: Volume V—Experimental Investigation of a Horizontal Axis Wind Turbine,* U. S. Dept. of Energy, *COO 3141-T1,* Sept. 1978.

[8] Dugundji, J. and Wendell, J. H. "Some Analysis Methods for Rotating Systems With Periodic Coefficients," *AIAA Journal, 21, No. 6,* June 1983.

[9] Friedmann, P. P. "Formulation and Solution of Rotary-Wing Aeroelastic Problems," presented at the International Symposium on Aeroelasticity, Nuremberg, Federal Republic of Germany, October 5–7, 1981.

[10] Friedmann, P. P. and Tong, P. "Nonlinear Flap-lag Dynamics of Hingeless Helicopter Blades in Hover and in Forward Flight," *Journal of Sound and Vibration, 31, No. 1,* 1973.

[11] Glauert, H. "The Analysis of Experimental Results in the Windmill Brake and Vortex Ring States of an Airscrew," *ARC R and M 1026,* 1926.

[12] Gray, R. B. et al. "Helicopter Hovering Performance Studies," *Princeton Univ. Aero. Eng. Report 313,* 1955.

[13] Ham, N. D. "Some Conclusions From an Investigation of Blade-Vortex Interaction," *Journal of the American Helicopter Society,* Oct. 1975.

[14] ————. Helicopter Individual-Blade-Control and Its Applications," presented at the 39th Annual Forum of the American Helicopter Society, May 9–11, 1983.

[15] Hodges, D. H. "An Aeromechanical Stability Analysis for Bearingless Rotor Helicopters," *Journal of the American Helicopter Society, 24, No. 1* (1979), 2–9.

[16] Hodges, D. H. and Ormiston, R. A. "Stability of Elastic Bending and Torsion of Uniform Cantilevered Rotor Blades in Hover," presented at AIAA/ASME/SAE 14th Structures, Structural Dynamics, and Materials Conference, Williamsburg, VA., March 20–22, 1973.

[17] ————————————. "Stability of Elastic Bending and Torsion of Uniform Cantilever Rotor Blades in Hover With Variable Structural Coupling," *NASA TN D-8192,* 1976.

[18] Hohenemser, W. G. "Hingeless Rotorcraft Flight Dynamics," *AGARDograph No. 197,* Sept. 1974.

[19] Johnson, W. *Helicopter Theory,* Princeton: Princeton Univ. Press, 1980.

[20] ————. "Comparison of Calculated and Measured Model Rotor Loading and Wake Geometry," *NASA TM 81189,* April 1980.

[21] ———. "A Comprehensive Analytical Model of Rotorcraft Aerodynamics and Dynamics," *NASA TM 81182,* June 1980.

[22] ———. "The Influence of Unsteady Aerodynamics on Hingeless Rotor Ground Resonance," *NASA TM 81302,* July 1981.

[23] Kendall, D. E. "Final Engineering Report on Phase I of the Development of Remotely Controlled Helicopter," *Kaman Aircraft Corp. Report G-59,* March 12, 1954.

[24] Lamb, H. *Hydrodynamics,* Dover, 1945.

[25] Landahl, M. T. "Roll-Up Model for Rotor Wake Vortices," *FFA (Sweden) HU-2262-Pt. 5-1981.* Also *MIT ASRL TR 194-4, 1981.*

[26] Landgrebe, A. J. "An Analytical and Experimental Investigation of Helicopter Rotor Hover Performance and Wake Geometry Characteristics," *USAAMRDL TR 71-24,* 1971.

[27] Landgrebe, A. J. and Cheney, M. C., Jr. "Rotor Wakes—Key to Performance Prediction," *AGARD Conference Proceedings No. 111 on Aerodynamics of Rotary Wings,* Sept. 1972.

[28] Landgrebe, A. J., Moffitt, R. C., and Clark, D. R. "Aerodynamic Technology for Advanced Rotorcraft—Part I," presented at the National Symposium on Rotor Technology, American Helicopter Society, Mideast Region, Essington, PA., August 1976.

[29] Liu, C. H., Thomas, J. L., and Tung, C. "Navier Stokes Calculations for the Vortex Wake of a Rotor in Hover," AIAA 16th Fluid and Plasma Dynamics Conference, July 1983.

[30] Loewy, R. G. "A Two Dimensional Approach to the Unsteady Aerodynamics of Rotary Wings," *Journal of Aerospace Science, 24* (1957), 82–98.

[31] ———. "Review of Rotary-Wing V/STOL Dynamic and Aeroelastic Problems," *Journal of the American Helicopter Society,* 1969. Presented at AIAA/AHS VTOL Research, Design, and Operations Meeting, Georgia Institute of Technology, Atlanta, GA., February 1969.

[32] McCroskey, W. J. et al. "Dynamic Stall on Advanced Airfoil Sections," *Journal of the American Helicopter Society, 26, No. 3* (1981), 40–50.

[33] Meyer, J. R. Jr. "An Investigation of the Bending Moment Distribution on a Model Helicopter Rotor Blade and a Comparison With Theory," *NACA TN 2626,* February 1951.

[34] Miller, R. H. "Helicopter Control and Stability in Hovering Flight," *Journal of the Aeronautical Sciences, 15, No. 8* (1948), 453–472.

[35] ———. "Unsteady Air Loads on Helicopter Rotor Blades," *Journal of the RAeS, 68, No. 640,* April 1964. (The Fourth Cierva

Memorial Lecture.)

[36] ———. "Rotor Blade Harmonic Air Loading," *AIAA Journal, 2, No. 7,* 1964.

[37] ———. "On the Weathervaning of Wind Turbines," *AIAA Journal of Energy, 3, No. 5,* Sept.-Oct. 1979.

[38] ———. "Simplified Free Wake Analyses for Rotors," *FFA (Sweden) TN 1982-7.* Also *MIT ASRL TR 194-3,* August 1981.

[39] ———. "A Simplified Approach to the Free Wake Analysis of a Hovering Rotor," 7th European Rotorcraft and Powered Lift Aircraft Forum, Garmisch-Partenkirchen, Federal Republic of Germany, Sept. 1981. Also *VERTICA, 6, No. 2,* May/June 1982.

[40] ———. "Free Wake Techniques for Rotor Aerodynamic Analysis: Volume I—Summary of Results and Background Theory," *NASA CR-166434.* Also *MIT ASRL TR 199-1,* Dec. 1982.

[41] Miller, R. H. and Ellis, C. W. "Blade Vibration and Flutter," *Journal of the American Helicopter Society,* July 1956.

[42] Miller, R. H. et al. "Wind Energy Conversion: Volume I—Design Manual, U. S. Dept. of Energy," *COO 3141-T1,* Sept. 1978.

[43] ———. "Wind Energy Conversion: Volume II—Aerodynamics of Horizontal Axis Wind Turbines," U. S. Dept. of Energy, *COO 3141-T1,* Sept. 1978.

[44] Murman, E. M. and Stremel, P. M. "A Vortex Wake Capturing Method for Potential Flow Calculations," *AIAA Paper 82-0947,* June 1982.

[45] Scheimann, J. "A Tabulation of Helicopter Rotorblade Differential Pressures, Stress, and Motions as Measured in Flight," *NASA TM-X-952,* March 1964.

[46] Scully, M. "Computation of Helicopter Rotor Wake Geometry and Its Influence on Rotor Harmonic Airloads," *MIT ASRL TR 178-1,* March 1975.

[47] Shaw, J. Jr. "Higher Harmonic Blade Pitch Control for Helicopter Vibration Reduction: A Feasibility Study," *MIT ASRL TR 150-1,* December 1968.

[48] Stremel, P. M. "Computational Method for Nonplanar Vortex Wake Flow Fields With Applications to Conventional and Rotating Wings," *M. S. thesis,* Dept. of Aeronautics and Astronautics, MIT, Feb. 1982.

[49] Sullivan, J. P. "Experimental Investigation of Vortex Rings and Helicopter Rotor Wakes Using a Laser Doppler Velocimeter," *ScD. thesis,* Dept. of Aeronautics and Astronautics, MIT, June 1973.

[50] Theodorsen, R. "General Theory of Aerodynamic Instability and

the Mechanism of Flutter," *NACA TR 496,* 1935.

[51] Van Dyke, M. *Pertubation Methods in Fluid Mechanics,* Academic Press, 1964.

[52] Viterna, L. A. and Janetzke, D. C. "Theoretical and Experimental Power From Large Horizontal-Axis Wind Turbines," 5th Biennial Wind Energy Conference and Workshop, October 5–7, 1981, Washington, D.C.

[53] Westwater, F. L. *Aero. Res. Council, R and M, No. 1962,* 1935.

[54] Wood, E. R. "On Developing and Flight Testing a Higher Harmonic Control System," presented at the 39th Annual Forum of the American Helicopter Society, May 9–11, 1983.

Special Opportunities in Helicopter Aerodynamics

*W. J. McCroskey**
NASA Ames Research Center
Moffett Field, CA 94035

Summary

Aerodynamic research relating to modern helicopters includes the study of three-dimensional, unsteady, nonlinear flow fields. A selective review is made of some of the phenomena that hamper the development of satisfactory engineering prediction techniques, but which provide a rich source of research opportunities: flow separation, compressibility effects, complex vortical wakes, and aerodynamic interference between components. Several examples of work in progress are given, including dynamic stall alleviation, the development of computational methods for transonic flow, rotor-wake predictions, and blade-vortex interactions.

1. Introduction

The flow fields of helicopters provide some of the most complex challenges to be found in the field of applied aerodynamics. For moderate flight conditions, a combination of practical experience, empiricism, static airfoil characteristics, and linear theory are usually adequate to estimate the overall performance and to assess the relative merits of various configuration changes. However, it is the boundaries of the flight envelope, which are typically set by vibrations, excessive power requirements, aeroelastic instabilities, and/or adverse handling characteristics, that often determine the operational success of modern helicopters. Expanding these boundaries, in turn, generally involves improving aerodynamic characteristics that are limited by very complex and nonlinear phenomena, such as boundary layer separation, shock wave formation, highly-distorted

*Senior staff scientist, NASA Thermo- and Gas-Dynamics Division and U. S. Army Aeromechanics Laboratory (AVRADCOM).

wakes, blade-vortex interactions, and aerodynamic interference between components of the helicopter. These sundry phenomena are indicated schematically in Figure 1.

This paper describes a few of the special research problems and opportunities that will be important if major improvements in helicopter characteristics are to be made in the future. The topics described in the following sections and the references are only intended to be representative, and perhaps provocative, rather than complete or exhaustive. Furthermore, the material presented tends to reflect the author's personal interests, biases, background, and research activities. In each of the three general categories of viscous flows, transonic flows, and vortical wake flows, the basic aerodynamic characteristics of the particular problem are described, and one or two examples of significant research in progress are cited. A number of other important categories and research issues are mentioned in passing in the final section of the paper.

2. Rotor Blade Boundary Layers and Stall

The phenomenon of wing stall is well-known for fixed-wing aircraft. It limits the *minimum* flying speed for takeoff and landing, and hence determines the minimum runway length that is required for aircraft operations. On the other hand, rotor-blade stall limits the *maximum* flight speed of helicopters. In blade-fixed coordinates, the difference in instantaneous local dynamic pressure on the retreating and advancing sides of the rotor disc increases monotonically with flight speed, resulting in asymmetric distributions of the sort illustrated in Figure 2, which is based on the experiment of Rabbott et al. (1966). To maintain some semblance of rotor equilibrium, the blade-element angles of attack must be increased on the retreating blade and decreased on the advancing blade. However, there is a limit to how far the retreating blade angle of attack can be increased before the boundary layer will separate, resulting in a loss of lift and an increase in blade-element drag and pitching moment. The variations in section lift and drag translate into limitations in rotor thrust and increases in power required, respectively; the blade torsion that is caused by the variations in pitching moment produce vibratory pitch-link loads and blade flutter.

Retreating-blade stall differs importantly in several respects from fixed-wing stall and for several different reasons, including the

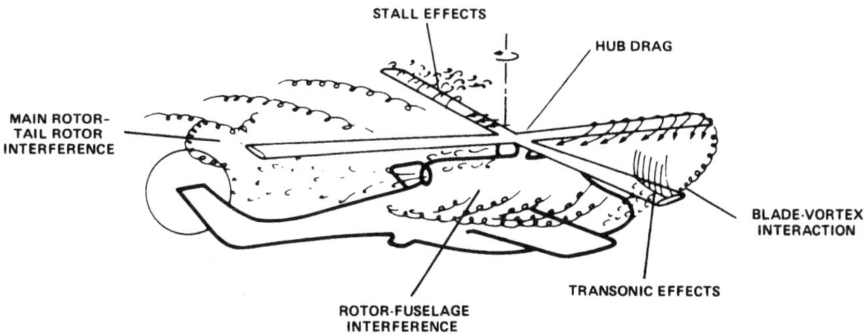

Figure 1. Major problem areas in helicopter aerodynamics.

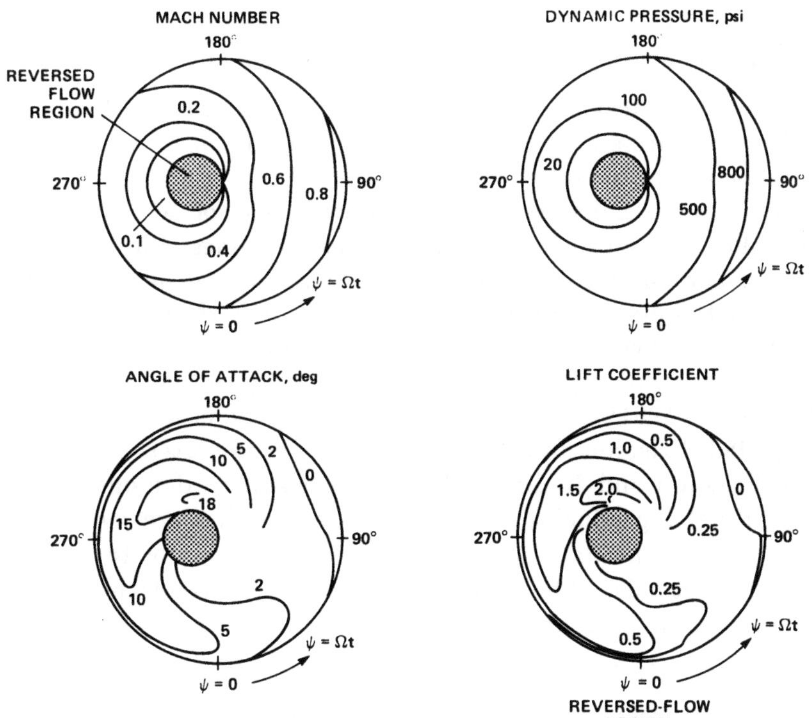

Figure 2. Blade-element conditions for a lightly-loaded rotor in high speed forward flight, (Rabbott et al., 1966); $V_\infty = 175$ knots, $\mu = 0.45$, $C_T/\sigma = 0.05$.

unsteady effects of rapidly-varying blade-element velocity and angle of attack, and the three-dimensional effects of yaw and rotational accelerations. Experiments indicate that all of these factors may be important, but the one that seems to be predominant is the time-dependent angle of attack, or pitch rate effect. The importance of the rate of change of $\alpha(t)$ is illustrated in Figure 3, which shows the lift and pitching-moment coefficient behavior for several experimental configurations. All of the unsteady cases are characterized by maximum airloads that exceed the static values and by large hysteresis in the flow behavior, according to whether the angle of attack is increasing or decreasing. The phenomenon of *dynamic stall* is characterized by strong vortical disturbances that are shed from the leading-edge region, convect over the upper surface of the airfoil, and induce highly nonlinear fluctuating airloads (McCroskey, 1981). Maximum lift is also known to be augmented by three-dimensional yaw effects (St. Hilaire et al., 1979), but the hysteresis evident in the model rotor measurements is only produced by unsteady effects.

To date, most of the research related to retreating-blade stall has been performed for the simpler model problem of a two-dimensional oscillating airfoil (Figure 3a), and for the associated unsteady boundary layer behavior. Furthermore, most of what is known today about the characteristics and various regimes of dynamic stall (Figures 4 and 5), has come from experiments, in two or three dimensions. Attempts to calculate the quantitative effects of dynamic stall have not been very successful up to now. Of the various two-dimensional semi-empirical correlation techniques that are available (McCroskey, 1981; Harris and Pruyn, 1968; Tran et al., 1982; Gangwani, 1983; Ericcson and Reding, 1983), only the latest (Gangwani, 1983; Ericcson and Reding, 1983) have begun to demonstrate the capability of reproducing experimental results that are significantly different from the data sets that were used originally to define or tune the methods. The three-dimensional aspects of the problem remain virtually untouched.

Three principal issues have thus emerged in the general area of retreating-blade stall. The first question is, how can the unsteady airloads be predicted with confidence over a wide range of unsteady flow conditions and blade geometries? An adequate data base already exists for the unsteady airloads on oscillating airfoils, although additional detailed measurements of boundary-layer separation characteristics would be useful. More importantly, a wide range of theoretical and numerical studies of unsteady viscous flows

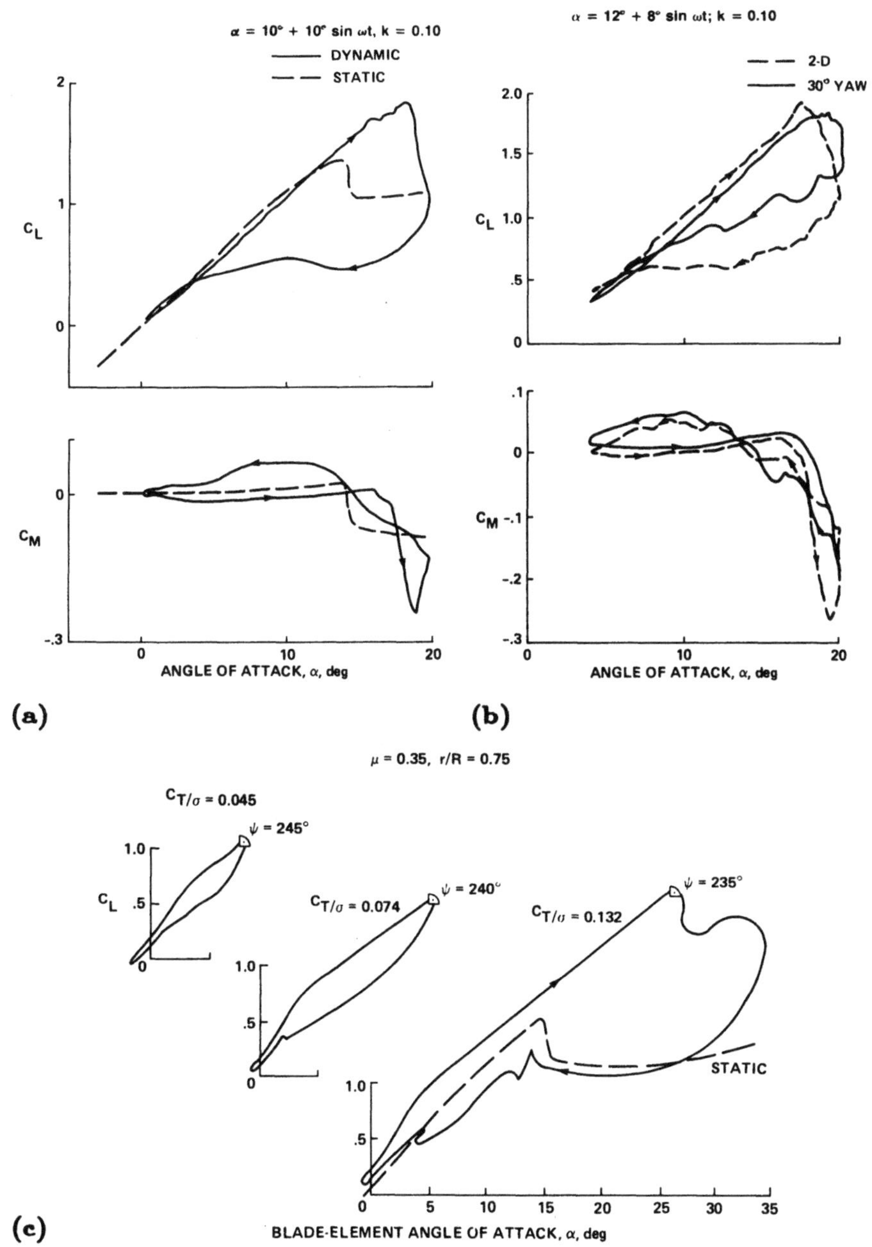

Figure 3. Lift and pitching-moment behavior vs. aerodynamic angle of attack, (a) unsteady two-dimensional flow, pitching oscillations (McCroskey, 1981); (b) unsteady yawed flow, pitching oscillations (St. Hilaire et al., 1979); (c) model rotor blade in forward flight (McCroskey and Fisher, 1972).

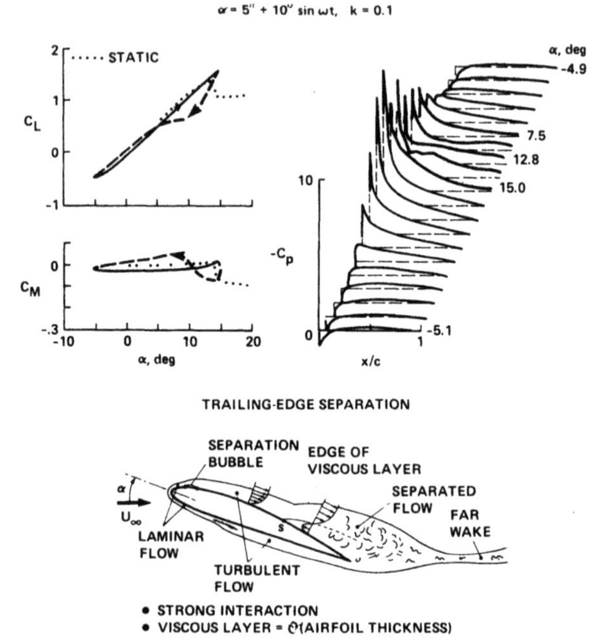

Figure 4. Light dynamic stall on an oscillating airfoil.

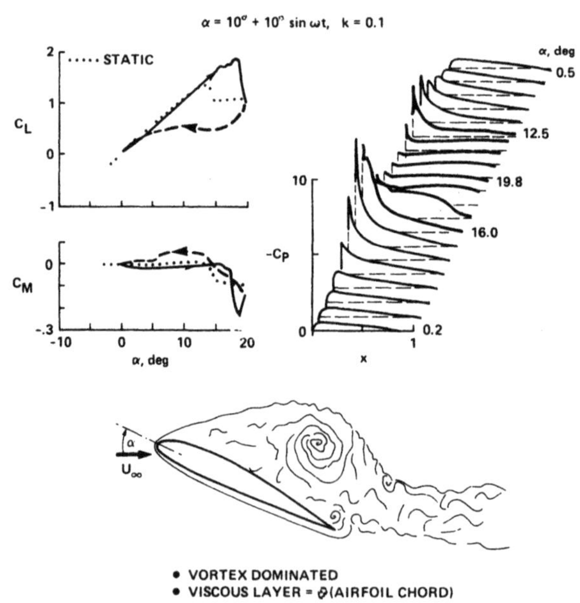

Figure 5. Deep dynamic stall on an oscillating airfoil.

is sorely needed to provide a better basic understanding of the phenomena, to improve the predictive capability, and to assess and guide the development of more general empirical correlation techniques.

The second question is, to what extent can information derived from either theoretical or experimental studies of dynamic stall in two dimensions be applied to the real three-dimensional problem of retreating-blade stall on helicopter rotors? This issue can probably only be addressed with new special-purpose experiments, on both rotating and nonrotating three-dimensional models.

Finally, perhaps the most important issue is whether practical means can be developed to delay retreating-blade stall or alleviate its adverse consequences. Since experiments have shown the qualitative features of dynamic stall to be surprisingly insensitive to airfoil geometry, major improvements in stall delay or control would seem to require a combination of creative new ideas for boundary-layer control or other unconventional devices, coupled with a series of careful experiments.

Figures 6 to 8 show three possibilities for affecting large changes in dynamic stall characteristics, by pneumatic or mechanical means. Although these sorts of devices have been used successfully in fixed-wing applications, only circulation control airfoils have been incorporated into an experimental helicopter rotor (Williams, 1976). However, much remains to be learned about the performance of either leading-edge or trailing-edge blowing under unsteady, three-dimensional, and separated flow conditions. The situation is likely to be further complicated by the development of transonic flow near the leading edge (see Section 3), if the blowing delays the onset of dynamic stall to significantly higher values of lift coefficient; that is, if it is successful.

With regard to mechanical devices, Carr and McAlister (1983) have found the leading-edge slat configuration shown in Figure 8 to be effective in suppressing dynamic stall on an airfoil undergoing sinusoidal oscillations in pitch, as shown in Figure 9. Their particular test configuration of a fixed-geometry slat suffered from high values of C_D at low angles of attack, but the dramatic improvement in C_L and C_M indicates that the concept is worthy of further exploration and refinement. As in the blowing cases, three-dimensional and transonic effects will also have to be studied before we can expect this device to be incorporated into an actual rotor. In any case,

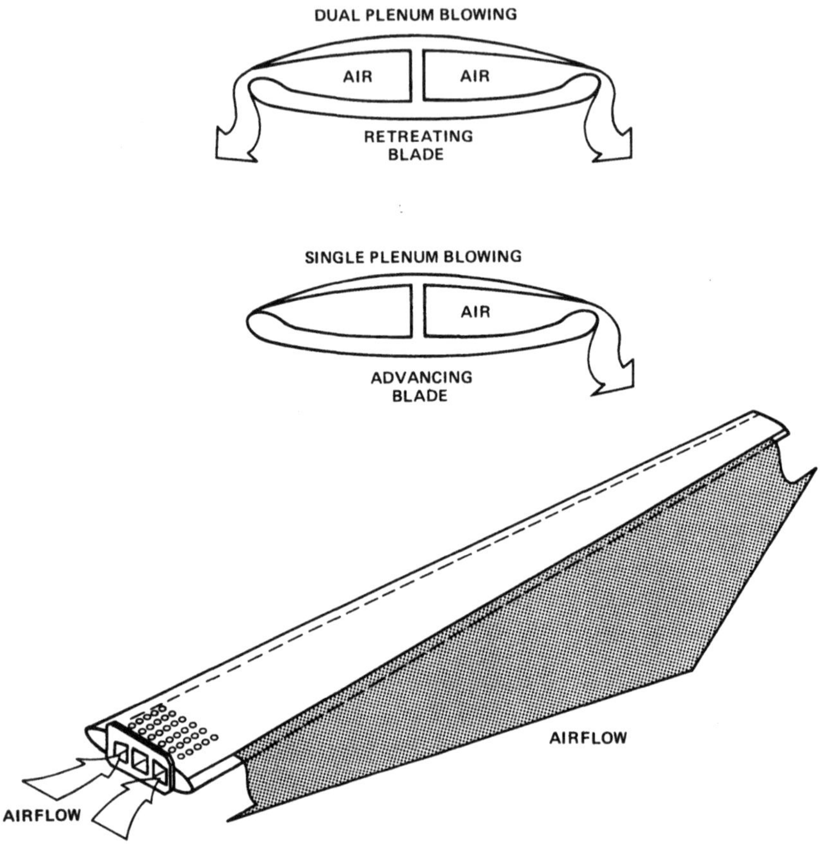

Figure 6. Circulation control to alleviate retreating blade stall (Williams, 1976). Reprinted with permission from Pergamon Press, Ltd.

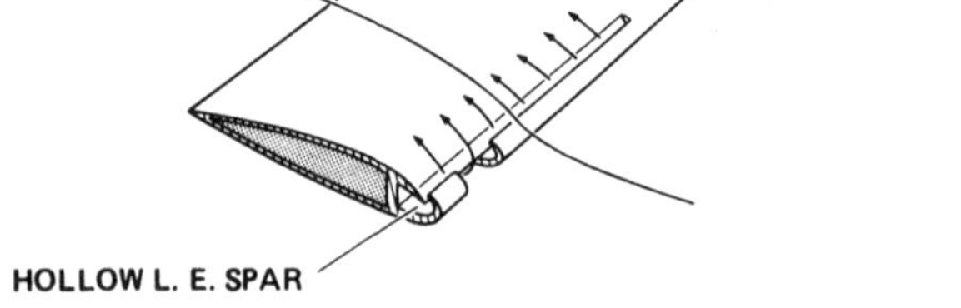

Figure 7. Leading-edge blowing to alleviate retreating blade stall (McCloud et al., 1960).

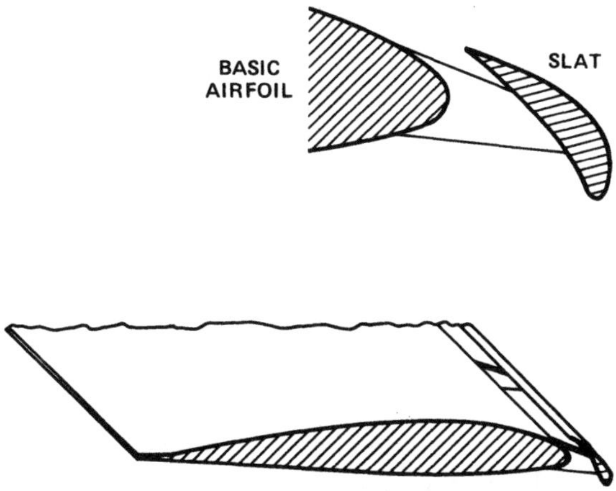

Figure 8. Leading-edge slat to alleviate retreating blade stall (Carr and McAlister, 1983). "Copyright c the American Institute of Aeronautics and Astronautics; reprinted with permission of the AIAA."

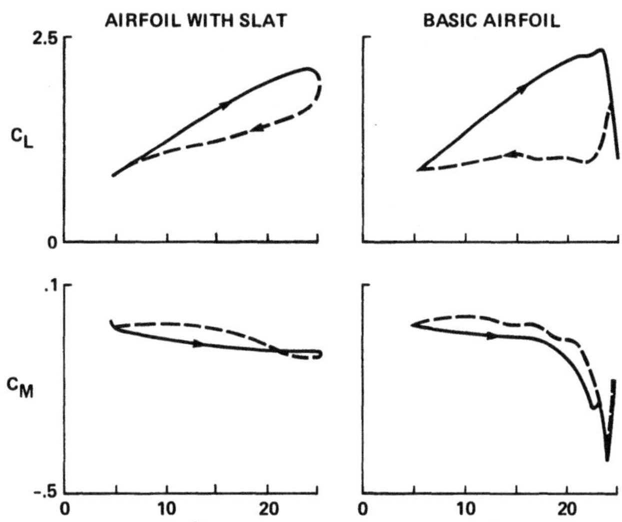

Figure 9. Comparison of lift, moment, and pressure distributions for a helicopter airfoil with and without a leading-edge slat; $\alpha = 15^\circ + 10^\circ \sin \omega t$, $M_\infty = 0.185$.

this or any other boundary-layer control device that could enable a rotor blade to approach the high-lift capabilities of modern airplane wings would pay handsome dividends for helicopter rotors.

3. Transonic Aerodynamics

Transonic flow phenomena commonly appear on the advancing blade tips of most modern helicopter rotors, where the blade-element Mach numbers are high; for example, as shown in Figure 2. It should be emphasized that shock waves may also appear on the retreating blade as well, when the blade-element angle of attack is sufficiently high to produce local regions of supersonic flow near the leading edge. Thus the transonic airfoil problem for rotors includes both of the facets illustrated in Figure 10. The periodic development and decay of shock waves on rotor blades in forward flight lead to vibratory airloads, degradations in performance and aerodynamic efficiency, and excessive noise.

In addition to the unsteady complications, the transonic flow on the advancing blade is highly three-dimensional, generally rendering two-dimensional information much less valuable than for the retreating-blade stall discussed above. Figure 11 shows representative spanwise variations in the pressure coefficient at the quarter-chord position on the advancing blade, along with an indication of the various flow regimes that develop. The change in the chordwise pressure distribution along the span is also indicated. In the figure, the transonic similarity parameter, following Isom (1974), is the following:

$$K = \frac{1 - M_T^2\,(r + \mu \sin\psi)^2}{[(\gamma+1)M_T^2\,(r + \mu\sin\psi)^2\,\tau]^{2/3}} \tag{3.1}$$

where M_T is the tip Mach number in hover, $\psi = \Omega t$ is the azimuthal position of the blade, r is the nondimensional radial dimension, $\mu = V_\infty/\Omega R$ is the advance ratio, τ is the airfoil thickness ratio, and γ is the ratio of specific heats. The inverse of K essentially represents the order of magnitude of the most important nonlinear term in the governing flow equation relative to the size of the linear terms. In practice, the relative importance of the various regions, especially Region 2 vs. Region 4 vs. Region 5, depends upon the flight conditions and the blade geometry. In all cases, however, the flow in the outboard regions is three-dimensional, and calculations

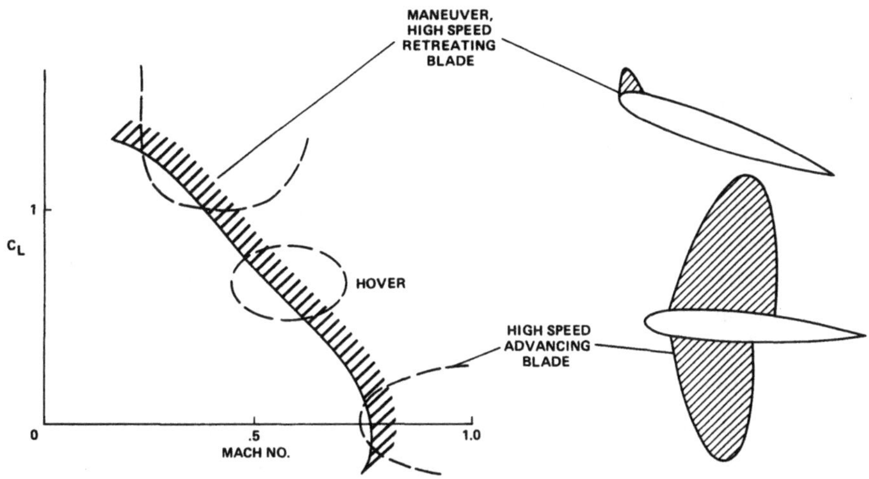

Figure 10. Flow regimes for a rotor airfoil.

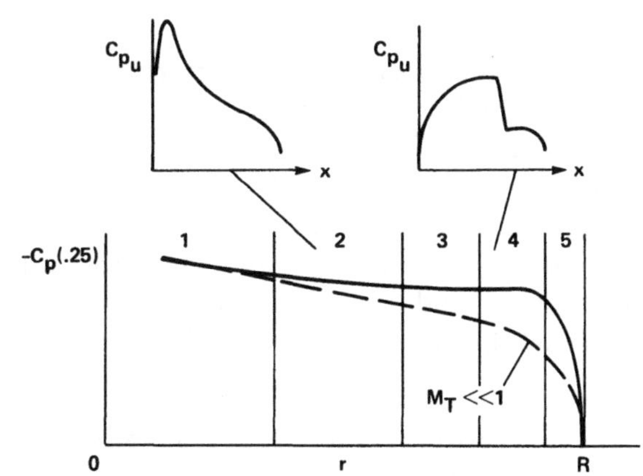

1: K → ∞ INCOMPRESSIBLE

2: K >> 1 COMPRESSIBLE, LINEAR, BLADE-ELEMENT

3: K ≃ 1.5 COMPRESSIBLE, NONLINEAR, NO SHOCKS

4: K ~ 1 TRANSONIC, NONLINEAR, SHOCKS

5: K ~ 1 TRANSONIC, 3-D TIP RELIEF, WEAKER SHOCKS

Figure 11. Spanwise variations and flow regimes for an advancing rotor blade.

also show important unsteady effects in forward flight, when the flow there is transonic.

It is important to recognize three essential differences between the transonic aerodynamics of fixed-wing aircraft and helicopter blades. The first is that the blade-element Mach number increases linearly with distance from the axis of rotation. This enhances transonic effects near the blade tip, in opposition to and in competition with the three-dimensional tip-relief effect there. This tip relief, of course, is basically the same as for a nonrotating wing. However, the behavior beyond the tip is different, because of the linearly-varying Mach number (Schmitz and Yu, 1983).

The second important difference is that in forward flight the blade-element Mach number and angle of attack vary with time, or rotor blade azimuthal position, as shown in Figure 2. This effect is relatively unimportant in the inboard, linear regimes, but the outboard transonic flow field takes longer to develop, and unsteady effects become essential features of the flow at high advance ratio. Because of the lag in adjusting to the time- and space-varying local Mach numbers, the flow in the second quadrant, $90° < \psi < 180°$, generally has stronger and more persistent shock waves than that of the first quadrant, $0 < \psi < 90°$.

The third difference, also of great importance, is the presence of concentrated tip vortices that trail from the tips of preceding blades. Depending on the flight conditions, these trailing vortices may remain near the path of the advancing blade, causing large disturbances to the shock-wave development and decay. This aspect of the problem cannot be uncoupled from the rotor wake structure, described in Section 4, and it is probably the most difficult and challenging aspect of the aerodynamics of the advancing blade.

Still another factor that is becoming increasingly important is the trend toward exotic tip geometries for advanced rotor designs. Figure 12, adapted from Philippe and Vuillet (1983), illustrates some examples which are already used on helicopters or are being developed. These various planforms are designed to alleviate shock-wave development by sweep effects, to reduce vibratory loads by either aerodynamic or structural dynamic changes, to increase hover performance by changes in spanwise circulation distribution, or to combine several of these factors. The systematic evaluation of such tip shapes is a difficult task beyond our present capabilities. In addition, each of the planforms in Figure 12 would require some adaptation of the computational grid in any numerical analysis of the tip region. Thus, the subject of grid generation becomes yet another fertile area for research.

The accurate treatment of the transonic regimes on rotor blades is, therefore, a formidable challenge. Fortunately, considerable assistance can be obtained from the large background of computational aerodynamic methods that have been developed over the past decade for fixed-wing applications, despite the aforementioned differences. Although it is by no means a trivial exercise to include both unsteady and three-dimensional effects, existing solution algorithms for the transonic small-disturbance, full potential, Euler, and even the Navier-Stokes equations, with sufficient effort, could be adapted to helicopter problems. Significant progress has already been made in recent years, and the level of complexity that has been implemented includes the following:

(1) The unsteady, three-dimensional small-disturbance equation for a lifting blade with an approximate wake model, e.g., Figure 13 from Caradonna et al. (1982).

(2) The quasi-steady full-potential equation for a highly-swept, nonlifting blade, e.g., Figure 14 from Tauber et al. (1984).

(3) The unsteady full-potential equation for a nonlifting blade, e.g., Figure 15 from unpublished work by I. C. Chang at the NASA Ames Research Center.

It should be mentioned that for realistic rotor calculations, the transonic field calculations have to be coupled with the rotor wake, which discussed in more detail in Section 4. At present it is not practical to do more than solve the governing equations by finite-difference methods on an isolated blade and within a limited computational volume around the blade tip. Therefore, the information about the vortex system in the wake and the associated rotor-induced inflow is accounted for either as prescribed modifications to the boundary conditions on the outer boundary of the computational box, or as an additional, interior-flow boundary condition that is prescribed within the computational box, or both. This is indicated schematically in Figures 16 and 17. As shown in Figure 17, the interior of the computational domain includes the usual trailing sheet of vorticity behind the blade, which is assumed to be undistorted, and one or more segments of line vortices representing the trailing tip vortex (or vortices) from the preceding blade(s). The line vortex segments are connected with the outer boundary by branch cut

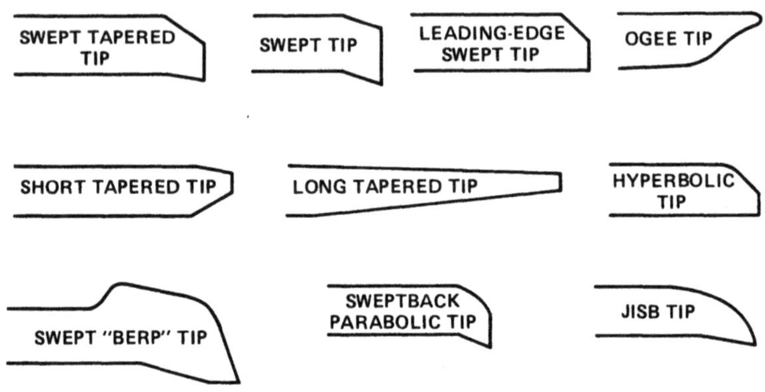

Figure 12. Examples of advanced rotor blade tips.

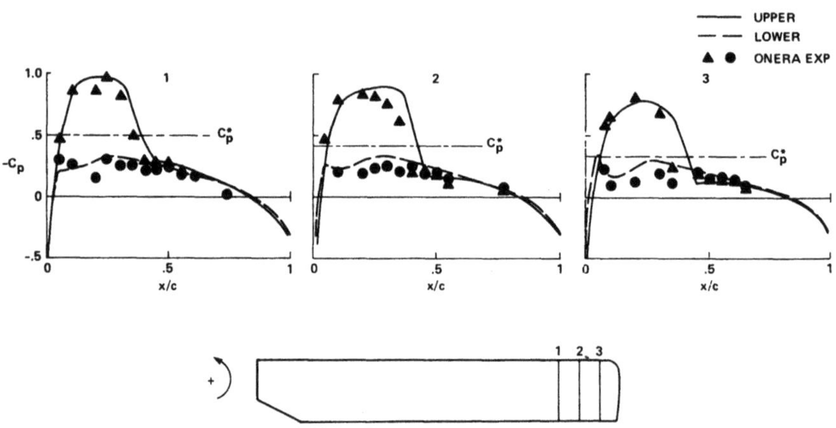

Figure 13. Pressure distributions on a lifting rotor at high advance ratio, using an unsteady small-disturbance code (Caradonna et al., 1982); $\mu = 0.36$, $C_T/\sigma = 0.05$, $\psi = 90°$.

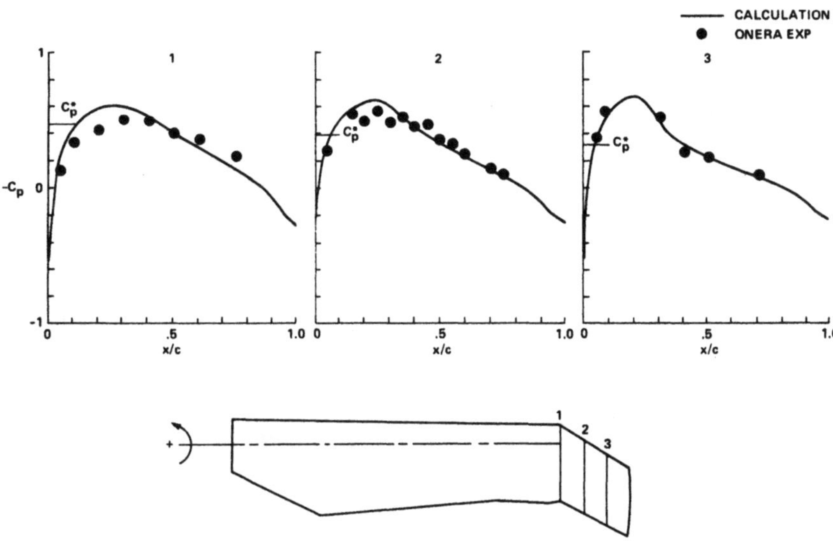

Figure 14. Pressure distributions on a nonlifting rotor at high advance ratio, using a quasi-steady full-potential code (Tauber et al. [24]); $\mu = 0.40$, $\psi = 90°$.

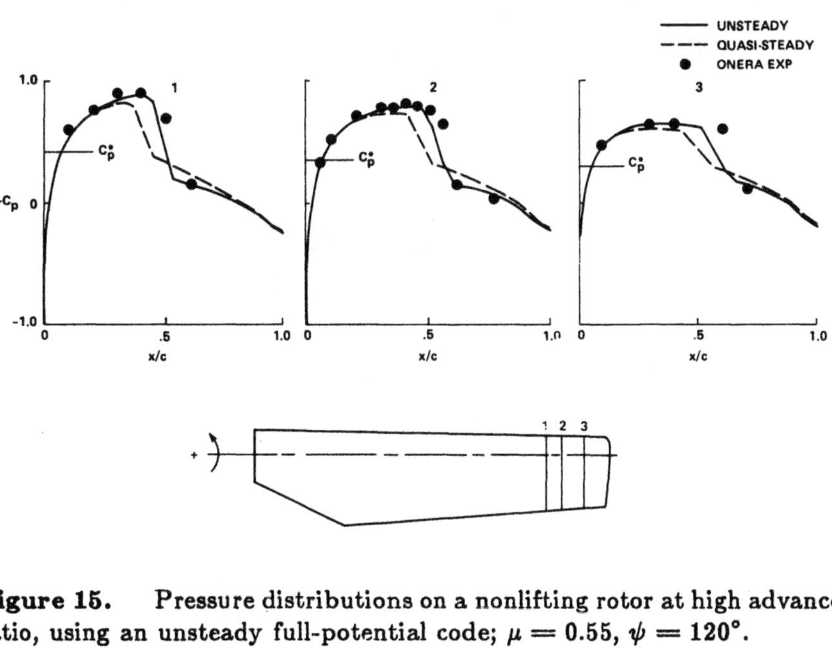

Figure 15. Pressure distributions on a nonlifting rotor at high advance ratio, using an unsteady full-potential code; $\mu = 0.55$, $\psi = 120°$.

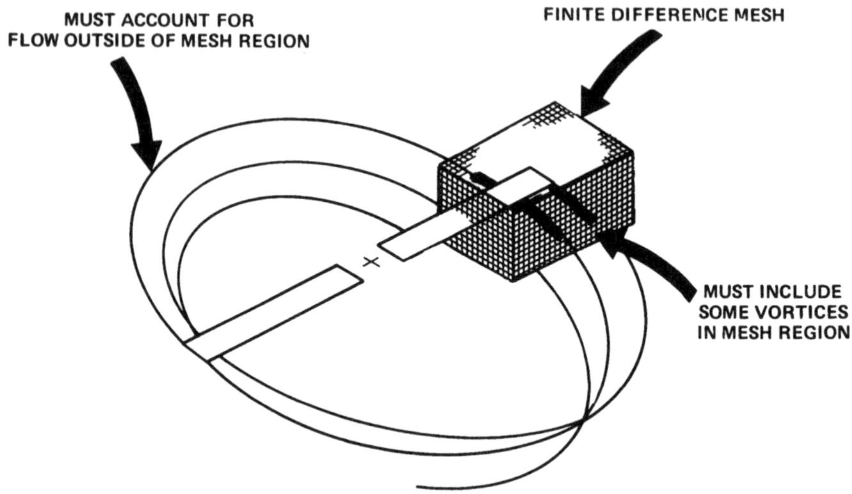

Figure 16. Aerodynamic modeling of the rotor for finite-difference methods.

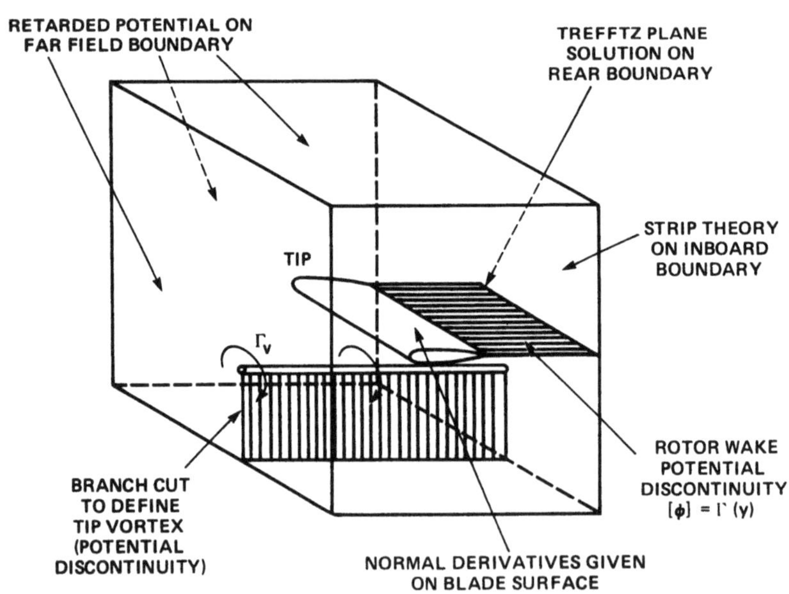

Figure 17. Boundary conditions for potential flow past rotor blades.

surfaces, across which jumps in potential are imposed that correspond to the strengths of the concentrated vortices.

The foregoing procedure makes it possible to perform meaningful transonic calculations for special cases. However, further improvements are required in two rather diverse directions. First, the existing codes are still too expensive and complex, in general, for routine use in engineering design. Therefore, serious efforts are needed to streamline the methodology and to improve the computational efficiency, without losing the essential features of the problem. On the other hand, the current assumptions and restrictions are still too severe, in general, to allow for accurate simulations of many of the most crucial operational conditions of advanced rotor systems. Therefore, improved transonic aerodynamics modeling is needed, but it must not add excessive burdens to the computational requirements. Clearly this area offers considerable opportunities and challenges for future research.

4. Rotor Wakes and Vortices

4.1 Global Wake Modeling

Accurate prediction of the vortical wake of a helicopter rotor is probably the most important, the most studied, and the most difficult aspect of helicopter aerodynamics; consequently, it can be argued that this topic provides one of the greatest opportunities for new improvements. Certainly the treatment of the wake is crucial in terms of performance, efficiency, structural vibrations, and aerodynamic noise. Current methods of analysis range from relatively simple momentum-theory applications to free-wake lifting-surface codes that consist of several thousand lines of Fortran statements, and that require approximately one million words of computer memory. In between these two extremes, there are a variety of so-called prescribed wake models, which rely on some degree of empiricism to determine the position of the wake vortices; then the Biot-Savart law is used to calculate the velocity field that is induced by these vortices at the plane of the rotor blades.

The complexity of the problem in comparison with fixed-wing aircraft is illustrated in Figures 18 and 19. The detailed structure

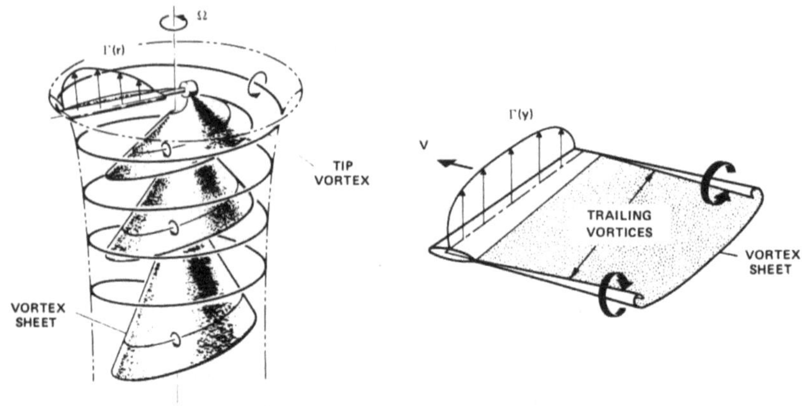

Figure 18. Wake of a rotor in hover vs. a fixed wing wake.

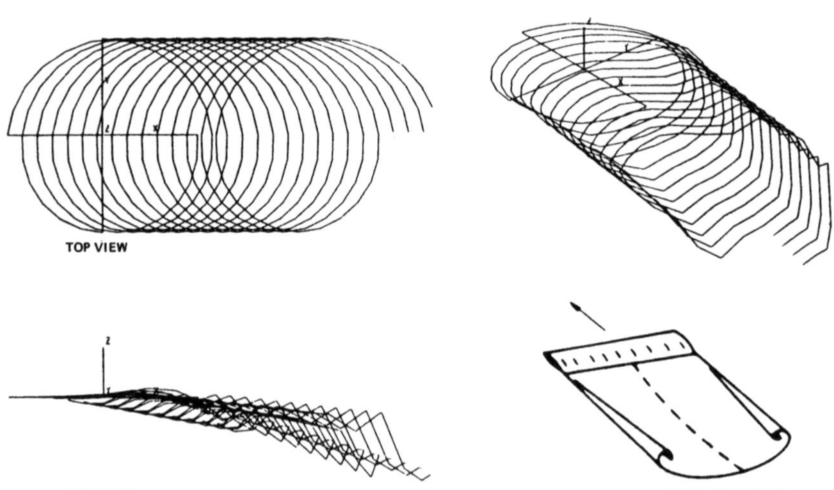

Figure 19. Wake of a rotor in forward flight (Egolf and Landgrebe, 1983); four blades, $\mu = 0.10$.

of an airplane wake is not all that simple, but because the wake elements trail rearward along approximately straight lines that are parallel to the flight direction, simple approximations to the tip-vortex rollup and downstream convection normally suffice. This is not at all the case for the helicopter rotor, whose blade tips trace out prolate cycloidal paths in space. This provides numerous opportunities for complex interactions between the vortices and the blades and between the vortices themselves.

It is beyond the scope and intent of this paper to review the state of the art in rotor wake modeling; a few examples are given to point out some features of the problem and to mention some of the limitations of existing methods. For example, Figure 19 from Egolf and Landgrebe (1983) is representative of the latest generation of prescribed wake analyses; this is an extension to forward flight of one of the better-known generalized prescribed-wake methods for hover (Landgrebe et al., 1977). This method is simpler and less expensive than the free-wake codes of comparable accuracy.

However, an inherent limitation of the prescribed-wake approach is that the empirical determination of the wake shape ignores some of the details of the particular case under consideration, including the mutual interaction between vortex elements. As a result, a prescribed-wake configuration is not, in general, a valid solution to an inviscid free-vortex flow (Bliss et al., 1983). In most cases where the wake geometry is not too different from the configurations that produced the original empiricism, this may not be a serious error. However, prescribed-wake models are unreliable for unusual blade planforms and/or twist distributions, and these are often the cases that are the most interesting to explore.

In some of these more challenging cases, the blade airloads depend strongly upon both the wake geometry and the method for calculating the flow. A relatively simple example of this is given in Figure 20. This figure, prepared by Dr. Chee Tung of the U. S. Army Aeromechanics Laboratory, shows a comparison of lifting-line and lifting-surface calculations with the hover data of Caradonna and Tung (1981) for a model rotor with two low-aspect-ratio blades of rectangular planform and zero twist. The computed results were obtained using the computer code described in Summa (1982). Here the local lift coefficient is nondimensionalized by the tip speed, ΩR.

For this example, the lifting-line calculations were in error, regardless of the wake model. The prescribed-wake option of the

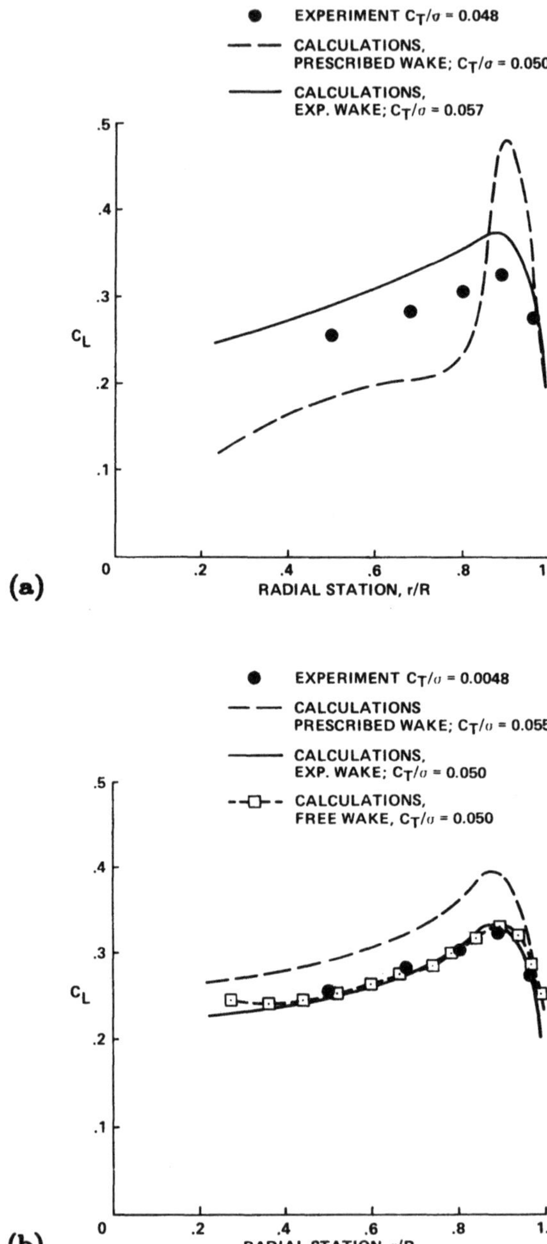

Figure 20. Spanwise lift distribution in hover; $AR = 6$, $\theta_c = 8°$, (a) lifting-line calculations for prescribed and experimentally-derived wake geometries; (b) lifting-surface calculations for prescribed, free-wake, and experimentally-derived wake geometries.

program also gave an erroneous spanwise load distribution, regardless of which representation was used for the surface of the blade, primarily because the prescribed wake geometry was not correct. On the other hand, the free-wake calculation predicted the actual wake geometry reasonably well. Consequently, the lifting-surface calculation gave essentially the same satisfactory results either when using the wake prescribed according to the experimental measurements or when using the free-wake option of the code. As a final footnote, however, Dr. Tung reports that this same free-wake, lifting surface code fails to predict adequately some cases with highly nonlinear twist distributions. This exercise illustrates the difficulty and importance of the wake geometry for the problem of a rotor in hover, and it indicates that further work is needed even without the complications of forward flight.

4.2 Blade-Vortex Interaction

The strong interaction between a segment of a rotor blade and the concentrated tip vortices in the wake is an important potential source of noise and vibration at low and moderate flight speed. The generic problem, sketched in Figure 21, is an unsteady, three-dimensional close encounter of a curved line vortex, at an arbitrary intersection angle, Λ_i, with a high-aspect-ratio lifting surface that is executing combined rotational and translational motion at transonic speeds. For many practical applications, it is almost impossible to separate the transonic aspects of the problem from the details of the vortex structure in the wake.

Insofar as basic research opportunities are concerned, the limiting cases of $\Lambda_i = 0$ and $90°$, as shown in Figure 22, are of particular interest. The former case is fundamentally unsteady but approximately two-dimensional, whereas the latter is essentially steady but highly three-dimensional. These two limiting cases are the basis of several ongoing research programs that are designed to explore the basic features of blade-vortex interactions, to develop methods of calculating such interactions, and to determine the minimum level of complexity that will be required in the future for adequate predictive capability in the complete rotor environment. The case of a vortex that is parallel to the free stream has received more attention in the past, as it is more relevent to fixed-wing applications and it is relatively easy to set up experimentally. On the other hand, the

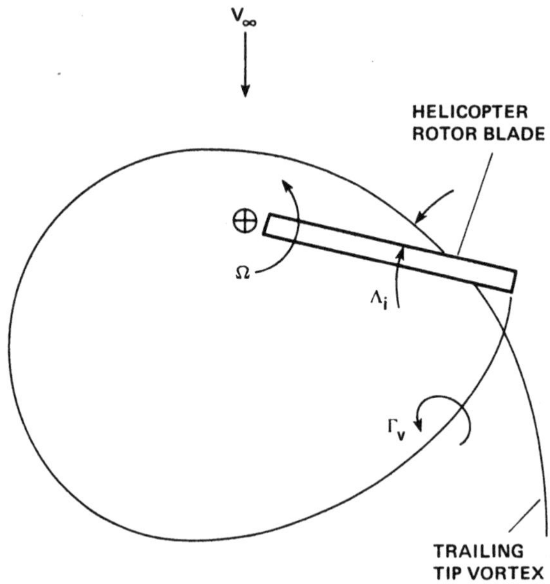

Figure 21. Sketch of helicopter blade-vortex interaction.

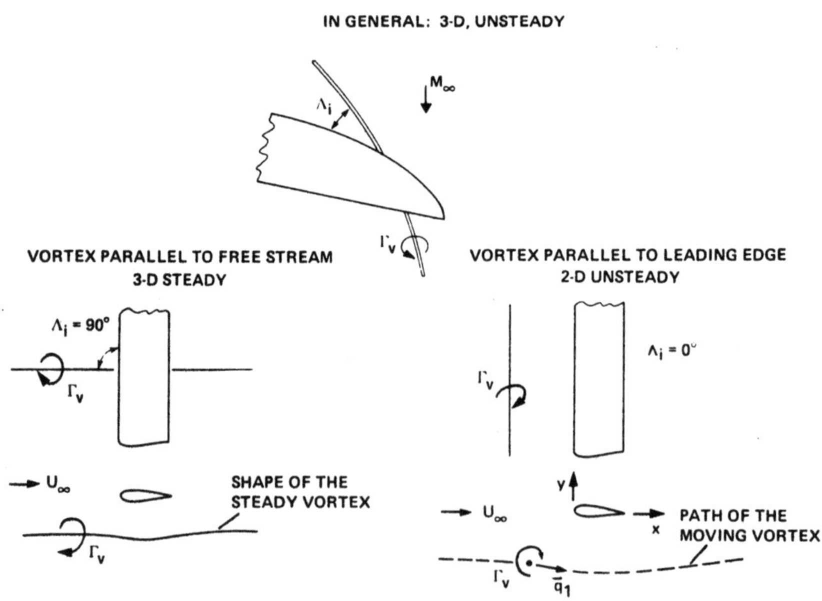

Figure 22. Limiting cases of blade-vortex interaction.

case of the vortex that is parallel to the leading edge is a simpler problem for theoretical and numerical analysis, and it requires less computer memory to store the solution variables.

For illustrative purposes, we shall briefly review some recent numerical work for the two-dimensional unsteady case, $\Lambda_i = 0$. Two of the basic issues are the introduction of the vortex into the numerical computations and the determination of its effect on the unsteady pressure distribution on the airfoil and nearby. The vortex can be introduced in two different ways, as indicated in Figure 23. Caradonna et al. (1982) and George and Chang (1983) have utilized the branch-cut method, as described in Section 3, and McCroskey and Goorjian (1983) employed the prescribed-disturbance method. All three groups solved the transonic small-disturbance equations by time-accurate ADI methods that had been developed and checked out previously, and the various results are in reasonable agreement on most points.

Representative results from McCroskey and Goorjian (1983) are shown in Figures 24 and 25 for two symmetrical airfoil sections. Figure 24 shows the distortions in the chordwise pressure distributions on the airfoil as the vortex passes underneath. Similar results are shown in Figure 25 for a thinner airfoil, including the time histories at individual locations on the airfoil. These results illustrate that strong gradients in pressure occur with respect to both time and space, due to the vortex encounter. These gradients can be especially significant in the leading-edge region of a thin airfoil.

Although the subject of airfoil-vortex interaction is just beginning to receive significant attention, several important features of the problem have already emerged. The first is that the effect of the vortex is felt primarily through the vertical velocity that it induces, which to first order appears as a time-dependent perturbation in the effective angle of attack, and secondarily through its horizontal induced velocity. Secondly, unsteady lag effects are very important, especially in the transonic case; calculations show enormous differences between quasi-steady and unsteady solutions and between the results for the vortex locations upstream and downstream of the airfoil. Thirdly, other strong nonlinear effects have been noted; even the qualitative pressure variations with respect to time and space differ markedly from the predictions of linear theory, and the vortex distorts the flow on the nearest surface of the body much more than

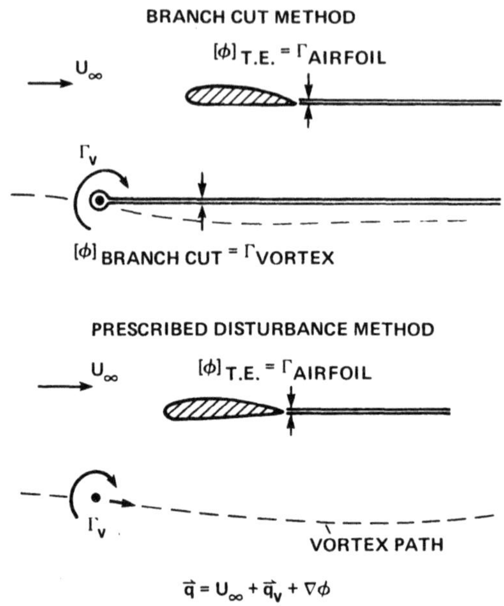

Figure 23. Two methods of introducing a vortex into the flow field.

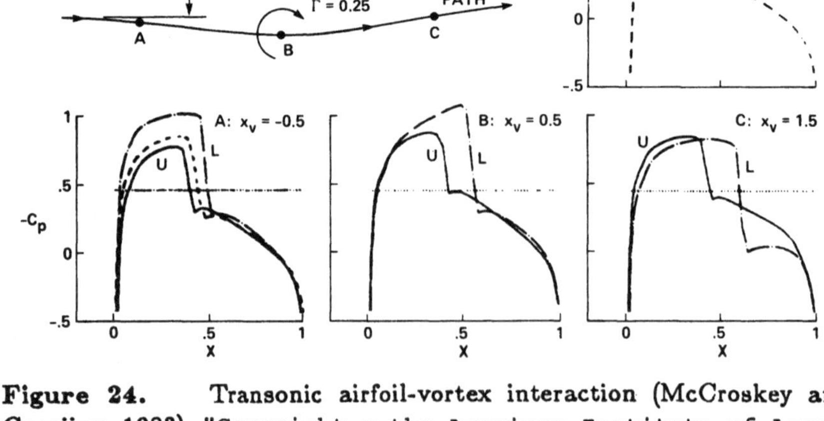

Figure 24. Transonic airfoil-vortex interaction (McCroskey and Goorjian, 1983). "Copyright c the American Institute of Aeronautics and Astronautics; reprinted with permission of the AIAA."

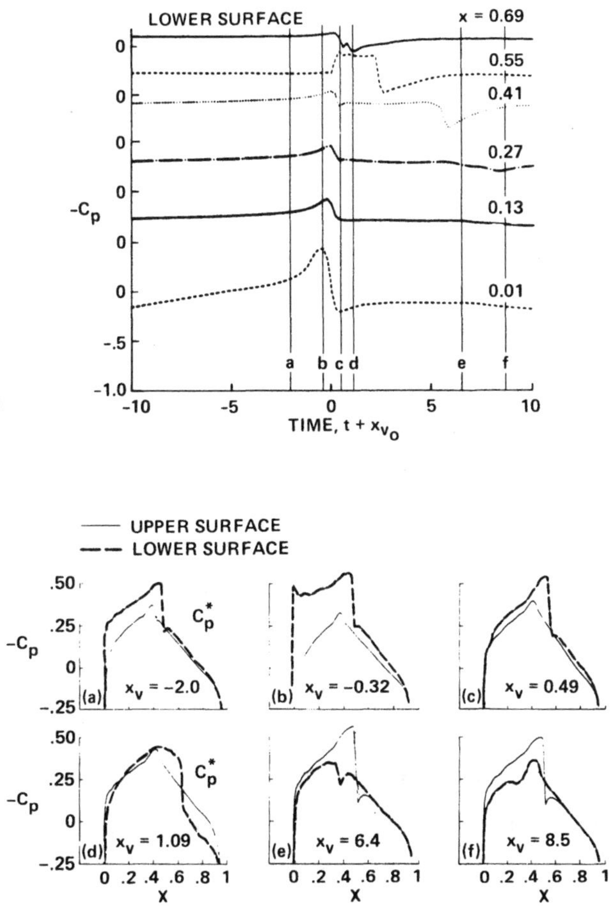

(a)

(b)

Figure 25. Pressure variations on the lower surface during an airfoil-vortex interaction (McCroskey and Goorjian, 1983); NACA 64A006 airfoil, $M_\infty = 0.85$, $\alpha = 0$, $\Gamma = 0.20$, $y_{v_o} = -0.26$, (a) time histories at specific points on the airfoil; (b) pressure distributions on the airfoil at specific times.

on the opposite surface. Finally, some of the pressure perturbations appear to leave the body as radiating waves; this phenomenon would seem to represent the far-field blade-vortex interaction noise of helicopter rotors.

The studies to date (Caradonna et al., 1982; George and Chang, 1983; McCroskey and Goorjian, 1983) have been performed with the small-disturbance equations, whose accuracy is questionable in the very region which seems to be important (Schmitz and Yu, 1983);

that is, near the leading edge. Therefore, it seems imperative to check the samll-disturbance codes with more accurate formulations and with experiments. Also, an ideal potential vortex with an invariant structure has been assumed thus far; but in reality, close encounters probably alter the vortex core significantly and may lead to vortex bursting. For rotor applications, future studies must be extended to include three-dimensions, where it will be even more important to establish what minimum level of complexity in the governing equations will suffice and to determine the most expeditious way of introducing the vortex into the computational domain. Finally, the severity of the blade-vortex interaction appears to be highly sensitive to the strength and position of the vortex, especially for transonic flow; therefore, the results of the computations will be no better than the predictions of the wake model used.

5. Concluding Remarks

The examples discussed in the preceding sections are but a few of the myriad of serious aerodynamic problems of helicopters that provide rich opportunities for future research. There are other examples which are not discussed in this selective review; however, there are several that are worth mentioning in passing.

The first is the general category of aerodynamic interference. As indicated in Figure 1, the flow field of the various components of the helicopter frequently influence other flow regions, and these interactions can be complex and nonlinear. Typical examples include the mutual interference and interaction between the main rotor and the fuselage, between the main rotor and the tail rotor, between the rotor hub and the fuselage, and between the tail rotor and the tail surfaces. Panel methods are being developed for predicting these flows; e.g., Clark and Maskew (1983), but the current industry approach is mostly empirical. A closely-related problem that has arisen recently is the rotor-induced *download* on the wing of the tilt-rotor aircraft, Figure 26. In hover, the downwash from the rotors impinges on the wing, causing a net loss in lifting capability of 10% or more of the gross weight of the aircraft. Predicting this effect accurately and designing flap configurations to minimize the download are particularly challenging problems, but ones which have great payoff.

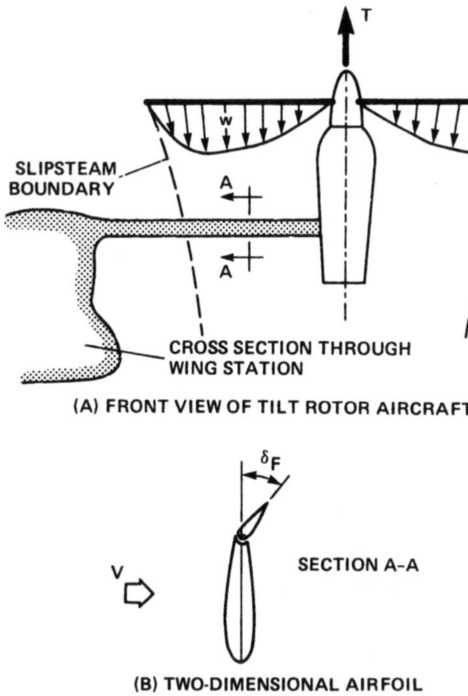

Figure 26. The tilt-rotor aircraft in the hover mode.

Another important area is that of simply improving the aerodynamic tools that are available to helicopter engineers. The need for simpler and more efficient computational techniques that can more accurately solve more complex problems has already been stressed. However, improvements in the coupling of aerodynamic prediction techniques with structural-dynamics codes is also essential. In addition, there would appear to be tremendous opportunities for coupling efficient aerodynamic codes with numerical optimization techniques to design improved blade configurations. This has already been demonstrated in two dimensions for transonic rotor airfoil sections (Hicks and McCroskey, 1980). The extension to three dimensions, that is, including planform and twist distributions, is the next logical step, although it is obviously a very large step.

Experiments will continue to play a major role in helicopter technology for the foreseeable future, and there is much to be done in the area of instrumentation, experimental methods, and testing techniques. One perennial issue is the gap between model and full-scale testing; the recent paper by Keys et al. (1983), for example, discusses some aspects of this problem. Mach-number scaling is clearly essential; but this is hardly a research issue anymore, even though constructing dynamically scaled, instrumented model rotors to operate at full-scale tip speeds can be a vexing engineering exercise. On the other hand, typical Reynolds numbers for model and full-scale helicopters span regimes where the boundary-layer transition and separation characteristics may be considerably different. Therefore, the model/full-scale issue will continue to be important for any aspect of helicopter aerodynamics in which viscous effects play a significant role.

A strong need for a wide variety of special-purpose experiments will continue to exist. For example, tests of complete rotor systems certainly have their place, but a full-scale helicopter blade is not the best experimental configuration to study the effects of unsteadiness on the Reynolds stresses of a separating turbulent boundary layer, nor to validate initially a new transonic code. Relatively simple experiments that are run under well-controlled and thoroughly-documented conditions are essential to sort out the dominant features of the various flow fields, to increase our fundamental understanding of the basic phenomena, to guide the development of new theoretical models, and to validate methodically the new computational algorithms and prediction techniques as they are being developed.

Finally, for both the global studies and for the special-purpose experiments, a wide variety of new instrumentation, measurement techniques, and data acquisition and processing methods need to be developed. These are required to cope with the complexity of the unsteady three-dimensional flow fields of the various helicopter components and with handling the large volume of data that are generated. It is as challenging to measure the right information and to eliminate the superfluous, as it is to develop the new predictive methodology.

In conclusion, the examples discussed in this paper illustrate the complexity of helicopter aerodynamics and the many important issues that remain to be studied. There are numerous challenges and a wide range of opportunities for the technologist and for the basic research scientist to explore. Their various contributions will assure continuing and major improvements to this unique flying machine.

References

[1] Bliss, D. B., Quackenbush, T. R., and Bilanin, A. J. "A New Methodology for Helicopter Free Wake Analysis," *Proc. Am. Hel. Soc. Forum, Paper No. A-83-39-75-0000,* May 1983.

[2] Caradonna, F. X. and Tung, C. "Experimental and Analytical Studies of a Model Helicopter Rotor in Hover," *NASA TM-81232,* Sept. 1981.

[3] Caradonna, F. X., Desopper, A., and Tung, C. "Finite Difference Modeling of Rotor Flows Including Wake Effects," *NASA TM 84280,* Aug. 1982.

[4] Carr, L. W. and McAlister, K. W. "The Effect of a Leading-Edge Slat on the Dynamic Stall on an Oscillating Airfoil," *AIAA Paper No. 83-2533,* Oct. 1983.

[5] Clark, D. R. and Maskew, B. "Aerodynamic Modeling of Helicopter and Tilt Rotor Configurations," *Proc. Am. Hel. Soc. Forum, Paper No. A-83-39-72-0000,* May 1983.

[6] Egolf, T. A. and Landgrebe, A. J. "Helicopter Rotor Wake Geometry and Its Influence in Forward Flight," *NASA CR 3726,* 1983.

[7] Ericcson, L. E. and Reding, J. P. "The Difference Between the Effects of Pitch and Plunge on Dynamic Airfoil Stall," *Proc. Ninth European Rotorcraft Forum, Paper no. 8,* Sept. 1983.

[8] Gangwani, S. T. "Synthesized Airfoil Data Method for Prediction of Dynamic Stall and Unsteady Airloads," *Proc. Am. Hel. Soc. Forum, Paper No. A-83-39-002-0000,* May 1983.

[9] George, A. R. and Chang, S. B. "Noise due to Transonic Blade-Vortex Interactions," *Proc. Am. Hel. Soc. Forum, Paper No. A-83-39-50-D000,* May 1983.

[10] Harris, F. D. and Pruyn, R. R. "Blade Stall—Half Fact, Half Fiction," *J. Am. Hel. Soc., 13, No. 2* (1968), 27–48.

[11] Hicks, R. M. and McCroskey, W. J. "An Experimental Evaluation of a Helicopter Rotor Section Designed by Numerical Optimization," *NASA TM-78622,* March 1980.

[12] Isom, M. P. "Unsteady Subsonic and Transonic Potential Flow Over Helicopter Rotor Blades," *NASA CR-2463,* Oct. 1974.

[13] Keys, C. N., McVeigh, M. A., Dadone, L., and McHugh, F. J. "Considerations in the Estimation of Full-Scale Rotor Performance From Model Rotor Test Data," *Proc. Am. Hel. Soc. Forum, Paper No. A-83-39-03-0000,* May 1983.

[14] Landgrebe, A. J., Moffitt, R. C., and Clark, D. R. "Aerodynamic Technology for Advanced Rotorcraft," *J. Am. Hel. Soc., 22, No. 2*

(1977), 21–27.

[15] McCloud, J. L., III, Hall, L. P., and Brady, J. A. "Full-Sclae Wind-Tunnel Tests of Blowing Boundary-Layer Control Applied to a Helicopter Rotor," *NASA TN D-335,* 1960.

[16] McCroskey, W. J. "The Phenomenon of Dynamic Stall," *NASA TM 81264,* 1981.

[17] McCroskey, W. J. and Fisher, R. K. "Detailed Aerodynamic Measurements on a Model Rotor in the Blade Stall Regime," *J. Am. Hel. Soc., 17, No. 1* (1972), 20–30.

[18] McCroskey, W. J. and Goorjian, P. M. "Interactions of Airfoils With Gusts and Concentrated Vortices in Unsteady Transonic Flow," *AIAA Paper 83-1691,* July 1983.

[19] Philippe, J. J. and Vuillet, A. "Aerodynamic Design of Advanced Rotors With New Tip Shapes," *Proc. Am. Hel. Soc. Forum, Paper No. A-83-39-04-0000,* May 1983.

[20] Rabbott, J. P., Jr., Lizak, A. A., and Paglino, V. M. "A Presentation of Measured and Full-Scale Rotor Blade Aerodynamic and Structural Loads," *US Army AVLABS Tech. Report 66-31,* 1966.

[21] St. Hilaire, A. O., Carta, F. O., and Fink, M. R. "The Influence of Sweep on the Aerodynamic Loading of an Oscillating NACA 0012 Airfoil," *NASA CR-3092,* 1979.

[22] Schmitz, F. H. and Yu, Y. H. "Helicopter Impulsive Noise: Theoretical and Experimental Comparisons," *Recent Advances in Aeroacoustics,* New York: Springer-Verlag, 1984; also *NASA TM 84390,* 1983.

[23] Summa, J. M. "Advanced Rotor Analysis Methods for the Aerodynamics of Vortex/Blade Interactions in Hover," *Proc. Eighth European Rotorcraft Forum, Paper No. 2.8,* Sept. 1982.

[24] Tauber, M. E., Chang, I. C., Caughey, D. A., and Philippe, J. J. "Comparison of Calculated and Measured Pressures on Straight and Swept-Tip Model Rotor Blades," *NASA TM,* to be published, 1984.

[25] Tran, C. T., Petot, D., and Falchero, D. "Aeroelasticity of Helicopter Rotors in Forward Flight," *la Recherche Aérospatiale, 1982, No. 4* (1982), 11–25.

[26] Williams, R. M. "Application of Circulation Control Rotor Technology to a Stopped Rotor Aircraft Design," *Vertica, 1, No. 1* (1976), 3–15.

The Rise and Demise of Trailing Vortices*

Turgut Sarpkaya
Naval Postgraduate School
Monterey, CA 93940

Abstract

Experiments were conducted with three Delta wings and two rectangular wings to investigate the evolution of trailing vortices in stratified and unstratified water. The vortex trajectories were determined as a function of the normalized time $t^* = V_o t/b_o$, stratification parameter $N^* = N b_o/V_o$, and an effective vortex core size $r^* = r_e/b_o$. The results have shown that the vortices rise only to a finite height as they demise gradually at first and rapidly thereafter under the influence of turbulence, sinusoidal instability, and core bursting. The effect of stratification is to reduce the lifespan of vortices and the maximum height attained by them.

1. Introduction

Vortices and vortex wakes have become a major theme of aerodynamics research since the advent of the large aircraft and the understanding of their evolution required an examination of many fundamental problems in fluid mechanics. Numerous studies have shown that there are several obstacles to the development of a better understanding of the important features of trailing vortices. The principal ones are as follows: roll-up process and its relation to the tip shape and the distributions of the initial velocity and turbulence; probe sensitivity of the vortices and the need to use nonintrusive means of measurement; large-scale instabilities, such as the sinusoidal instability (Crow instability, (Crow, 1970)) and the vortex breakdown; scale effects (even the highest Reynolds numbers, based on wing chord, reached in wind tunnels or towing basins, are an order

---

*This is an extended summary of the paper which was published in the *Journal of Fluids Mechanics, 136* (1983), pp. 85–109, under the title "Trailing Vortices in Homogeneous and Density Stratified Media."

of magnitude lower than what is possible for an aircraft); and the ambient conditions, such as shear, turbulence, and stratification which play major roles in the evolution of vortices. The quantification of these effects offers exceedingly complex difficulties.

Mathematical and numerical dynamic modelling of a two-dimensional vortex pair has been attempted, sometimes with discordant results (Widnall, 1975) regarding the vortex spacing, migration speed, and the maximum height attained by the vortices.

2. Observations and Measurements

The experiments were performed in a towing tank with five lifting surfaces (three Delta wings and two rectangular wings, with a NACA-0012 cross section). The interior of each model was hollowed and used as a dye reservior to seed the vortex cores. All trailing vortices and vortex rings were recorded on high-speed film until the time they have completely dissipated.

The data were analyzed according to

$$H/b_o = f\,(V_o t/b_o, D/b_o, Nb_o/V_o, V_o^2/gb_o, V_o b_o/\nu, r_e/b_o) \qquad (2.1)$$

in which H represents the displacement of the vortex; b_o, the initial separation of the vortex pair; V_o, the initial mutual-induction velocity of the vortex pair; D, the initial depth of the vortex pair; N, the Brunt-Vaisala frequency; and r_e, the core radius. The results have shown that for D/b_o larger than about 7, H/b_o is found to depend only on $t^* = V_o t/b_o$, $N^* = Nb_o/V_o$, and $r^* = r_e/b_o$, i.e., $H^* = f(t^*, N^*, r^*)$.

In a weakly stratified medium (N^* less than about 0.3), linking of the vortices and/or the cascade of core bursts are primarily responsible for the destruction of the vortex pair. In this process, the role of the ever present turbulent diffusion becomes paramount after the onset of the said instabilities. In a strongly stratified medium, the amplitude of the sinusoidal instability remains relatively small and the vortices do not link. In this case, the level of initial turbulence generated by the vortex pair, additional turbulence generated by core bursting, and the countersign vorticity generated at the boundaries of the recirculation cell determine the lifespan of the vortices even in a medium otherwise free from turbulence. The numerical simulations of Hecht et al. (1981) for $N^* = 0.67$ tend to support this view. A precise demarcation of the boundaries of weak and strong stratification in terms of N^* and the initial level of turbulence is not yet possible. There may even exist an intermediate stratification for which the sinusoidal instability, occasional linking, core bursting, and turbulent diffusion play equally important roles throughout the rise and demise of the vortices.

The suitability of a towing tank in conducting controlled laboratory experiments with trailing vortices and the importance of t^*, N^*, and the tip-shape of the test models (in determining the initial turbulence level and the core size) have been shown either directly or indirectly throughout the paper (Sarpkaya, 1983).

References

[1] Crow, S. C. "Stability Theory for a Pair of Trailing Vortices," *AIAA J., 8* (1970), 2172–79.

[2] Hecht, A. M., Bilanin, A. J., and Hirsh, J. E. "Turbulent Trailing Vortices in Stratified Fluids," *AIAA J., 19* (1981), 691–98.

[3] Sarpkaya, T. "Trailing Vortices in Homogeneous and Density Stratified Media," *Journal of Fluid Mechanics, 136* (1983), 85–109.

[4] Widnall, S. E. "The Structure and Dynamics of Vortex Filaments," *Annual Review of Fluid Mechanics, 7* (1975), 141–65.

MIX
Papier aus verantwortungsvollen Quellen
Paper from responsible sources
FSC® C105338

If you have any concerns about our products,
you can contact us on
ProductSafety@springernature.com

In case Publisher is established outside the EU,
the EU authorized representative is:
Springer Nature Customer Service Center GmbH
Europaplatz 3, 69115 Heidelberg, Germany

Printed by Libri Plureos GmbH
in Hamburg, Germany